COMPUTER AND COMPUTING TECHNOLOGIES IN AGRICULTURE II, VOLUME 3

IFIP – The International Federation for Information Processing

IFIP was founded in 1960 under the auspices of UNESCO, following the First World Computer Congress held in Paris the previous year. An umbrella organization for societies working in information processing, IFIP's aim is two-fold: to support information processing within its member countries and to encourage technology transfer to developing nations. As its mission statement clearly states,

> *IFIP's mission is to be the leading, truly international, apolitical organization which encourages and assists in the development, exploitation and application of information technology for the benefit of all people.*

IFIP is a non-profitmaking organization, run almost solely by 2500 volunteers. It operates through a number of technical committees, which organize events and publications. IFIP's events range from an international congress to local seminars, but the most important are:

- The IFIP World Computer Congress, held every second year;
- Open conferences;
- Working conferences.

The flagship event is the IFIP World Computer Congress, at which both invited and contributed papers are presented. Contributed papers are rigorously refereed and the rejection rate is high.

As with the Congress, participation in the open conferences is open to all and papers may be invited or submitted. Again, submitted papers are stringently refereed.

The working conferences are structured differently. They are usually run by a working group and attendance is small and by invitation only. Their purpose is to create an atmosphere conducive to innovation and development. Refereeing is less rigorous and papers are subjected to extensive group discussion.

Publications arising from IFIP events vary. The papers presented at the IFIP World Computer Congress and at open conferences are published as conference proceedings, while the results of the working conferences are often published as collections of selected and edited papers.

Any national society whose primary activity is in information may apply to become a full member of IFIP, although full membership is restricted to one society per country. Full members are entitled to vote at the annual General Assembly, National societies preferring a less committed involvement may apply for associate or corresponding membership. Associate members enjoy the same benefits as full members, but without voting rights. Corresponding members are not represented in IFIP bodies. Affiliated membership is open to non-national societies, and individual and honorary membership schemes are also offered.

COMPUTER AND COMPUTING TECHNOLOGIES IN AGRICULTURE II, VOLUME 3

The Second IFIP International Conference on Computer and Computing Technologies in Agriculture (CCTA2008), October 18-20, 2008, Beijing, China

Edited by

Daoliang Li
China Agricultural University
China

Chunjiang Zhao
National Engineering Research Center for Information Technology in Agriculture
China

Springer

Library of Congress Control Number: 2009921742

Computer and Computing Technologies in Agriculture II, Volume 3

Edited by Daoliang Li and Chunjiang Zhao

p. cm. (IFIP International Federation for Information Processing, a Springer Series in Computer Science)

ISSN: 1571-5736/1861-2288 (Internet)
ISBN: 978-1-4419-0212-2
eISBN: 978-1-4419-0213-9

Printed on acid-free paper

9 8 7 6 5 4 3 2 1

springer.com

Contents

Foreword

The papers in this volume comprise the refereed proceedings of the Second IFIP International Conference on Computer and Computing Technologies in Agriculture (CCTA2008), in Beijing, China, 2008.

The conference on the Second IFIP International Conference on Computer and Computing Technologies in Agriculture (CCTA 2008) is cooperatively sponsored and organized by the China Agricultural University (CAU), the National Engineering Research Center for Information Technology in Agriculture (NERCITA), the Chinese Society of Agricultural Engineering (CSAE) , International Federation for Information Processing (IFIP), Beijing Society for Information Technology in Agriculture, China and Beijing Research Center for Agro-products Test and Farmland Inspection, China. The related departments of China's central government bodies like: Ministry of Science and Technology, Ministry of Industry and Information Technology, Ministry of Education and the Beijing Municipal Natural Science Foundation, Beijing Academy of Agricultural and Forestry Sciences, etc. have greatly contributed and supported to this event. The conference is as good platform to bring together scientists and researchers, agronomists and information engineers, extension servers and entrepreneurs from a range of disciplines concerned with impact of Information technology for sustainable agriculture and rural development. The representatives of all the supporting organizations, a group of invited speakers, experts and researchers from more than 15 countries, such as: the Netherlands, Spain, Portugal, Mexico, Germany, Greece, Australia, Estonia, Japan, Korea, India, Iran, Nigeria, Brazil, China, etc. are gathering Beijing to review the new advancement of Information and Communication Technology (ICT) applications for sustainable agriculture and food quality and safety control, to present new research findings, and to look for the new challenges and opportunities in the future.

Information technology, the convergence of computing and communication technologies, has had an enormous impact on all aspects of socio-economic development and human life in the past 30 years. Powered by the unprecedented and continuous advances in microelectronics and photonics, the power and capacity of our expanding information infrastructure has risen exponentially, while simultaneously its cost has fallen also exponentially. At least for the foreseeable future, the exponential pace of technology improvement is likely to be continued. The modern ICT is playing increasingly important roles in every facet of agricultural and biological system improvement. While traditional sectors of agricultural technology are being constantly updated, the new sectors, such as biological informatics, information network services, information & knowledge-based precision farming system are bringing new concepts and contents into the agricultural & food chain management. The arrival of new requirements for agricultural system sustainability is accompanied by greater challenge in our profession. The goals for farming productivity, resources conservation & environmental sustainability require to develop intelligent equipment, technologies & services in extension of ICT for agriculture. The automated data acquisition is the fit way to provide spatial and temporal high-resolutions and safe documentations. The huge amount of raw data needs to be processed by a easy-to-use and safe data processing systems. A well-founded documentation will be the base of many agricultural applications in the future. A web based data management and information system are able to provide safety and effective information management for the farmers-avoiding problems with local installed software, time and costs. To promote ICT for agriculture, we need "Simplicity Theory", that is to find the simplest method to solving real problems in farming management. To develop a low-cost with high technologies are the future of innovation activities of ICT engineers for agriculture.

The main subjects of this conference are:

■ Exploitation of the strategic problems on ICT for agricultural resources, environment & production system management, web-based technology & agro-information and knowledge service system;

■ Spatial information technologies (GPS, GIS, RS) for agriculture, modeling of resources, ecological and biological systems; Precision Agriculture; advanced sensors and instrumentation for farm use; & process automation; expert system and knowledge system & DSS development;

■ Applied software development for farm users and macro management;

■ Intelligent & virtual technology for agriculture, knowledge dissemination and remote education, etc.

More than 432 academic manuscripts have been received by this organizing committee. After review process by a group of experts, 244 English papers are accepted and published by Spring IFIP US. Taking this opportunity, We would like to express our gratefulness to the hard word by all the contributors and members of Academic Committee.

Finally, we would like to extend the most earnest gratitude to our organizers, College of Information and Electrical Engineering (CAU), EU-China Centre for Information & Communication Technologies (CAU), also to Beijing Eu-Chi Technology Co., Ltd., all members and colleagues of our preparatory committee, for their generous efforts, hard work and precious time!

This is the Second series of conferences dedicated to real-world applications of computer and computing technologies in agriculture around the world. The wide range and importance of these applications are clearly indicated by the papers in this volume. Both are likely to increase still further as time goes by and we intend to reflect these developments in our future conferences.

Daoliang LI 

Chunjiang Zhao 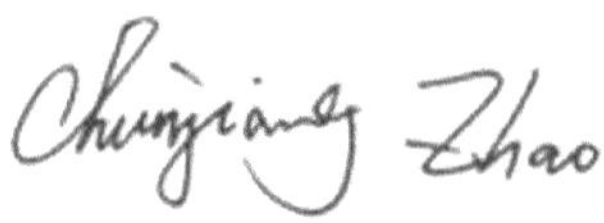

Co-Chairs of CCTA2008

Organizing Committee

Co-Chairs

Prof. Daoliang Li
China Agricultural University, China
Director of EU-China Center for Information & Communication technologies in Agriculture

Prof. Chunjiang Zhao
Director of National Engineering Research Center for Information Technology in Agriculture, China

Members [in alpha order]

Baozhu Yang, Professor of National Engineering Research Center for Information Technology in Agriculture, China

Dehai Zhu, Professor of College of Information and Electrical Engineering, China Agricultural University, China

Haijian Ye, Professor of College of Information and Electrical Engineering, China Agricultural University, China

Jianing Cai, Official of Department of International Cooperation, Ministry of Science and technology, China

Ju Ming, Official of Department of science and technology, Chinese Ministry of Education, China

Qingshui Liu, Secretary-general of China Agricultural University Library, China

Rengang Yang, Professor of College of Information and Electrical Engineering, China Agricultural University, China

Renjie Dong, Professor of Office of International Relations, China Agricultural University, China

Songhuai Du, Professor of College of Information and Electrical Engineering, China Agricultural University, China

Wanlin Gao, Professor of College of Information and Electrical Engineering, China Agricultural University, China

Weizhe Feng, Professor of International College at Beijing, China Agricultural University, China

Xinting Yang, Associate Professor of National Engineering Research Center for Information Technology in Agriculture, China

Program Committee

Chair

Maohua Wang
Professor of China Agricultural University, Academician of Chinese Academy of Engineering, China

Members [in alpha order]

Baoguo Li, Professor of College of Resources and Environmental Sciences, China Agricultural University , China
Béatrice Balvay, Professor of Institut de l'Elevage, France
Benhai Xiong, Professor of Institute of Animal Science, Chinese Academy of Agricultural Sciences, China
Chunjiang Zhao, Professor of National Engineering Research Center for Information Technology in Agriculture, China
Daoliang Li, Professor of College of Information and Electrical Engineering, China Agricultural University, China
Deepa Thiagarajan, Doctor of Michigan State University, USA
Dehai Zhu, Professor of College of Information and Electrical Engineering, China Agricultural University , China
Fangquan Mei, Professor of Agricultural Information Institute, Chinese Academy of Agricultural Sciences, China
Fanlun Xiong, Professor of Hefei Institute of Intelligent Machines, Chinese Academy of Sciences
Fazhong Jin, Professor of Center for Agro-food Quality & Safety, Ministry of Agriculture, China
Fernando Bienvenido, Professor of Universidad de Almeria, Spain
Gang Liu, Professor of College of Information and Electrical Engineering, China Agricultural University , China
Guohui Gan, Professor of Institute of Geographic Sciences and Natural Resources, Chinese Academy of Sciences, China
Guomin Zhou, Professor of Agricultural Information Institute, Chinese Academy of Agricultural Sciences, China
Heinz-W. Dehne, Professor of University of Bonn, Germany
Jihua Wang, Professor of National Engineering Research Center for Information Technology in Agriculture , China

Jinsheng Ni, Doctor of Beijing Oriental TITAN Technology Co., LTD, China
Joanna Kulczycka, Doctor of Polish Academy of Sciences Mineral and Energy Economy Research Institute, Poland
João Cannas da Silva, Vice President of European College of Bovine Health Management
K.C. Ting, Professor of University of Illinois at Urbana-Champaign
Koji Sugahara , Professor of National Agricultural Research Center, NARO, Japan
Kostas Komnitsas, Professor of Technical University of Crete, Greece
Liangyu Chen, Professor of Rural Technology Development Center , Ministry of Science & Technology , China
Louise Marguin, Professor of Institut de l'Elevage, France
Max Bramer, Professor of University of Portsmoth , UK
Michele Genovese, Director of Unit Specific International Cooperation Activities, International Cooperation Directorate, DG Research , UK
Minzan Li, Professor of College of Information and Electrical Engineering, China Agricultural University , China
Nick Sigrimis, Professor of Agricultural University of Athens , Greece
Nigel Hall, Professor of Harper Adams University College , England
Rohani J. Widodo, Professor of Maranatha Christian University , Indonesia
Shihong Liu, Professor of Agricultural Information Institute, Chinese Academy of Agricultural Sciences, China
Theodoros Varzakas, Doctor of Technological Educational Institution of Kalamata , Greece
Weixing Cao, Professor of Nanjing Agricultural University , China
Xiwen Luo, Professor of South China Agricultural University, China
Yanqing Duan , Professor of University of Bedfordshire, UK
Yenu Wan, Professor of Taiwan Chung Hsing University , China
Yeping Zhu, Professor of Agricultural Information Institute, Chinese Academy of Agricultural Sciences, China
Yibin Ying, Professor of Zhejiang University , China
Yiming Wang, Professor of College of Information and Electrical Engineering , China Agricultural University , China
Yud-Ren Chen, Professor of Instrumentation and Sensing Laboratory, Department of Agriculture, USA
Yuguo Kang, Professor of China Cotton Association, China
Zetian Fu, Professor of China Agricultural University , China

Zhujun Zhu, Professor of College of agriculture and food science, Zhejiang Forestry University, China
Zuoyu Guo, Professor of Information Center , Ministry of Agriculture , China

Secretariat

Secretary-general

Baoji Wang , China Agricultural University, China
Jihua Wang, National Engineering Research Center for Information Technology in Agriculture, China
Liwei Zhang , China Agricultural University, China

Secretaries

Bin Xing, China Agricultural University, China
Chengxian Yu, China Agricultural University, China
Dongjun Wang, China Agricultural University, China
Liying Xu, China Agricultural University, China
Miao Gao, National Engineering Research Center for Information Technology in Agriculture, China
Ming Li, National Engineering Research Center for Information Technology in Agriculture, China
Ming Yin, National Engineering Research Center for Information Technology in Agriculture, China
Rui Guo, China Agricultural University, China
Xiaochen Zou, China Agricultural University, China
Xiaohong Du, National Engineering Research Center for Information Technology in Agriculture, China
Xin Qiang, China Agricultural University, China
Xiuna Zhu, China Agricultural University, China
Yanjun Zhang, China Agricultural University, China
Yingyi Chen, China Agricultural University, China
Zhenglu Tao, China Agricultural University, China

THE RESEARCH OF COMPUTER AIDED FARM MACHINERY DESIGNING METHOD BASED ON ERGONOMICS

Xiyin Gao*, Xinling Li, Qiang Song, Ying Zheng
Department of Mechanical and Electronic Engineering, Hebei Agricultural University, Baoding, Hebei Province, P. R. China 071001
* *Corresponding author, Address: Department of Mechanical and Electronic Engineering, Hebei Agricultural University, Baoding 071001, Hebei Province, P. R. China Tel: +86-0312-7526576, Email: gaoxiyin@163.com*

Abstract: Along with agricultural economy development, the farm machinery product type Increases gradually, the ergonomics question is also getting more and more prominent. The widespread application of computer aided machinery design makes it possible that farm machinery design is intuitive, flexible and convenient. At present, because the developed computer aided ergonomics software has not suitable human body database, which is needed in view of farm machinery design in China, the farm machinery design have deviation in ergonomics analysis. This article puts forward that using the open database interface procedure in CATIA to establish human body database which aims at the farm machinery design, and reading the human body data to ergonomics module of CATIA can product practical application virtual body, using human posture analysis and human activity analysis module to analysis the ergonomics in farm machinery, thus computer aided farm machinery designing method based on engineering can be realized.

Keywords: farm machinery, ergonomics, CATIA, human model

Please use the following format when citing this chapter:

Gao, X., Li, X., Song, Q. and Zheng, Y., 2009, in IFIP International Federation for Information Processing, Volume 295, *Computer and Computing Technologies in Agriculture II, Volume 3*, eds. D. Li, Z. Chunjiang, (Boston: Springer), pp. 1527–1532.

1. COMPUTER AIDED ERGONOMICS SOFTWARE HAS LIMITATION WHEN USING IT TO DESIGN FARM MACHINERY

Along with technology development, farm machinery is playing vital role to agricultural economy development, the farm machinery user not only sets request to the product working efficiency, the structure performance, but also has its security, comfortableness. But many farm machinery products more considered realization itself use function did not treated the person as a design goal, did not consider the person's factor, but more considered how to let the human adapt the machine. For example, some threshers, grinders have created many casualties because the feeding board was reduced in no safe protective measure situation in order to save the material and reduce the cost also; some agricultural wastage processing mechanical produces the dust, but no dustproof installment; the agricultural machines and tools structural style and size not too adapt person's custom and needs and so on. In no consider the ergonomics situation, the designed machines and tools are unable to cause the person comfortable in operating process very much, very inaccessibility machine operation efficiency optimum condition (Rychtyckyj et at, 2004).

Along with the technical unceasing development, people further knew the human, machine and environment in product design, the ergonomics application gradually moved towards the practical stage, specially the automobile design, home environment design aspect. Along with the computer technology specially the computer-aided design technology (CAD/CAE/CAM) development, virtual reality technical and high performance graph technology breakthrough, the ergonomics gradually from the theoretical formula computation, the experience material accumulation as well as the simple application computation moved towards the computer assistance ergonomics design technology. Through researching domestic and foreign project technological development condition, we discover that although the overseas ergonomics simulation technology is quite mature, but its development simulation method mostly aims at the western nation crowd, when use these software, our country farm machinery designer must choose the data close our country human body size, then according to our country national condition approximately adjusts(FeiXue Mei at, 2006), in this kind of situation, the design inevitably has error, which causes inappropriate factor between product and user(Rychtyckyj et at, 2004). Therefore, it is important for us to develop the assistance ergonomics design system, which suits our country human body characteristic, and supports the farm machinery product design in china(FeiXue Mei at, 2006).

2. THE HUMAN BODY SIZE DATA NEEDED IN FARM MACHINERY DESIGN

2.1 Regional human body size data

Various countries, various areas, various nationalities as well as the different sex person's human body size data all are different. In our country different provincial capital area human body size also is different. In order to design machines and tools that conform to person's physiology and psychological characteristic, cause the operator to be comfortable and suitable, reduce energy consumption, enhances the labor efficiency, it requests the farm machinery designer to fully understand the product use scope and human body each size scope, massively use the body measurement parameter(GUANGYAN LI et at, 1999). In our country various regional, the difference climate causes the crops planter to have a stronger region, which had also decided the farm machinery is regional, that is to say some machines and tools only be used in some region. Therefore, we should determine the machines and tools most superior design parameter according to the various difference local human body sizes.

Our country human body size standard GB/T10000-1988 divides the human body data into 5 kinds: Northeast and the North China area, the South China area, the China central area, the Southwest area, Northwest area. We may choose the appropriate human body size data according to the machines and tools use region scope when design farm machinery product.

2.2 Seasonal human body size data

In order to guarantee the data scientific and accuracy, the human body data obtains from person measured, which did not put on shoes, was naked or was short in clothes. But the product is for the normal users, which put on the clothes and shoes, therefore we should revise the obtained data, namely if the prime data was taken as product design basis, we must add on the shoe heel altitude and clothes thickness. In addition, people clothing thickness is different along with the changed season; therefore its revision value also should be different. Speaking of the farm machinery, its use season and temperature are certain, then the operator's clothing revises value also is may determine. The designer may adjust the human body size according to product use season.

3. BUILT UP VIRTUAL BODY NEEDED IN FARM MACHINERY DESIGN BASED ON CATIA

The CATIA V5 is a new generation high-end CAD/CAM software system, has provided many kinds of highly effective ergonomics analysis tool and method, may comprehensively analyze the complete factor in the man-machine engineering interaction process, provide the ergonomics engineering design detailed solution(Wei Zhang at, 2006). The CATIA human model is simple, practical, real. But it does not contain Chinese human body data in the CATIA basic model storehouse. This article key work is to establish the virtual body using the open database interface procedure of CATIA, which has the Chinese human body characteristic, conforms to the farm machinery product design.

The CATIA V5 allows the designer to produce the high-quality user definition human model through the massive advanced anthropometrics tool. The farm machinery product user has the strong localization; therefore we should establish our country five regions human body size database files. The body constitution size are quite many, when manufactures each kind of data size the digitization human model, in order to simplify the input data work, we may use the relational formula between human body's structure metrical data and the height, the body weight, after inputting certain data, may direct calculate size value through the specific formula. Constructs the database file according to the CATIA establishment rule, then increases the file directly to the CATIA human body size database. Reads in the appropriate human body size data to the CATIA ergonomics module, through software's human body roll-in function the virtual body can be completed which may support farm machinery product human-machine analysis.

4. THE OPERATION FLOW OF COMPUTER AIDED FARM MACHINERY DESIGN SYSTEM BASED ON ERGONOMICS

4.1 Establish farm machinery product three-dimensional model

Establishing the product model that needs ergonomics analysis is the first step. Its analysis object three-dimensional model may establish in the CATIA software, may induct to the CATIA ergonomic analysis module also from other three dimensional software.

4.2 Join the virtual body in the product model

If we carry on the ergonomics project analysis certainly to have the human body in the analysis system. Before product virtual body, we first should confirm the farm machinery product use area and the crowd, because we had already joined our country five wide range human body databases in CATIA, might find the most appropriate person size from the database to take the analysis directly the object. Then according to the product use choice human body percentage scope, inputs the percentage and the sex in the human model structure module, then carries on the function revision according to the national standards and the product use season to the human body size(Wei Zhang at, 2006). Read the revised human body data to the CATIA human-machine module, through software's human body roll in function, the production virtual body and product user basic consistent.

4.3 Simulate human body operation product posture

The CATIA human posture analysis module (HPA) may analysis ergonomics each kind of posture qualitatively. Human body entire and each kind of posture may examine repeatedly systematically comprehensively from each aspect and analyze, evaluates user's comfortableness, and may compared with the comfortable database data which exists in the announced comfortable database(Bullock, M. I. 1984). In this module, when operates the product according to the human operation product movement request and the human body joint's movement scope dummy, the contact surface dialog box has supplied the human model each spot posture information, the color coding technique may discovery the question posture spot use different color code fast, then makes the analysis, optimize the posture(Greenstein,E.1999).

5. CONCLUSION

The computer aided farm machinery product design method based on Ergonomics mainly use the CATIA ergonomics analysis module to analysis whether the product does meet the human body size requirement, whether the product operation be reasonable, then optimize the product design. Says in view of the Chinese farm machinery design, in order to guarantee ergonomics analysis accuracy, we need to post virtual body in the CATIA ergonomics analysis module, which and the farm machinery user are the same. Using the open database interface procedure in CATIA to establish human body database is the article important work. The farm machinery product has the regional and the seasonal characteristic, so we make the essential revision to the human body size standard GB/T10000-1988, and establish five wide ranges human body databases, reading the human body data to ergonomics module of CATIA can built up practical application virtual body, use human posture analysis and human activity analysis module to analysis the ergonomics in farm machinery, thus computer aided farm machinery designing method based on engineering can be realized.

REFERENCES

Bullock, M. I. Pruct development of an ergonomic keboard, Behaviour and information Technology. 1984(in Chinese)

FeiXue Mei.The Research of Computer Aided Machinery Designing System Based on Ergonomics;2006(in Chinese)

Greenstein,E., Hunting,W. and Nishiyama,K.,1984. PerferredVDT workstation settings, bodyposture and physical impairments, Applied ergonomics. 1999(in Chinese)

GUANGYAN LI, PETER BUCKLE. Current techniques for assessing physical exposure to work-related musculoskeletal risks with emphasis oposture-based methods [J].Ergonomics. 1999

Rychtyckyj, Nestor. Ergonomics analysis for vehicle assembly using artificial intelligence [J]. Proceedings of the National Conference on Artificial Intelligence. 2004

Wei Zhang .A Dissertation Submitted to Shanghai Jiao Tong University for Master Degree of Engineering 2006(in Chinese)

ESTIMATING THE PADDY SPIKE YIELD USING FRACTAL DIMENSION

Hongju Gong [1], Hua Li [1], Haiming Yu [1], Changying Ji [1,*]

[1] *College of Engineering ,Nanjing Agricultural University, Nanjing, Jiangsu Province, P. R. China 210031*

[*] *Corresponding author, Address: College of Engineering ,Nanjing Agricultural University, Nanjing 210031, Jiangsu Province, P. R. China, Tel: +86-25-58606570, Fax: +86-25-58606699, Email:gonghongju@gmail.com*

Abstract: Paddy is one of major crops grown in China. Fractal dimension is an important parameter used to describe geometrical characteristics of many natural objects. The objective of this study is to investigate mathematical relationships between the fractal dimension and the yield of the paddy spike based on the fractal theory and the image processing technologies. The color images of paddy spikes were taken with a digital camera, then they were transacted to the gray using image processing. Fractal dimensions of the spikes were computed using the Box-Counting Method based on the gray images. The results indicate that there is a significant linear relationship between the fractal dimension and the yield.

Keywords: Fractal Dimension (FD), yield, paddy spike, Box-Counting Method (BCM)

1. INTRODUCTION

The fractal theory, brought forward by Mandelbrot who is considered as the father of fractal, has found wide applications in areas ranging from material science, geology, power technology to micro-electronic(Biswas et al., 1998; Valdez et al., 1998), as well as in areas of Agricultural Engineering (Li Yan et al., 2000; Liu Taoju et al., 2002; Zhang Shirong et al., 2002; Ren Xianzhong et al., 2005). Simple objects can be described by ideal shape primitives, such as cubes, cones, and cylinders. But most of the natural objects are so complex and erratic that they cannot be described in terms of

---

Please use the following format when citing this chapter:

Gong, H., Li, H., Yu, H. and Ji, C., 2009, in IFIP International Federation for Information Processing, Volume 295, *Computer and Computing Technologies in Agriculture II, Volume 3*, eds. D. Li, Z. Chunjiang, (Boston: Springer), pp. 1533–1541.

simple primitives (Kenneth, 1991). On the other hand, the essence of fractal is self-similarity, and the concept of self-similarity seems to play an important role in the description of nature.

Paddy is one of major crops grown in China. LiuTaoju (Liu Taoju et al., 2002) considered that paddy spikes have the self-similarity, which is the primary characteristics of the fractal geometry and they base it simulating the rice morphology with fractal theory. Currently, the further researches on fractal characteristic of paddy spike, such as Fractal Dimension (FD) and the relationship between FD and yield, have been not reported.

In this paper, fractal analyses of paddy spikes and measurement of FD of paddy spikes based on the Box-Counting Method are presented; The correlativity between FD and the yields of paddy spikes and a mathematical model for estimating the yields using FD are also discussed.

2. BOX-COUNTING METHOD(BCM)

Fractal Dimension (FD) can be defined in many ways, in which, the Hausdorff Dimension, the Self-similarity Dimension and the Box-counting Dimension are the leading definition. By comparison, the Box-counting Dimension is easier in mathematical computation and measurement, therefore it was widely used (Chaudhuri et al., 1995; Biswas et al., 1998; Cheng Kunjie, 2005; Ren Xianzhong et al., 2005; Wan Yaonan et al., 2006; Zhang Tao et al., 2006). In the research, it is also used to measure the FD of paddy spikes.

Box-counting Dimension can be explained as follows. If X is a bounded set in Euclidean n-space. And it is the union of $N_e(X)$ distinct (no overlapping) copies of itself, each of which is similar to X and scaled down by a ratio e . Then fractal dimension D of X can be derived from the following equation when the limit in the equation exists,

$$D = \lim_{e \to 0} \frac{\log N_e(X)}{\log(1/e)} \qquad (1)$$

It is difficult to compute the parameter D directly using Eq.(1). An approximate method was used extensively in many literatures. Just as its name implies, our work is counting the numbers of the boxes. There is a digital image with M × M pixels, in which the target information is contained and it could be divided into boxes with $r \times r$ pixels, where $M/2 \geq r \geq 1$, and r is an integral number. Then we count the boxes in which portions of the target information are contained. Let $N(r)$ be the total number of countable boxes, an array of r- $N(r)$ data can be obtained by varying the values of r . If the data satisfies following equation,

$$N(r) = \left(\frac{1}{r}\right)^D \tag{2}$$

It indicates that the target information possesses fractal characteristics.

In practice, the Eq.(2) could be transformed into a logarithmic form, i.e.,

$$log(N(r)) = D\,log\left(\frac{1}{r}\right) \tag{3}$$

With that, according to the array of r- $N(r)$ data obtained above, a plot of $log(1/r)$ versus $log(N(r))$ could be drawn in a log-log grid. Afterwards, a fitted straight line can be got with the method of least squares. Then the slope of the line is the Box-Counting Dimension.

3. MATERIAL AND METHOD

3.1 Material and Equipments

Digital Camera-Olympus U400D(DC), Constant Illuminating Source Box(CISB), Electronic Weighing Scale-JY12001(EWS) are the Hardware and equipments used in the experiments. The Software used are VC++ and MATLAB6.0.

The paddy species is Sujing-45, it was grown in the Jiangpu Experimental Farm of Nanjing Agricultural University ,Jiangsu Province of China which is located at latitude of 32° 3' 4.96" N, and longitude of 118° 36' 38.78" E. 27 paddy spikes picked up randomly from the field with visible difference in yields are weighed with the EWS and pictured in the CISB with a black velvet background using the DC. When picturing, each spike was positioned just as it was in the original field. Each image is of sides 512 × 512 pixels(Fig. 2).

3.2 Digital Image Processing

Fig.1 Source image of paddy

Fig.2 Grayscale image of paddy

The source color images (Fig.1) were converted to grayscale images (Fig.2) according to the following equation (Yang Shuyin, 2003)

$$R_1 = G_1 = B_1 = Max(R, G, B) \tag{4}$$

Where R,G,B represent respectively the gray values for each pixel of the origin color image, R1,G1,B1 represent the gray values for each pixel of grayscale image. Afterwards, using a threshold value 100, grayscale images were binarized into binary images(Fig3).

Fig.3 Binary images of paddy spike

3.3 Fractal Dimension Measurement

In the binary images such as the Fig.3, the part of paddy spike is white (where pixel value equals 1), whereas the background is black (where pixel value equals 0).

Divide the binary image into boxes sides $r^{\prime}\ r$, avoiding the singularity where $1 < r \quad 216$ and r is an integer.

Let $r = 2,3,4,L\ ,256$, and then count the number of grids where pixel value equals 1 and note it as N_r. With the arrays of r - N_r, a log-log grid of $log(1/r)$ versus $log(N(r))$ could be constructed. So the Box-Counting Dimension of the paddy spike, i.e., the slope of the fitted straight line could be obtained, which is noted as DBC.

4. EXPERIMENTAL RESULTS AND DISCUSSION

A scatter plot of $log(1/r)$ versus $log(N(r))$ of the image (for example the sample 1) is shown in Fig.4.

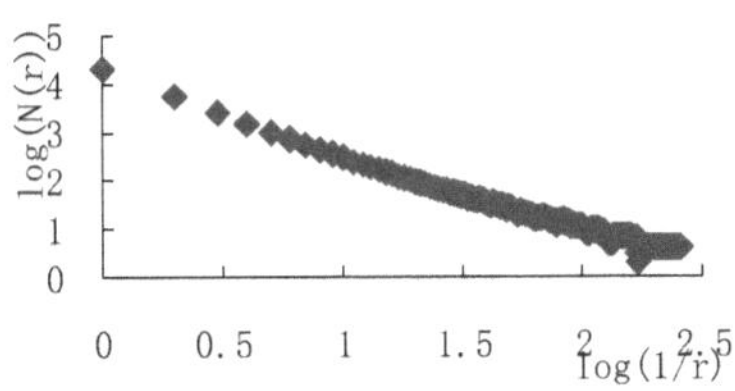

Fig.4 Relationship between $log(1/r)$ and $log(N(r))$

According to the definition of the Box-counting Dimension, if there is a linear interval in the scatter plot, it means that the target information bears fractal characteristic. It could be seen from Fig.4 that when r varies from 2 to 103 pixels, there do exist distinct linear relationship between $log(1/r)$ and $log(N(r))$ Fitting the scatter plots in the region from $r = 2$ to $r = 103$, the slop of fitting straight line, so the DBC of the paddy spike, can be calculated. The results for every sample are presented in Table 1.

Table 1. D_{BC} and yields of paddy spikes

sample	1	2	3	4	5	6	7
yield(g)	3.55	3.96	3.97	4.17	4.46	4.5	4.76
D_{BC}	1.5465	1.5221	1.4707	1.4478	1.5317	1.5005	1.5191
sample	8	9	10	11	12	13	14
yield(g)	5.1	5.12	5.46	5.54	5.72	5.8	5.83
D_{BC}	1.5165	1.521	1.5104	1.5122	1.5716	1.511	1.5944
sample	15	16	17	18	19	20	21
yield(g)	5.84	5.86	6.08	6.11	6.28	6.31	6.45
D_{BC}	1.5487	1.5582	1.5102	1.565	1.5421	1.601	1.561
sample	22	23	24	25	26	27	
yield(g)	6.67	6.79	6.83	6.88	7.83	8.5	
D_{BC}	1.5481	1.5674	1.5798	1.5539	1.6105	1.5852	

A regression analysis between the FDs and the yields of paddy spikes was performed. By means of calling the Regression (Y,X,a) (He Runbing, 2001)function of MATLAB 6.0. fitting equations can be easily obtained. when the confidence level being $1-\ a = 99.9\%$, the fitting equation can be described as follows,

$$y = 21.4017D_{BC} - 27.2623 \tag{5}$$

with, $F = 23.7419, R^2 = 0.4871, p = 0.0001$ Since $F > F_{0.001}(1,100), p > a$, It can be concluded that there is a distinct linear relationship between the FDs and the yields of corresponding paddy spikes, at the confidence level of 99.9%.

With rcoplot(r,rint) (He Runbing, 2001) function of MATLAB6.0, the residuals of regression were plotted as shown in Fig.5. From the figure, it

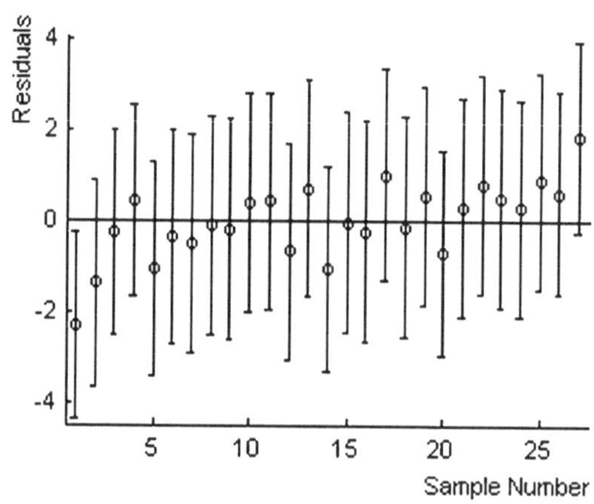

Fig.5 Residuals of regression

can be seen that the FD of sample 1 is an outlier in the regression. After getting rid of the outlier, the renew regression equation was founded as follows

$$y = 21.7397D_{BC} - 27.6951 \tag{6}$$

With $R^2 = 0.5798$, $F = 33.1145$, $p = 0.0000$. The renew Residuals were plotted in Fig.6. It is seen that the coefficients of the regression equation are equable, but the value of R^2 and F were increased clearly,

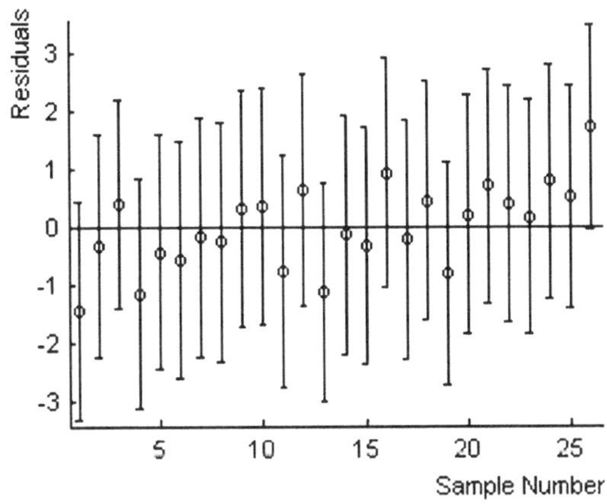

Fig6. Renew Residuals of regression

and the value of p was diminished to zero, and the fitting results were optimized clearly.

5. MODEL CHECKING

To check the adequacy of the mathematical model which describes the relationship between FD and the yield of each paddy spikes, 20 spikes of same species were collected as a checking specimen. Calculated values getting from the model and real values were plotted in Fig.7, and the error distribution of the model checking was shown in Fig.8.

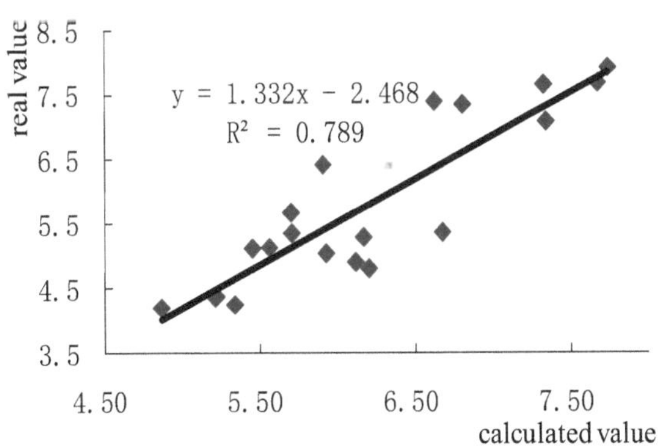

Fig.7 Model checking

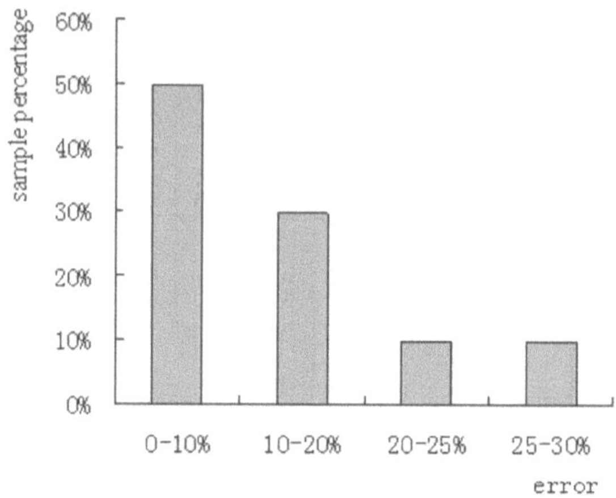

Fig.8 Error Distribution of Samples

As shown in Fig.7, the correlation coefficient $R^2 = 0.7897 > (r_{0.01}^{19})^2 = 0.4425$, the conclusion can be made that the correlation between the calculated values and the real values is significant at $a = 0.01$. It is also shown in the plot of error distribution that 80% of errors in the specimen are limited in 20%. Further investigation shows that when the yield of all, single spikes in specimen is higher than 5g, the maximum error is 24.57%, and 93% of errors in the specimen are limited in 20%.

6. CONCLUSION

From the investigation on fractal properties of paddy spikes and the analyses of the relationship between the yields of paddy spikes and their fractal dimensions, the following conclusions can be drawn.

(1)The paddy spikes bear fractal characteristic.

(2)The correlation relationship between the FDs and the spike yields is significant and it can be described as:

$$y = 21.7397 D_{BC} - 27.6951 \quad (7)$$

(3)Although the accuracy of the model is not very satisfactory, the reserch on the FDs and estimating yields of paddy spikes should be helpful for the further study in relevant domain.

ACKNOWLEDGEMENTS

Funding for this research was provided by the department of Jiangsu agricultural machinery management and the College of Engineering , Nanjing Agricultural University (P. R. China). The first author is grateful to the two departments for their supports.

REFERENCES

Biswas, M. K. and T. Ghose, et al. (1998). "Fractal dimension estimation for texture images:A parallel approach." Pattern Recognition Letters 19 : 309-313.

Chaudhuri, B. B. and N. Sarkar (1995). "Texture Segmentation Using Fractal Dimension." Transactions on Pattern Analysis and Machine Intelligence 17 (1): 72-77.

Kenneth, F. (1991). Fractal Geometry-Mathematical Foundations and Applications. Shenyang, Shenyang University Publishing Company.

kunjie, C. (2005). Study on Beef Automatic Grading Based on Fractal Dimension & Machine Vision. Nanjing, Nanjing Agricultural University. Doctor.

Runbing, H. (2001). Calculation and Application in Engineering using MATLAB6.0. Chongqin, Chongqin Publishing Company.

Shi-rong, Z. and G. Guosu (2002). "Fractal analysis of patch patterns on plant diseases." Plant Protection 28 (6): 9-13.

shuyin, Y. (2003). Digital Image Processing Using VC++. Beijin, TsingHua publishing company.

Tao, Z. and Y. Zhibiao, et al. (2006). "Improved Extracting Algorithm of Fractal Dimension of Remote Sensing Image." Journal of Ordnance Engineering College 18 (5): 61-65.

Taoju, L. and T. Jianjun, et al. (2002). "A Study on the Fractal Characters and the Visual Simulation of Rice Morphology." Acta Agricultural University Jiangxiensis 24 (5): 583-586.

Valdez-Cepeda, R. D. and E. Olivares-Sanz (1998). "Fractal analysis of Mexico's annual mean yields of maize,bean,wheat and rice." Field Crops Research 59 : 53-62.

Xianzhong, R. and W. Chun, et al. (2005). "Study on Fractional Characteristics of Wheat Kernel Shape Parameter and Its Application on Agricultural Engineering." Transactions of the Chinese Society for Agricultural Machinery 36 (10): 85-87.

Yan, L. and P. Jiaxiong (2000). "The Target Segmentation and Detection Based on Fractal Dimension Feature." Journal of Huazhong University of Science and Technology(Nature Science) 28 (8): 1-5.

Yaonan, W. and W. Shaoyuan, et al. (2006). "Image Texture Analysis Based on Fractal Dimension." Journal of Hunan University(Natural Sciences) 33 (5): 67-72.

STUDY ON WEB-BASED TOOL FOR REGIONAL AGRICULTURE INDUSTRY STRUCTURE OPTIMIZATION USING AJAX

Xiaodong Huang[1], Yeping Zhu[1,*]
[1] *Agricultural Information Institute of CAAS, Beijing 100081, China;*
[*] *Corresponding author, Address: Agricultural Information Institute of CAAS, 12 South ZhongGuanChun Road Beijing 100081, China, Tel:+86-10-68919652, Fax:+86-10-68919652, Email: zhuyp@mail.caas.net.cn*

Abstract: According to the research status of regional agriculture industry structure adjustment information system and the current development of information technology, this paper takes web-based regional agriculture industry structure optimization tool as research target. This paper introduces Ajax technology and related application frameworks to build an auxiliary toolkit of decision support system for agricultural policy maker and economy researcher. The toolkit includes a "one page" style component of regional agriculture industry structure optimization which provides agile arguments setting method that enables applying sensitivity analysis and usage ot data and comparative advantage analysis result, and a component that can solve the linear programming model and its dual problem by simplex method.

Keywords: Agriculture, Industry structure optimize, AJAX, Linear programming

1. INTRODUCTION

China is a large agricultural country, and now she has enter a new era of agriculture industry structure strategic adjustment that composed of improving the quality of agricultural products, optimizing the agriculture structure and increasing the income of farmers(Soft Science Committee on the Ministry of Agriculture Research Group. 2000).

Please use the following format when citing this chapter:

Huang, X. and Zhu, Y., 2009, in IFIP International Federation for Information Processing, Volume 295, *Computer and Computing Technologies in Agriculture II, Volume 3*, eds. D. Li, Z. Chunjiang, (Boston: Springer), pp. 1543–1550.

Many researchers have studied the application of regional agriculture industry structure adjustment theory, and have built a certain number of information system or decision support system to support this kind of adjustment. By study on these researches, we have found a weakness in those information systems. Many of these information systems include agriculture industry structure optimizing modeling tools. These tools have a common character that they all based on external tools such as MATLAB or SPSS (Xiong. 2006). This character makes a separation between information system and modeling tool that the modeling tool can't make use of data in information system. To address this issue, this paper proposed a solution that makes use of AJAX technology, builds a flexible web-based tool for regional agriculture industry structure optimization modeling and solving.

2. AJAX TECHNOLOGY AND ASP.NET AJAX FRAMEWORK

AJAX stands for Asynchronous JavaScript and XML and includes a large set of technologies such as DOM, XMLHttpRequest, XML, XSLT, JSON etc (Garrett. 2005). AJAX technology has been widely implemented in many web applications since it is first brought out in 2005, and it has triggered the revolution of web application that so called trend of Web2.0, and is now supported by many industry giants such as Google and Microsoft.

AJAX brings a new interaction mode between users and web applications which is called asynchronous mode. In a tradition interaction mode, user need to submit each action to web server and wait the page reloaded to see the result. Ajax overcomes the page loading requirements of HTML/HTTP-mediated web pages and creates the necessary initial conditions for the evolution of complex, intuitive, dynamic, data-centric user interfaces in web pages. Fig.1 shows the working mechanism of AJAX:

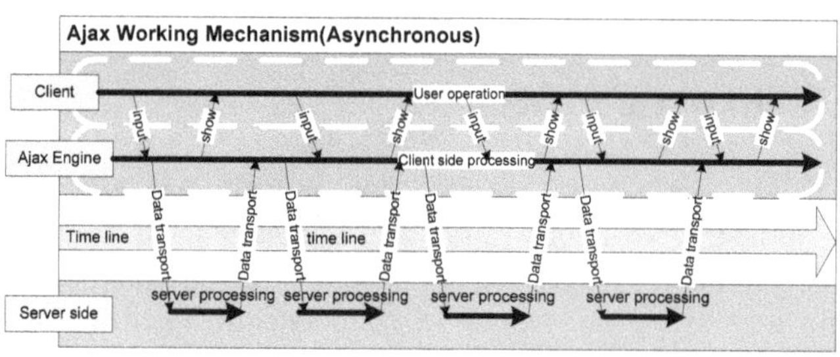

Figure 1. Working Mechanism of AJAX

In this paper, we use ASP.NET AJAX framework as foundation of our web-based tool. ASP.NET AJAX is the name of the solution that provided by Microsoft. It is the most powerful and most completely encapsulation of AJAX techniques currently, and provide full support of client side object oriented programming, a large number of client side and server side components, web service integration and ASP.NET components extension.

3. LINEAR PROGRAMMING FOR REGIONAL AGRICULTURE INDUSTRY STRUCTURE OPTIMIZATION

According to regional agriculture industry structure adjustment theory, there are two ways to modeling agriculture industry structure optimization problem, one is linear programming, and another is non-linear programming. Linear programming is most used method. Here is the common form of a linear programming model (Hillier et al. 1999):

$$Max(Min)Z = \sum_{i=1}^{n} c_i x_i$$

Subject to

$$\begin{aligned} & a_{11}x_1 + a_{12}x_2 + \cdots + a_{1n}x_n \le (=,\ge) b_1 \\ & a_{21}x_1 + a_{22}x_2 + \cdots + a_{2n}x_n \le (=,\ge) b_2 \\ & \ldots \\ & \qquad \ldots \\ & a_{m1}x_1 + a_{m2}x_2 + \cdots + a_{mn}x_n \le (=,\ge) b_m \\ & x_1 \ge 0, x_2 \ge 0, x_3 \ge 0 \cdots x_n \ge 0 \end{aligned} \tag{1}$$

This model includes one objective function, several decision variables, and several constraint functions. By research of a large number of empirical study cases, we found that the objective function of the regional agriculture industry structure optimization model often refers to maximize the agricultural economy benefits, or maximize the social benefits, or maximize the environment benefits. We also found the linear programming model mainly includes two type constraint functions, one is constraint that reflects resource restrictions, such as farm land, labors, agriculture machines, and another is constraint that shows the target or objective which need to be achieved, such as security of food, keep advantage and etc.

By summarize the characteristic of regional agriculture industry structure optimization model, we found that the modeling tool is composed of some basic element block. The objective function contains a type and a multi-monomial, the constraint function contains a multi-monomial and a type and a right hand side coefficient. Each multi-monomial contains several monomials which composed of one coefficient and one decision variable. Because of actual meaning of objective function and constraint functions, coefficients in model objective function and constraint functions often come from agricultural economic data. So the modeling tool can be built on a certain set of agriculture economic data and make use of them.

4. USER INTERFACE OF MODELING TOOL

According to the AJAX design principles, the modeling tool is designed as a "one page" style application (Mahemoff et al. 2005). A "one page" style application means user can perform all actions that application provides in one user interface, just like Word or Excel. User can create a model then edit it, save it, and edit it again. With this kind of web application, we can provide the same user experience as desktop application. Fig.2 shows the user interface of regional agriculture industry structure optimization modeling tool:

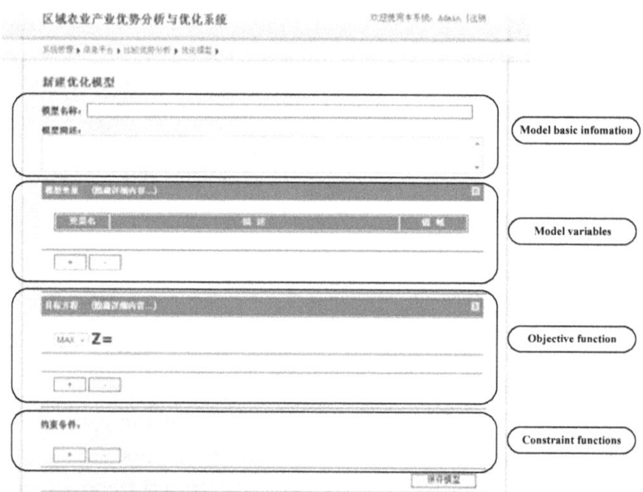

Figure 2. User Interface of Modeling Tool

The whole page is divided to 4 parts; first one is for filling model's basic information such as model name, model description; second part is variables setting panel, user can add or remove decision variable in this part; third part is objective function editing panel; and the last part is for editing constraint functions. The last three parts each contains an add button and a remove button, respectively response for adding or removing element in model. Each part with a title bar can be collapsed by clicking the title bar in order to save the work space and prevent come out with a long page.

A monomial is designed as a single widget, by clicking the add button in objective function editing panel or constraint function editing panel, user can add a monomial to target function like in fig.3:

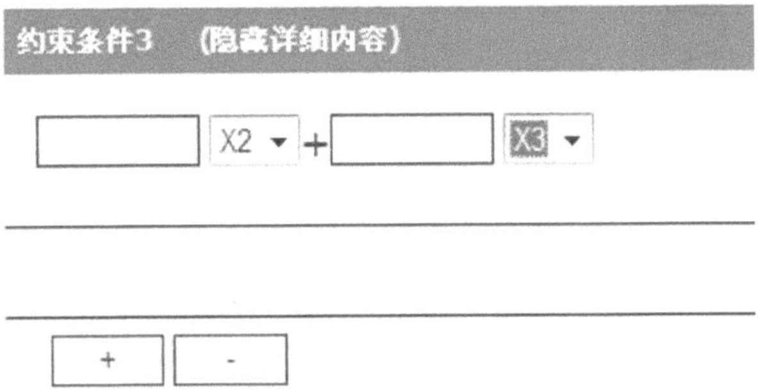

Figure 3. Add a Monomial to Target Function

In order to provide flexibility and make full use of agriculture data and comparative advantage analysis result, regional agriculture industry structure optimization modeling tool provide four ways to set the coefficient of monomial:

■ Const coefficient: user input the coefficient with a number when create model, and this number stays unchanged during model calculation.

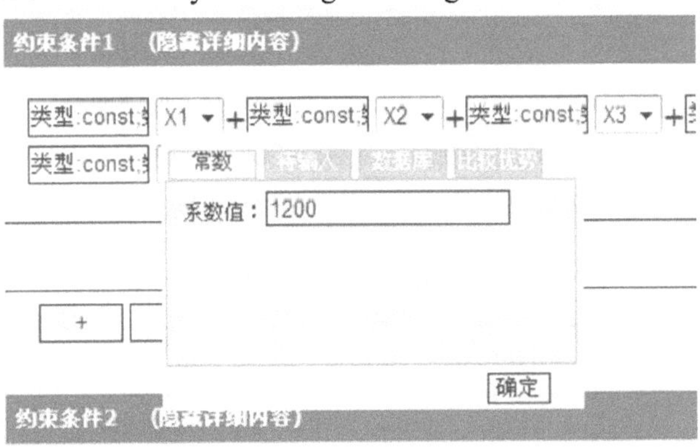

Figure 4. Model Coefficient Setting-Const Coefficients

■ Input coefficient: user provides a description of this coefficient, the modeling tool will automatically generate an input textbox before model calculation, and user can input different value of this coefficient in every other calculation.

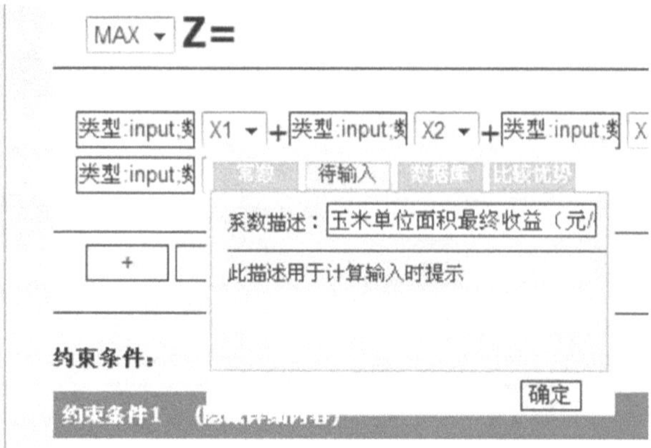

Figure 5. Model Coefficient Setting-Input Coefficients

■ From database coefficient: user can set some criteria to locate a data in database, modeling tool will get the actual data when model calculation.

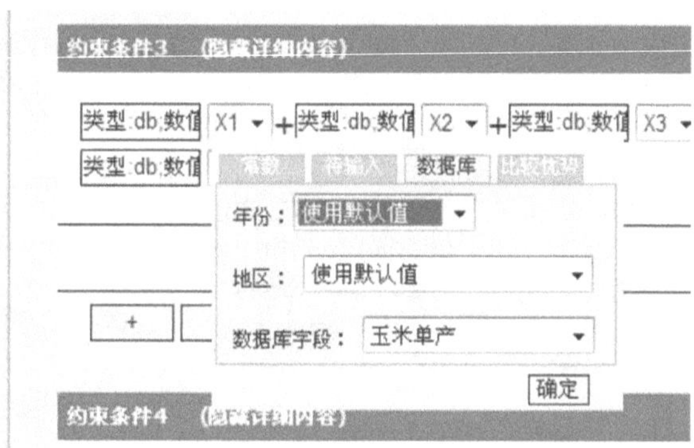

Figure 6. Model Coefficient Setting-From Database Coefficients

■ Compare index coefficient: user provides some criteria to find a compare index in the result of comparative advantage analysis, modeling tool will get the actual data when model calculation.

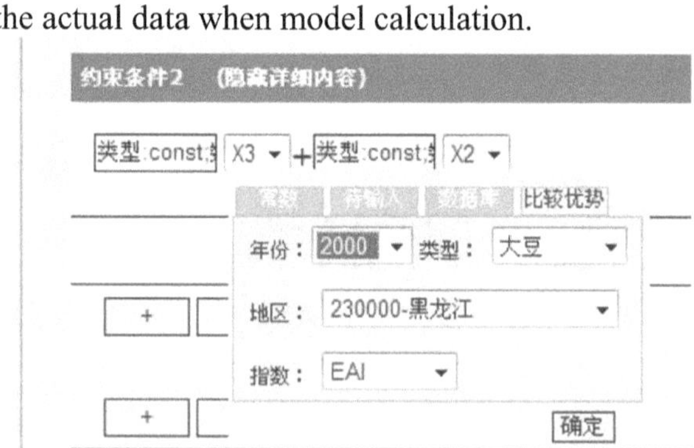

Figure 7. Model Coefficient Setting-Compare Index Coefficients

5. COMPONENTS DESIGN OF REGIONAL AGRICULTURE INDUSTRY STRUCTURE OPTIMIZATION MODELING TOOL

We design the components of regional agriculture industry structure optimization modeling tool according to the N-tiers design principle (Wikipedia. 2007), fig.8 shows the structure of components:

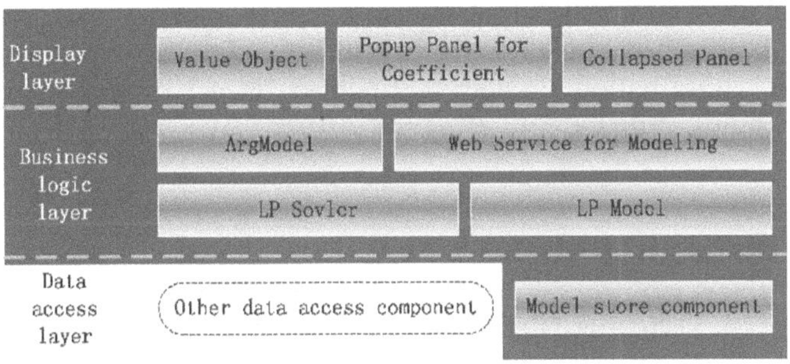

Figure 8. Components in Optimization Modeling tool

■ Value object: Value object is used for transporting complex data type between client-side JavaScript and server-side component (Gibbs et al. 2007). ASP.NET AJAX framework can perform this transform automatically.

■ Popup panel for coefficient: used for setting coefficient type and related criteria and value. This panel is design according to the popup pattern (Popup pattern. 2006), the popup pattern is one of the most important AJAX design pattern which allows user perform a quick task without distracting from the main flow. In this modeling tool, popup panel allows user set the coefficient without distracting from model building.

■ Collapsed panel: after editing a block of model, user can make that block collapse, this action makes higher usability of page space

■ Web services for modeling: a set of web services that response directly to client-side requests and dispatch those requests to components in business logic layer.

■ ArgModel: a ArgModel object stores all information that user entered in model creation, includes model name, model description, variables name and description, every coefficient's value or find criteria. ArgModel also provides method that returns a LPModel object.

■ LPModel: represents a linear programming model that generated from ArgModel. LPModel just contains data and function, and has nothing to do with actual meaning of the model.

■ LPSovler: provide algorithm that can solve the linear programming problem and its dual problem by simplex method.

■ Model store component: responsible for reading ArgModel information from and writing to xml files.

6. DISCUSSION

The AJAX technology can bring web application a better user experience and make use of client-side computing power. In this paper, we design and implement a flexible web-based tool for regional agriculture industry structure optimization modeling and solving by using AJAX technology. This modeling tool is a relatively independent component that can be use in agricultural information system, provides linear programming model building, solving abilities.

ACKNOWLEDGEMENTS

The work described in this paper was funded by National Scientific and Technical Supporting Programs Funded by Ministry of Science and Technology of China (2006BAD10A06), Beijing Municipal Natural Science Foundation (4042026), and Special Fund of Basic Scientific Research and Operation Foundation for Commonweal Scientific Research Institutes (2008J-1-06).

REFERENCES

Frederick S. Hillier, Gerald J. Lieberman. Introduction to Operation Research(sixth edition). China Machine Press. 1999

http://www.adaptivepath.com/ideas/essays/archives/000385.php

Jesse James Garrett.A New Approach to Web Applications.

Mahemoff, Michael. Ajax Design Patterns. Oreilly & Associates Inc. 2005

Matt Gibbs, Dan Wahlin. Professional ASP.NET 2.0 AJAX. Wrox Press. 2007

Popup pattern. http://ajaxpatterns.org/Popup

Soft Science Committee on the Ministry of Agriculture Research Group. New Stage of China Agriculture Development. 2000

Wikipedia http://en.wikipedia.org/wiki/Multitier_architecture

Xiong Shuanglin. A Research on Decision Support System for Agriculture Structure Adjustment at County Level--A Case Study in Hua County of Henan Province. Chinese Academy of Agricultural Sciences. 2006

DEVELOPMENT OF A MODEL FOR DISSEMINATING AGRICULTURAL TECHNOLOGY INFORMATION BASED ON SIGNALING THEORY

Xiaohong Wang, Baofeng Chen *

Economic and Management College, China Agricultural University,17#, Qinghua East Road, Haidian District, Beijing, P. R. China 100083

* *Corresponding author, Address: Economic and Management College, China Agricultural University,17#, Qinghua East Road, Haidian District, Beijing, P. R. China 100083,Tel: +86-10-62736284, Fax: +86-10-62736284,E-mail: cbfisu@163.com*

Abstract: The effective information dissemination model is greatly meaningful to rural economic development. Applying the theory of signaling games, it's concluded that farmer's demand centered technology supply is essential. On this basis, a new agricultural technology information dissemination model was built in the paper.

Key words: signaling, information, dissemination, model

1. INTRODUCTION

Agricultural technology information dissemination plays an important role in the improvement of rural development and new Socialist Countryside Construction. It is necessary to set up fluent spread and feedback channels in order that the advanced and useful technology information will be available to farmers (Qu, 2001). The agricultural technology information transmit systems in our country ignored the farmers' information demands. There was a gap between information demand and supply (Sun,2003). In the processes of technical information transmission and spread, farmers lack the effective information feedback channels. It will be not fluent for

Please use the following format when citing this chapter:

Wang, X. and Chen, B., 2009, in IFIP International Federation for Information Processing, Volume 295, *Computer and Computing Technologies in Agriculture II, Volume 3*, eds. D. Li, Z. Chunjiang, (Boston: Springer), pp. 1551–1558.

information dissemination unless the problem is solved. And the result of agricultural technology spread will be negatively affected by the problem. For one thing, the risks of both resource waste and economic losses would be increased. Furthermore, the risk of new technology adoption of farmer will also rise. Therefore, for solving the problem, it is essential to set up a new agricultural technology dissemination model. Based on the theory of signaling games, an agricultural technology information dissemination model from the angle of farmers' demand will be discussed in the paper.

2. THE ORETICAL BACKGROUND

2.1 The meanings of agricultural technology information demand for farmers

Agricultural technology information demand refers to farmers' demand for agricultural technology information under certain conditions, which implicates two meanings: 1, farmers' desirableness for informational purchase or adoption. 2, the payment capacity of technical information. The payment capacity is not just refers to finance but time, perception and so on.

2.2 Assumptions and model development

There are often asymmetry of information between the organization of agricultural technology spread (hereinafter referred to A) and farmers (hereinafter referred to B).A does not understand fully about B's demand, receptivity and the way they deal with information in the process of receiving. According to the information dissemination model they chose, whether B could get desirable information or not depends on if the problem about asymmetry can be solved. A would bring about wastes in finance and recourses if they do not understand B's demand in the process they provide agricultural technology information to B. For B, it would also lead losses if the technology information not fit to them.

There are differences in quality of information demand among farmers. In the process of agricultural technology information dissemination, A always hopes for more Bs who are higher quality of information demand as their targets customer, because the higher quality farmers can realize themselves' information demand and are able to search, evaluate and use them efficiently. They have high abilities in using information provided by A and they will bring great benefits in information technical spread to A (social benefit for governments and financial benefits for enterprises). It is not easy thing to distinguish farmers' quality of information demand because A often needs certain signal mechanisms such as B's incoming, education level and experience in informational application. It will get different results according to Bs who have different quality levels. In this way, A can effectively distinguish farmers with different information demand, thereby enhancing the average quality of information demand of farmers who apply information technology.

We assume that the standard distribution range of B's quality of information demand is [0,1].Generally, the farmers who have higher quality of information demand have the better ability to use technology information, and the farmers who have lower quality of information demand could use less. Here we assume that it is linear function between the ability of using the technology information and the quality of information demand, expressed by the line pp', and assuming that the externality of information dissemination can get compensation or controlled through certain means. We use Fig.1 to express the quality of information demand of farmers and the corresponding ability of using the technology information. If A doesn't use any signal mechanism, and selects the target customers B randomly, then the average expect-quality of the chosen target customers is 0.5, and the average expect-ability of using information technology is b, both of which belong to the average level. A are always would like to use signal mechanisms to assess B's income level, education level and past experience using information technology, thereby finding the target customers with high-quality of information demand. It is assumed that the cost which B met A's requirements is related to its own quality of information demand, or further assume that the cost is linear and decreasing to the quality of information demand, cc' showing that in the Fig.1. A further assumption is that the interest of getting technology information from A is d. It is cost-effective to send signals for farmers who have quality of information demand above e and have a lower cost of sending signals than their benefits. The farmers who have the quality of information demand bellow e would face higher cost than benefits. It has no sense for them to send signals. In this way, the signal mechanism of A could exclude the farmers with the lower quality of information demand, and the average expect-quality of B can be achieved (e +1) /2, thus the average expect-ability of using technology information of B would be higher than a. When A further increase their demand, its average expect-ability of using information technology would be further enhanced, but it is unrealistic to over demand the quality of information demand of farmers. Because on the one hand, B which has high quality of information demand maybe require the higher quality of technical information, on the other hand, it is difficult for A to find sufficient target customers when B having a lower quality of information demand.

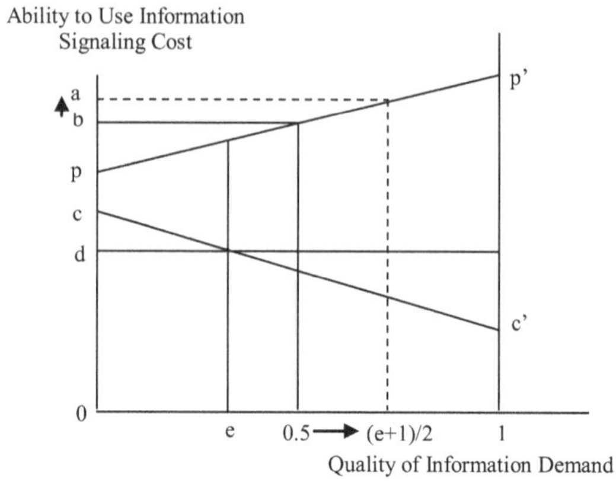

Fig.1: Relation between Quality of Information Demand and Ability to Use Information

From the analysis above, we can get that it is conducive to raise target customers' quality of average expectant information demand and enhancing the adoption rate of technology information for the agricultural technology information provider. For farmers, their quality of information demand should meet the technology providers' requirements. This would improve the applying efficiency of agricultural technology information to some extent, but two issues must be satisfied: firstly, B can convey their real and effective signals of technology demand quality to the technology information providers, namely, the existence of effective channels for feedback is needed; The second is in the request of A's signal mechanism, A could obtain enough target customers to maintain their investment costs. Therefore, there are two problems at present: first, whether B has condition to convey a real and effective technology demand signals to A? It is effective that A disseminate information to B only when conditions permit, if the conditions are not full or possessive, the informational disseminating from A to B would be invalid or even failure. The second problem is how to access to adequate number of target customers for A in the signal mechanism circumstances. It can meet A's request better only when B possesses of high quality of information demand. On the contrary, it would be difficult for A to obtain sufficient number of target customers.

Based on the analysis above, we should think about that whether the two problems exist in reality. In the current process of agricultural technology dissemination, we can conclude that the answer is no.

Firstly, the farmers are short of channels for information feedbacks. Recently it is mainly by unidirectional transmission of technical information. For instance, the information providers spread their information through newspapers, broadcast, TV, websites, and so on. The farmers can not fully realize the actual useful effect and real value of the information they received because lack the professional informational knowledge. The information supply lacks pertinence and interactivity. The relevant surveys indicate that presently the fluent information communicational channels have not popularized in the wide countryside in China. Especially pointed, what the farmers need is effective channels for information feedbacks to solve the difficulties they encounter. And otherwise it will result in informational lying idles and wastes. The existing methods of interactive information spread by telecoms, short massagers' platforms, the clerks for information spread, enterprise-farmer household ordered cooperation, expert consulting are still not fully popularized. Generally speaking, the agricultural technical information suppliers do not spread information pertinently and the feedbacks from farmers are usually not available to the suppliers. Therefore it is resulting in transplacement between supplies and demands. Now farmer-feedback channel should be provided urgently to make the information spread more pertinently and increase the efficiency of technical information dissemination.

Secondly, the whole quality of the farmers is not high enough for agricultural technical information demands at present. The whole educational level of the farmers is still on the low side now. With the widespread use of the current techniques of production in agriculture, however, farmers rely increasingly on technical information to help them and Meanwhile it is request the qualities of farmers should be higher. Agricultural technology information dissemination requires of the farmers having a certain scientific and technological knowledge and skills. Under the market economy conditions, farmers only have physical couldn't adapted to the development of the times.

3. MODEL DEVELOPMENT

Based on the analysis above and the experience of foreign countries, this paper will build an appropriate model of agricultural technology information dissemination (as shown in Fig.2). An effective model of agricultural technology information dissemination should be able to fully mobilize all the elements' positivity. The elements include the disseminator, the recipients, as well as resources, media, and so on. The analysis about the model is from the following two aspects: Firstly, the elements involved in the model and their functions. Secondly, the driving force and the way of the dissemination model act, and feedback mechanisms.

3.1 Elements and their functions

Generally, The information dissemination process includes five elements such as the disseminator, the recipients, the message, the media and feedback. For information dissemination of agricultural technology, "disseminator" mainly refers to government-oriented Agricultural Techniques Extension Organizations, Rural Special Technique Association, agricultural enterprises, agricultural technical education and scientific research institutions, and so on. They are the main force of agricultural technology information dissemination; "recipient" refers to the broad masses of farmers which contain core of farmers who have high recognition, judgment, application and evaluation capacity and general farmers;" Message" is that agricultural technology information itself; "media" means the channels, means or tools of agricultural technology information dissemination including oral media, print media, electronic media, and so on.

3.2 Operating Way

The process of agricultural technology information dissemination is divided into two stages. The first stage is before the agricultural technology information transmitted. We should enable the broad masses of farmers to get the information demand feed backed to the hands of information disseminators through certain channels and means (known as "External Feedback", or "EF" for short).Ordinary farmers may rely on the key information personal in the village to reflect their desire or needs, this process is called "Inner Feedback"("IF" for short).Then the disseminators should dispose the farmers' information about the demand of agricultural technology information, and transmit the required and appropriate information the farmers need through certain channels. The second stage is after the agricultural technology information dissemination. Information disseminator should have certain channels to track and investigate farmers' response in time, including information technology awareness, the existing problems in the application process and evaluation (also "EF") in order that the disseminators can timely adjust the contents and dissemination channels of the information and transmit the re-adjusted information to the farmers again. After the re-transmission of information, Both the disseminator and the recipients can get better effect in supplying and using the agricultural technology information.

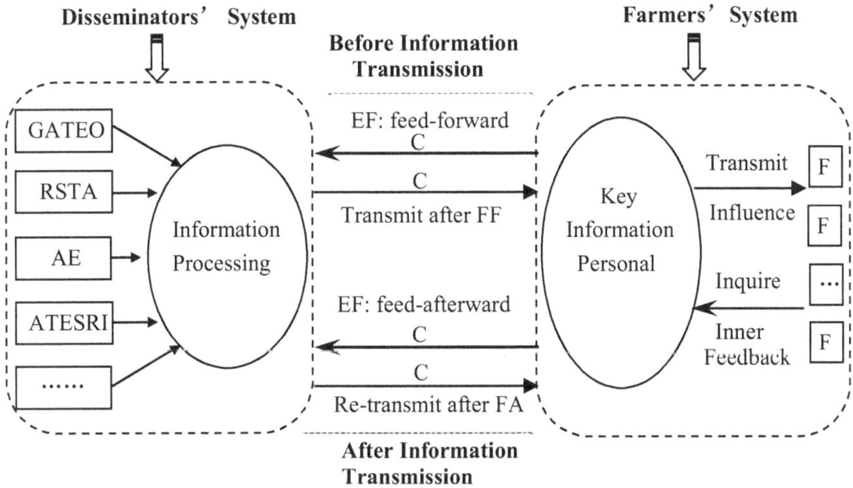

Fig.2: Agricultural Technology Dissemination Model

4. CONCLUSION

Through analyzing the signaling theory, it aims at establishing the feedback channel of information demand for farmers, which is of great significance to improve the farmers' overall quality, improve the adoption rate of agricultural technology, and promote the dissemination of agricultural technology information. The new dissemination model of agricultural technology information puts forwards the two-stage information dissemination model which includes dissemination and feedback of information, and considers the differences within key information personal in the village with higher quality of information demand and ordinary farmers adapting China's reality. It would have given some attention and support by forming the feedback and dissemination, and been contribute to the spread of agricultural technology information.

Farmers today need the guidance and service of agricultural technology information more pressingly than ever before to guide their production and management. The disseminator of agricultural technology information should master farmers' information demand comprehensively and dynamically, and regard them as the important subject developing rural information technology, with orient to meet the farmers' information demand by taking measures in a targeted strategy. It provides the accurate, reliable and useful information to farmers timely, and promotes the development of building the socialistic new rural.

ACKNOWLEDGEMENTS

Thanks Miss Xiaihua, Renyan, and Zhangnini for giving good advice to this paper. Thanks Liu Xutan for the help of translation.

REFERENCES

Jing Jipeng, Zhang Xiangxian.Economics of Information [M].Beijing: Science Press, 1996

Lei Na, Zhao Banghong. Farmers' Information demand of and Agricultural Information Dissymmetry Between Supply and Demand - an Empirical Research Based on the Farmers' Information demand in Hebei Province, The Agricultural Economy[J], 2007(3)

Li Zhengliang.The Principles of Communication. China's Communication University Press.2006

Quzhou Committee of People's Communist Party of China, the People's Government of Quzhou City. Reform the Spreading System of Agricultural Technology to Satisfy the Farmers' demands to Technological Information. Review of China Agricultural Science and Technology 2001(3)

Ryan Womack. Information intermediaries and optimal information distribution. Library & Information Science Research. 2002, 24:137

Sun Lianhui.Studies on the Mechanism of Chinese Agricultural Extension. Northwest Sci-tech University of Agriculture and Forestry,2003

Tan Ying, Wang Dehai. Analysis of Information Demand of Farmers at Different Level in Depressed Areas Journal of China Agricultural University(Social Science Edition) [J].,2003(3)

Wathmanel Seneviratne. Local information upload strategy in building up rural community information pages on the Internet: An approach to attending the information needs of rural Sri Lanka. The International Information & Library Review. 2004, 241~252

Zhang Weiying. Game Theory and Information Economics [M].Shanghai: San Lian Bookshop of Shanghai, People's Press of Shanghai.

Zheng Yegang, Jin Kuiwen. The Information Demand of Farmers and the Existing Problems. Agricultural Library and Information Science Journal [J].1999（1）

TRANSIENT ANALYSIS OF A BLDC STARTER/GENERATOR SYSTEM USED IN ELECTRIC VEHICLES

Xinli Xu [1], Yan Shi [2,*], Anbo Liang [3], Ming Zhang [4], Qian Liu [1]

[1] *College of Engine & Electronic Engineering, Qingdao Agricultural University, Qingdao, China., 266109*

[2] *College of constructional engineering, Qingdao Agricultural University, Qingdao, 266109,China*

[3] *International Intercommunion School, Qingdao Agricultural University, Qingdao, China, 266109*

[4] *College of Electric and information Engineering of Jiangsu University of Science and Technology Zhen Jiang 212013，China*

[*] *Corresponding author, Address: Foodstuff Science and Engineering* college, *Qingdao Agricultural University, 700 Changcheng Road, Qingdao, 266109, P. R. China, Tel:0532-88030446, Email: shiyan517@163.com*

Abstract: The brushless DC motor used in ISAD of Electric and Hybrid Electric Vehicles is introduced in this paper. Magnet field distribution and static characteristic of permanent magnet brushless DC motor are calculated by finite element methods.

Keywords: Integrated Started Alternator Damper, Static and dynamic characteristic, Permanent brushless DC, Finite element method

1. INTRODUCTION

Integrated Started Alternator Damper is a new technique which has been used in auto's framework in recent years. The electric motor of ISAD integrates starting motors and generators, so it can lighten auto's weight, save space and reduce the cost. Not only can this system realize Regenerative Braking, but also can provide accessional start/brake function,

---

Please use the following format when citing this chapter:

Xu, X., Shi, Y., Liang, A., Zhang, M. and Liu, Q., 2009, in IFIP International Federation for Information Processing, Volume 295, *Computer and Computing Technologies in Agriculture II, Volume 3*, eds. D. Li, Z. Chunjiang, (Boston: Springer), pp. 1559–1567.

what's more, it can boost engine's torque when necessary. All this system can promote the efficiency of the system, export more electrical energy, reduce the cost of auto's energy, and tail gas and noise(Yin Wei et al.2006).

Instead of the mechanical commutate of normal direct current electrical motor's brush and commutator, brushless DC motor makes use of electrical commutator. It has many advantages, such as higher running efficiency, none excitation cost, good timing function, and so on(Jianzhong Sun et al.2005). And the rotor of brushless motor is made of permanent magnet steel. As the rare-earth permanent material has high magnetic energy product, it can reduce the weight of the motor, spare the volume of the motor, which can do good to auto and motorcycle's start/generating system which has limit weight and volume for the installation's place and using method. The usual magnetor of start/generating system is permanent magnet synchronous motor, which can be reformed easily to be the start/generating motor for its framework as similar as the brushless DC motor(A Permanent Magnet Brushless DC Motor Problem). And the block diagram of the framework of brushless DC motor is as follows (Fig.1):

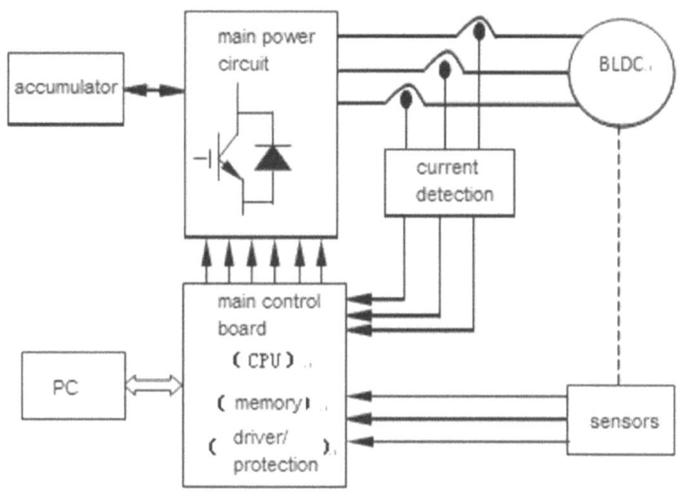

Fig.1. The block diagram of the framework of brushless DC motor

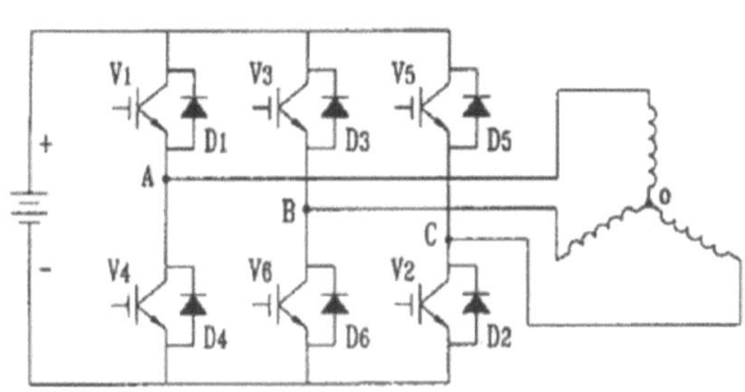

Fig.2. The main circuit of brushless DC motor

2. THE BASIC EQUATION OF BRUSHLESS DC MOTOR

This paper makes use of star connection-three-phase bridge main circuit in which each two phase bridge energizes to each other. That's to say, there are two power pipe being broken over every moment, and phase changes every 1/6 cycle, each phase changes one power pipe and every power pipe breaks over 120° electrical angle. As Fig.2 shows. The voltage equation of brushless DC motor can be simplified as:

$$U_s - 2U_T = E + 2rI_a \tag{1}$$

U_T——the voltage drop of switch apparatus

E——the back elevation of motor

I_a——Armature Current

Any has the two-phase winding to lead at any time all,

As for three phase-six mode brushless DC motor, two phase winding will be lead at any time , so the back elevation of the motor is:

$$E = 2E_m = \frac{2pW_\varphi}{15a_I}\phi_\delta n = C_e\phi_\delta n \tag{2}$$

C_e——electromotive force constant of the motor, $\frac{2pW_\varphi}{15a_I}$

ϕ_δ——flux of each electrode, Wb, $\phi_\delta = B_\delta \alpha_i \tau L_a$

W_φ——Number of turns of each phase's cascade of armature winding

the electric current of armature winding:

$$I_a = \frac{U_s 2U_\tau - E}{2r} \tag{3}$$

At each moment, motor's electromagnetic torque is formed by the interact of two-winding's composite magnetic field and rotor field:

$$T_e = \frac{2E_m I_a}{\Omega} = \frac{EI_a}{\Omega} = \frac{4pW_\varphi}{\pi\alpha_i}\phi_\delta I_a = C_T\phi_\delta I_a \tag{4}$$

Ω——mechanical angular velocity of rotor, $\Omega = 2\pi n/60$

C_T——Torque constant of the motor, $C_T = \frac{4pW_\varphi}{\pi\alpha_i}$

3. ESTABLISH THE MODE

Firstly, we should design the framework of the motor and import the basic data and main size to Ansoft's Rmxprt, then we can get the two-dimension Finite Element Model of the motor(Guoqiang Liu et al.2005). Main parameters of brushless DC motor's mode are as follows: voltage rating:220V, power rating 1.5KW, rotate speed 3176r/min, load 4.5N.m, moment of inertia 0.00112825Kgm^2 . The Finite Element Model of the motor is as Fig.3 shows. The Skewed Stator Slot I use can reduce the cost of brushless DC motors, lower the noise and vibration and improve the stationarity of motors(Renyuan Tang.1997).

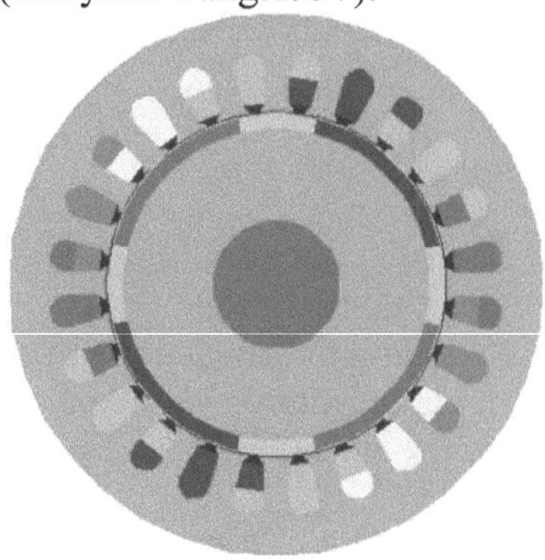

Fig.3. Finite Element Model of brushless DC motor.

The most valid method to study the motor's Transient Behavior is time-stepping finite element method, which has high degree of accuracy for it can involve every factor of the motor, such as: slot effect, skewed slot effect, saturation effect, vortex, skin effect and so on. Then definiting the characteristic of each subdivision part of the mode and subdividing the grid are done. When frequency changes, system can change the part automatically and realize adaptive subdivision. If the grid is broken, and the broken is big enough, then the air gap can subdivide the grid anew. After subdividing the gird, we should assure proper Boundary Condition and load. Boundary Condition adopts the natural boundary, and excitation source adopts the additional voltage source which is controlled by rotor's place. As Fig.4 shows, S1-S12 are switches controlled by V1-V6, and V1-V6 are functions of place, whose delay angle is 60° electrical angle one by one. Then definiting the parameter of count process is done, such as counting march, time, accuracy and the process parameter.

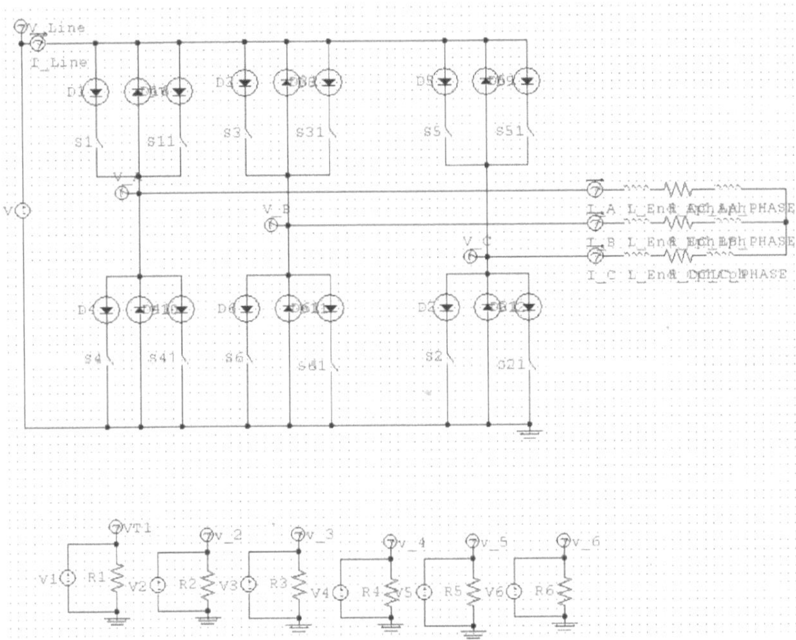

Fig.4 Simulation by the Canadian circuit

4. PROCESS OF EMULATION ANALYSIS

4.1 Analysis of static character

Through analyzing the motor's magnetic field, Ansoft can test the shape, size and Rotor Laminations' size. What's more, it can make the motor's function be the best through adjusting some parameter of the motor. Fig.5 shows the distribution of rotor's magnetic line of force when motor idle loads. Fig.7 shows circle distributing curves of flux density. We can see that the flux density and magnetic line of force is well-distributed along the circle, which indicated that the design of the motor is reasonable. Fig.6 and Fig.8 shows the distribution of rotor's magnetic line of force when motor loads, which is different from Fig.5 and Fig.7, that's because the magnetic field of armature winding affect the main magnetic field when motor loads, which is also called armature reaction. The main effect of q-Axis Armature's magnetomotive force to airspace magnetic field is changing the distribution of airspace magnetic field. Armature reaction lies on the degree of saturation of magnetic circuit, the changing of motor, connecting style of armature winding and electrifying mode of inverter.

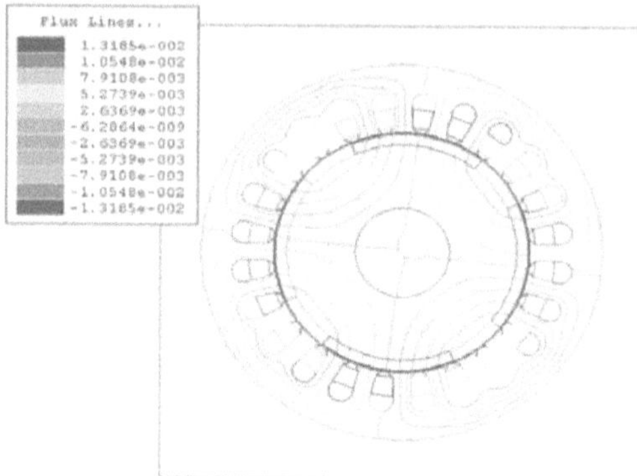

Fig. 5 the distribution of motor's magnetic field when idle loads

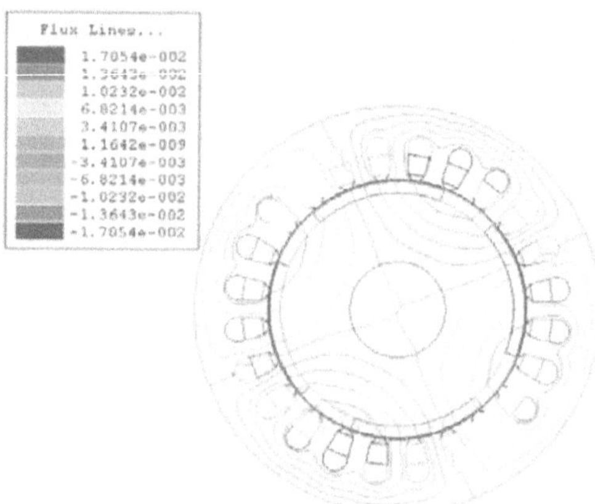

Fig. 6 Distribution of motor's magnetic field when loads

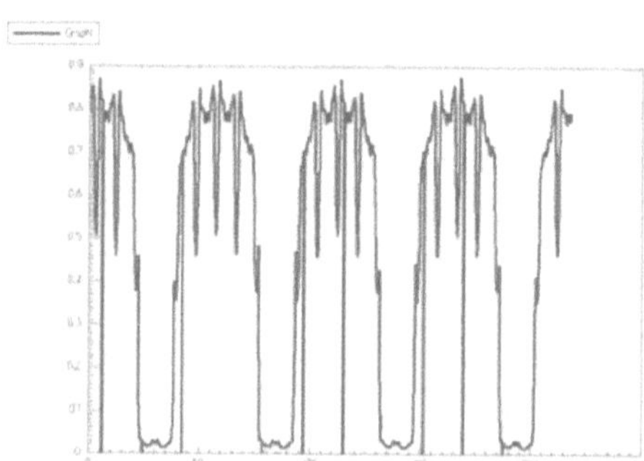

Fig.7 Distribution of motor's air gap flux density when idle loads

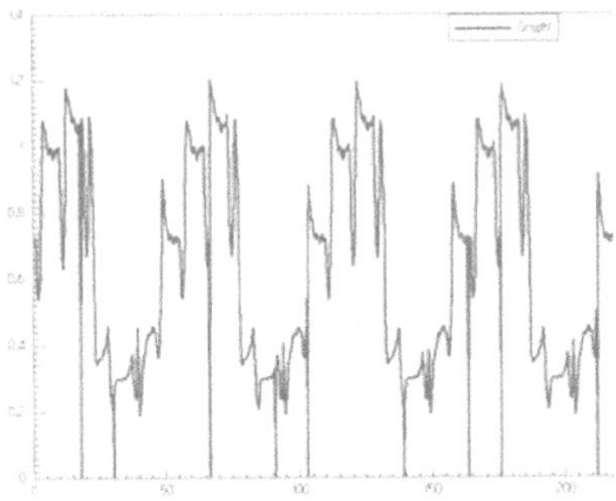

Fig.8 the distribution of motor's air gap flux density when loads

4.2 Analysis of dynamic character

ISAD system requires the smallest synthetic starting torque when the motor starts should be greater than the biggest load torque of the motor's engine. The author makes use of Ansoft's transient soft and counts the motor's starting, loading and other transient process. The results are as the diagram below, including electric current, torque, speed and power cost's varying curves along with time. As the turning inertia of the motor is small, so the motor will be stable after starting 0.04s when loading running. At 0.3s, the motor's loading is adjusted suddenly, when t<0.3s, $T_L = 4.5$ N• m; when t>0.3s, $T_L = 6$ N• m; when t=0.3s, the load change suddenly. At this time the motor overloads, the rotate speed declines and the torque increases, however, the motor which the the author designs can still run stably, so the function of the motor which the author designs is stable.

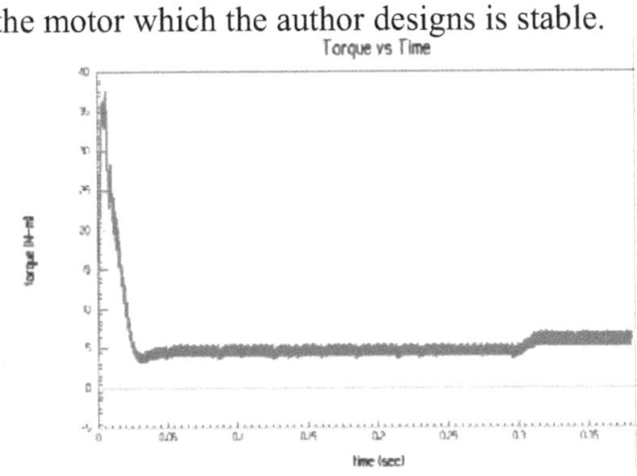

Fig. 9 Changing curves of the torque when the motor starts.

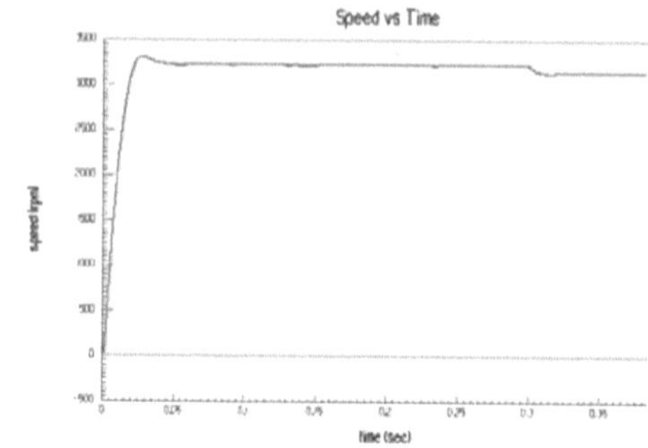

Fig. 10 Changing curves of rotate speed when the motor starts.

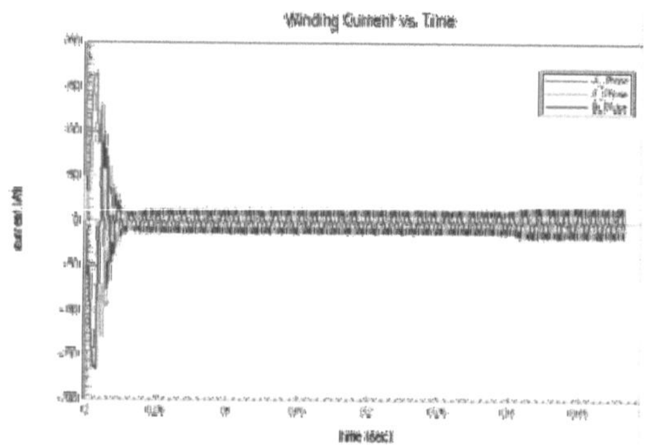

Fig. 11 Changing curves of three-phase winding current when the motor starts

5. CONCLUTION

We can see from the diagrams that the brushless DC motor designed by the author which has application in mixed-power auto's starting generating system has many advantages, such as: torque's pulsation is small, the starting torque is big which is about 10times of the rating torque, the efficiency is 90% and it is stable at low speed. But one point should be concerned, when the motor starts, as for its low speed, the electromotive force which caused by rotor flux cutting stator winding is very little, so the starting electric current may be over the rating current. To solve this problem, we should add current-limiting circuit. We can see from Fig.10 that there is torque ripple caused by commutation current delay which is affected by winding inductance when communicating. We can adopt commutation method by overlap and optimal current control to reduce torque ripple.

REFERENCES

A Permanent Magnet Brushless DC Motor Problem [Z]. Ansoft RMxprt Application Note
Analysis software.Beijing: Electronics Industry Publishing House,2005 (in Chinese)
Conservancy and hydropower Press,2005 (in Chinese)
Generator system Starting Performance Analysis and Simulation[J]. Micro-motor 2006,(1) (in Chinese)
Guoqiang Liu,Lingzhi Zhao,Jiya Jiang. Ansoft electromagnetic engineering finite element
Jianzhong Sun,Fengxian Bai. Special motor and control[M].Beijing: China's water
Renyuan Tang. Modern permanent magnet motor theory and design ［M］.Beijing: Machinery Industry Press,1997(in Chinese)
Yin Wei,Li quan,Zhen xu,Qiang liu. Six of the 12/10 vehicle is switched reluctance starter

DESIGN OF SINGLE-PHASE POWER METER BASED ON ETHERNET

Huimei Yuan [1,*], Yan Wang [1]
[1] *College of Information Engineering, Capital Normal University, Beijing, 100089, China*
[*] *Corresponding author, Address: College of Information Engineering, Capital Normal University, No. 56 Northern Road of Xisanhuan, Haidian District, Beijing, P. R. China, 100048. Tel: +86-10-68902762-230, Fax: +86-10-68416830, Email: yuanhmxxxy@263.net*

Abstract: According to the related technology about power meter, single-phase power meter based on Ethernet is designed and researched. Single-chip processor MC9S12NE64 is used as system's CPU and in charge of the system control and information processing. Power measurement chip CS5460A is applied to calculate electrical parameters. EMAC and EPHY modules which are MC9S12NE64's inter parts are adopted to accomplish Ethernet communication, and how to design Ethernet protocol and data processing are illuminated in detail. The system has realized real-time settlement of electric charge and other network-controlled functions such as power stoppage and supply. It overcomes current problems about electric charge's disagreement due to offline running of power meter. The power meter is mainly used in the environment where Internet network has been realized, and it is unnecessary to establish wired communication network in addition. Resources of network sharing can be achieved and cost can be saved greatly. Keywords: Ethernet; single-phase power meter; single-chip processor; communication protocols, power measurement.

Keywords: ethernet; single-phase power meter; single-chip processor; communication protocols, power measurement

1. INTRODUCTION

Current domestic market of the power meter is entering a stable period of development, but power meter still hold a large market capacity because of

Please use the following format when citing this chapter:

Yuan, H. and Wang, Y., 2009, in IFIP International Federation for Information Processing, Volume 295, *Computer and Computing Technologies in Agriculture II, Volume 3*, eds. D. Li, Z. Chunjiang, (Boston: Springer), pp. 1569–1577.

resident's increased demand, power girds' reconstruction and external demand for export. At the same time, the functions of the power meter have became long life, high-precision, and multi-functional compared with the ordinary functions of original power meter (CHEN Chi-pei et al., 2002).

Traditional power meters usually use infrared and 485 serial transmission as their communication methods when it transcribes. Distance of the infrared communication is not more than 20 meters, but for 485 serial communications, it is required to lay cable separately. These traditional communication methods are time-consuming, laborious and unreliable (ZHU Xiao-hong et al., 2007; WANG Yue-zhi et al., 2007;SHA Zhan-you et al.2006).

With the wide use of the internet, technology of the computer and network develop rapidly, and all these impel the related technology become mature. The Ethernet technology is widely used in various computer networks in order to join the different network equipments with each other. The equipment-control technology based on Ethernet develops more rapidly, and it has a deep impact on instrument industry. So, there appears a strong tendency of the network instruments.

In recent years, the related research about the network power meter has been done by many (LI Wei-dong et al., 2000; ZHOU Yao-yi et al., 1999), and how to realize it has been discussed in these papers. However, these researches are basically concentrated on the possibility of achieving it with laboratory network communication technology. Some researches such as cost constraints, anti-jamming capability and the choice of the whole system scheme are not done deeply, and the practical significance is not enough.

With the requirements of communications network, there exist too many problems to deal it with 8-bit microcontrollers which are the mainstream MCU products (DU Qiong et al., 2005). Now, ARM32-bit microprocessor is very common on the market, it's network function is so strong that it can meet requirements of network-based instruments. But as a commercial chip, ARM32-bit microprocessor takes too many market factors into account, so it integrated many on-chip and out-chip equipments(GUO Shu-jun et al., 2004), some of them are unnecessary, and it is unreasonable for consumers to pay for these additional costs .On the other side, for Ethernet network communications, ARM32-bit microprocessor need external related devices such as Ethernet Token ring physical layer protocol/ Media Access Control (PHY/ MAC) Control Chip, network electromagnetism transformer, RJ45 and so on. The further cost is unacceptable for users.

According to the problems above, the single-phase power meter based on Ethernet is designed and researched in this paper, so that the power meter's network can de realized. The power meter is faced to the environment where the internet connection has been achieved, so that establishing the wired communication again can be avoided, resources of network sharing can be achieved, and cost can be saved greatly too.

2. DESIGN OF HARDWARE EQUIPMENTS

Block diagram of single-phase power meter based on 16 bits single chip MC9S12NE64(freescale semiconductor,2004) has been put forward in this paper. If methods of wires connecting in the hardware and the corresponding program modules in the system are adjusted, the block diagram can be applied to the design of three-phase power meter too.

The block diagram of single-phase power meter's hardware system based on Ethernet is showed in Fig.1

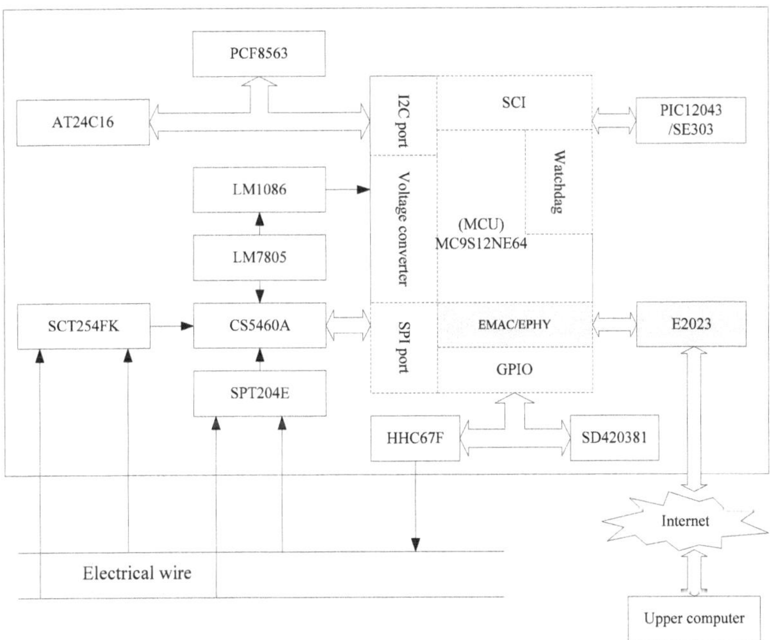

Fig.1: block diagram of hardware

In our system, single-chip processor MC9S12NE64 is used as the system's CPU and in charge of the system's control and information's process.

LM7805 is used to stable 5V DC voltage produced by transformer and supply power to CS5460A. LM7805's output is connected to the input of LM1086, whose function is transforming 5V DC voltage to 3.3V DC voltage to provide MC9S12NE64 with working voltage.

Voltage from electrical wire is transformed to small current signal and small voltage signal through current transducer SCT254FK and voltage transducer SPT204E, then power measurement chip CS5460A(HUAN Ling et al. 2002; GUO Song-lin et al. 2002; LIAN Xiao-qin et al.2005; ZHU Yun et al.2004; WANG Yi-jun et al.2001) takes charge of calculating all kinds of parameters such as the effective values of voltage and current, values of active power, reactive power and apparent power, values of active energy, reactive energy and apparent energy with these signals. MC9S12NE64 can read all parameters calculated by CS5460A through Serial Peripheral Interface (SPI) and process them with software. These parameters are stored in EEPROM chip AT24C16 by I2C bus and exported to the LED or LCD module by general pin input and output（GPIO）in order to shown, and whether energy power is send by relay HHC67F is determined by operating conditions. In our design, the special power meter nixietube-SD420381 which is made by Wuxi Ark Technology Electronic Co., Ltd. is used as LED module. Real time chip PCF8563 is used to provide reference time to system by IIC bus pin SCL and SDA. The system mainly communicates with the upper computer through Ethernet. When these communicate method do not work, workers should take advantage of infrared communication to read meter and check it in order to recover the communication. In our design, infrared communication module is composed of infrared transmitting tube SE303 and infrared receiving tube PIC12043.

MC9S12NE64 contains internal module of Ethernet Media Access Controller (MAC)and Ethernet Physical Layer (EPHY)(LI Wei-dong et al., 2000), so its Ethernet interfaces to the outside just need electromagnetic isolation transformer(in our design E202 is selected), RJ45,resistance and capacitance. The internal EPHY module can also be replaced by exterior PHY chip, but doing that not only adds cost of the system, increases area of PCB but also influences the system's stability. The hardware diagram of communication circuit which uses internal EMAC and EPHY module is shown in Fig.2.

In Fig.2, PHY_RXP and PHY_RXN are positive and negative pins in MC9S12NE64 for Ethernet communications to transmit differential signals, while PHY_TXP and PHY_TXN are positive and negative pins to receive differential signals. The internal electromagnetic isolation transformer have 8 signal wires followed as J1....J8, in which J1 and J2 are the positive and negative pins separately to transmit differential signal, J3 and J6 are the positive and negative pins separately to receive differential signal, and other four signal wires are unused.

In Ethernet communications, data transmitted by MC9S12NE64 are transformed into differential level signal and then delivered by PHY_TXP and PHY_TXN, and they are sent into internet by J1 and J2 after passing internal electromagnetic isolation transformer of RJ45. When received, the data on internet enter J3 and J6 of RJ45 as the differential signal data, and they are sent to PHY_RXP and PHY_RXP of MC9S12NE64 after passing through electromagnetic isolation transformer, at last they are received by EPHY of MC9S12NE64.

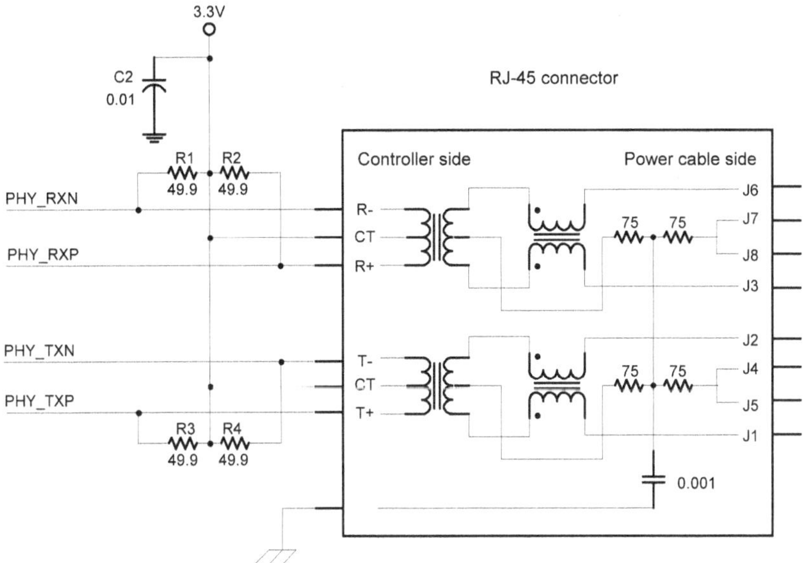

Fig.2: Circuit diagram of Ethernet communication

3. DESIGN OF ETHERNET PROTOCOL

On the basis of realization of hardware equipment, the functions of Ethernet communications should be as: responding to the ping request of upper computer on internet, so as to inform upper computer that network connection is normal; accomplishing the communication with upper computer, including timing proof requisition of upper computer, request of transmitting electro-parameter data and requisition of current failure.

According to system requirements, Ethernet communications only need to accomplish parts of transmission control protocol/internet protocol (TCP/IP)l stack, such as Address Resolution Protocol(ARP),IP, Internet Control Message Protocol (ICMP)and User Datagram Protocol (UDP).It is enough to finish the system. TCP is not selected because it is more complicated than UDP and its realization occupies too many resources. By contrast, UDP is simpler and requires fewer resources. Although reliability of the data transmitted by UCP is uncertain, it can be vouched by upper software. At the same time, considering functions of the whole system and limitation of hardware resources, project of transplant Embedded Operating System is not adopted too.

Structure of TCP/IP protocol stack in our design is shown in Fig.3.

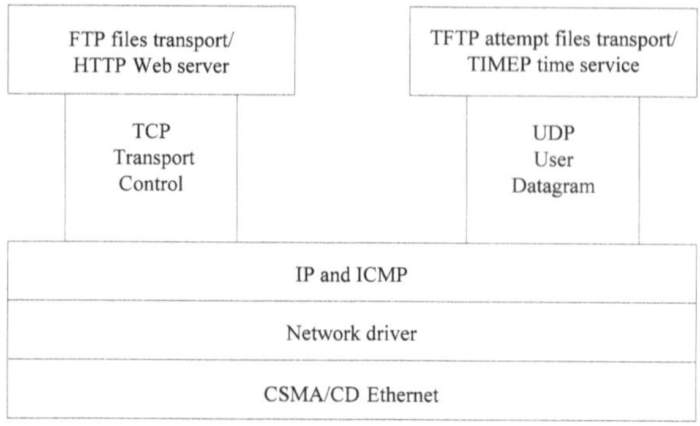

Fig.3: structure of TCP/IP protocol stack

When system begins, it must complete initialization of Ethernet communication, which is the most necessary work of communication. These works include initial configurations such as address of EMAC、EPHY module, size of receiving and transmitting buffer, data type of communication, and enable of interrupt mode. After initialization, the system comes into the state of waiting for receiving information.

When system receives request for receiving interruption, the program judges and handles the data transmitted by Ethernet. If received data are true, the system will begin to analyze them.

If received data is ARP frame, the system will judge whether the other side is searching for it. If the answer is right, ARP response sign will be established and interruption will be quitted, otherwise, this frame will be discarded and interruption will be quitted. If received data is IP frame, the system will continue to judge if it is ICMP datagram or UDP datagram.

The system only needs the simplest Ping function to test condition of the network. If received data is ICMP datagram, the system will judge if it is Ping requisition or not. If the answer is right, Ping response sign will be established. Otherwise, if the checkout is incorrect, this frame will be discarded, error sign will be established and interruption will be quitted.

If received data is UDP datagram, the system will check it. If checkout is wrong, error sign will be established and interruption will be quitted. If checkout is right, data will be read into EMS memory, corresponding signs will be set up for the master program to deal with, and then the system will be quitted.

Transmitted data of Ethernet communications in the system are carried out in master program. When corresponding signs are read by master program, the data will be dealt with according to them. If the sign is ARP responses sign, the corresponding data frames will be sent out by EMAC and EPHY; If the sign is Ping responses sign, a response frame will be sent out; If the sign is UDP datagram and it's function is to inform the system adjusting time, register settings of the timepiece chip PCF8563 are amended according to the time received; If the sign is UDP datagram and it's function is to inform the system sending electrical parameters of local user to upper computer of network, the system will transmit power consumption of this month and last month, peak value voltage and current.

The transmitted data need to be packaged in the format of all kinds of protocol frames. If needed, some padding bytes and Cyclic Redundancy Check (CRC) should be added.

4. DESIGN OF DATA PROCESSING MODULE

After the power meter terminal receives interruption of Ethernet communications, it will finish request of upper computer by inquiring all signs in master program, so that it could accomplish these functions such as response、timing, tariff readjust, transmission of electrical parameters (power consumption, peak value of voltage and current and so on) and power-off while communication network is connective.

Master program read the sign named EtherFlag which is produced by Ethernet communications circularly. If the sign is zero, it means that Ethernet request of upper computer have not been received, and then master program will jump over other signs produced by Ethernet communications. If the sign is '1', it means that request have been received, and then master program will judge whether it is Ping request whose sign named PingFlag, ARP request whose sign named ArpFlag, system timing request whose sign named TimeFlag, electrical parameters transmission request whose sign named ElecFlag or power on/off request whose sign named RelayFlag/Relay1Flag. If one of them is set to 1, it means that upper computer has produced request, and then master program will call corresponding functions to meet the request and clear the sign at the same time. The respondent data will be packaged into corresponding format of data frames and sent out by Ethernet, and then EtherFlag sign will be cleared to zero in order to prepare for accepting the next Ethernet communications request of upper computer.

After this Ethernet communications is achieved, master program will continue to check the other signs or to complete display function of the system.

The system also set up the infrared communication to avoid communication broken and guarantee normal communication. If the power meter unable to connect with Ethernet, worker could go to local to gather data and recover Ethernet communications network in time.

Fig.4 shows data processing of Ethernet communication in the master program.

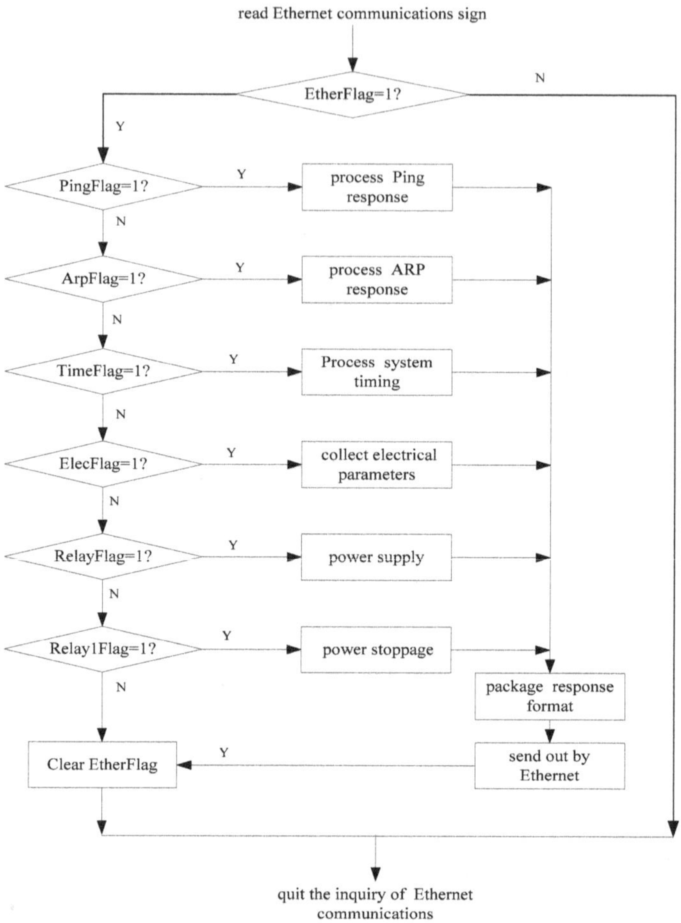

Fig.4: flow of data processing

5. CONCLUSIONS

The single-phase power meter based on Ethernet is a digital power meter with Ethernet interface and communication protocols. A network power meter based on 16 bits single-chip is designed in this paper, and how to design the hardware configuration, Ethernet protocol and data processing module of the meter have been illustrated in detail. The system has realized real-time settlement of electric charge and other network-controlled functions such as power stoppage and supply. It overcomes current problems about electric charge's disagreement due to offline running of power meter. The power meter is mainly used in the environment where Internet network has been realized, and it is unnecessary to establish wired communication network in addition.

If this kind of power meter is used in power grid, resources of network sharing can be achieved and cost can be saved greatly, so the system has good practical value.

ACKNOWLEDGEMENTS

Funding for this research was provided by Beijing Municipal Science & Technology Commission (P. R. China). The first and second author is grateful to thc Capital Normal University for providing support to this research.

REFERENCES

CHEN Chi-pei. The application and development of electric energy meter[J].Electric Power Automation Equipment,2002,22(4):55-57.

DU Qiong, ZHOU Yi-jie. Motorola SCM and its application in power meter[J]. Electric Power Automation Equipment. 2005,25(2):76-79.

freescale semiconductor. MC9S12NE64 Data Sheet. 2004,9.

GUO Shu-jun,WANG Yu-hua.Theory and Application of embedded processors[M].Beijing: Tsinghua University Press,2004

GUO Song-lin, LIN Hai-jun, ZHANG Li-yong. Classifing and Theory of Electrical Power measurement chip[J]. Electrical measurement and Instrumentation,2002,39(10):5-7,52.

HUAN Ling, GAO Zhong-wen. An electronic single-phase power meter based on the chip of CS5460[J]. Automation and Instrumentation, 2002,(5):38-40.

LI Wei-dong,FANG Zi-li, WANG Hai-jin et al. R & D of multiuser electric energy acquisition terminal. Electric Power Automation Equipment,2000,20(4):49-51.

LIAN Xiao-qin,BAI Li-ping,JIN Liang.Design of Three-phase Multi-functional Power Meter[J].Chinese Journal of Scientific Instrument,2005,26(8):287-289,295.

SHA Zhan-you. New dedicated digital instrumentation and application of principles [M].Beijing. China Machine Press,2006

WANG Yi-jun, REN Yan, WANG Ke-ying. Appliaction of power/energy chip CS5460 in the power system parameter measurement[J], 2001,28(11):51-54.

WANG Yue-zhi The electric energy measure(The second edition)[M]. China Electric Power Press 2007

ZHOU Yao-yi, BAO Bin-shou. Remote automation meter date reading system for low voltage electric power user. Electric Power Automation Equipment, 1999,19(2): 42-43.

ZHU Xiao-hong. The electric energy measure[M].China Electric Power Press 2007.

ZHU Yun, LE Xiu-fan, GAO Xuan. Application of high-precision single-phase energy chip in electric parameter measurement [J]. Electric Power Automation Equipment.2004,24(1):50-53.

AUTOMATIC ASSEMBLY OF COMBINED CHECKING FIXTURE FOR AUTO-BODY COMPONENTS BASED ON FIXTURE ELEMENTS LIBRARIES

Jingtao Jiang [1] , Rendong Sui [1] , Yan Shi [2] ,Furong Li [1] , Caiqi Hu [1,*]

[1] *College of Mechanical and Electrical Engineering, Qingdao Agricultrual University, Shandong Qingdao, 266109, Email: jjtao_2518@163.com*

[2] *Foodstuff Science and Engineering college, Qingdao Agricultural University, Qingdao, China, 266109;*

[*] *Corresponding author, Address: Corresponding author, Address: College of Engine & Electronic Engineering, Qingdao Agricultural University, 700 Changcheng Road, Qingdao, 266109, P. R. China, Tel:0532-86080844, Email: hucaiqi@163.com*

Abstract: In this paper 3-D models of combined fixture elements are designed, classified by their functions, and saved in computer as supporting elements library, jointing elements library, basic elements library, localization elements library, clamping elements library, and adjusting elements library etc. Then automatic assembly of 3-D combined checking fixture for auto-body part is presented based on modularization theory. And in virtual auto-body assembly space, Locating constraint mapping technique and assembly rule based reasoning technique are used to calculate the position of modular elements according to localization points and clamp points of auto-body part. Auto-body part model is transformed from itself coordinate system space to virtual assembly space by homogeneous transformation matrix. Automatic assembly of different functional fixture elements and auto-body part is implemented with API function based on the second development of UG. It is proven in practice that the method in this paper is feasible and high efficiency.

Keywords: combined checking fixture; automatic assembly; auto-body part

Please use the following format when citing this chapter:

Jiang, J., Sui, R., Shi, Y., Li, F. and Hu, C., 2009, in IFIP International Federation for Information Processing, Volume 295, *Computer and Computing Technologies in Agriculture II, Volume 3*, eds. D. Li, Z. Chunjiang, (Boston: Springer), pp. 1579–1584.

1. INTRODUCTION

During auto-body parts and components are assembled and wielded, the spatial shape of parts and components are complex, meanwhile the match precision of matching parts is required to be high, so the quality of parts and components need to be inspected comprehensively after the process of stamping and forming and before the process of welding and assembling, then rejects are avoided to flow into next production process. Conventional Equipments for checking quality of auto-body parts and components are special checking fixtures, the data obtained through special checking fixtures are relative values based on some special datum elements, and those data can not be compared with the data obtained through Coordinate Measuring Machine measuring parts of white auto-body in the auto-body coordinate system, so engineers can not judge that the checking error is because of parts manufacturing error or fixtures benchmark error. Thus to certain measuring points based on special checking fixtures only the qualitative appraisal can be given. Meanwhile as special-purpose checking fixtures, they can only be suitable for some specific components, when product remodeled, those special fixtures would be discarded.

In this paper according to the auto-body components' shape characters, clamp characters, localization characters and the checking technological requirement, the assembly space of automobile body coordinate system is simulated, solid models of combined fixture elements are created, based on the CAD automatic assembly technology, automatic assembly of combined fixtures are studied based on UG/API, the feasible scheme of virtual assembly can instruct operator to assemble fixture in practice. Those combined fixtures obtained through that method can overcome the above special-purpose fixture's shortcomings, and realize the special-purpose fixture's function completely and can meet requirement of product update.

2. ESTABLISHMENT OF MODULAR FIXTURE ELEMENTS LIBRARIES

Solid models libraries of combined fixture elements are established based on auto-body part characteristic parameters. According to their different functions those libraries can be divided into: ①Foundation elements library E_f; ②Supporting elements library E_s; ③Adjusting elements library E_a; ④Location elements library E_l; ⑤Connection elements library Ec; ⑥Clamping elements library E_c; ⑦Auxiliary elements library E_a. When elements solid models are created, two principles (one is easy to describe the geometry characters of solid models and one is easy to solve the assembly direction of parts) should be ensured. So the assembly direction of models and the direction of datum axle should be parallel. In order to realize automatic assembly of combined fixtures, besides massive 3D parameterizations models of elements, assembly rule databases adapting to product design should be constructed, such as installment way database, installment antecedent database, related connection database, disassembly antecedent database and so on.

3. CALCULATION OF PARTS POSITION IN THE ASSEMBLY SPACE OF AUTO-BODY COORDINATE SYSTEM

solid models of elements and parts are created in the original themselves coordinate system at first, in order to assemble each element solid model of part should be transformed to the assembly space of auto-body coordinate system through transferring matrix, and then 3D model of part transformed is saved as .prt files, meanwhile that model of part is loaded into the assemble space of UG, thus the virtual assembly space of auto-body coordinate system is constructed. In this virtual space, all elements assembly position of fixture is determined by auto-body part position.

Size parameters of fixture bottom plate are determined according to auto-body parts 2D size in the first location plane, then those size parameters are used as input, corresponding fixture elements libraries are accessed based on input parameters, and interrelated fixture elements are identified, transferred and loaded into the virtual assembly space of auto-body coordinate system from elements libraries. And all other fixture elements are called, installed, localized, and connected by above method. At last combined checking fixture of special function is formed. After the auto-body have produced for a period of time it maybe need to be remodeled, corresponding combined checking fixtures can not be used any more, but they can be disassembled into individual fixture elements, and which can be reassembled for new type of auto-body.

During design of combined checking fixtures modularization idea is followed, locating function or clamping function of one point is carried out by combining different types fixture elements, this combination is a function module of checking fixture. Position of every function module in 2D plane of bottom plate is determined by location restriction mapping, in other words the assembly position of each fixture function module in bottom plate is determined by mapping of part location point in space to fixture bottom plate. The mapping flow is showed in Fig1.

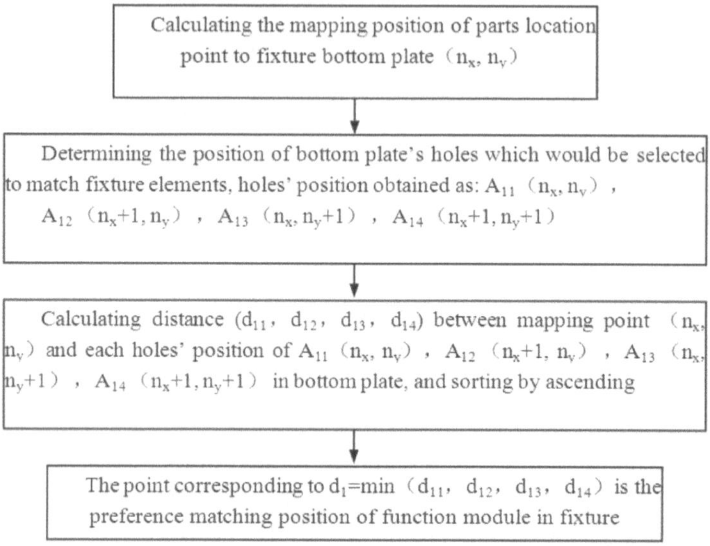

Fig1. Location restriction mapping flow

4. PARAMETERIZED DESIGN OF EACH FUNCTION MODULE IN COMBINED CHECKING FIXTURE

In this paper each locating or clamping module is approximately composed of 3 portions: supporting elements e_s, adjusting elements e_a and locating elements e_l respectively. After the install position of function module in fixture bottom plate is determined, parameters of each element of the function module should be determined, thus construction space of checking fixture can be expanded.

the following is the steps of determining element parameters for function module:

Step 1: position point is described as $P_l(x_l, y_l, z_l)$, location type is judged to be location plane or location hole;

Step 2: if the location type is location plane(the value range of master parameter S for location elements: 10mm$\leq$S$\leq$60mm);

Then the type of adjusting elements e_a is A, and the master parameter of A is H=50mm;

the parameter H_s (H_s=N$\times$d) of support element e_s is determined as following:

N= (H`-H) /d (d is the distance of two nearest holes in fixture bottom plate);

If $(10 \leqslant (H`-N \times d) \leqslant 60)$ then select $H_s = N \times d$

Else if $((H`-N \times d) \leqslant 10)$ then select $H_s = (N-1) \times d$;

the high parameter of location element e_l is $S = H`-H- H_s$;

Step 3: if the location type is location hole, the type of adjusting elements e_a is B, the master parameter of B is H=50mm; the determining method of parameters for supporting elements e_s and location elements e_l is same as step2.

5. REALIZATION OF AUTOMATIC ASSEMBLY FOR COMBINED CHECKING FIXTURE

Automatic assembly of combined checking fixture could be carried out by programming and calling function of UG/OPEN API based on the above algorithm. The transformation matrix $[T]_i$ of each matching element and part should be generated before solid models are assembled, and so it is easy to determine the position of each element and part in virtual assembly space. $[T]_i$ is generated by searching and identifying and saving the features of the assembly solids. Then the DLL application program module of UG is activated by calling API function and the position ($[T`_O]_i$) of element in the assembly model can be calculated.

The following are some functions called many times in the program mentioned above:

UF_ASSEM_mating_condition_t ftf;

Int UF_ASSEM_ask_component_data(part, name, refset, instance, orig, csys_matrix, trans);

Int UF_ASSEM_solve_mc(&ftf, &status, &dof, transform);

Ret=UF_ASSEM_apply_mc_data(&ftf, &struct_status, &status);

Among parameters in functions above, Ftf is a structure type data parameter of assembly relation, at first ftf is defined, the parameter of part passes path and file name of user-defined object, features of solid model are searched by UF_ASSEM_ask_component_data () and the transformation matrix is saved in the parameter of trans, which is 4×4 array, then match relationship is solved by UF_ASSEM_solve_mc () , if it successes for solving, status=1. UF_ASSEM_apply_mc_data ( ) apply restriction relation to assembly solid model by parameters passed, in order to obtain assembly position matrix of assembly elements in assembly model, the data is passed by the parameter of transform and automatic assembly is realized.

So the elements libraries are called, identified and read. Different components have different position features, if some special location elements do not exit in location elements library, they must be created by calling UG/API function and according to position features.

6. CONCLUSION

Along with speeding up of automobile update, in order to control the manufacture quality of auto-body parts, the quality of corresponding checking fixtures are request more highly. Combined checking fixture is one of the best choices for remodeling frequently of automobile. In this paper the method of 3-D combined checking fixture automatic assembly for auto-body part is presented based on modularization theory. And in virtual auto-body assembly space, Locating constraint mapping technique and assembly rule-based reasoning technique are used to calculate the position of modular elements according to location points and clamp points of auto-body part. The model of auto-body part is transformed from itself coordinate system space to virtual assembly space by homogeneous transformation matrix. Automated assembly of different functional fixture elements and auto-body part is implemented with API function based on the second development of UG. It is proven in practice that the method in this paper is feasible and high efficiency.

REFERENCES

Kim YeoKeun, Kim YongJu and Kim Yeongho. Genetic algorithms for assembly line balancing with various objectives[J], Computers & Industrial Engineering, 2004, 30(3):397-409

Michael Yu Wang. Optimum 3D Fixture Layout Design. Proceeding of the 3rd World Congress on Intelligent Control and Automation. P.R.China. 2000

Wang Y. A time finite element method for dynamic analysis of elastic mechanisms in link coordinate systems[J], Computers and Structures, 2001, 79(2):223-230

RESEARCHES OF OPTIMUM LEAF AREA INDEX DYNAMIC MODELS FOR RAPE(BRASSICA NAPUS L.)

Hongxin Cao[1,*], Chunlei Zhang[2], Guangming Li[2], Baojun Zhang[3], Suolao Zhao[3], Baoqing Wang[3], Zhiqing Jin[1], Dawei Zhu[1], Juanjuan Zhu[3], Xiufang Wei[1]

[1] *Institute of Agricultural Resources and Environment Research/Engineering Research Center for Digital Agriculture, Jiangsu Academy of Agricultural Sciences, Nanjing 210014, Jiangsu Province, P.R.China*

[2] *Institute of Oil Crops Research, Chinese Academy of Agricultural Sciences, Wuhan 430062, Hubei Province, P.R.China*

[3]*Northwest Sci-Tech University of Agriculture and Forestry, Yangling 712100, Shaanxi Province, P.R.China*

[*] *Corresponding author, Address: Institute of Agricultural Resources and Environment Research/Engineering Research Center for Digital Agriculture, Jiangsu Academy of Agricultural Sciences, Nanjing 210014, Jiangsu Province, P.R.China, Tel: +86-25-84390125, Fax: +86-25-84390248, Email: caohongxin@hotmail.com*

Abstract: The objectives of developing optimum leaf area index dynamic models for rape (OLAIDM) was to develop Rape Cultivation Simulation-Optimization-Decision Making System(Rape-CSODS) , to design its planting , to regulate and control its growth and development, and to fulfill its high yield, good quality, high benefits and standard production eventually. The OLAIDM were developed based on field experiments with 3 cultivars, 6 sowing dates, 2 types of plant pattern and 4 sites from 2002 to 2007 in middle and lower valley of Yangtze river in China and relative data from references of rape researches, employed ideas of R/WCSODS (Rice/Wheat Cultivation Simulation-Optimization-Decision Making System), and in the same time, the OLAIMR and its parameters also were assessed, calibrated and tested. The average absolute deviation(de), correlation coefficients(r) and the standard errors of their absolute deviation(Sde) of between the observed and simulated values for LAI of two cultivars in Wuhan and Nanjing were -0.03~0.1533, 0.9707~0.9997 and0.1332~0.4032, respectively. 1:1 line of them were in Fig. 1 to 4. Multi-factors such as the ramification types, cultivars, and light et al. were taken into account in this study, therefore, the OLAIDM with general

---

Please use the following format when citing this chapter:

Cao, H., Zhang, C., Li, G., Zhang, B., Zhao, S., Wang, B., Jin, Z., Zhu, D., Zhu, J. and Wei, X., 2009, in IFIP International Federation for Information Processing, Volume 295, *Computer and Computing Technologies in Agriculture II, Volume 3*, eds. D. Li, Z. Chunjiang, (Boston: Springer), pp. 1585–1594.

adaptability, clear yield aim, mechanism, and dynamic characteristic can simulate optimum LAI dynamic for rape under different sites, cultivars and ramification types, and yielding levels.

Key words: rape; optimum LAI dynamic; models

1. INTRODUCTION

Rape is one of four main oil crops in the world, whose plant area is about 14,100,000 ha in general. In the same time, it also is main oil crops in China, and its plant area is from 4,000,000 to 5,000,000 ha.

The crop simulation and the technology of 3S (GPS, GIS, and RS) are core and foundation of digital cultivation technique systems (Cao et al., 2005). The crop simulation technology mainly included the crop growth simulation, the optimum models in cultivation, and the simulation-optimization-decision making. The optimum models in rape cultivation mainly included models of the optimum season, population dynamic, seed rate, fertilizer rate, and soil water, etc., which ware foundation and goals of the simulation-optimization-decision making in rape, and had an important significant in promoting digital cultivation and realizing management with objective, dynamic, quantitative and optimum.

In the world, the researches of rape models can be divided into two stages obviously, i.e. experiential models as main (70-80's of 20 century) and mechanism models as main (since 90's of 20 century). In the later stages, rape growth and development and ecological system models such as EPR95, DAR95, LINTUL—BRASNAP, CERES, APSIM—Canola, and CECOL etc. were developed by Kiniry et al., Peterson et al., Habekotté, Gabrielle et al., Robertson et al., Husson et al.(Kiniry et al.,1995; Peterson et al.,1995; Habekotté, 1997; Gabrielle et al.,1998; Robertson et al.,1997; Husson et al.,1997), respectively, and they can simulate in real time. However, in China, the researches of rape simulation model were not more, and Liu Hong et al.(Liu et al., 2004), Liu Qiemei et al.(Liu et al.,2004) set up simulation model of rape phenophase etc.. Cao Hongxin et al.(Cao et al.,2006) carried out the studies of simulation models of phenophase, leaf age, dry matter, leaf area, and ramification numbers.

The objectives of this study were to establish the OLAIDM based on field experiments in Yangtz river middle and lower valley of China.

2. MATERIALS AND METHEODS

2.1 Research Methods

2.2 Referring to references

The aims were to apprehend and master rape growth and development rules and principles of growth regulation and control for rape whole, to obtain initial parameters on varieties, soil, climates etc., through referring to the references on papers and books of rape physiology, ecology, cultivation etc.

2.3 Development models

The optimization models of rape cultivation were established based on referring to references and apprehending rape growth and development rules fully, in terms of rape physiology and ecology principles and optimization principles of rape cultivation.

2.4 Experiments

In order to decide the parameters of the models and verify the models, the field experiments in multi-sites and multi-years were carried out adopting different rape cultivar types

Experiment 1: Two rape genotypes "No.9 Zhongshuang"(conventional) and "No.2 Zhongyouza"(hybrid) were grown in the field from 2002 to 2005 on Red loam soil near Wuhan(The total nitrogen was at 0.12%, organic matter at 2.07%, pH 7.79, and volume weight at 1.50g/cm^3 in pre-planting in soil), Hubei Province, respectively. The six planting dates included the first planting date on 15 September, the other planting date every 7 days each year. 12 treatments, 3 replications, and 36 plots with 8m long, 2.5m width were included in the experiments. The plots were arranged random. The varieties were breed by Institute of Oil Crops Research, China Academy of Agricultural Sciences (CAAS). Fertilizing and other field management in plots were the same. Nitrogen at 180 kg ha-1, phosphorus at 120 kg ha-1, potassium at 180 kg ha-1, and boron at 15 kg ha-1 were incorporated.

Experiment 2: The rape genotypes "No.16 Ningyou" was grown in the field from 2006 to 2007 on Yellow umber soil with higher fertility in pre-planting in soil in Nanjing, Jiangsu Province. 2 treatments, 3 replications, and 6 plots with 3m long, 2.5m width were included in the experiments. The plots were arranged random. The varieties were breed by Institute of Economic Crops Research, Jiangsu Academy of Agricultural Sciences. Fertilizing and other field management in plots were the same.

Phenophase, LAI, the total shoot numbers, biomass, leaf age, leaf photosynthesis, plant characters, and meteorological and soil data etc. were observed during rape growth or after harvest.

3. DEVELOPMENT OF MODELS

The Optimum LAI(OLAI) was basis of forming the optimum pod area index (OPAI), if the OLAI had been gotten, the OPAI also was gotten(Liu,1987). In terms of definition of the optimum LAI, if light intensity of down part leaves of population just equal to compensating light intensity for photosynthesis, and the most biomass had been gained, the LAI of the population was the OLAI (Wang,1991).

3.1 The optimum maxmum LAI (OMLAI) at early anthesis

The OMLAI in whole life occurred generally from early anthesis to middle anthesis(Ling,2000). The essence of deciding the OMLAI at this stage were to realize the optimum accumulation of photosynthesis in 40 days around early anthesis. Therefore, according to Monsi equation, the OLAI in this phenophase (L_{AO}) of rape can be expressed as follows:

$$L_{AO} = -（1/E_A）\ln(B \cdot I_{bA}/ I_{0A}) \qquad (1)$$

The parameter B can be gained using the definition of light compensating point for rape and relationships between photosynthesis rate and light intensity of single leaf, i.e., the relationships among the gross photosynthesis rate(D_{GP}), the respiration rate(D_R) and the net photosynthesis rate (D_{NP})of rape in all day can be expressed as follows. When equation (2) was satisfied, the LAI closed to the optimum.

$$D_{GP} = D_R \text{ 或 } D_{NP} = D_{GP} - D_{LR} = D_{DR} \qquad (2)$$

When light intensity(I) was enough small, photosynthesis rate for rape equals to b·I, the D_{NP} can be computed as follows(wang, 1991).

$$D_{NP} = D_L \cdot b \cdot I_C \cdot B \qquad (3)$$

$$D_{DR} = D_{AYR} + N_{IR} \qquad (4)$$

$$D_{AYR} = D_L \cdot b \cdot I_C \tag{5}$$

$$N_{IR} = (24 - D_L) \cdot m \cdot b \cdot I_C \tag{6}$$

$$m = Q_{10}^{-TR/20} \tag{7}$$

$$B = ((24 - D_L) \cdot m + D_L) / D_L \tag{8}$$

where I_{0A} is the horizontal nature light intensity above the population at this phenophase in the local, which equals to product of daily average solar radiation by coefficient (C). $(B \cdot I_{bA})$ is the light intensity of basic part of population at this phenophase. E_A is the extinction coefficient of population in this phenophase. I_{bA} is the compensating light intensity of basic part of population at this phenophase. D_{LR} and D_{DR} are the photo respiration rate and dark respiration rate, respectively. D_L is the daily sunlight length(h). b is the parameter of photosynthesis for rape. I_C is the instantaneous compensating light intensity for rape leaf. $b \cdot I_C$ is the hourly net photosynthesis rate at light compensating point. D_{AYR} and N_{IR} are the day and night respiration rate for rape. $(24 - D_L)$ is the dark length(h). m represents the ratio of day respiration rate to night respiration rate resulted by difference in temperature between day and night, and T_R represents daily difference between the maximum temperature and the minimum temperature. Q_{10} is temperature coefficient of respiration for rape, which equals to 2.0(Wang,1991). The formula (1) showed that the OLAI for rape at early anthesis can be decided by the following factors: (a) the total radiation in 40 days around early anthesis, (b) the local average day length around early anthesis, (c) the local average daily difference between the maximum temperature and the minimum temperature around early anthesis, (d) light compensation point of rape cultivars, and (e) the plant types of cultivars and extinction coefficient, etc.(responsible to yielding levels) .

3.2 The OLAI at enlongation

The principle of estimating the OLAI at this phenophase was that middle and down part of population had some extent light distribution when this phenophase was occurring the OLAI. Therefore, the OLAI at this phenophase, L_{EO}, can expressed as follows.

$$L_{EO} = -1/E_E \cdot \ln(B \cdot I_{bE} / I_{0E}) \tag{9}$$

where E_E is extinction coefficient of population at this phenophase, I_{bE} light compensation point of basic part of population at this phenophase and I_{0E} horizontal nature light intensity above the population at this phenophase in the local. The formula (9) showed that the OLAI for rape at enlongation can be decided by the following factors: (a) the plant types of cultivars and extinction coefficient, (b) the local total solar radiation at this phenophase, and (c) plant density, etc.

3.3 The OLAI at ten leaves(LTO) and reviving(LPO)

They can be computed as follows, respectively.

$$L_{TO} = B_{SO} \cdot L_{TS}/10000 \qquad (10)$$

$$L_{PO} = B_{SO} \cdot L_{PS}/10000 \qquad (11)$$

where B_{SO} is the optimum plant numbers($10^4 \cdot hm^{-2}$), and L_{TS} and L_{PS} represent the average leaf area per plant (main stem leaf) at ten leaves and reviving, respectively. The formula (10) and (11) showed that the optimum LAI for rape at the phenophase can be decided by the following factors: (a) the optimum basic seedling number, the stem and the first ramification number, and the available stem and ramification number, which dealt with yielding levels, and (b) the average leaf area per plant at ten leaves and reviving, which dealt with yielding levels and growth types of rape.

3.4 The OLAI at end anthesis (LDO)

It generally was around 25% that of early anthesis according to experiment and reference (Ling,2000). Therefore, it can be computed as follows:

$$L_{DO} = b_A\, L_{AO} \qquad (12)$$

where b_A represents percent of the OLAI at end anthesis accounting for that of at early anthesis in a yielding level.

4. PARAMETER ASSESSING AND CASES OF MODEL VERIFYING

4.1 Parameter Assessing

The conversion coefficient, $C=5007.6 \mu mol \bullet S^{-1} \bullet m^{-2}$(corresponding to $1 MJ \bullet m^{2} \bullet h^{-1}$).

The population extinction coefficient: Through mensuration of light distribution and reverse deducing, the extinction coefficient of population at the enlongation, E_E was 0.68, and the extinction coefficient of population at the early anthesis, E_A was 0.48 at 2400kg/hm^2 yield planted in Wuhan area; the extinction coefficient of population at the enlongation, E_E was 0.55, and the extinction coefficient of population at the early anthesis, E_A was 0.55 at 2550kg/hm^2 yield planted in Nanjing area.

The light compensate points of population basis part: ($B \cdot I_{CE}$) was 30.943μmol•S^{-1}•m^{-2} at the enlongation, and I_{CH} was 6.872μmol•S^{-1}•m^{-2} at 2400kg/hm^2 yield planted in Wuhan area; ($B \cdot I_{CE}$) was 30.943μmol•S^{-1}•m^{-2} at the enlongation, and I_{CH} was 5.065μmol•S^{-1}•m^{-2} at 2550kg/hm^2 yield planted in Nanjing area.

The percent of the OLAI after end anthesis account for that early anthesis: b_A was 25% around at 2400kg/hm^2 yield planted in Wuhan area; b_A was 20% around at 2550kg/hm^2 yield planted in Nanjing area.

4.2 The cases of model verification

The OLAI for "zhongshuang 9" planted in Wuhan and "Ningyou 16" in Nanjing area were simulated, which were compared with the observed values at the corresponding sites and yielding levers(Fig.1 and 2) using meteorological data(the monthly average temperature, the monthly average maximum temperature, the monthly average minimum temperature, the monthly average precipitation, the monthly average rain days, as well as latitude) from Wuhan and Nanjing area, and the corresponding cultivar parameters.

The average absolute deviation(d_e), correlation coefficients(r) and the standard errors of their absolute deviation(S_{de}) of the observed(Over-year 1.5, Enlongation 2.0, Early anthesis 4.75, End anthesis 1.18) and simulated values for the OLAI for "zhongshuang 9" in Wuhan were -0.03, 0.9707 and 0.4032, respectively, and r was significant at 0.05 level(Fig. 1).

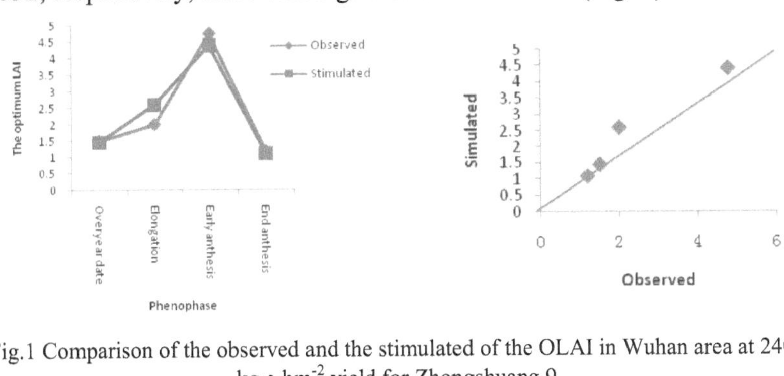

Fig.1 Comparison of the observed and the stimulated of the OLAI in Wuhan area at 2400 kg • hm^{-2} yield for Zhongshuang 9

The average absolute deviation(d_e), correlation coefficients(r) and the standard errors of their absolute deviation(S_{de}) of the observed(Over-year 2.04, Early anthesis 5.9, Yellow mature 1.02) and simulated values for the OLAI for "Ningyou 16" in Nanjing were 0.1533, 0.9997 and 0.1332, respectively, and r was significant at 0.05 level($r_{0.05(1)}$=0.997)(Fig.2).

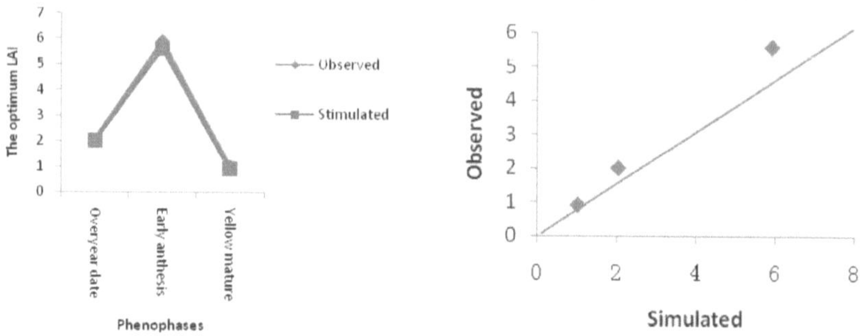

Fig.2 Comparison of the observed and the stimulated of the OLAI in Nanjing area at 2550 kg • hm^{-2} yield for Ningyou 16

5. CONCLUSIONS AND DISCUSSION

The conventional methods of deciding OLAI in rape cultivation were that the OLAI at various phenophase under different areas, cultivars and yielding levels were obtained by researches of the relationship between LAI and yielding formation, which had important roles in directing rape cultivation with high-yielding, good quality and high benefit. However, the method lacked of widely utility. This paper developed the OLAIDM according to the principle of high-yielding cultivation and optimization and the definition of OLAI, which taken in account multi-factors such as cultivars and light, etc., can estimate the OLAI dynamic with universal suitability, objective, mechanism and dynamic aimed at different areas, cultivars, and yielding levels.

The key factors to assure the OLAI at rape various phenophase under different areas, cultivars and yielding levels can be found out theoretically through analysis of models developed in this paper. e.g., the area with high solar radiation rate and large difference in temperature between day and night had larger OLAI dynamic, and in the same area, the OLAI dynamic could generally had close range when they had the same yielding levels, which had a slightly change in different cultivars. But the OLAI dynamic were clearly different in the same area with different yielding levels.

The OLAIDM had mechanism of physiology and biology and the principles of feedback controlling for yielding levels. But the verification data were less, which showed that the models developed in this paper needed more verification and continuous perfection.

ACKNOWLEDGEMENTS

The authors would like to express their appreciation to Prof. Dr. Cunkou Qi and deputy Prof. Xinjun Chen from Economic Crop Institute of Jiangsu Academy of Agricultural Sciences for helping of field experiments in this work, and for funding of national projects in the tenth five-year plan(2001BA507A-09-04), 948 projects of Agricultural Ministry of China(2004- Z 30), and Jiangsu Province projects(06-G-169)

REFERENCES

Cao H X, Zhang C L, Jin Z Q, Shi C L, Ge D K, et al. Discussing of Frame and Technological Systems for Digital Cultivation. J Farming and Cultivation, 2005, （3）:4-7

Cao H X, Zhang C L, Li G M, Zhang B J, Zhao S L, Wang B Q, Jin Z Q. Researches of simulation models of rape(Brassica napus L.) growth and development. Acta Agron Sin, 2006,32(10):1530-1536

Farré M J, Robertson G H, Walton S, Asseng. Simulating response of canola to sowing data in Western Australia. In: Hobart Proceedings of the 10th Australia Agron Conference, 2001. pp36—40

Gabrielle B, Denoroy P, Gosse G, Justes E, Andersen M N. Development and evaluation of a CERES-type model for winter oilseed rape. Field Crop Res, 1998, 57(1):95—111

Gao L Z, Jin Z Q, Huang Y, Li B B, Chen H. Rice Cultivational Simulation-optimization-decision Making System. Beijing: Chinese Agri Sci & Tech Press, 1992. pp21—40 (in Chinese)

Gao L Z, Jin Z Q, Zheng G Q, Feng LP, Zhang LZ, Shi CL, Ge D K. Wheat cultivational simulation-optimization-decision making system, J Jiangsu Agric, 2000, 16（2）: 65—72 (in Chinese with English abstract)

Guan C Y. Cultivational Techniques of Good Quality and High Yielding for Rape. Changsha: Hunan Sci & Tech Press, 1992. (in Chinese)

Habekott B.é. A model of the penological development of winter oilseed rape (Brassica napus L.). Field Crop Res, 1997, 54:137—153.

Hu L Y, Liu T M, Zheng X L, Cao C G, Cao W X, Yan M C. A mechanistic of phasic and phenological development in rape. II. Verification and evaluation of the model. Chin J of Oil Crop Sci , 2004, 26(2):51—54 (in Chinese with English abstract)

Husson F, Wallach D, Vandeputte A. Evaluation of CECOL, a model of winter rape (Brassica napus L.). Europe J of Agron, 1997, 18: 205-214

Kiniry J R, Major.D J, Izaurralde R C, Williams J R, Gassman P W, Morrison M, Bergentine R, Zentener R P. EPIC model parameters for cereal, oilseed, and forage crop in the north Great Plain region. Can J Plant Sci, 1983, 63:1063—1065

Liao G P, Guan C Y,Wu Q Y, Li A P, Chen S Y, Wang G H. Development of knowledge-base of the expert system for rapeseed production. Crop Res,2002, （3） :118—121 (in Chinese with English abstract)

Liao G P, Guan C Y. Study on characteristics of dry matter accumulation, distribution and transfer of winter rapeseed (Brassica napus). Acta Agron Sin, 2002,28(1):52—58 (in Chinese with English abstract).

Ling Q H. Crop population quality， 1st ed. Shanghai: Shanghai Sci and Tech Press, 2000. 217-286(in Chinese)

Liu H L. Rape Culture Science for Application. 1st ed. Shanghai: Shanghai Sci & Tech Press, 1987. pp220—221 (in Chinese)

Liu H, Jin Z-Q. The simulation models of rape development dynamic. J of Appl Meteo, 2003, (5):634—639 (in Chinese with English abstract)

Liu T-M, Hu L-Y, Zhao Z-H, Cao C-G, Cao W-X, Yan M-C. A echanistic of phasic and phenological development in rape. I. Description of the model. Chin J of Oil Crop Sci, 2004, 26(1): 27—30 (in Chinese with English abstract)

Petersen C T, Svendsen H, Hansen S, Jensen H E, Nieksen N E. Parameter assessment for simulation of biomass production and nitrogen uptake in winter rape. Europe J of Agron, 1995, (4):77—89

Wang Y R. Population Physiology of Crop High-yielding, 1st ed. Beijing: Sci. and Tech. literature Pub. House, China.1991.(in Chinese)

Zhu Y, Cao W X, Shen W X,Tian Y C. A dynamic knowledge model for fertilization management in rapeseed. Chin J Eco, 2005 ,24 (2) :209～213

STUDY ON AGING AND DAMAGE EVALUATION SYSTEM OF SEEPAGE CONTROL CANAL WITH CONCRETE

Chengbin Yuan*, Mingyao Zhou, Dingjiu Yang, Susheng Wang, Shuang Song

College of Hydraulic Science and Engineering, Yangzhou University, Yangzhou, Jiangsu Province, P. R. China 225009

* *Corresponding author, Address: College of Hydraulic Science and Engineering, Yangzhou University, Yangzhou, 225009, Jiangsu Province, P. R. China, Tel: +86-0514-87978650, Email: ycbyx999@126.com*

Abstract: Based on knowing clearly about the goal and principle of the system development, the paper formulated its basic methods and tools as well as the environment of development, studied the structure and the operation procedure of the system. The system also established the database and the evaluation system of the aging and damage situation of seepage control canal with concrete, and therefore improved the evaluation quality and efficiency. It also will play a positive role in maintaining and transforming canals, and raising the efficiency of water-distribution.

Keywords: seepage control canal with concrete; aging and damage; evaluation system

1. INTRODUCTION

Seepage control canals with concrete which were important parts of building in irrigation district, had played an outstanding role in saving water, seepage control, anti-scour, fixing canals and distributing water safely. However, after long using and various factors influencing, these seepage control canals with concrete would inevitablely be damaged and cannot work normally. Currently, the study of the evaluation system of the aging and damage situation mainly concentrated on those buildings like: pumping

Please use the following format when citing this chapter:

Yuan, C., Zhou, M., Yang, D., Wang, S. and Song, S., 2009, in IFIP International Federation for Information Processing, Volume 295, *Computer and Computing Technologies in Agriculture II, Volume 3*, eds. D. Li, Z. Chunjiang, (Boston: Springer), pp. 1595–1602.

stations, sluices, dams and aqueducts, but paid less attention to the aging and damage situation of seepage control canals with concrete which had been widely used (Jian Gong, 2001; Li Anguo, 2000; Xu Yunxiu et at., 2004; Zhang Haiyan et at., 2004). Therefore, it was necessary to evaluate the aging and damage situation of seepage control canals with concrete through reasonable evaluation systems, as well as to get scientific evaluation conclusions which would greatly support the further transformations.

2. THE MAJOR AIM AND PRINCIPLES OF THE SYSTEM DEVELOPMENT

2.1 The major aim of the system development

(1)By establishing the database of the aging and damage situation of seepage control canals with concrete, the basic information about the aging and damage would be inputted, searched, maintained and exported.

(2)The results of investigation about seepage control canal with concrete will be stored, and the aging and damage situations would be evaluated. Moreover all of these information will be searched and exported by Visual Foxpro software.

2.2 The principle of system development

(1)Practicability. After analyzing all demands in detail and grasping the practicable needs of clients, a complete and useful evaluation software and detailed and reliable materials will be submitted to them. What's more, it designed a rolling development that once some new problems happened in practical using; it would communicate with the clients and take their reasonable suggestions, to perfect the functions of the system.

(2)Reliability. It is essential for the system to work stably and accurately during evaluation. As a result, it has to consider fully about its reliability when designing the structures and functions. By using some mature theories and techniques to analyze the probable uncommon conditions, it can make the corresponding solutions, debug the system's trouble, and avoid the unknown disruptions.

(3)Advance. It should try best to keep the top level of softwares and hardwares, user interface preferences, system operation mechanisms, service processing methods and so on. Especially, when selecting the softwares and hardwares of system, it should not only think about the development limitation, the current management, the practical level and the economic factors, but also leave some space for the further developments.

3. THE BASIC METHODS OF SYSTEM DEVELOPMENT

The system tries to use the object-oriented technique to develop. The methods of the object-oriented system, which based on the objects, utilized the specific softwares to transform directly from object descriptions to software structures. This approach which shortened the developing cycle and resolved the complicated process of many transfer maps from analyses and designs to software module structures, had surpassed many other ways and been considered as one of the most potential and promising methods in reality.

The object-oriented system development had the following merits, compared with the traditional system development:

(1)All aging and damage present situations were sealed by object form, which would make the system organizations and structures good and clear, but not been saved in the complex relational forms;

(2)The system took the object as the foundation and eliminated the concept of stratification;

(3)The structure of object-oriented, such as "general-special" structure and "the whole-part" structure, enabled the system development to directly define and process complex aging and damage conditions;

(4)According to the thought of the object-oriented's late binding, users may define their needed data type and the space operating procedure based on the existing abstract data types and the spatial operations, which would enhance its exploitability and extensibility;

(5)Based on the icon object-oriented user interface, it was advantageous for the user to operate and use.

Therefore, the softwares which used the object-oriented method were easy to revise, compared with the traditional methods. The procedure was more reliable, maintainable, reusable, suitable and the intelligible. Certainly, the object-oriented system development also had some problems which still needed to be further studied, for example the hardware limitation which the big object operations still suffered, the object independent question, and so on.

4. SYSTEM DEVELOPMENT TOOL AND THE DEVELOPMENT ENVIRONMENTS

4.1 Development tool

Visual FoxPro, a kind of relational database management system, which was promoted by Microsoft Corporation, was selected for system development tool. It used the object-oriented programming thought and the visual operating procedure, which had many characteristics, such as the friendly user interfaces, the sophisticated technology of object-oriented programme, the fast establishment of application programme, the simple database operation, the timely updating of former editions, etc.

Visual FoxPro 7.0, as the database system development software, had many new characteristics compared with the former edition: true object-oriented programming ; providing many visual programming objects including toolbar, grid, three dimensional buttons and so on; giving the better guide tool; helping users easily to complete the work, etc.

4.2 Development environments

The system development environment was the Windows XP Chinese operating system, and the object-oriented data manipulation environment in VFP. It formed a completed set of database and analysis product, which can rapidly provide solutions for the various services, the data warehouse and the evaluation model in current as well as the next generation. Its unique characteristics, such as usability, reliability and advanced, ensure the excellent development environment for the system opening.

This system development used the pattern of Visual FoxPro, and the overall system used the medium and small scale database management system of Visual FoxPro for storing data. The platform of the system development was the programming language provided by VFP. Fig.1 was the integrative overall of the system.

Fig.1: Integrative overall of the system

5. SYSTEM COMPOSITION STRUCTURES AND OPERATION PROCESS

5.1 Request of system

According to the related requests and the investigations, the system should renew, maintain, research and evaluate the basic information of the seepage control canals with concrete, and enable the engineering data and the evaluation results to show in the form of images directly.

The system should have two functions, evaluating the aging and damage of seepage control canals with concrete and querying information. The system composition should be with organization and modulation, have the relative independence, and be convenient to use, maintain and renew, meanwhile different parts can relate mutually and combine organically.

5.2 System composition

The aging and damage evaluation system of seepage control canals with concrete had a man-machine contact interface, a database and a model repository. The database provided the basic data for the aging evaluation, including the basic document data about channels, the data from investigations, and the final outcome of computation. The model repository was the evaluation and the computation parts of the system, also was the core of the entire software, which had the rational distribution in structure. It would enable the database and the model repository, the evaluation and analysis of channel project aging and damage to separate and unify organically (Hong XiaoLin et al., 2007). Fig.2 showed the composition schema of the system.

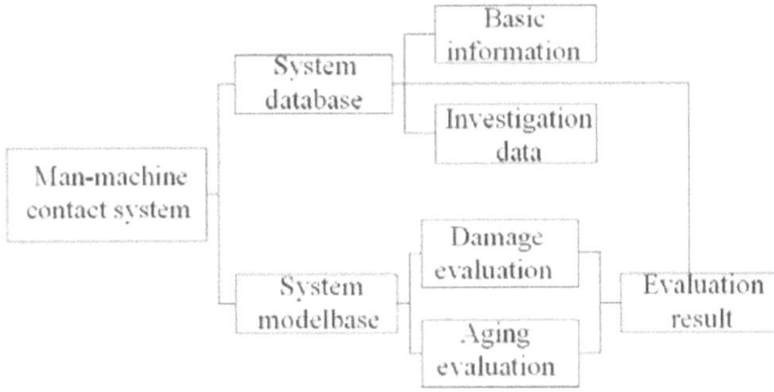

Fig.2: System composition schematic drawing

5.3 System operation procedure

After the objectively investigating of seepage control canals with concrete according to the requirements, the date from investigations will be input and store in the system. Then it used the established evaluation model of aging of concrete channel to evaluate (Hong Xiaolin et at., 2001; Liu Baiqing et at., 1998; Xu Yunxiu et at., 2004; Zhang Zhijun et at., 1998), drew the evaluation conclusions, and gave the processing advises. The system detailed operation procedure showed in Fig.3.

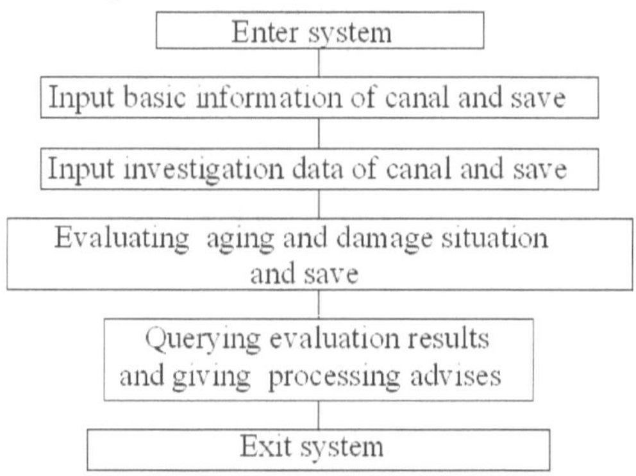

Fig.3: System operation flow chart

6. SYSTEM IMPLEMENTATIONS

6.1 Maintenance of the information database of the seepage control canals with concrete

Information database maintenance mainly referred to revise, add and delete the existing database.

The revised and the added information included the name of the concrete channel, the construction time, the capacity of designing capacity, the shape of cross section, the construction methods, the related sizes, etc. (SL18-2004). Information of doubtless mistake or impractical information should be deleted from the database permanently. In the database, the character information was mainly used to locate and distinguish the seepage control canals with concrete, while the digital information was mainly used to help to evaluate the aging and damage situation of the seepage control canals with concrete.

6.2 Implementation of the evaluating system of aging and damage of the seepage control canals with concrete

The evaluation may follow the steps below:

(1) Inputting the basic document of the channel. The basic document database of concrete channels was first opened, inputting the channel's name. The querying database was clicked to find whether this channel had been recorded. It would go to the next step if existing; otherwise it must store the data of this channel.

(2) Investigating the aging and damage situation of the seepage control canals with concrete. In accordance with the evaluation requirements, the conditions about surface destructions, cracks, dislocations, cappings, seams, expansion joints, etc., should be investigated, then some related examination tables be written down truthfully.

(3) Inputting above investigation data. In the examination tables, the correspondence aging and damage situation information should be inputted. After checking, clicking to preserve and calculate.

(4) Querying, browsing and printing the evaluation results. After evaluating, the results can be queried, searched and printed when needed.

7. CONCLUSIONS

This paper described the general goals and the principles of the system development, which took the software VFP, a kind of related database, as the system development tool, and used the object-oriented system method for development. The paper also established the information database and evaluation system of aging and damage, which contained the man-machine contact interface, the database and the model repository. Thus, it can evaluate the damage conditions of the concrete channels efficiently and accurately based on it. In addition, the rationality of the evaluation model would influence the final conclusions remarkably, so the model should be studied further.

ACKNOWLEDGEMENTS

This research was funded by National key Technology R & D Program of China (accession number 2006BAD11B03-02).

REFERENCES

Hong Xiaolin, Ke Minyong and Jin Chuyang. Safe Inspection and Appraisal Analysis of Sluice [M]. China Water Power Press, 2007, 172-225

Jian Gong. Cause of Damage of Old Seepage Control Canals with Concrete and Measures to Their Repair and Reconstruction[J].Technique of Seepage Control, 2001, 7(2): 1-7

Li Anguo. Summarization about Seepage Control of Concrete Channels [J]. Technique of Seepage Control, 2000, 6(1): 1-5

Liu Baiqing, Zhou Suzhen, Lei Shenglong and Du Jinping. The Study on Destruction Rating Index and Standard of Concrete Building in Irrigation District [J]. China Rural Water and Hydropower, 1998, (5): 16-18

National Center of Irrigation and Drainage Development.Standard for Engineering Technique of Seepage Prevention on Canal [SL18-2004]. Beijing: China Water Power Press, 2004

Xu Yunxiu, Fang Kunhe. The Inspection and Evaluation for Destruction of Building in Irrigation District [M]. China Water Power Press, 2004, 1-20, 380-430

Zhang Haiyan, Du Yingji, Zhang Jinkai. Some Problems on Seepage Control of Concrete Channels. Advances in Science and Technology of Water Resources [J], 2004, 24(6): 52-54

Zhang Zhijun, Wu Taipin and Shan Li. A Method of Rating Deterioration With Fuzzy Sets Theory for Sluice [J]. Journal of Nanjing Hydraulic Research Institute, 1998, (3): 249-254

THE MECHANISMS OF AGRICULTURAL RISKS MANAGEMENT AND ITS RESEARCHING FRAME

Jian Wang [1], Xuejun Zheng [2], Zhengjia Wang [3]

[1] *Corresponding author, Adddres: Dept. of Economics, Agricultural University of Hebei, Baoding City, Hebei, P.R. China, 071000. Tel(Fa):0312-7528631, Email: wangjian@hebau.edu.cn*

[2] *College of Foreign Language, Agricultural University of Hebei, Baoding City, Hebei, P.R. China, 071000.*

[3] *College of Computer Science, Agricultural University of Hebei, Baoding City, Hebei, P.R. China, 071000.*

Abstract: This paper started from the understanding of the concept of the agricultural risk. It takes asymmetric information and existed uncertainty as the risk occurring mechanisms. Therefore, through out some related theoretical analysis to setup a frame on the agricultural risk management study. It gives a rational knowledge on agricultural risk management.

Key words: Risk and uncertainty, Agricultural risk management, Risk decision-making

The development of agriculture is directly under the mutual influences of the natural risk and the market risk. When we enter into the new century, we can find our lives are facing more and more risks that are brought by the economic and advanced technologic development in uncertainty situations. It almost becomes the general knowledge that "The high technology with high risk, the high profit with high risk" in the real life. However, the natural disaster, as the risk background factor, is becoming more and more complex that has enhanced people's welfare loses. Therefore, the risk exists universally, if there is less or asymmetric information we will be difficult to treat uncertainty, which should be increasing in risks.

Please use the following format when citing this chapter:

Wang, J., Zheng, X. and Wang, Z., 2009, in IFIP International Federation for Information Processing, Volume 295, *Computer and Computing Technologies in Agriculture II, Volume 3*, eds. D. Li, Z. Chunjiang, (Boston: Springer), pp. 1603–1611.

1. THE CONCEPT OF RISK

Different person has different understanding about the risk. Even today still, we cannot find a unified conception about risk. Regarding the risk wording significance, the Weber dictionary defines the risk as "the faced damage or the possibility of losses." In the agricultural economic management, the risk refers to that the operators make a variety of judgments and probabilities which will happen in the future production according to their formers experiences and related knowledge·(H· Knight, 1921； Peter H. Calkins, 1983).

In the field of insurance, risk means "disaster or possible loss". It is usually refers to the insured object, the causes of the loss and the loss opportunity in the risk management research. Therefore, usually the risk defined by the relative changes between expected results and the actual result (Harold D. Skipper, Jr, 1998).

The statistician Wald (1950) proposed that, the risk is the sum of anticipated experimental cost and the anticipated losses because of when using a special policy-making function to make the wrong final decision-making (Ding Yiming et al., 2001).

In addition, some scholars thought (Chen Chijun et al.,1999; Zhang Jikang， 1991), the risk refers to the uncertainty of loss happening. It is a function between the probability of the disadvantageous event or the loss occurs and the related consequence, which we can use a mathematical formula to express: R=f (P， C); R stands for the risk; P stands for the probability and C stands for the loss of the consequence when the disadvantageous event occurs. Therefore, we think the risk is the synthesized negative deviations of the consequence from the anticipated objective, because of people decisions for future behavior and the reality with the uncertainties. The risk refers to the negative deviations of the possible consequence from the main body. The negative deviations are several of paradigms and different importance. In the complex agricultural production and management activities, we need to perform the concrete analysis and the synthesis judgment according to the special details. Although the risk emphasizes the negative deviation, the positive deviation also exists in the reality. The positive deviations are what people want so they belongs to the risk revenue, which should be paid attention to in the risk decision-making in order to arouse people's willing to take risks, and get the risk revenue. The theory of risk utility is considered from this viewpoint.

2. RISKS AND UNCERTAINTIES

There are different opinions about the conceptions of the risk, which are always used to describe the results of something is uncertain. Uncertainty roots in the inability of forecasting the future, so, under the complete uncertain circumstances, we do not know the probability that some things will occur. As a result, when producing a new agricultural product there are much more uncertainties, because when producing a new product the producer usually lack of production management experiences than his producing an old one. Especially when coupled with the variation of unknown natural environmental factors, these uncertainties will get more.

However, the approaches of decision-making to fight with risks and uncertainties can be divided into two kinds. One is risk decision and the other is decision with uncertainty. There are many research achievements in risk decision. While, the research on decision with uncertainty is just on a starting position because without information the uncertainty is hard to measure and forecast. Along with the advancements in scientific experiments, as well as the human production and rich life experiences, research on uncertainties can transform into research on risks. The typical example is about the climate change, now the people may through collection of rainfall amount and illumination materials and so on to understand the meteorological target in probability distribution situation, which turn the problem of uncertainty to the problem of risk. So far, people have little command on the occasional things such as earthquakes. Actually, uncertainties are everywhere in our daily lives because of a limitation of people's cognition and reaction ability. Both kinds of decision-making are involved in modern risk management, which carries on the classification, the analysis model, to study on causes and effects, and to change the features of uncertainty, in order to choose the suitable countermeasures. There is no clear boundary between the risk and the uncertainty in some works. Such as "the uncertainty comes from the risk's existence." "The existence of risk leads to the existence of uncertainty, and causes the decision-making process more complex" (Harold D. Skipper, Jr, 1998) (Harold D. et al., 1998). In Fact, the risk is another side of the uncertainty, namely probability, not the certainty. Davidson P. (1988) applied the stochastic process of the probability theory to carry on the explanation, which may explain this question better. He said, "If the situation has been faced is universality that could be considered as risks. If not, it is uncertainty." (Davidson P et al., 1998)

3. EXPECTED VALUE AND RISK UTILITY

The risk can be the probability's expectancy of the wealth losses. It relates to the value of a certain asset and its losses probability, and it is the algebra sum of the product of the two. If the value of the asset is too lower, it will not bring big losses or harm even if the loss probability is relative higher. Hence, if the value of the asset is very higher, when the risk comes true, (although the probability is small) the losses must be huge. Moreover, the risk has the relativity. First, it relates to the owner's asset size and its lost value. Second, it relates to the asset owner's attitude to losses. This is because every people's attitude or stand for various losses and the harms is different, and their abilities of taking risks are also varied that usually called "preference". This has caused the study on the risk utility and the risk value theories. Moreover, the bold speculators are tending to gain the opportunity benefit from the other side of the risk. This is, at the same time, the attitudes of the people who buy lotteries.

According to people's attitudes to the risks, who can be divided into different types of risk-aversion to affect people's choice when they facing the risks. In real lives, some people are willing to take risks by themselves; others are willing to pass the risk to the society. We are used to divide them into three types: risk preferential, risk neutral and risk averted people. For a society, what types of risks need to avoid or keep lookout by the social members and through market to transform are general decision-making choices. Answers to these questions relate to the policy, law, and the depth of the market, leading to different decision-making models and approaches to different risks. Therefore, in personal significant, risk utility and risk expected value may be not equivalent, as the former relies on individual preference choice and the later is general a social value judgment (such as market price). However, takes a society as a whole, risk utility and risk expected value should be equivalent, different societies may have their own choice.

4. INFORMATION, RISKS AND UNCERTAINTIES

Information has built a bridge from uncertainty to probability. C. E. Shannon (1948) as the founder of information theory, who defined the information entropy refers to the measurement of information, choice and uncertainty. Knowing to increase the probability means to reduce the uncertainty. Information entropy definite as the single value reducing function of probability, the bigger the entropy is and the smaller the probability is, and which means to need more information to deal with the uncertainty.

Therefore, to reduce uncertainty means to increase probability, which makes the entropy decrease and we obtain more information. Qualified (accurate, on time and effective) information thus turns leads to increase expected value and reduces the risk, which becomes a basic condition to avoid and prevent from, and a key of the decision-making on risk. We have to know, Information is the necessary and sufficient condition for the recognition of risks, but it is only the necessary condition for risks aversion, because we also need efficient investment on material, energy, workforce modern tech, and build an efficient economic management system to transfer, disperse or reduce the risks. That is, if there is not enough information, means incomplete information or asymmetric information, the Game Theory should be used to prevent the uncertainty and the emergence, at same time a prevented infrastructure platform as the sufficient condition.

In this study, the applications of information theory and information management approaches have been paid more consideration. Especially the development of modern information technology and the study on the precise agriculture and modern management have shown that people's methods to reduce the uncertainties strengthened. We must notice the relevant scientific definition on the uncertainty, the probability, and the information entropy. If investment theory in economics has proved that square deviation is equivalent with the expected utility value, and from the information theory we can also find that information entropy is positive related with its distributed deviation, and negative related with its distributed probability. That shown us a new equivalent relations, "the mean value ~ the square deviation ~ the information entropy" (Wang Jian, 2002). Therefore, these can help us to find the necessary operation mechanisms of the agricultural risk management. At the same time, to build a relation among risks, uncertainties and information with both on theoretical and cognitive in order to probe a new direction on methodology and models, and put up new ideas on the decision-making of the risks.

Humans are seeking the approaches to reduce the uncertainties, which not only have promoted the formation of all kind of communities or social organizations but also have developed the information technology. The famous economist Kenneth J. Arrow (1988) (Kenneth J. Arrow, 1984) who has won the Nobel Prize with the research on the information economics said that, "once the existence of the uncertainty can be analyzed in the paradigm; the economic function of the information will become very important. People can spend labor and money in changing the uncertainty in the economy and other aspects of lives and this kind of change is just the gaining of the information." Therefore, gaining the information means reducing uncertainties, and thus uncertainties means the economic cost. Hence, reducing uncertainties become a kind of revenue, which, to some extent, can reduce the risks or provide the prerequisites for preventing the inevitable natural disasters.

In a system view, when we try to find the relations among the concepts of risks, the uncertainties and the information, we can turn to the explanation of modern system science. According to the theory of dissipative structure (I. Prigogine 1967, 1969), the exchange entropy can form a negative entropy flow in an open system, which is far away from the balancing state and exchanges materials and energies with the surrounding of the system. The increase of negative entropy flow can produce a new structure of the system, making the system better organized. The better organizing means more information can be got, and the more systematic uncertainties or the higher entropy will be reduced. Seeing from the system theoretical point of view, system disorder related to the chaos of the system. The more randomness means the more entropy, when the system reaches balance entropy is the biggest. In this significant, entropy has two explanations, one is uncertainty related to information, another is disorder related to system structure, and thus we can understand those conceptions in equivalent: loss information~ uncertainty ~ complex ~ disorder ~ chaos.

We can see that system complex is relatively, it is linked the people intelligent system and surrounding physics system, where the entropy can be as a unified measurement. Einstein once said that, "The theory of entropy is the first law of the entire natural sciences." In a thermodynamics physical system, the rule that the natural process displays is the second law of thermodynamics. That is, entropy always in increasing (R. J. Clausius, 1822-1888) (Raymond A.et al., 1982). Energies have been dispersed in the natural evolution process from the order to disorder. Therefore, it is only can depend upon people's wisdom to forecast and deal with the flood of natural disasters. Through fixing and using the energy, artificially join the energy to the negative entropy flow in a closed natural process, making it open and forming a dissipative structure means that the people can change the nature and themselves in order to get rid of the natural shackles. The human beings, who survive in the natural environment, have been reproducing, surviving and developing up today by fighting with the negative factors of the environment. Karl Marx once explicitly pointed out that, "can only consist in socialized man, the associated producers, rationally regulating their interchange with Nature, bringing it under their common control, instead of being ruled by it as by the blind forces of Nature;... ..." ("Marx, Engels Complete works" 25th volume, page 926-927, People's Publishing House, Beijing, in 1979). Facing bigger uncertainties and universal risks in today's world, the humanity must adopt combined actions and resist the common disasters. The United Nations "the 21st Century Agendas" was a common manifesto in a sustainable development. Therefore, global risks management information systems need to be reconstructed.

5. AGRICULTURAL RISKS MANAGEMENT COUNTERMEASURES AND STUDY FRAME

Risk management related a serial processes, which include risk recognition and measurement, risk evaluation and forecasting, risk programs and decision, risk control and effective appreciation. The countermeasures to deal with agricultural risks we can take a wide view about risk management nowadays. There are several major countermeasures, generally to be shown in following six aspects.

(1) The risk dissolve, that means turn peril into safety, cancellation risk, if want to exterminate risk still have to remove risk concealed suffer from, make it become naught, this is a kind of exhaustive risk treatment. In practice, the first, when we find the risk occurring can escape from it and avoid the risk directly harm us. The second, we usually use futures market for some agricultural products to evade risk.

(2) The risk dispersion, that means to transfer the risk and carry out its total load in which people can bear of the scope. This is a kind of measure of coordination. General risk market has this kind of function, such as insurance and futures markets are usually used for risk dispersion, which can transfer the risks to society.

(3) The risk control, that means limit the risk in the certain scope, or open up a path to the risk, limits its diffusion proliferation and changing its moving direction. It is also a rules and regulations risk, but do not let it injure people's basic benefits. A precondition is that recognizes the risks, then measures and analyzes them to make risk programming, to realize optimal risk decision.

(4) The risk reduction, that means to make the strong risk become small risk, but not only is to carry on dispersion to the risk, the key wants to contract the whole scale of risk existence, namely reduce have already existed of uncertainty factor. A basic work that needed is risk information statistic and makes risk forecasting, finally using the methods of statistical analysis, programming on risk, stochastic dominance analysis etc help us to realize the decision of the risk reduction.

(5) The risk prevention, that is subjective to strengthen the defense and away from risk and uncertainty, taken measures to avert the damages. Based on the universal theory of stochastic process, risk occurring is a random in subjective to prevent it is very important. However, uncertainty outside of a universal stochastic process used to be an inevitable event. In fact, for a society generally requires great invest in workforce, material and finance to build preventing mechanisms. To setup the prevention system is a giant system engineering, it ask for mastering the game theory under the every

cases of uncertainty, to complete market system, to enforce social infrastructures and build modern information system.

(6) The risk utilization, including turn bad luck into good of plan, exactly recognized that risk is either an opportunity or a threat, thus tend benefit to avoid harms can be named as risk utilization. Pursuing a risk speculation income in bond and stock markets are many peoples preference. In modern society, become natural disasters into benefits is also an idea, such as become water inundation into water conservation that means to create a reasonable irrigation system and to go in for large-scale hydraulic construction.

About agricultural risk management, the paper tries to give a study frame in following figure 1. It illustrates that under the guiding theories and methodology of modern science, to combine market economics with complex system approaches, we are able to study agricultural risk management. A real process of the researching will be risk recognition, risk measuring, risk managing decision making and the implementation of the every plans. Only if the risk is inescapable, we have to apply the strategies of transferring, dispersing or reducing the risks.

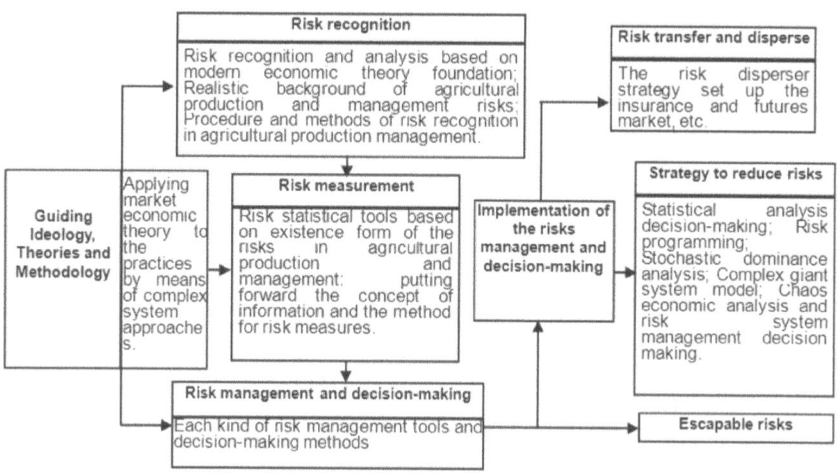

Figure 1. A Frame of Agricultural Production and Operational Risk Management

In conclusion, the agricultural risk management is complex system engineering, except to spend a great quantity of social resources, which it also needs a complete market system. Applying the modern information technology to collect and process data need to setup special agricultural risk management information system. The risk decision-making methodology has formed a frontier knowledge set that is to treat uncertainty and complex things. Especially to find the important disaster occurring rules, it generally asks through out the analysis of nonlinearity, chaos and uncertainty.

Through the analysis of above, we find by means of information and its technological application, establishes the related social economic organization that forms a dissipative structure, which both are the guaranties of the agricultural risk management mechanisms operation.

ACKNOWLEDGEMENTS

This Study Based on the Soft Scientific Program in Hebei Province ①(No: 01457224D, 2001-2003): "Study on Agricultural Production and Operation Risk Management Theory and Methods", AUH Program (2003-2005); ②(No. 07457201D-6) "Study on the strategy of Counties Regional Economic Development and Cities and Towns Construction in Hebei Province.

REFERENCES

Chen Chijun, Ye Huanting, "The Risk Aversion In Enterprise Production and Development", Chinese Management Science [J]. Sept. 1999, P1-10.

Davidson P, Financial Markets, Investment and Employment [M], London: Macmillan, 1998, P73-92.

Ding Yiming, Fang Fukang, "Analysis on the Risk Conceptions", System Engineering [J]. (Vol.16, No.5), Oct. 2001, P402-406.

Harold D. Skipper, Jr. "International Risk and Insurance: An Environmental Managerial Approach", [M]. Copyright © 1998 by McGraw_Hill. Inc.

Kenneth J. Arrow. The Economics of Information [M], Basil Blackwell Limited, 1984.

Peter H. Calkins, Agricultural Enterprise Management [M]. New York: MacGraw-Hill, 1983.

Raymond A. Serway, Physics: For Scientists and Engineers [M], Printed in the United States of America, 1982. P380-398.

Wang Jian. "Economic Theoretical Foundation of Agricultural Production and Business Risk Analysis". China Business Review [J], Feb. 2002, Vol.2, No.5, P54-65.

Zhang Jikang, "Business Risk Management" [M]. Lixin Accounting Publishing House, Jan. 1991, P74~84.

FOREST PEST OCCURRENCE PREDICTION USING CA-MARKOV MODEL

Fangyi Xie[1], Xiaoli Zhang[1,*], Xiaoyan Chen[1]
[1] *Key Laboratory for Silviculture and Conservation, Ministry of Education, Beijing Forestry University, 100083, P. R. China*
[*] *Corresponding author, Address: Department of cartography and GIS, Beijing Forestry University, Beijing 100083.P.R.China,Tel:+86-010-62337830, Email: zhang-xl@263.net*

Abstract: Since the spatial pattern of forest pest occurrence is determined by biological characteristics and habitat conditions, this paper introduced construction of a cellular automaton model combined with Markov model to predicate the forest pest occurrence. Rules of the model includes the cell states rules, neighborhood rules and transition rules which are defined according to the factors from stand conditions, stand structures, climate and the influence of the factors on the state conversion. Coding for the model is also part of the implementations of the model. The participants were designed including attributes and operations of participants expressed with a UML diagram. Finally, the scale issues on forest pest occurrence prediction, of which the core are the prediction of element size and time interval, are partly discussed in this paper.

Key words: e-government, knowledge management, frameworks, e-governance

1. INTRODUNCTION

In cellular automata models, information from different sources can be translated into a set of transition rules which define the behavior of the system (Sergio A. Cannas et al., 1999). Thus, it was widely used in prediction and simulation of landscape dynamics combined with geographic models to increase their own realism, such as forest-landscape evolution(Cannas et al., 1999; Sprott et al., 2002), land use and cover changes (Li Xia et al., 2005), urban development simulation and analyzed

Please use the following format when citing this chapter:

Xie, F., Zhang, X. and Chen, X., 2009, in IFIP International Federation for Information Processing, Volume 295, *Computer and Computing Technologies in Agriculture II, Volume 3*, eds. D. Li, Z. Chunjiang, (Boston: Springer), pp. 1613 1618.

(Ke Chang Qing et al., 2006). While Markov models can show us a set of conditional probability states of different kinds of conversion with a transition matrix which is calculated by analyzing a pair of objects of different kinds. The CA-Markov model is a kind of enhanced cellular automata model, of which the knighthood rules and transition rules are combined with Markov models. Thus, the influence of subjective factors when modeling can be decreased to some extent.

From the perspective of insect population ecology, as an important property of population, spatial pattern is determined by biological characteristics and habitat conditions (Xu Lumei et al., 2005). Forest disease and pests prediction is not only quantitative calculation, but also spatial features analysis, such as location of the disease, spreading ranges and distribution patterns. So, with the research on the formation and change of forest pest spatial distribution, it's practicable to predict the forest pest occurrences from the perspective of spatial patterns. Obviously, the work would not be carried on without spatial models incorporated in. while it's just what CA-Markov can help us: transition factors from cellular automata and transition probability from Markov predicate the forest pest occurrence (Hou Xiyong et al., 2004).

In this paper, a CA-Markov model based on GIS was constructed to analyze the prediction of forest pest occurrence for GIS system. Here is something worth to noting that this model is not for some specific forest pest, such as Fall Webworm (Hyphantria cunea), but for major forest diseases and insect pests.

2. MODEL: RASTER-BASED CA-MARCOV

There are three key points when modeling with Raster-based CA-Markov: cell states rules, neighborhood rules and transition rules. Cell states represent the occurrence of forest pest, and transition rules express the likelihood of the cell state.

Five states should be included in this model: unchanged, healthy, slight, moderate and serious. If what the original cell state represent is not kind of vegetation, such as urban land, or kind of plants the pest never harm, the cell state should be unchanged, and should not change after the transition. While, other four states' determination depends on the probability calculated by transition rules integrated with Markov model's analyzing.

Generally, the occurrence of forest pest depends on these factors: date, climatic, stand conditions, stand structures; among those, climatic includes climatic type, temperature, humidity, rainfall; stand conditions includes slope, aspect, slope position and altitude; stand structures includes tree species (which and pure or mixed), stand layer structure, stand age, biodiversity and distance from the forest edge. Those factors are different from different kinds of forest pests, depended on the different biological characteristics (Zhang Xingyao et al., 2003).

The neighborhood rules should be made combined with those factors. Besides the four-neighbor method or eight-neighbor method, some distance, such as distance from the nearest road or from the nearest river, could also be used in the transition probability calculations, because some pest spread by transportation or water flow.

The core of CA models is how to define transition rules that control the conversion of states in the simulation. There are lots of methods to define the rules, such as the Multi-Criteria Evaluation (MCE), Logistic Regression, Principal Component Analysis (PCA) and Artificial Neural Network (ANN). Here we define the transition probability as

$$P = \sum_{i=1}^{m} \omega_i K_i + e \tag{1}$$

Where P is the primary transition probability from the original state to the transform state, m is the number of the factors the model use, e is the error term. For each factor, i is the index, K_i is the value and ω_i is the weight.

This primary probability is calculated with the transition rules in cellular automata, which should be combined with the random probability from the matrix outputted by the Markov model's analyzing a pair of forest pest image. The two kinds of probability can be calculated as

$$PT = f(P, M) \tag{2}$$

Where PT is the final conversion probability, while M is the random probability from Markov model's analyzing. The function f should be defined to decrease the influence of subjective factors.

In the constrained simulation, only the cells of which the conversion probability meets certain criteria can change those states.

3. CODING FOR THE MODEL

According to the designing of the model, we draw a UML diagram to descript the class structure. In Fig.1, it is shown that the member variables and member functions of classes and the relationship among the classes, which include Markov, StandCondition, StandStructure, DayClimate, MonthClimate, YearClimate, CAObject, CAMask, CAGroup and CommonFunction.

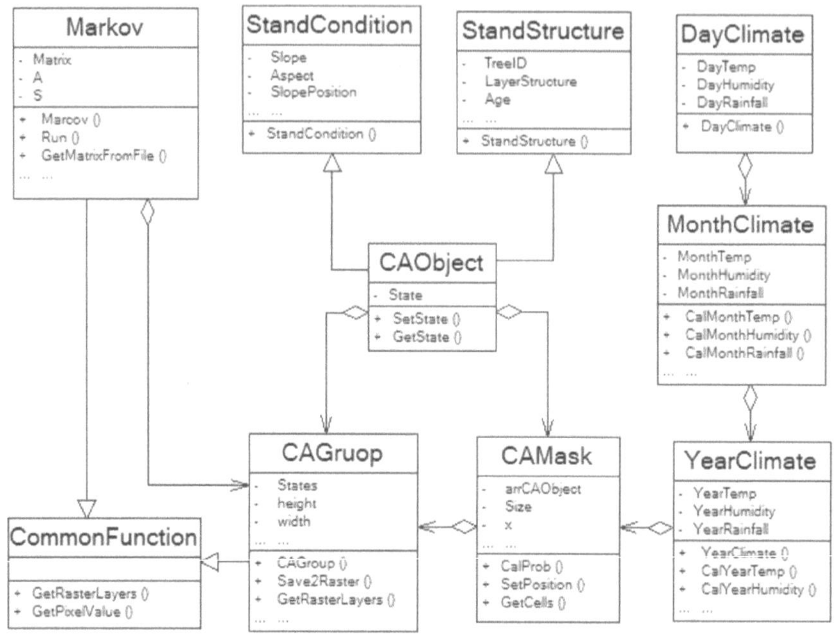

Figure 1. The UML diagram of CA-Markov model

Here are the Participants as follow:

Markov, which defines the matrix according to CAGroup, provides functions to calculate the transition probability matrix, and it also reads and saves the matrix from and to an existing file so that it's not necessary to calculate the transition probability matrix repeatability.

StandCondition defines the factors of stand conditions: slope, aspects, slope position, altitude.

StandStructure defines the factors of stand structures: tree species, stand layer structures, stand age, stand biodiversity and kinds of distance.

DayClimat, defines the factors of climate on the scale of the day: temp, humidity, rainfall.

MonthClimate provides functions to calculate the average, maybe maximum or minimum, of instances of DayClimate.

YearClimate provides functions like MonthClimate on the scale of the year.

CAObject which is the cell or element of the study scale, with the same size of the pixel size of raster layer data defines the state of the cell.

CAMask, which defines a gliding window to collect data around the core cell with the traversal of a pixel block, provides functions to calculate the primary probability according to the neighborhood rules.

CAGroup, which defines variables of CAObject instances, with the same size of a giving pixel block, provides functions to calculate the final probability, with the primary probability and the random probability, and determinate the convert states.

CommonFunction provides the common functions which could be used frequently, such as reading the raster layer data and getting value from a certain pixel from a raster layer data.

In all the calculating work, values of cells should be got from the raster data layers. It's not suggested to develop the analysis application only using advanced language. Using ArcGIS Engine, we can embed GIS functions into our existing applications for pre-processing vector and grid data. Especially, spatial extension, one of the extensions available for ArcGIS Engine Runtime, provides a set of functions to create, query, and analyze cell-based raster data. Factors such as slope and aspect can be easily calculated with the help of the extension.

The raster data should be calculated by pixel blocks, which conducive to the reading and writing data, using the interfaces such as IPixelBlock, IRasterCursor, IRasterEdit and so on. So that large raster can be divided into smaller pieces by creating a pixel block and reading portions of the raster sequentially.

4. DISCUSSIONS

Here is an aspect we could not neglect in the study of geological spatial analysis: scale. It's available to comprehend scale as the particularities of spatiotemporal characteristics of study objects (Zhang Tong et al., 2004). Generally, it can be repressed by spatial resolution and temporal sequence.

The scale issues, in which there are two core points, the prediction of element size and time interval, should be paid attention to in the study of forest pest occurrence prediction. It's more easily on the determination of time interval than that of element size, as most forest pest disaster occurs periodically because of the biological characteristics of the pest.

Obviously, the determination and acquisition of factors of neighborhood rules depends on the scale, and would change as the change of scale, so that it can adapt to the particularities. Factors act differently on the difference scale. The local scale incarnates the diversity of soil, plants and microclimate, which are influenced by biological effects, such as competitions. The regional scale incarnates the diversity of upper structures, which are influenced by geomorphic structures and soil combinations. And the global scale incarnates the diversity of macroclimate, which is influenced by terrestrial environment and atmosphere changes (Yu Xingxiao et al., 2005).

On the same scale, all the data should be raster data with the same cell size, not only the maps, but also the remote sensing data. All the data should be georeferenced to the same coordinate system. The choosing of the date should be paid attention to. The date of the data should be closed to that of the occurrence of forest pest, and the spatial resolution should not be lower than that of the cell size set in the model.

ACKNOWLEDEGMENTS

Funding for this research was provided by National Science and Technology Support Project in the 11the Five Years (Projcct No. 2006BAD23B04-4), and NCET (Programme for New Century Excellent Talents in University, Project No. NCET-06-0122).

REFERENCES

Cannas S A, Pɑ́ez S A, Diana E M. Modeling plant spread in forest ecology using cellular automata. Computer Physics Communications, 1999, 121~122

Hou Xiyong, Change Bin, Yu Xinfang. Land use change in Hexi corridor based on CA-Markov methods. Transactions of the CSAE, 2004, 20(5):286~291

Ke Chang Qing, Ouyang Xiao Ying. The Advances in Modeling Urban Spatial Change Based on Cellular Automata. Journal of Nanjing University (Natural Sciences), 2006:42(1): 103~110(In Chinese)

Li Xia, Anthony Gar-On Yeh.. Celluar automata for simulating complex land use systems using neural networks. Geographical Research, 2005, 24(1): 19~27(in Chinese)

Liu Xiaoping, LI Xia. Retrieving CA Nonlinear Transition Rule from High-dimensional Feature Space.Acta Geographica Sinica

Sergio A. Cannas, Sergio A. Pɑ́ez, Diana E. Marco. Modeling plant spread in forest ecology using cellular automata. Computer Physics Communications, 1999, 121(122):131~135

Sprott J C, Bolliger J, David J M. Self-organized criticality in forest-landscape evolution. Physics Letters A, 2002, 297: 267~271

Xu Lumei, Cheng Xinyue. Insert Population Ecology: Foundation and Front, 2005

Yu Xingxiao, Niu Shuzhi, Guan Wenlin, Feng Zhongke. Landscape Ecology, 2006

Zhang Tong, Cai Yongli. Scale in Ecological Research. Ecologic Science, 2004, 23(2):175~178

Zhang Xingyao, Luo Youqing. Major Forest Diseases and Insect Pests in China, 2003

Zhou Guo-Fa, XU Ru-Mei, LI Tian-Sheng. Spatio-temporal explicit interactive model for forecasting the pine caterpillar population dynamics. Acta Ecological Sinica, 1999, 19(4):464~467(in Chinese)

ECONOMIC METHODS OF GINGER PROTEASE'S EXTRACTION AND PURIFICATION

Yuanyuan Qiao [1], Junfeng Tong [2], Siqing Wei [3], Xinyong Du [1,*], Xiaozhen Tang [1]

[1] *Collage of Food Science and Engineering, Shan dong Agricultural University, Tai'an, China, 271018*

[2] *Collage of Food, Zhe Jiang University, Hangzhou,China,310000*

[3] *Shandong External Economy Trade Technical School, Taian,China, 271000*

[*] *Corresponding author, Address: Collage of Food Science and Engineering, Shan dong Agricultural University, Tai'an, China, 271018,Tel: 0538-6164139, Fax: 0538-8242850,, Email: txz@sdau.edu.cn*

Abstract: This article reports the ginger protease extraction and purification methods from fresh ginger rhizome. As to ginger protease extraction, we adapt the steps of organic solvent dissolving, ammonium sulfate depositing and freeze-drying, and this method can attain crude enzyme powder 0.6% weight of fresh ginger rhizome. The purification part in this study includes two steps: cellulose ion exchange (DEAE-52) and SP-Sephadex 50 chromatography, which can purify crude ginger protease through ion and molecular weight differences respectively.

Keywords: ginger protease, extraction, purification, DEAE-52, SP-Sephedax 50

1. INTRODUCTION

There are several kinds of vegetal protease used in food industry nowadays, such as papain, bromelin etc, and all of them have similar proteolytic activity. It is proved that, compared with papain, ginger (Zingiber officinale roscoe) protease has 10-folds activity in meat tenderization. And furthermore, it can significantly improve not only the flavor but also the

Please use the following format when citing this chapter:

Qiao, Y., Tong, J., Wei, S., Du, X. and Tang, X., 2009, in IFIP International Federation for Information Processing, Volume 295, *Computer and Computing Technologies in Agriculture II, Volume 3*, eds. D. Li, Z. Chunjiang, (Boston: Springer), pp. 1619–1628.

quality of meat by increasing its nutritious value (Naveena, B.M., 2001; Yan Zhou et. al.,1996; Zhang Pingping et. al.,2001; Jin-neng Lin,1991), which makes it a valuable meat tenderization. It also can significantly improve the clarifying degree of wine and beer (Tang Xiaozhen et.al. 2002). Ginger protease can improve the quality of food processing on the ground of not reducing nutritious value. With the proteolytic of ginger protease, those proteins which are difficult to digest, such as casein, bean separate protein (Yuan-lin Song et.al., 2001; Ming-fa Zhang et.al., 1991), can be decomposed into small molecules as peptides or amid acids. Some of those small molecule materials have bio-activity, and belong to functional food. Also, ginger protease can be used in the proteolysis of animal and vegetal proteins, or as dough adjuster in baking, or milk solidification factor, or addictive in cosmetic. Our Chinese traditional food, Ginger Freezing Milk, is solidified by ginger juice, and the milk solidification factor here is proved to be ginger protease (Zhang Pingping et.al., 1999). Ginger protease consists of some complicated components, which have similar structures and properties, and make it difficult to attain highly purified ginger protease. Hai-yan Cai et, al have researched the distributing of ginger protease in fresh ginger and its activity changing tendency during storing, and the method of crude ginger protease extracting by using AOT-isooctane and CTAB heptane/octanol anti-micelle extracting (Cai Haiyan et, al., 2004), Jing-quan Dai have purified ginger protease by cellulose ion exchange chromatography (DEAE-52), by the following steps: dissolve ginger juice in acetone, then dry it into powder. Before purification, first dissolve the powder again by buffer, then deposit it by various concentrations of ammonium sulfate, then dialysis, and at last purified those materials by chromatography column (Dai Jingquan et al., 2003). Researchers throughout the world had tried many methods to purify this enzyme (Hompson ,E.H et al.,1973; Zhao Zhiping et.al.,1995; Ohtsuki,Kozo; et al.,1995; Kyung H. Choi et.al.,2000), and had got the crystal structure and X-ray structure as well (Kyung H. Choi et.al.,2000). But the process of extraction and purification are too complicated to be extended the scale from laboratory to industry. Therefore, finding some economic methods to market it needs further researches, especially on how to extract and purify it by a simple and effective method which can be used in industry has been highlighted.

2. MATERIAL AND METHODS

2.1 Material and samples preparation

Ginger rhizome is high quality yellow ginger of LaiWu, Shangdong.

Tyrosine, supplied by Bio-Chemical College of Science Institution of China; Casein and Bovine serum albumin, supplied by Shanghai Chemical Reagents Station (with N: 14.5—15.5%); Commassie Blue G-250, repacked import by Shanghai Chemical Reagents Station. All the above are G. R. grade.

The chromatography column was obtained from Shanghai Ya-Long Bio-chemical Apparatus Company (16mm*20cm).

2.2 Methods of assays

2.2.1 Ginger protease activity determination (Qian Yuanze 1991)

The definition of protease activity is under a certain temperature (40℃) the mass of protease which is used to produce 1μg tyrosine by hydrolyze casein. The steps of determining the activity of ginger protease is as following:

Mix the solutions of enzyme (1.0ml) and casein (1.0ml 0.5%), and incubated for 10min at 40 ℃. The assay is stopped by adding 2.0ml trichloroacetic acid (0.4mol/l), and then the solution is filtered after 10min standing. The color reaction is to mix 1.0ml filtrate with 5.0ml sodium carbonate (0.4mol/l) and 1.0ml Flionhydroxybenzene, incubated for 20min at 40℃. The comparision reaction was same as the assay above except for the adding sequence was reverse between casein and trichloroacetic acid. And then using the solution of the blank group to adjust the extinction, and measure the absorbance at 280nm.

The formula of enzyme activity calculation is:

U=A*K*N

U: the unit of enzyme activity

A: the value of absorbance

K: calculated by dividing 100 with the corresponding value of absorbance of 280nm at the 100μg/ml tyrosine standard curve.

N: the multiple of enzyme dilute

2.2.2 Extraction of crude ginger protease

Ginger rhizomes are cleaned and extracted juice, and then the juice is deposit in low temperature and centrifuged at 4800 r/min for 20min to get rid of starch. Then it is suction filtrated to remove some large contaminants. The filtrate is mixed with acetone which is pre-cooled in refrigerator (1:1), and then the sediment is collected after centrifuged at 30000r/min for 10min. The sediment is dissolved in phosphoric acid buffer and centrifugated again. This time, the supernatant is collected and mixed with ammonium sulfate to saturation of 30%. The solution is place in low temperature for 24h, and then centrifuged and filtrated. The supernatant is collected again and mixed with ammonium sulfate to saturation of 60%, and again place in low temperature for 24h, and centrifugated to get the sediment. Then it is dissolved by phosphoric acid buffer (pH=6), and dialyzed in dialysis bag. At last, crude ginger protease powder is attained after freeze-drying.

2.2.3 DEAE-52 Sepharose assay

According to the result of primary assays, the crude ginger protease was dissolved in Tris-HCl buffer (pH 7.5) with the concentration of 30 mg/ml, and then was absorbed on a column of DEAE-52, which was first washed by Tris-HCl buffer (pH 7.5) with sodium chloride (0.1mol/l) in order to get rid of the un-combined protein. Then, a gradient concentration eluting buffer began with an initial speed of 1ml/min with 0.2~0.5mol/l sodium chloride and Tris-HCl buffer (pH 7.5). Then the optimization wash concentration and volume were determined by measuring the OD value of protein in washing, and the optimization wash speed also could be determined by comparing the protein curve in the washing under the optimization condition that mentioned above. The efficiency of purification of DEAE-52 was measured by testing the enzyme activity in the washing. The cellulose ion exchange chromatography purify ginger protease through the difference of ion between compounds, and because of the hydrophilic nature of cellulose, it has little tendency to denature proteins which is very important for enzymes.

2.2.4 SP-Sephadex 50 assay

The enzyme purified through DEAE-52 is taken into bag to be dialyzed in order to get rid of the ions and condense it. Then diluted it into20mg/ml by phosphate buffer (pH6.0, with 0.1mol/l sodium chloride), and then refined it by SP-Sephadex 50. During this process, the volume of the dissolve was 4ml, and the wash speed 0.4ml/min, collect speed 7ml per tube which meant 18 ml/min. The Sephadex gel chromatography column purifies ginger protease from the difference of molecular weight of components, which can get rid of contaminants with different molecular weight.

3. RESULTS AND DISCUSSION

3.1 Extraction of crude ginger protease

In this study, the method used to extract crude ginger protease can get higher enzyme activity compared with former ones (extracted by acetone powder), the higher rate is 43% according the experimental data. After freezing and drying, the crude enzyme powder weight is 0.6% of fresh ginger rhizome, with the activity of 1.2×105U/g.

3.2 Cellulose ion exchange chromatography(DEAE-52 Sepharose) (Jian Lu et,al.,2005)

3.2.1 Solution concentration and volume determination

It is shown in fig1 that when the solution concentration is lower than 30mg/ml, nearly all of the protein is absorbed when the volume is 0.1ml or 0.3ml. While the volume becomes 0.5ml, protein appears in the supernatant as the concentration higher than 30mg/ml, and has a positive correlation trend of the volume and concentration of the solution. While the volume becomes 0.7ml, the same trend appears at the concentration of 20mg/ml. According to the assay, two groups of optimization of concentration and volume are 30mg/ml, 0.5ml and 20mg/ml, 0.7ml respectively. Concerned the length of the column, the former group is determined as the best one. (According the column contains 40ml DEAE-52 homogenized solution, a result of volume to it can be calculate as 4ml)

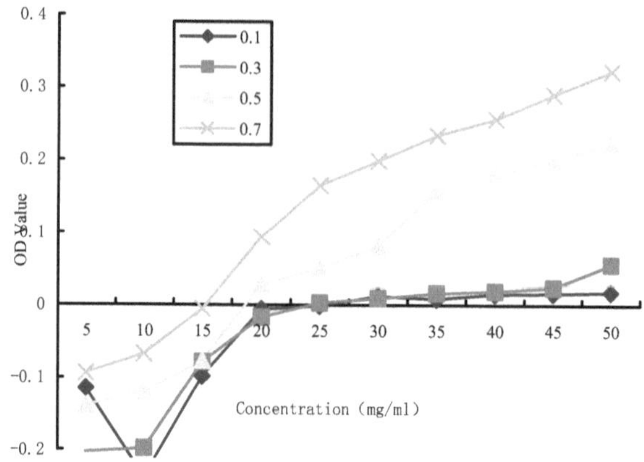

Fig 1. Effects of solution concentration and volume on the protein concentration in supernatant

3.2.2 Wash speed determination

There is hardly any protein available in the washing while the wash speed is 0.5ml/min as shown in fig2. It is probably that the wash speed is too slow to wash down the enzyme which is firmly absorbed by DEAE, and an incompletely wash process is made. However, when the wash speed rises to 1.5ml/min, the protein concentration in the washing becomes very high, but with no obvious peak. The reason for this may lies in the too fast speed, which makes the enzyme cannot exchange completely with exchanger in the column and cannot be absorbed, therefore, the enzyme is washed out into the foregoing tubes. With the wash speed of 1.0ml/min, three distinct protein peaks are attained, with OD values of 0.495, 0.549, and 0.771 respectively, and this speed is determined as the best wash speed.

3.2.3 Purifying effects of DEAE

There are three protein peaks in fig2, but compared by fig3, no enzyme activity is found in the peaks except of the second one. After measuring the purified enzyme solution in the second peak, the OD value is 0.955, and compared with the crude lyophilization ginger protease, the effects of DEAE-52 purification are listed below in table1.

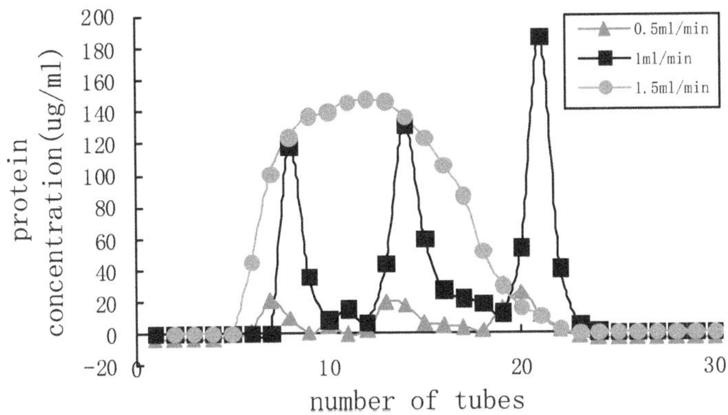

Fig 2. Effects of wash affected by speed

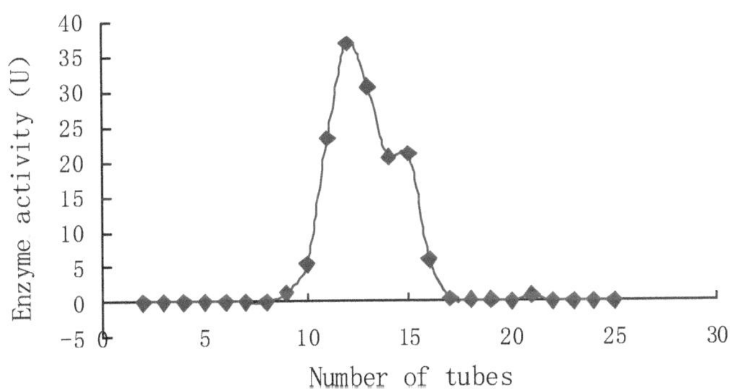

Fig 3. The curve of enzyme activity

Table 1. Effects of DEAE-52 purification

Items	Protein μg/mL	Enzyme activity U	Specific activity U/μgPr	Multiple Purification
Lyophilization protease	58.84	21.19	0. 36	——
DEAE-52	45.00	26.42	0. 59	1.64

3.3 SP- Sephadex (Ming-fa Zhang et.al., 1991)

Collecting the washing of the second peak, refine it by SP-Sephadex, and protein concentration and enzyme activity are shown in fig4 and fig5 respectively.

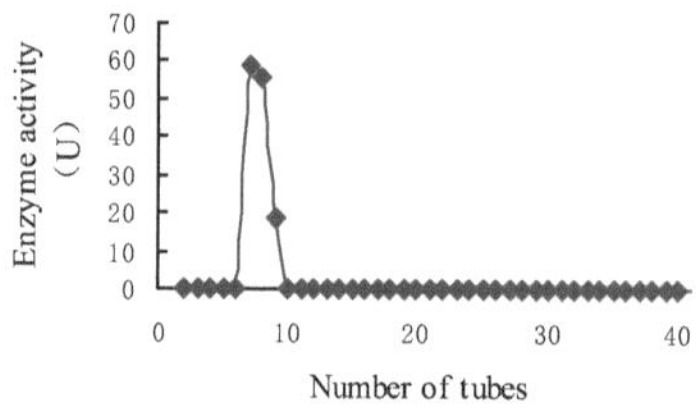

Fig4 Washing Curve of protein concentration by SP-Sephadex

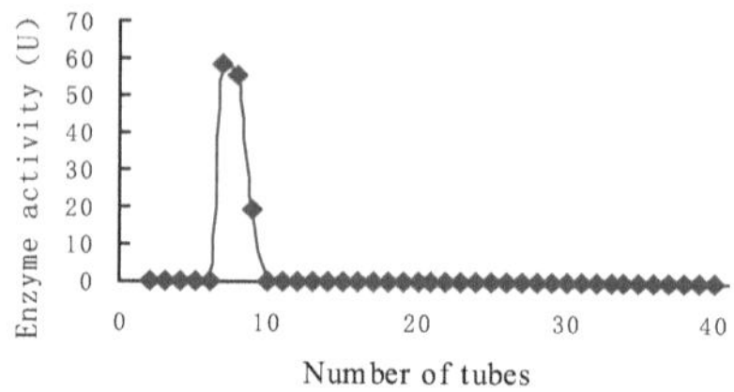

Fig5 Enzyme proteolytic activity curve

It is obvious that there is a solo protein as well as enzyme peak, and the tube with high protein concentration also with high enzyme activity and purifying effect. The effects of SP-Sephadex are recorded in table2.

Table 2. Effects of SP-Sephadex purification

	Protein μg/mL	Enzyme activity U	Sepecific activity U/μgPr	Multiple of purification
Crude enzyme	58.84	21.19	0. 36	1
DEAE-52	45.00	26.42	0. 59	1.64
SP-Sephadex50	43.00	31.91	0.74	2.06

It is shown in table2 that refined by DEAE and SP-Sephadex, the enzyme solution with high activity and low protein concentration, and the multiple is 2.06.

4. CONCLUSION

4.1 Extraction of ginger protease

There are many methods of enzyme extraction, such as organic solvent deposit, super filtrate concentration, dialysis, adsorption and freeze-drying. Those mentioned here are methods of crude enzyme extraction, used to concentrate and enrich crude proteases. Refining methods used to purifying proteases include polypropylene amide gel, dextran gel, DEAE-cellulose, DEAE-gelose, and electric chromatography etc. The methods such as spray-drying and freeze-drying adapt to large extent extraction, for example industrial extraction. The method used in this study can not only attain a comparable higher yield but also transfer to industrial procession.

Enzyme additives used in food industry are always crude enzymes (National Standard of PR, 1986), because the high rate of purification is just a luxury in common application.

4.2 Purification of ginger protease

One enzyme washing peaks after DEAE, and the multiple of purification is 1.64. This result is different from the report of Jing-quan Dai (Zhang Pingping et.al.,1999)that two enzyme washing peaks was attained. The reason may lies behind the different working condition and outer environment. According to the experimental result of this study, the purification of SP-Sephedax only attains a non-significant effect. If it were the industrial extraction, this step can be omitted.

4.3 Components differences of Crude enzyme and Purified enzyme

The term of crude enzyme is defined by OD value in 280nm, which indicate the material is protein. After the testing of enzyme activity, the crude enzyme is attained which includes a lot of impurities such as contaminated proteins, organic and inorganic salt etc. After purification of DEAE and SP-Sephadex, most of the contaminants are got rid, which is exhibited by the higher enzyme activity.

ACKNOWLEDGEMENTS

Funding for this research is provided by Nature Science Project Funds of Shandong Province (Q2004B01); and Scientific Technology Research Projects of Shandong Province (2007GG30009002).

REFERENCES

Cai Haiyan et.al.Primary research of preparation and anti-micelle extracting about ginger protease (GP-Ⅱ) [J]. Natural Product Research and Development， 2004， 16(6)： 543~547,551.3.

Dai Jingquan Huang Xuesng.Purification of Ginger Protease [J]. Food Science，2003,24 （2）:73~79.

Hompson ,E.H., Wolf,L.D. Ginger Rhizome :A new source of proteolytic enzyme. [J）.Food Scince.1973,38:652~655

Jian Lu et,al Technology and Application of protein purification, Chemical Industry Publication, 2005: 53~119

Jin-neng Lin, Natural edible flavor's producing and application, [M] Beijing: Light industry Publication, 1991:435

Kyung H. Choi and Richard A. Laursen. Amino-acid sequence and glycan structures of cysteine proteases with proline specificity from ginger rhizome Zingiber officinale . Eur. J. Biochem,2000, （267）： 1516~1526.

Ming-fa Zhang et.al Dry ginger's pharmacological research of Warming catamenia and acesodyne [J] Chinese Medical Research, 1991, (1) :41~43

National Standard of PR. China [S] GB2786—86.

Naveena，B.M.，& Mendiratta，S.K.Tenderizationof spent hen meat using ginger extract. British Poultry Science，2001,42，344-350.

Ohtsuki,Kozo; et al. Purification of ginger protease by DEAE-sepharose and isoelectric focusing. Biophys. Acta. 1995, 1423(2):181~184

Qian Yuanze translate.Method for determination of enzyme activity[M].Beijing: China Light Industry Press, 1991.

Tang Xiaozhen et.al. Effect of zingibain on beer clarification. Science and Technology of Food Industry,2002.23(8):12~14

Yan Zhou et. al Research of the ginger protease's usage in several foods digestion. Food Science, 1996,17 (7):6~9

Ya-shan Han: Instruction of Food Chemist Experiments, [M] Beijing: Beijing Agricultural University Publication, 1992.42~50.

Yuan-lin Song et.al Series books of Export vegetables: High yield planting and processing technology of Ginger, Yam and dasheen, Shandong Scientific Technology Publication, 2001:55

Zhang Pingping et.al. Studies on the Milk Clotting of Ginger Juice[J]. Chian Dairy Industry，1999,27 (5) :17~ 19

Zhang Pingping，Liu Xian hua.Study on Ginger Proteases Hydrolysis of Isolated Soy Protein[J].Journal of Tianjin Agricultural College，2003,10(3):26~29

Zhao Zhiping et.al. Studies on Purification and Glycosidic Linkage of Zingibain. Journal of ANHUI University (Natural Sciences), 1995，2：93~97

DETERMINATION OF IRON IN MILK POWDER BY MICROWAVE DIGESTION AND FLAME ATOMIC ABSORPTION SPECTROPHOTOMETRY

Guangyuan Zhao[1,*], Bo Li[2]

[1] *School of Food and Biological Engineering, Zhengzhou University of Light Industry, Dongfeng Road, Zhengzhou, 450002, China.*

[2] *College of Food Engineering, Henan Institute of Science and Technology, Hualan Road, Xinxiang, 453003, China*

[*] *Corresponding author, Address: School of Food and Biological Engineering, Zhengzhou University of Light Industry, Dongfeng Road, Zhengzhou, 450002, P.R.China. Tel: +86-37-63627116, Fax: +86-371-63556627, Email: guangyuan-zhao@163.com*

Abstract: To investigate the conditions of microwave digestion for determining Iron in milk powder by flame atomic absorption spectrophotometry(FAAS), the content of iron in milk powder was determined by flame atomic absorption spectrophotometry after the samples were microwavely digested under different conditions. The optimum parameters for microwave digestion were obtained by the orthogonal test at last. The best optimum parameters for microwave digestion was that, the volume of digestion solution was 8mL, the reagent proportion for HNO_3 and H_2O_2 was 4:1, the digestion time was 8min, the digestion pressure was 2.6 Mpa and the digestion power was 1000 W. The content of Iron in assayed milk powder was 0.0560mg/g. Microwave digestion was a time-saving and practical pretreatment of samples.

Key words: microwave digestion; flame atomic absorption spectrophotometry(FAAS); milk powder

1. INTRODUCTION

Due to its superior sensoury and nutritional properties, the milk powder has a considerable market potential and people pay more and more attention

---

Please use the following format when citing this chapter:

Zhao, G. and Li, B., 2009, in IFIP International Federation for Information Processing, Volume 295, *Computer and Computing Technologies in Agriculture II, Volume 3*, eds. D. Li, Z. Chunjiang, (Boston: Springer), pp. 1629–1636.

to the nutritional properties of it such as the content of trace element. So, the determination of trace element in milk powder is more important. Flame atomic absorption spectrophotometry (FAAS) is a preferred method for the derermination of trace element (Wang et al., 2004). The advantages of FAAS are that it is highly sensitive and quick. The traditional dry ashing and acid dissolution standard methord for the samples of FAAS is time-consuming and carries risk of loss and contaminative. Microwave digestion with acid mixtures in closed teflon vessels avods these problems and ensures that the total mental content can be analysed(Li., 2007). Chen Wenting et al determined the metal elements in milk powder byultrasonic extraction(Chen et al., 2007). Zhang Limin and Zhou Xuemei et al determined the metal elements in milk powder by flame atomic absorption spectrophotometry after the samples were microwavely digested(Zhang., 2007; Zhou et al., 2005), but they did not investigate the optimum parameters for microwave digestion.

Iron is a essential trace element for people because it could inactivate the enzyme in people body. The objective of the present work was to investigate the optimum working condition for microwave digestion used in assay the amount of Iron in milk powder by FAAS, and this results also could be used in other trace element in milk powder by FAAS.

2. EXPERIMENTAL

2.1 Reagents and standard solutions

HNO_3 used was purified by quartz sub-boling distillation. Stock standard of Fe was prepared by dissolving 9.99%pure metal in HNO_3. Working standard was prepared daily by subsequent dilution.

2.2 Apparatus

AA140/240 atomic absorption spectrophotometry（VARIAN,USA）; MDS-200 microwave digestion equipment（CEM,USA）

2.3 Decomposition of sample by conventional heating

About 2.5 g of samples were taken in PFA containers and 15 mL HNO_3 was added. Once the initial reaction ceased, the screw cap was tightened and the sample was digested on the hot plate for 1 h at 80-90℃. The sample solution was evaporated near to dryness and made up to 50 mL using ultrapure water. The time required was about 3 h.

2.4 Decomposition of sample by microwave digestion

About 2.5 g of samples were taken in PFA containers and digestion solution (contained HNO_3 and H_2O_2) was added. The sample was pre-digested on the hot plate at 100℃ some time until white smoke rised. When the temperature of containers decreased, they were put into the digestion vessel and the microwave digestion was producted according to the prosedures designed before. After microwave digestion the sample solution was made up to 50 mL using ultrapure water.

2.5 Standard curve and samples measurements

Working standard was prepared daily by diluting stock standard of Fe to some solution levels of 0.020mg/L， 0.040mg/L, 0.060mg/L, 0.080mg/Land 0.100 mg/L.

The absorption value of working standard solution were obtained according to the conditions shown in table 1 by FAAS and the standard curve equation (drawn automaticly by the computer) was

$Abs=0.03734\times C+0.00143$($r^2=0.9995$). Then, the samples and the blank were assayed in the conditions shown in table 1 by FAAS .

Table1. Instrument working conditions

Element	Wave/nm	Electricity mA	Slit nm	Highth of burning nm	Flow of acetylene $L.min^{-1}$	Flow of air $L.min^{-1}$
Iron	248.3	5.0	0.2	7.5	2.00	13.50

3. RESULT AND DISCUSSION

3.1 Effect of microwave digestion power on the determination of Iron in milk powder

When other factors were fixed, only the microwave digestion power was a variation to determine Iron in milk powder by flame atomic absorption spectrophotometry(FAAS) and the results were shown in fig1.

The content of Iron in milk powder increased as the microwave digestion power increased, and when the microwave digestion power was 1000W the content of Iron in milk powder was the most (Fig 1). So, the the microwave digestion power was fixed at 1000 W in the following experiments.

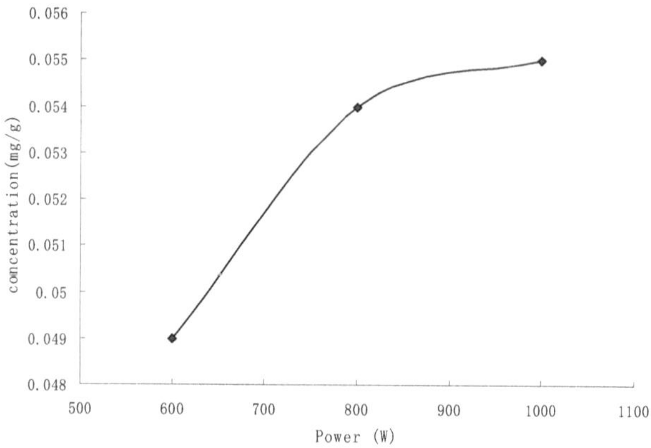

Fig 1 Effect of microwave digestion power on the determination of Iron in milk power

3.2 Effect of the reagent proportion on the determination of Iron in milk powder

When the reagent proportion(HNO_3:H_2O_2) was 4:1, the content of Iron in milk powder was 0.552 mg/g and was the most(Fig 2). So, a proportion of 4:1 was the best reagent proportion(HNO_3:H_2O_2).

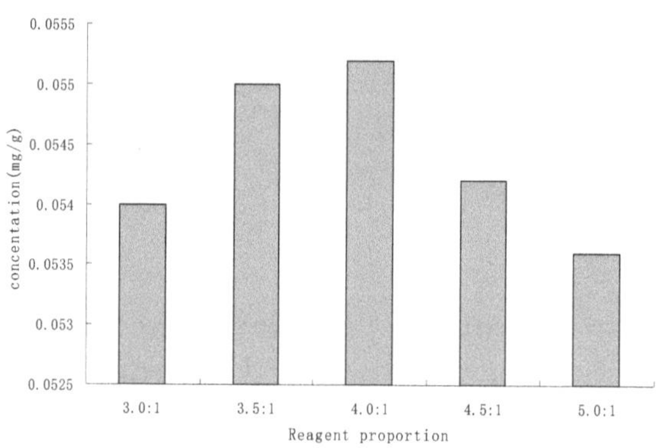

Fig 2Effect of reagent proportion on determination of Iron in milk power

3.3 Effect of the digestion pressure on the determination of Iron in milk powder

The content of Iron in milk powder increased as the microwave digestion pressure increased from 2.1 to 2.4 MPa, and when the microwave digestion pressure was 2.4MPa the content of Iron in milk powder was the most(Fig 3). Then, the content of Iron in milk powder decreased as the microwave digestion pressure decreased. If the microwave digestion pressure was lower, not only decreased the digestion efficacy, but also cost more digestion time.

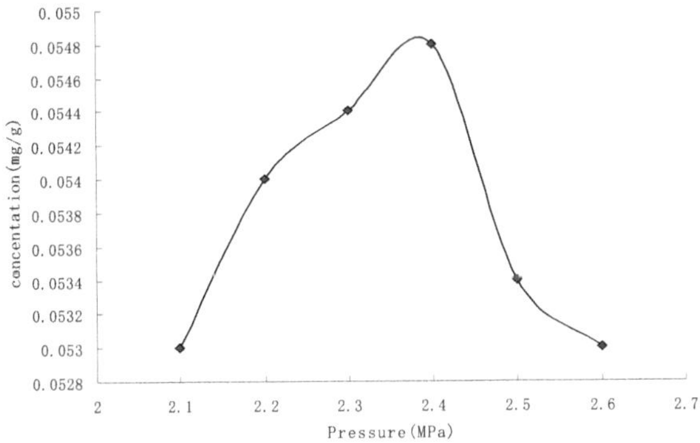

Fig 3 Effect of digestion pressure on the determination of Iron in milk power

3.4 Effect of the digestion time on the determination of Iron in milk powder

The content of Iron in milk powder increased as the microwave digestion time increased from 6 to 8 min, and when the microwave digestion time was 8 min the content of Iron in milk powder was the most and was0.0550mg/g (Fig 4). Then, the content of Iron in milk powder decreased as the microwave digestion time decreased. The microwave digestion could destroy the surface of the samples and this could provide more new surface of samples to touch with the HNO_3 to increase the speed of reaction. If the microwave digestion time was too long, the microwave digestion pressure would be decreased to low the the determination of Iron in milk powder.

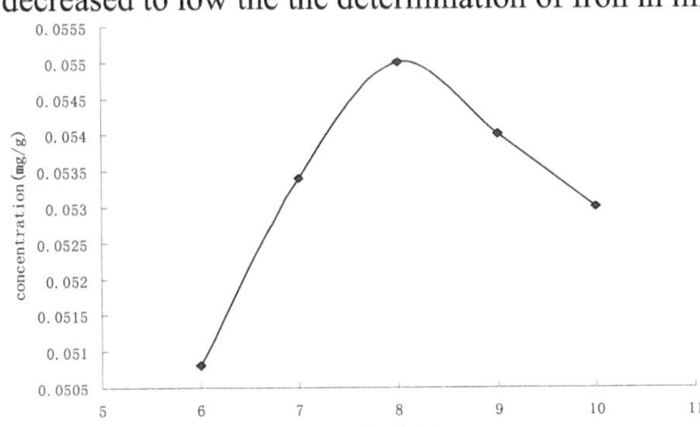

Fig 4 Effect of digestion time on the determination of Iron in milk power

3.5 Effect of the digestion solution volume on the determination of Iron in milk powder

The content of Iron in milk powder increased at first and then decreased when the volume of digestion solution increased from 8 to 12 mL. The content of Iron in milk powder was the most (0.0548mg/g, Fig 5) when the volume of digestion solution was 10 mL. The samples surface could not touch with the acid enough if the digestion solution was less, and more acid left after digestion would effect the accuracy of the dermination of Iron in milk powder. So, the digestion solution volume should be control at a optimum level.

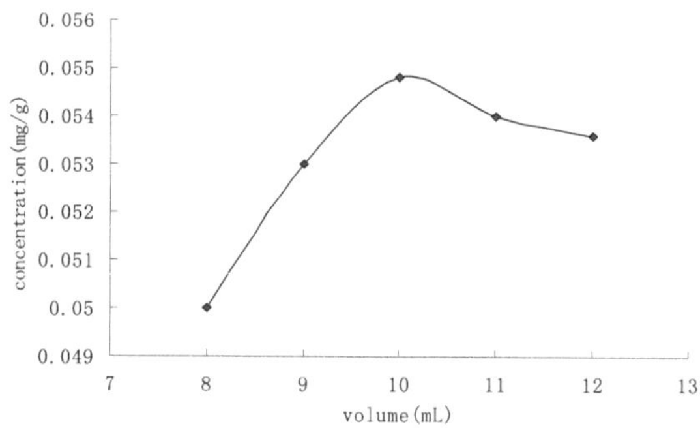

Fig 5 Effect of solution volume on the determination of Iron in milk power

3.6 Results of orthogonal test of microwave digestion

A orthogonal test for microwave digestion of four factors(time, solution volume, pressure and power) and three levels was designed according to the single factor experimenta above, and the results were shown in table 2.

Table 2 Results and analysis of orthogonal test of microwave digestion

Number	Factors				Concentration (mg/g)
	(A) Time (min)	(B) Solution volume (mL)	(C) Pressure(Mpa)	(D) Power(W)	
1	7	8	2.2	600	0.0476
2	7	10	2.4	800	0.0474
3	7	12	2.6	1000	0.0512
4	8	8	2.4	1000	0.0552
5	8	10	2.6	600	0.0504
6	8	12	2.2	800	0.0532
7	9	8	2.6	800	0.0550
8	9	10	2.2	1000	0.0540
9	9	12	2.4	600	0.0486
K1	0.1462	0.1578	0.1548	0.1466	
K2	0.1598	0.1518	0.1512	0.1556	
K3	0.1576	0.1530	0.1566	0.1604	
R	0.0126	0.0060	0.0054	0.0138	

The power was the most important and the pressure was the lest important for the microwave digestion of milk powder(Tbble 2). The optimum parameters for microwave digestion were $D_3A_2B_1C_3$ and that were, the volume of digestion solution was 8mL, the reagent proportion for HNO_3 and H_2O_2 was 4:1, the digestion time was 8min, the digestion pressure was 2.6 Mpa and the digestion power was 1000 W. The content of Iron in assayed milk powder was 0.0560mg/g according to these parameters for microwave digestion.

3.7 Comparison of decomposition of sample by microwave digestion and by conventional heating

For the same milk powder sample assayed by FAAS, the content of Iron was more in sample decomposed by microwave digestion than it by conventional heating. So, for FAAS, decomposition by microwave digestion was time-saving and more accurate than by by conventional heating.

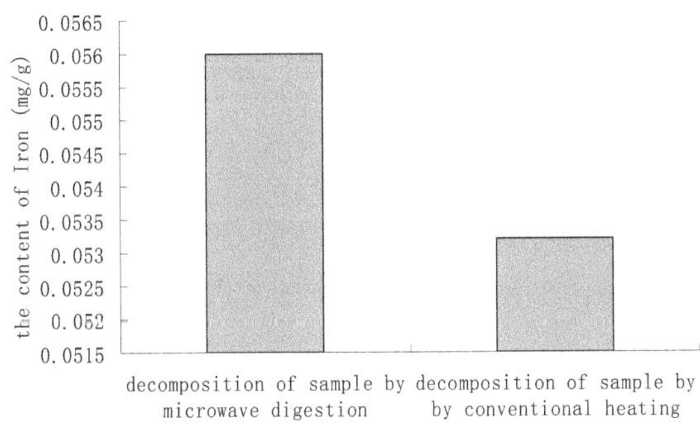

Fig 6 Comparision of the result by different pretreatment methods

4. CONCLUSION

The best optimum parameters for microwave digestion was that, the digestion power was 1000 W, the digestion time was 8min, the volume of digestion solution was 8mL and the digestion pressure was 2.6 Mpa. The content of Iron in assayed milk powder was 0.0560mg/g assayed by FAAS. Microwave digestion was a mordern ,advantageous and future pretreatment of samples.

REFERENCES

Chen W T and Wen C Z. Determination of metal elements in milk powder by ultrasonic extraction. Chemical Industry Times, 2007,21(1):48-51

Li Q X. Determination of laed in health products by microwave digestion-GFAAS. Chinese journal of health laboratory technology. 2007 ,17(7): 1214-1216

Wang X M, Chen Y C, Xie L Q, et al. Flame atomic absorption spectrophotometry for measurement of iron. manganese and zinc contents of wheat seeds. Spectroscopy and spectral analysis, 2004,24(11):1467-1469

Zhang L M. Application of microwave digestion-flame atomic absorption method in the determination of 5 nutritional elements in milk powder. Occupation and Health, 2007,23(22):2048-2049

Zhou X M, Zheng M and Wu X J. Application of microwave digestion-flame atomic absorption method in the determination of 9 trace elements in milk powder. Occupation and Health, 2005, 21(7):1005-1007

EFFICIENT BAR-CODE WATERMARK SYSTEM TO PROTECT AGRICULTURAL PRODUCTS INFORMATION AND COPYRIGHT

Lin Deng[1,*], Xiaoming Wen[2]

[1] *China Patent Information Center, State Intellectual Property Office of P.R.China, Beijing, P. R. China, 100088*

[2] *Shandong University Library, Jinan, Shandong Province, P. R. China, 250100*

[*] *Corresponding author, Address: Hui Zhong Bei Li 116-1-401, Chao Yang District, Beijing, P.R.China, 100012, Tel: +86-13439892596, Email: denglin1021@hotmail.com*

Abstract: In order to protect agricultural product information and copyright, this paper proposes an efficient bar-code watermark system with digital signature. The proposed system adopts digital signature to prevent a buyer from unauthorized copies and to prevent a seller from forged unauthorized copies. The proposed system also encodes the signature with bar-code and embeds the bar-code image into the original image. As long as the similarity of watermark extracts from the damaged image over a threshold, the signature can be fully recovered. It is a novel idea to bring the bar code concept into watermark system to protect agricultural product information and copyright. Detailed simulation results show that the proposed system gets much better results than that with error correcting code scheme, and prove that the proposed system can protect agricultural product information and copyright effectively.

Keywords: bar-code, digital signature, agricultural product, copyright protection

1. INTRODUCTION

At the global level, significant progress has been made towards improved agricultural product safety and information (Fernando, et al. 2006; Clark, et al. 2005). Since then, world gross agricultural production has grown. In order to manage agricultural production effectively, many efforts have been made, such as the scheme (N., et al. 2008; Konde, et al. 2005; Anon, 2005)

Please use the following format when citing this chapter:

Deng, L. and Wen, X., 2009, in IFIP International Federation for Information Processing, Volume 295, *Computer and Computing Technologies in Agriculture II, Volume 3*, eds. D. Li, Z. Chunjiang, (Boston: Springer), pp. 1637–1645.

for information and its management for differentiation of agricultural products, and a knowledge-based intelligent e-commerce system for selling agricultural products called KIES (W., et al. 2007; Kresic, 2005; Buf, 2007). The KIES system not only provides agricultural products sales, financial analysis and sales forecasting, but also provides feasible solutions or actions based on the results of rule-based reasoning. The authors proposed a method of adopting a bar code technology which is successfully applied to a production processing information management of a grocery enterprise (X., et al. 2007). The method can real-timely be analyzed the devotion, the output and wearing off of the original materiel, burden, equipment and manpower, and the individual client cost and individual product cost by the management systems. The cryptography and the digital watermark have begun to attract more attention from the agricultural field along with the rapid growth of the Internet and electronic businesses. Robust watermark can be seen as a mean for the declaration of copyright (S. et al., 1998; R., et al., 2007), and at the same time offers a way to hide data safely. Bar code technology has widely been applied in materiel management of production processing (X. et al., 2007; R., et al., 2007), however, such schemes did not consider protecting agricultural product information and copyright.

In this paper, we show how digital signature protects agricultural product information and copyright traded in the Internet. Copyright protection is known to be a buyer-seller problem (Nasir, et al. 2001) and must meet the three following requirements. First, when we find an unauthorized copy image, we should be able to detect where the original source comes. Second, the message embedded in the image should include buyer information that the seller has no access to. This will prevent the seller from making the signature himself. The last and the most important point is that the signature should be accessed to anyone to verify the validation, but not only to those who own special information. In our method, we encode a long message into a bar-code. The maximal length of the message in our simulation is 64 bytes. It is long enough for the inclusion of identification, the object name, the object number, the price, the timestamp, and so on. Another point that obviously differentiates the method introduced in the paper from those in other watermark references is that we use bar-codes to replace meaningful logos or random sequences. We find that bar-codes can sustain various kinds of attacks and therefore are much better than any other error correcting codes. This is very useful in recovering the damaged watermark.

2. PROPOSED METHOD

2.1 Bar-code scheme

The bar-code scheme is similar to that of code 39. We expand the number of lines to present each byte from nine to twelve. The twelve lines are composed of three bold black, three thin black, two bold white and four thin white lines. So there are 300 combinations and they are enough to express all ASCII codes and the remaining 44 combinations are reserved. Assumption a black line as a start line so that it will end with a white one. The bold line must be presented in two pixels and the thin one in one pixel. That is each byte will be presented in 17 pixels.

A bar-code scheme can not only declare the ownership but also can record various information that includes the buyer, the seller, the object for sale, the object price and so on. Because the bar-code is made follow some orders, we can recover the damaged bar-code image with some recovery rules so that the bar-code has high recoverability against various attacks.

The recover rules of the proposed bar-code scheme are listed below: According to the majority rule and the average gray level, we judge the line is black or white. By the assumption, the start line representing each byte is a black one and its ending line is a white one. When we judge the line is black or white, we record the error pixels of each line and the number of error pixels can be viewed as the severity of attack. If one line and its neighboring line are of the same color, we view these two lines as a bold line. We can check if there are three bold black, three thin black, two bold white and four thin white lines for each byte. If not, we change the color of the line that has the most error pixels until the number of bold and thin lines for each byte is correct. We group every twelve lines as a set for the bar code and map a set of twelve lines to an ASCII code. If we can not map it to any ACSII code, we mark this byte to be a '?' to indicate an error. Every Chinese character is denoted as two ASCII codes.

To show the high recoverability of the bar-code scheme we do several simulations. The original bar-code image and the damaged bar-code images of these simulations are shown in Table 1. We encode a trade message to be a bar-code image. We attack the bar-code images with several kinds of image processing. In the extraction process, we use the recovery rules described above to fix the damaged bar-code images and map the recovered bar-code images to corresponding characters.

The extracted messages and the number of error characters of these simulations are listed in Table 2. The trade message is: "Richard Marks=>Tom 张 2008/3/29 $2000". It means that the seller is Richard

Marks and the buyer is Tom Smith. The transaction of $2000 is done in 2008/3/29. These simulations prove the high recoverability of the bar-code scheme. Ever if the damage is great enough to cause some errors, we can still retrieve the rest part of the trade message from the damaged bar-code image.

Table 1. Original bar-code image and damaged bar-code images.

Processing	Image
Original bar-code image	
100% uniform noise	
JPEG compression	
Sharpening	
100% uniform noise and blurring	
120% uniform noise	

Table 2. Extracted messages and number of error characters.

Processing	Similarity	Extracted message	Error
Original bar-code image	1.0	Richard Marks=>Tom 张 2008/3/29 $2000	0
100% uniform noise	0.503	Richard Marks=>Tom 张 2008/3/29 $2000	0
JPEG compression	0.595	Richard Marks=>Tom 张 2008/3/29 $2000	0
Sharpening	1.0	Richard Marks=>Tom 张 2008/3/29 $2000	0
100% uniform noise and blurring	0.0006	Richard Marks=>Tom 张 2008/3/29 $2000	0
120% uniform noise	0.497	R?chard Marks=>?om 张 2008/3/29 $2000	2

2.2 Verifying procedure

The proposed bar-code watermarking system is composed of three stages: the verifying procedure, the embedding procedure and the extraction procedure.

Suppose buyer B wants to buy an object from a seller A. He must make a trade message M which includes the buyer's name, the seller's name, the kinds of objects, the trade price and the trade date. The message is signed with B's private key and buyer B encrypts the signature with A's public key to avoid the message is known to the third party. Then he tells A who he is and transmits the cipher-text to A:

$$
\begin{aligned}
&B: M=ID_A||ID_B||Objects||Price||Date \\
&B: S=E_{Bs}(M) \qquad (1) \\
&B \rightarrow A: T=ID_B||E_{Ap}(S)
\end{aligned}
$$

where E_{Bs} means encrypt with B's private key.

When A receives the cipher-text from B, he can decrypt it to $E_{Bs}(M)$ with his private key and decrypt $E_{Bs}(M)$ to M with B's public key. So, he can confirm the content of the trade message. If the seller A accepts the deal, he calculates SHA-1(Bp) and concatenates it with $E_{Bs}(M)$ to be the information encoded into a bar-code, otherwise the seller would reject the trade and notify the buyer.

A: S= EBs(M)=DAs(EAp(S))
A: M=DBp(EBs(M))
If A accepts the deal, (2)
A: I=SHA-1(Bp)||EBs(M);
Otherwise the trade is rejected.

2.3 2.3 Watermark embedding procedure

The second stage of the proposed bar-code watermarking system is the embedding procedure. We encode the information I to be a bar-code image. Each byte takes 17 pixels width of bar-code image. The default bar-code height is 21 pixels. So, the bar code image is a bitmap, whose size is 64*17 by 21, equal to 22848 pixels. We decompose the original image with 8*8 block Integer DCT and obtain 4096 blocks of 8*8 pixels from the images of 512*512 pixels. In our simulation, we hide 21 pixel height bar-code image into the 4096 blocks. In average, we must embed each pixel height bar-code image into every 195 block and give up the last one block. One pixel height bitmap includes 64*17=1088 bits. As a result the front 82 blocks out of 196 blocks should choose 5 positions to embed the watermark and the rest 113 blocks should choose 6 positions. We show the relationship in the equation (3) and (4).

$$\lfloor 4096/21 \rfloor = 195,\ 4096 \text{ mode } 21 = 1 \quad (3)$$

$$\lfloor (64*17)/195 \rfloor = 5,\ 64*17 \text{ mode } 195 = 113 \quad (4)$$

For the reason we want the watermark embedded algorithm can be public and the watermark still can maintain its robustness. We choose the embedded position according to the coefficients in the frequency domain and the seller's private key. Therefore, each image will have different embedded point list.

Suppose we want to choose five positions to embed the watermark each block. We sort the 63 AC terms by their absolute values in a descending order first. Then we pick out the top ten terms and encrypt them with A's private key. We sort the value after encryption in a descending order again. The five large ones are the positions where we want to embed the watermark.

Expression (5) is the watermarking equation.

$$\begin{cases} C_i' = C_i(1+\alpha), if \quad W_i = 255 \\ C_i' = C_i(1-\alpha), if \quad W_i = 0 \end{cases} \tag{5}$$

Where C_i is the original coefficient and C_i' is the embedded coefficient. W_i is a sequence of gray levels of bar-code bitmap composed of 0 and 255. α is an unfixed scaling factor varying with the DCT coefficients. The rule of adjusting a value is shown in equation (6).

$$\alpha = \begin{cases} 0.15, if \quad C_i > 35 \\ 0.3, if \quad C_i \in (20,35] \\ 0.6, if \quad C_i \in (10,20] \\ 0.8, if \quad C_i \in (5,10] \\ 0, C_i = C_i + 12, if \quad C_i < 5 \quad and \quad W_i = 255 \\ 0, C_i = C_i - 12, if \quad C_i < 5 \quad and \quad W_i = 0 \end{cases} \tag{6}$$

After watermark embedded, we do the inverse DCT to get a watermark embedded image.

2.4 Watermark embedding procedure

The third stage of the proposed bar-code watermarking system is the extraction procedure. In the embedding procedure, we decompose the original image by using the 8*8 block DCT. Then we choose the embedded positions according to the coefficients in the frequency domain and the seller's private key. In the extraction procedure, we use the same two steps to find the positions where the watermark is embedded.

Expression (7) denotes to retrieve the watermark sequences.

$$\begin{cases} T_i = C_i' / C_i \\ If \quad T_i \le 1, W_i' = 0 \\ If \quad T_i > 1, W_i' = 255 \end{cases} \tag{7}$$

Where C_i is the original coefficient and C_i' is the watermarked image coefficient. W_i' is the extracted watermark sequence.

We use the watermark sequences gotten in (7) to reconstruct the bar-code image and fix the bar-code with the recovery rules describing in section 2.1. Then we translate the content of the recovered bar-code image into bytes. These bytes can be taken to compose an information which equals to SHA-1(Bp)||E_{Bs}(M). These steps are shown in Fig.1.

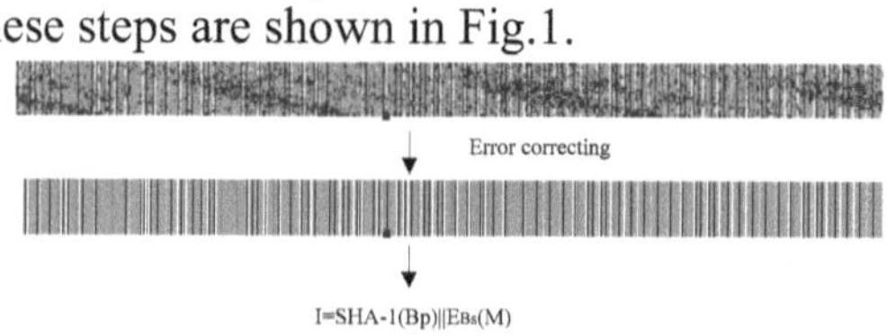

Fig.1: Steps of reconstructing bar-code and using to compose a signature

The decryption process is shown in expression (8). The information I can be clipped into two parts. The first part is SHA-1(Bp) and the second part is EBs(M). SHA-1(Bp) can help us quickly find the buyer in the masses of the buyers. And we can decrypt the cipher-text EBs(M) with B's public key to obtain the trade message M.

$$\begin{aligned}&I=SHA\text{-}1(Bp)||EBs(M)\\&DBp(EBs(M))=M\\&M=IDA||IDB||Objects||Money||Date\end{aligned} \tag{8}$$

3. EXPERIMENT RESULTS

In the simulation, we encrypt a trade message with 1024-bits RSA [6] and confirm the ciphertext by both the buyer and the seller. Then we encode it as a signature with the buyer's private key. Then, the signature is encoded into a bar-code to be a watermark to embed in the original image. The embedded point list is decided by SHA-1 hash function with the seller's private key as a random seed. In the watermark detection, only the seller can extract the watermark because nobody has his private key. However, after the signature is extracted everyone can verify the message with the buyer's public key. So we can make certain who the real buyer and the real seller are. In this experiment, four 512*512 test images are used to be the original images. The trade message hided into each original image is different. We encode these messages to the bar-code images that are watermarks we will embed into the original images. The PSNR of the watermarked image is computed to show its image quality.

The similarity is to measure the degree of damage between the original bar-code watermark W and the extracted bar-code watermark W*. It equals to the number of pixels that are the same gray level between two bar-code watermarks divided by the number of total pixels. These image processing include JPEG compression, blurring, sharpening, uniform noise and cropping. We should define some keywords we will use lager. "Jpeg4" means that we do Jpeg compression with image quality level 4(low) to the image and "Jpeg6" means compressing with image quality level 6(medium). We should list the image sizes after compressing below the tables. "Noise5" means we add 5% uniform noise into the watermarked image and so on. The number of errors in the result table presents the different characters between the original trade message and the extracted message. The trade message we presuppose for the test image is: &Alice &Bob &2007\5\18 &4,000,000. When the ciplier text the seller receive from the buyer, the seller will decrypt the cipher text with B's public key and verify the trade message. If the message is correct, the seller produces the bar-code watermark image as shown in Fig.2.

Fig.2: Bar-code watermark image embedded in test images

The image quality, the extracted message and the number of errors of the test images are shown in Table 3 and Table 4.

Table 3. Experiment results of image "Scene".

Processing	PSNR	Similarity	Extracted message	Error
WMKed	39.28	1	&John &K.J.陈&2007\6\1 &12,345	0
Jpeg6	34.68	0.921	&John &K.J 陈&2007\6\1 &12,345	0
Jpeg4	33.00	0.869	&John &K.J. 陈&2007\6\1 &12,345	0
Blur	35.41	0.755	?J??? ?K????? ???0??6??????,345	21
Sharpen	29.05	0.905	&John &K.J. 陈&2007\6\1 &12,345	0
Noise5	29.81	0.860	&John &K.J. 陈&2007\6\1 &12,345	0
Noise10	24.30	0.716	&John &K.J. 陈&200??6\1 &12,345	2
Crop0.25	10.49	0.739	&John &K.J. 陈&2007\6\1 &12,345	0

Table 4. Experiment results of image "Lena".

Processing	PSNR	Similarity	Extracted message	Error
WMKed	39.08	1	&Catherine &Serlina &2007\6\11 &5,000	0
Jpeg6	36.53	0.942	&Catherine &Serlina &2007\6\11 &5,000	0
Jpeg4	35.11	0.870	&Cather??e &Serlina &2002\???1? &5,000	6
Blur	38.07	0.869	&Catherine &?erlina &?007\6\11 &5,000	2
Sharpen	31.16	0.955	&Catherine &Serlina &2007\6\11 &5,000	0
Noise5	29.93	0.861	&Catherine &Serlina &2007\6\11 &5,000	0
Noise10	24.31	0.718	&Ca??erine &Ser?ina &2007\6\11 &5,000	3
Crop0.25	11.64	0.726	&Catherine &Serlina &2007\6\11 &5,000	0

4. CONCLUSION

In this paper, we have successfully combined the applications of cryptology and watermark to protect agricultural product information and copyright traded in the Internet. Cryptography is used not only for choosing embedded positions but also for encrypting and decrypting to protect the trade messages. We use asymmetrical encryption algorithm, RSA, to avoid the buyer from distributing the unauthorized copies and to prevent the seller himself from generating the signature. This is the key point in protecting agricultural product information and copyright. The bar-code watermark also offers a high recoverability to against the damage caused by image processing. For the future studies, we will select another watermark embedding method to replace the integer DCT used in our study. The spatial distribution and the neighboring pixels will be taken into considerations to adjust the weights of embedding watermark. We believe that it may get better results to protect agricultural product information and copyright.

ACKNOWLEDGEMENTS

The authors would like to thank the anonymous reviewers for the invaluable comments that have significantly improved the paper quality.

REFERENCES

Anon. Water-efficient agriculture key to China's food supply security. International Water and Irrigation, 2005, 25(3): 44-46.

Buf J. M. H.. Improved grating and bar cell models in cortical area V1 and texture coding. Image and Vision Computing, 2007, 25(6): 873-882.

Clark J. Peter. Food plant security. Food Technology, 2005, 59(11): 66-68.

Fernando P. Carvalho. Agriculture, pesticides, food security and food safety. Environmental Science & Policy, 2006, 9(7): 685-692

Konde Victor. Industrial biotechnology applications for food security in Africa: Opportunities and challenges. International Journal of Biotechnology, 2005, 7(1): 95-112.

Kresic-Juric S. Edge detection in bar code signals corrupted by integrated time-varying speckle. Pattern Recognition, 2005, 38(12): 2483-2493.

N. Niederhausera, T. Oberthu, S. Kattnigb, et al. Information and its management for differentiation of agricultural products: The example of specialty coffee. Computers and Electronics in Agriculture, 2008, 61(2): 241-253

Nasir Memon, Ping Wah Wong. A Buyer-Seller Watermarking Protocol. IEEE Transactions on Image Processing, 2001, 10(4): 643-649

R. Rivest, A. Shamir, and L. Strawczynski. A method for Obtaining Digital Signature and Public-Key Cryptosystem. Communication of the ACM, 1978, 21(2)

R. Shams, P. Sadeghi. Bar Code Recognition in Highly Distorted and Low Resolution. IEEE International Conference on Images Acoustics, Speech and Signal Processing, 2007, 1: 737-740

S. Craver, N. Memon, B.L and M. M Yeung. Resolving rightful ownerships with invisible watermarking techniques: limitations, attacks, and implications. IEEE Journal of Selected Areas in Communications, 1998, 16(4): 573-586

W. Wen. A knowledge-based intelligent electronic commerce system for selling agricultural products. Computers and Electronics in Agriculture, 2007, 57(1): 33-46

X. Liu, C. Lin. Information Management System of Grocery Production Processing Based on a Bar Code Identification Technology. IEEE International Workshop on Anti-counterfeiting, Security, Identification, 2007, 164-168

EGG VOLUME AND SURFACE AREA CALCULATIONS BASED ON MACHINE VISION

Ping Zhou, Wengang Zheng*, Chunjiang Zhao, Changjun Shen, Gang Sun
National Engineering Research Center for Information Technology in Agriculture, Beijing, China, 100097
* *Corresponding author, Address: Room 507, Building A, Beijing NongKe Masion, 11# Shuguang Huayuan Middle Road, Haidian District, Beijing, China,100097,Tel:+86-10-51503590, Email: zhengwg@nercita.org.cn*

Abstract: Egg volume(V) and surface area(S) are two important geometrical calculations for the poultry industry and in biological studies, as they can be used in research on population and ecological morphology, and to predict chick weight, egg hatchability, shell quality characteristics, and egg interior parameters. With machine vision technology, the objective of the research herein was to develop a much more fast and available method to predict them. Besides, the researcher gave the definition of Volume in Pixels (V_p) and Surface area in Pixels (S_p) for the first time. Finally, new linear models were established. The R value for volume model is 0.88 and surface area is 0.86. The prediction ability of the models are: for volume, the correctness reaches 52% with $\pm 1cm^3$ statistical errors, 89% with $\pm 2cm^3$, for surface, the correctness reaches 49% with $\pm 1cm^2$ statistical errors , 87% with $\pm 2\ cm^2$.

Keyword: Egg, Volume,Surface area, Machine vision

1. INTRODUCTION

Egg volume and surface area are important geometrical calculations in poultry studies, which were reported by Narushin(2001b,2002,2002a,2002b). So how to measure them quickly and accurately is considered to be a hot topic. As far as so on, several models, based on the measurements of the egg length (L) and the maximum breadth (B), have been suggested to calculate the parameters. The basic formula for calculating egg volume was:

$$V = k_v LB^2 , \tag{1}$$

Please use the following format when citing this chapter:

Zhou, P., Zheng, W., Zhao, C., Shen, C. and Sun, G., 2009, in IFIP International Federation for Information Processing, Volume 295, *Computer and Computing Technologies in Agriculture II, Volume 3*, eds. D. Li, Z. Chunjiang, (Boston: Springer), pp. 1647–1653.

in which k_v is a coefficient for volume calculation. The coefficient values were firstly reported by Romanoff(1949). Later, data gave k_v=0.523 (Ayupov,1976), and k_v=0.496 (Narushin, 1997a), and a number of values of k_v were proposed by Hoyt(1979). Narushin(2001a) showed that egg volume can also be estimated by means of a theoretically deduced formula:

$$V = \frac{2\pi L^3}{3(3n+1)}, \qquad (2)$$

in which $n = 1.057(\frac{L}{B})^{2.732}$.

Later, Narushin(2005) gave a more accurate and available formula than ever before:

$$V = (0.6057 - 0.0018B)LB^2 \qquad (3)$$

Similar to equation(1) and taking into account the changes of the dimensionalities, the lasted formula for calculating S were reported by Narushin(2005):

$$S = (3.155 - 0.0136L + 0.0115B)LB \qquad (4)$$

However, there are still several problems: firstly, the supposed curve in the reported models(taking Narushin,1997's for example) will not always best resemble all eggs' shapes; secondly, if thus, the measurements of L and B with a vernier caliper can't be fast and automatic, which will not be acceptable in poultry industry.

Thus, with machine vision technology, the objective of the research herein was to develop a much more fast and available method to solve the above problems. Characteristic parameters firstly defined by the researcher were calculated from the edge information detected from the egg image. Besides, new linear models were built to predict the volume and surface area based on the calculated parameters.

2. MATERIAL AND METHODS

2.1 Egg samples

In this experiment, 100 fresh eggs were collected from hens of 70 weeks old raised by Hubei Xiantao Food Corporation in China. They were labeled in order after cleaning and air-drying.

2.2 Experimental system

As shown in Fig.1, the test system was composed of lamp(DPS-24W30-4T), CCD camera(SRC-500HP), Additional Lamp, Image Card(DH-VRT-CG200), Computer, and etc.

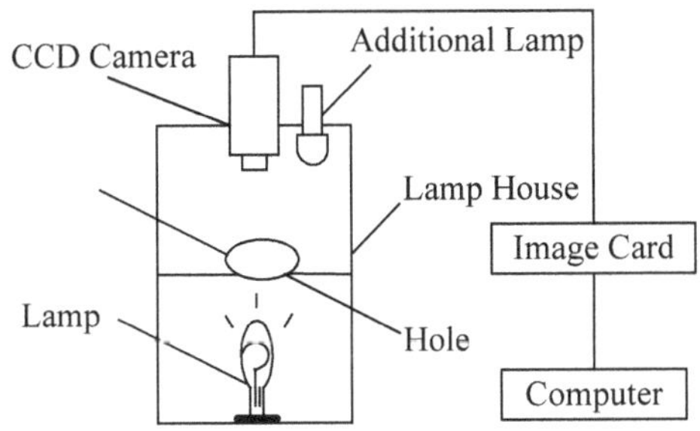

Fig.1 the test system

2.3 Experimental Method

2.3.1 Specific procedure

The experiment should be excuted through these steps: (i) put each egg on the hole, through where the light can shine out, and (ii) adjust the direction of egg till the projection of longitudinal axis in the image was vertical in the screen(manual viewing the screen while adjusting), then capture the image and store it in the harddisk, and (iii) with the self-developed programme, detect the edge information, later, calculate the characteristic parameters of volume in pixels(V_p,unit:p_{ix}^3,p_{ix}=pixel) and surface area in pixels(S_p,unit: p_{ix}^2), which were explained in the next paragraph, and (iv) calculate the exact volume V and V_s using the difference between the measurements of egg weight in air and water, and finally calculate the exact surface area S from the formula:

$$S = {}^{V_s}\!/_{T_s} \tag{5}$$

where V_s is the volume of the eggshell, T_s the average thickness of the eggshell, measured by vernier caliper.

2.3.2 The algorithm for characteristic parameters

As practice shows, if en egg may be considered as a solid of revolution about the longitudinal axis, it is possible to consider the contour of the cross section to be circular. Fig.2 was the profile of an egg, L is the longitudinal axis, O is the peak, Line MN is perpendicular to longitudinal axis, z is the perpendicular distance from the cross section to the peak of the profile, M is the left intersection with the profile, and N is the right, $\theta(z)$ is the angel between long axis and the tangent at point M. Fig.3 is the cross section at the position of Line MN, $A(z)$ is the area and $C(z)$ is the circumference length.

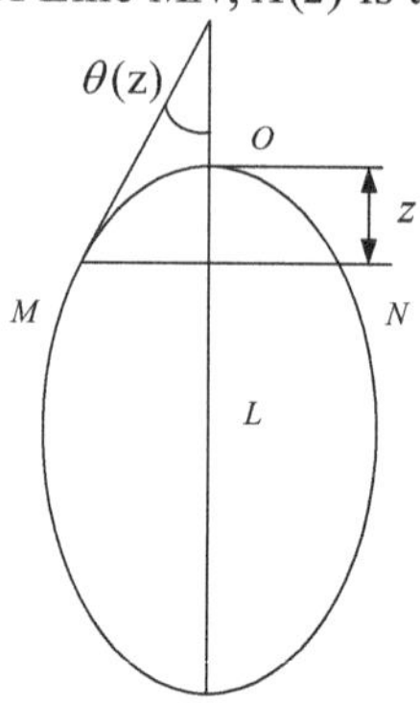

Fig.2 The profile of egg

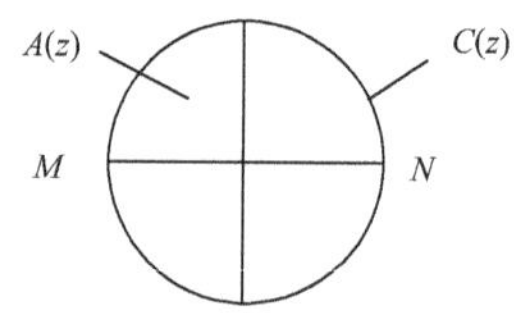

Fig.3 *MN* cross section

Hence, from the perspective of the differential calculus and integral calculus, the volume V and surface area S of the solid of revolution about the long axis can be found from the equations

$$V = \int_L A(z)dz \tag{6}$$

and

$$S = \int_{S'} C(z)ds \tag{7}$$

Where s' is the left contour of the egg's half-projection, ds is the differential operator along s', with the following relation to dz .

$$ds = {dz}/{\cos\theta(z)} \tag{8}$$

Thus, equation (7) can be rewritten as

$$S = \int_L \frac{C(z)}{\cos\theta(z)} dz \tag{9}$$

Egg edge found out from the image was similar to the profile shown in Fig.2. Considering the discrete edge information, Volume in pixels (V_p) and Surface in pixels (S_p) are defined as followings:

$$V_p = \sum_{i=0}^{L} A(z_i) \tag{10}$$

and

$$S_p = \sum_{i=0}^{L} \frac{C(z_i)}{\cos\theta(z_i)} \tag{11}$$

Where L (unit: p_{ix}) is long axis in the image, $A(z_i) = {}^{\pi d^2(z_i)}/_4$, $d(z_i)$ (unit: p_{ix}) is the diameter of the cross section which has a distance of z_i to the peak of the edge in the image, $C(z_i)$ is the corresponding circumference length, $\theta(z_i)$ has the similar meaning to the above $\theta(z)$, which can also be calculated from the edge information.

3. RESULTS AND DISCUSSION

The results of the measurements and calculations are shown in Fig.4 and Fig.5. Fig.4 is the scattergram of samples about the V and V_p, from which we can see that the volume(V) seems to increase linearly accompanying with the increasing of V_p. Hence, with SAS V9.0, we can easily find the model to predict egg volume:

$$V = 3.447E-05 \times V_p + 23.296991 \tag{12}$$

with $R = 0.88$. The proportion of samples with statistical errors in domain of $\pm 1cm^3$ reaches 52%, and 89% in $\pm 2cm^3$.

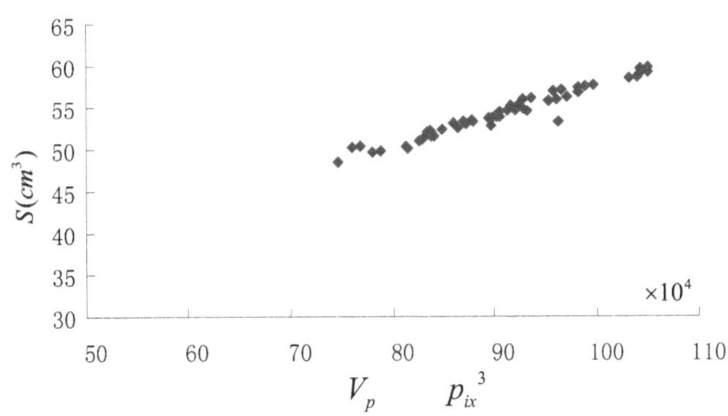

Fig.4 Scattergram of samples about V and V_p

And Fig.5 is the scattergram of samples about S_p and S, from which we can see that the surface area(S) tends to become large as long as the increasing of S_p linearly. Thus, we can easily get the model for surface area:

$$S = 5.985E - 04 \times S_p + 44.684279 \qquad (13)$$

with $R = 0.86$. The proportion of samples with statistical errors in domain of $\pm 1cm^2$ reaches 49%, and 87% in $\pm 2cm^2$.

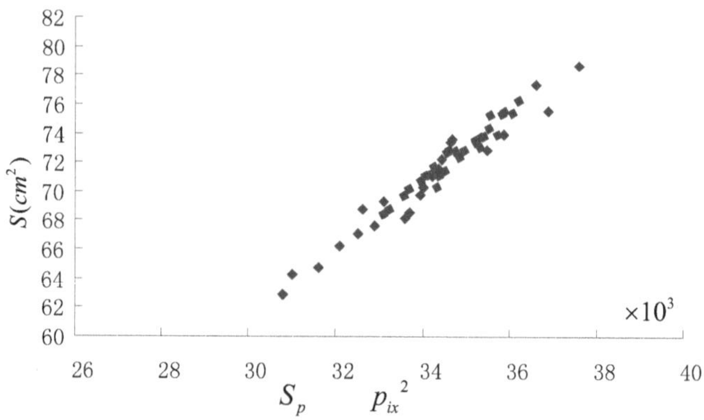

Fig.5 Scattergram of samples about S and S_p

There are may several factors working on the models' precision: (1) the projection of egg's longitudinal axis in the image may not be strictly vertical,(2) the machine vision system may have dimension distortion. However, the results show that machine vision technology can also be used to predict volume and surface area instead of traditional models in practice. Besides, the new method can improve the calculation efficiency, which will be welcome in poultry industry and biological studies. Furthermore, the linear models can be explained much better in geometrics.

ACKNOWLEDGEMENTS

This study has been funded by China National 863 Plans Projects (Contract Number: 2006AA10Z202)

REFERENCES

Ayupov, F. G. 1976. On the egg mathematical model. Advanced Scientific-Industrial Experience in Poultry Breeding. Express Information. 9:14-16.(In Russian)

Hoyt, D.W. 1979. Practical methods of estimating volume and fresh weight of bird eggs. AuK96:73-77.

Narushin, V. G. 1997a. The avian egg: Geometrical description and calculation of parameters. J. Agric. Eng. Res. 68:201-205.

Narushin, V. G. 2001a. Shape geometry of the avian egg. J. Agric. Eng. Res. 79:441-448.

Narushin, V. G. 2001b. What egg parameters predict best its shell strength? Pages 349-355 in IX European Symposium on the Quality of Eggs and Egg products. Kusadasi, Turkey.

Narushin, V. G. 2005. Egg Geometry Calculation Using the Measurements of Length and Breadth. Poultry SCI. J. 84:482-484.

Narushin, V. G., and M. N. Romanov 2002a. Physical characteristics of chicken eggs in relation to their hatchability and chick weight. Paper #026066 in CD-ROM Proceedings of ASAE Annual International Meeting/CIGR World Congress, Chicago, IL.

Narushin, V. G., and M.N. Romanov. 2002b. Egg physical characteristics and hatchability. World Poult. SCI. J. 58:297-303.

Narushin, V. G., M. N. Romanov, and V. P. Bogatyr. 2002. Relationship between pre-incubation egg parameters and chick weight after hatching in layer breeds. Biosystems Eng. 83:373-381.

Romanoff, A.L., and A.J.Romanoff. 1949. The Avian Egg. John Wiley, New York, NY.

ADVANCES IN STUDIES ON NATURAL PRESERVATIVES FOR FRUITS AND VEGETABLES

Haisheng Gao*, Pengbao Shi, Yuhua Zhao

Department of Food Engineering, Hebei Normal University of Science & Technology, Changli, Hebei Province, P.R. China 066600

* *Corresponding author, Address: Department of Food Engineering, Hebei Normal University of Science & Technology, Changli, 066600, Hebei Province, P.R. China, Tel: +86-0335-2039374,Fax: +86-0335-2039374, Email:spxghs@163.com*

Abstract: The author introduced general research and application situations of natural preservatives for fruits and vegetables all over the world these years, and summarized application of vegetation of Murraya in Rutaceae, Cinnamomum in Lauraceae, Artemisia in Compositae and other families and genera on fruits and vegetables preservation and fresh-keeping. Decoction or extraction of Chinese traditional medicine, such as Alpinia Officinarum, Amarphalus Konjac K., stemona etc, could be used in fresh-keeping for orange, apple, strawberry, edible fungi and so on. Garlic could be used in fresh-keeping for orange. Phytic acid and fresh-keeping agents compounded with Phytic acid could extend storage periods of easily rotting fruits and vegetables, such as strawberry, banana, cantaloup, edible fungi and so on, and better keep original fresh condition. Extraction of Snow Fresh, Semper Fresh, Arthropod shell extraction, and halite also had better effect on preservation and fresh-keeping for fruits and vegetables. Main problems exsited in the application of natural preservatives for fruits and vegetables were showed in this article and the applying prospect were discussed too.

Keywords: fruits, vegetables, fresh-keeping agent, fresh-keeping effect, applying prospect

Fruits and vegetables fresh-keeping agents could be divided into two categories, i.e. compound and natural fresh-keeping agents. For a long time, people mainly used compound substance as fresh-keeping agent to keep stored fruits and vegetables fresh. Although this method had better effect on

Please use the following format when citing this chapter:

Gao, H., Shi, P. and Zhao, Y., 2009, in IFIP International Federation for Information Processing, Volume 295, *Computer and Computing Technologies in Agriculture II, Volume 3*, eds. D. Li, Z. Chunjiang, (Boston: Springer), pp. 1655–1669 fresh-keeping of stored fruits and vegetables and preventing them from rotting, many compound substance had bad effects to man's health, and might lead to cancer, distortion and virulence mutation(Gao Haisheng,2002; Yi Ming,2006). So people began to pay attention to developing and researching natural fresh-keeping agents, and had made a great progress.

1. ANTISEPSIS AND FRESH-KEEPING OF EDIBLE FLAVOR VEGETATION

The ancients had applied antisepsis and fresh-keeping function of edible flavor vegetation for a long time. Hoffman and Evans found that mustard seeds, clove and cassia all had the effects of preventing fruits and vegetables from rotting. Adding 0.5g mustard seeds in 100g apple juice, this could keep its quality for 4 months. The reason why edible flavor vegetation had antisepsis function was that there was active matter—essential oil in it(Lin Jinneng,1991;GaoHaisheng,1993). Ripptter and Wise approved that essential oil in mustard seed, clove, cassia bark, elettaria cardamomum, caraway seed, allspice and thyme all had antisepsis function to a certain degree. Essential oil existed in vegetation root, stem, leaf, flower and fruit. For instance, clove bud contained essential oil by 15%～20%, and 85%～92% of it was eugenol. It was eugenol that had main antisepsis function. Similarly，Chinese cassia contained essential oil by 1.0%～3.5%, and 80%～95% of it was cinnamal; Sri Lanka cassia contained essential oil by 0.5%～3.5%, and 42%～75% of it was cinnamal. Cinnamal was the active ingredient.

1.1 Vegetation of murraya in rutaceae

Essential oil in branch and leaf of Murraya almost could control and kill familiar mould. They were ideal raw materials of natural fresh-keeping agents for fruits. Vegetation of Zanthoxylum were widely used as food anticorrosive, had strong controlling and killing function to fruit malignant disease-producing germ. Essential oil(especially not mature essential oil) in Citrus vegetation peel had very strong controlling and killing function to Penicillium digitiatum and Penicillium italicum causing orange blue and green mould. This discovery led to success in researching natural fresh-keeping agent for orange, and a national patent was obtained.

1.2 Vegetation of cinnamomum in lauraceae

Most of vegetation of Cinnamomum were arbores, its root, stem, branch and leaf contained rich essential oil. In essential oil contained a great deal of cinnamal, borneo camphor, 1,8～cineole, eugenyl methyl ether, linalool and so on, and these could control and kill microbe. Eugenyl methyl ether, cinnamal, and linalool could be used in fruit fresh-keeping experiments.

1.3 Other vegetation

Besides the vegetation mentioned above, many other flavor vegetation had active matter of controlling and killing function to fruit malignant disease-producing germs. For instance, ascitriodora, eucalyptus, blue gum of Eucalyptus in Myrtaceae; tsaoko, curcuma of Amomum and Curcuma in Zingiberaceae were widely used as food anticorrosive and fresh-keeping agents. In addition, there were citronella, lemongrass, rue grass in Poaceae; acorus gramineus soland, anise calamus in Araceae; perilla, sweet basil in Labiatae.

2. EFFECTS OF EXTRACTION FROM CHINESE HERBAL MEDICINE VEGETATION ON FRUITS AND VEGETABLES FRESH-KEEPING

Many vegetation could be used as medicine. The useful ingredients extracted by decocting and immersing were used to treat fruits and vegetables combining with other drugs, and got better effect on fresh-keeping.

2.1 Amarphalus Konjac K. Extraction

Tuber of Amarphalus Konjac K. contained rich konjaku flour, the main ingredients in it were konjaku mannoside, konjaku mannoside content was 50%, and konjaku mannan could be obtained after extracted using organic solvent. konjaku mannan had no color, innocuity and no peculiar smell. It had certain function to fruit fresh-keeping and antisepsis of fish and meet. One treatment was as follows: fresh strawberries were putted in 0.05% konjaku mannan solution and doused for 10 minutes, taken them out, redundant sugar fluid was dripped, and dried in the air. The strawberries used this treatment would lose a little gloss after stored 1 week under room temperature, but they would not mildew and rot, and even not mildew after stored for 3 weeks. On the contrary, untreated strawberries would lose their gloss after stored 2 days under room temperature, and start to mildew and rot after 3 days(Zhu Liangfeng et al.,1991).

2.2 Decoction of Alpinia Officinarum

The main antisepsis component in Alpinia Officinarum was naphtha, occupying 0.5%～1.5% and including 1,8～cineole, methyl cinnamate, clove oil, phenol pinene, cubeben etc. Its main piquancy component was galangol, and main yellow components were galangin, galangin, and kaempferide.

Preparation and application of Alpinia Officinarum decoction were as follows: 1kg Alpinia Officinarum and 10kg water were put together and decocted for 45 minutes. Water was being supplemented at any time to supplement evaporated water during boiling. Be sure that the mixture was approximately being kept 10kg. The obtained mixture was filtered when it was hot, then cooled for using. The obtained decoction and bleaching (decoction : bleaching=1：1.5) were put together and adjusted to coating. Oranges were painted on their surfaces and encased in wicker baskets, and stored in rooms under normal temperature. Or the painted oranges were dried in the air, wrapped in 0.01mm thickness plastic films with little holes(0.25mm^2), and stored under normal temperature. After stored for 95 days, the treated oranges rotting rate was only 7.8%, and untreated oranges rotting rate was 37%(Yi Ming,2006;Zhu Liangfang et al.,1991).

2.3 Allium Sativum extraction

Allium Sativum contained garlic naphtha， the main component in garlic naphtha was garlicin. It came into being through garlic amino acid in fresh garlic being decompounded by garlic enzyme. Allicin had controlling and killing function to fungus, such as anthracnose fungi, damping-off fungi and rhizopus nigricans. Preparation of garlic extraction was as follows: Fresh garlic pieces were soaked in cold water for 12 hours， then decocted to boiling, finally 10% garlic extraction was made for using. Or 1 portion of triturated garlic and 10 portions of 80～90℃ water were put together and cooled to normal temperature. One using method was that picked oranges were soaked in garlic extraction, taken out after 10～15 minutes, dried in the air, put in fir boxes with empty cases and hardboards, and stored in ventilated storehouses or common rooms. Good fruits rate after stored for 70 days was over 92.4%(Gao Haisheng,1990a).

2.4 Phytic Acid fresh-keeping agent

Phytic Acid was a sort of organic acid widely existing in vegetation seeds. Fruits and vegetables anticorrosive made using Phytic Acid could be used in antisepsis and fresh-keeping of easily rotting fruits and vegetables and edible fungi. Some melons and fruits, such as grape, strawberry, Hami melon, banana, ananas and litchi could be painted with the agents. This could not only maintain their faint physiological activity, ideal air permeability and water permeability, but also raise melon and fruit shine, enhance resisting ability of outside germs invading, and observably inhibit enzyme activity. Edible fungi could be treated with Phytic Acid, this could prevent edible fungi from changing color, solve the problem of SO_2 remaining, extend edible fungi fresh period from 2～3 days to 5～7 days. This method also worked in fresh cherry, which was difficult in storing(Wu Fan,2003).

According to L zuyou's experiments in strawberry fresh-keeping(Luo Zuyou et al.,2002), the obtained parameter constitution through orthogonal selection was 0.1%～0.15% Phytic Acid ＋ 0.05% sorbic acid ＋ 0.1% acetic hyctro peroxide. Strawberries treated with the mixture fluid could be kept fresh for 1 week under normal temperature, and 15 days under low temperature, good fruit rate was 90%～95%. In addition， Phytic Acid could keep fruits quality. Dissolubility solids, organic acid and Vc content in Strawberries would be raised after treated with 0.1% or 0.15% Phytic Acid. The reason was that Phytic Acid had anti-oxidation effect. Another report was that Phytic Acid could efficiently prevent fruits and vegetables from brown changing and Vc losing in processing.

2.5 Chinese traditional medicine compound semipermeable film fresh-keeping agent

Stemona, giant knotweed rhizome, galangal rhizome, berberine and so on could be extracted under exceed critical condition, the extraction was mixed with starch, konjaku and lecithin, finally Chinese traditional medicine compound semipermeable film fresh-keeping agent was obtained. Some formulas were as follows:

In formula Ⅰ(Zhu Xuan,2001;Kong Qiulian et al.,2002)，included 0.5～2.0 g/L starch, 0.5～1.0 g/L galangal rhizome, 0.5～1.0 g/L stemona, 0.5～1.0 g/L giant knotweed rhizome, 0.5～1.5 g/L konjaku, 0.5～1.0 g/L berberine and 0.1～2.0 g/L lecithin. The apples treated with the mixture could be preserved for 6 months; tomatoes, mad apples and cucumbers treated with it could be stored for 2 months under 10～15℃.

In formula Ⅱ(Song Xiaogang et al.,1996)，included 50g Chinese prickly ash, 50g cassia bark, 50g clove and 300ml starch oxide. 20g starch, 0.12g $KMnO_4$, 0.15g borax, 1g NaOH and 100ml water were put together, heated to 90～95℃, continuously stired for 20～30minutes，cooled it, and then starch oxide was obtained. At first, the mixture was even painted on common packing paper to form a continuous and even Chinese herbal medicine painting layer. The layer was baked under 60～80℃ to form compound fresh-keeping paper. The Ya pears stored in it under room temperature for 35days would be kept better fresh condition and nourishment.

In formula Ⅲ(Song Xiaogang,1997)，included 33g stemona, 33g giant knotweed rhizome, 54g galangal rhizome, 14g licorice and 270ml starch oxide. Compound fresh-keeping paper was made according to the method mentioned above. Tomatoes, sweet peppers and green cucumber were wrapped in pieces of this paper, then stored under 26～34℃, relatively humidity 75%～98%， and deposited in no direct sun light places. This could diminish nutrient losing. The tomatoes wrapped in this films would be basically kept fresh after stored for 18～24 days， except their color became a little dark, and untreated tomatoes were completely rotted. Sweet peppers stored in this films for 12～18 days didn't rot, except their color had a changing period of green, yellow and red. Green cucumber wrapped in this films would be kept its fresh for about 6～12 days.

Formula Ⅳ (Wei Qin,1998)， i.e. NFP fresh-keeping agent, included modified 0.3% konjaku glucan, 0.2% CMC-Na and 0.1%garlic extraction. Scatheless oranges were put in NFP fresh-keeping extraction and soaked for 1～2 minutes， then took them out and stored under room temperature. The orange rotting rate was 25% after stored for 60 days， weight losing rate was 3.0%, and they kept yellow and shining surface. Untreated oranges rotting rate was 50%, weight losing rate was 3.7%, their surface was lackluster.

2.6 Application of Nicandra Physaloides(L.) Gaertn seed extraction

Nicandra Physaloides(L.) Gaertn seed extraction had good film forming character. The film had no color and no smell, dissolved in water rapidly and washed easily. So we can use it in fruit and vegetable fresh-keeping.

Jinguan apple, early ripening variety, painted with 0.5% ～ 1% the extraction would not change its surface color and basic quality after stored for 40 days under 25～30℃ room temperature in open places. But the untreated apples would became yellow in 2 weeks， and when stored to 40 days, they would be wizened seriously and lose perfume and merchandise value. Using the same extraction to treat middle or late ripening varieties would had better effects. Ya pears painted with 0.25%～0.5% the extraction and stored for 40 days in open places under room temperature would not be changed in quality, and the untreated pears would become wizened, and quality and merchandise value decreased. Similarly， using 0.5%～1% the same extraction to paint fruits and vegetables, such as sweet peppers, cucumber and tomatoes could efficiently prolong their shelf life for over 10 days. Vc content of the sweet peppers painted with 1% the extraction and stored for 15 days was 1.364g/kg, still at high level, and Vc content in untreated ones was only 1.047g/kg。

Fresh-keeping mechanisms of Nicandra Physaloides(L.) Gaertn seed extraction were as follows: after painted with the extraction, there would form a even and colorless film on fruit surface. This provided a small air conditioning environment. It efficiently controlled fruit respiration metabolizing, postponed cell becoming decrepit and pigment degrading， and extended fruit fresh-keeping time. Different varieties of fruits had different tolerance to CO_2, so different varieties of fruits had different ideal treating concentration. The experiments results showed that apple ideal treating concentration was 0.5%～1%, Ya pear and Xuehua pear ideal treating concentrations were 0.25%～0.5%. Sweet pepper, other fruit and vegetable ideal treating concentrations were 0.5%～1.0%(Zhu Xiangqiu et al.,1997).

After painted with the extraction, the fruit surface gloss was enhanced. It had glazing and waxing function, so the extraction as a fresh-keeping agent was very suitable in fruits and vegetables fresh-keeping in shelf period.

3. PHOSPHOPROTEIN HIGH MOLECULAR PROTEIN FRESH-KEEPING MEMBRANE

Phosphoprotein high molecular protein widely existed in animals and vegetation. In protein molecular contained a lot of hydrophilic group, such as —NH—, —O—, —COOH, —OH and —NH. After forming film, phosphoprotein high molecular protein fresh-keeping membrane had suitable air permeability, water permeability and gas selectivity. After soaked in phosphoprotein high molecular protein, in fruit surface would form even film, and its thickness could be easily adjusted among tens micron according to fruit physiological development. The film could observably control respiration intensity of fruits and vegetables(Guan Wenqiang et al.,2006). Hygienic toxicity experiments result showed that little albino rat oral LD of phosphoprotein high molecular protein was over 10000mg / kg, it belonged to no poison medicines.

Its using concentration should be 3%～7%, pH value was 2.7～8.5. On fruits and vegetables surface would form films in 3 hours after they were soaked. General dosage was 0.15～0.5kg / t. Storing experiments result showed that Good fruit rate of Jingguan apples stored for 5 months was 95%, and Good fruit rate of Guoguang apples stored for 6 months was 98%. It had better effects in orange and banana. The experiment result in strawberry showed that strawberry treated using concentration of 1%～2% would be prolonged its store time for 2 days under normal temperature, and 15～20 days under 4～8℃(Guan Wenqiang et al.,2006).

4. NATURAL FRUIT AND VEGETABLE FRESH-KEEPING AGENT STUDIED ABROAD AND STARTED

4.1 Snow fresh

Snow Fresh was a new-style high effect multiple function fruit and vegetable fresh-keeping agent made in MONSANTO chemical company in America. It could defer oxidization and brown changing of fruits and vegetables. It had better effect on fresh-keeping of fruit and vegetable, semi-manufactured goods, which had no peel and pit. According to experiments, Snow Fresh could at least keep semi-manufactured goods color and organization shape for 5 days, and the effect was better than that of sulfite which commonly used at present. Snow Fresh contained four safe and avirulent components, i.e. sodium pyrophosphate, citric acid, ascorbic acid and CaCL. Snow Fresh was different from sulfite, which had blanching function. Fruits and vegetables treated with sulfite and stored for some time not only had obvious peculiar smell, but also had rudimental sulfur in them. Fruits and vegetables treated with Snow Fresh had no peculiar smell.

Snow Fresh using methods were simple and convenient. Under room temperature, the fixed quantity of Snow Fresh (white powder) was added in water, mixed for 1min, and 1%～3% solution was compounded. Fruit and vegetable soaking time was 0.5～3 min. Snow Fresh solution should be made and used in one day, and supplemented Snow Fresh powder in it to keep its concentration to get optimal effects(Gao Haishengb,1990).

4.2 Semper fresh

Semper Fresh was a sort of fruit and vegetable fresh-keeping agent made in SEMPER Genetic Engineering Company in England. It had such characteristics as no color, no smell, no virulence, no pollution and no side effect. The edible Semper Fresh was widely used in fruit and vegetable fresh-keeping, and succeeded in flower fresh-keeping. Semper Fresh was a chemical compound, containing vegetable oil and sugar, its active component was sugar ester, other components were cellulose, edible oil etc. Its fresh-keeping mechanism was that it could control fruits and vegetables respiration, prevent water evaporating in order to make them rest， slow them aging speed. Its using concentration was 0.8%～1.0% for smooth peel melon and fruit, proper higher concentration for rough peel fruit, and proper lower concentration for strawberries and leaf vegetables. The concrete using concentrations were 1.0%(variation between 0.8%～1.0%), 0.8%, 1.0%～

1.5% and 1.2% for apples, pear (no stipe), orange and banana respectively. In addition, the fresh-keeping experiments in strawberry, cherry and apricot using Semper Fresh were done in Shanxi agricultural science academy, and the effects were ideal. Grape fresh-keeping was usually carried out in grape trees (when the grapes were 70～80% ripeness) with 0.8%～1.0% concentration Semper Fresh, then picked after 3～4 days. The using concentration was 0.6%～1.0% for melon, about 0.8% for tomato, and 0.8%～1.0% for fresh kidney beens and pepper. Semper Fresh could be stored in cool and dry places for a long time. Blended solution could be stored for 5 days. According to different proportions of agent and water, Semper Fresh was dissolved in water and deposited over one night for using. This could assure the powder pervading completely. Selected fruits and vegetables were soaked in Semper Fresh solution for 30 seconds, then taken them out and dried in the air. In a general way, 1kg Semper Fresh powder could treat 28～35t apples(Gao Haisheng,1990; Wu Fan,2003).

4.3 Arthropod shell extraction

Arthropod shell extraction was generally called chitosan. It was reported(Gao Haisheng,1990c;Zhu Dongxing,2003) that its main component was derivative of deacetylated chitin, a sort of cation high molecular polysaccharide. The extraction was safe and innoxious. It could be washed away by water, degaded by creature. It had no rudimental toxicity problem. The semipermeable film forming on fruit surface could adjust fruits physiological metabolizing(Ghaouth,1992a), and control many microorganisms(Ghaouth,1992b). Using the compound membrane to control fruits and vegetables air condition and slow them physiological metabolizing had been already reported(Asahi,1981,1982,1983). The compound film segregating gas factors could be expressed as α a/b=Pa/Pb. Among them "p" stands for pervious ability coefficient, "a" and "b" stand for different gases. For the film containing hydroxyl molecular, $\alpha\ CO_2/O_2<1$, it had better effect on fruits and vegetables fresh-keeping. For general compound film, $\alpha\ CO_2/O_2>1$, it had no fresh-keeping function to fruits and vegetables. Chitosan cames into being from chitindeacetylation, so its molecular contained amino-group as well as hydroxyl group. Its forming compound film segregating gas factors could be expressed as, $\alpha\ CO_2/O_2<1$. If metal ions were added in chitosan, α a/b value would be much smaller(Bai,1988), so it was more suitable to fruits and vegetables fresh-keeping. Du chuanlai et al.(2004)reported that fixed quantity of chitosan (50% NaOH deacetylation) was dissolved in 2% acetic acid solution, added in 0.05% metal chloride salt solution, stired for 30 minutes, finally chitosan painting agent containing metal ion was prepared. It had better effect on grape fresh-keeping under normal temperature. Among the treatments, the extraction containing Ca^{2+} had the best effect on fresh-keeping, grape rotting rate and fresh weight losing rate were both the lowest.

EL-Ghaouth et al.(1983)found that using 0.1% deacetylated chitin to paint strawberries could greatly inhibit rotting of them stored under 13℃. After 21 days, rotting rate was about 1/5 that of untreated strawberries, the effect was better than germicide. It had no harm to strawberries, and better kept them hardness. Yang yuan'er(2002)uesd 2% chitosan to treat cucumbers, the fresh-keeping period lasted 20 days long, and the color was not changed. Sweet peppers treated with chitosan could be stored for 25 days or much longer. The experiments in amaranth, Chinese chive, greens also displayed fresh-keeping effects at different degrees. Although it worked to bananas and apples, but not to wild peaches, even accelerated them deteroprating. Neutral solution of N, O-carboxymethylated-chitosan extracted from shrimp and crab was used to treat Huaguang 2 Kivifruit, single fruit was painted to form membrane on it, wrapped in polyethylene bag, in the bag $KMnO_4$ ethylene absorbefacient was added. Yangtao used this kind of treatment could be stored for 3 months under normal temperature, good fruit rate was 100%, and weight losing rate was only 3%.

chitosan dissolving capability was better in thin acid, and H^+ concentration was 1～100nmol/L. In order to easily control acidity, some organic acids, sush as Ctiric Acid and Acetic Acid, were used as menstruum. Chitosan doesn't dissolve in water, so its using is not convenient. Chitosan water-solubility ramification A was developed in Hebei Agricultural University, it had more powerful inhibiting bacterium function. The lowest inhibiting concentrations to sporeine, penicilla, rhizopus nigricans, main various food harmful microorganisms were all 1%. The product had stable inhibiting function under high temperature and high pressure(Qi Guohui et al.,2004).

4.4 Compound Vc ramification fresh-keeping agent

American scientists found that Vc ramification and its chemical compound could keep opened apples not to become brown for 48 hours. Components of this kind of compound Vc ramification included Vc ramification (for instance, Ascorbic acid -2- Phosphate or Ascorbic acid-6-aliphatic acid), cassia acid, β -cyclodextrin, sodium phosphate and so on. It could be used as fresh-keeping agent after fruits were flayed and before processing, replacing sulfite in can, preserved fruit, confection, and fruit juice drink producing. It was widely used as a agent of no social effects of pollution and not influencing products flavor(Gao Haisheng,1990d).

4.5 Halite extraction

Halite extraction was a sort of substance extracted from mineral in rock layer. It was white powder, mainly containing metal-salt of Ca, P, Mg, Na, Mn and so on, and used as fresh-keeping agent for fruits and vegetables. When using, put 10g halite extraction in 75L water and wait until it dissolved. Experiment results indicated that the agent had significant effects in easily rotting fruits and vegetables fresh-keeping, such as strawberry, arbutus, mushroom and so on. Its another advantage was that halite extraction could decompose chemical fertilizer and agricultural chemicals on the surface of fruits and vegetables, and kill parasite eggs(Wu Fan,2003).

5. PROBLEMS AND THE VIEW OF APPLICATION

5.1 Problems

5.1.1 The lack of systematic theoretical guide and fundamental studies

There are several steps in the study of natural preservatives for fruits and vegetables: the distillation and separation of substance that can effectively restrain germ growth, the characteristic of the substance in medium, the application test in fruits and vegetables preserving of the substance, the toxic evaluation, etc. So the development of new natural preservatives requires multidisciplinary knowledge as its basis, such as food science, microbiology, biochemistry, medicine, Chinese traditional medicine, etc. However, researchers in these fields seldom collaborate. On the other hand, there is a general lack of enthusiasm for developing new preservatives among enterprisers because of their unwillingness to invest in fundamental studies. Thus, the development of new natural preservatives for fruits and vegetables is limited in some degree.

It's known that the extractives of many plants possess the property of antisepsis, and many of them have already been marketed as natural preservatives. However, most of these preservatives are crude products, and concentrations of the effective species vary with season and geographical environments. For some natural preservatives, the identification of chemical structures for its compositions has not been carried out thoroughly. Consequently, the active species are still unknown and its separation and toxicology evaluation can not be performed. As for the action mechanisms, antimicrobial spectrum, application domains, and the relationship between molecular structure and its activity, it is still unclear for most new natural preservatives.

5.1.2 Problems in film coating for fresh-keeping

Because of its own structural limitations, polysaccharides have defects in several aspects of film coating such as the moisture retention and antibacterial action. Thus, it is necessary to add functional additives and active substances or to alter the characteristics of polysaccharides, so as to make them capable of antisepsis.

It is essential to take care on each step during application of the film coating reagents, for factors such as film concentration and thickness, marinating time and apparatus are all influencing for the final results. It is difficult to determine the thickness of the film and the concentrations of the reagents, because the surface roughness vary with breeds and batches.

On the other hand, film coating is not suitable for all produces. Some produces are capable of self-protection because there is a layer of fuzz on their surfaces. It is likely that the fuzz be destroyed by film coating and the storage time shortened.

5.1.3 Problems in natural products applications

The essential oils have been industrialized as pesticide, but they are still not industrialized as antibacterial agent and they are barely used in preservatives for fruits and vegetables. The studies of mechanism and applied technology are not thorough yet. And so much studies of these fields are in the test period and the mature products are absence.

The reports about natural products for fresh-keeping of the bacteriostasis test are all carried through in medium. The results showed that the bacteriostasis concentration should be increased in the production application. While the essential oils are all hydrophobic compound, they can influence the flavor of fruits and vegetables. Besides the studies mostly processed in the conditions that environment invariety and single microorganisms exist. Lack in the studies that the combination and application of natural extractives on antisepsis characteristic.

There are still some problems though the major studies are about bacteriostasis test of plant crude extract. One of the problems is that the extraction and seperation do not proficiency. The second problem is that the extraction methods are variable and the structure of active ingredient can not be ascertained easily. The last one is that the bacteriostasis mechanism still dimness and studies on applied technology are not enough yet.

5.2 View of applying prospect

Rencently years, peoples are focuse on the studies and developments about organism polysaccharides and then green fresh-keeping methods became the hot point in the domain. The development of polysaccharides film will centralize in the following aspects in future. One of it is to confirm the coat structure ,moisture content, polysaccharides concentration and marinating time by mathematical model.The second one is presentation fresh-keeping mechanism of the polysaccharides film-coating. The third one is combining the research results with production to enhance the industrialization degree.

Internationally, the trend in development of preservatives for fruits and vegetables is natural and functionality. What is more, studies and application about it will more and more. But it is still in start-up periods in china. Resently years, many natural preservatives have been empoldered under the active endeavor of scientific research institutions and manufacturers. Some of the products are spread the application positively and achieved good results.

With the development of food industry and advance in living standard and awareness raising, the security of some synthetic preservatives have been questioned. For these reasons, hunting safe and efficient natural preservatives became the focus point in food industry domain. Developing botanical and natural antiseptic for fruits and vegetables turns into the first choice according to the security , resource and economics because there are extensive territory and abundance on plant resource in china. Most of thc materials are good taste, abundant nutrition and health function that can be used as food and medicinal materials or condiment in the same time normaly. They can endow special aroma to fruits and vegetable, moreover, they are low mammalian toxicity. Such antiseptics easier to be accepted and they appease the demand for green food. So it is not only the mainstream development in abroad but also the major aspect in china. Though botanical and natural antiseptic impossible replace the chemosynthesis one entirely because the restrain of price and bacteriostasis effect , the chemosynthesis one must be substitute in future for the technology advancement, living standard improving and safety consciousness gather head more.

REFERENCES

Asahi Dow Ltd.,J. P. Patent,1981,78,968.
Asahi Dow Ltd.,J. P. Patent,1982,63,25.
Asahi Dow Ltd.,J. P. Patent,1983,415,925.
Bai R. K. Polymer Bulletin 1988,20,83.

Du Chuanlai, Wang Jiahong, Yu Zhifang. Research and application of natural preservrtive of fruits and vegetables. Science and Technology of Food Industry,2004,25(5):135-138(in Chinese)

EL-Kazzaz, M K et al . Effect of different atmospheres on postharvesy decay and quality of fresh strawberries. Photopathology,1983,73:282-285

Gao Haisheng. Edible spice plant and application of it[J]. World Agriculture,1993,(5):23-26(in Chinese)

Gao Haisheng. Natural preservatives for fruits and vegetables—snow fresh. Shan Xi Fruit Trees,1990b,(3):57(in Chinese)

Gao Haisheng. New method to prevent fruits broening. Chinese Agricultural Science Bulletin,1990c,(2):9(in Chinese)

Gao Haisheng. New methods for fruits fresh-keeping.[J] Herbei Agricultural Science and Technology,1990d,(1):31(in Chinese)

Gao Haisheng. Production and application of natural fresh-keeping agents for fruits and vegetables[J]. The Farmers Consultant,2002,(8):19(in Chinese)

Gao Haisheng.Introductions about several natural fresh-keeping agents for fruits and vegetables[J]. World Agriculture,1990a,(4):44-46(in Chinese)

Ghaouth A E,Hortscience,1992a,27(9):1016.

Ghaouth A. E. Phytopathnology,1992b,82(4):398.

Guan Wenqiang, Li Shufen. Research advances in application of natural plant extracts to post-harvest preservation of fruits and vegetables[J]. Transactions of the Chinese Society of Agricultural Engineering,2006,22(7):200-204(in Chinese)

Kong Qiulian, Zhang Mingrong, Zhang Lili. Inhibition on aspergillus niger and penicillium expansun of Chinese medicinal herb[J]. Storage and Process,2002,(3):17-18(in Chinese)

Lin Jinneng.Edible spice plant and application of it[M]. Beijing: Light Industry Press,1991,28-30. (in Chinese)

Luo Zuyou, Luo Shunhua. Preliminary probe into the applying organic acids compound fresh-keeping agents to strawberry storage[J]. Hubei Agricultural Sciences,2002,(3):67-69(in Chinese)

Qi Guohui,Nan Yan, Li Baoguo,etc.Effect of free-polluntant fresh-keeping agents on Taihang chestnut (Castanea mollissima Bl.)[J]. Journal of Agricultural University of Hebei,2004,27(1):53-56(in Chinese)

Song Xiaogang, Chen Min, Wu Yahong. Studies on the fresh-keeping effect of several Chinese medicinal herb and compound preservative paper for juicy pear [J]. Food Science,1996,17(2):67-69(in Chinese)

Song Xiaogang, Chen Min, Wu Yahong. Studies on the fresh-keeping effect of small packing of Chinese medicinal herb compound preservative paper[J].Food Science,1997,18(2):54-57(in Chinese)

Wei Qin.Manufacture of natural preservative for orange preservation and its effect[J]. Science and Technology of Food Industry,1998,(2): 32-33(in Chinese)

Wu Fan. Application of natural fresh-keeping agents for fruits and vegetables in abroad[J].Shan Dong Food Science and Technology,2003,(4):28-30(in Chinese)

Yang Yuan'er, Luo Faxing.A new crude food preservative-chitosan[J]. Journal of Wuhan Polylechnic University, 2002,21(3):22-25(in Chinese)

Yi Ming. Studies on natural fresh-keeping agents for fruits and vegetables[OL]. China Net on Applying of Food Additives,2006.1.10(in Chinese)

Zhu Dongxing, Li Shengyin. Summarization of applying studies on preservatives for fruits and vegetables[J].Shan Xi Jurnal of Agricultural Science,2003,(1):30-34(in Chinese)

Zhu Liangfeng, Lu Biyao. Studies on natural fresh-keeping agents for fruits[J].Letters in China Food Additives,1991,4:32-35(in Chinese)

Zhu Xiangqiu, Wei Jianmei, Mao Zhenchuan.The coloid characteristc of seeds of saloides (L). Gaertn and its application on food industry.Colloquium collection of essays about international food process and packing technology[C].Beijing:Engineering Industry Publishing House,1997(in Chinese)

Zhu Xuan. Advance in using natural Chinese herb as fruits and vegetables preservatives[J]. Xinjiang Agricultural Sciences,2001,41(Special Issue):110-111(in Chinese)

A COMPARISON OF LINEAR REGRESSION METHODS FOR THE DETECTION OF APPLE INTERNAL QUALITY BY NEAR INFRARED SPECTROSCOPY

Dazhou Zhu[1], Baoping Ji[1], Chaoying Meng[2], Bolin Shi[3], Zhenhua Tu[1]
Zhaoshen Qing[1,*]

[1] *College of Food Science and Nutritional Engineering, China Agricultural University, Beijing, P. R. China 100083*

[2] *College of Information and Electrical Engineering, China Agricultural University, Beijing, P. R. China 100083*

[3] *Institute of Food and Agriculture Standardization, China National Institute of Standardization, Beijing, P. R. China 100088*

[*] *Corresponding author, Address: College of Food Science and Nutritional Engineering, China Agricultural University, Beijing 100083, P. R. China, Tel: +86-010-81906784, Fax: +86-010-62347334, Email: qingzhaoshen@cau.edu.cn*

Abstract: Hybrid linear analysis (HLA), partial least-squares (PLS) regression, and the linear least square support vector machine (LSSVM) were used to determinate the soluble solids content (SSC) of apple by Fourier transform near-infrared (FT-NIR) spectroscopy. The performance of these three linear regression methods was compared. Results showed that HLA could be used for the analysis of complex solid samples such as apple. The predictive ability of SSC model constructed by HLA was comparable to that of PLS. HLA was sensitive to outliers, thus the outliers should be eliminated before HLA calibration. Linear LSSVM performed better than PLS and HLA. Direct orthogonal signal correction (DOSC) pretreatment was effective for PLS and linear LSSVM, but not suitable for HLA. The combination of DOSC and linear LSSVM had good generalization ability and was not sensitive to outliers, so it is a promising method for linear multivariate calibration.

Keywords: linear regression, partial least-squares, hybrid linear analysis, least square support vector machine, near infrared spectroscopy, apple

---

Please use the following format when citing this chapter:

Zhu, D., Ji, B., Meng, C., Shi, B., Tu, Z. and Qing, Z., 2009, in IFIP International Federation for Information Processing, Volume 295, *Computer and Computing Technologies in Agriculture II, Volume 3*, eds. D. Li, Z. Chunjiang, (Boston: Springer), pp. 1671–1680.

1. INTRODUCTION

The soluble solids content (SSC) of apple is one of the main factors that influence the consumer's decision on purchasing. The nondestructive determination of the internal quality of apple, such as SSC, is of great importance for quality evaluation. As a fast, green, and nondestructive technique and its ability of on-line detection, near infrared (NIR) spectroscopy is widely used for the determination of the internal quality of apple (Ventura et al., 1998; Lammertyn et al., 1998). Until now, various chemometrics algorithms have been introduced to improve the robustness of model. For analyzing the quality of apple, traditionally used linear multivariate calibration methods were multiple linear regression (MLR) (Ventura et al., 1998; Murakami et al., 1994), principle component regression (PCR) (Lu et al., 2000), and the partial least-squares (PLS) regression (Peirs et al., 2000; Zude et al., 2006). In order to develop the relation between SSC and the spectra by taking into account some non-linear variations, artificial neural network (ANN) was used to construct the model (Ying et al., 2005). However, some problems still exist for these methods. The determination of the optimal number of PLS factors may be a difficult problem, particularly when an automatic analysis process is concerned. In some cases, the empirical criteria (Malinowski, 1986) for factor selection may fail to achieve a satisfying result (Xu & Schechter, 1997). In addition, ANN is based on the empirical risk minimization principle, whereby robust calibration modeling requires massive representative samples to be analyzed, and ANN can easily lead to local minimization. Therefore, new excellent methods are needed to construct the calibration model for SSC of apples.

Berger et al. (1998) have introduced a new method, which combines the advantages of different standard methods and is therefore called hybrid linear analysis (HLA). HLA incorporates the spectrum of the desired species into the calibration procedure, and it is based on the theory of net analysis signal (NAS) (Lorber, 1986). Berger et al. have used HLA to analyze the near infrared Raman spectra of aqueous mixtures of glucose, lactic acid, and creatinine, and found that HLA produced lower prediction errors than PLS. However, this original HLA can only be used when the pure spectrum of the desired species can be obtained. Two variant HLA algorithms were introduced which did not require the pure spectrum to be known. The hybrid linear analysis developed by Xu & Schechter was called HLA/XS (Xu & Schechter, 1997), and the hybrid linear analysis developed by Goicoechea & Olivieri was called HLA/GO (Goicoechea & Olivieri, 1999a). HLA was used to determinate the tetracycline in blood serum (Goicoechea & Olivieri, 1999a), the bromhexine in cough-cold syrups (Goicoechea & Olivieri, 1999b), and several components of ophthalmic solutions (Collado et al.,

2000). Overall, the performance of HLA seems to be better than that of PLS. Therefore, HLA was used for the determination of SSC of apple.

Support vector machine (SVM) is a relatively new learning algorithm based on the statistical learning theory (Vapnik, 1995). Least square support vector machine (LSSVM) (Suykens et al., 2002) is the reformulation of the principles of SVM. Compared with SVM, LSSVM is much easier due to its solution of only a set of linear equations. In addition, LSSVM has fewer parameters to be tuned. LSSVM was applied to NIR analysis for discrimination (Belousov et al., 2002; Goodacre, 2003) and quantitative predictions (Cogdill & Dardenne, 2004; Chauchard et al., 2004). In all of these cases, LSSVM was used as a non-linear modeling technique, where the RBF kernel function was mostly used. Delightfully, LSSVM performed well when the spectra had non-linear information, and it had some advantages over other non-linear multivariate regression techniques such as ANN. For analyzing apple quality by NIR, our work before have proved that both ν-SVM and LSSVM were powerful for developing the non-linear relation between spectra and chemical values (Zhu et al., 2007; Zhu et al., 2008). Unfortunately, a fact has been ignored in the chemometrics: The biggest difference between SVM and ANN is the use of a regularized parameter which determines the trade-off between minimizing the training error (or empirical risk) and the model complexity in SVM. Similarly, Bayesian regularization that uses a regularization parameter to modify the performance function is a useful method to improve the generalization ability of ANN. Therefore, linear LSSVM may have advantages over PLS and other linear methods due to the use of regularization parameter.

In this study, PLS, HLA, and linear LSSVM were used to construct the linear calibration models between the SSC of apples and the NIR spectra, the performance of these three linear regression methods was compared. In order to improve the prediction error, direct orthogonal signal correction (DOSC) (Westerhuis et al., 2001) was used to preprocess the spectra.

2. EXPERIMENTAL

2.1 Samples

A total of 113 'Fuji' apples coming from the Shandong province of China was used. The weights of all the apples were larger than 200 g. Apples were stored at room temperature (~25°C) for at least 12 hours before NIR measurements. Three outliers were eliminated according to the plots of studentized residuals versus leverage value (Otto, 2003), and the retained

110 samples were divided into a calibration set and a prediction set. Samples were sorted according to SSC. Every third sample was included in the prediction set, while all remaining samples made up the calibration set. The calibration set was used to construct the model, and the prediction set was used to test the model performance for external samples. The statistical characteristics of the SSC for two data sets are summarized in Table 1.

Table 1. Statistic SSC values of calibration and prediction sets of apple[a]

Sample Set	n[b]	Range	Mean	SD[c]
Calibration	73	10.5-18.0	13.18	1.75
Prediction	37	10.5-17.7	13.14	1.74

[a] Unit used, °Brix. [b] n=number of samples. [c] SD=standard deviation.

2.2 Spectra measurement

FT-NIR spectra were recorded on an ANTARIS FT-NIR spectrometer (Thermo Nicolet Corporation, USA) equipped with a NIR fiber-optic probe. It has a spectral range of 3800~12000 cm^{-1}. The resolution was set to 2 cm^{-1}, and the number of spectrum scans was set to 64. The reference was measured every 30 minutes. All the reflectance spectra of apples were measured in the lab at room temperature (~25 °C). For each apple, the spectra were measured on four evenly distributed equatorial positions, and the mean of these four spectra was calculated. Since the lower and higher parts of the spectra had noise, only the spectral range of 4500-9500 cm^{-1} was selected for latter analysis. Thus, each spectrum had 2593 data points.

2.3 SSC Measurement

The SSC of apples was measured by a refractometer (WAY, Shanghai Precision & Scientific Instrument Co., Ltd, China). About 100 g eatable flesh of each apple was cut and the juice was obtained by a juice extractor (AF2000, Dondxing desheng food mechanicl factory, Shantou, China). Then the juice was centrifuged for ten minutes, and a little pellucid juice was dropped onto a refractometer to record the °Brix.

2.4 Data processing

DOSC was used to preprocess the spectra, the optimal number of DOSC components removed from the spectra and the tolerance factor was selected by leave-one-out (LOO) cross-validation. The cross-validation process was applied on the calibration set. The calibration models of SSC were constructed by PLS, HLA/XS, HLA/GO, and linear LSSVM, respectively. The optimal number of PLS factors and HLA factors were selected by LOO cross-validation and the criterion proposed by Haaland and Thomas (Haaland & Thomas, 1988). In this study, the value of F corresponding to a probability smaller than 0.75 yielded the optimum number of factors for PLS and HLA. For linear LSSVM, the regularization parameter C was selected by two-step LOO cross-validation. First, C increased with an exponential sequence and a cursory range of C was obtained; and then, the optimal C was selected with an equal increasing sequence within the cursory range.

The quality of the calibration model was evaluated by the determination coefficient (R^2) for the model of calibration set, the standard error of cross-validation (SECV), and the relative standard deviation (RSD) for the prediction set. All the calculations were performed in the software system MATLAB 2006a (The Math Works, Inc., Natick, MA, USA).

3. RESULTS AND DISCUSSION

3.1 Influence of parameters on the calibration model

The number of factors for PLS and HLA, as well as the parameter C of LSSVM, had significant influence on the quality of models. SECV decreased sharply with the increase of PLS factor numbers and then reached a local minimum (SECV=0.7102). When the factor number was very large (>20), SECV kept constant (SECV=0.6946) (Fig. 1a). It should be noted that the model was over-fitted with a large factor number (R^2=0.9996 with a factor number of 20). In this case, the traditional cross-validation with minimum PRESS criterion was invalidated. Therefore, the criterion proposed by Haaland and Thomas (1988) was used to select the optimal number of PLS factors. Similar result was obtained for HLA. For both HLA/XS and HLA/GO, SECV decreased with the increase of HLA factors first and then kept constant (Fig. 1a). This was consistent with the result reported by Xu and Schechter (1997), where RMSEP reached a constant value. Xu suggested that all the m (the sample number of calibration set) factors can be used. In fact, R^2 was very close to one when the number of HLA factors was larger than 30 (R^2=1.0000 with a factor number of 73), suggesting that the model was over-fitted. Therefore, we selected the optimal HLA factors by cross-validation for both HLA/XS and HLA/GO. For the linear LSSVM model, SECV decreased with the increase of C when C was small (Fig. 1b), suggesting that the increase of model complexity gave better result. When C was very large, the model was too complicated, and the model was over-fitted. Therefore, SECV increased subsequently.

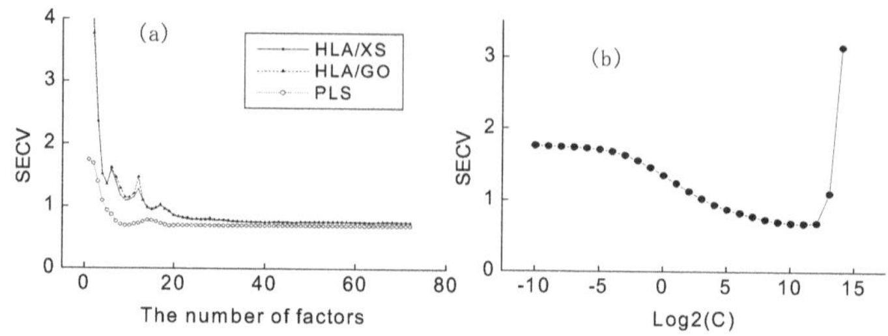

Fig. 1: (a) SECV values versus the numbers of factors for PLS, HLA/XS and HLA/GO models; (b) SECV values versus the parameter C for LSSVM models with linear kernel.

3.2 Calibration models obtained by three regression methods

The optimal models constructed by the three kinds of regression methods were obtained. The results of both HLA/XS and HLA/GO were comparable to PLS (Table 2). Compared with PLS, HLA had higher SECV while their RSD showed no obvious difference. In fact, HLA is a NAS based calibration method, which decomposes the total signal into analyte contribution and contributions from interferences. However, the decomposition may be too radical, thus could not exactly extract useful spectra information. In addition, the optimal number of factors for HLA was relatively large. This was different from the result of Collado et al. (2000); Collado reported that HLA used fewer factors than PLS for the determination of Tetracycline in Blood Serum. Maybe the components of serum were simple or the content of the main components of serum were relatively high, therefore, only few factors could define the projection matrix for HLA. However, the component of apple is very complicated, thus more factors were needed. Among all the three kinds of regression methods, linear LSSVM gave the best results (with R^2 of 0.9304, SECV of 0.6627, and RSD of 4.41%). Due to the foundation of structural risk minimization, linear LSSVM has excellent generalization ability.

Table 2. The calibration models of SSC constructed by three kinds of linear regression methods with DOSC pretreatments.(^{N} The number of factors for PLS and HLA.)

Method	Parameters	R^2	SECV (°Brix)	RSD (%)
PLS	N=8	0.8930	0.7102	5.13
HLA/XS	N=23	0.9635	0.7978	5.17
HLA/GO	N=24	0.9739	0.7954	5.02
LSSVM	C=4800	0.9304	0.6627	4.41
DOSC-PLS	N=1	0.8989	0.5731	4.05
DOSC-HLA/XS	N=18	0.9409	0.8584	5.54
DOSC-HLA/GO	N=19	0.9488	0.8718	5.53
DOSC-LSSVM	C=0.05	0.8988	0.5732	4.01

3.3 Influence of DOSC pretreatment on the three regression methods

DOSC was used to preprocess the spectra. For DOSC-PLS, the optimal number of DOSC components removed from the spectra and the tolerance factor selected by leave-one-out (LOO) cross-validation were 2 and 0.01, respectively. The parameters of DOSC in other regression methods were set the same as that of PLS for latter comparison. After DOSC preprocessing, the model quality of PLS and LSSVM improved, both SECV and RSD decreased (Table 2). The factors of PLS model decreased from eight to one, making the model easier to interpret. Similar results were obtained for HLA, where the factors decreased from 23, 24 to 18, and 19, respectively (Table 2). However, the combination of DOSC and HLA gave even worse results then single HLA. For both HLA/XS and HLA/GO, SECV and RSD increased. Since DOSC had removed the information that was linearly unrelated (orthogonal) to the SSC of apples, the regression methods that directly built the relation between spectra and SSC, such as PLS and LSSVM, would obtain better results. As for HLA, all other components were used to define the projection matrix of HLA and to calculate the NAS. The use of DOSC would remove part of the information of these other components, thus influence the accuracy of NAS. Moreover, the principle of DOSC and HLA was similar, that is, dividing the spectra into two parts: useful signal and interference. However, DOSC is a "mild" one, and HLA is a rather "stringent" one. So the mildness of DOSC is lost on combining it with HLA. This indicated that it may not be appropriate to apply DOSC and HLA simultaneously. Among all the calibration models of SSC, DOSC-LSSVM gave the best results.

3.4 Sensitivity to outliers for the three kinds of regression methods

Outliers are data that have a rather large influence on the regression solution and the occurrence of such data points can lead to considerable deviations from normality (Philips & Eyring, 1983). To investigate the sensitivity to outliers for the three kinds of regression methods, three outliers that had been eliminated were again put into the calibration set, while the prediction set kept constant. The models constructed with and without DOSC pretreatment are shown in Table 3. Both SECV and RSD increased compared with the models without outliers (compare Table 2 and Table 3), suggesting that the existence of outliers had affected the model robustness. Among all of the four models, linear LSSVM had the smallest increase extent for RSD. In addition, the C value of linear LSSVM model decreased from 4800 to 750, this indicated that linear LSSVM gave a smaller weight to the prediction error (see equation 10), thus minimized the influence of outliers. As a result, linear LSSVM performed better than PLS, HLA/XS, and HLA/GO, indicating the latter three methods were more sensitive to outliers than linear LSSVM. Specially, the results of HLA models were very bad, suggesting that outliers must be eliminated when HLA was used to construct a regression model. After DOSC pretreatment, the RSD of PLS and linear LSSVM decreased, while the RSD of HLA/XS and HLA/GO increased. This again indicated that DOSC preprocessing was not suitable for HLA. In addition, PLS and linear LSSVM obtained almost the same result, with RSD of 5.33% and 5.28%, respectively. The same phenomenon has appeared in Table 2, where the RSD of DOSC-PLS and DOSC-LSSVM were 4.05% and 4.01%, respectively. It seems that linear LSSVM performed better than PLS for the original spectra; while the use of DOSC pretreatment helped PLS to achieve a result as good as that of linear LSSVM.

Table 3. The calibration models constructed by three kinds of linear regression methods with three outliers in the calibration set. ([N] The number of factors for PLS and HLA.)

Method	Parameters	R^2	SECV (°Brix)	RSD (%)
PLS	N=7	0.7449	1.1216	6.87
HLA/XS	N=27	0.9585	1.4785	7.93
HLA/GO	N=27	0.9697	1.4497	8.02
LSSVM	C=750	0.7817	1.0504	5.66
DOSC-PLS	N=1	0.7715	0.8695	5.33
DOSC-HLA/XS	N=24	0.8950	1.7085	8.91
DOSC-HLA/GO	N=22	0.8795	1.8652	9.21
DOSC-LSSVM	C=0.03	0.7714	0.8695	5.28

4. CONCLUSIONS

For the determination of SSC in apples, linear regression methods of PLS, HLA/XS, HLA/GO, and linear LSSVM all obtained satisfying results. HLA was sensitive to outliers, thus the application of HLA requires the elimination of outliers first. The predictive ability of SSC model obtained by HLA was comparable to that of PLS. Among the three kinds of methods, linear LSSVM performed better than PLS and HLA, since it uses a regularized parameter to determine the trade-off between minimizing the training error and the model complexity. DOSC pretreatment was effective for PLS and linear LSSVM, but seems to be not suitable for both HLA/XS and HLA/GO. The combination of DOSC and linear LSSVM had good generalization ability and was not sensitive to outliers, thus gave the best calibration model for the SSC of apples.

REFERENCES

A. Belousov, S. Verzakov, J. von Frese, Applicational aspects of support vector machines, J. Chemom. 2002, 16:482-489

A. J. Berger, T. W. Koo, I. Itzkan, M. S. Feld, An enhanced algorithm for linear multivariate calibration, Anal. Chem. 1998, 70: 623-627

A. Lorber, Error propagation and figures of merit for quantification by solving matrix equations, Anal. Chem. 1986, 58: 1167-1172

A. Peirs, K. Ooms, J. Lammertyn, B. Nicolaï, Prediction of the optimal picking date of different apple cultivars by means of VIS/NIR spectroscopy, Postharvest Biol. Tec. 2000, 21:189–199

D. M. Haaland, E. V. Thomas, Partial Least- Squares Methods for Spectral Analyses, Anal. Chem. 1988, 60: 1193-1208

D. Zhu, B. Ji, C. Meng, B. Shi, Z. Tu, Z. Qing, The application of direct orthogonal signal correction for linear and non-linear multivariate calibration , Chemom. Intell. Lab. Syst. 2008, 90:108-115

D. Zhu, B. Ji, C. Meng, B. Shi, Z. Tu, Z. Qing, The performance of ν -support vector regression on determination of soluble solids content of apple by acousto-optic tunable filter near-infrared spectroscopy, Anal. Chim. Acta. 2007, 598:227–234

E. R. Malinowski, Factor Analysis in Chemistry, Wiley, New York (1986)

F. Chauchard, R. Cogdill, S. Roussel, J. M. Roger, V. Bellon-Maurel, Application of LS-SVM to non-linear phenomena in NIR spectroscopy: development of a robust and portable sensor for acidity prediction in grapes, Chemom. Intell. Lab. Syst. 2004, 71: 141-150

G. R. Philips, E. M. Eyring, Comparison of conventional and robust regression in analysis of chemical data, Anal. Chem. 1983, 55:1134-1138

H. C. Goicoechea, A. C. Olivieri, Determination of bromhexine in cough–cold syrups by absorption spectrophotometry and multivariate calibration using partial least-squares and hybrid linear analyses. Application of a novel method of wavelength selection, Talanta. 1999b, 49: 793–800

H. C. Goicoechea, A. C. Olivieri, Enhanced Synchronous pectrofluorometric Determination of Tetracycline in Blood Serum by Chemometric Analysis. Comparison of Partial Least Squares and Hybrid Linear Analysis Calibrations, Anal. Chem. 1999a, 71: 4361-4368

J. A. K. Suykens, T. Van Gestel, J. De Brabanter, B. De Moor and J. Vandewalle, Least Squares Support Vector Machines. World Scientific Publishing, Singapore, 2002

J. A. Westerhuis, S. de Jong, A. K. Smilde, Direct orthogonal signal correction, Chemom. Intell. Lab. Syst. 2001, 56:13-25

J. Lammertyn, B. Nicolaï, K. Ooms, V. D. Semedt, J. D. Baerdemaeker, Non-destructive measurement of acidity, soluble solids, and firmness of Jonagold apples using NIR-Spectroscopy, Trans. of the ASAE. 1998, 41: 1089–1094

L. Xu, I. Schechter, A calibration method free of optimum factor number selection for automated multivariate analysis. Experimental and theoretical study, Anal. Chem. 1997, 69: 3722-3730

M. Murakami, J. Himoto, K. Itoh, Analysis of apple quality by near infrared reflectance spectroscopy, J. Fac. Agr. Hokkaido Univ. 1994, 66: 51–61

M. Otto wrote; X. G. Shao, W. S. Cai, X. J. Xu translated, Chemometrics: Statistics and Computer Application in Analytical Chemistry. Scientific Publishing Company, Beijing, China, 2003

M. S. Collado, V. E. Mantovani, H. C. Goicoechea, A. C. Olivieri, Simultaneous spectrophotometric-multivariate calibration determination of several components of ophthalmic solutions: phenylephrine, chloramphenicol, antipyrine, methylparaben and thimerosal, Talanta. 2000, 52: 909–920

M. Ventura, A. de Jager, H. de Putter, F. P. M. M. Roelofs, Non-destructive determination of soluble solids in apple fruit by near infrared spectroscopy (NIRS), Postharvest Biol. Tec. 1998, 14: 21–27

M. Zude, B. Herold, J.-M. Roger, V. Bellon-Maurel, S. Landahl, Non-destructive tests on the prediction of apple fruit flesh firmness and soluble solids content on tree and in shelf life, J. Food Eng. 2006, 77: 254-260

R. F. Lu, D. E. Guyer, R. M. Beaudry, Determination of firmness and sugar content of apples using near-infrared diffuse reflectance, J. Texture Stud. 2000, 31: 615–630

R. Goodacre, Explanatory analysis of spectroscopic data using machine learning of simple, interpretable rules, Vib. Spectrosc. 2003, 32:33-45

R. P. Cogdill, P. Dardenne, Least-squares support vector machines for chemometrics: an introduction and evaluation, J. Near Infrared Spectrosc. 2004, 12:93-100

U. Thissen, B. Üstün, W. J. Melssen, L. M. C. Buydens, Multivariate Calibration with Least-Squares Support Vector Machines, Anal. Chem. 2004, 76:3099-3105

V. Vapnik, The nature of statistical learning theory, New York, Springer, 1995

Y. Ying, Y. Liu, X. Fu, H. Lu, Application of principal component regression and artificial neural network in FT-NIR soluble solids content determination of intact pear fruit, in Optical Sensors and Sensing Systems for Natural Resources and Food Safety and Quality, Ed by Y.-R. Chen, G.E. Meye, S.-I. Tu, Proc. of SPIE, 2005, Vol. 5996, p. 292

COST–BENEFIT ANALYSIS OF HACCP IMPLEMENTATION IN THE CHINESE SLAUGHTERING AND MEAT PRODUCT PROCESSING INDUSTRY

Shunlong Gong [1] , Chenglin Ma [2] , Yinsheng Yang [2] , Li Bai [2,*] , Linyi He [3]

[1] *College of Management, Jilin University, 5988 Renmin Street, Changchun 130025, PR China*

[2] *College of Biological and Agricultural Engineering, Jilin University, 5988 Renmin Street, Changchun 130025, PR China*

[3] *College of Management, Beijing Normal University, Beijing 100875, PR China*

[*] *Corresponding author: College of Biological and Agricultural Engineering, Jilin University, 5988 Renmin Street, Changchun 130025, PR China, Tel: +86 13504334103; E-mail: bl@jlu.edu.cn (L. Bai)*

Abstract: The paper reports the results of a study of the costs and benefits associated with the implementation and operation of HACCP in the Chinese slaughtering and meat product processing industry. The research results suggest that although some kinds of intangible costs are more regularly referred to, the major costs of implementing and operating HACCP in the industry are still relatively tangible, such as investment in new equipments and product testing. And although most respondents indicated that the costs of implementing and operating HACCP were approximately in accordance with their prior expectations, still a significant majority indicated that some costs exceeded their expectations. The results also suggest that the slaughtering and meat product processing enterprises do derive benefits from implementing and operating HACCP, and some of them have derived distinct benefits. The results have implications for the further adoption of HACCP not only in the industry itself but also in the Chinese food industry as a whole. Policy makers should take account of these research results and make more quantitative researches to offer more comprehensive and classified information to help food enterprises make decisions on HACCP implementation and operation.

Keywords: cost–benefit analysis; HACCP; slaughtering and meat product processing industry

---

Please use the following format when citing this chapter:

Gong, S., Ma, C., Yang, Y., Bai, L. and He, L., 2009, in IFIP International Federation for Information Processing, Volume 295, *Computer and Computing Technologies in Agriculture II, Volume 3*, eds. D. Li, Z. Chunjiang, (Boston: Springer), pp. 1681–1689.

1. INTRODUCTION

Hazard Analysis and Critical Control Point (*HACCP*) system was designed to focus on prevention of possible hazards and to improve the process (Swanson and Anderson), and has been developed and promoted in many countries to obtain better food safety. Although *HACCP* do help enterprises improve food quality (Gillespie et al., 2000) and may support inspection of regulatory authorities (WHO, 1993), the implementation of the *HACCP* system requires additional resources for *staff training*, *investments in buildings and equipments*, *extra purchase of supplies* as well as *technical support* (Henson, Holt, & Northen, 1999). Like other kinds of enterprises, food enterprises are in pursuit of profits as well. Therefore, inadequate information on costs and benefits have prevented food enterprises from implementing *HACCP* to a certain extent since it is important for them to possess necessary information to concretely evaluate the *economic* impacts prior to implementation.

And the published studies showed that the costs and benefits for food enterprises to implement and operate *HACCP* varied from country to country. In the companies of Northern Greece, the major costs of implementing and operating *HACCP* were respectively *staff training* and *product testing*, and the benefits were ascribed to 3 headings: *clientele benefits*, *product improvements* and *improvements in production procedures* (Semos and Kontogeorgos, 2007). And in the UK dairy processing plants, the major cost of implementing and operating *HACCP* was *staff time required to document the system*, and the most important benefit was the *enhanced ability to retain existing customers* (Henson, Holt & Northen, 1999). However, in the Mexican meat plants, *investment in new equipment* and *microbiological tests of products* accounted for most of the implementation and operational costs, and the main benefit is *reduction in microbial counts* (Maldonado et al., 2005).

However, there is little published information on the status of adoption of *HACCP* system in the food industry of China, and the resulted costs and benefits that come with the system. Bai L. (2005), et al. depicted the features of the China's food enterprises which have implemented the *HACCP* system, which laid a foundation for the further researches in the field. Liu et al (2006). reported that the large-sized food enterprises in China knows *HACCP* system well and have implemented either *ISO, GMP*, *SSOP* or *HACCP* system; however, the small-sized enterprises in China know *HACCP* system little and most of them lack enough motivation to implement the *HACCP*. Although Wang Z.G., et al (2006). verified the effectiveness of *HACCP* on the firms' economic performance, their verification is based on the financial data in the first year of *HACCP* implementation rather than the real costs and benefits coming from the implementation of the system.

Moreover, Wang et al. didn't identify the major costs and benefits of the implementation of *HACCP* in the food industry.

The aim of this paper is to explore the process of *HACCP* implementation in the food industry of China. The paper reports the initial results of a survey that addresses the following questions in turn: What are the major costs of implementing and operating *HACCP*? What are the major benefits?

2. MATERIALS AND METHODS

In September, with the help of *Jilin Food and Drug Administration* in *Jilin* Province, northeast of China, questionnaires were sent by faxes and e-mails to the forty-nine slaughtering and meat product processing enterprises who wish to be involved. Referring to previous researches in this field, the questionnaire was comprised of three sections. The first section investigated the status of the quality assurance systems, the second examined the costs and benefits derived by the *HACCP* implementation and operation, and the last one dealt with the general data of the surveyed enterprises.

The structure of the questionnaire and the list of costs and benefits referred to several researchers in the specific field of cost-benefit analysis of *HACCP* or *ISO* implementation and operation (see for example Bredahl and Holleran, 1997; Henson, Holt and Northen, 1999; Maldonado et al., 2005). And personal interviews, based on a pilot questionnaire, were used to form the final questionnaire. Personal interviews were conducted in three food companies located in the city of Changchun (Northeast China).

The final form of the questionnaire was sent, with the help of *Jilin Food and Drug Administration* in *Jilin* Province, to the forty-nine slaughtering and meat product processing enterprises in September. And a reminder was mailed to non-respondents at the end of October, 2007. Finally, a total of twenty-six questionnaires were returned, and only twenty-one of them, locating in eighteen provinces in China, were fully completed and accepted for survey purposes. Some enterprises refused to return questionnaires for the following reasons of "confidential information", "incapable of answering", "persons concerned unavailable" or "no responsibility for offering information to scientific institution", etc.

Although only twenty-one questionnaires were available, it was considered sufficient enough to serve the objectives of such a survey for three reasons. Firstly, the top 3 of the slaughtering and meat product processing enterprises, Shineway (Shuang Hui), Yurun (Yu Run) and Jinluo (Jin Luo), were involved in the research, and their sales revenue occupied more than 10 percent of the total revenue of all the Chinese slaughtering and meat product processing enterprises in 2006. Secondly, the sales revenue of the 21 respondents almost occupied one fourth of the total revenue of the industry. Thirdly, the paper is still a tentative descriptive analysis aiming at inspiring further exploration and quantitative investigation in the field.

The Statistical Package for the Social Sciences – *SPSS* (Release 12.0, September 2003) for Windows was used to analyze the data. Descriptive statistics was used to describe the *HACCP* implementation status and the general data of the respondents.

3. RESULTS AND DISCUSSION

According to the Interim Measures for Statistical Definitions of Large, Medium and Small Enterprises (the State Economic and Trade Commission, No. 143, 2003) promulgated by the State Economic and Trade Commission, the State Development Planning Commission, the Ministry of Finance and the National Bureau of Statistics, eight (38.1%) of the twenty-one surveyed respondents were large-sized (employees over 2000, sales over *RMB* 300 million Yuan and total assets over *RMB* 400 million Yuan), ten (47.6%) were medium-sized (employees ranged from 300 to 2000, sales ranged from *RMB* 30 million to 300 million Yuan and total assets ranged from *RMB* 40 million to 400 million Yuan) and 14.3% were small-sized (employees less than 300, sales less than *RMB* 30 million Yuan and total assets less than *RMB* 40 million Yuan). Considering the higher proportion of small-sized food enterprises in China, the results showed that more large-sized and medium-sized food enterprises implemented the *HACCP* system than small-sized ones did.

The product types of the survey respondents are frozen meat, cooked meat and iced fresh meat. And more than one thirds of the respondents also deal in the fields of livestock and poultry breeding, Biological drugs and concentrated feedstuff. Only five of the respondents just distribute products in domestic market. And the top five exporting countries of the sixteen respondents are Hong Kong, Russia, Singapore, Japan and Korea.

100% of the respondents claimed that they had full operation of *HACCP* system in place. And eighteen (85.7%) of them had a fully operational *HACCP* system for more than twelve months, and only one (4.8%) less than six months. Respondents were asked to indicate how long it took from their starting to implement *HACCP* system to fully operating it. Sixteen (76.2%) of them indicated that the whole process took 6-12 months. However, four (19.0%) indicated that it took more than 18 months.

3.1 Costs of implementing and operating HACCP

Respondents were presented with a list of costs which previous studies have suggested can be incurred when implementing and operating quality control/assurance systems such as *HACCP* or *ISO 9000* (for example Vanguard Consulting, 1993; Bredahl and Holleran, 1997; Maldonado et al., 2005). They were then asked to rank these costs, a score of one meaning major importance and of zero if a cost not incurred. Respondents were allowed to give a score of one to none, one or more items.

3.1.1 Costs of implementing HACCP

There was great variation in the costs of implementing *HACCP* among individual respondents (Table 1). For example, although 33.3% of the respondents judged *investment in new* equipments to be the most important cost associated with the implementation of *HACCP*, 37.7% of them had not incurred that cost at all. In a whole, more medium-sized and small-sized enterprises took the cost of *investment in new equipment* as the most important cost, while more large-sized ones took *staff training* as the most important one. As showed in table 1, *staff time in documenting system* was the most frequently incurred cost, being ranked zero by none of the respondents, and one by 19.0% of the respondents.

Table 1. Rank score for the costs of implementing *HACCP*

Costs	Proportion of respondents giving rank of zero (%)	Proportion of respondents giving rank of one (%)
External consultants	48.3	8.3
Investment in new equipments	37.7	33.3
Staff training	10.0	28.6
Managerial changes	50.0	14.2
New employment of Personal	66.7	8.3
Staff time in documenting system	0.0	19.0

Furthermore, sixteen (76.2%) of the twenty-one respondents had conducted a formal cost analysis before implementing *HACCP* in their enterprises. eleven (66.7%) of the 16 responded that the actual implementation costs were almost the same with their expectations, while five (23.8%) underestimated the implementation costs. And among the lists of underestimated costs, the costs of *staff training* and *investment in new equipments*, which was reported by medium-sized and small-sized enterprises, were referred to more regularly.

3.1.2 Costs of operating HACCP

The respondents were asked to rank the individual costs of operating *HACCP* in a manner similar to that described above. Like the costs of implementing *HACCP*, there was still great variation among the respondents (Table 2). Considering the results as a whole, *record keeping* was the most frequently incurred cost (ranked zero by none of the respondents), but *product testing* was ranked as the most important cost (ranked one by 41.7% of the respondents).

Table 2. Rank scores for the costs of operating *HACCP*

Costs	Proportion of respondents giving rank of zero (%)	Proportion of respondents giving rank of one (%)
ecord keeping	0.0	23.8
roduct testing	19.0	41.7
taff training	28.6	19.0
lanagerial/supervisory time	52.4	14.3

Seven (33.3%) of the respondents indicated that they underestimated the operating costs and the most widely underestimated cost were *product testing* and *record keeping*, while the others indicated that the actual operating costs were almost same with their expectation. And twelve (52.4%) of the respondents indicated that their total production cost had increased after implementing and operating the system. Three (14.3%) indicated that their total production cost decreased after implementing and operating the system. And after implementing and operating *HACCP* system, more small-sized respondents indicated an increase in total production cost than middle-sized and large-sized respondents did. 64.3% of the respondents with employees less than1000 indicated an increase in total production cost, while only 37.6% of the respondents with employees over 1000 indicated an increase. Although the costs of implementing and operating *HACCP* varied greatly among the respondents, the most regularly occurred and important costs were relatively tangible costs, like the cost of *product testing*, rather than the intangible costs as reported by some other researchers.

However, as mentioned by many researchers (for example Maldonado et al., 2005), the reliability of these results on cost analysis depends on whether the respondents could clearly identify and distinguish the costs of implementing and operating *HACCP*.

3.2 Benefits of implementing and operating HACCP

Respondents were presented with a list of benefits which referred to previous studies and were adapted to the Chinese food enterprises based on a pilot survey in three enterprises. The respondents were asked to rank these benefits, a score of one meaning major importance and of zero if a cost not incurred. Respondents were allowed to give a score of one to none, one or more items.

A relatively higher proportion of the respondents allocated zero ranks to each of the possible benefits, indicating significant differences in the types of benefits derived from implementing and operating *HACCP* among different enterprises. As showed in table 3, according to the one scores allocated by the respondents, the top three benefits were *increased ability to retain existing customers*, *reduced product microbial counts* and *increased ability to access new overseas market.*

Table 3 Rank scores for benefits of implementing and operating *HACCP*

Benefits	Proportion of respondents giving rank of zero (%)	Proportion of respondents giving rank of one (%)
icreased ability to improve production procedures	38.1%	14.3%
nproved staffs' consciousness of food safety	47.6%	9.5%
icreased product sales	47.6%	9.5%
icreased ability to access new overseas market	42.9%	19.0%
icreased ability to attract new customers	38.1%	14.3%
icreased ability to retain existing customers	33.3%	23.8%
educed product microbial counts	33.3%	19.0%
educed product wastage	52.4%	9.5%
icreased product prices	61.9%	4.8%
icreased product shelf life	75.0%	9.5%
icreased motivation of supervisory/managerial staff	80.9%	0.0%

Results also showed that all of the respondents indicated that they did derive benefits and nine of them (42.9%) indicated having derived distinct benefits. And the majority (76.2%) of the survey respondents indicated that the benefits they had experienced from implementing and operating *HACCP* were approximately in accordance with their prior expectations, and 28.6% of the survey respondents indicated that some intangible benefits, such as *increased ability to retain existing customers* and *increased ability to access new overseas market* were even higher than their expectation. However, a significant minority (23.8%), especially small-sized enterprises, indicated that some benefits, like *increased product prices* and *increased sales* were lower than they had expected.

However, as mentioned by many researchers, these results on benefits presented above should be interpreted with some caution. Given the intangible nature of many of the potential benefits of *HACCP* and the fact that a higher proportion of Chinese food enterprises do not formally monitor the costs and benefits of *HACCP*, many respondents may not have been aware of the benefits and their degree that *HACCP* had on their operations.

4. CONCLUSIONS AND SUGGESTIONS

4.1 Conclusions

The research results on the costs of implementing and operating *HACCP* showed that although most slaughtering and meat product processing enterprises indicated that the costs they had experienced from implementing and operating *HACCP* were approximately in accordance with their prior expectations, still a significant proportion of the survey respondents indicated some costs exceeded their expectations.

And the research results on the benefits of implementing and operating *HACCP* showed that: firstly, slaughtering and meat product processing enterprises can derive benefits from implementing *HACCP* system and some of them have derived distinct benefits; secondly, the benefits from implementing the system are relatively intangible; thirdly, still a significant minority, especially small-sized enterprises, indicates that some benefits, like *increased product prices* and *increased sales* are lower than their expectation.

4.2 Suggestions

Firstly, policy makers should take account of these research results and make more quantitative researches on the costs and benefits of implementing and operating *HACCP* to offer more comprehensive and classified information to help food enterprises make decisions on *HACCP* implementation and operation.

Secondly, policy makers should also take due regard of these costs and benefits when making *HACCP* relative regulation policies or offering directions and destructions to support food enterprises to implement the system.

Thirdly, policy makers can effectively support food enterprises by providing interest subsidy to purchase necessary quality control facilities and offering free manpower training since the major costs of implementing and operating *HACCP* in slaughtering and meat product processing enterprises were still relatively tangible, such as investment in new equipments and product testing.

Fourthly, the research results also suggest that policy-makers should take due regard of the differences of costs and benefits among different sized enterprises when formulating regulation and/or support as part of measures designed to promote *HACCP* adoption since it seems that the costs of implementing and operating *HACCP* of small-sized enterprises are different from those of large-sized ones.

ACKNOWLEDGEMENTS

The research is sponsored by Humanities and Social Science Project of Ministry of Education of China (06JC630013). The opinions expressed here do not necessarily reflect those of the funding agency. We also thank the Editor and the anonymous referees for helpful comments and suggestions.

REFERENCES

Bai, L., Ma, C. L., Gong, S. L., & Yang, Y. S. Implementation of HACCP system in China: A survey of food enterprises involved. Food Control, 2007, 18: 1108 - 1112.

Bredahl, M., Holleran, E. Transaction costs and institutional innovation in the British food sector. Food Safety, 1997, 3: 403 - 419.

Gillespie, I., Little, C. and Mitchell, R. "Microbiological examination of cold ready-to-eat sliced meats from catering establishments in the United Kingdom", Journal of Applied Microbiology, 2000, 88: 467 - 474.

Henson,S.J., Holt,G. and Northen, J. Cost and benefits of implementing HACCP in the UK dairy processing sector, Food Control, 1999, 10: 99 - 106.

Liu X., Zheng F. T. and Liu, J. H. Industry cognition, implementation and orientation of diffusing HACCP: Demonstrating study on survey samples. Business economics and administration, 2006, 15 (10): 39 - 43.

Maldonado E. S., et al. Cost-benefit analysis of HACCP implementation in the Mexican meat industry. Food Control, 2005, 16: 375 - 381.

Seddon, J., Davis, R., Loughran, M., Murrell, R. Implementation and value added: A survey of registered companies. 1993, Vanguard Consulting Ltd, Buckingham.

Semos, A. and Kontogeorgosn, A. HACCP implementation in northern Greece - Food companies' perception of costs and benefits. British Food Journal, 2007, 109 (1): 5 - 19

Swanson, K. M. J. and Anderson, J. E. Industry perspective on the use of microbial data for HACCP validationand verification. Journal of Food Protection, 2000, 63 (6): 815 - 818.

Wang, Z. G., Weng, Y. Z., Yang, Z. G. and Zheng, F. T. The effectiveness of adopting HACCP system certification in food industry: From the questionnaire of 482 food manufacturing enterprises in China. China Soft Science, 2006, 9: 69 - 76.

WHO (1993), Training considerations for the application of the hazard analysis critical control points system to food processing and manufacturing, WHO Document, WHO/FNU/FOS/93.3, World Health Organization, Division of Food and Nutrition,

STUDY OF APPLICATION OF THZ TIME DOMAIN SPECTROSCOPY IN FOOD SAFETY

Liying Lang[1,*] , Na Cai[2]

[1] Hebei University of Engineering, Handan, China, 056038;

[2] College of Information and Electrical Engineering, Hebei University of Engineering, Handan, 056038, China

[*] Corresponding author, Address: Hebei University of Engineering, Handan,Hebei province, Guangming South Street 199, 056038 , P. R. China, Tel: 0310-8579551, Fax:0310-8578746, Email: langliying@126.com

Abstract: In this paper, a new spectroscopy technology named terahertz time-domain spectroscopy (THz-TDS) is introduced, which is used in food safety. We describe a coherent subpicosecond THz spectroscopy system based on non-resonant optical rectification for the generation of THz radiation. As an example, we measured absorption spectrum of water vapor by THz-TDS in frequency from 0.5 to 2.5 THz, The experiment demonstrated that the spectroscopy resolution of system was up to 0.0001THz, which can be measured vegetable pesticide residual, for it neither need sample pretreatment nor cause pollution.

1. INTRODUCTION

In recent years, food quality safe has attracted more and more attention. According to statistics, there are 15 billion of diarrhea cases a year in the whole world, which makes more than 3 million children death. And 70 percent of these result from food and water polluted by many kinds of pathogenic microorganism. There are approximately 76 million food-borne disease cases every year in the US, bringing on 325 thousand persons hospitalization treatment and 5000 deaths, which directly or indirectly causes tens billions dollars loss every year. In particular, the frequently breaking out of International food security incidents, O17 (Japan, 1996), bse (Britain, 2000), Swine streptococcus (China, 2006) and so on, has attacked

Please use the following format when citing this chapter:

Lang, L. and Cai, N., 2009, in IFIP International Federation for Information Processing, Volume 295, *Computer and Computing Technologies in Agriculture II, Volume 3*, eds. D. Li, Z. Chunjiang, (Boston: Springer), pp. 1691–1697.

the alarm bell to the common people again. In order to avoid the occurrence of these accidents, the first thing we should do is to carry on the safe examination to the food, which is the base of reducing the food-borne disease. Whether one product has the unsafe factor and what is the unsafe degree both rely on the examination. Therefore, the examination is the most essential method of guaranteeing food safety. In agricultural production, the pesticide which is one kind of organic mixtures has been universally using in pest control. Because of toxic and high residual, while promoting socio-economic development, the pesticide can cause the harm to the human health and the ecological environment either. Presently, the environmental pollution created by the pesticide residue has already aroused widespread social attention. Therefore, it has important meaning to carry on the high sensitive detection aiming at the pesticide residue.

How to detect pesticide residue in food quickly and efficiently is a key problem in food safety controlling system. The conventional methods need complicated sample pretreatment, consume biochemical reagents and last a long measuring period. So we attempt to achieve the pesticide residual measurement by optical method. This paper is concentrated on the methodology study for direct measurement on vegetable pesticide residue based on terahertz time domain spectroscopy, for it neither needs sample pretreatment nor causes pollution.

2. THZ TIME DOMAIN SPECTROSCOPY

THz time domain spectroscopy (A. Rice, 1994) has recently emerged as an extremely useful and powerful probe of charging transport in materials, owing to the fact that it provides a probe of the complex conductivity over a wide frequency range, with sub-picosecond time resolution. As such, it has many advantages over conventional techniques such as resistivity and hall measurements: it is an all-optical technique (and therefore contactless) and it allows the study of non-doped (pristine) samples. Moreover, the experiments are ideally suited for pump-probe experiments that can access charge dynamics such as recombination and trapping. Inorganic and organic semiconductors and non-polar liquids are some of the materials to which THz spectroscopy has been successfully applied. At AMOLF we are currently looking at charge transport in novel materials such as semi-conducting polymers and quantum dots. These materials have important technological applications in solar cells and other electronic devices (D. Dragoman, 2004).

THz generation via optical rectification can be accomplished in two different configurations characterized by slight and tight focusing of femtosecond laser beams. Slight focusing means that the beam size in the emitter is much larger than the extension of the pulse in the longitudinal direction while for tight focusing they are comparable. When the laser beam is slightly focused on the velocity mismatch, it reduces the effective thickness of the emitter contributing to THz generation. In this situation the THz radiation originates from two thin layers near the front and the back surfaces of the emitter. In the case of tight focusing, the smaller beam waist of the pump thus the increased intensity results in high conversion efficiency of THz radiation.

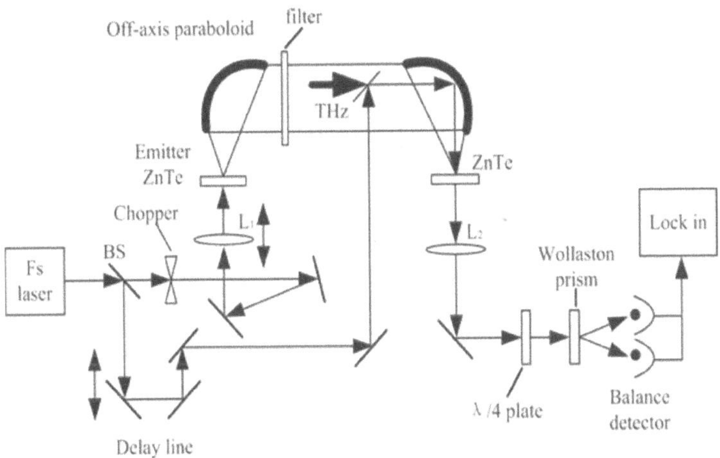

Fig.1: Schematic drawing of the experimental setup

In a wideband coherent THz radiation system using ZnTe crystals, the generation of THz radiation is based on optical rectification, and THz radiation detection is via electro-optic effects in ZnTe. Optical rectification allows producing much broader THz bands, and the simplicity of the experimental realization is another advantage. Figure 1 schematically illustrates the experimental setup of THz radiation. A homemade self-mode-locked Ti: sapphire laser delivered 40fs pulses with an average power of 300mw at 810nm (Qirong Xing, 1995), repetition rate of 100MHz. The linearly polarized femtosecond laser beam with polarization direction along the table plane was split into two parts, in which 70% was used as the pump beam, and the remains 30% was used as the probe beam. The pump beam chopped into 1050Hz was focused by lens L_1 (f=5cm) on the ZnTe emitter. Lens L_1 was mounted on a translation stage, which was carefully set so that it could be moved along the pump beam direction. Two 1mm thick <110> ZnTe crystal was used as the emitter and the detector of the THz radiation respectively. The polarization of both the THz and optical probe beams were aligned parallel to the [1, -1, 0] direction of the ZnTe detector. The THz radiation was collimated with two off-axis parabolic mirrors and focused on the detector for electro-optic sampling of the THz radiation. A teflon filter, F, was used to remove the residual pump beam. The probe beam modulated by

THz radiation was collimated with lens L_2 and optically biased at a quarter-wave plate, which allows the system to be operated in the linear range. A Wollaston polarizer was used to convert the THz field induced phase retardation of the probe beam into an intensity modulation between the mutually orthogonal linearly polarized beams. A balanced photoelectric detector was used to detect the optical intensity modulation, which wired to give zero current for balanced power. The signal current was sent to a lock-in amplifier for amplifying the signal and filtering noise. The optical delay line worked through a computer-controlled translation stage (Physikinstrumente, M-405.DG) with a resolution of 0.1μm. Via this setup, we succeeded in detecting the spatial profile of a THz radiation. Figure 2 shows the THz temporal signal and the corresponding frequency spectrum (insert) obtained from the emitter ZnTe beyond focus.

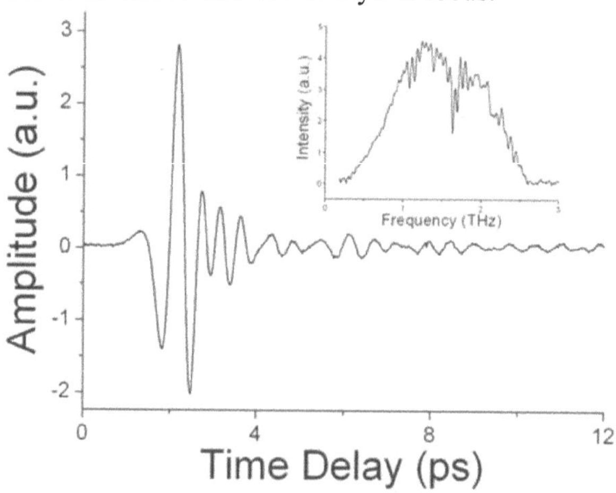

Fig.2: Typical THz waveform and its frequency spectrum

As an application example of our THz system to TDS, we measured the absorption spectrum of water vapor by test THz radiation pulses signal through water vapor. We performed the experiments as follows; firstly, we detected THz radiation pulses as reference background after only propagating through atmosphere in the lab. And then by putting a cup of hot water under the THz radiation beam between two collimating mirrors, we measured THz radiation pulses through water vapor the amplitude absorption spectrum of the atmosphere and water vapor are given from Fourier transform of corresponding to the THz time-domain signals. By subtracting the amplitude absorption spectrum of the atmosphere from that of sample with water vapor the amplitude absorption spectrum of water vapor in frequency range from 0.5 to 2.5 THz shows in Figure 3 with a measurement accuracy as high as reported in the Ref.8. The data in table 1 indicate the comparison of the water vapor spectrum between in Ref (R. A. Cheville, 1999) in our experiment.

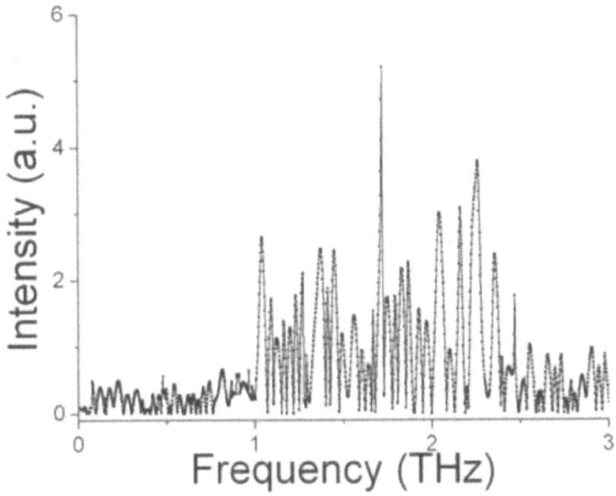

Fig.3: the water vapor absorption spectrum

Table .1 Water vapor spectrum

D1	1.09922	1.1156964	1.16257	1.20938	1.23055	1.41094	1.60323	1.67109
D2	1.09367	1.1256175	1.16686	1.20078	1.23272	1.41312	1.60480	1.66869
D1	1.71808	1.79766	1.87031	1.91965	2.04147	2.16563	2.19860	2.25703
D2	1.72131	1.79272	1.86788	1.92801	2.04452	2.16479	2.19861	2.26251

* *D_1-the water vapor spectrum of in Ref (R. A. Cheville, 1999), D_2-the spectrum of water vapor in our experiment。 Units： THz*

As an application example of our system to TDS, we measured the amplitude absorption spectrum of water vapor in frequency range from 0.5 to 2.5 THz. The measured spectrum data show that our system possesses a measured accuracy as high as reported in the literature.

3. ANALYSIS OF MEASURED THEORY

Taking into account these important advantages it was believed that TDS will surmount in many ways the conventional spectroscopic method used in the THz range of frequencies, i.e. the Fourier-transform infrared (FTIR) spectroscopy. The FTIR spectrometer has as a source of radiation an incoherent lamp and is based on the two-beam interferometry principle. Contrast to TDS, the output signal of the FTIR spectrometer is proportional to the intensity of the FIR field.

However, recent results have demonstrated that the SNR of TDS is better than that of FTIR for frequencies in the range 10GHz-4THz. In this range the field amplitude SNR of THz TDS is 104 while that of FTIR is about 300.

Beyond this frequency range (the FTIR method is able to work from FIR up to visible spectral range), the SNR of FTIR is better. The SNR of TDS is drastically decreased when its bandwidth is increased up to 40 THz. Both methods show a similar spectral resolution≈$0.1cm^{-1}$.

Thus, TDS is an extremely powerful method of spectroscopy, working at room temperature with an impressive sensitivity. TDS is performed in two steps. Firstly, a reference signal $E_r(t)$ is detected in the absence of the sample. Then, a signal in the presence of the sample, $E_s(t)$, is detected. The transmission function of the TDS is given by (L. Xu, 1992)

$$\mathrm{T}(\omega) = E_s(\omega) / E_r(\omega) = |T(\omega)| e^{i\varphi(\omega)} \tag{1}$$

The index of refraction of the sample $\mathrm{N} = n + iK$ is related to the transmission function through the equation:

$$T(\omega) = 4N \exp[i\omega(N-1)d/c] \sum_{i=0}^{P} \{[(N-1)/(N+1)] \exp(i\omega Nd/c)\}^{2i} \tag{2}$$

Where d is the thickness of the sample, Expression (2) is the transfer function resulting from the passage of the THz field through the sample, the THz field being reflected P times inside the sample. These repeated reflections inside the sample, which can be modeled as a Fabry-Perot-like effect, are seen in the time domain as distinct pulses accompanying the main time domain response. Thus, P is easily determined at least in the case of thick samples.

When p=0, Eq. (2) allows a simply determination of the real and imaginary parts of the refractive index of the material as:

$$|\mathrm{T}(\omega)| = 2(n^2+k^2)^{\frac{1}{2}} \exp(-\omega k d/c) / [(n+1)^2 + k^2] \tag{3}$$

$$\varphi(\omega) = \omega(n-1)d/c + \arctan\left\{-k(n^2+k^2-1)/[n(n+1)^2 + k^2(n+1)]\right\} \tag{4}$$

Solving the system formed by the last two equations we can extract the real and imaginary part of the index of refraction from the known amplitude and phase of the transmission function ate a certain frequency. The procedure is repeated for any frequency with the bandwidth of the TDS system. Thus only two time-domain measurements are necessary to determine $\mathrm{T}(\omega)$, from which the frequency behavior of the complex index of refraction of the material is determined. Both analytical methods and numerical algorithms are used to determine simultaneously.

The index of refraction and the thickness of the sample, a procedure useful mainly for thin samples, is difficult to estimate the sample thickness.

4. CONCLUSION

In this paper, a coherent subpicosecond THz spectroscopy system is introduced based on non-resonant optical rectification for the generation of THz radiation. As an example, we measured absorption spectrum of water vapor by THz-TDS in frequency from 0.5 to2.5 THz, The experiment demonstrated that the spectroscopy resolution of system was up to 0.0001THz. With the installation of the system continuously improving, it will certainly have the widespread application prospect.

REFERENCES

A. Rice, Y. Jin, X. F. Ma, and X. C. Zhang, Appl.phys.Lett.1994,64:1324-1326
D. Dragoman, M. Dragoman, Progress in Quantum Electronics. 2004, 28:1-66
L. Xu, X. C. Zhang and D. H. Auston, Appl.phys.Lett.1992,61:1784-1786
Qirong Xing, Weili Zhang and K. M. Yoo, Opt Commun.1995,119:113-116
R. A. Cheville, D. Grischkownsky. J. Opt. Soc. Am. B.1999,16:317-321

NONDESTRUCTIVE DETECTION OF THE INTERNAL QUALITY OF APPLE USING X-RAY AND MACHINE VISION

Fuzeng Yang[1,2,*], Liangliang Yang[1], Qing Yang[1], Likui Kang[1]
[1] *College of Mechanical and Electronic Engineering, Northwest Agriculture and Forestry University, Yangling, China, 712100*
[2] *Department of Computer Science and Technology, Northwestern Polytechnical University, Xi'an, China, 710072*
[*] *Corresponding author, Address: College of Mechanical and Electronic Engineering, Northwest Agriculture and Forestry University, Yangling, 712100, P. R. China, Tel: +86-29-87092913, +86-13772025795, Email: yfz0701@163.com*

Abstract: The internal quality of apple is impossible to be detected by eyes in the procedure of sorting, which could reduce the apple's quality reaching market. This paper illustrates an instrument using X-ray and machine vision. The following steps were introduced to process the X-ray image in order to determine the mould core apple. Firstly, lifting wavelet transform was used to get a low frequency image and three high frequency images. Secondly, we enhanced the low frequency image through image's histogram equalization. Then, the edge of each apple's image was detected using canny operator. Finally, a threshold was set to clarify mould core and normal apple according to the different length of the apple core's diameter. The experimental results show that this method could on-line detect the mould core apple with less time consuming, less than 0.03 seconds per apple, and the accuracy could reach 92%.

Keywords: X-ray; machine vision; detection; apple; internal quality.

1. INTRODUCTION

Mould core of apple could be aroused by *Trichothecium roseum* and other factors (Zhang et al., 2008). The *fumaric acid* in apple juice increases

---

Please use the following format when citing this chapter:

Yang, F., Yang, L., Yang, Q. and Kang, L., 2009, in IFIP International Federation for Information Processing, Volume 295, *Computer and Computing Technologies in Agriculture II, Volume 3*, eds. D. Li, Z. Chunjiang, (Boston: Springer), pp. 1699–1706.

dramatically from 0.8 to 2.6 due to the mould core (Yi et al., 2001). Therefore, it is necessary to detect the internal quality to guarantee the quality of apple reaching market.

In the past few years, there are many methods which have been proposed to detect internal quality of fruits and vegetables. Multi spectroscopy has been investigated for detecting the brown heart of pear (Zerbini et al., 2002; Han et al., 2006). Near-Infrared Spectroscopy (NIR) has been used for predicting sugar content of Apple (Ying et al., 2006). Ultraviolet (UV) fluorescence has been reported for detecting freeze damage of citrus (Slaughter et al., 2008). Researches has been done at using Magnetic resonance imaging (MRI) (Thybo et al., 2004), laser technique methods to measure fruit tissue texture and internal disorders (Muramatsu et al., 1999). X-ray technology has also been adopted for detecting the agricultural quarantine materials (Toyofuku et al., 2007), seed weevil-infested mango fruits (Thomas et al., 1995) and the edible ratio and sugar content of rambuta (Zhang et al., 2005). However, there is little report on the detection of mould core apple using X-ray. It is necessary to do research on detecting the apple's internal quality using low-cost X-ray instrument.

Furthermore, digital image processing algorithm is vital in detecting and extracting the image features, such as brown heart in pear (Zerbini et al., 2002; Han et al., 2006), stem-end injury in citrus (Blasco et al., 2007) and edible ratio and sugar content of rambuta (Zhang et al., 2005). The algorithms were focused on spatial domain in former researches. In recent two decades, Wavelet Transform (WT) from Multi Resolution Analysis (MRA) has been successfully used in image processing, which can detect the signal catastrophe point in spatial and frequency domain. The WT is known as the mathematic microscope (Daubechies, 1992; Yang, 2007). Ying et al. (2006) has used the WT de-noising NIR image of apple; the results showed WT can improve the de-noising effort. Yang et al. (2005, 2007) has done many researches in agricultural image de-nosing and enhancement, the results showed the WT is an effective algorithm in agricultural image processing. In 1990's Sweldens et al. provided the Second Generation Wavelet Transform – Lifting Wavelet Transform (LWT), which provides a quick and less memory needed algorithm in frequency domain (Sweldens, 1995, 1996). However, there are not any researches on agricultural image processing using LWT.

The objectives of this study are to:

(1) Construct a real time detection system to detect the internal quality of apple using X-ray.

(2) Determine a suitable image processing algorithm to extract image's features using lifting scheme wavelet in frequency domain, which could efficiently detect the mould core apple.

(3) Determine a processing method, using the (2) algorithm, can efficiently classify the apples with or without mould core, and determine its accuracy.

2. MATERIALS AND METHODS

2.1 Apple samples

One hundred apple samples were harvested in Mid-September 2007 from an orchard in Yangling, Shaanxi province of China and stored without any special instruments. The apples were transported to the laboratory in March 2008, and then stored in the room condition.

2.2 Instrumentation

The real-time internal quality detection system consists of two parts. The first part is an X-ray radiator (BJI-1U, Bo Jin Electronic Instrument, Inc., Shanghai, China) and a machine vision system, which includes a black and white (BW) camera (ECB-1793, Hong Tian Zhi Electronic, Inc., Shenzhen, China) and an image grabber (BS-602A, Bao Shi, Inc., Taiwan, China). The other is an X-ray image processing workstation (Itellistation Z Pro MT: 9228, IBM, New York) with the image processing program programmed by ourselves using the Matlab (MathWorks, Inc., Natick, Mass).

2.3 Detecting the mould core of apple

The prime purpose of the system is to determine whether the apple is a mould core apple. The success of the detection depends mostly on the correct image feature extraction. The following sections will focus on how to extract image features.

2.3.1 X-ray image processing algorithm

We proposed an effective algorithm that is used to detect the mould core apple in frequency domains inspired by the LWT.

2.3.2 Lifting scheme wavelet

A canonical case of lifting consists of three stages: split, predict, and update (Sweldens, 1995, 1996). It is simply described as follows.

Firstly, the signal λ_0 is divided in to two parts λ_{-1} and γ_{-1}.

Secondly, we use the λ_{-1} subset to predict the γ_{-1} subset based on the correlation present in the original data. If a prediction operator P can be found,

$$g_{-1} = P(l_{-1}). \tag{1}$$

If the prediction is reasonable, the difference will contain much less information than the original γ_{-1} set. We can denote this abstract difference operator as

$$g_{-1} := g_{-1} - P(l_{-1}) \tag{2}$$

In this way, the wavelet subset encodes how much the data deviates from the model on which P was built. Then split λ_{-1} into λ_{-2} and γ_{-2}, and replace γ_{-2} with the difference between γ_{-2} and $P(\lambda_{-2})$. After n steps, the original data is replaced by the wavelet representation $\{\lambda_{-n}, \gamma_{-n}, \cdots, \gamma_{-1}\}$. In order to solve the aliasing problem, the third step - update is introduced. Using an operator U and the already computed wavelet set γ_{-1} update λ_{-1} so that a certain scalar quantity Q could be obtained,

$$Q(g_{-1}) = Q(g_{-0}). \tag{3}$$

Therefore, we use the already computed wavelet set γ_{-1} to update λ_{-1} so that the latter preserves Q. In this way, an operator U can be constructed and updating λ_{-1} as

$$l_{-1} := l_{-1} + U(g_{-1}) \tag{4}$$

The original data could be decomposed as

$$l_0 = \sum_i l_i + \sum_i g_i, (i = -n \sim -1) \tag{5}$$

The above steps illustrate the basic idea of lifting. One property of lifting is that the inverse wavelet transform can immediately be found by undoing the operations of the forward transform. In practice, this comes down to simply reversing the order of the operations and changing each + to a - and vice versa.

2.3.3 Detection algorithm and its implementation

This section would analyze the image to determine the mould core apple. The original RGB (Red, Green, Blue; the native space for color representation in computers) image was converted to grayscale image before processing. The X-ray image could be transformed into a low frequency and three high frequency images by LWT. The low frequency image always represents the approximate character of the image, which can represent the original image with low noise, while the high frequency image always represents the noise and the points of discontinuity.

The procedure of the detection method is described as,

(1) Extract the low frequency image using LWT that represents the feature of the apple's image (Fig.1 (b), Fig.2 (b)).

(2) Enhance the image using histogram equalization in order to recognize the mould core (Fig.1 (c), Fig.2 (c)).

(3) Detect the apple's edge using Canny operator to get the apple core's diameter (Fig.1 (d), Fig.2 (d)), from which we can clearly obtain that the diameter of mould core apple is longer than the normal one's.

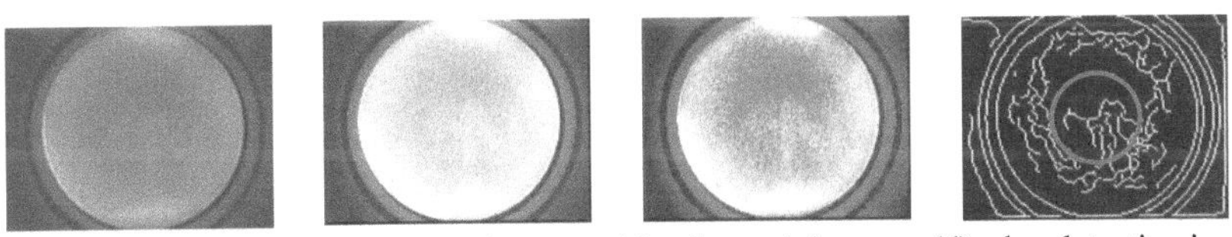

(a) original image (b) low scale image (c) enhanced image (d) edge detection image

Fig.1: X-ray images of a mould core apple

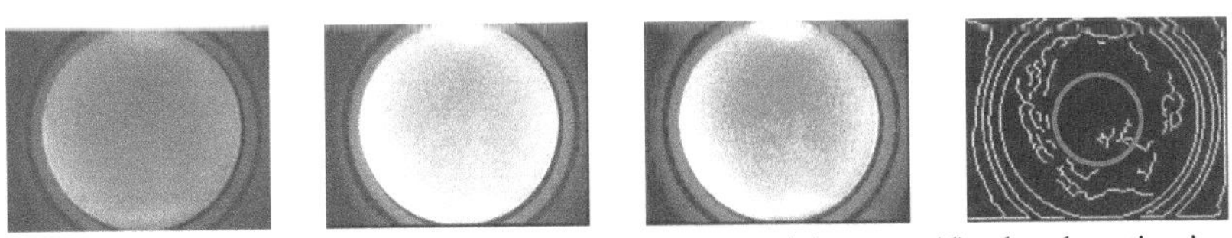

(a) original image (b) low scale image (c) enhanced image (d) edge detection image

Fig.2: X-ray images of a normal apple

(4) Calculate the diameter of the apple core by the statistical method - counting the pixels corresponding to the mould core (the pixel value is one in the binary image) in four directions (Fig.3), the algorithm is showed as the equation (6).

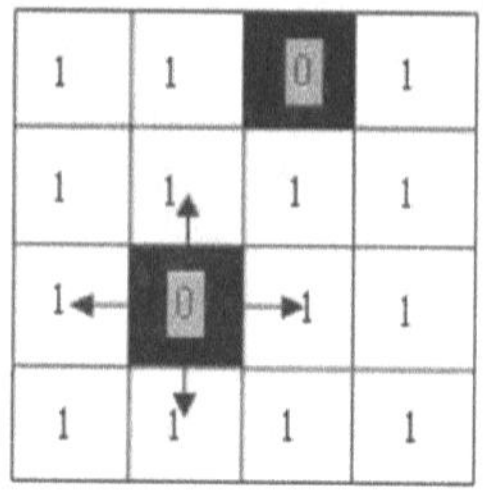

Fig.3: Four directions in the detection

$$m = \begin{cases} m+1, & pixel\ value=1 \\ m, & pixel\ value=0 \end{cases} \tag{6}$$

(5) Compare the value of m with the threshold that has been set to sort the apple, if m>threshold we can classify the apple to mould core apple, otherwise to good one.

3. RESULTS AND DISCUSSION

The results of this research are the development of a new online feature detection algorithm based on LWT, construction a detection system using machine vision and X-ray that can obtain the internal apple's image.

Table 1 shows the time consuming using different algorithms to detect the mould core. It illustrates that LWT is less-consuming than space domain method (contrast enhancement). We can clearly notice that the time consuming is less than 0.03 seconds by the frequency domain algorithm (LWT), which is quick enough to detect the quality in the on-line system.

Table 1. Time cost in different algorithms.

Time cost	Contrast enhancement	LWT (Wave name)				
		Haar	Db2	Db4	Lazy	9.7
1st	0.1424	0.0235	0.0245	0.0287	0.0225	0.0269
2en	0.1427	0.0229	0.0295	0.0276	0.0225	0.0265
3rd	0.1427	0.088	0.0238	0.0274	0.0224	0.0263
Average	0.1426	0.0251	0.0259	0.0279	0.0225	0.0266

The unit of time cost is s.

Table 2 illustrates the detection results. The apple could be sorted at the accuracy of 92%. Generally, the failed detection is aroused when the apple did not sharply face the X-ray radiator.

Table 2. Detection results of the experiment.

Sample	Number	Detected	Undetected	Accuracy rate
Normal	72	69	3	92%
Mould core	28	23	5	

Overall, the results reveal that X-ray image is available as a non-destructive method used for detecting the apple's internal quality. Although this paper focused only on apple, the algorithm could also be used for pear, peach, pineapple and kiwi fruit. Moreover, the algorithm can be polished by other methods that need more study to promote its robust and accuracy.

4. CONCLUSIONS AND FUTURE WORKS

From the experimental results, the following conclusions can be drawn.

(1) The X-ray can obtain clear internal image of apple, which can be processed to detect the quality.

(2) Lifting wavelet transform is an efficient way to process the X-ray image, which can easily reduce the noise and enhance the image.

(3) The detection algorithm can quickly (less than 0.03 seconds per apple) and accurately detects the mould core apples, with an accuracy of 92% in the experiment.

The results of this research demonstrate the feasibility of an on-line X-ray based sensor for detecting internal quality of apple. In addition, the other works can be done at the automatic conveyor to push the internal quality detection to a new stage.

ACKNOWLEDGEMENTS

This research was supported by the Excellent Young Scientists Fund of Northwest A&F University (01140303) and Postdoctoral Fund Project of China (20060401012). The authors would like to thank Mr. Fengtao Chen for his assistance in the preparation of the apple samples.

REFERENCES

Blasco J, Aleixos N, Gómez J, Moltó E. Citrus Sorting By Identification of The Most Common Defects Using Multispectral Computer Vision. Journal of Food Engineering, 2007, 83: 384-393 (in Chinese)

Daubechies I. Ten Lectures on Wavelets. SIAM, 1992

Han D, Tu R, Lu C, Liu X, Wen Z. Nondestructive Detection of Brown Core in the Chinese Pear 'Yali' by Transmission Visible-NIR Spectroscopy. Food Control, 2006, 17: 604-608

Muramatsu N, Sakurai N, Wada N, Yamamoto R, Takahara T, Ogata T, Tanaka K, Asakura T, Ishikawa-Takano Y, Nevins D J. Evaluation of Fruit Tissue Texture and Internal Disorders by Laser Doppler Detection. Postharvest Biology and Technology, 1999, 15: 83-88

Slaughter D C, Obenland D M, Thompson J F, Arpaia M L, Margosan D A. Non-destructive Freeze Damage Detection in Oranges Using Machine Vision and Ultraviolet Fluorescence. Postharvest Biology and Technology, 2008, 48: 341-346

Sweldens W. Lifting Scheme: A New Philosophy in Biorthogonal Wavelet Constructions. Wavelet Applications in Signal and Image Processing III. 1995: 68-79

Sweldens W. The lifting scheme: A Custom-Design Construction of Biorthogonal Wavelets. Applied and Computational Harmonic Analysis, 1996, 3: 186-200

Thomas P, Kannan A, Degwekar V H, Ramamurthy MS. Non-destructive Detection of Seed Weevil-infested Mango Fruits By X-ray Imaging. Postharvest Biology and Technology, 1995, 5: 161-165

Thybo A K, Jespersen S N, Lærke E, Stødkilde-Jørgensen H J. Nondestructive Detection of Internal Bruise and Spraing Disease Symptoms in Potatoes Using Magnetic Resonance Imaging. Magnetic Resonance Imaging, 2004, 22: 1311-1317

Toyofuku N, Schatzki T F. Image Feature Based Detection of Agricultural Quarantine Materials in X-ray Images. Journal of Air Transport Management, 2007, 13: 348-354

Yang F Z, Wang H B, Yang Q, Wang Z. Wavelet Transform and Its Application in the Processing of Fruit Image. Transactions of The Chinese Society of Agricultural Machinery, 2005, 36: 61-64 (in Chinese)

Yang F Z, Wang Z, Yang Q, Zhang Y N. Application of Wavelet Transform-based Wiener Filtering Method to Denoise in Agricultural Product images. Transactions of the Chinese Society of Agricultural Engineering, 2007, 23: 145-150 (in Chinese)

Yi J H, Qiu N X, Hu B, Zhu Z B. Research of the Fumaric Acid in Concentrated Apple Fuice. China Fruit and Vegetable, 2001: 24-24 (in Chinese)

Ying Y B, Liu Y D, Fu X P. Sugar Content Prediction of Apple Using Near-Infrared Spectroscopy Treated by Wavelet Transform. Spectroscopy and Spectral Analysis, 2006, 26: 63-66 (in Chinese)

Zerbini P E, Grassi M, Cubeddu R, Pifferi A, Torricelli A. Nondestructive Detection of Brown Heart in Pears by Time-resolved Reflectance Spectroscopy. Postharvest Biology and Technology, 2002, 25: 87-97

Zhang C H, Liu C Q, Liu M H, Wang Q. A Study on Inspecting Internal Quality of Rambuta Using X-ray CT Imaging. Acta Agriculturae Universitis Jiangxiensis, 2005,27,06 : 939-942 (in Chinese)

Zhang J, Yuan X L, Zhao L, Cheng S L. The Rule and Control of Mould Core of Apple. Northwest horticulture: fruit trees, 2008, 1: 52-52 (in Chinese)

VALIDATION OF AN HPLC-DAD-ESI/MS/MS METHOD FOR THE CLASSIFICATION OF GREEN TEAS

Jingbo Yu, Nengsheng Ye, Xuexin Gu*, Ni Liu
Department of Chemistry, Capital Normal University, Beijing, China 100048
* *Corresponding author, Address: Department of Chemistry, Capital Normal University, No.105, Xisanhuan North Road, Beijing 100048, P. R. China, Tel:+86-10-68902490, Email: guxuexin@263.net*

Abstract: A reversed phase high performance liquid chromatography (RP-HPLC) separation coupled with diode array detection (DAD) and electrospray ionization mass spectrometer (ESI/MS) was developed and optimized for the classification of green teas. Five catechins [epigallocatechin (EGC), epigallocatechin gallate (EGCG), epicatechin (EC), gallocatechin gallate (GCG), epicatechin gallate (ECG)] had been identified and quantified by the HPLC-DAD-ESI/MS/MS method. The limit of detection (LOD) of five catechins was within the range of 1.25-15 ng. All the analytes exhibited good linearity up to 2500 ng. These compounds were considered as chemical descriptors to define groups of green teas. Chemometric methods including principal component analysis (PCA) and hierarchical cluster analysis (HCA) were applied for the purpose. Twelve green tea samples originating from different regions were subjected to reveal the natural groups. The results showed that the analyzed green teas were differentiated mainly by provenance; HCA afforded an excellent performance in terms of recognition and prediction abilities. This method was accurate and reproducible, providing a potential approach for authentication of green teas.

Key words: green tea, catechins, classification, HPLC

Please use the following format when citing this chapter:

Yu, J., Ye, N., Gu, X. and Liu, N., 2009, in IFIP International Federation for Information Processing, Volume 295, *Computer and Computing Technologies in Agriculture II, Volume 3*, eds. D. Li, Z. Chunjiang, (Boston: Springer), pp. 1707–1719

1. INTRODUCTION

Green tea is the most popular beverage in China. Recent studies have proved that green tea confers beneficial effects to human's health, especially for the effects of anti-oxidation, anti-cancer and prevention of cardiovascular disease. It is generally believed that catechins (Fig.1), the principal bioactive compounds of green tea, are responsible for the claimed therapeutic activities (Sakanaka et al., 1989; Blentine, 1997; Yang, 1997). Therefore it is essential to establish routine quality control method for the analysis of catechins in green tea.

Several analytical methods for analyzing catechins had been developed (Lee et al., 2000; Bonoli et al., 2003; Zhao et al., 2006; Shaghaghi et al., 2008; Tsukagoshi et al., 2008). Among the methods, HPLC was widely applied due to its excellent reproducibility and accuracy. Although most of HPLC methods were practicable, many established strategies were rarely available for simultaneous analysis of the content of catechins (Goto et al., 1996; Dalluge et al., 1998; Wang et al., 2000; Wang et al., 2003; Yao et al., 2004; Nishitani et al., 2004; Pelillo et al., 2004; Zhu et al., 2004; Liang et al., 2005; Owuor et al., 2007), no available method was suitable for authentication of tea. However, tea authenticity was vital to tea quality. Therefore, methods to guarantee tea authenticity, based on chemical analysis and sophisticated data analysis procedures, were demanded by consumers and tea producers. Chemometric techniques are commonly employed in order to develop system for the differentiation of geographical origin and fraud detection.

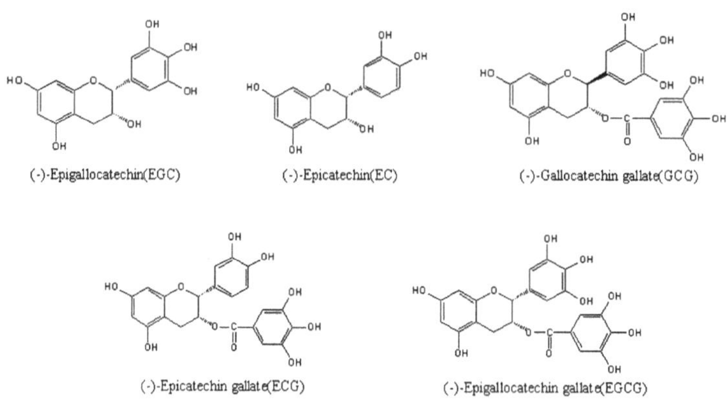

Figure 2: Chemical structures of catechins

This paper reported an application of chemometric techniques to HPLC data for the classification of green teas based on chemical characteristics. In this study, twelve green tea samples of three species originating from Henan, Jiangsu and Zhejiang province had been investigated to define groups. Five catechins in samples had been simultaneous analyzed and utilized as chemical descriptors to apply chemometric methods including principal component analysis (PCA) and hierarchical cluster analysis (HCA) for grouping.

2. MATERIALS AND METHODS

2.1 Materials

The standard chemicals of (-)-epigallocatechin [(-)-EGC], (–)-epigallocatechin gallate [(-)-EGCG], (-)-epicatechin [(-)-EC], (-)-gallocatechin gallate [(-)-GCG] and (-)-epicatechin gallate [(-)-ECG] were purchased from Sigma Chemical Co. (St. Louis, MO, USA). Methanol (HPLC grade) was purchased from Fisher Scientific (Fairlawn, NJ, USA). The water was prepared with a Millipore Milli-Q purification system (Bedford, MA, USA). Other reagents were analytical grade.

Chinese green tea samples were purchased from local tea shops (Beijing, China).

2.2 Preparation of samples

1.00 g tea leaves were accurately weighed and extracted with 200 mL boiling water in 100℃ water bath. After 30 min extraction, the extraction mixture was cooled to room temperature and filtered into a 250 mL volumetric flask and made to volume with water. Approximately 2 mL sample solution was filtered through a 0.45 μm polytetrafluroethylene (PTFE) filer prior to HPLC analysis.

2.3 Preparation of catechin standard solutions

Stock solution, 2.5 mg of each standard chemical was accurately weighed into a 5-mL volumetric flask, dissolved and made to volume with methanol, avoided light at -20℃. Working standard solutions were prepared by 2-1000 fold dilution of the stock solutions with methanol prior to HPLC analysis.

2.4 Analytical determinations

An HPLC-DAD-ESI/MS/MS system comprising an Agilent 1200 series liquid chromatography system(Palo Alto, CA, USA) equipped with a

G1322A degasser, a G1311A quaternary pump, a G1329A autosampler, a G1316A thermostat column compartment, a G1315B diode array detector and a Micromass Q-TOF mass spectrometer (Manchester, UK) was used. The column was a C18 reversed phase Agilent Zorbax SB-C18 5 μm (250 × 4.6 mm) with a Zorbax SB-C18 5 μm (30 ×4.6 mm) guard column. Mobile phases consisted of 0.5% acetic acid in water (v/v) (eluent A) and methanol (eluent B). The gradient elution system was: 0-5 min, 13% B; 5-20 min, linear gradient from 13 to 20% B; 20-25 min, 20% B; 25-35 min, linear gradient from 20 to 25% B; 35-40 min, 25% B; 40-50 min, linear gradient from 25 to 40% B; 50-60 min, linear gradient from 40 to 80% B. Elution was performed at a solvent flow rate of 0.8 mL/min. Detection was accomplished with a diode array detector and chromatograms were recorded at 278 nm. The column was maintained at 35 ℃. The injection volume was 5 μL.

Mass spectra of catechins were operated in negative mode using an electrospray ionizing source with nitrogen as desolvation gas. Spray chamber parameters: ion spray voltage, 3.5 kV; desolvation gas temperature, 300 ℃; desolvation gas flow, 400 L/h. The full-scan mass was acquired over the range 50-1500 m/z. Cone voltage was 10 V in full-scan mode, and the voltage for selected ion monitoring (SIM) was set at 30 V.

2.5 Data analysis

2.5.1 Principal component analysis

Principal component analysis (PCA) is a statistical tool commonly used for classification of data. PCA can reduce the dimensionality of the data considerably, enabling effective visualization regression and classification of multivariate data (Qian et al., 1994). PCA compresses a large number of variables to a much smaller number of principal components (PCs) that capture the majority of variance in the data by means of mathematical transformation. The PCs can be displayed in a scatter diagram, presenting the individual samples as points in a lower-dimensional (generally 2-D or 3-D) space.

2.5.2 Hierarchical cluster analysis

Hierarchical cluster analysis (HCA) is performed to classify samples on the basis of the similarities of their measured properties. Objects are grouped in clusters in terms of their nearness in the multidimensional space. The elements or clusters are joined with the criterion that the sum of heterogeneities of all clusters shall increase as little as possible. In this work, the distance matrix was calculated using Euclidean distances. From the distance matrices the dendograms were created using the Ward algorithm.

3. RESULTS AND DISCUSSIONS

3.1 Optimization of HPLC-DAD-ESI/MS/MS conditions

3.1.1 Column temperature

Column temperature was critical for the separation of catechins. The effect of column temperature on catechins was investigated in the range of 20~45℃ (Fig.2). The experimental results indicated that with the increase of column temperature, the retention times of analytes decreased.

At lower column temperature, the chromatograms showed baseline merging of analytes and other components. Baseline separation could be obtained upon 30℃. As could be seen from Figure 2, the difference of capacity factor (k') between adjacent analytes reached a maximum at 35℃. Hence, column temperature of 35℃ was chosen for further studies.

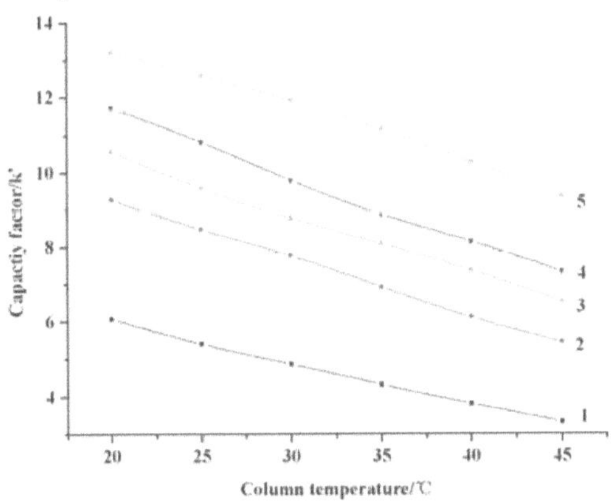

Figure 2: Effect of column temperature on capacity factor on Zorbax-SB C18 column; 1.EGC 2.EGCG 3.EC 4.GCG 5.ECG

3.1.2 Acidity of mobile phase

Acidity of the mobile phase was also essential for the separation and peak shape of catechins (Dalluge et al., 1998). In this paper, acetic acid was utilized as modifier of the mobile phase. To optimize the acidity of mobile phase, the analysis was done at different volume fraction of acetic acid in eluent A viz. 0, 0.25%, 0.5%, 0.75%, 1.0%, 1.5% (Fig.3). It was found that with increasing the acidity of mobile phase, the retention time of catechins decreased and peak shape became sharper.

The complete separation and well-resolved, symmetrical, sharp peaks could be obtained at 0.5%, above which the separation effect had no obvious change. However, higher acidity of mobile phase could have negative impact on the function of stationary phase. Therefore, the volume fraction of acetic acid in eluent A was fixed at 0.5%.

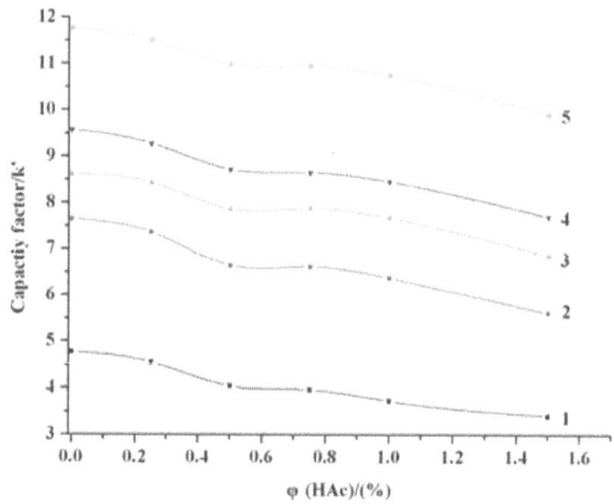

Figure 3: Effect of acidity of mobile phase on capacity factor on Zorbax-SB C18 column; 1.EGC 2.EGCG 3.EC 4.GCG 5.ECG

3.1.3 Detected wavelength

Wavelengths between 210 and 220 nm or between 270 and 280 nm were ordinarily applied to the detection of catechins (Dalluge et al., 1998; Goto et al., 1996). Catechins showed maximum absorbance at 210 nm. However, the baseline shifted seriously at 210 nm due to the change of the mobile phase. As the baseline was smooth during the gradient elution at 278 nm, the chromatograms were recorded at 278 nm.

3.2 Optimization of sample preparation conditions

3.2.1 Extraction temperature

Extraction temperature could obviously affect the extraction efficiency due to the increase in solubility of catechins in water with temperature. In this work, the effect of extraction temperature on the extraction efficiency over the temperatures interval of 50℃ to 100℃ with an extraction time of 30 min was studied (Fig.4). It was demonstrated that the extraction efficiency for all target analytes except EGCG increased with temperature and maximum extraction efficiency was obtained at 100℃. The most significant effect was observed between 90℃ and 100℃. To obtain better extraction efficiency, 100℃ was selected as extraction temperature for subsequent experiment.

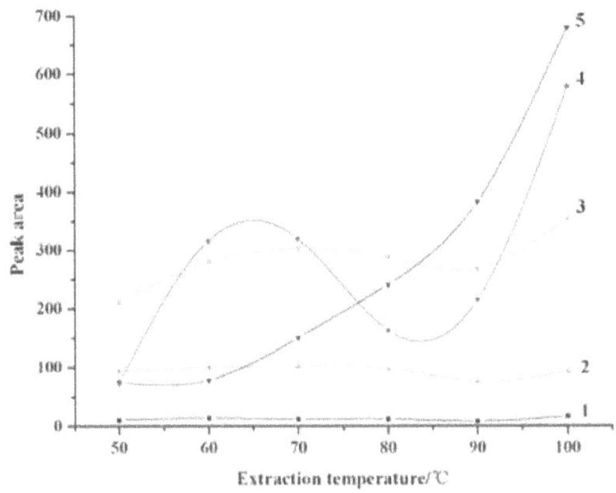

Figure 4: Effect of extraction temperature on extraction efficiency; 1.EGC 2.EC 3.ECG 4.EGCG 5.GCG

3.2.2 Extraction time

Extraction time was another factor that influenced the extraction efficiency. The effect of extraction time on the extraction efficiency was investigated in the range of 10~120 min (Fig.5). As could be seen from Fig.5, the peak area for each analyte increased sharply within 30 min. It appeared that the time effect faded after 60 min, which might be caused by oxidation of catechins. Therefore, 30 min was most suitable for extraction.

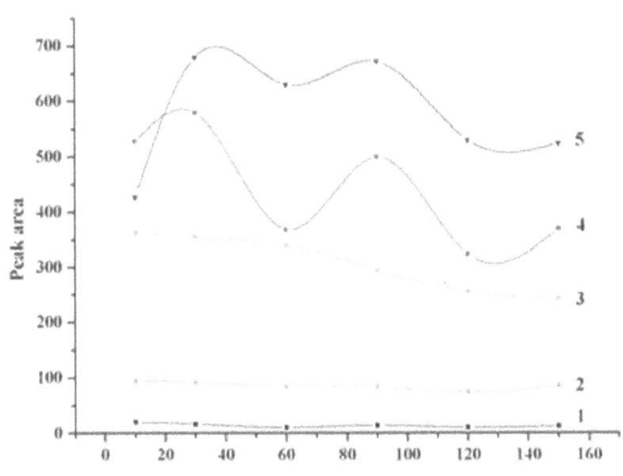

Figure 5: Effect of extraction time on extraction efficiency; 1.EGC 2.EC 3.ECG 4.EGCG 5.GCG

3.3 Method validation

3.3.1 Linearity and limit of detection

Calibration curves of the catechins were constructed using six levels of concentration which covered the concentration ranges expected in samples (Fig.6). The characteristics of calibration curve, including the range of linearity, the square of correlation coefficient (R^2) and limit of detection (LOD) was listed in Table 1. The LOD was evaluated as a signal equals three times of noise (S/N=3). It could be seen that an excellent linearity was observed for each analyte in the range studied.

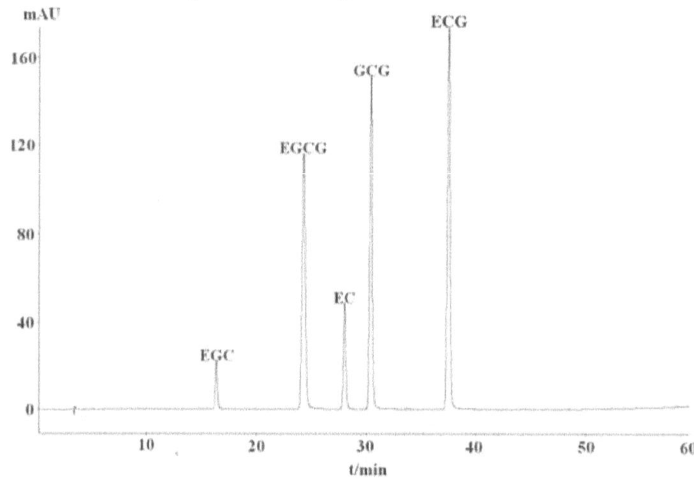

Figure 6: Chromatogram of catechin standards monitored at 278 nm.

Table 1: Characteristics of the calibration curve

Compound	*Calibration curve*	*Correlation coefficient (r^2)*	*Liner range (ng)*	*LOD (ng)*
EGC	Y = 0.5828x + 11.062	0.9986	50~2500	15
EGCG	Y = 4.5871x + 83.4	0.9978	50~2500	0.75
EC	Y = 1.496x + 28.661	0.9982	50~2500	5
GCG	Y = 4.721x + 82.72	0.9983	50~2500	0.75
ECG	Y = 5.7258x + 112.42	0.9977	50~2500	1.5

3.3.2 Precision

To test the precision of the assay method, one of the samples to be analysed was injected five times under the HPLC conditions described above. The relative standard deviations (RSD) of relative peak area for all analytes were within 1.15%.

3.3.3 Stability

The stability of the catechins in the sample solution was determined by analyzing the same sample after 2, 4, 6, 12, 24, 30, 48 h of preparation. The experiment showed stability within 24 hours when the sample was kept at room temperature.

3.3.4 Reproducibility

The reproducibility was evaluated by the RSD of relative peak area of five analytes in various batches of samples. It was found that the maximum RSD was less than 4.20%.

3.4 Identification of catechins by HPLC-DAD-ESI-MS/MS

Catechins were identified by comparing retention time, UV spectra, m/z of their quasi-molecular ions, and MS^2 fragmentation patterns of unknown peaks to the standards. The typical HPLC chromatogram and total ion current chromatogram of Biluochun, a well-known green tea, was shown in (Fig.7 and Fig.8), while Table 2 showed the retention times, MS and MS^2 data and the identification results for the peaks numbered in the chromatogram.

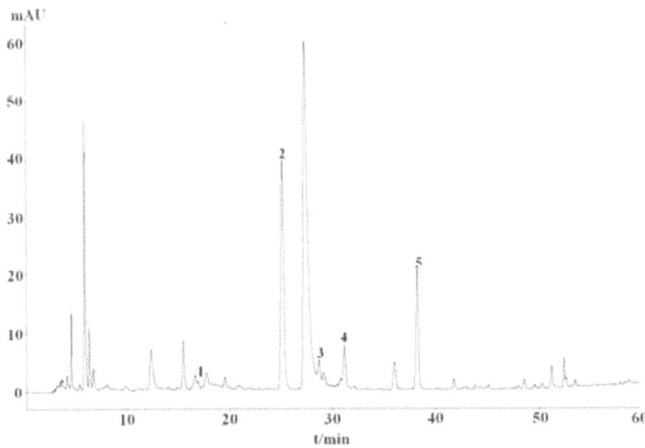

Figure 7: Chromatogram of Biluochun monitored at 278 nm. Peak identification: 1.EGC 2.EGCG 3.EC 4.GCG 5.ECG

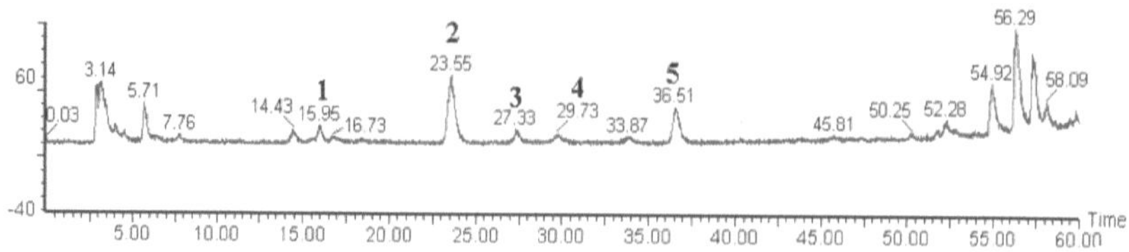

Figure 8: Total ion current chromatogram of Biluochun; Peak identification: 1.EGC 2.EGCG 3.EC 4.GCG 5. ECG

Table 2: Retention time, molecular weight and the *m/z* values of selective ions for identification of catechins

No	*Rt*	*Compound*	*Molecularweight*	*Selectiveion (m/z)*	MS^2(m/z)
1	16.7	EGC	306	305	219,167,137,125
2	25.0	EGCG	458	457	305,169,125
3	28.5	EC	290	289	203,151,109
4	31.0	GCG	458	457	305,169,125
5	38.2	ECG	441	441	289,245,169,125

3.5 Analysis of green tea samples

Twelve Chinese green tea samples of three species (Xinyang maojian, Biluochun, Xihu longjng) were analyzed for catechins by employing the developed method. The content of five analytes were calculated as milligram per gram of dry weight (Tab.3)

Table 3: Content of five catechins for 12 tea samples (mg/g of dw)

No	*Sample**	*Grade*	*Source*	*EGC*	*EGCG*	*EC*	*GCG*	*ECG*
1	MJ	4	Hennan	1.3	12.3	12.6	19.7	4.0
2	MJ	3	Hennan	4.6	21.3	9.3	29.0	6.6
3	MJ	2	Hennan	1.2	17.7	8.4	25.9	7.0
4	MJ	1	Hennan	1.5	15.6	7.7	25.0	6.6
5	BLC	4	Jiangsu	4.3	24.1	8.7	27.2	11.9
6	BLC	3	Jiangsu	1.2	15.2	7.7	20.2	9.8
7	BLC	2	Jiangsu	1.4	23.1	6.9	25.1	10.6
8	BLC	1	Jiangsu	2.1	19.7	7.2	20.3	9.5
9	LJ	4	Zhejiang	1.5	24.2	10.8	14.6	6.0
10	LJ	3	Zhejiang	1.8	7.5	10.2	10.5	3.5
11	LJ	2	Zhejiang	1.9	6.8	10.9	11.4	2.4
12	LJ	1	Zhejiang	1.9	12.2	12.5	12.7	3.5

* *MJ: Xinyang Maojin; BLC: Biluochun; LJ: Xihu Longjin*

3.6 Hierarchical cluster analysis

HCA applied the information obtained from the measured variables to reveal the natural clusters exiting between the studied samples. The

Euclidean distance was used as similarity measurement and the Ward's method was utilized as amalgamation rule to obtain the hierarchical associations. The data matrix used for the multivariate analysis was composed of a 12×5 data matrix (12 samples and 5 variables). The result of HCA was presented as a dendogram (Fig.9). The resulting dendogram was interpreted to classify the twelve Chinese green tea samples in three groups based on the similarity of the five studied parameters. Group A was composed of 4 Xinyang maojian. Group B comprised 4 Biluochun. Group C contained 4 Xihu longjing. This meant that 100% correct classification was achieved by HCA.

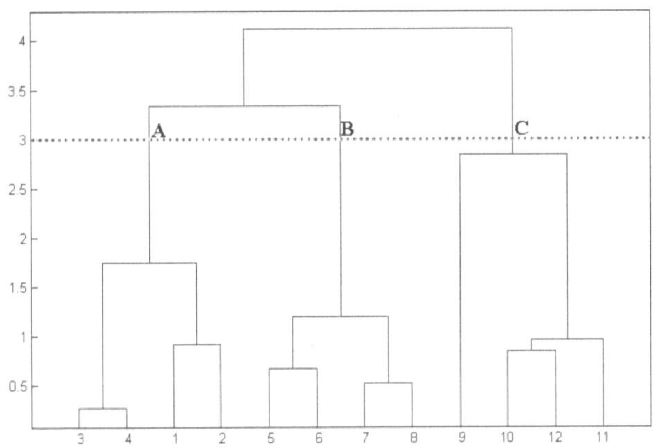

Figure 9: Hierarchical cluster analysis dendgram of 12 green tea samples using Euclidean distance; 1-4 Xinyang maojian, 5-8 Biluochun, 9-12 Xihu longjing

3.7 Principal component analysis

PCA technique was applied to reduce the number of dimensions present in the data matrix (reducing 5 variables to 2 PCs in this study), to select the most discriminating parameters, and to investigate the overall variation of data. Rotation of principal components was carried out using the Varimax normalized method and Kaiser Criterion. Varimax normalized procedure for eigenvector rotation resulted in two principal components (PC1 and PC2), which explained 85% of the total variance. Figure 10 showed the result of the PCA analysis of the 12 green tea samples. The first principal component (PC1) contained 43.9% of the total variance and the second component (PC2) represented 41.1% of the total variance. The twelve green tea samples with various species and origins were grouped into three clusters. Representation of the green tea employing PCA was effective and reasonable as the first two components accounted for 85% of the total variance.

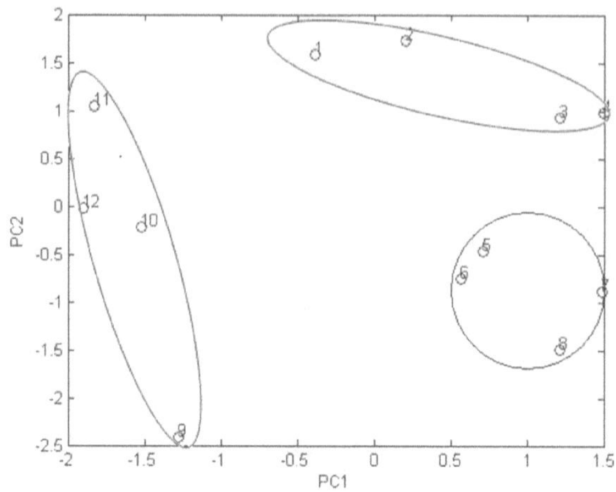

Figure 10: Samples in the space defined by first and second principal components; 1-4 Xinyang maojian, 5-8 Biluochun, 9-12 Xihu longjing

4. CONCLUSIONS

In this study, an HPLC-DAD-ESI/MS/MS method was developed and optimized for analyzing green teas. Five catechins had been simultaneously identified and quantified. The HPLC tandem with DAD and ESI-MS/MS provided multi-dimension qualitative information comparing with ordinary HPLC analysis of catechins.

Chemometric methods (PCA and HCA) had proved to be available approaches to the classification of green teas based on the content of catechins. Twelve Chinese green tea samples originating from three regions were subjected to reveal the natural grouping among the source. PCA provided information on the overall components of green teas and the overlap of the clusters. HCA rendered three significant green tea groups on the basis of similarities in their chemical properties with percentage of correct classification of 100%. Variations in the chemical compositions of green teas were mostly caused by environments (climate, geologic formation).

The result suggested that PCA and HCA were available for the classification of green teas. This study confirmed a potential of HPLC in connection with chemometric method for authentication of green tea.

REFERENCES

Balentine D. Tea and health. Critical Reviews in Food Science and Nutrition, 1997, 37: 691-692

Bonoli M, Pelillo M, Toschi TG, Lercker G. Analysis of green tea catechins: comparative study between HPLC and HPCE. Food Chemistry, 2003, 81: 631-638

Dalluge JJ, Nelson BC, Thomas JB, Sander LC. Selection of column and gradient elution system for the separation of catechins in green tea using high performance liquid chromatography. Journal of Chromatography A, 1998, 793: 265–274.

Goto T, Yoshida Y, Kiso M, Nagashima H. Simultaneous analysis of individual catechins and caffeine in green tea. Journal of Chromatography A, 1996, 749: 295 - 299.

Lee BL, Ong CN. Comparative analysis of tea catechins and theaflavins by high-performance liquid chromatography and capillary electrophoresis. Journal of Chromatography A, 2000, 881: 439-47.

Liang YR, Zhang LY, Lu JL. A study on chemical estimation of pu-erh tea quality. Journal of the Science of Food and Agriculture, 2005, 85: 381-390

Nishitani E, Sagesaka YM. Simultaneous determination of catechins, caffeine and other phenolic compounds in tea using new HPLC method. Jornal of Food Composition and Analysis, 2004, 17: 675-685

Owuor PO, Obanda M. The use of green tea (Camellia sinensis) leaf flacan-3-ol composition in predicting plain black tea quality potential. Food Chemistry, 2007, 100: 873-844

Pelillo M, Bonoli B, Biguzzi B, Bendini A, Toschi TG, Lercker G. An investigation in the use of HPLC with UV and MS-electrospray detection for quantification of tea catechins. Food Chemisty, 2004, 87: 465-470

Qian G, Gabor G, Gupta RP. Principal components selection by the criterion of the minimum mean difference of complexity. Journal of Multivariate Analysis, 1994, 49: 55 - 75

Sakanaka S, Kim M, Taniguchi M, Yamamoto T. Antibacterial substance in Japanese green tea extract against Streptococus mutans, a carcinogenic bacterium. Agriculture Biology and Chemistry, 1989, 53: 2307-2311

Shaghaghi M, Manzoori JL, Jouyban A. Determination of total phenols in tea infusions, tomato and apple juice by terbium sensitized fluorescence method as an alternative approach to the Folin - Ciocalteu spectrophotometric method. Food Chemistry, 2008, 108: 695-701

Tsukagoshi K, Taniguichi, Nakajima R. Analysis of antioxidants using a capillary electrophoresis with chemiluminescence detection system. Analytica Chimica Acta, 2007, 589: 66-70.

Wang H, Helliwell K, You X. Isocratic elution system for the determination of catechins, caffeine and gallic acid in green tea using HPLC. Food Chemistry, 2000, 68: 115 - 121.

Wang HF, Provan GJ, Helliwell K. HPLC determination of catechins in tea leaves and tea extracts using relative response factors. Food chemistry, 2003, 81: 307-312

Yang CS. Inhibition of carcinogenesis by tea. Nature, 1997, 389: 134-135

Yao LH, Jiang YM, Datta N , Singanusong R, Liu X , Duan J, Raymonte K, Lisle A, Xu Y. HPLC analyses of .avanols and phenolic acids in the fresh young shoots of tea (Camellia sinensis) grown in Australia. Food Chemistry, 2004, 84: 253-263

Zhao JW, Chen QS, Huang XY, Fang CH. Qualitative identification of tea categories by near infrared spectroscopy and support vector machine. Journal of Pharmaceutical and Biomedical Analysis, 2006, 41: 1198-1204

Zhu XL, Chen B, Ma M, Luo XB, Zhang F, Yao SZ, Wan ZT, Yang DJ, Hang HW. Simultaneous analysis of theanine, chlorogenic acid, purine alkaloids and catechins in tea samples with the help of multi-dimension information of on-line high performance liquid chromatography/electrospray-mass spectrometry. Journal of Pharmaceutical and Biomedical Analysis, 2004, 34: 695-704

IDENTIFICATION OF TRACEABILITY BARCODE BASED ON PHASE CORRELATION ALGORITHM

Liying Lang [1,*], Xiaofang Zhang [2]

[1] *Hebei University of Engineering, Handan, China, 056038;*

[2] *College of Information and Electrical Engineering, Hebei University of Engineering, Handan, 056038, China*

[*] *Corresponding author, Address: Hebei University of Engineering, Handan,Hebei province, Guangming South Street 199, 056038 , P. R. China, Tel: 0310-8579551, Fax:0310-8578746, Email: langliying@126.com*

Abstract: In the paper phase correlation algorithm based on Fourier transform is applied to the traceability barcode identification, which is a widely used method of image registration. And there is the rotation-invariant phase correlation algorithm which combines polar coordinate transform with phase correlation, that they can recognize the barcode with partly destroyed and rotated. The paper provides the analysis and simulation for the algorithm using Matlab, the results show that the algorithm has the advantages of good real-time and high performance. And it improves the matching precision and reduces the calculation by optimizing the rotation-invariant phase correlation.

Keywords: traceability barcode recognition, Fourier transform, phase correlation, image registration

1. INTRODUCTION

At present in our country, the major food output such as cereals, fruits, meat, eggs and aquatic products occupy first in the world. In order to guarantee the people's food security and effectively control of food-borne disease outbreak, the work to establish food tracking and retrospect will have a tremendous impact on the development of the food industry in China.

Please use the following format when citing this chapter:

Lang, L. and Zhang, X., 2009, in IFIP International Federation for Information Processing, Volume 295, *Computer and Computing Technologies in Agriculture II, Volume 3*, eds. D. Li, Z. Chunjiang, (Boston: Springer), pp. 1721–1727.

Compared the way of log file retrospect (Tillett R D, 1991), it is more efficient, real-time and convenient to make use of the barcode superiority characteristic to manage the agricultural product security and the retrospect. This way has many merits (Paulsen, 1997). It not only provides the completely transparent managed capacity in the agricultural product supply chain, ensures safety of agricultural products throughout visual control, monitoring and retrospect, but also may monitor the planter and cultivation source pollution, the chemical additive and deleterious substance in the process of production produce, the security risks in the circulation link. However, there will inevitably appear the attrition and the distortion of the traceability barcode in the course of agricultural product circulation. For this purpose, the phase correlation algorithm which has powerful anti-geometric deformation capacity to the barcode identification is applied to deal with this problem.

2. BASAL PRINCIPLE OF ALGORITHM

Phase correlation algorithm is non-linear in frequency domain, which is based on Fourier transform (Stone, 2001). It makes use of a number of important properties of the image in frequency domain, such as translation, rotation, zooming invariance characters (Srinivasa R B, 1996), that is to say there exists corresponding relation between the airspace and the frequency domain when there is translation, rotation, zooming between the images. Fast Fourier Transform a large extent improves the operation efficiency of the algorithm, and it behaves a good robust to the noise of the frequency correlation. Therefore phase correlation algorithm is a more effective method to achieve the image registration.

2.1 The theory estimation of translating parameter

Hypothesize $f_1(x,y)$ and $f_2(x,y)$ are two one-dimensional barcodes which have definitely translating relation (Kobayashi, 1996). They satisfy with formula (1). (That is to say, $f_2(x,y)$ results from simply translating by $f_1(x,y)$)

$$f_2(x,y) = f_1(x - x_0, y - y_0) \tag{1}$$

The character according to Fourier transform can may:

$$F_2(u,v) = F_1(u,v)e^{-j(ux_0 + vy_0)} \tag{2}$$

In above formula, $F_1(u,v)$ and $F_2(u,v)$ are neutralized respectively Fourier transform drawing from $f_1(x,y)$ and $f_2(x,y)$ .Their cross-power spectrum is that:

$$\frac{F_1^*(u,v)F_2(u,v)}{\left|F_1^*(u,v)F_2(u,v)\right|}=e^{-j(ux_0+vy_0)} \tag{3}$$

$F_1^*(u,v)$ is complex conjugation of $F_1(u,v)$ in style. The formula (4) carries out Fourier transform on the contrary alternation but gets a two-dimensional pulse function. That is:

$$F^{-1}(e^{-j(ux_0+vy_0)})=\delta(x-x_0,y-y_0) \tag{4}$$

It can find that, phase correlation function at the surface all is zero apart from δ function, which is the location of their displacement value (x_0, y_0)(Harold S S,2002).The function peak value should be 1 under ideal circumstances, but due to noise and other reasons, its peaks often less than 1. The peak value depends on whether two images mate the degree goodness or badness.

2.2 The theory estimation of rotating parameter

Hypothesize $f_1(x,y)$ and $f_2(x,y)$ are two one-dimensional barcode images. $f_2(x,y)$ results from simply rotating by $f_1(x,y)$.That is to say:

$$f_2(x,y)=f_1(x\cos\theta_0+y\sin\theta_0,-x\sin\theta_0+y\cos\theta_0) \tag{5}$$

According to the rotation of Fourier transform, the relationship between the two images after shift as follows:

$$F_2(u,v)=F_1(u\cos\theta_0+v\sin\theta_0,-u\sin\theta_0+v\cos\theta_0) \tag{6}$$

From above formula it can see that the image of the space domain in the rotation θ_0 perspective, the corresponding Fourier transform in the frequency domain is the same rotation angle θ_0 . Due to Cartesian coordinates corresponding to the rotation angle of the polar coordinate translation (Wolberg, 2000), it will be (7) converted to the form of polar coordinates as follows:

$$F_2(\rho,\theta)=F_1(\rho,\theta-\theta_0) \tag{7}$$

In the formula, $\rho = \sqrt{u^2 + v^2}$, θ = arctan $\frac{v}{u}$, from formula (7) it can find that they satisfy a simple translation relation between $F_2(\rho, \theta)$ and $F_1(\rho, \theta)$. Make use of above-mentioned phase correlation method to be ok to gain the revolution angle.

3. EXPERIMENT AND RESULT ANALYSIS

According to situation which the barcode possibly appears in the actual circulation, we divide the experiments into two following parts: (1) When the traceability barcode is partly frayed or masked, translation parameters are estimated based on phase correlation algorithm; (2) Using the rotation-invariant phase correlation algorithm to estimate the rotation parameters when the barcode revolves slightly.

3.1 Translation experiment

In the process of algorithm simulation, the paper first pretreats the images using median filter to filter out pulse interference and image scanning noise, and uses MATLAB tools to simulate the image of one-dimensional barcodes. The results show that the phase correlation algorithm only has good effect on translation image detection, even if only in the relevant content of 1 / 4, the success rate is still over 90%. There are two barcode images of mutual translation. The two figures with relevant contents of half and quarter were compared. The results clearly show that with the relevant content reducing gradually the peak values reduce sharply, as shown in figure 1 and figure 2. The relevant peak soon declines when there exists rotating in the images .However, it has a better match rate within only 1.5 degrees, more than 3 degrees the algorithm completely ineffective.

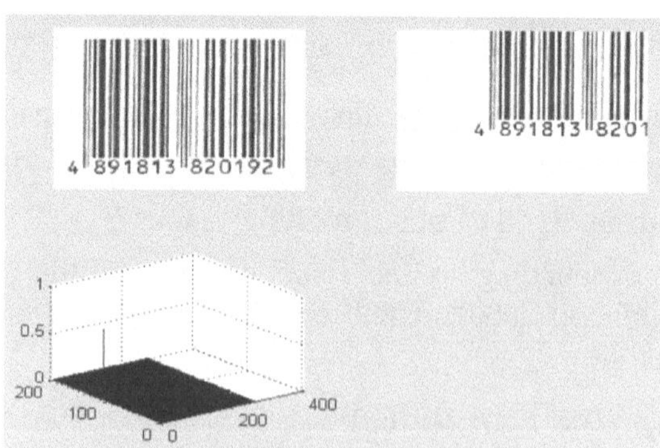

Fig.1: Correlation result of half related

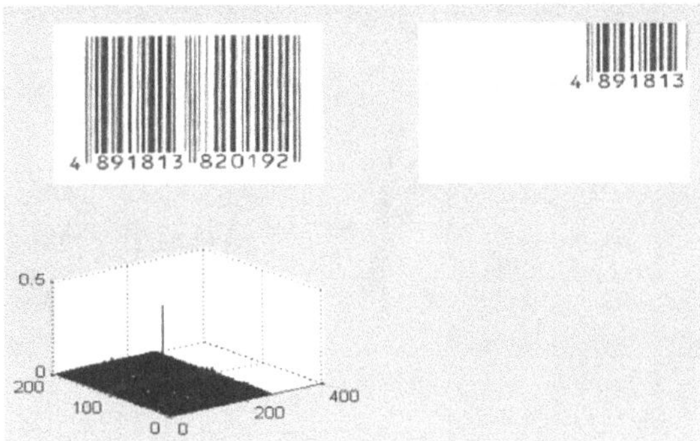

Fig.2: Correlation result of quarter related

3.2 Rotation experiment

The algorithm is carried out a certain amount of optimization when in the process of rotation-invariant phase correlation algorithm simulation. In the polar coordinate mapping, it first seeks admission points in a very homogeneous coordinates in turn to obtain the value in the Cartesian coordinate system mapping, so that it may sample around the neighborhood part of origin point in the Cartesian coordinate system. Because after the Fourier transform and centering the image energy mainly concentrates in the neighborhood of the origin, it can reduce the mapping error from Cartesian to polar coordinates. The experiment results prove that it can have great correlation peak growth. So that it will not only improve the matching accuracy but also greatly reduce the computation of following stage. Below are the barcodes with different rotational angle carried out the simulation. Figure 3-a and figure 4-a are polar coordinate transform of the original and rotating barcodes, corresponding correlation result in figure 3-b and figure 4-b. The results show that the smaller the angle of deformation registration the results more accurate.

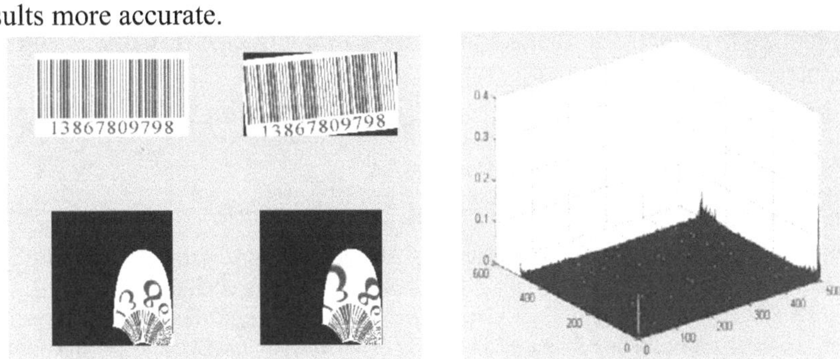

(a) polar coordinate transforms (b) correlation result

Fig.3: Result of rotating angle of 5

(a) polar coordinate transforms (b) correlation result

Fig.4: Result of rotating angle of 30

From above translation experiment and rotation experiment, in order to make it clearly we sum up the comparison as following table:

Table 1. Comparison results of different related content

Translation peak values		Rotation peak values	
Half related	Quarter related	Rotating 5°	Rotating 30°
0.791	0.493	0.3852	0.2103

4. CONCLUSIONS

In the paper, it takes one-dimensional barcodes for example, and makes use of MATLAB tool to the phase-correlation algorithm simulation. The two experimental results prove the feasibility and effectiveness of the method. The difficult issues such as rotation and defaced which traceability barcode may arise are well solved. So that it will ensure that the food industry can be promptly identified and recall the products which may exist security risks, enhancing the confidence of consumers. Therefore, it will have a very wide application prospect, and make contribution to the traceability of agricultural products

REFERENCES

G.Wolberg,S.Zokai. Roubust image registration using log-polar transform[A]. Proc. IEEE Int. Conf. Image Processing[C].Vancouver, BC,Canda:IEEE，2000:493-496.

Harold S S, Michael T Orchard, Ee2Chien Chang, et al. A fast direct Fourier based algorithm for subpixel registration of images [J]. IEEE Transactions on Geoscience and Remote Sensing , 2001 ,39 (10) : 2235-2243.

Harold S S, Robert Wolpov. Blind cross-spectral image registration using prefiltering and Fourier-based translation detection [J]. IEEE Transactions on Geosciences and Remote Sensing,2002 , 40 (3) : 637-650.

K.Kobayashi,H.Nakajima.T.Aoki.M.Kawamata.T.Higuchi. Principals of phase only correlation and applications. ITE Technical Report, 1996, 20(41):2-4

Ni B, Paulsen MR, Reid J F. Corn kernel crown shape identification using image processing[J] . Trans of the ASAE, 1997, 40(3) :833-838.

Srinivasa R B, Chatterji B N. An FFT-based technique for translation , rotation , and scale-invariant image registration [J] . IEEE Transactions on Image Processing,1996, 5 (8) : 1266-1271.

Tillett R D. Image analysis for agricultural processes: A review of potential opportunities[J] . J Agric Engng Res, 1991, 50: 247-258.

STUDIES ON THE RAPID METHODS FOR EVALUATING SEED VIGOR OF SWEET CORN

Guangwu Zhao[1,*], Linlin Yang[2], Jianhua Wang[3], Zhujun Zhu[1]

[1] *School of Agriculture and Food Science, Zhejiang Forestry College, Lin'an, China, 311300*

[2] *College of Foreign Languages, Wenzhou University, Wenzhou, China, 325027*

[3] *Department of Seed Science and Technology, China Agricultural University, Beijing, China, 100094;*

[*] *Corresponding author, Address: School of Agriculture and Food Science, Zhejiang Forestry College, 88 Huancheng North Road, Lin'an, 311300, P. R. China, Tel:+86-571-63742087, Fax:+86-571-63741276, Email:gwuzhao@126.com*

Abstract: Since the germination test is still the main method for evaluating seed vigor of sweet corn, it is necessary to study the rapid methods for evaluating their vigor. First, two vigor levels of 9 super sweet corn and 9 sugar enhanced corn hybrids were distinguished by accelerated ageing treatment. Next, their vigor statuses were tested by rapid methods such as electrical conductivity (EC), soluble sugar (SS), volatile aldehyde (VA), dehydrogenase activity (DA) and Q2 oxygen sensing technology. Correlation relationship was analyzed between the above tested values and vigor performances. The results showed that EC, SS, VA, DA, oxygen metabolism rate (OMR), critical oxygen pressure (COP) were all suitable to evaluate seed vigor of sweet corn. Furthermore, VA was optimal for evaluating seed vigor of super sweet corn. DA was optimal for evaluating sugar enhanced corn.

Keywords: super sweet corn, sugar enhanced corn, seed vigor, rapid evaluation methods

1. INTRODUCTION

Sweet corn (*Zea mays* L.), especially super sweet corn and sugar enhanced corn are widely planted. Seed vigor is a comprehensive characteristic that is an important index of seed quality and is closely correlated with field performances. Field emergence percentage (FEP) and field emergence speed

---

Please use the following format when citing this chapter:

Zhao, G., Yang, L., Wang, J. and Zhu, Z., 2009, in IFIP International Federation for Information Processing, Volume 295, *Computer and Computing Technologies in Agriculture II, Volume 3*, eds. D. Li, Z. Chunjiang, (Boston: Springer), pp. 1729–1738.

(FES) are the most direct indicators of seed vigor. FEP is the only index in most reports about seed vigor evaluation, whereas FES has been often ignored (Ching et al., 1977; Waters and Blanchette 1983; Wilson et al., 1992; Zhang 1997; Woltz and TeKrony 2001; Wang et al., 2004; Zhao and Wang 2005; Zhao et al., 2007). In our study, FEP and FES or laboratory germination percentage (LGP) and laboratory germination speed (LGS) were together considered.

The germination test is still the main method for evaluating seed vigor of sweet corn. Germination test is time consuming. Too often seed lots are sent either to the next step in processing or to the client without having been thoroughly analyzed. Due to time constraints and a competitive market, too many times seeds are processed without adequate quality information needed to get the best from the seeds. Therefore, it is necessary to study the rapid methods for evaluating their seed vigor.

Some tests such as electrical conductivity (EC), soluble sugar (SS), volatile aldehyde (VA) and dehydrogenase activity (DA) are faster than the germination test. EC can indirectly determine seed membrane integrity of different quality seed lots and is an easy, fast and objective vigor testing method. It has been listed in the International Rules for Seed Testing as a seed vigor testing method for pea. Sweet corn is famous for high sugar content. SS in the soaked solution has not been reported for evaluating seed vigor of sweet corn. In addition, VA is toxic to germinating seeds. VA released during seed germination was negatively correlated with seed vigor, which was validated in vigor testing of soybean, rice and wheat (Wilson and McDonald, 1986; Chen et al., 1996; Zhu and Zeng, 1999). It might be a potential method for vigor testing (Tao and Zheng, 1991). DA could be used to evaluate seed vigor of rice and maize (Akram et al., 2002; Qiao et al., 2003).

The Q2 oxygen sensing technology is revolutionary for seed testing in basic research and commercial operations alike. It provides a fast and accurate measurement of the germination level of a seed lot. In addition, Q2 data is more robust and defining than traditional germination test. Depending on the species, the estimated time needed will be between 5 and 96 hours and will easily determine dead, dormant or actively germinating seeds. Although it currently does not provide specific details on seedling abnormalities, the Q2 can quickly give more accurate indications of the vigor and the homogeneity of a seed lot.

The Q2 instrument, developed by Netherlands ASTEC Global Corporation, measures the percentage of oxygen in closed microtiter plates, consisting of 24-48-96 wells, in regular time intervals. The purpose of the measurements is to relate the oxygen consumption of seeds over time, to the quality of the seed lot. Q2 stands for Quality and Quick, as well as O2 (oxygen), the element the Q2 instrument is based upon. The Q2 instrument can provide results far more quickly than a germination test. Moreover, due to the latest developments in optical technology for measuring oxygen, the Q2 instrument is able to detect seeds' activity in the first stages of germination (imbibition). Imbibition is characterized by an increased amount of oxygen consumption. In addition to very fast detection of germinating seeds, the Q2 instrument provides data concerning the dormancy of non-germinating seeds, the homogeneity of a seed lot, as well as information concerning different stress related factors, and so much more.

Q2 test can be used as fast and accurate indications of germination under certain conditions. They can be used to detect maximum germination (correlating to an International Seed Testing Association (ISTA) test) under various environmental conditions, such as temperature, moisture, and substrates. In the second case, Q2 test can be used as a stress test, telling something about the vigor of seeds. Certain abnormalities may be detected if oxygen consumption is affected. The accuracy of this test depends on the protocols used and of course, the integrity of the sample. The instrument enables testing of a large number of samples (from 1 seed to several microtiter plates) and the duration of the test will influence the quality of the results (from a minimum of a single scan to an indefinite period; 48 hours for the most common seeds). The development of protocols for specific seed testing objectives is important and the precision of executing these protocols, such as the application of water, or the substrate application, will obviously influence the results as well.

The key innovation to the Q2 oxygen sensing technology is that oxygen consumption is directly and proportionately related to energy use. Thus, a seed's energetic potential can be determined by measuring its oxygen consumption in a simulated field environment. Because of this proportional relationship between oxygen consumption (respiration) and caloric energy consumption (metabolism), robust inferences for a seed's performance in the field can be made. To characterize a greater population or seed lot, measuring respiration and metabolism seed by seed is thus very precise and accurate using the Q2 instrument. Measuring energy use can give us new insights on germination and vigor in one single, quality controlled test, quickly. Up to now, Q2 test is only successfully applied to evaluate seed vigor of a few crops such as sugar beet. Q2 test depends on the specific protocol according to specific species. Therefore, the Q2 oxygen sensing technology provides a great room to seed vigor testing of so many species.

However, whether these vigor indices are suitable for evaluating seed vigor of sweet corn needs further study. As for the Q2 oxygen sensing technology, we will focus on the initial research in super sweet corn "Green Superman". First, the specific protocol for sweet corn is developed. Then, oxygen percentage was tested at an interval of 30 min by Q2 instrument. Next, five Q2 values were calculated including increased metabolism time (IMT), oxygen metabolism rate (OMR), critical oxygen pressure (COP), relative germination time (RGT), homogeneity, or uniformity of the RGT (HOM) developed by ASTEC. Finally, correlation relationship was analyzed between the five Q2 values and laboratory germination performances and optimal indices were confirmed. The focus of our study was to develop the rapid methods for evaluating seed vigor of sweet corn.

2. MATERIALS AND METHODS

2.1 Seed materials

Seeds of 18 sweet corn hybrids were collected from 9 provinces in China. Seeds of super sweet corn "Chaotian 204", "Chaotian 43", "Green Superman", "Nongtian 2", "Nongtian 3", "Huatian 1", "Xiangtian 1", "Yangtian 1", "Mitian 8" were bought from Zhejiang, Beijing, Guangdong, Hubei, Hunan, Jiangsu. Seeds of sugar enhanced corn "Jingketian 115", "Jingketian 116", "Tiandan 8", "Tiandan 10", "Zhongtian 2", "Nongtian 1", "Yuetian 3", "Sutian 8", "Jiatian 16" were bought from Beijing, Guangdong, Jiangsu, Shanghai. Except for 10 lots of "Green Superman" seeds, the other hybrid seeds were only bought for 1 lot.

2.2 Accelerated ageing treatment

To distinguish seed vigor to two levels (high vigor and low vigor), all seeds were aged in a container with 100% relative humidity at 45℃ for 48 h. Then, seeds were re-dried.

2.3 Electrical conductivity test

The EC test was conducted on three replicates of 50 seeds each for each treatment. The test referred to the international rules for seed testing (ISTA, 2007). EC was tested by a DDS-11A conductivity meter (Shanghai, China) and was calculated by the formula of the international rules for seed testing.

2.4 Soluble sugar test

The imbibition solution (1 mL) from the EC test was added to 5 mL anthrone, then vibrated in a test tube to mix them thoroughly. After the test tubes were stoppered, they were placed in boiling water for 10 min and then cooled on a shelf. The solutions were poured in colorimetric cups to determine the OD value at 620 nm. The test was conducted on three replicates.

2.5 Volatile aldehyde test

Two layers of adequately moistened filter paper were placed on the bottom of a 1000 mL flask and 100 seeds per treatment planted on the paper until germination occurred. Then, a test tube with 5 mL 0.2% 3-methyl-2-benzothiazolinone hydrazone hydrochloride (MBTH) was put in the flask to imbibe VA released from the seeds. Next the orifice was stoppered and the flask was put into a case at 20℃ in the dark. The imbibition solution (1 mL) was taken out after 48 h, blended with 2.5 mL 0.2% FeC13 and reacted for 5 min. Finally, 6.5 mL acetone was mixed with the solution and the OD value determined at 635 nm. 1mL 0.2% MBTH was used for the control colorimetric solution. The test was conducted on three replicates.

2.6 Dehydrogenase activity test

Ten embryos were extracted from seeds germinated for 1 d and placed into a test tube with 10 mL 0.1% triphenyl tetrazolium chloride (TTC) and the tube stoppered to allow staining for 3 h at 35℃. Seed embryos were moistened with distilled water 3 times and 10mL anhydrous ethanol added and placed at 35℃ for 24 h. Finally, the extraction solution was used to determine the OD value at 490 nm. The test was conducted on three replicates.

2.7 Field emergence test

The field emergence test was conducted on three replicates of 50 seeds each for each treatment. Seeds were sown in Zhejiang and Beijing. The number of seedling was recorded after emergence until no emergence. FES was denoted by the number of days when 50% FEP was reached.

2.8 Validation of vigor indices

To validate the veracity of VA, DA as optimal index for evaluating seed vigor of super sweet corn and sugar enhanced corn, seeds of super sweet corn “Green Superman” and sugar enhanced corn “Zhongtian 2” were bought in 2005. A part of seeds were stored at 4℃ and the others were stored at room temperature. VA of “Green Superman”, DA of "Zhongtian 2” and their field performances were tested.

2.9 Q2 test and laboratory germination test

Ten lots of super sweet corn “Green Superman” seeds were used for Q2 test. The Q2 test conducted on four replicates of 100 seeds per treatment quantified energy use by measuring oxygen consumption of individual seeds, planted in 16 microtiter plates of 96 wells with 200 μl water and relating the oxygen consumption seed-by-seed to laboratory germination performances of the seeds. The Q2 oxygen measurement is temperature dependent because the measurement is based on partial oxygen pressure, which is temperature dependent. Thus, for sweet corn oxygen measurement, the temperature was adjusted to 20°C. The test was performed at an interval of 30 min until 72 h. These information was set by Q2 software program. The seeds were airproofed and oxygen content was measured by Q2 instrument. Five Q2 values including IMT, OMR, COP, RGT, HOM were calculated. The laboratory germination test was conducted on four replicates of 100 seeds per hybrid between two layers of filter paper. Seeds were incubated for 7 days at 20℃ with alternating 12 h periods of darkness and light. Normal seedlings were judged according to the ISTA rule (ISTA, 2007) and the number of germinated seedlings was recorded at the 7th day. Germination speed was denoted by the number of germinated seedling days at the 4th day.

2.10 Statistics

Variance analysis, correlation analysis and regression analysis were carried out with SAS V8.02. χ^2 test was conducted according to the following formula: $\chi^2 = \Sigma(AV - PV)2/PV$ (AV is for actual value and PV is for prediction value).

3. RESULTS AND DISCUSSION

3.1 Evaluation indices for seed vigor of sweet corn

It was seen from Table 1 that EC, SS, VA and DA of sweet corn were significantly correlated ($P \leq 0.01$) with both FEP and FES. The results suggested that all the four indices were suitable for evaluating seed vigor of sweet corn.

It is known that there are different genetic backgrounds between super sweet corn (*sh2*) and sugar enhanced corn (*se*). Therefore, we considered that there might be some differences in evaluating seed vigor of super sweet corn and sugar enhanced corn.

The correlation analysis results showed that EC ($P \leq 0.05$), SS ($P \leq 0.05$) and VA ($P \leq 0.01$) significantly correlated with both FEP and FES. However, there was a higher correlation of VA than the former two indices. Therefore, VA was optimal index for evaluating seed vigor of super sweet corn.

The correlation analysis results showed that EC ($P \leq 0.05$) and DA ($P \leq 0.01$) significantly correlated with both FEP and FES. However, there was a higher correlation of DA than EC. Therefore, DA was optimal index for evaluating seed vigor of sugar enhanced corn.

Wang et al. (2004) considered EC was optimal index for evaluating seed vigor of legumes. Furthermore, it was also suitable for evaluating seed vigor of super sweet corn (Wilson et al., 1992), which was consistent to our results. In our results, SS could be used for evaluating seed vigor of super sweet corn, whereas it was unsuitable for evaluating seed vigor of sugar enhanced corn. The difference could be related with higher SS of super sweet corn than that of sugar enhanced corn. Our results showed that VA was an optimal index for evaluating seed vigor of super sweet corn while it was unsuitable for evaluating seed vigor of sugar enhanced corn, and DA was an optimal index for evaluating seed vigor of sugar enhanced corn while it was unsuitable for evaluating seed vigor of super sweet corn.

Table 1. Correlation coefficient between testing index and field performance of sweet corn.

Testing indices	Sweet corn (n=36)		Super sweet corn (n=18)		Sugar enhanced corn (n=18)	
	FEP (%)	FES (d)	FEP (%)	FES (d)	FEP (%)	FES (d)
EC ($\mu\Omega.cm^{-1}.g^{-1}$)	-0.678**	0.543**	-0.548*	0.571*	-0.519*	0.571*
SS ($mg\ g^{-1}$)	-0.603**	0.610**	-0.533*	0.586*	-0.447	0.421
VA (μg)	-0.529**	0.564**	-0.713**	0.725**	-0.408	0.203
DA (OD_{490})	0.585**	-0.566**	0.368	-0.214	0.702**	-0.668**

*, ** *denote significant correlation at $P \leq 0.05$ and $P \leq 0.01$, respectively. FEP, field emergence percentage; FES, field emergence speed; EC, electrical conductivity; SS, soluble sugar; VA, volatile aldehyde; DA, dehydrogenase activity; OD_{490}, optical density at the wave length of 490 nm.*

3.2 Validation for evaluation indices of seed vigor of sweet corn

According to the above results for correlation analysis, regression equations were established by VA or DA as independent variable x and FEP or FES as dependent variable y. The regression equations of $y=-16.074x+86.355$ and $y=1.940x+4.050$ were used to forecast FEP and FES of super sweet corn, respectively. The regression equations of $y=196.830x+15.634$ and $y=-27.315x+14.513$ were used to forecast FEP and FES of sugar enhanced corn, respectively.

It was seen from Table 2 that VA of "Green Superman" at room temperature was higher than at 4℃, which was consistent to their vigor status. DA of "Zhongtian 2" at room temperature was lower than at 4℃, which was also consistent to their vigor status. To test the consistency of the actual value and the value of field emergence performance predicted by the above formulas, χ^2 test was performed. χ^2 test results showed that the four values of χ^2 were all far lower than the value of $\chi^2_{0.05,\ 1}$ (3.84), which suggested that the predicted value was consistent to the actual value. The verification experiment sufficiently indicated that VA and DA was very reliable for evaluating seed vigor of super sweet corn and sugar enhanced corn, respectively.

Table 2. The validation result of VA for seed evaluating seed vigor of super sweet corn and DA for evaluating seed vigor of sugar enhanced corn.

Hybrid	Storage condition	VA (μg)	DA (OD_{490})	FEP (%)			FES (d)		
				AV	PV	χ^2	AV	PV	χ^2
Green Superman	RM	2.566	-	48.0	45.1	0.28	9.5	9.0	0.07
	4℃	1.995	-	52.0	54.3		8.5	7.9	
Zhongtian2	RM	-	0.151	42.0	45.2	0.35	11.0	10.4	0.04
	4℃	-	0.176	48.0	50.5		10.0	9.7	

FEP, field emergence percentage; FES, field emergence speed; RM, room temperature; OD_{490}, optical density at the wave length of 490 nm; AV, actual value; PV, prediction value; $\chi^2_{0.05,1}$=3.84.

3.3 The application of Q2 oxygen sensing technology to seed vigor testing

It was seen from Table 3 that OMR of super sweet corn "Green Superman" was positively correlated with both LGP ($P \le 0.01$) and LGS ($P \le 0.05$). Furthermore, COP was negatively correlated with both LGP ($P \le 0.05$) and LGS ($P \le 0.01$). However, the other three Q2 values were not significantly correlated with LGP or LGS. The results suggested that OMR and COP were suitable for evaluating seed vigor of super sweet corn "Green Superman". The above results were the initial ones, which need further study and validation.

Table 3. Correlation coefficient between Q2 value and laboratory germination performance of super sweet corn "Green Superman".

Q2 value	LGP (%)	LGS (%)
IMT (h)	-0.321	-0.339
OMR (%.h^{-1})	0.582**	0.549*
COP (%)	-0.487*	-0.570**
RGT (h^{-1})	0.395	0.264
HOM (h^{-2})	0.213	0.378

*, ** *denote significant correlation at $P \le 0.05$ and $P \le 0.01$, respectively. LGP, laboratory germination percentage; LGS, laboratory germination speed; IMT, increased metabolism time; OMR, oxygen metabolism rate; COP, critical oxygen pressure; RGT, relative germination time; HOM, homogeneity, or uniformity of the RGT.* $r_{0.05,18}$=0.444, $r_{0.01,18}$=0.561.

4. CONCLUSIONS AND FUTURE WORKS

In conclusion, EC, SS, VA, DA, OMR and COP were all suitable to evaluate seed vigor of sweet corn. Furthermore, VA was optimal for evaluating seed vigor of super sweet corn. DA was optimal for evaluating sugar enhanced corn.

For Q2 test, however, laboratory germination performances were used as the vigor indices, which was lack of stringency to response the vigor status. Furthermore, only super sweet corn "Green Superman" was used. Therefore, to validate the reliability of OMR and COP, in the future works, field test will be conducted and other sweet corn hybrids will be used for Q2 test.

ACKNOWLEDGEMENTS

We give our thanks to Johan van Asbrouck, vice president of research & technology and Freek Schreurs, director of new business, ASTEC Global, Netherlands for Q2 oxygen sensing technology information.

REFERENCES

Chen Ruizheng, Zhang Hongwei, Fu Jiarui, et al.. Studies on the association of rice seed vigor witth volatile aldehydes, Acta Scientiarum Naturalium Universitatis Sunyatseni, 1996, 35: 54-57 (in Chinese)

D. O. J. Wilson, J. C. Alleyne, B. Shafii, et al. Combining vigour test results for prediction of final stand of shrunken-2 sweet corn seed, Crop Science, 1992, 32: 1496-1502

International Seed Testing Association (ISTA). International rules for seed testing edition 2007

J. M. Woltz, D. M. TeKrony. Accelerated ageing test for corn seed, Seed Technology, 2001, 23: 21-34

L. J. Waters, B. L. Blanchette. Prediction of sweet corn field emergence by conductivity and cold tests, Journal American Society for Horticultural Science, 1983, 108: 778-781

M. Akram, M. Munir, S. U. Ajmal, et al. Seed and seedling vigour in rice: Association among the traits, Pakistan Journal of Seed Technology, 2002, 1: 25-30

Qiao Yanxiang, Gao Pingping, Ma Junhua, et al.. Study on physiological characteristics and seed activity changes of two maize inbreds in ageing course, Acta Agronomica Sinica, 2003, 29: 123-127

S. O. J. Wilson, M. B. J. McDonald. A convenient volatile aldehyde assay for measuring soybean seed vigour, Seed Science and Technology, 1986, 14: 259-268

T. M. Ching, S. Hedtke, M. C. Boulger, et al. Correlation of field emergence rate and seed vigor criteria in barley cultivars, Crop Science, 1977, 17: 312-314

Tao Jialing, Zheng Guanghua. Seed vigor. Beijing: Science Press, 1991 (in Chinese)

Y. R. Wang, L. Yu, Z. B. Nan, et al. Vigour tests used to rank seed lot quality and predict field emergence in four forage species, Crop Science, 2004, 44: 535-541

Zhang Shourui. Preliminary study on corn seed vigour testing and relationship between it and field emergence percentage, Gansu Agricultural Science and Technology, 1997, 2: 13-15 (in Chinese)

Zhao Guangwu, Sun Qun, Wang Jianhua. Improving seed vigour assessment of super sweet corn and sugar enhanced sweet corn, New Zealand Journal of Crop and Horticultural Science, 2007, 35(3): 349-356

Zhao Guangwu, Wang Jianhua. Seed vigor testing of sweet corn and evaluation of its field establishment ability, Plant Physiology Communications, 2005, 41: 444-448 (in Chinese)

Zhu Cheng, Zeng Guangwen. A convenient method for measuring the volatile aldehydes released from seed, Plant Physiology Communications, 1999, 35: 39-40 (in Chinese)

FACTORS ASSOCIATED WITH THE ADOPTION OF FOOD SAFETY CONTROLS BY THE MEXICAN MEAT INDUSTRY

Ema Maldonado-Simán [1,*], Pedro Arturo Martínez-Hernández [1], José G. García-Muñiz [1], José Cadena-Meneses [1]

[1] *Posgrado en Producción Animal. Universidad Autónoma Chapingo. Km. 38.5 Carretera México-Texcoco. 56230. Texcoco, Edo. de México.*

[*] *Corresponding author, Address: Posgrado en Producción Animal, Universidad Autónoma Chapingo. Km. 38.5 Carretera México-Texcoco. 56230. Texcoco, Edo. de México, Tel: +52-(595)95216-21, Fax: +52-(595)95216-89, Email: mamaldonado@correo.chapingo.mx*

Abstract: Food marketing at international and domestic markets has focused on processing systems that improve food safety. The objective of this research is to determine the factors influencing the implementation of the HACCP system in the Mexican meat industry, and to identify the main marketing destination of their products. Only 18.5% of enterprises reports fully operational HACCP in their plants. The main destination of their production in the domestic market is supermarkets, suppliers and distributors and specific niches of the domestic market. Exports are to USA, Japan, Korea and Central America and some niches of the domestic market with requirements of higher quality. The four principal factors that motivate enterprises to adopt HACCP are associated with improvement of plant efficiency and profitability, adoption of good practices, improvement of product quality and waste reduction. It is concluded that Mexican enterprises adopt HACCP to successfully remain and face competition by foreign enterprises in the domestic market and to a lesser extent to compete in the international market.

Keywords: Factors, HACCP adoption, Mexican meat industry

Please use the following format when citing this chapter:

Maldonado-Simán, E., Martínez-Hernández, P.A., García-Muñiz, J.G. and Cadena-Meneses, J., 2009, in IFIP International Federation for Information Processing, Volume 295, *Computer and Computing Technologies in Agriculture II, Volume 3*, eds. D. Li, Z. Chunjiang, (Boston: Springer), pp. 1739–1746.

1. INTRODUCTION

Food safety is the common ground for developing and developed countries in the food commercial trade (Maldonado et al., 2005). International and domestic markets experienced important changes due to the jointly application of government and private regulatory requirements and standards along all the food chain, in order to satisfy the increased demands for food safety. (Caswell et al., 1998). Meat industry in Mexico includes Federal Inspection Type (TIF) enterprises that slaughter domestic animals, and others that carry out transformation processes as cuts, boners, sausages and stuffers, meat patties, meat dehydration, preparation of diverse meat foods and tamales (SAGARPA, 2001). There is growing interest in food safety by local consumers that are better informed about disease outbreaks (OECD, 1999), which in turn has promoted more demanding market (Martin et al., 1993).

Along with a stiffer domestic market there is the increased competition of meat plants from countries that have Commercial Agreements with Mexico. Foreign enterprises offer food products processed under tighter safety controls (Caswell and Henson, 1997; Caswell et al., 1998), as Hazard Analysis Critical Control Points (HACCP) and international certification by the International Organization for Standardization (ISO 9000) (Caswell et al., 1998; Herrera, 2004). Government standards related to the food sector may be classified into two groups: one group is the set of standards that define type and final quality of the food product (Caswell and Johnson, 1991; Henson and Caswell, 1999; Henson y Caswell, 1999) and the other is the nutrimental information that must be provided to consumers by the enterprises (Henson, 1997). In Mexico, the National Plant and Animal Health and Food Safety and Quality Services (SENASICA), is the federal agency in charge of regulations and surveillance of food quality and safety (Constitución Política de los Estados Unidos Mexicanos, 2005; Ley de la Administración Pública Federal, 2003). Following political and economical demands of certain countries, food processing industries are required to apply effective food safety controls such as the HACCP System (Caswell and Henson, 1997), in addition to voluntary systems such as ISO 9000 and ISO 14000 (ISO, 1996; Noelke and Caswell, 2000). Food processing enterprises must face different economic and social factors, which determine the pressure in adopting these controls. These factors may be similar for enterprises that operate within the same sector, while among others, they arise from particular demands derived from the need to comply with regulations, avoid lawsuits related with food quality and safety, reduce internal and external costs or achieve a clear differentiation of product brand (Hooker and Caswell, 1999).

Therefore, the objective of this study is to define the nature of the motives that lead Mexican Federal Inspection Type (TIF) enterprises in the adoption of the HACCP system in their plaints and the main market destination of their products.

2. MATERIALS AND METHODS

Field information for this study came from questionnaires (Henson et al., 1999; Maldonado et al., 2005; Martin et al., 1993) sent to the 160 TIF enterprises registered in the Official Listing of TIF Enterprises of the Ministry of Agriculture, Livestock, Rural Development, Fisheries and Food (SAGARPA) (SAGARPA, 2001). Questionnaires were filled out by direct interview, this field phase was from 2000 to 2001; in 2003, and the enterprises identified with a fully operational HACCP system were revisited to validate the accuracy of the information given in the questionnaire.

The questionnaire included a list of factors that in previous studies (Carlsson and Carlsson, 1996; Tompkin, 1994) were proposed as determinants for the enterprises in the adoption of food safety protocols. In order to assure a positive evaluation of the main motives that lead meat processing plants to the implementation of HACCP system, the first step taken was the classification of enterprises by the level of adoption of this system. Then, further analyses were performed only within the group of enterprises that showed the HACCP fully implemented and functional in their plants; these enterprises were then classified by size following the Food Safety and Inspection Service (FSIS) criteria (FSIS, 1996). Enterprises were asked to grade the importance of each deciding factor based on the Likert scale, grades were from 1(very important factor) to 7(not-important factor). After that, with the purpose of better understanding the incentives that the enterprises had to adopt HACCP, a Principal Components Analysis (PCA) was used (Henson and Holt, 1999).

3. RESULTS

Ninety-two (57.5% of all registered Mexican TIF enterprises) answered and filled out the questionnaire. Within these enterprises four groups were identified in relation to the extent of adoption or interest in adopting the HACCP system (Table 1). Only 18.5% of the enterprises had fully operational HACCP system in their plants. Of the Mexican TIF enterprises that had an operational HACCP system 76.5% were small and medium size, while less than a third (23.5%) of this group was represented by large size enterprises (Table 2). In general, fully implementation and operation of HACCP was relatively recent, since at least 75% of the enterprises did not have more than two years with HACCP system operating in their plants and approximately; 10% of them that had an operational HACCP at the time of the interview had it for about four years (Figure 1). These last enterprises could be considered the pioneers in the sector.

Table 1. Grouping of Mexican TIF enterprises based on HACCP implementation status

HACCP status	Number of enterprises	Proportion(%)
Fully operational	17	18.5
Being implemented	27	29.3
Planned but not implemented	30	32.6
No plans to implement	18	19.6

In relation to participation in foreign markets, 47% of the enterprises send approximately 17% of their production to export, and the domestic market took the 83% of the production. Distribution of production within the domestic market was: 63.3% went to national supermarket chains and nation-wide distributors; 16.9% went to the major caterers; 10.6% to meat plants for further processing; 5.3 and 3.9% to local wholesalers and final consumers, respectively. Therefore, the incorporation of the Mexican meat industry to international markets, in number and sale volumes, was incipient. Export destinations were USA, Japan, Korea, Central America and specific niches of the domestic market, such as transnational enterprises (restaurants and fast food chains).

Table 2. Size of Mexican TIF enterprises with fully operational HACCP

Size (number of employees)	Number of enterprises	Proportion (%)
Small (10 to 200)	7	41.2
Medium (201 to 500)	6	35.3
Large (over 500)	4	23.5
Total	17	100.0

Factors that the enterprises considered of greater importance to make a decision in favor of implementing HACCP were related with legal requirements compliance as demanded by domestic and international markets, having access to international markets, and to be certified by a third party (Table 3).

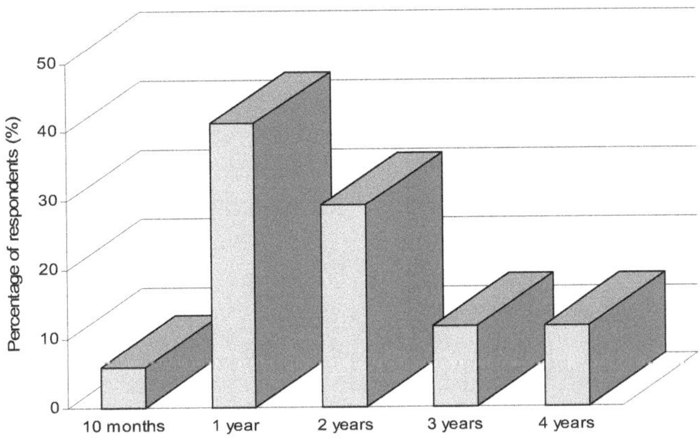

Fig. 1: Time intervals of Mexican TIF enterprises with fully operational HACCP

Seventy-one percent of the enterprises indicated that their current customers were asking them to have an operational HACCP system for all the products generated in their plants. This fact could indicate the importance that market regulations and incentives have on pressing Mexican meat enterprises to adopt food safety protocols.

Table 3. Mean importance scores in ascending order for factors influencing decision to develop/implement HACCP.

FACTOR	GRADE*
Legal requirements	1.24
Have access to international markets	1.41
Need of plant certification by a third party	1.71
Improve product quality	1.76
Attract new customers for their products	1.88
Comply with demands of most of their customers	1.94
Improve production process control	2.59
Maintain existing clientele	2.65
Improve plant efficiency / profitability	2.71
Because it was considered a good practice	3.53
Reduce need for quality auditing by customers	3.71
Reduce product waste	4.41
Recommended by commerce organization	5.06
Reduce client complaints	5.24

** Grade scale was from 1 (very important deciding factor) to 7 (not-important factor in the decision making).*

Four factors explained 69.8% of the variance in the selection of the 14 reasons based on the need to have the HACCP system implemented (Table 4). To enable the advantages derived by the implementation of the HACCP system in their plants, the largest variation proportion factor (22.9%) included the following reasons: "plant efficiency and productivity improvement" and "maintaining existing clientele". Efficiency improvement is an internal factor, while clientele responds to a condition external to the plant." This suggests that this factor is associated firstly with the plant efficiency and then through clientele benefits as a major benefit associated to the HACCP system implementation and operation.

The second factor (20.5%) included internal reasons, such as: "being a good practice" and "improving process control". This suggests that only internal factors are influential to this group of enterprises in order to implement HACCP in their plants. Third and fourth important factors with a total variance of 13.7 and 12.7%, respectively, are also associated with internal reasons: "improving product quality" and "reducing product waste", respectively. Both suggest that these factors are associated with the product improvement as a major benefit derived from the HACCP system implementation and operation.

Table 4. Factor loadings for motivation to implement HACCP derived from principal components analysis.

Reason for implementation of HACCP	Factor 1	Factor 2	Factor 3	Factor 4
Comply with demands of their customers	0.60610	0.04607	-0.43834	0.55344
Legal requirements	-0.07564	-0.31224	-0.02201	-0.58168
Improve production process control	0.37980	0.67632	-0.25184	-0.15528
Reduce product waste	-0.09597	0.43271	0.28037	0.66871
Reduce client complaints	0.44497	0.61000	0.38825	-0.12471
Improve plant efficiency / profitability	0.87042	0.05919	0.21612	0.15484
Improve product quality	0.48784	-0.04600	0.67501	-0.17888
Recommended by commerce organization	-0.31905	-0.36915	0.49419	0.47458
Reduce need for quality auditing by customers	0.61004	-0.51796	-0.18694	0.39521
Need of plant certification by a third party	-0.05413	-0.59356	-0.45168	0.29219
Because it was considered a good practice	0.03376	0.77908	-0.29799	0.07104
Maintain existing clientele	0.77150	-0.04119	-0.30656	-0.20216
Attract new customers for their products	0.54403	-0.53632	0.43462	-0.18414
Have access to international markets	-0.35003	0.29732	0.17489	0.19548
Proportion of variation explained	22.9	20.5	13.7	12.7

4. DISCUSSION

This analysis indicates that the adoption of the HACCP system is motivated by a wide range of factors, also reported in other studies (Caswell et al., 1998; Mortlock et al., 1999), which are related with internal and external factors to the enterprises. The most relevant internal factors for the TIF enterprises to make a decision to adopt the HACCP system are the fact of increasing plant efficiency and good practices during processing. To maintain existing clientele and comply with legal requirements are the most important external factors for this group of Mexican enterprises.

Probably the size of the enterprise is another factor that influences the incentives for implementing HACCP, since the largest ones are the first in adopting and operating the HACCP system in Mexico. Evidences found in other studies (Holleran and Bredahl, 1997; Mortlock et al., 1999), suggest that the largest enterprises adopted this system before the smallest enterprises because of their own benefit expectations.

Mexican enterprises with HACCP system are placed as the most efficient ones in Mexico, with a strong presence in the domestic market and strength to cope with transnational enterprises that are trying to gain customers in this Mexican market on the bases of low number of complaints from unsatisfied customers. Evidence suggests that enterprises adopted food safety protocols as a mean to avoid lawsuits due to defective or unhealthy products placed in the market (Buzby et al., 2001; Henson and Holt, 1999) and with this they keep their local markets (Buchanan and Whiting, 1998).

It is concluded that the adoption of HACCP and ISO 9000 is associated with the goal of increasing sales due to a certified improvement in product quality, in addition to improvements in efficiency and yields of all internal processes. Mexican TIF enterprises adopted HACCP and ISO 9000 to comply with international market regulations and in order to maintain domestic customers with higher level demands on food quality and safety food products.

ACKNOWLEDGMENTS

We acknowledge the financial support of the Consejo Nacional de Ciencia y Tecnología (CONACYT); the assistance of Asociación Nacional de Empresas TIF (ANETIF) and their Managers and the Official TIF Veterinarians of Federal Inspection Type (TIF) enterprises.

REFERENCES

A.G. Herrera. The hazard analysis and critical control point system in food safety. Journal of Methods in Molecular Biology, 2004, 268:235-280.

C. M. Noelke, J. A. Caswell. A Model of the Implementation of Quality Management Systems for Credence Attributes. PhD dissertation. Department of Resource Economics. University of Amherst, Massachusetts. 2000.

Constitución Política de los Estados Unidos Mexicanos. Porrúa. México. 2005.

E. Holleran, M. E. Bredahl. Food Safety, Transaction Costs, and Institutional Innovation in the British Food Sector. Working Paper 97-1, Center for International Trade Studies, Missouri University. Columbia, MO. 1997.

FSIS. Food Safety Inspection Service. Pathogen reduction: Hazard Analysis and Critical Control Point (HACCP) Systems; Final Rule. Federal Register. 61(144):38805-38889. United Sates Department of Agriculture, Washington. 1996.

ISO. ISO Standards Compendium-ISO 9000 Quality Management. International Organization for Standardization Geneva, Switzerland. 1996.

J. A. Caswell, G. V. Johnson. Firm Strategic Response to Food Safety and Nutrition Regulation. In: Economics of Food Safety, Julie A. Caswell, (ed.). pp. 273-297. New York: Elsevier Science, 1991.

J. A. Caswell, M. E. Bredahl, N. H. Hooker. 1998. How Quality Management Metasystems are affecting the Food Industry. Review of Agricultural Economics, 1998, 20: 547-557.

J. A. Caswell, S. J. Henson. Interaction of Private and Public Food Quality Control Systems in Global Markets. Paper presented at the conference Globalization of the Food Industry: Policy Implications. September. The University of Reading, United Kingdom, 1997.

J. C. Buzby, P. D. Frenzen, and B. Rasco. Product Liability and Microbial Foodborne Illness. Washington, DC. U.S. Department of Agriculture. Agr. Econ. Rep. No. 799. USDA/Economic Research Service, Food and Rural Economics Division, Washington, DC, 2001.

Ley de la Administración Pública Federal 2003. Porrúa. México.

M. Carlsson, D. Carlsson. Experiences of Implementing ISO 9000 in Swedish Industry. International Journal of Quality & Reliability Management, 1996, 13:36-47.

M. P. Mortlock, A. C. Peters, C. J. Griffith. Food hygiene and hazard analysis critical control point in the United Kingdom food industry: practices, perceptions, and attitudes. Journal of Food Protection, 1999, 62:786-92.

N. H. Hooker, J. A. Caswell. 1999. Two Case Studies of Food Quality Management Systems. Journal of International Food & Agribusiness Marketing, 11:57-71.

OECD. Food Safety and Quality: Trade Considerations. Organisation for Econ17461746omic Co-operation and Development. Paris. 1999.

R. B. Tompkin, HACCP in the Meat and Poultry Industry. Food Control, 1994, 5:153-161.

R. L. Buchanan, R. C. Whiting. Risk assessment: a means for linking HACCP plans and public health. Journal of Food Protection, 1998, 61:1531-1534.

S. A. Martin, B. J. Bowland, B. Calingaert, N. Dean. Economic Analysis of HACCP Procedures for the Seafood Industry. North Carolina: Research Triangle Institute. 1993.

S. E. Maldonado, S. J. Henson, J. A. Caswell, L. A. Leos, P. A. Martinez, G. Aranda, and J. A. Cadena. Cost–Benefit Analysis of HACCP Implementation in the Mexican Meat Industry. Food Control, 2005. 16:375-381.

S. J Henson, J. A. Caswell. Food Safety Regulation: An Overview of Contemporary Issues. Food Policy, 1999, 24:589-603.

S. J. Henson, G Holt, J. Northen. 1999. Cost and Benefits of Implementing HACCP in the UK Dairy Processing Sector. Food Control, 1999. 10:99-106.

S. J. Henson, G. Holt. Exploring Incentives for the Adoption of Food Safety Controls: HACCP Implementation in the U.K. Dairy Sector. Review of Agricultural Economics, 1999, 22(2):407-420.

S. J. Henson, J. A. Caswell. 1999. La regulación de la seguridad alimentaria: Perspectiva general de las cuestiones actuales. Revista Asturiana de Economía, 22:7-26.

S. J. Henson. Costs and Benefits of Food Safety Regulations. Paris: Organization for Economic Co-operation and Development. 1997.

SAGARPA. Directorio de Empresas Tipo Inspección Federal (TIF). Dirección de Rastros TIF. Dirección de Sanidad Animal. México. 2001.

LEVEL OF ADOPTION OF QUALITY MANAGEMENT SYSTEMS INTO THE MEXICAN PORK INDUSTRY

Ema Maldonado-Siman [1,*], Agustín Ruíz-Flores [1], Rafael Núñez-Domínguez [1], Mariano González-Alcorta [1], Bertha Alicia Hernández-Rodríguez [1]

[1] *Posgrado en Producción Animal. Universidad Autónoma Chapingo. Km. 38.5 Carretera México-Texcoco. 56230. Texcoco, Edo. de México.*

[*] *Corresponding author, Address: Posgrado en Producción Animal, Universidad Autónoma Chapingo. Km. 38.5 Carretera México-Texcoco. 56230. Texcoco, Edo. de México, Tel: +52-(595)95216-21,Fax:+52-(595)95216-89, Email: emamaldonado@correo.chapingo.mx*

Abstract: This research studies the characteristics of the Mexican pork sector; adoption status of quality management systems, and product destinations. Ninety six percent of fifty enterprises have fully answered a questionnaire. Ninety percent are small and medium-sized, the rest are large-sized firms. Nineteen percent of them have totally adopted HACCP, sixty three percent are implementing or planning to do it, the rest have no plans to adopt it. Thirteen percent of the enterprises had ISO 9000. Thirty four percent of their sales go to supermarkets, 57% to other retail chains and 9% to exportation. Product destinations are mainly Central America, United States of America, Asia and Mexico. To improve efficiency and the quality of process it is necessary to implement HACCP. Besides, customers and legal requirements are the external factors, which result in this adoption. In the process of implementing, there are some problems, such as staff motivation and training. The results suggest that HACCP system operating is important for the Mexican pork industry. It also has relevant implications in domestic trade. It is necessary to encourage adoption of quality management systems in the sector.

Keywords: Mexican Pork industry, HACCP, ISO 9000

Please use the following format when citing this chapter:

Maldonado-Siman, E., Ruíz-Flores, A., Núñez-Domínguez, R., González-Alcorta, M. and Hernández-Rodríguez, B.A., 2009, in IFIP International Federation for Information Processing, Volume 295, *Computer and Computing Technologies in Agriculture II, Volume 3*, eds. D. Li, Z. Chunjiang, (Boston: Springer), pp. 1747–1756

1. INTRODUCTION

Food safety is the major focus of consumers, producers, industry groups and government agencies in the whole world. Sockett (1996) indicates that meat is an important agent of food-borne infections. However, consumers assume that safety in pork meat is crucial to reduce outbreak trends of human illness in domestic and foreign countries. The field of pork production has recognized the importance of having consumer's confidence in the pork product by offering a safe product. In order to obtain this consumer's recognition, food safety is an essential coordination of efforts throughout the food chain (Lautner, 2005). This means that research programs in production, processing, distribution, retail and consumer, need to be developed.

In general, it is accepted that Quality Management Systems (QMS) are the most effective means of ensuring the safety and quality of food that identify a series of food attributes (Caswell et al., 1998). The most important QMS in the food industry is HACCP (Hazard Analysis and Critical Control Points), which is now mandatory in some industries and countries, and ISO 9000 which is no mandatory. HACCP focuses to reduce food borne hazards throughout the food supply chain (FSIS, 1996; Hathaway and Cook, 1996) and ISO 9000 aims to standardize quality management and to facilitate communication in a set of international quality assurance standards that guarantee a consistent production process (Bredahl et al., 1997).

It is evident that implementation of HACCP within the Mexican pork industry sector has become widespread over last several years (Gallardo, 2005), and it is beneficial to the firm. Mexican meat industry includes Federal Inspection Type (TIF) enterprises that carry out slaughter domestic animals and/or diverse transformation processes. Food quality in these Mexican TIF enterprises is supervised by the Ministry of Agricultural, Livestock, Rural Development, Fisheries and Food (SAGARPA). These plants incorporate new control points for materials used to handle residues and to detect possible causes of ecological contamination (DGG-CEA, 2000). In addition, the National Food Safety and Quality Program (Programa Nacional para la Inocuidad y Calidad Alimentaria- PRONINCA) promote the implementation of HACCP in the processing and handling of fresh vegetables and meat processing and handling (SAGARPA, 2000).

On the other hand, Holleran and Bredahl (1997) developed a conceptual outline in order to analyze food safety. They observed that firms were motivated by internal and external factors when implementing ISO 9000. It was also found that ISO 9000 was particularity adopted by large firms due to internal operating efficiency and cost reduction. Alternatively, intra – and inter-firm factors are registered in adopting QMS such as HACCP and ISO

9000 (Fouayzi et al., 2005). Price premiums for better quality products, efficiency and improved management are included into the intra-firms factors, whilst inter-firm factors incorporated the improvement of the ability to identify other firms into the supply chain and better opportunities of having trade with them. The objectives of this paper are to characterize the Mexican pork sector, to explore the adoption status of Quality Management Systems (QMS) such as HACCP and ISO 9000, and to identify the reasons and problems faced by the enterprises to implement the systems and product destination.

2. MATERIALS AND METHODS

The data used in this study comes from a survey of firms in pork processing industry (Henson et al., 1999; Maldonado et al., 2005). The survey includes a series of questions intended to collect information about firm characteristics, customers and products, and the status of adoption of QMS as well as other relevant elements, such as the importance of factors in the decision to develop/ implement these two QMS and problems faced by the enterprises during the implementing process. Survey data is collected by direct interview of the General Manager and the Veterinarian assigned of each Pork Meat Federal Inspection Type (TIF) enterprises registered in the Official Listing of TIF Enterprises of the Ministry of Agriculture, Livestock, Rural Development, Fisheries and Food (SAGARPA) from 2000 to 2001. The enterprises were revisited or phoned in 2003 to validate the accuracy of the information given in the questionnaire.

Respondents are presented with a list of factors that previous studies suggested can motivate the decision to implement QMS such as HACCP and ISO 9000 (Martin et al, 1993; Vanguard Consulting, 1993). Then, they are asked to indicate how important each of them have been in their decision, using a seven-point Likert scale ranging from "very important" (1) to "very unimportant" (7).

For the problems faced in the implementation and/or operation of these two QMS, enterprises are also asked to identify the relative importance on a seven-point Likert scale. The scale ranges from "major problem" (1) to "minor problem" (7) of a series of variables that previous studies have suggested can be difficulties in the implementing process of these two QMS (Institute of Quality Assurance, 1991; Tompkin, 1994). Chi-square tests are used to analyze the data using the Statistical Analysis System package (SAS, 2001).

3. RESULTS AND DISCUSSION

All the 50 firms that have answered the survey are members of the National Association of Enterprises Federal Inspection Type (ANETIF - the Trade Association of Meat Industry). Forty-eight (96.0%) of the questionnaires were fully completed. Twenty enterprises (41.7%) had between 51 and 200 employees that represent the size of most of the Mexican Pork TIF processing firms; 33.3% of the enterprises had between 201 and 500 employees; seven enterprises (14.6%) had less than 50 employees; and the remaining five (10.4%) registered more than 500 employees.

Fifty-two percent of the firms are established at the Center region of the country near the cities with heavy domestic trade markets. Forty six percent are settled at the north part where the other main market destination of pork meat products is located. Only 2.1% are found at the south region. Plants carry out four types of processes, sausage-making activities registered the major production level (37.5%), which clusters most of the enterprises; 35.4% of them were slaughterhouses. The rest of the plants concentrate their activities in prepared meals (6.3%), and cold cuts and packing (20.8%).

Additional details about the status of QMS, HACCP and ISO 9000 adoption by firms are presented in Fig. 1. Thirteen of the surveyed firms have two QMS, nine (18.8%) with HACCP under fully operation and four (8.3%) with ISO 9000 certification. It seems relevant the adoption of QMS for this group of processing industry. It may perceive benefits from QMS by dealing with other firms with adopted QMS, and having HACCP seems to be common business practice between them. As suggested in the literature, some firms require their suppliers to have QMS in place to reduce transaction costs associated with hesitation about the quality of the product they buy. Five out of nine enterprises with HACCP fully operating (55.6%) require suppliers to have QMS. Fouayzi et al., (2005) suggested that depending on enterprise position in the chain, it might be required to use specific QMS. Commonly, the leader in the supply chain has the most responsibility for assuring the quality of a product.

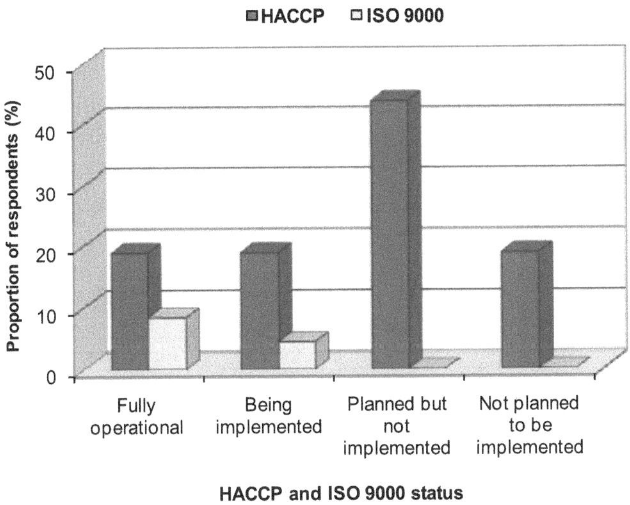

Fig. 1: HACCP and ISO 9000 status of plants that responded the survey

Besides, two more enterprises (4.2%) had the voluntary standard ISO 9000 in full operation, although the group is also formed for those nine with the status still of being operating HACCP (18.8%). This second group declared its interest to be included in the list of regular domestic and foreign suppliers. On the other side, 30 enterprises of pork meat TIF firms have Mexican Official Norms adopted/operating, but any QMS, nevertheless 21 (43.8%) were on the path of beginning to adopt HACCP system. They were mainly pulling off with structural changes of their plants, optimum staff training, and investment in new equipment. The last group of nine firms pointed out that they have no interest in adopting any QMS. The main factors for the decision are related with high costs of implementing/operating, not being a legal requirement in domestic market and not required by major customers. However, capacity for food safety remains far below international standards, and food safety requirements are considered as a significant barrier (Henson et al., 2000; Henson and Wilson, 2002) to markets of developed countries or even in domestic developing countries.

In the food industry, the supply chain can be represented by producer of raw good, as pork producers, who had recognized the importance of producing a product in which their domestic and international market and consumers could have the highest confidence. Then, the processors followed in this segment of chain can be more than one type. Subsequently, the wholesalers, retailers or caterers may follow and are in charge of selling the final product to consumers.

Additional details about the percentage of Mexican meat pork production and main market destination of the 48 TIF enterprises surveyed are registered in Table 1. They reported seven destinations for their production. Supermarkets had the highest proportion (35.0%), 30 firms participated of this proportion of market. This segment of sale chains frequently moves faster in the direction of certification and traceability requirements (Henson and Caswell, 1999), thus the quality control systems are highly required.

Table 1. Percentage of Mexican meat pork production and main marketing destination of the 48 TIF enterprises surveyed.

	Frequency	Percentage	Production (%)
Sales destination			
Supermarkets	30	62.5	35.0
Other retailers	11	22.9	7.5
Major caterers	34	70.8	32.3
Wholesalers	9	18.8	7.0
Final consumers	6	12.5	1.8
Other food processors	9	18.8	7.5
Exportation	17	35.4	8.9
Cumulative Frequency/Percentage			100.0
Exportation			
USA	9	18.8	14.4
Asia (Japan, Korea, China)	10	20.8	35.1
Puerto Rico	4	8.3	10.0
Central America	10	20.8	25.8
Cuba	1	2.1	5.3
Fast food chains and restaurants	2	4.2	9.3
Cumulative Frequency/Percentage			100.0

Major caterers consumed 32.3% of production, and the largest group of enterprises is working with them. Although 17 firms export, less than 10% of production is exported. That is, because capacity to implement effective food safety controls is of vital importance to food exports from developing countries. Henson and Wilson (2002) indicate that the exporting country must be able to comply with food safety requirements and to demonstrate that compliance has been achieved.

Six export destinations are related with this pork processing companies (Table 1). They reported that most of their production (35.1%) is shipped to Japan, Korea and China, however only ten firms share this trading relationship. There may be a couple of reasons to endorse the fact; it could mean the number of TIF pork slaughterhouses have improved their infrastructure and processing technology in 2001 due to financial support. However, through 2002 the increased trends of ten years of exportation went down because major foreign marketing competence from USA and Brazil, and higher food safety standards took place in Japan (Gallardo, 2005).

Central America is the second destination for the Mexican pork meat (25.8%). It has lower quality standards and requirements. Less than 10% of the production goes to USA market; only nine Mexican enterprises participated of this market share. Henson et al., (2000) stated the necessity of the application of HACCP and other hygienic practices along the meat production chain when the goal is to export to most developed countries such as the USA. Those countries need a proof that food safety requirements are achieved. Domestic market with specific high pricing niches has 9.3% of the production and only two enterprises report relationship with them. Local markets in Latin America countries with these characteristics affect local producers by imposing quality and safety standards on them (Reardon and Berdegué, 2002).

Concerning the reasons to implement QMS, the overall analysis showed a difference ($P<0.0001$) whether they were regarded as of major or minor importance by the industries. Improvement of product quality, the need of the plant to be third party accredited, and attraction of new customers for products were considered the major incentives for the enterprises (Table 2). The enterprises are interested in improving their product quality because the increased competence both at domestic and international levels. In addition, the firms third party accredited are able to export, and as consequence, they might attract new customers for their products. The results of this study are similar to those found by Deodhar (2003) for the Indian Food Processing Industry and by Semos and Kontogeorgos (2007) in northern Greece. On the other side, reduce product wastage, reduce customer complaints, and recommended by trade organization were reasons considered of minor importance. Since most of the enterprises are domestic market oriented, the compliment with a trade organization recommendation is not in their priorities: In a study in India, Deodhar (2003) reported that recommended by trade organization, and consignment rejection/retention were rated the lowest among all food categories and among all reasons in motivating adoption of HACCP. Similarly, Semos and Kontogeorgos (2007) found that reduced product wastage, reduced production costs, and increased product prices were the least important benefits of operating HACCP.

The overall analysis for problems perceived by the industries in implementing QMS showed a difference ($P\leq0.0038$) whether they were regarded as of major or of minor importance (Table 3). Need to retrain both production staff and supervisory/management staff, and reduced staff time available for other tasks were considered as major problems. The low level of formal education, staff labour desertion, and the perception of being low-paid are cited among the reasons for these problems.

Table 2. Classification of incentives in implementing Quality Management Systems into the Mexican pork industry.

Incentives	I[a]	II[b]	Percentage[a]	Percentage[b]
To meet the needs of major customers	21	17	55	45
To meet legal requirements	22	10	69	31
Improve control of your production process	24	10	71	29
Reduce product wastage	11	25	31	69
Reduce customer complaints	11	23	32	68
Improve efficiency/ profitability of the plant	26	9	74	26
Improve product quality	31	6	84	16
Recommended by trade organization	14	23	38	62
Reduce need for quality audits by customers	17	19	47	53
Needed for plant to be third party accredited	28	7	80	20
Generally regarded as good practice	25	10	71	29
Hold onto existing customers for products	28	10	74	26
Attract new customers for products	27	7	79	21
Access new overseas markets	22	15	59	41

[a]I= major incentive, [b]II=minor incentive

Table 3. Problems faced in implementing Quality Management Systems into the Mexican pork industry

Problem	I[a]	II[b]	Percentage[a]	Percentage[b]
Need to retrain production staff	8	0	100	0
Need to retrain supervisory/managerial staff	5	1	83	17
Attitude/motivation of production staff	5	4	56	44
Attitude/motivation of supervisory/managerial staff	2	4	33	67
Reduced flexibility of production staff	5	3	63	38
Reduced staff time available for other tasks	6	3	67	33
Recouping cost of implementing HACCP	3	3	50	50
Reduced flexibility to introduce new products	3	5	38	63

[a]I= major problem, [b]II=minor problem

The results of this research are similar to those found by Deodhar (2003) in a study in India. In an another study, Semos and Kontogeorgos (2007) found that need to retrain production staff, attitude/motivation of production staff, and reduced flexibility of production process were the main difficulties faced during HACCP operation in northern Greece. Meanwhile, attitude/motivation of supervisory staff, and reduced flexibility to introduce new products were considered as minor problems by the firms surveyed. However, in general, there were no incentives or formal programs for increasing staff motivation. Similarly, in India Deodhar (2003) found that attitude/motivation of supervisory/managerial staff, and reduced flexibility to introduce new products were regarded as less important by the enterprises surveyed. Semos and Kontogeorgos (2007) reported similar results in a survey to northern Greece food companies.

4. CONCLUSION

The study reports the particular characteristics of the Mexican pork TIF industry and destination of their production, and provides indications of internal and external incentives and problems associated with the implementation of HACCP and ISO 9000. The HACCP operation has been a respond to requests of international markets and very specific domestic niches, and less for local public demand. However, this industry needs to go ahead, in order to compete with foreign markets with higher quality products into a globalization trade holding higher rigorous standards. This kind of information is useful for understanding the key factor of QMS adaptation in Mexican pork industry. They can be a tool for improving management and production efficiency along the meat production chain, in order to compete onto domestic and foreign markets, where the demand of quality products is increasing.

REFERENCES

A. Semos, A. Kontogeorgos. HACCP implementation in northern Greece: Food companies´ perception of costs and benefits. British Food Journal, 2007, 109:5-19.

B. Lautner. HACCP- Its Application for the Pork Industry. National Pork Producers Council. 2005-.Available at: http://www.cvm.uiuc.edu/HACCP/Symposium/LAUTNER.HTM

DGG-CEA (Dirección General de Ganadería y Centro de Estadística Agropecuaria). Situación actual y perspectiva de la producción de carnes en México 1990-2000. Elaborado por el Centro de Estadística Agropecuaria y la Dirección General de Ganadería y el apoyo del Departamento de Agricultura de los Estados Unidos de América (USDA). México, D. F. 2000.

E. Holleran, M. E. Bredahl. Food safety, transaction costs, and institutional innovation in the British food sector. Working Paper 97-1, Center for International Trade Studies. Missouri University. Columbia, MO. 1997.

FSIS. Pathogen reduction: Hazard Analysis and Critical Control Point (HACCP) systems; Final Rule. Federal Register. 61(144), 38805-38889. Food Safety Inspection Service. United States Department of Agriculture, Washington, D. C. 1996.

H. Fouayzi, J. A. Caswell, N. H. Hooker.. Motivations of Fresh-Cut Produce Firms to Implement Quality Management Systems. Review of Agricultural Economics, 2005, 28:132-146.

Institute of Quality Assurance (IQA). An IQA survey on the use and implementation of BS 5750 standards by third party assessment bodies as seen by the end user". Institute of Quality Assurance. London, U. K. 1991.

J. A. Caswell, M. E. Bredahl, N. H. Hooker. How Quality Management Metasystems are affecting the Food Industry. Review of Agricultural Economics, 1998, 20:547-557.

M. E. Bredahl, E. Holleran, L. Zaibet. ISO 9000 in the UK Food Sector. Working Paper 97-2. Center for International Trade Studies, Missouri University, Columbia, MS. 1997.

N. J. Gallardo. Situación actual y perspectivas de la producción de carne de porcino en México 2005. Coordinación General de Ganadería. Secretaría de Agricultura, Ganadería,

Desarrollo Rural, Pesca y Alimentación. 2005. Available at: http://www.sagarpa.gob.mx/Dgg/estudio/sitpor05.pdf

P. N. Sockett . The epidemiology and costs of diseases of public health significance, in relation to meat and meat products. Journal Food Safety, 1995, 15: 91-112.

R. B. Tompkin. HACCP in the meat and poultry industry. Food Control, 1994, 5:153-161.

S. A. Martin, B. J. Bowland, B. Calingaert, N. Dean. Economic analysis of HACCP procedures for the seafood industry. North Carolina. Research Triangle Institute. 1993.

S. E. Maldonado, S. J. Henson, J. A. Caswell, L. A. Leos, P. A. Martinez, G. Aranda, J. A. Cadena, 2005. Cost–benefit analysis of HACCP implementation in the Mexican meat industry. Food Control, 2005, 16:375-381.

S. Hathaway, R. L. Cook. A regulatory perspective on the potential uses of microbial risk assessment in international trade. Paper presented at the Second International Conference on Predictive Microbiology Power with Precision. Hobart, Australia. February 18-22, 1996.

S. J. Henson, G. Holt, J. Northen. Cost and benefits of implementing HACCP in the UK dairy processing sector. Food Control, 1999, 10:99-106.

S. J. Henson, J. A. Caswell. Food safety regulation: An Overview of Contemporary Trends. Food Policy, 1999, 24:589-603.

S. J. Henson, J. Wilson. Understanding the nature of sanitary and phytosanitary capacity. World Bank. Washington, D. C. 2002.

S. J. Henson, R. Loader, A. Swinbank, M. Bredahl, N. Lux. Impact of Sanitary and Phytosanitary Measures on Developing Countries. University of Reading. Centre for Food Economics Research. Reading, U. K. 2000.

S. Y. Deodhar. Motivation for and cost of HACCP in Indian food processing Industry. Indian Journal of Economics and Business, 2003, 2:193-208.

SAGARPA (Secretaría de Agricultura, Ganadería, Desarrollo Rural, Pesca y Alimentación). Programa nacional para la inocuidad y calidad (PRONICA). México, D. F. 2000.

SAS. 2001. SAS/STAT User´s Guide (Release 8.2). Cary NC, USA: SAS Inst. Inc.

T. Reardon, J. A. Berdegué. The Rapid Rise of Supermarkets in Latin America: Challenges and Opportunities for Development. Development Policy Review, 2002, 20(4): 371–388.

Vanguard Consulting. BS 5750 Implementation and Value Added. Vanguard Consulting, Ltd. Buckingham, U. K. 1993.

IMPLEMENTATION OF HACCP IN THE MEXICAN POULTRY PROCESSING INDUSTRY

Ema Maldonado-Siman [1,*], Pedro Arturo Martínez-Hernández [1], Agustín Ruíz-Flores [1], José G. García-Muñiz [1], José A. Cadena-Meneses [1]

[1] *Posgrado en Producción Animal. Universidad Autónoma Chapingo. Km. 38.5 Carretera México-Texcoco. 56230. Texcoco, Edo. de México.*

[*] *Corresponding author, Address: Posgrado en Producción Animal, Universidad Autónoma Chapingo. Km. 38.5 Carretera México-Texcoco. 56230. Texcoco, Edo. de México, Tel: +52-(595)95216-21, Fax: +52-(595)95216-89, Email: mamaldonado@correo.chapingo.mx*

Abstract: Hazard Analysis and Critical Control Point (HACCP) is a safety and quality management tool used as major issue in international and domestic trade in food industry. However, detailed information on costs and benefits of HACCP implementation is needed to provide appropriate advice to food processing plants. This paper reports on the perceptions of costs and benefits by the Mexican poultry processing plants and sale destinations. The results suggest that the major costs of implementing and operating HACCP within poultry processing plants are record keeping and external technical advice. The main benefit indicated by the majority of processing plants is a reduction in microbial counts. Over 39% of poultry production is sent to nation-wide chains of supermarkets, and less than 13% is sent to international markets. It was concluded that the adoption of HACCP by the Mexican poultry processing sector is based on the concern to increase and keep the domestic market, rather than to compete in the international market.

Keywords: HACCP implementation, Mexican Poultry Industry, costs and benefits.

1. INTRODUCTION

Meat and poultry products are sensitive to biological contamination and after then, they provide an excellent environment for growth of bacteria. This kind of contamination may result in foodborne illness (Northcutt and

---

Please use the following format when citing this chapter:

Maldonado-Siman, E., Martínez-Hernández, P.A., Ruíz-Flores, A., García-Muñiz, J.G. and Cadena-Meneses, J.A., 2009, in IFIP International Federation for Information Processing, Volume 295, *Computer and Computing Technologies in Agriculture II, Volume 3*, eds. D. Li, Z. Chunjiang, (Boston: Springer), pp. 1757–1767.

Scott, 2003). For that reason, food sector trends focus on food safety regulations. Besides, consumers demand safer food because of their better knowledge and handling of food, affluence and longevity (Unnevehr and Jensen, 1998). They also recognize the potential seriousness of foodborne diseases and the problem involved because of lack of information on safe handling and storage of food products (Unnevehr and Jensen, 1998).

As a consequence, food safety becomes one of the priorities of countries as foodborne diseases are a large public health problem (Engler et al., 2005). Moreover, stricter regulations apply for importing and exporting countries involved on world trade (Dohlman, 2003). Exporting developing countries center on the effect of rigorous food safety standards set in import markets of developed countries and their ability to export to these markets. In contrast, principal issue for importing countries is to meet the demand of consumers on food safety. As a result, a new approach to ensure safety of the food is the mandatory use of the Hazard Analysis Critical Control Point (HACCP) system in food industries (Unnevehr and Jensen, 1998).

A complex interrelation among governments, industry and consumers is required to implement HACCP (Salay and Caswell, 1998). Private and public incentives are implicated in food industry's response to regulations in order to improve food safety. It involves a complex regulatory analysis of costs (Unnevehr and Jensen, 2001). Segerson (1999) mentioned that enterprises may adopt HACCP system to satisfy incentives or in anticipation of more rigorous regulation. Nevertheless, cost and time for full HACCP implementation vary widely in food processing enterprises (Martin and Anderson, 1999). On the other side, studies reported that many enterprises may not have reliable costs and benefits estimation of HACCP implementation beforehand and it could be major restrain for planning HACCP (Calatore and Caswell, 1999). Thus, essential information is necessary to evaluate concretely the magnitude of costs in each type of plant prior to adoption (Maldonado et al., 2005). The aim of this paper is to explore the process of HACCP implementation in the Mexican poultry meat sector, to identify product destinations, and to offer information about the costs and benefits of HACCP implementation and major problems involved into the process.

2. MATERIALS AND METHODS

A total of 32 plants listed under the poultry meat sector were sent a questionnaire (Henson et al., 1999) during the period 2000-2001 and revisited in 2003. A total of 21 questionnaires were fully completed, 9.5%

were returned by mail and 90.5% by direct interview. These enterprises form part of Federal Inspection Type (TIF) meat processing plants of the Ministry of Agriculture, Livestock, Rural Development, Fisheries and Food (SAGARPA) which is in charge of direct control of food quality on meat and poultry processing plants in Mexico.

In addition, the survey included a series of questions intended to collect information about firm characteristics, customers and products and the status of implementation of HACCP and asked about aspects related to costs, benefits and difficulties of implementation. The respondents ranked difficulty factors in order of importance according to their own conditions, using a seven-point Likert scale ranging from "very important" (1) to "very unimportant" (7); and a set-up cost and benefits in HACCP implementation and operation (Deodhar, 2003), identifying "major" or "minor" cost and benefits by the enterprises. Data was analyzed using SAS (SAS, 2001).

3. RESULTS AND DISCUSSION

Fully answered questionnaires represented 65.6% of poultry meat TIF plants. Table 1 shows basic sample characteristics of those enterprises. Plants carry out seven types of poultry processing. Slaughter and packing activities (38.1%) registered the major production level, following by cold cuts (19.0%). The rest five processes (42.9%) were recorded by nine plants.

One third of the sampled plants had between 51 and 200 employees, and other third of respondents had more than 401 employees on plants. Forty-three percent of the firms have been in operation no more than five years, whilst less than 10.0% were operating on market for more than 21 years. On the other side, only 33.3% of the enterprises reported having more than one plant. Three of them had two plants operating, and four of them operated between three to nine plants. In addition, 71.4% of the products of the enterprises were branded under their own company's brand name, 47.6% under a retailer's brand name and only 19.0% under another manufacturer's brand name.

Further details of the seven destinations for the production of the 21 TIF poultry meat enterprises surveyed and main marketing destination are reported in Fig. 1. Fifteen firms were involved in having commercial contracts with supermarkets and send them 39.6% of their production. The quality control systems for this segment of sale chains frequently are a primary target, so it moves faster in the direction of certification and traceability requirements (Henson and Caswell, 1999).

Table 1. Number of survey respondents by characteristics of enterprises

	Frequency	Percentage	Cumulativepercentage
Type of processes			
Cold cuts	4	19.0	19.0
De-boning	2	9.5	28.6
Hamburgers and nuggets	2	9.5	38.1
Packing	1	4.8	42.9
Pasteurized liquid and dehydrated eggs	2	9.5	52.3
Prepared meals	2	9.5	61.9
Slaughter and packing	8	38.1	100.0
Cumulative Frequency	**21**		
Number of employees			
1-50	4	19.0	19.0
51-200	7	33.3	52.3
201-300	3	14.3	66.7
401-500	5	23.8	90.4
More than 501	2	9.5	100.0
Cumulative Frequency	**21**		
Years of operating			
1-5	9	42.9	42.9
6-15	7	33.3	76.2
16-20	3	14.3	90.4
More than 21	2	9.5	100.0
Cumulative Frequency	**21**		
Additional plants per enterprise			
1-2	3	14.3	42.9
3-4	2	9.5	71.4
5-9	2	9.5	100.0
Cumulative Frequency	**7**		

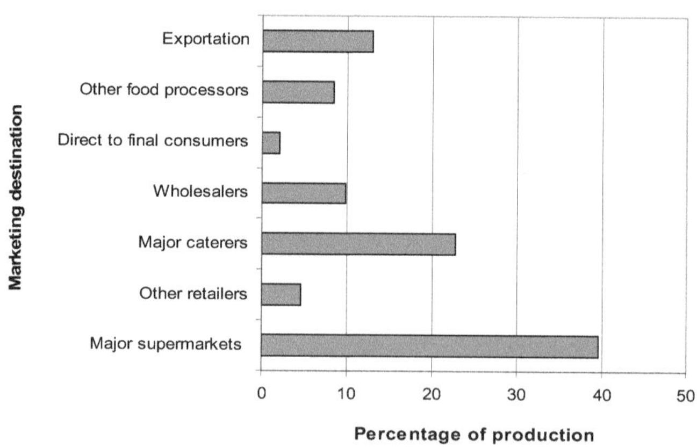

Fig. 1: Percentage of Mexican poultry meat production and main marketing destination of the 21 TIF enterprises surveyed.

Major caterers received approximately 22% of the poultry meat production and a large group of enterprises formed part of their customer list. In contrast, eleven enterprises exported 12.9% of the poultry meat production. As it is indicated (Henson and Wilson, 2002), this low exporting percentage is because enterprises must be able to comply with food safety requirements and to demonstrate that compliance has been achieved. Thus, this goal to implement effective food safety control is of vital importance to food exports from developing countries.

There were six reported export destinations. They are shown in Fig. 2. The firms mentioned that the major part of their production, almost 50.0%, was shipped to Japan; however, only three firms shared with the trading relationship. Northcutt and Scott (2003) indicated that it was possible because these firms operated HACCP system with other control systems to improve product safety.

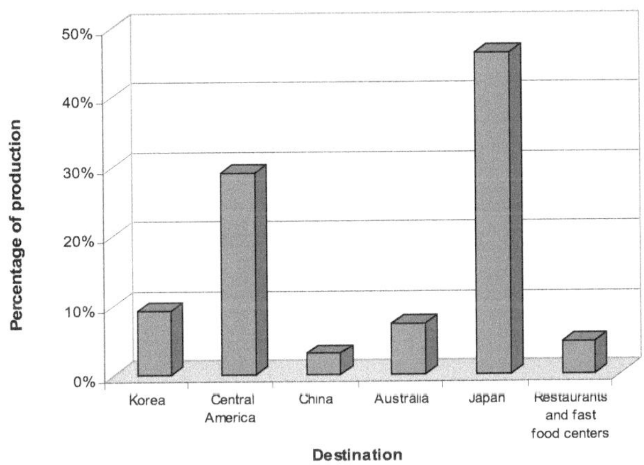

Fig. 2. Exporting poultry meat production and country destination of TIF enterprises surveyed.

Central America is the second destination for poultry meat (29.1%), with lower quality standards and requirements. Each of the other four destinations acquires less than 10% of the production, three of them are of foreign origin and only one firm reported domestic trade relationship (4.8%) with specific high pricing destination as restaurants and fast food centers. It is necessary to operate HACCP and other hygienic practices along production chain since they facilitate the exportation to most industrialized countries where food safety requirements should be warranted (Henson et al., 2000).

3.1 HACCP Status

In reference to the status of HACCP adoption, in total, 38.1% of respondents had fully operational HACCP systems in place (Table 2). Of these, all firms claimed to have full documentation. Furthermore, only 37.5% of them declared that the HACCP system operating in their plants has been verified and audited by a third party and only one firm had operational ISO 9000. Consequently, it can be possible that most of respondents were on path of having fully operational HACCP systems. Besides, they may perceive benefits from HACCP by dealing with other firms with adopted and having HACCP as well as a common business practice between them. Over 50.0% had a fully operational system for at least 24 months, whilst the remaining 50.0% had a fully operational HACCP system for 12 months.

Table 2. HACCP status of meat poultry plants responding survey

HACCP Status	Frequency	Percentage	Cumulative percentage
Fully operational	8	38.1	38.1
Being implemented	3	14.3	52.4
Planned but not implemented	7	33.3	85.7
No plans to implement	3	14.3	100.0
Cumulative Frequency	***21***		

The second group of firms involved those with the status still of being operating HACCP (14.3%) and one of them had fully operational ISO 9000. They were interested to be included in the list of regular domestic and foreign suppliers.

On the other side, 33.3% were on the path to begin the adoption of the HACCP system. They were mainly pulling off with structural changes of plants, optimizing staff training and investing in new equipment.

Finally, the last group of three firms pointed out that they had no interest to adopt HACCP. The main reasons for this decision were high costs to implement/operate, lack of knowledge of HACCP and the fact that in Mexico is not a legal requirement in domestic market. However, capacity for food safety remains far below international standards (Henson et al., 2000; Henson and Wilson, 2002) and food safety requirements are considered as barrier.

3.2 Motivating factors for HACCP implementation

Previous studies have registered motivations for implementing HACCP systems on firms (Vanguard Consulting, 1993; Deodhar, 2003; Fouayzi et al., 2005). Respondents were presented a list of reasons to adopt HACCP. They were asked to rank them by the importance of 14 reasons on a scale of 0-7. Results of the average responses of TIF poultry processing sector are described in Table 3.

Among these poultry processing firms, the attraction of new customers for their products had the highest score, following by improving of product quality. This probably reflects the fact that domestic market is the most important destination for their products. Moreover, the lowest rated was the reason of being recommended HACCP implementation by trade organizations. This could be an indication that trade organizations were either not having close communication, or HACCP was not yet of importance for domestic market.

Table 3. Average scores for HACCP adoption reasons by TIF poultry meat processing enterprises (scale 0 to 7)*

Concept	Mean Score*	Std Dev
Attract new customers for products	1.2	0.5
Improve product quality	1.9	2.0
Reduce need for quality audits by customers	2.2	2.1
Improve control of your production process	2.2	2.0
Access new overseas markets	2.3	2.4
Improve efficiency/profitability of the plant	2.3	1.7
Needed for plant to be third party accredited	2.4	2.3
Hold onto existing customers for products	2.5	1.7
Generally regarded as good practice	2.8	1.8
To meet the needs of major customers	2.9	2.4
To meet legal requirements	3.1	2.3
Reduce customer complaints	4.2	2.6
Reduce product wastage	4.9	2.2
Recommended by trade organization	5.7	1.9

**Where 1=very important and 7=unimportant.*

3.3 Costs and benefits

Respondents were asked to rank the individual costs of implementing and operating, as well as the benefits involved in adopting HACCP process, which previous studies have suggested they result from implementation and/or operation of quality control assurance systems (Vanguard Consulting, 1993). In order to describe the importance of set-up cost in HACCP implementation (Deodhar, 2003), identified a number of one-time fixed cost concepts as "major cost" or "minor cost" by the enterprises (Table 4).

An important fact is that more than 85.5% of enterprises considered "external consultants" as the major cost to adopt HACCP. It was registered as a major cost by 37.5% of slaughter and packing, and none nuggets in poultry processing plants. On the contrary, 50.0% of them considered "managerial changes to the plant" as a minor cost of adopting HACCP system, and 25.0% of slaughter and packing plants recorded as minor cost as well. These results agreed with those presented by Deodhar (2003) in the Indian Food Processing Industry. The difference between the responses by processes for adopting costs considered "major" or "minor" was statistically significant (P<0.0001).

Table 4. Sector-specific importance of HACCP Set-up cost of adopting and operation (I=Major Cost; II = Minor Cost).

	Set-up Costs of Adopting											
	Cold cuts		Nuggets		Pasteurized liquid and dehydrated eggs		Prepared meals		Slaughter and packing		Total	
	I	II	I	II	I	II	I	II	I	II	I	II
External consultants	1	0	0	1	2	0	1	0	3	0	7	1
Investment in new equipment	1	0	1	0	0	1	1	0	3	0	6	1
Staff training	1	0	1	0	2	0	0	0	2	1	6	1
Managerial changes to the plant	0	1	0	0	0	1	1	0	0	2	1	4
Structural changes to plant	0	1	1	0	0	1	1	0	0	1	2	3
Staff time in documenting system	1	0	0	0	2	0	0	0	3	0	6	0

In reference to four operating costs involved in HACCP implementation, enterprises agreed that all of them were considered of major importance to their cost estimation. However, costs generated by record keeping activities and staff training programs were considered of major relevance into their formal estimating plans. Then, all costs involved for product testing and payments to managerial and/or supervisory time were followed in significance for these enterprises.

Cost item with all activities involved in record keeping was considered a critical cost by slaughter and packing poultry processing plants in Mexico, this is also reported by small-scale food processing industry (SSI) in India and in United Kingdom (UK) in the dairy processing plants (Deodhar, 2003; Henson et al., 1999). Finally, it is of relevance that there were not minor costs registered by any of these TIF poultry processing enterprises. However, the consistency of these results is directly correlated by the ability of firms to identify these two types of costs. Given that simultaneously many respondents may have other changes on plants, so it could be difficult for them to identify the two types of costs clearly.

Regarding main benefits of implementing and operating the HACCP system in the Mexican poultry processing sector, those concerning product and customer were of key significance for enterprises, Fig. 3. Reduction of microbial counts and the possibility of increased price for products were the major benefits pointed out. The following benefits in order of importance were those involved in having higher ability to retain existing customer and to attract new customers, as well as the fact of increasing shelf life of products. In the opposite, benefits of minor importance for poultry meat processing industries were related with reduction of product wastage and the possibility of increasing the supervisory/managerial staff's motivation.

The quantitative impact of HACCP into poultry industrial processes was difficult to assess because it was complicated to isolate the impact of HACCP from other changes that were carried out simultaneously in their plants. Therefore, the higher than expected levels of certain benefits, such as the ability to attract new customers and retain existing clients, should be regarded with caution. In addition, there is evidence that a high proportion of the meat-processing enterprises interviewed do not formally monitor costs derived from insuring food safety and quality (Maldonado et al., 2005).

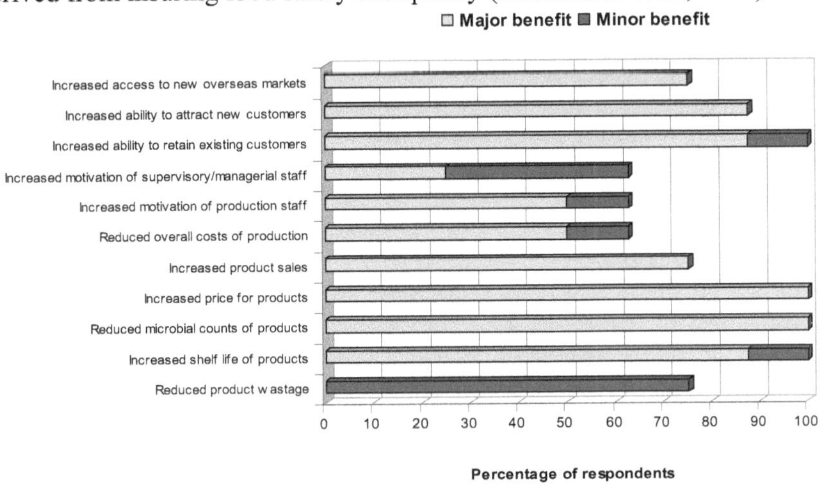

Fig. 3: Specific-sector of major and minor importance of benefits of implementing/operating HACCP in the poultry meat processing industry.

4. CONCLUSIONS

It is already affirmed that HACCP system is based on scientific, systematic and rational approach. It also identifies, assesses and controls hazards not only from production to distribution, but also on transportation and preparation of food. This is integration into the design of the food process that is applied to the whole food chain. For that reason, this study provides one of the first assessments of costs and benefits associated with the implementation of HACCP in the Mexican poultry processing sector. For instance, those costs of adopting, as consulting and operating services involved with record keeping and staff training are of major concern for the enterprises surveyed.

Some of the major benefits of HACCP are primarily intended to consumers, industry and governments as well. Specific domestic markets has been the major reason of the implementation of HACCP system in the Mexican poultry processing sector, joined with requests of international trade. Therefore, HACCP offers a good chance of attracting new customers for enterprises' higher quality products. It seems that, the trend towards higher food safety standards will be a continuous demand to both developing and developed countries, thus HACCP system for poultry meat industries ensures food safety production.

ACKNOWLEDGMENTS

We acknowledge the assistance of Managerial Directors and the Official Veterinarians of the TIF enterprises.

REFERENCES

A. Engler, H. Guy, D. H. Iglesias, A. A. Florindo, G. Gutiérrez, E. Salay, Actor-organization for QAS along agro-supply-chains: the case of mycotoxins reduction in Sourthern Cone Grains. 92 Seminar EAAE, Göttingen, Germany. 2005.

A. S. Martin, W. D. Anderson. HACCP Adoption in the U.S. Food Industry. In: The Economics of HACCP: Costs and Benefits. Laurian J. Unnevehr (Eds.) Eagen Press. Minnesota. USA. 15-28. 1999.

C. Calatore, A. J. Caswell. The Cost of HACCP Implementation in the Seafood Industry: A Case Study of Breaded Fish. In: The Economics of HACCP: Costs and Benefits. Laurian J. Unnevehr (Eds.) Eagen Press. Minnesota. USA. 1999, 45-68.

E. Dohlman. Mycotoxin Hazards and Regulations: Impacts on Food and Animal Feed Crop Trade. In: Buzby Ed. 2003. International Trade and Food Safety: Economic Theory and Case Studies, Agricultural Economic Report 828. USDA, ARS. 2003.

E. Salay, J.A. Caswell. Developments in Brazilian Food Safety Policy, International Food and Agribusiness Management Review, 1998, 1: 167-177.

H. Fouayzi, J. A. Caswell, N. H. Hooker.. Motivations of Fresh-Cut Produce Firms to Implement Quality Management Systems. Review of Agricultural Economics, 2005, 28: 132–146.

J. K. Northcutt, M. R. Scott. General Guidelines for Implementation of HACCP. The University of Georgia, Cooperative Extension Service College of Agricultural and Environmental Sciences Department of Poultry Science. Bulletin 1155. 2003.

K. Segerson. Mandatory Versus Voluntary Approaches to Food Safety. Agribusiness, 1999, 15: 53-70.

L. J. Unnevehr, H. J. Jensen,. Industry Compliance Costs: What Would They Look Like in a Risk-Based Integrated Food System?. Working Paper 01-WP 278. Center for Agricultural and Rural Development. Iowa State University. Ames, IA, USA. 2001.

L. J. Unnevehr, H. J. Jensen. The Economic Implications of Using HACCP as a Food Safety Regulatory Standard. Working Paper 99-WP 228. Center for Agricultural and Rural Development Iowa State University. Ames, IA, USA. 1998.

S. E. Maldonado, S. J. Henson, J. A. Caswell, L. A. Leos, P. A. Martinez, G. Arand, J. A. Cadena. Cost–Benefit Analysis of HACCP Implementation in the Mexican Meat Industry. Food Control, 2005, 16: 375-381.

S. J. Henson, G. Holt, J. Northen. Cost and benefits of implementing HACCP in the UK dairy processing sector. Food Control, 1999, (10): 99-106.

S. J. Henson, J. Wilson. Understanding the Nature of Sanitary and Phytosanitary Capacity. World Bank. Washington, D. C. 2002.

S. J. Henson, J.A. Caswell. Food Safety Regulation: An Overview of Contemporary Trends. Food Policy, 1999, 24: 589-603.

S. J. Henson, R. Loader, A. Swinbank, M. Bedahl, N. Lux Impact of Sanitary and Phytosanitary Measures on Developing Countries. University of Reading. Centre for Food Economics Research. Reading, UK. 2000.

S. Y. Deodhar. Motivation for and cost of HACCP in Indian food processing Industry. Indian Journal of Economics and Business, 2003, 2:193-208.

SAS. SAS/STAT User's Guide (Release 8.2). Cary, NC, USA: SAS Inst. Inc. 2001.

Vanguard Consulting. BS 5750 Implementation and Value Added. Vanguard Consulting Ltd. Buckingham. 1993.

DEVELOPMENT AND IMPLEMENTATION OF PRODUCTION AREA OF AGRICULTURAL PRODUCT DATA COLLECTION SYSTEM BASED ON EMBEDDED SYSTEM

Lei Xi [1] , Wei Guo [1] , Yinchao Che [1] , Hao Zhang [1] , Qiang Wang [1] , Xinming Ma [1,2,*]

[1] *College of Information and Management Science, Henan Agricultural University, Zhengzhou, He Nan, China, 450002*

[2] *College of Agriculture, Henan Agricultural University, Zhengzhou, He Nan, China, 450002*

[*] *Corresponding author, Address: College of Information and Management Science ,Henan Agricultural University, 63 Agricultural Road, Zhengzhou, 450002, P. R. China, Tel:+86-371-63558388, Fax:+86-371-63558090, Email: xinmingma@126.com*

Abstract: To solve problems in detecting the origin of agricultural products, this paper brings about an embedded data-based terminal, applies middleware thinking, and provides reusable long-range two-way data exchange module between business equipment and data acquisition systems. The system is constructed by data collection node and data center nodes. Data collection nodes taking embedded data terminal NetBoxII as the core, consisting of data acquisition interface layer, controlling information layer and data exchange layer, completing the data reading of different front-end acquisition equipments, and packing the data TCP to realize the data exchange between data center nodes according to the physical link (GPRS / CDMA / Ethernet). Data center node consists of the data exchange layer, the data persistence layer, and the business interface layer, which make the data collecting durable, and provide standardized data for business systems based on mapping relationship of collected data and business data. Relying on public communications networks, application of the system could establish the road of flow of information between the scene of origin certification and management center, and could realize the real-time collection, storage and processing between data of origin certification scene and databases of certification organization, and could achieve needs of long-range detection of agricultural origin.

Keywords: embedded system, data collection, middleware, Web Services

---

Please use the following format when citing this chapter:

Xi, L., Guo, W., Che, Y., Zhang, H., Wang, Q. and Ma, X., 2009, in IFIP International Federation for Information Processing, Volume 295, *Computer and Computing Technologies in Agriculture II, Volume 3*, eds. D. Li, Z. Chunjiang, (Boston: Springer), pp. 1769–1781.

1. INTRODUCTION

The collection of agricultural production environment information is an important link to guarantee the quality and safety of agricultural products, and these information on the safety of agricultural products are also main source of data and parameters of pre-warning system and decision support systems(Meng, 2006). Some domestic and abroad research institutions have developed convenient terminal equipment of field environmental data collection and the corresponding software based on embedded systems(Joann, 1998), such as the United States StarPal company's HGIS system, which based on the PocketPC, could collect vector and attribute information of GPS location and soil sampling. China's Xiao Chunhua(Xiao, et al. 2007) has developed Embedded Wireless Remote Environmental Monitoring System based on the ARM7, which apply to multi-source analog signal acquisition. Zhou Guoxiang (Zhou, et al. 2005), a specific remote monitoring controller (ReMoC) based on GSM was developed for this situation to communicate field equipment or apparatus with the monitoring computer on a farm. Meng Zhijun (Meng, et al. 2005), based on DGPS, using embedded GIS development components, realized the integration of GPS and GIS functions under Pocket PC. Currently embedded data collection system emphasis more on integration capabilities: multi-source data collection, data transmission based on variety of physical link and providing data support for business systems (Kevin, et al. 2002; Meng, et al. 2003; Wang, et al. 2006; FAN, et al. 2007; Qiu, et al. 2005).

The research is for environmental monitoring origin of agricultural products. It uses embedded data terminals to construct data collection node, to complete the data collection of multi-sensor analog signals and multi-channel serial signals. Based on GPRS / CDMA / Ethernet physical links, it realizes rapid data transmission. The data are preserved in the data center node. The data center node uses Web Services technology to provide data service interface, and to achieve loose coupling with business systems.

2. SYSTEM DESIGN

The system uses C / S model, which is constructed of data collection node and a data center node, the system physical structure show in Figure 1. Data collection nodes take embedded data terminal NetBoxII (NetBoX-II,2003) as the core, be constructed of data acquisition interface layer, controlling information layer and data exchange layer, complete the data reading of different front-end acquisition equipment, pack the TCP data package, and achieve data exchange between the data center nodes based on Physical link

(GPRS / CDMA / Ethernet). Data center node is constructed of the data exchange layer, the data persistence layer, and the business interface layer, which achieves durable storage of collection data, and provides standardized data through the business interface layer for business systems.

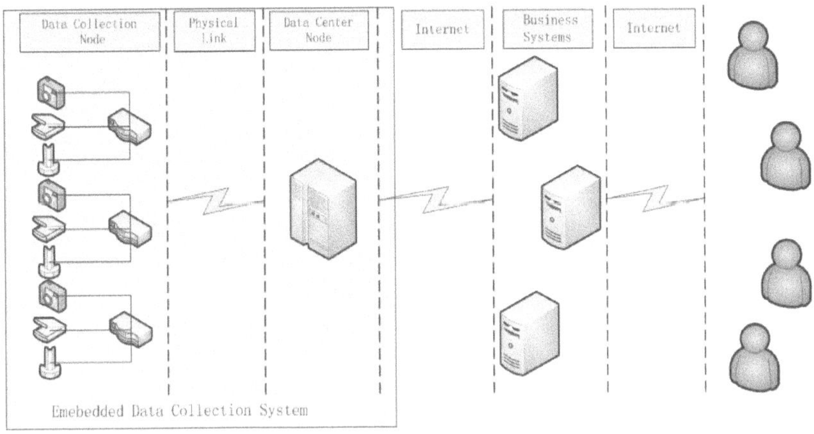

Figure 1. the system physical structure

3. DATA CENTER NODE DESIGN

3.1 Module structure and implementation technology

The main modules function as follows: (1) application modules, the system's main modules, make system run into the cycling of information, (2) configuration modules, complete configuration of network parameters, database parameters, security parameters, and the auto-launch parameters, (3) service launching Module, launch services, monitor the connection request of data collection nodes and complete the initialization of database interface, network interface and system buffering zone (4) database interface module, in charge of connection and lasting of system database; (5) network interface module, is responsible for monitoring network, establishing / releasing connect, send / receive TCP messages, (6) Costumer interactive module, responsible for receiving data acquisition request of data collection node, packaging / sending messages for data collection nodes, receiving / analyzing information from data collecting nodes and retrieving system database; (7) data service module, responsible for providing data accessing interface for the operational system. Data center node module structure shown in Figure 2.

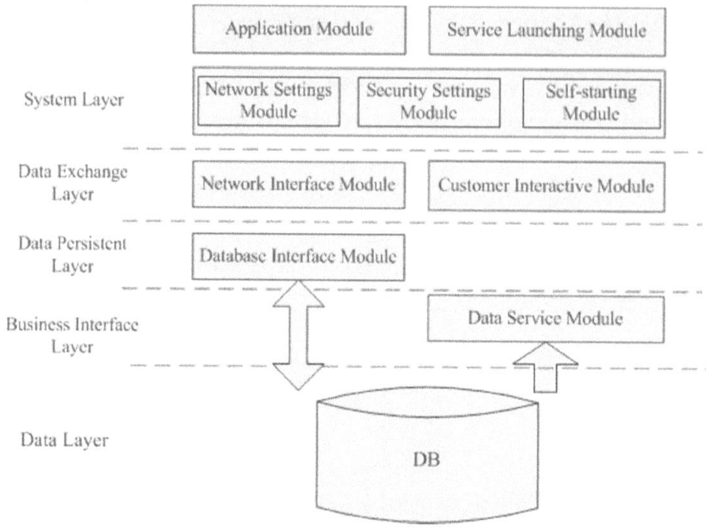

Figure 2. Data center node module structure

Data center node using windows 2003 server, oracle 10g as platform for data, VC.NET 2003 developing the data exchange layer, data lasting layer and system layer. J2EE technology is used to achieve business interface layer. The application services and web services containers adopt BEA Weblogic 9.2.

3.2 Data Model

Core data is constructed of metadata and special topic data, the relationship model of the main data tables show in Figure 3. Special topic data are actually measured data of various types of objects, such as the latitude and longitude data of sampling points and soil nutrient data. Metadata are used to describe the relationship between data, including (1) Data Acquisition Category Table, describing the type of the system data acquisition. Data collection domain is the way of describing the initiation of data collection from the point of the data center node: the active approach, launched by the central node of data, data collection nodes to receive data acquisition orders; passive approach, initiated by the joint data collection, data center node passively receive data acquisition request, (2) table of thematic data categories, is the description of thematic data, and the entity table specifying the data storage. (3) Table of data acquisition categories and thematic data relationship, describing the relevance of the thematic data.

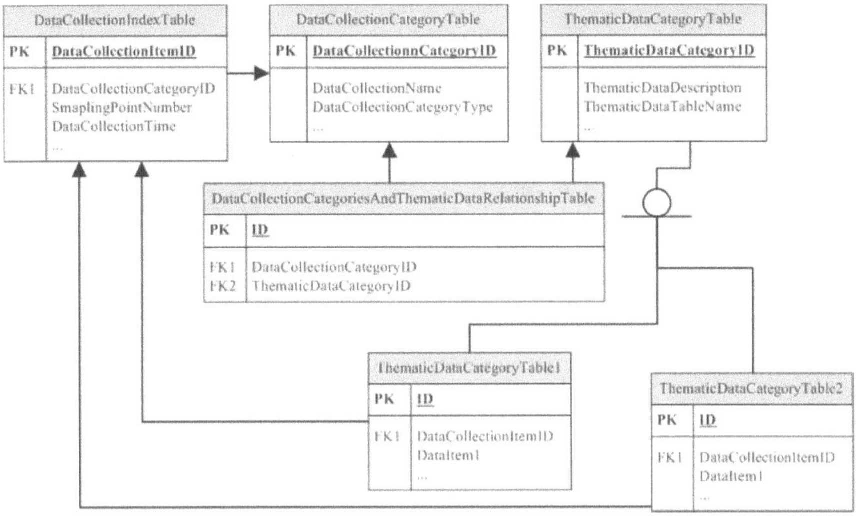

Figure 3. the relationship model of the main data tables

3.3 Workflow and messaging format

3.3.1 Workflow

Passive mode's data centers and data collection node's working process described as follows:

(1) launch and initialization;

(2) in the 4000 TCP port monitoring connection requests of data collection node;

(3) receive connecting requests of data collection nodes, establish TCP connection;

(4) wait for data collection nodes' next round request of a data collection;

(5) receive data collection nodes' next round request of a data collection, send a welcome message;

(6) search database, send data acquisition options to data collection node; wait for feedback;

(7) receive information containing data from data collection node, analyze and order;

(8) continue to perform from (6) until the completion of an acquisition;

(9) save the data of the current round of collection into the database, go to (4);

(10) in this process, if received request of closure from data collection nodes, close the connection, go to (2).

3.3.2 Message format

(1) there are three types of information from data central node to data collection node: welcome information, information of data option and finished information. Welcome information is the first information given to clients at the beginning of the acquisition, which contains tips for users' option and data option is options of the available data; finished information is the instruction of the end of an acquisition. Message format in Figure 4:

\r	Message Type	Message Subject	Options 1	Options content 1	Options 1	Options content 2	… … … …	\n

News head A Data option News end

Figure 4. Data central node message format

'\ r' and '\ n'. Respectively, stands for the head and the end.

Information types, 0: welcome information; 1: data options; 9: Acquisition completed.

Information theme, a summary of the content of the message. Type 1, the theme of the content of data options.

Options identifier, only for type 1, is the identifier for users to choose an option of data.

Options content, when type 1, this is for users to choose an option of a data content.

For example, data collection nodes in the choice of "data collection categories," the feedback information is "soil nutrient acquisition," then the data center node retrieves "data acquisition categories and thematic data relationship Table" and " thematic data categories table ",there is an information in Figure 5. In this case options identifier is "thematic data Category ID" in "thematic data categories table".

\r	1	Soil Nutrient Acquision	5	Sampling Point Latitude And Longitude Data	16	Soil Nutrient Data	\n

Message Head A Data Options Message End

Figure 5. Data central node message format example

(2) The client gives server two kinds of information: requests to begin information collection and data contained information. Message format see Figure 6. When type is 0, that is requesting begin information; when it is non-0, there is the data information which is the option identifier for users to

select. Only when “information type" is non-0, the data type is effective. If its value is 00, then the information is the collected data.

\r	Message Type	Data Type	Message	\n

Figure 6. Data collection node message format

3.4 Business interface layer

The data center node provide business systems the following functions: (1) to provide data collection log list based on "data acquisition catalog" (2) to provide associated data according to the business system’s choice of the data items of "data acquisition catalog table". Release these functions to Web services to form the business interface layer of data center node.

3.4.1 Data service components

The main components of data services include: (1) DataServiceDAO, to abstract bottom data to access. (2) DataServiceXML, to provide data object XML mapping for DAO, and to construct XML documents to a string format. (3) DataServiceValueObject, to provide the value objects of table of entities. (4) DataServiceSession, is no-condition conversation Bean, which packages mutual operations with other data service components, acts as the agent of business systems. Remote interface DataServiceSession defines the interface of Web services, interface definition is as follows.

```
/ **
* this is the remote interface for the System Session EJB
* /
public interface DataServiceSession extend EJBObject {

  public String getDataCollectionList () throws RemoteException;
  public String getSpecialData (int dataCollectionItemID) throws RemoteException;
}
```

3.4.2 Web services

Use BEA WebLogic servicegen to make DataServiceSession EJB form interface of services based on JAX-RPC (Naveen , et al. 2007; Mikhail,2007) and deployment documents, and package it as the deployable EAR files. Take business system’s data acquisition directory as an example, data services components’ interactive sequence show in Figure 7.

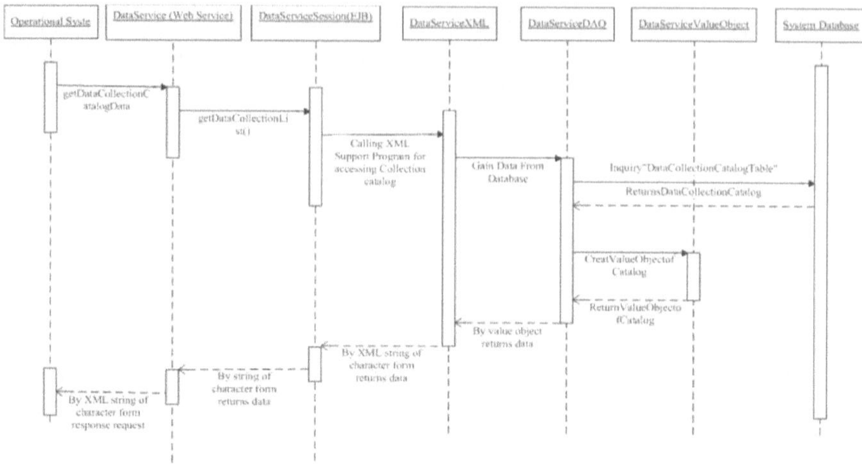

Figure 7. Data services components' interactive sequence

4. DATA COLLECTION NODES DESIGN

Node adopts embedded system and uses Borland C + +3.1 integrated development environment. Hardware includes: NetBox-II development board, 4 × 5 digital keypad and 320 × 240 LCD screen. Through the board integrated serial interface, it can realize the multi-source data collection to increase ETA197 data acquisition module (ETA197,2006). To increase Siemens MC39i GPRS communications module or WAVECOM Q2358C CDMA communications module and the board integrated Ethernet interface can achieve the support of variety of links. In addition NetBox-II provides a TCP / IP and PPP-TCP/IP protocol stack, which can be convenient link to variety of network-based application development.

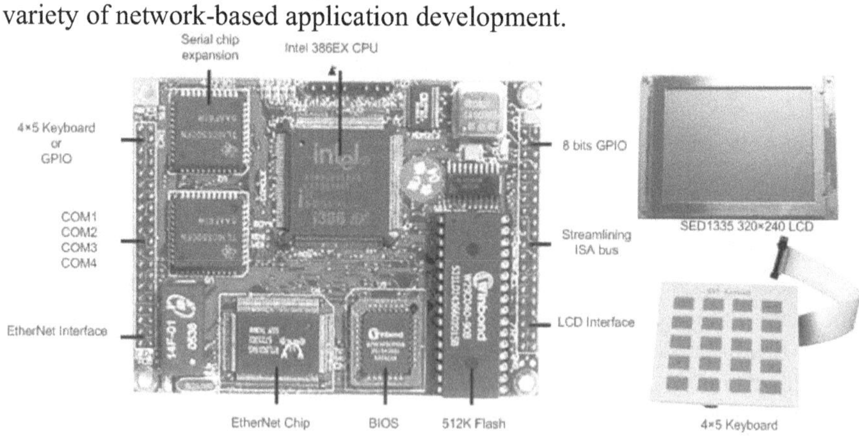

Figure 8. NetBox-II development board, LCD and Keyboard

4.1 Module structure

The main function modules include : (1) keyboard interface module, responsible for receiving the user's keyboard input, making filtering process, sending it into the keyboard buffer zone, and returning to the LCD screen. (2) the interface showing module, due to the NetBox-II of the screen's showing problem, users can not directly use development tools' base functions to screen out Chinese characters, they need to use the specific driver function of specific LCD to dealing with characters' display, in which character dealing Module is used to initialize LCD screen, load English and Chinese characters to allow the system to support the Chinese characters. (3) Network processing module, responsible for processing network transmission-related work, including start-up / shut down Ethernet / GPRS / CDMA interface, create / release of TCP connection, receiving / sending TCP packet, and so on. (4) Information processing module, responsible for processing information interacted with data center node. Data collection node module structure show in Figure 9.

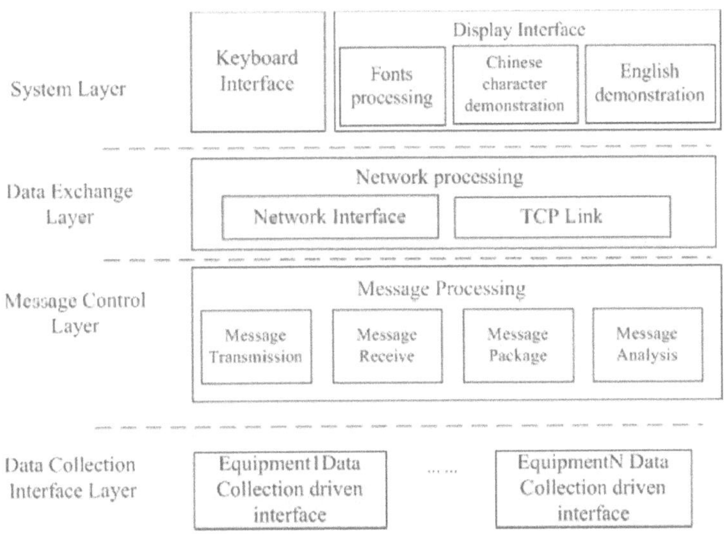

Figure 9. Data collection node module structure

4.2 Workflow and status conversion

The main cycle of data collection node complete receiving and processing data center node information, processing keyboard input and sending messages. The workflow show in Figure 10. In each cycle it deals with this three-part work orderly. To carry out mutually exclusive three-part work, a state variable 'ns' is introduced to record the client running state, and to give different treatment in accordance with different states.

Data collection node running states are showed as the main and sub-state, the main state has four kinds of circumstances: EP_IDLE, EP_SVR, EP_FDBK and EP_QUIT, the sub-state has two kinds of circumstances: ING and ED, the status conversion of data collection node see Figure 11.

EP_IDLE: idle. When the state is ING, wait for users to start a keyboard order of collection or a keyboard order to withdraw from the system, deals only with the keyboard input, not to receive information. If received user's acquisition orders at this time, construction begins requesting information and sent into the buffer zone, the sub-status changes into the ED, after sending messages has been completed, the main state changes into EP_SVR; If user presses the exit button, the main state changes to preparing for Exit status EP_QUIT.

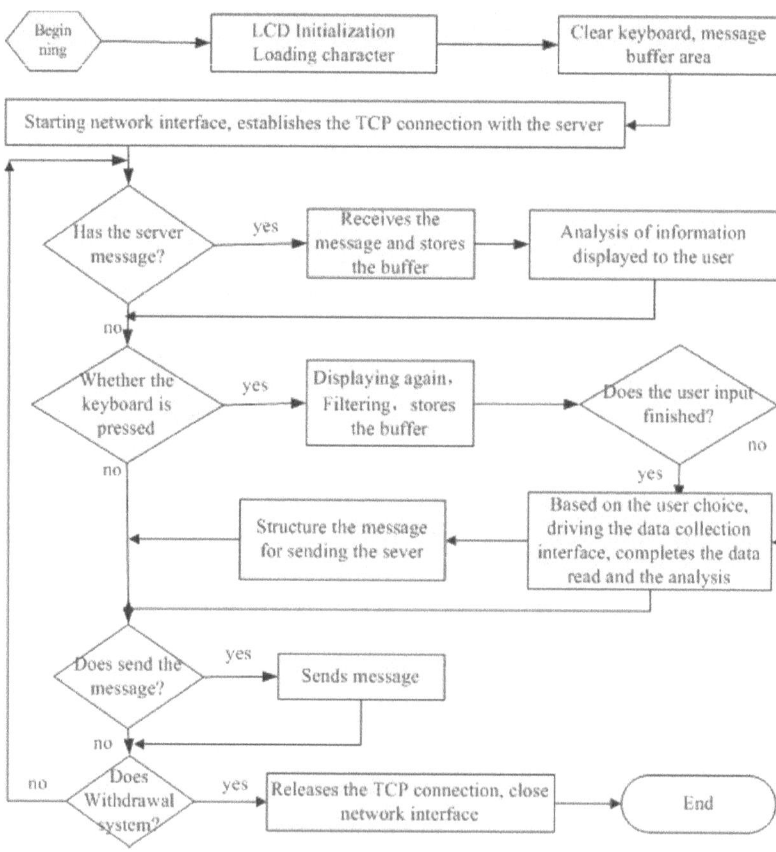

Figure 10. workflow chart

EP_SVR: waiting for receiving information of data center node status. At this time, a collection has already begun; it can receive information from the data central node or the suspension keyboard orders of user. If it receives data option information of server data center node, analyzes and outputs to the LCD screen, the main state changes into EP_FDBK, waiting for users to input data from the keyboard; if the user presses withdraw, the main state will change into preparing for exit status EP_QUIT.

EP_FDBK: waiting for users to input data from the keyboard. If the sub-state is ING, to receive the user's keyboard input and sent into the keyboard buffer zone, when the input has been completed, the status will change into the ED. At this time, construct the data sending to the central node of data sources, sent them into the buffer zone, wait for sending.

EP_QUIT: Preparing for withdraw. If the main status is EP_IDLE, and the sub-status is ING, the user can press the ESC key to lead this system into this state, wait for the after-treatment and withdraw from the system.

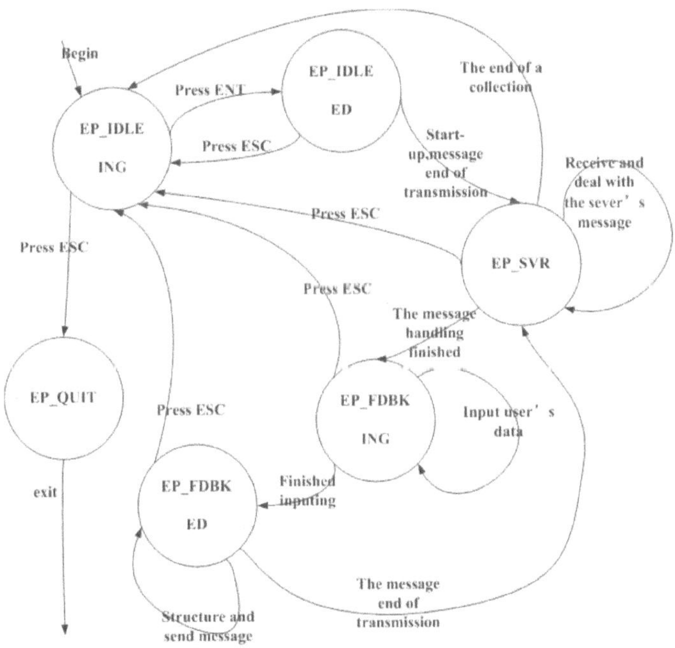

Figure 11. The status conversion of data collection node

4.3 Serial signal acquisition

To send and receive serial communication data of RS232/RS485, in the course of system development to see the serial signal providers as serial signal acquisition targets, to abstract a serial port on each object to a serial signal acquisition drive category. Such a package: (1) int InitUART (int Comldx, int Baudldx, int Parity), completed serial initialization; (2) int InstallISR (int Comldx), serial port interrupt settings, (3) int PutOutputData (int Comldx, char abyte), the data will be sent into output data buffer zone. (4) void StartSend (int Comldx), launching interrupted, the sending data process begins (5) int GetInputData (int Comldx), to retrieve data from the receiving data buffer zone, analyze the collection of data, form a format data. In addition, could add ETA518 expansion boards (ETA518,2006) to achieve timer interrupt serial operation.

5. CONCLUSION

The embedded agricultural origin data acquisition system developed in this research has the following characteristics (1) The realization of multi-source data collection, data transmission based on a variety of physical link and the opening of the data service interface integration, can provide the technical support for monitoring pollution-free agricultural products or green Agricultural production (2) it has a good scalability and adaptability. With proper adjustments to hardware and software of the system, it can be easily applied to other areas of industry data' collection work. In practical application, the system provides the monitoring data of agricultural production environment for pollution-free agricultural products digital authentication system, and realizes the combination of the survey and directions for soil expert system (Xi, et al.2005), which could provide the sampling point's space data and soil nutrient data.

ACKNOWLEDGEMENTS

This study has been funded by China National 863 Plans Projects (Contract Number: 2006AA10Z270－4) and. It is also supported by the Education department of Henan province Scientific and technological projects (Contract Number: 200510466005).

REFERENCES

ETA197. http://www.emtronix.com.cn/download/ETA197.pdf
ETA518. http://www.emtronix.com.cn/download/ETA518.pdf
Fan Li, Ding Zhuyu, Yu Guangwei, Huang Jie, Zhou Yinghua. Research of the Application of Embedded GPRS Long-Range Transmission Technology in Accurate, Agriculture Journal

of Southwest University(Natural Science Edition) , 2007,05 Vol.29,NO 5.pp137-140 (in Chinese)

Joann Wilcox. Embedded systems. Successful Farming. (Iowa edition). Des Moines: Apr 1998. Vol. 96, Iss. 6; pg. 19, 1 pgs

Kevin B. Mayer, Dr Ken Taylor CPEng, MIEAust. An Embedded Device Utilizing GPRS for Communications. ICITA2002.

Meng Hong.Trend of the technology of agricultural information collection, Agriculture Network Information,2006,12 ,NO 12.pp35-38 (in Chinese)

Meng Zhijun, Wang Xiu, Zhao Chunjiang, Xue Xuzhang. Development of field information collection system based on embedded COM-GIS and pocketPC for precision agriculture , Transactions of The Chinese Society of Agricultural Engineering 2005,04 Vol.21,NO 4.pp 91-96 (in Chinese)

Meng Zhijun, Zhao Chunjiang, Wang Xiu, Chen Liping, Xue Xuzhang. Field multi-source information collection system based on GPS for precision agriculture, Transactions of The Chinese Society of Agricultural Engineering,2003,07 Vol.19,NO 4. pp13-18 (in Chinese)

Naveen Balani, Rajeev Hathi. Design and develop JAX-WS 2.0 Web services. https://www6.software.ibm.com/developerworks/education/ws-jax/index.html, 2007,09

NetBoX-II. http://www.emtronix.com.cn/download/NetBox-II_low_price.pdf

Qiu Huanguang, Deng Xiangzheng, Zhan Jinyan, Guo Zaocheng. Design of a distributed data acquisition and spatial decision making system for precision agriculture. Transactions of The Chinese Society of Agricultural Engineering 2005,10 Vol.21,NO 10.pp109-113 (in Chinese)

Wang Qian, Sun Zhongfu, Li Xiuhong, Zhang Hongtao, Wang Yingchun. Agricultural Environment Monitoring Based on Embedded System,Control & Automation, 2006,Vol.23,NO 8-2.pp38-41 (in Chinese)

Xi Lei, Ma Xinming, Hu Xiaohong, Zhao Hao, Li Jing. Agriculture Intelligent Decision System Development Platform Based On Knowledge And Growth Model, Journal of Shanxi University of Science & Technology, 2005,12 Vol.23,NO 6.pp113-117 (in Chinese)

Xiao Chunhua, Li Xiuhong, Sun Zhongfu, Zhang Hongtao, Song Zikai.Implementation of a wireless and remote environment monitoring system, Microcomputer Information, 2007,11 Vol.23,NO 4-2. pp 4-6 (in Chinese)

Zhou Guoxiang, Zhou Jun, Miao Yubin, Liu Chengliang. Development and application on GSM-based monitoring system for digital agriculture, Transactions of The Chinese Society of Agricultural Engineering, 2005, 06. Vol.21, NO 6. pp87-91 (in Chinese)

APPLICATION OF ISO22000 AND FAILURE MODE AND EFFECT ANALYSIS (FMEA) FOR INDUSTRIAL PROCESSING OF POULTRY PRODUCTS

Theodoros H. Varzakas [1,*], Ioannis S. Arvanitoyannis [2]

[1] *Technological Educational Institute of Kalamata, School of Agricultural Sciences, Department of Agricultural Products Technology, Hellas (Greece) E-mail: theovarzakas @yahoo.gr*

[*] *Whom all correspondence should be addressed to*

[2] *University of Thessaly, School of Agricultural Sciences, Department of Agriculture, Animal Production and Aquatic Environment, Fytoko Street, 38446 Nea Ionia Magnesias, Volos, Hellas (Greece)*

Abstract: Failure Mode and Effect Analysis (FMEA) model has been applied for the risk assessment of poultry slaughtering and manufacturing. In this work comparison of ISO22000 analysis with HACCP is carried out over poultry slaughtering, processing and packaging. Critical Control points and Prerequisite programs (PrPs) have been identified and implemented in the cause and effect diagram (also known as Ishikawa, tree diagram and fishbone diagram).

However, the main emphasis was put on the quantification of risk assessment by determining the RPN per identified processing hazard. Evisceration, reprocessing, immersion chilling and cooling/freezing/ distribution were the processes identified as the ones with the highest RPN (324, 252, 240, 210, respectively) and corrective actions were undertaken. Following the application of corrective actions, a second calculation of RPN values was carried out leading to considerably lower values (below the upper acceptable limit of 130). It is noteworthy that the application of Ishikawa (Cause and Effect or Tree diagram) led to converging results thus corroborating the validity of conclusions derived from risk assessment and FMEA. Quantification of risk assessment was also carried out for a sliced poultry sandwich and for refrigerated chicken salad. RPN was determined before and after corrective actions. Therefore, the incorporation of FMEA analysis within the ISO22000 system of a poultry processing industry is considered imperative.

Please use the following format when citing this chapter:

Varzakas, T.H. and Arvanitoyannis, I.S., 2009, in IFIP International Federation for Information Processing, Volume 295, *Computer and Computing Technologies in Agriculture II, Volume 3*, eds. D. Li, Z. Chunjiang, (Boston: Springer), pp. 1783–1795.

Keywords: FMEA; ISO22000; HACCP; Ishikawa diagrams; Poultry manufacturing.

1. INTRODUCTION

The L. monocytogenes surveillance programme has shown an increase in the incidence of this micro-organism in meat products, especially poultry (unpublished data). It is thus possible that some of the sporadic cases of listeriosis observed in Italy for which the vehicle of infection was not identified may have been associated with the consumption of foods of animal origin.

Sumner et al. (2004) reported that their study had parallels with the findings of Schlosser et al. (2000) and Sarwari et al. (2001). Firstly, there was no great concurrence between the suite of serovars isolated most frequently from red meat and those most commonly associated with salmonellosis in Australia. Secondly, S. sofia was isolated from 50% of poultry samples and yet caused only 0.3% of salmonelloses, pointing to a very low ability to cause human illness.

In the case of poultry meat, however, phage types of S. typhimurium isolated from poultry were responsible for a significant proportion of salmonelloses, a situation which pertained both before and after regulatory change.

Poultry carcasses usually have very high levels of contamination on the skin. They can present microorganisms that cause food-born illness as well as food spoilage. There are a series of microorganisms on the surface of the carcasses which can be analyzed in order to indicate the microbiological quality, the level of hygiene in production and handling, and the correct maintenance of the cold chain (Sandrou and Arvanitoyannis, 1999). Furthermore, they can help to predict products' shelf-life. The flow diagram of slaughter and dressing of chicken broilers is given in Figure 1.

Three mechanisms exist for the attachment of bacteria on poultry carcasses (NACMSF, 1997). Retention occurs when carcasses come into contact with water containing bacteria. Entrapment occurs when exposed tissue surfaces (skin, collagenous connective tissue layers of muscle) absorb water and begin to swell. Adhesion occurs when microorganisms adhere to surface tissues. Only some strains of salmonellae are capable of adhesion. It looks certain that bacteria are firmly attached to carcasses before the start of processing and hence it is very difficult to get eliminated in the processing premises.

Stunning and bleeding: After leaving the holding areas, the animals are located in a stunning or immobilization area where they are rendered unconscious. Cattle stunning in the United States are usually done by a bolt pistol or electric shock. The anesthetized animals are then shackled and hoisted (hind quarters up) onto a overhead rail or dressing trolley. Bleeding (exsanguination) or sticking is conducted with the blood collected in a trough or floor drain for disposal or further processing.

Scalding is a process by which the bird is subjected to moist heat for a short time to loosen hair and facilitate the removal of feathers. Immersion scalding using a single or two-stage scalder usually occurs in New Zealand. Soft scalds (<=55°C) and hard scalds (>55°C) are being used for up to 3 minutes. When birds are immersed in the scalding tank, some of the dirt, faecal material and other contaminants on the surface of the bird are removed and contaminate the scald water, hence scalding could be a means of cross-contamination.

The number and incidence of Salmonella and Campylobacter in raw poultry carcass are greatly affected by the operating conditions of scalding and defeathering, evisceration, washing, skinning, boning, portioning, chilling, decontaminating (irradiation) and freezing. In commercial scalding, temperature is set in a range of 50 to 60°C for 2 to 2.5 min. One study found that increasing of scalding water temperature from 50 to 60°C reduced the number of C. jejuni and S. Typhimurium by approximately 6 log CFU/ ml (Yang et al., 2001). In contrast, Humphrey (2004) found that hot water (up to 80°C) treatments were not much more effective in removing campylobacters as compared to cold water. So, he suggested that heat treatment is not a significant control measure, particularly when chicken carcasses are to be sold whole.

Treatment of wash water has been found to be a potential processing control to reduce contamination. Alternatives are treatment with chlorine water, electrolyzed water, trisodium phosphate, cetylpyridinum chloride, hydrogen peroxide etc.

The main objection to immersion scalding is the possible inspiration of contaminated scald water by the birds, with subsequent contamination of air sacs, lungs and other internal organs and edible tissues by pathogenic bacteria (Bailey et al. 1987). Hard scalding at about 58-60°C, followed by mechanical plucking, results in removal of the outer epidermal layer of the bird's skin. This cuticle-free skin of hard-scalded broilers serves as a more suitable substrate for bacterial attachment (Kim et al., 1993).

Counter-current scalders and multi-stage scalders have a greater impact on reducing the level of microorganisms on the carcass (Bolder, 1998). Multi-tank also reduce both the total aerobic and enterobacterial counts (Stals, 1996).

Defeathering is considered as a major site of cross-contamination for poultry carcasses by pathogens such as Camplylobacter and Salmonella and E. coli. (NACMCF, 1997).

During the process, feathers, dirt and bacteria from carcasses are removed, however, creation of aerosols is being carried out that spread bacteria, water and solid matter, contaminating other carcasses (Tinker et al., 1996). Staphylococcus aureus might also colonise the machinery by getting associated with the rubber fingers used to remove the feathers. Despite all these bacteria and their possible presence, defeathering can result in a reduction on carcass contamination by 1000-fold as reported by Hinton et al. (1996). Cross-contamination could be reduced if defeathering took place on a carousel.

Dressing and hide removal: The bled carcasses are conveyed to the slaughter area where dressing (cleaning) and evisceration occurs. Dressing is carried out from the overhead position or by placing the carcass in a cradle. The skin is removed from the head and the head separated from the body. The fore and hind feet are then removed to prevent contamination of the carcass with manure and dirt dropped from the hooves (shanking or legging). Each leg is then skinned. The remaining hide is removed from the carcass with electric or air-powered rotary skinning knives. The hides are preserved by salting or chilling on ice before being sent to a tannery for processing into leather (http://wrrc.p2pays.org/p2rx/subsection.cfm?hub=449andsubsec=11andnav=11andCFID=1245734andCFTOKEN=43813664).

Evisceration can be a major source of faecal contamination on carcasses and edible offal, particularly if the intestines are cut or broken. This will result in an increase in contamination by mesophilic bacteria including pathogens such as Salmonella, Campylobacter, Clostridium perfringens, and Listeria.

New Zealand uses small processing premises whereas large premises use autmoamtic equipment (MAF, 2000).

The skinned carcasses are opened to remove the viscera (internal body organs). The abdomen is opened from the top to bottom where the internal organs are loosened and removed from the body. The abdominal organs are inspected and the stomach and intestine are emptied of manure and cleaned for further processing. A handsaw is used to halve the remaining carcass by cutting through the center of the backbone. The inedible materials are collected and sent to a rendering plant for manufacture of feed materials. The beef sides are washed to remove any remaining blood or bone dust and the carcasses are physically or chemically decontaminated. The simplest physical decontamination method involves spraying with high pressure water or steam. Chemical decontaminants include acetic and lactic acids, and aqueous solutions of chlorine, hydrogen peroxide and inorganic acids.

Operator skills influence the levels of gut breakage in smaller plants. Moreover, full crops and intestinal tracts increase the risk of gut breakage, hence feed should be withdrawn carefully prior to processing (Izat et al.,

1989). In addition , shorter withdrawal periods may have more advantages in the reduction of bacteria in the crop and the caeca.

Finally, spray cleaning of carcasses during evisceration could prevent the occurrence of Salmonella and Enterobacteriaceae. Moreover, rinsing with chlorinated water could aid the minimization of cross-contamination.

Proper cleaning and effective sanitation is an essential component of processing poultry, as it contributes significantly to the prevention of product contamination with microorganisms that cause food-borne disease and spoilage. Rapid expansion of production volume, increased further-processing and introduction of diverse ready-to-cook and ready-to-eat products, sophistication of the processing equipment, implementation of HACCP and microbial finished-product standards, and, more importantly, expensive product recalls have necessitated greater control over the cleaning and sanitation process.

The main emphasis of HACCP in slaughter operations is to minimise the potential for contamination of the finished raw products (Untermann, 1993). Current technology cannot assure a pathogen-free product (Baumgart, 1993), but it can minimise contamination from any slaughter operation (Goodfellow, 1995). Two fundamental concepts must be considered during the slaughtering process. The first one is related to the use of procedures which will minimise the degree of contamination on carcasses, such as training workers, providing adequate work space and time, providing a plant layout that flavours microbial control, and selecting equipment. For this purpose, cost-effective technology is needed (Mackey and Roberts, 1993). The second concept is to include procedures which can reduce or destroy pathogens which inadvertently contaminate the carcass during slaughtering (Tompkin, 1994).

Immersion chilling (immersion in ice or chilled water) or a combination of immersion chilling and wet air chilling (blast air chillers) are the methods used to reach an internal carcass temperature of 4°C within 24 hours of dressing. It takes approximately 50 minutes to achieve this temperature and the use of chlorinated water is essential. However, immersion chilling is an area of cross-contamination with C. jejuni, Salmonella and Clostridium perfringens (Lillard, 1990). Chlorine can facilitate the hygienic operation of these systems. Chlorine dioxide could also be added to reduce Salmonella in chiller water.

Four hazards are likely to occur: physical contamination with fecal material and potential pathogen contamination at evisceration/presentation, pathogen contamination at reprocessing, pathogen cross-contamination and proliferation at chilling, and pathogen proliferation at finished products storage (cold). Four CCPs address these four hazards: proper evisceration/presentation, proper reprocessing, and proper chilling of product and proper maintenance of finished product temperatures during storage

(Table 1, 2). The ISO 22000 Analysis worksheet for the determination of prerequisite programs for poultry slaughtering and synoptical presentation of HACCP plan are given in Table 1 and Table 2, respectively. Moreover in Fig.1 the main processing stages of poultry slaughtering are presented.

2. RESULTS OF FAILURE MODE AND EFFECTS ANALYSIS

In FMEA analysis, risk of contamination and its presence at Hazardous Fraction in the final product, is expressed with the risk Priority Number (RPN) which is defined as follows:

$$RPN=S\times O\times D \quad (1)$$

Where S: Severity of contamination risk, O: supply probability occurrence of contaminated ingredient, D: Detection probability of contaminated ingredient.

FMEA tables were constructed where Hazardous Elements (ingredients) are assessed and RPN was calculated as can be seen in Table 3.

The classification of hazardous elements occurs according to the RPN assessment as can be seen in Table 5 and corrective actions are proposed per identified hazard. The new RPN is calculated following undertaking of corrective actions.

The limit for RPN is 130 and if this number exceeds 130 corrective action should be implemented.

Ishikawa or Fishbone diagrams analyse all the dangers at all processing stages where CCPs are incorporated. These diagrams consist of five basic axes: man, machine, materials, methods and environment. In these axes the danger is described. Only one diagram is presented here and that refers to the cooling, storage and distribution of poultry processing.

In Figure 2 the Ishikawa is applied to detect the causes that might be due to the physical, chemical and microbiological dangers occurring at the storage and distribution stages. Regarding man the main problem is bad distribution due to the inadequate training of workers. Regarding machines the problem might be due to the wrong placement of packaging, or the bad adjustment of the temperature of the transportation vehicle due to inadequate personnel training.

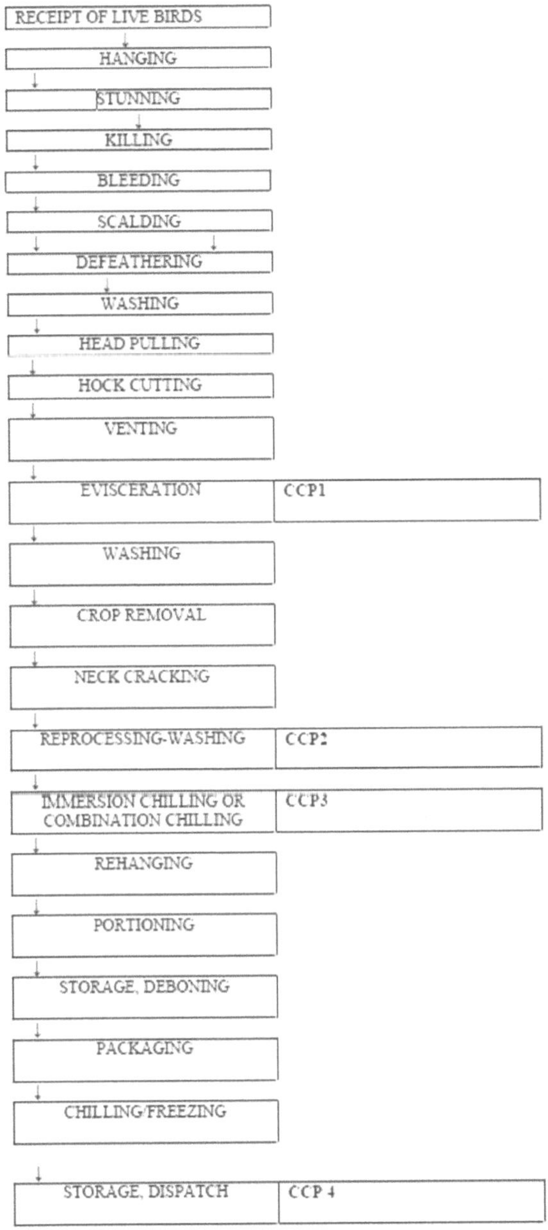

Figure 1. Flow Diagram of Slaughter and Dressing of Chicken Broilers (Adapted from MAF New Zealand, 2000)

Table 1 Determination of Critical Control Points for poultry slaughtering

A/A	STEP	Q1: Do control measure(s) exist	Q2: Is the step specifically designed to eliminate or reduce the likely occurrence of a hazard to an acceptable level	Q3: Could contamination with identified hazard(s) occur in excess of acceptable level(s) or could these increase to acceptable levels	Q4: Will a subsequent step eliminate identified hazards or reduce likely occurrence to an acceptable levels	CCP/PrP
1	RECEIVING OF LIVE BIRDS	YES	NO	NO		PrP
2	HANGING	YES	NO	NO		PrP
3	STUNNING	YES	NO	NO		PrP
4	KILLING	YES	NO	NO		PrP
5	BLEEDING	YES	NO	NO		PrP
6	SCALDING	YES	NO	NO		PrP
7	DEFEATHERING	YES	NO	NO		PrP
8	WASHING	YES	NO	NO		PrP
9	HEAD PULLING	YES	NO	NO		PrP
10.	HOCK CUTTING	YES	NO	NO		PrP
11	VENTING	YES	NO	NO		PrP
12	EVISCERATION	YES	YES			CCP1
13	WASHING	YES	NO	NO		PrP
14	CROP REMOVAL	YES	NO	NO		PrP
15	NECK CRACKING	YES	NO	NO		PrP
16	REPROCESSING	YES	YES			CCP2
17	IMMERSION CHILLING	YES	YES			CCP3
18	REHANGING	YES	NO	NO		PrP
19	PORTIONING	YES	NO	NO		PrP
20	DEBONING	YES	NO	NO		PrP
21	PACKAGING	YES	NO	NO		PrP
22	CHILLING/FREEZING	YES	NO	NO		PrP
23	STORAGE/DISPATCH	YES	YES			CCP4

Table 2 Synoptical presentation of HACCP plan (Adapted from USDA, 1999)

CCP/Processing step	Critical Limits	Monitoring CCPs	HACCP Records	Verification	Corrective Actions
1P/B Evisceration	Zero visible fecal contamination after processing; equipment kept properly adjusted; no gut breakage due to improper equipment adjustment; range of 20-50 ppm chlorine.	Visible check; check chlorine at start up and every 2 hours using documented random sampling procedures to demonstrate control.	Plant Finished Product Standards Form Antimicrobial Sheet, Equipment Maintenance Sheet, Corrective Action Sheet	The QA supervisor will review the Plant Once per shift. Antimicrobial Sheet and observe chlorine level testing. Maintenance Supervisor will review Equipment Maintenance Log twice per shift.	QA will reject or hold product until zero fecal tolerance is achieved. Equipment will be properly adjusted to assure zero contamination. All suspect product will be visually examined between evisceration and after final wash. Contaminated product will be condemned or reconditioned.
2P/B Reprocessing	Zero visible fecal contamination after re-processing; equipment kept properly adjusted; range of 20-50 ppm chlorine.	Visible check on each lot (at least once per hour of production); check chlorine at start up and every 2 hours using documented random sampling procedures. Designated Quality Assurance employee will record results in appropriate Log.	Reprocessing Log (using Plant Finished Product Standards) Antimicrobial Log Equipment Maintenance Log Corrective Action Log	Once per shift the QA supervisor will review the Reprocessing Log and Antimicrobial Log. Twice per shift Maintenance Supervisor will review Equipment Maintenance Log	QA will reject or hold product until zero fecal tolerance is achieved. Product will be reworked and reinspected by QA for fecal contamination. Any equipment adjustments will be made.
3B Chilling (All Products)	Temperature of 5° C or less will be reached within 4 hours on all product. Chlorine level in chiller will be maintained at > 20 ppm.	Product temperature check monitored by QA technician at end of chilling procedure (every hour of production). Chill water will be tested for Chlorine level every 2 hours by QA.	Chilling Log Carcass Chiller Recording Chart Neck/Giblet Chiller Recording Chart Thermometer Calibration Log Corrective Action Log Antimicrobial Log	Once per shift the QA supervisor will review the Chilling Log, Plant Post Chill Product Standards Form, and Antimicrobial Log. Maintenance supervisor will verify accuracy of the carcass chiller and neck/giblet chiller temperature recording charts once per shift. QA will verify the chlorine concentration in the chiller. Thermometers will be calibrated.	QA will reject or hold product depending on time, temperature and or antimicrobial level deviation. QA will identify the cause of the deviation and prevent reoccurrence. Maintenance will check chiller circulation and water exchange rate and make adjustments as required.
4B Finished Product Storage (Cold)	Finished product will not exceed 5° C.	Maintenance personnel will check product temperature on carcasses every two hours.	Chilling Log Thermometer Calibration Log Corrective Action Log	Maintenance supervisor will verify the accuracy of the product temperature log once per shift. QA will check all thermometers used for monitoring and verification activities for accuracy daily and calibrate.	If a deviation from a critical limit occurs, the following corrective actions will be taken: 1. The cause of the temperature exceeding 5° C will be identified and eliminated. When the cause of the deviation is identified, measures will be taken to prevent it from recurring 2. The CCP will be monitored hourly after the corrective action is taken to ensure that it is under control.

Table 3 FMEA of hazardous processing methods for poultry processing

Defective Products											Estimated corrective Actions result
Production step-CCP	**Hazards**	**Causes**	**S**	**O**	**D**	**RPN**	**Corrective actions**	**S**	**O**	**D**	**RPN**
Evisceration	Pathogens, parasites	Wrong chlorine level	9	9	4	324*	Check chlorine level	8	3	3	72
Reprocessing	Fecal contamination	Wrong chlorine level	9	7	4	252*	Check chlorine level	9	2	2	36
Immersion chilling		Wrong temperature and wrong chlorine level	8	6	5	240*	Correct temperature and chlorine level	8	2	2	32
Cooling/freezing/Storage & Distribution	Contamination from wrong temperature or time remaining at room temperature	Improper control of storage conditions and distribution	7	5	6	210*	Check storage rooms and distribution vehicles	7	2	2	28

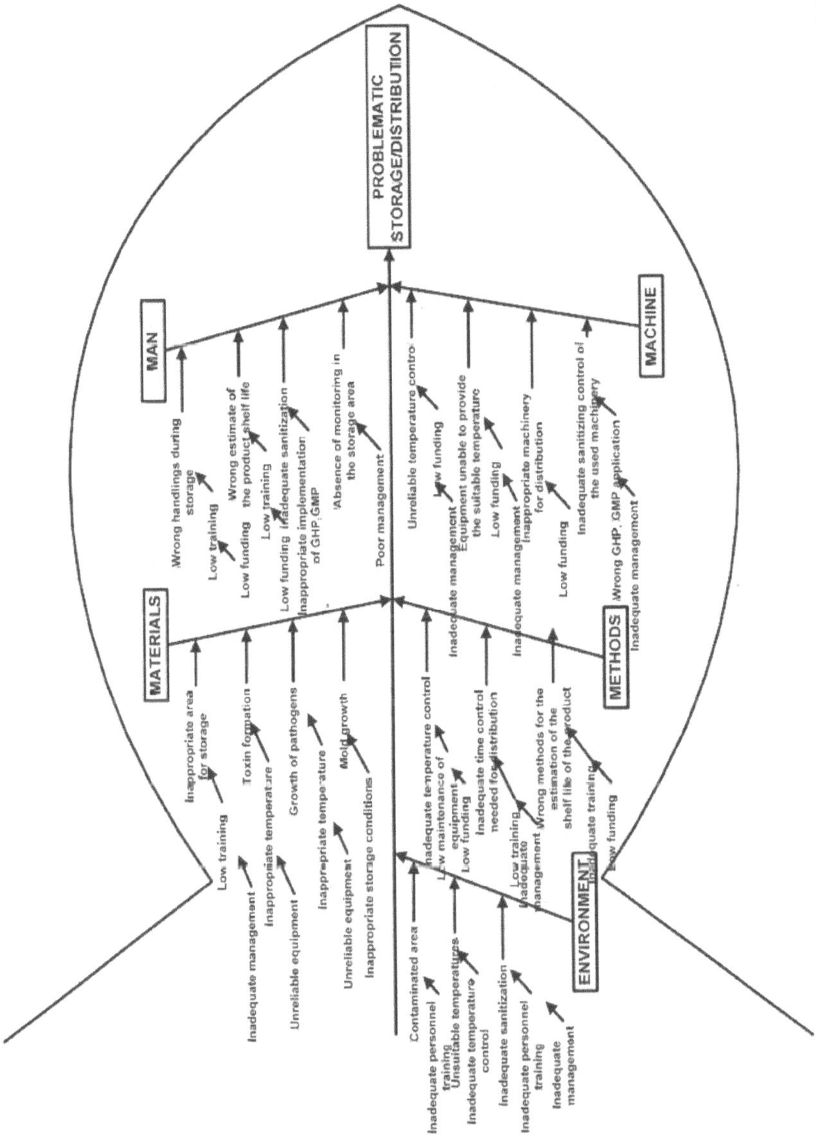

Figure 2. Application of the Cause-and-Effect diagram (Ishikawa diagram) to ready to eat vegetables (CCP problematic storage/distribution of poultry).

REFERENCES

Bailey, J.S., Thomson, J.E. and Cox, N.A. (1987). Chapter 5: contamination of poultry during processing. In: "The Microbiology of poultry meat products".

Baumgart, J. (1993). Lebensmitteluberwachung und Qualitatssicherung. Fleischwirtsch, 73(4), 392-396.

Bolder, N.M. (1998). Chapter 5: The microbiology of the slaughter and processing of poultry. In: "The Microbiology of meat and poultry". Davies, A., and Board, R. (EDS.). Blackie Academic and Professional, London.

Goodfellow, S.J. (1995). Implementation of the HACCP program by meat and poultry slaughterers. In: "HACCP in meat, poultry and fish processing", eds. A.M. Pearson and T.R. Dutson, Blackie Academic and Professional, London, pp. 58-71.

Hinton, M.H., Allen, V.M., Tinker, D.B., Gibson, C. and Wathes, C.M. (1996). The dispersal of bacteria during the defeathering of poultry. In: "Factors affecting the microbial quality of meat". 2. Slaugher and dressing. Hinton, M.H., and Rowlings, C. (eds.). University of Bristol Press, Bristol, U.K. pp. 113-121.

Humphrey T. (2004). Control of Campylobacter spp. in the food chain: a far from simple task. Culture, 25, 6-9.

Izat, A.L., Colberg, M., Driggers, C.D. and Thomas, R.A. (1989). Effects of sampling method and feed withdrawal period o n recovery of microorganisms from poultry carcasses. Journal of Food Protection, 52, 480-483.

Kim, J.W., Slavik, M.F., Griffis, C.L. and Walker, J.T. (1993). Attachment of Salmonella typhimurium to skins of chicken scalded at various temperatures. Journal of Food Protection, 56, 661-665, 671.

Lillard, H.S. (1990). The impact of commercial processing procedures on the bacterial contamination and cross-contamination of broiler carcasses. Journal of Food Protection, 53, 202-204.

Mackey, B.M. and Roberts, T.A. (1993). Verbesserung der Schlachthygiene durch HACCP und Uberwachung. Fleischwirtsch, 73(1), 34-43.

MAF Food Assurance Authority. (2000). Generic HACCP plan for slaughter, dressing, portioning and deboning of chicken (broilers).

NACMSF (National Advisory Committee on Microbiological criteria for foods) (1997). Generic HACCP application in broiler slaughter and processing. Journal of Food Protection, 60, 579-604.

Sandrou, D.K. and Arvanitoyannis, I.S. (1999). Implementation of Hazard Analysis Critical Control Point in the meat and poultry industry. Food Reviews International, 15(3), 265-308.

Sumner, J., Raven, G. and Givney, R. (2004). Have changes to meat and poultry food safety regulation in Australia affected the prevalence of Salmonella or of salmonellosis? International Journal of Food Microbiology, 92, 199– 205.

Sarwari, A., Magder, S., Levine, P., McNamara, A., Knower, S., Armstrong, G., Etzel, R., Hollingsworth, J., Morris, J., 2001. Serovar distribution of Salmonella isolates from food animals after slaughter differs from that of isolates found in humans. Journal of Infectious Diseases 183: 1295-1298.

Schlosser, W., Hogue, A., Ebel, E., Rose, B., Umholtz, R., Ferris, K., James, W., 2000. Analysis of Salmonella serovars from selected carcasses and raw ground product sampled prior to implementation of the pathogen reduction: hazard analysis and critical control point final rule in the US. International Journal of Food Microbiology 58: 107–111.

Stals, P. (1996). Slaughter and dressing of poultry. In: "Factors affecting the microbial quality of meat. 2. Slaughter and dressing". Hinton, M.H., and Rowlings, C. (eds.). University of Bristol Press, Bristol, U.K. pp. 99-105.

Tinker, D.B., Gibson, C., Hinton, M.H., Wathes, C.M. and Allen, V.M. (1996). Defeathering-engineering developments. In: "Factors affecting the microbial quality of meat. 2. Slaughter and dressing". Hinton, M.H., and Rowlings, C. (eds.). University of Bristol Press, Bristol, U.K. pp. 123-131.

Tompkin, R.B. (1994). HACCP in meat and poultry industry. Food Control, 5(3), 153-161.

Untermann, F. (1993). Hygienanforderungen an die Verarbeitung von Fleisch. Fleischwirtsch, 73(4), 389-392.

Yang, H., Li, Y. and Johnson, M.G. (2001). Survival and Death of Salmonella Typhimurium and Campylobacter jejuni in Processing Water and on Chicken Skin during Poultry Scalding and Chilling. Journal of Food Protection, 64, 770-776

TRACEABILITY OF BEEF PRODUCTION AND INDUSTRY IN FRANCE

Unique identification number, passport, movements, slaughtering number...

L. Marguin – B. Balvay

INSTITUT DE L'ELEVAGE 23 rue Baldassini F- 69364 LYON cedex 7beatrice.balvay @inst-elevage.asso.fr

Abstract: The French cattle tracing system results from a long evolution, which began in the mid sixties for cattle selection purposes. In addition to its main objective which is cattle tracing and sanitary uses, the system is widely used by many breeder organisations for very different uses: parentage recording, performance recording, herd book keeping, breeding value calculation and animal marketing.

The organisation of this system consists on two levels: the breeder organisations level and the level of the national identification database, which is run by the Ministry of Agriculture.

The Ministry of Agriculture commits some of the breeder organisations to perform some of the tracing operations: ordering the official eartags, data collection and animal passport printing. The system is designed in order to allow the breeders to provide data only once, independently of their future utilisation as well as to facilitate access to data for other utilisation than cattle tracing. Consequently, the official animal passport gives the official parentage, and electronic data interchange between the information systems of the breeder organisations allows easy access to data of the official tracing system.

At the entry of slaughterhouses, the passport corresponding to the identification number of the animal on the eartags is recorded in the local database with the unique number of slaughtering and sent to the national identification database. This number of slaughtering follows all the parts of the carcass, till the pieces of meat.

Keywords: Identification, Tracing system, French cattle

Please use the following format when citing this chapter:

Marguin, L. and Balvay, B., 2009, in IFIP International Federation for Information Processing, Volume 295, *Computer and Computing Technologies in Agriculture II, Volume 3*, eds. D. Li, Z. Chunjiang, (Boston: Springer), pp. 1797–1805.

1. CATTLE INDIVIDUAL IDENTIFICATION

The traceability of beef meat began with the individual identification in many big breeding countries. It has been compulsory in all the European Union since 1998, in Australia and New Zealand one year after, then in Canada, and after Japan, Brazil, Uruguay, Mexico, Argentina, Chili...

Traceability is a necessity for four reasons developed under:

1.1 To eradicate some animal diseases and to manage sanitary crisis

For eradication or management plans relative to designed diseases, control veterinarians from the Ministry have to check that all cattle get compulsory vaccinations and regularly realize blood tests.

These tests may give a positive result. In such cases, the veterinarian responsible for the test needs to find very quickly all animals which could have been in contact with the sick one. Indeed, depending on the disease (ESB for example), all these animals will have to be destroyed.

For this purpose, he must have access to all movements of the sick animal since he was born but also to the list of those who were present in the same holding at the same moment.

Therefore, farmers and databases need to get the list with all alive cattle, updated with each movement of animal, in or out.

1.2 For safety of human food

To protect consumers from any sanitary risk in food, some technical ways of breeding are forbidden: for example hormones or some medicine given to cattle.

From the birth where the identification number is fixed to the animal till the agreed slaughterhouse where the animal will be transformed into pieces of meat, breeding systems are followed to make sure that animal hasn't been present in any holding which wouldn't have respected this rules.

In any case of sanitary doubt, the cattle concerned is destroyed.

1.3 For herd management

The same official individual identification can be used by the farmer or technicians who manage animals of the herd.

For example, choosing the insemination bull to inseminate a particular cow, depends on its performance characteristics: it's important that both performance record an insemination technicians use the same identification number for the animal.

These crossed uses of official identification by many users have also great favorable consequences on reliability of data: the more users use a number the less mistakes will remain in database because they'll be detected and corrected, while an identification used only yearly for the administration may have failure.

1.4 For cattle trade

When some alive cattle or carcass is to be sold abroad, importer asks for identification number and data about the animal, on one hand to guaranty on the sanitary quality and on the other hand to get technical data about animal as breed, birth place or even parentage.

2. THE FRENCH SYSTEM OF TRACEABILITY

The traceability of beef means that all cattle get individual identification numbers just after the birth, and then all the movements of all cattle are recorded so that trace back is possible to find all the animals in contact anytime. And this is anywhere till the slaughterhouse or the natural death.

2.1 The French history of identification

In France the first law on breeding was in 1966. It defined rules of a permanent individual identification for volunteers breeders who needed it for technical reasons: when you want to select mothers and fathers for genetic improvement and to sell bulls and heifers, private working number is no more sufficient.

After this genetic purpose, the sanitary crisis pushed the official identification complete itself after 12 years. In 1978, this permanent individual identification was widespread and became compulsory for all breeders and all animals, this time on the official purpose of traceability of cattle. At this date, all farms were recorded with a unique number of premises, all births were notified very soon after born with an individual unique number and all movements between premises were also notified and recorded.

This compulsory identification was widespread to the whole European Union by the regulation of 1998 which makes compulsory the traceability of the beef meat till the pieces of meat with a system of labels for the consumers and specifies the necessity of a national database for alive cattle data.

2.2 General French organization

From the beginning of the national system, French Ministry of Agriculture delegates all the work of identification to local breeders organizations, named "EDE" which are responsible each of them for a defined geographical zone. There are around 70 EDE to manage with 250 000 cattle herds and 20 millions of cattle.

A technical organization, named Institut de l'Elevage (French Livestock Institute), supports the Ministry, to coordinate the implementation of the identification:

- to define methods which have to be the same for each level
- to help for the implementation of each level
- to control each level

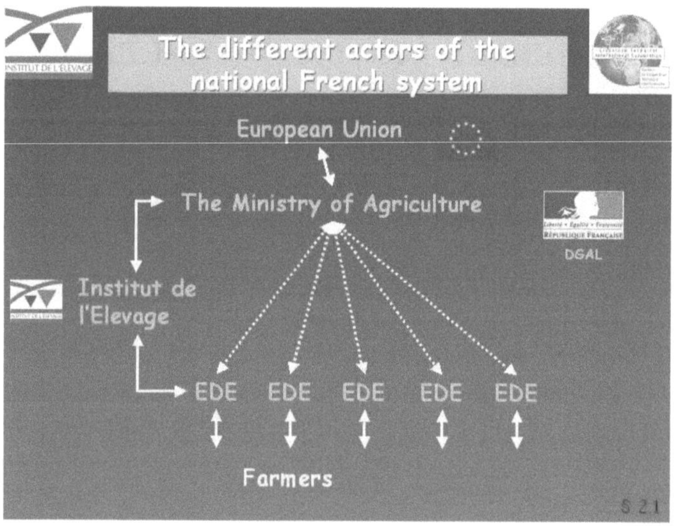

Figure 3. The national French organization

2.3 What is identification from the birth till the slaughtering, and who realizes it?

2.3.1 Identification of the calves

For a long time, it has been a technician from the EDE, who applied the eartag to the calves, when he passed once each three months on each farm.

In 1998, the European Commission decided that the calves have to be identified before 21 days of age, so the farmers were charged to put the two eartags themselves.

The farmer orders to the EDE only the number of eartags he needs for the calving year : the uniqueness of the numbers granted is the responsibility of the EDE (and the State behind). When some eartags are not used or damaged, the farmer has to give back to the EDE.

To trace the data of any system (genetic, herd management, trading…) the same unique identification number has to be used by anybody for the animal.

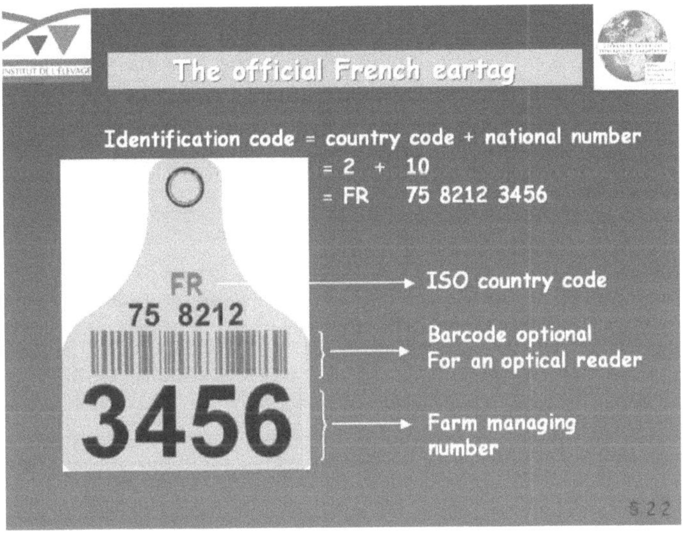

Figure 2. An official French eartag

Some electronic (RFID) eartags can be used now, but it doesn't change a lot in the identification system: it only permits an automatic reading of the identification number of the animal with a receiver, instead of a visual reading by the farmer and reporting to a paper, and then recording it in a database. This can avoid many human mistakes on the number.

When RFID is used to identify animals, the ISO standards 11784 and 11785 have to be taken in account.

2.3.2 The notification of all events on the farm

The farmer has to declare all events on the annual book of the herd:

- the births with the identification number, birth date, sex…
- the entries on the farm, identification number of new animals
- the exits from the farm with the date of departure to slaughterhouse, other farms or death.

2.3.3 Record of data and controls

The record of all information sent by the farmer is done by the EDE (or directly by the farmer with specific software and sent by an electronic net) after some controls on the data in the local database.

2.3.4 The French cattle passport

After recording a birth, the data are sent to the national database to check them at this level, and errors or OK are coming down to the EDE before to print the passport. The passport has to follow the animal all his life, till the slaughterhouse and after they are recovered to be destroyed.

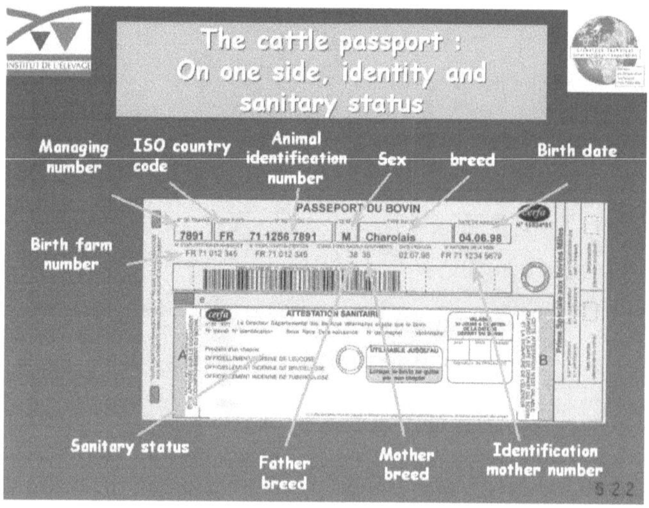

Figure 3. A face of a cattle French passport

2.3.5 Controls before slaughtering the animal

Before the animal have been slaughtered, the identification number between his passport and his eartags are checked : if there is not the right passport, the animal is not slaughtered.

If the passport corresponds to the animal at the entry of the slaughterhouse, a unique slaughtering number is associated to the identification number: a number with slaughtering date and an order number on the chain.

This short number is easier to use for the work on the chain, but one eartag stays on the carcass till the cold chamber, in case of any doubt on the identification number.

2.3.6 Controls on the carcass and pieces of meat

A label is printed with many barcodes to stick on each piece of carcass. On the label you find with the identification number, all the data from the passport (birth farm, birth date, breed, sex, type of animal) and the slaughtering data (slaughter date, place, weight).

The European regulations and French standards ask for 3 types of checking :

- Controls from the slaughter
- Controls from private firms agreed for that (EN 45011)
- Controls from veterinaries of the Ministry of Agriculture

Figure 4. The label on the carcass

2.4 The roles of the Ministry of Agriculture

All the technical work is supported by French Livestock Institute and validated by the Ministry.

That means the Ministry has the responsibility of the EDE and agrees them, the EDE have to help and check the operations of the farmers and other stakeholders. The Ministry agrees the official eartags on the results of laboratory and field tests and the official documents.

The Ministry validates the field procedures on the proposal of the Institute.

The same for the data exchanges between agreed computer centers.

The Ministry manages the national database himself.

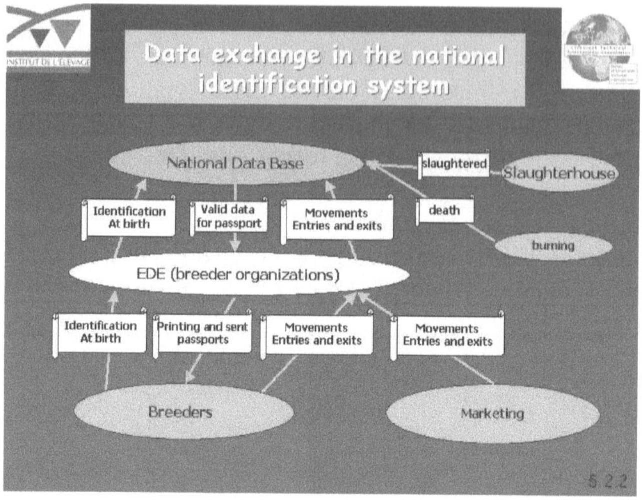

Figure 5. Data exchanges in the French identification system

Identification and genetic data are managed in the same information system.

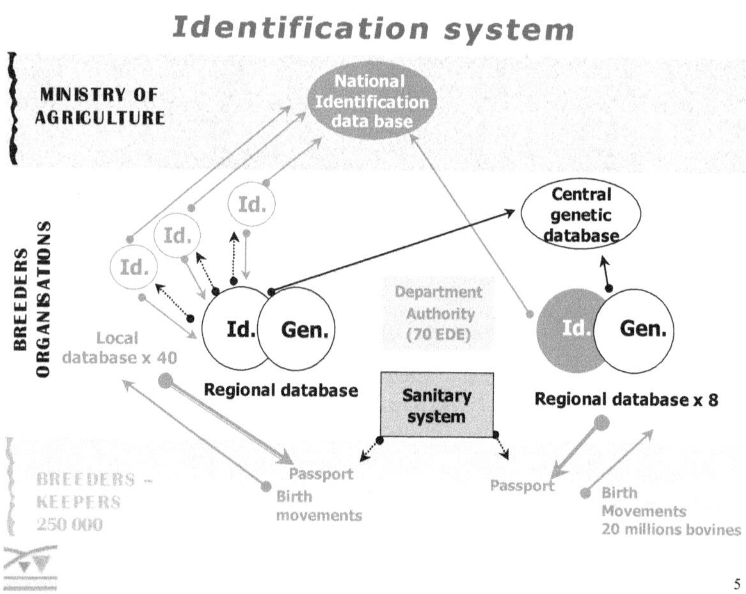

Figure 6. French information system

3. CONCLUSIONS

It takes long time to get an identification and traceability system like this one to be reliable. That's why it is necessary to plan several steps till the final aim and first with only some of the more interested farmers and stakeholders.

The global databases system is not the most important first, the main question is to collect the real data on the field and record them perhaps only locally but with a good reliability. And this point needs a lot of explanations and interest to be well done.

REFERENCES

COMMISSION REGULATION (EC) No 911/2004 of 29 April 2004 implementing Regulation (EC) No 1760/2000 of the European Parliament and of the Council as regards eartags, passports and holding registers (Text with EEA relevance)

Council Directive 92/102/EEC of 27 November 1992 on the identification and registration of animals

Council Directive 97/12/EC of 17 March 1997 amending and updating Directive 64/432/EEC on health problems affecting intra-Community trade in bovine animals and swine

Council Directive 98/99/EC of 14 December 1998 amending Directive 97/12/EC amending and updating Directive 64/432/EEC on health problems affecting intra-Community trade in bovine animals and swine

French decree No 2006-376 of 23 March 2006 on identification of bovine animals and modifying rural code

French decree No 98-764 of 28 August 1998 on identification of bovine animals

French departmental order of 10 February 2000 creating the national database of identification and traceability of bovine animals and their products

French departmental order of 9 May 2006 on identification of bovine animals modalities

Norm ISO 11784: 1996 Radio frequency identification of animals -- Code structure

Norm ISO 11785: 1996 Radio frequency identification of animals – Technical concept

Regulation (EC) No 1760/2000 of the European Parliament and of the Council of 17 July 2000 establishing a system for the identification and registration of bovine animals and regarding the labelling of beef and beef products and repealing Council Regulation (EC) No 820/97

CONTROL SYSTEMS FOR IMPROVEMENT QUALITY OF HUMAN RESOURCES

R. J. Widodo[1], K.C. Ting[2]

[1] *President of MASDALI/ICSS (Masyarakat Sistem Kendali Indonesia/Indonesian Control Systems Society) rjwidodo@ yahoo. com & masdali@yahoo.com*

[2] *Department of Agricultural and Biological Engineering, University of Illinois, Urbana-Champaign, USA. kcting@uiuc.edu*

Abstract: This paper presents Control Systems (CS) to support the Programs of Education & Training (E & T), Research & Development (R & D) for improvement quality of Human Resources (IQHR).

The programs of the IQHR must be related with the modern Art, Science, Engineering and Technology (ASET). The prosperity of the Human Resources will be approached from prosperity approach which is mainly emphasized with three universes of aspects: Physical, Intellectual, Emotional and Spiritual (PIES) aspects. There are three basic requirements for developing the Physical aspect: Material, Energy and Information (MEI). Social welfare and increase of life quality are desired output that must be attained from the MEI aspects. Current standard of living of human being should be reported as feedback information in the programs of the IQHR. In global communication, developed countries and developing countries should build several attractive and sound symbiosis bridges, to prevent loss of universe balances. High quality of human resources as products of IQHR activities, have social impacts not only in developed countries but also in developing countries. A new work force strategy without denying the existing of high quality of human resources is established by retooling the work forces, thus the challenges of social impacts should be answered wisely and would be bright opportunities to improve human standards of living.

Keywords: CS, E&T, R&D, ASET, PIES, IQHR, MEI, human standards of living, social impacts, work force.

Please use the following format when citing this chapter:

Widodo, R.J. and Ting, K.C., 2009, in IFIP International Federation for Information Processing, Volume 295, *Computer and Computing Technologies in Agriculture II, Volume 3*, eds. D. Li, Z. Chunjiang, (Boston: Springer), pp. 1807–1815.

1. INTRODUCTION

The development of Control Systems (CS) to support Programs of the IQHR represents forces that strongly affect man, life, and civilization. It means that those can be good and can also be evil. The goodness or evil is determined by the purpose and utilization.

The purpose of CS is to support the programs of the IQHR efforts, thus they constitute parts of the development while they also can be the driving force for speeding up development.

The objectives of the programs of the IQHR are to increase the quality of human life. Sustainable development must be supported by health economics. In an expose on health economics, the heath industries are highly labor-intensive. Therefore the development is closely tied with the redistribution of income seen from two different aspects, namely with regard to equitable distribution of the results of development, and with the expansion of employment opportunities. These cannot be separated from the programs of the IQHR particularly in the rural area. In this connection, one should keep in mind that in developing CS to support the programs of the IQHR, one must pay heed to the local condition, such as tradition, customs and usages, standard of ASET and quality of infrastructure.

It is now generally accepted that the principle of the appropriate ASET comprises:

1. Low-cost, in proportion with the income level of the local population.
2. Providing job opportunities.
3. The tools and equipment used must be only a few and simple maintenance.
4. The equipment and material should be locally available or only small quantities need to be imported.
5. In harmony with the local socio-cultural conditions.
6. Not needing any intricate infrastructure.
7. A high added value compared with ASET being applied.

The increase of added value shall be achieved through application of ASET in the widest sense. ASET developed must be adapted to social conditions, natural resources and environment.

The main constraint of ASET development in developing countries is the fund, material and manpower. In this connection, it will be necessary to adapt the existing ASET and bring them into harmony with the local conditions. That is why it is necessary to emphasize the main managerial aspect that must cover ASET for existing tradition, customs, and desires of community with regard to the local skills.

2. IMPROVING THE QUALITY OF HUMAN RESOUCES (IQHR)

There are two approaches that must be considered in the IQHR namely: 1. Prosperity and 2. Security.

The prosperity approach will be mainly discussed in this paper. There are three aspects that must be accounted for. They are Physical, Intellectual, Emotional and Spiritual (PIES) aspects.

The basic need for the physical aspect are MEI:

1. Material,
2. Energy,
3. Information.

The Subsystems of the Social System are:

1. Demography (human resources),
2. Natural Resources,
3. Art, Science, Engineering and Technology (ASET),
4. Law and order,
5. Infrastructure.

The IQHR should be related with:

1. Education & Training (E&T),
2. Research and Development (R&D,
3. Industry and Manufacturing,
4. Implementation and Application of ASET,
5. Repair, Maintenance, and Rehabilitation.

The programs of E&T and R&D in developed countries are needed for innovation of new ASET, but in general, in developing countries they are for importation and utilization of ASET only.

2.1 Basic Requirements and Development

2.1.1 Considerations.

The first human basic need is material: food, clothing, housing, health, and education. We need to greatly increase the productivity of food production. We need to use the latest and modern ASET for modern production, so that marketable surplus far exceeds the amounts by farmers, for their own consumption and production.

The second large area for further investment in ASET infrastructure, namely for energy, information, infrastructure development and natural resources utilization.

In industrial R&D, for development and expansion of ASET facilities, efforts are being made to make the industrial ASET appropriate solutions to the various problems. The needs for communication and transportation infrastructure includes sea, air and transportations, while postal and telecommunication services and a network of radio and television are also important to be concerned.

To implement of progressive manufacturing plans requires that the industries adopt an active and even aggressive attitude towards the acquisition of ASET. They cannot remain passive and static waiting for ASET to be passed on to them. In order to help develop this dynamics and active posture they must be sustained and supported by ASET institutes, providing ASET services relevant to the production processes. This will become the focus of planned efforts to transfer, adapt and develop ASET appropriate and useful for the creation and expansion of productive processes throughout the nation.

Defense-related ASET is the third area in which additional investments are to be made. The need to strengthen ASET, support national's defense and

security capability is obvious given her strategic location, her abundance of resources and her unique geographic features. One nation must be able to control her area of jurisdiction and defend it against actions endangering her sovereignty, the achievement of objectives for her national development, such as equitable distribution, growth, stability, and the progress of her efforts for nation building. One nation has to be ready to commit resources to utilize defense technologies relevant to the strengthening of her people's defense system adapted to the national cultural, geographic, and ASET conditions.

Lastly, a people committed to transfer, adapt, and further develop even the most modern ASET and apply these in already existing cultural environments must take measures to prevent the application of the new ASET from eroding and destroying those cultural values upon which its national identity rests. As is the case in other societies, a national's identity is exemplified by her own unique way of working and living together and by distinctive ways social order is created and maintained. For some developing countries, they already possess ancient cultures long before she became a modern state. For the purpose of maintaining an orderly-balance between established cultural values and the new values inherent in the modern ASET, in the social sciences, in culture and in it's philosophies of life has been designated as the fifth but perhaps the most important area in which infrastructure is to be strengthened. The development of ASET infrastructure in this area is geared towards the maintenance of the ability of a nation's people to preserve the compatibility of its cultural values with the progress of ASET in the first four areas discussed previously.

To achieve this stage of development, ASET and skills are indispensable. Without ASET, no one country will be able to develop her economic potentiality.

Successful transfer and development of ASET can only take place through the vehicle of consistently executed realistic and concrete programs of production which incorporate a systematic, step by step increase in the degree of comprehension and mastery of successively the more sophisticated ASET. Only through these kinds of progressive manufacturing plans can a society develop itself into a productive power and hence a credible and respected member of the family of nations existing in the world today and in the future.

2.2 Policies on ASET

The policies on ASET shall be directed toward developing national capabilities in ASET as necessary for the national development in accordance with the needs and priorities of national development.

The global policies on ASET can in broad outlines reach their targets, when the definite ASET programs can be implemented according to plan.

However, it will still be necessary to study and formulate the supportive ASET programs, to achieve a convergence between the plans and the results we hope for. Special attention must be given to the following:

1. Inter-institutional coordination and interaction, and particularly the optimal distribution and utilization of the available information on ASET.

2. The provision of funds and of skilled staff in accordance with ASET, whereby the development and participation of the regions in accordance with the development and participation of the regions in accordance with the development conditions must be further increased.

3. Proper inducements and remuneration for ASET staff that show superior performance must be planned and formulated as a policy which shall be increasingly introduced in our society.

4. Supportive policies are also needed to distribute and popularize the modus operandi for ASET transfer from overseas, through involving national manpower and materials, in a programmed, systematic, and effectively phased program.

5. In taking benefit of the foreign ASET capacities through bilateral, regional or multilateral collaboration, a mutually beneficial institutional must be pursued.

6. In determining the objectives of ASET programs, the results and utilization should be institutionalized harmoniously in stages.

This is appropriate, as the development of ASET has been funded from taxes and other government revenues raised by the community, and therefore the interests of the community should be given proper attention.

Relations between industry and university

Educational and Training (E&T) purposes, are as follows:

1. Have obtained a firm foundation in ASET basis to their fields.
2. Have begun to acquire a working knowledge of the current ASET in their area of interest.
3. Have begun to understand the diverse nature and history of human societies as well as their literary, philosophical and artistic traditions.
4. Have acquired the skills and motivation for continued self-education.
5. Have had an opportunity to exercise ingenuity and inventiveness on a research project.
6. Have had an opportunity for engineering synthesis on a design project.
7. Have developed oral and written communication skills.
8. Have begun to understand and respect the economic, managerial, political, social and environmental issues the surrounding ASET development.

A good vision for the universities-industries cooperation should have some distinctive features including the followings:

1. Innovative and future-oriented.
2. Utopian enough to lead to a clearly better future for the organization.
3. Fitting in with the organization's culture, values and history.
4. Reaching out for new dimensions.
5. Setting standards of excellence and reflecting high ideals and aspirations.
6. Clarifying purpose and direction, including measurable objectives.
7. Inspiring enthusiasm and encouraging Commitment.
8. Reflecting the uniqueness of the organization, its identity and core competencies.

Industry can get the following advantage from its cooperation with research institutions:

1. A redefinition of industry's role, causing it to be increasingly oriented towards need and demand.
2. Consultation and counseling services on industrial property laws, design and development of products, quality control and suchlike.
3. Testing facilities.
4. Assistance in negotiations with projects and equipment suppliers.
5. Techno-economic forecasting.
6. Exchange of ASET information
7. Establishing communications with the overseas ASET institutions, etc.

The factors, which encourage cooperation between ASET institutions, and industrial world include:

1. The improvement of communication lines, contributing to better harmony between resources and needs.
2. The involvement and responsibility of personnel, supported by a satisfactory remuneration system.

3. The mobility of personnel between the ASET institutions and the industrial world.
4. The beneficial effect of the development projects.

2.3 Challenges & Opportunities of ASET

2.3.1 Industrial Robot Case

In manufacturing plants in several countries, there has been a large-scale increase in the usage of industrial robots, which are programmable machine tools designed in many cases to accomplish arduous or complex tasks. Although there has been some opposition to the fact that robots often replace human labor, the author explains why he believes the trend toward robotics will continue, and on balance, be beneficial to the national economy.

ASET developments of the last decade are likely to have as profound a potential impact on productivity, labor markets, working conditions, and the quality of life in the developed countries as the introduction of robot into workplace. According to this bleak scenario, in time, workers either will have to accept lower wages or increase their output if the wholesale loss of jobs to an army of robots is to be prevented.

This pessimistic viewpoint has little basis in fact. The conclusions can be reached based on four factors:

First, the estimate of the number of jobs that could be performed by is relatively small.

Second, almost all of these workers would be spared forced unemployment because of retraining and in some cases the job attrition that occurs through normal retirement.

Third, total employment is a function of real economic growth: robots can have a positive effect on real economic growth and, therefore, a positive effect on total employment.

Fourth, in 10 years, retraining programs can adequately shift displaced workers to new careers. In fact, the main challenge posed to policymakers by increased use of robots is not unemployment but need for retraining.

History shows that labor-saving techniques have led to improved living standards, higher real wages, and employment growth. In large measure, the robotics revolution is merely a continuation of a centuries-long trend that has resulted in enormous material progress.

New forms of employment can be created to offset any jobs directly lost to robots. Protection from job loss can come through retraining programs. Working condition and job safety will improve as robots take over dangerous and undesirable forms of works.

Technological advances in computers and microprocessors are increasing the sophistication of robots, giving them some "thinking" capacity that increase potential uses.

The key to usage in such area as office work depends in large part on the ability to develop "intelligent" robots capable of performing tasks that vary somewhat over time. Some industry observers believe breakthroughs may allow for extensive introduction of robotics in non-manufacturing tasks within a few years.

Three important dimension of the growth of robotics are subjects to economic analysis. *The first* is the determinants of the magnitude of the growth of the robotics industry.

The second is the impact robotics unemployment.

The third is the impact that robots will have on wages, profits and prices.

There are two reasons for the growth in the use of robots, one related primarily to supply and the second primarily to demand.

In the long run, robots will be increasingly utilized because the cost of traditional labor-intensive techniques is rising over time, while the cost of the capital-intensive robotic techniques is falling relative to prices generally. These costs decline because the technological advances in robotics lower the capital costs of robots per unit output.

On the other hand, some government policies may speed robotic introduction. For example, where environmental regulations lower worker productivity or raise capital costs associated with the traditional ASET, the traditional technique cost line will shift upward, advancing the date at which robotic adoption becomes profitable. The actual threshold date will vary with the individual job involved, depending on wage levels of human labor, the capital cost of robots, the number of workers the robot displaces, the impact of robotics on product quality, the nature of union contracts, and other factors previously mentioned.

2.3.2 Retooling the Work Force.

Important changes in the composition of the work force have occurred over the past four decades, and, in some opinions, even more massive changes lie ahead as many thousands of low-skill jobs are eliminated while at the same time large numbers of new jobs are created to meet the demands of technological advance. If serious employment displacement effects are to be avoided, development of broad-scale training programs in which the private sector plays a key role, in concert with various governmental bodies, will be required, he explains.

It already is clear that robotics will do for manufacturing what mechanization did for agriculture. The robotics of jobs in basic industries such as automobiles is only foretaste of similar changes likely to occur in virtually every manufacturing industry.

3. CONCLUDING REMARKS

It is concluded that :

1. The development of ASET for the IQHR represents forces that strongly affect man, life, and civilization.
2. With ASET, most countries will be able to develop her economic potentiality.
3. The policies on ASET shall be directed toward developing national capabilities in ASET, as necessary for the national development in accordance with the needs and priorities of national development.
4. The spirit of the cooperation among scientists/academics inter-universities should very high.
5. The exchange programs are very useful for scientists, academics and engineers in order to discuss and share the latest development in ASET.

REFERENCES

Choate, P. , Retooling The American Work Force, Northeast-Midwest Institute, 1982.

Dato' Lee Yee Cheong, The President of World Federation of Engineering Organizations, Keynote Address : Current Activities of the World Federation of Engineering Organizations, CAFEO-21, Jogjakarta, Indonesia, 22-23 October 2003.

Habibie, B.J., Science, Technology and Nation Building, vol. I, Technology Indonesia & The Agency for The Assessment and Application of Technology, Jakarta, 1991.

Habibie, B.J., Science, Technology and Nation Building, vol. II, Technology Indonesia & The Agency for The Assessment and Application of Technology, Jakarta, 1991.

Widodo, R.J, .Automatic Control for Reducing Energy Consumption and Improving Energy Conservation, (CAFEO-10), Manila, Phillipines, 5-6 November 1992.

Widodo, R.J. Control Education at Bandung Insitute of Technology,(CAFEO-11), Singapore, 18-19 November 1993.

Widodo, R.J., Development of Control Applications in Electrical Power Systems ,PSDC'95, Bandung Institute of Technology, Bandung, 14-16 March 1995.

Zuboff, S., Computer-Mediated Work : A New World, The President and Fellows of Harvard College, 1982.

KEES: A PRACTICAL ICT SOLUTION FOR RURAL AREAS

Xiaoye Dai[1], Sabin Tabirca[2], Eamon Lenihan[3]

[1] *Corresponding author, Address: Room 223, O'Rahilly Building, Centre for Sustainable Livelihoods, University College Cork, Ireland. Tel: +353 21 490 3548 Fax: +353 21 490 3540 Email: x.dai@cs.ucc.ie*

[2] *Department of Computer Science, University College Cork, Ireland*

[3] *Centre for Sustainable Livelihoods, University College Cork, Ireland*

Abstract: This paper introduces a practical e-learning system, identified as Knowledge Exchange E-learning System (abbr. KEES), for knowledge distribution in rural areas. Particularly, this paper is about providing a virtual teaching and learning environment for small holders in agriculture in those rural areas. E-learning is increasingly influencing the agricultural education (information and knowledge learning) in all forms and the current e-learning in agricultural education appears in informal and formal methods in many developed countries and some developing areas such as Asian Pacific regions. KEES is a solution to provide education services including other services of information distribution and knowledge sharing to local farmers, local institutes or local collection of farmers. The design of KEES is made to meet the needs of knowledge capacity building, experience sharing, skill upgrading, and information exchanging in agriculture for different conditions in rural areas. The system allows the online lecture/training materials to be distributed simultaneously with all multimedia resources through different file formats across different platforms. The teaching/training content can be contextless and broad, allowing for greater participation by more small holders, commercial farmers, extension workers, agriculturists, educators, and other agriculture-related experts. The relative inconsistency in content gives farmers more localised and useful knowledge. The framework of KEES has been designed to be a three-tier architecture logic workflow, which can configure the progressive approach for KEES to pass on and respond to different requests/communications between the client side and the server.

Keywords: e-learning, multimedia synchronisation, knowledge dissemination, video

---

Please use the following format when citing this chapter:

Dai, X., Tabirca, S. and Lenihan, E., 2009, in IFIP International Federation for Information Processing, Volume 295, *Computer and Computing Technologies in Agriculture II, Volume 3*, eds. D. Li, Z. Chunjiang, (Boston: Springer), pp. 1817–1824.

1. INTRODUCTION

Knowledge – and related information, skills, technologies, and attitudes – will play a key role in the sustainable intensification of agriculture and success for rural development investment (Alex, 2002). However, the dissemination of knowledge in rural areas is not easy. Low level of education hampers the adoption of new technologies, farming skills and innovations. The literacy rate is usually low in these under-served areas and it is a crucial problem for farmers to learn and absorb new knowledge. Somehow, farmers are reluctant to accept new development unless it is the fact that the development would bring them more benefits when compared with their traditional practices.

Agricultural knowledge is often difficult to reach local farmers. The agricultural knowledge is often protected by the interests of chemical and seed manufacturers. As for marketing information, it is always cut off partially or even fully by the middle-men.

Colleges and institutes are the main forces for agricultural skills upgrading and innovations. Research students, fellows and lecturers can be the experts in farming fields. They pass over many agricultural skills, new research and development results, advanced technologies, and even diagnose some issues. However, the cost of field trips and the amount of time required are enormous.

The traditional means of passing on agricultural knowledge and skills from one generation to the other and/or among community members is becoming less effective. This is a result of several factors, such as fewer experts in rural areas, agricultural innovations are increasingly coming from outside the community, fewer traditional courses are being offered, and much information is time-sensitive and/or needed quickly (Agriculture and Agri-Food Canada, 2003). To solve these problems, it requires that an online agricultural education should provide a powerful means of overcoming these constraints and allow the individual student/learner to obtain up-to-date information by clicking their mouse.

When comparing the global non-agriculture e-learning trends versus global agriculture e-learning trends, report shows that "e-learning in Canadian, European, American and Australian agriculture falls significantly behind the current adoption rates seen in non-agriculture sectors." and "each of these countries are linking learning to marketable skills and selling e-training programme for specific agricultural training purposes." (Berge & Leary, 2006)

The introduction of ICT into rural areas could be a possible solution for achieving the improvement of farmers' agricultural knowledge. However, it is more practical to develop a compatible online platform for the entire rural areas, including those in developed and developing countries or regions.

Bearing this practical solution design in mind, researches are driven to find out a suitable system for farmers participatory learning.

2. METHODOLOGY

The research work is divided into two stages. The first stage was to collect requirements from agricultural educators and agriculture-related workers. It began with field trips selected in agricultural oriented countries such as Ghana and China. The main method of collecting information and data requirements is to interview agricultural educators in colleges and institutes, such as CARD centre (The Center for Agricultural and Rural Development) in Zhejiang University, and also some agriculture-related workers or researchers in agricultural research organisations, such as members of FARA (Forum for Agricultural Research in Africa). Through interviewing these experts in agriculture and agricultural education, it was found that users require a low-cost and easy-to-manage system.

Case studies proved that the constrains imposed due to lack of infrastructural facilities for the local farmers are the determinant factors for applying the special techniques needed in rural areas. However, the infrastructural conditions vary in different rural areas, and the research should find out the benchmark for grouping the rural areas generally. Based on rural area infrastructure situation, those developing areas could be divided into two experimental levels: mature level and beginning level. Regions or areas in mature level are usually equipped with TV, telephone or even Wi-Fi Internet connection. Those in beginning level are worse hit in facilities. They probably do not have TVs and/or telephones in the whole village nor the neighbouring villages. In this case, the difference in system dissemination should also be taken care of. DVD-based hard-copy dissemination is a supplementary to the Internet-based distribution in the newly developed system. It broadens the accessibility of the online courses, and also reduces costs to a realistic level.

The second stage was to develop a compatible system to meet the requirements of the agricultural educators and workers. During the research period, the designed system was tested at every milestone in rural areas in Ghana, Bosnia and Herzegovina, Czech Republic and China for its feasibility and robustness as well as for evaluation purposes. Questionnaires were carried out by agricultural educators and other agriculture-related researchers or workers in order to get feedback and comments to improve the system at each developing stage.

3. THE CHARACTERISTICS OF VIDEO CONTENT

The characteristic of video content is a special issue to be addressed in this research. Here, farmers are the research target objects, they are various in their educational background, but the video based approach will help users to access the needed agricultural knowledge freely. The teaching or training content made in video format is a new, but efficient way of acquiring agricultural knowledge for most farmers regardless their educational background. After all, they are going to adopt a new practice instead of replacing their conventional operations completely. However, different areas have their localised situations and knowledge demanding. Farmers are also difficult to engage with during their leisure time. Therefore, the system should allow the video content to be simple, diverse and flexible. Video presenters should avoid using any terminology or jargons so as to make the content easy to understand while the topics could be various. Due to the informal community-based study in rural areas, the video content should be contextless and irrelevant to each other; this is to ensure farmers participation and also allow greater participation by other entities such as commercial farmers, extension workers, agriculturists, educators and other agriculture-related experts.

4. TECHNICIAL DESIGN

Knowledge Exchange E-learning System (abbr. KEES) is designed based on a three-tier structure (see Figure 1). The first layer on the left side is composed of HTML and SMIL files. They form simple and easy to control interfaces for system clients to use. Clients make requests and get feedback from SMIL embedded HTML pages. The middle layer is a client server comprised of Java and XML files. On this server, requests from client side are preceded and all the resources will be located or pointed to the physical links through XML file to the Streaming server. The third layer is the streaming server where all the educational resources are physically stored.

The design of the three-tier architecture work flow through the entire system is for the purpose of separating users from the complicated back end coding, and help them visualise their final production of multimedia synchronised presentations. The process improves the system's usability and also enhances the efficiency of content management characteristic of KEES.

Experts and agriculturists are intermediary users at one client side, but system remote users are the other main users which consist of local farmers, agricultural educators and researchers or local communities. Thus the topic raised by the intermediary users would be distributed and responded to.

KEES works as a common platform for the two different levels of users by connecting their communication. (See Figure 2)

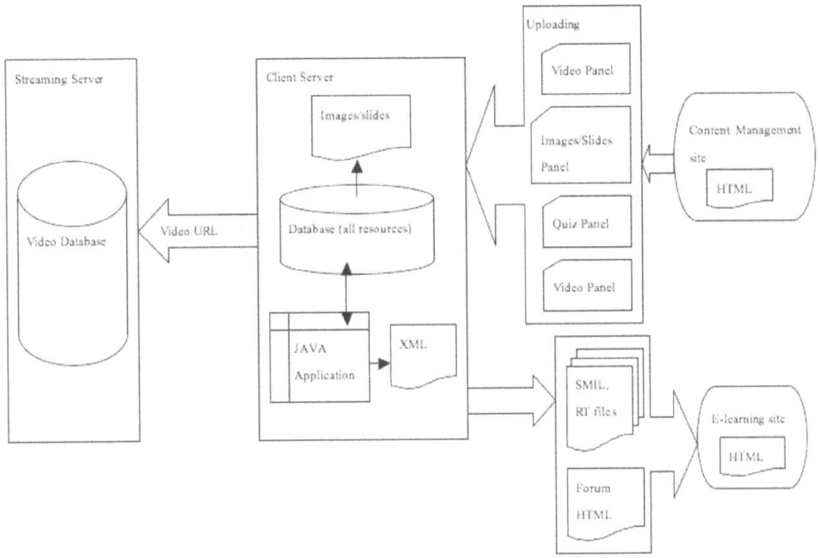

Figure 1 Three-tier Architecture Logic Workflow (Source: Xiaoye Dai, 2006)

As there are two level users in this system, KEES is designed with two interfaces for different users. KEES is focusing on simplifying the IT knowledge requirements of the educational content preparation for intermediary users and allows them take full advantage of all multimedia resources and distribute their presentations online.

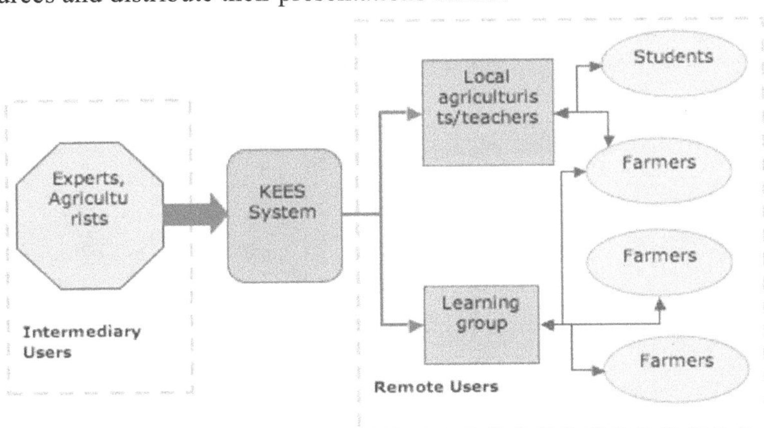

Figure 2 KEES user distribution diagram (Source: Xiaoye Dai, 2006)

For the intermediary users, they would use Java Assisted SMIL (abbr. JAS) interface to deal with the video editing and teaching content synchronisation. JAS is programmed to be executed as a presentation-producing factory. To edit/produce a multimedia presentation requires all basic functions for editing videos and other additional features for generating the streaming files (see Figure 3).

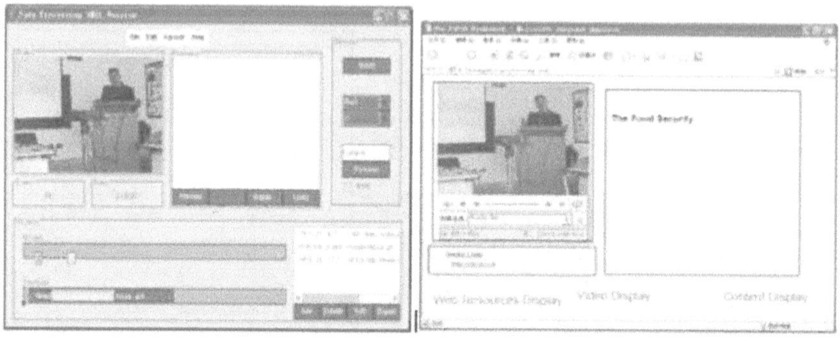

Figure 3 JAS interface and its Real Time based output layout (Source Xiaoye Dai, 2005)

JAS interface allow users to load video clips and synchronise text content, images and PowerPoint slides afterwards. External web resources can also be edited as URLs embedded into the synchronized outputs. Therefore, JAS takes the responsibilities of supplying end users options for editing and preparing the video for synchronisation performance.

For most remote users, the outputs are formatted as shown in Figure 3. Some computers or streaming servers might not be equipped with Real Time player so that another possible way to guarantee the distribution is to develop the other layout for Quick Time player (see Figure 4).

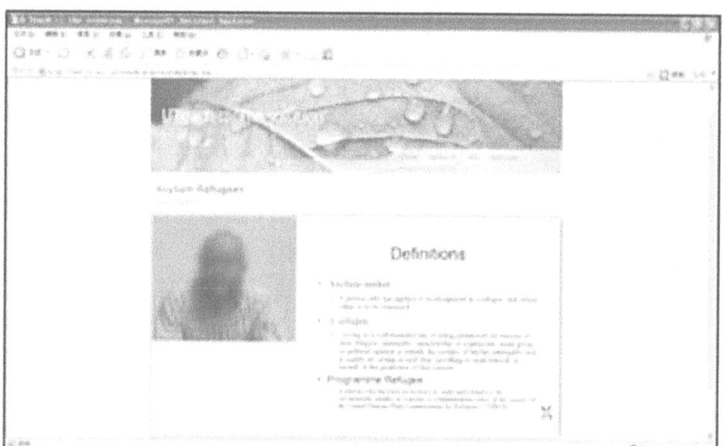

Figure 4 Quick time based layout (Source: Xiaoye Dai, 2006)

5. DISSEMINATION

Consequent upon infrastructural conditions, mature level areas will have access to connect to the website to access the online video archives. Users in those relatively advanced areas can watch the training course or knowledge information via the web pages as shown in Figure 3 and Figure 4. As for the beginning level areas, TVs and DVDs should be equipped in some selected collecting points for local farmers to gain access to the information. TV and DVD equipments could be shared within several villages by providing gathering places to play the video. With funds, each village could be equipped with one set of TV and DVD to repeat playing the video so that farmers would have more optional time to join the participant learning process.

6. CONCLUSION

In farming communities, literacy rate is substantially low. There is no doubt that teaching from books and leaflets would meet certain difficulties. The benefits of traditional learning would be long-term and intangible rewards to local farmers. However, farmers rely on their farming products for living, and the short-term profits to improve their life would be the only motivation for them to keep learning after their daily agricultural activities. Watching and listening to the demonstration is a faster approach for them to acquire knowledge. Video creation tends to be the way forward for those particular participant learners, which is less expensive compared to other types of media.

KEES builds up and improves the ICT roles of communication between the experts and the local farmers. Through applying different teaching and learning materials, farmers' trust and confidence in learning from virtual environment will be formed and enhanced.

KEES specifies the template of multimedia presentation outputs and the regulation of synchronising multimedia educational resources. Under certain infrastructure conditions, KEES has been proved to be capable of improving access to agricultural education, helping in knowledge capacity building, enabling educational information distribution, skill upgrade, information exchange and offering a platform to exchange teaching and learning experience.

REFERENCES

Agriculture and Agri-Food Canada. (2003, August 29). An overview of e-Learning in Canadian agriculture and Agri-business. Retrieved from:

Alex G, Zijp W, Byerlee D, et al. 2002. Rural extension and advisory services: new directions. Rural Development Strategy Background Paper No. 9. Washington, D.C. (USA): Agriculture and Rural Development Department, World Bank. Retrieve from:

Berge, Z., & Leary, J. (2006, June 12). Trends and Challenges of eLearning in National and International Agricultural Development. International Journal of Education and Development using ICT [Online], 2(2). Retrieved from: http://ijedict.dec.uwi.edu/viewarticle.php?id=179.

CSL E-Learning Online Archive. (2006, January). http://student.cs.ucc.ie/~xd1.

http://www.agr.gc.ca/ren/serv/elearn_e.pdf.

http://www.worldbank.org/wbi/sdruralpoverty/ag_extension1/Materials/additional/Rural_extension.pdf.

Xiaoye, D., Sabin, T. (2007, March). A Theoretical Multimedia Synchronization Framework for E-Learning System. Vol. 10 No. 2 pp 225 – 230.

Xiaoye, D., Sabin, T., Eamon, L. (2005, June). JAS - An E-Learning Tool for Building Multimedia Presentations. IMSCCS'06 Vol. 1 pp 743 – 746.

A REFLECTIVE PERSPECTIVE TOWARDS AGRICULTURAL INFORMATION-BASED SYSTEM DEVELOPMENT IN GENERAL RURAL CHINA AND FARMERS' ECONOMIC ASSOCIATION AS THE ENTRY POINT OF ICTS

Miao Ji[1,*], Ting Zuo[1], Eamon Leninhan[2]

[1] *College of Humanities and Development, China Agriculture University, Beijing, 100094*

[2] *Center for Sustainable Livelihoods, University College Cork, Ireland*

[*] *Corresponding author, Address: College of Humanities and Development, China Agriculture University, Beijing, 100094, P. R. China, Tel: +353876194121, +8613010774671, Fax: +353 214903358, Email: watermiao2003@hotmail.com*

Abstract: The information-based agricultural system plays a profound role in agricultural economic and social development in long term. There are various practices worldwide, both successful and unsuccessful. In recent years new technologies blossomed, if those new technologies succeed in the developed world, does it hold true for China? The alternative choices in information-based system development should be noticed in China. This article argues that China's development policy should choose the way that adapts to China's circumstances, not totally adopt the western way. In the current situation of general rural China, the theme of information-based agricultural system needs to be further addressed. It is suggested that the Farmers' Economic Association can be considered as one of the most accessible entry-points for further development way.

Keywords: Farmers' Economic Association, ICTs, Information-based system

---

Please use the following format when citing this chapter:

Ji, M., Zuo, T. and Leninhan, E., 2009, in IFIP International Federation for Information Processing, Volume 295, *Computer and Computing Technologies in Agriculture II, Volume 3*, eds. D. Li, Z. Chunjiang, (Boston: Springer), pp. 1825–1834.

1. SMALL-SCALE FARMERS, AGRICULTURE DEVELOPMENT AND INFORMATION

Small-scale farming is a typical feature in rural China. Agriculture development policies should aim at small-scale farmers. In research on small-scale farmers in Africa (Wapakala,1982), several agriculture development areas have impacts on the productivity and on the income of small-scale farmers; these areas are covering agriculture extension, local farmer participation, credit, marketing, social services, project administration and training. Hence, development in these areas will be meaningful for small-scale farmers.

In agriculture development, information plays active roles. Many researches have demonstrated the profound contributions of information towards agriculture development and poverty reduction. (Kalusopa, 2005) pointed out agricultural information has become core dynamic factor for rapid rural development. Agricultural development activities are based on the utilization of information. (Williams & Williams, 1987) highlighted the importance of the information dissemination, and the relationships between information transfer and the agricultural development. Referring to information utilization, (Richardson, 1997) deemed when information or knowledge was harnessed by strong organizations of small producers, strategic planning could be used to provide members with lower cost inputs, better storage facilities, improved transportation links and collective negotiations with buyers. (Mchombu, 2003), described a rural development information centre model, based on broad information strategy, competent information facilitators, a multi-channel approach to information dissemination, and the participation of different rural groups.

Though different researchers have different research perspectives, the above research demonstrates implicit linkages across small-scale farmers, agriculture development and information. Farmers, as a dynamic factor, play roles in information dissemination and agriculture development; information, as a new emerging factor, contributes sufficiently to rapid agriculture development; agriculture development will turn out to be favorable to farmers. In this process, infrastructure is not the absolute affecting factor in terms of information-based system establishment; institution building is an alternative option for small-scale farmers except for infrastructure building. The relationships can be shown in Fig.1:

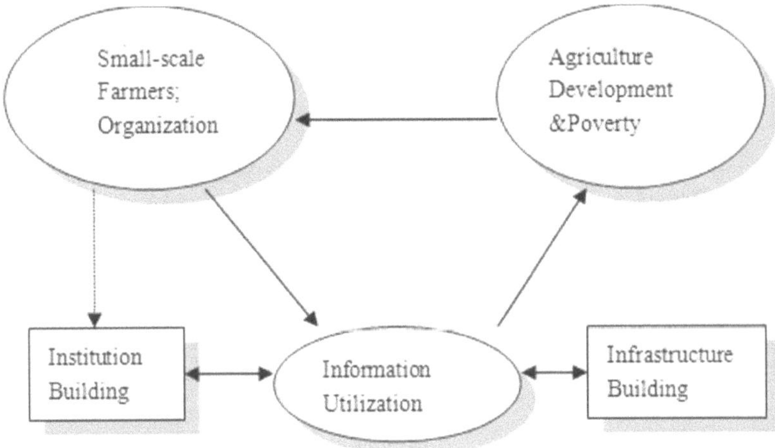

Fig. 1: Relationships between small-scale farmers, information and agriculture development

2. THE APPLICATIONS OF ICTS WORLDWIDE

In the process of information utilization, there are a lot of experiences of ICTs (information and communication technologies) worldwide. The experiences of ICTs have shown that many new ICTs technologies emerge in the form of e-techniques, including group discussion via Internet forums and bulletin boards, voter guides, e-mail, instant messaging, blogs, CDs, video, film, PC-produced publications, community broadcasting, and other new media; rapid group mobilization via e-mail, mobile cell phones, instant Internet messaging; controls on spam, access to Web sites and other content; interactive voting and polling via the Internet, digital TV, text messaging; government, public agencies, NGOs and the politicians using all ICTs to inform, communicate and consult with the public (Cfr. UNESCO,2004). Though it seems these technologies contribute to the efficiency of information dissemination among different people and different organizations. While an important point which needs to be identified is whether human skills can feed the technologies, and whether these technologies can feed local people real needs.

The usage of ICTs in both developed and developing countries have confronted some barriers. According to (Kalusopa, 2005; Obijiofor, 1998; Agriculture and Agri-Food Canada, 2003), in terms of the information development, several challenging factors can be summarized, these factors are social culture and local basic needs, appropriate institution and systems, users' capacity building and skills improvement, infrastructure establishment, management skills, and the appraisal method improvement, etc.

3. CHALLENGES IN INFORMATION-BASED SYSTEM IN CHINA: FROM A GENERAL REFLECTIVE PERSPECTIVE

Generally speaking, in the process of development in China, main policies have been based on the 'giving priority to efficiency with due consideration to fairness' principle, the same happened in the information development policy design. An industrialization characterized developing path depends more on advanced sciences and technologies. Hence, referring to what is information-based system, (Daoliang, 2007) has given a definition that it is 'a process characterized by application of mobile and computer technologies in agriculture production, rural life and social management', 'In this process, human beings can employ high technologies to exploit information resources sufficiently, in order to promote agricultural economic development and social development'. In the current developing path mentioned above, this definition of information-based system seems feasible, while when considering the development path from a reflective perspective, such definitions need to be reconsidered.

During the process of information-based system establishment, five main challenges emerge in the general picture:

First, there is a challenge of infrastructure limitation in poor rural areas. In the current situation in rural China, limited by the infrastructure constraints, especially in many poor rural areas, human factors are still the main means of collecting and disseminating information (Miao, 2008). Hence, it will be meaningful to investigate the human factors' roles, as the basis for further information policy design. Computer and mobile infrastructure shouldn't be simply set as the indicators to measure the developing level of information-based agricultural system.

Second, there is a dramatic gap of information infrastructure building between east and west areas in China. In this research, the comparison data on rural informationize index can be divided into two groups, one is typical east developed areas constituted by Beijing, Shanghai and Zhejiang, the other group is typical west underdeveloped areas which are comprised by Yunnan, Guizhou and Tibet. The index of national average level is set in the middle. From the comparison, it shows the extreme big gap between these two groups. Take Beijing and Tibet for example, Beijing's informationize index is three times Tibet's. From the anti-poverty function of information, olicies should be adjusted more to the west characterized poor and main agriculture areas. It is the time to implement the development principle, that is, 'consideration for fairness'.

Table 1: Development gap between east area and west area in China

Areas	Beijing	Shanghai	Zhejiang	Average	Yunnan	Guizhou	Tibet
Index	68.25	67.8	60.69	40.53	28.33	27.7	20.74

Source of Data: Daoliang, 2007

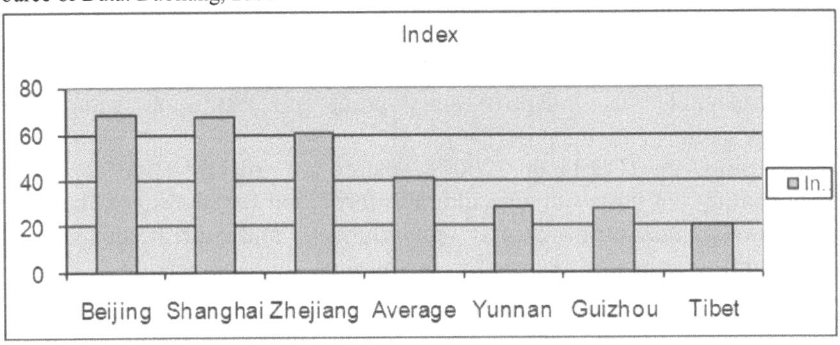

Third, according to information supply system, this is characterized as scattered, poorly developed and unfocused in the current supply system. It seems lack of co-ordinate efforts in the establishment process leading to the duplication of efforts among different information developing entities.

Fourth, the traditional top-down development model burdens the government and it usually causes neglect of farmers' real and diverse needs.

Fifth, cultural factor is always neglected in policy planning and implementation. In rural areas, compared with traditional information and communication ways, which are formed in the long history, characterized by certain cultural background, to create transformative space for new information and communication technologies is not easy. The main point is to consider the cultural and history factors.

Information establishment is a systematic process. These five challenges mentioned above are just summarized from a general perspective. There are still other diverse challenges in practice, which need to be addressed and shared further in practice.

4. OPTIONS AND DIRECTIONS FOR THE FUTURE: FARMERS' ECONOMIC ASSOCIATION AS ENTRY POINT

The general development of ICTs in rural areas is still at an original stage with many challenges in current period. While from the worldwide experiences of ICTs application in rural areas, it seems that two ways can be figured out for information-based system establishment in the future, first, developing totally along with the Western way without considering any influences; second, developing according to certain local circumstances in China, at the same time adapting appropriate technologies to local development. From my point of view in this paper, the second choice seems more reasonable. The adaptive entry point and institution arrangement should be explored in this process.

According to (Daoliang, 2007), there are mainly five kinds of implementing entities in an agricultural information-based system, these are, government, including central government and local government, agricultural companies, scientific institutions and universities, farmers' associations and farmers' households, different entities play different important roles in information-based system building. In this paper, when referring to the development options of the information-based systems establishment in China, Farmers' Economic Associations have been decided as an entry point for a long-term information-based agricultural system development strategy, especially in the underserved rural areas.

Farmers' Economic Association is an association of persons, who work together to achieve certain commercial objectives. As an interest organization, usually is created from below by individuals to satisfy their common needs, the objective of which can cover many aspects, mainly reflecting social values. Besides, farmers' economic association is usually understood as an economic activity owned and managed by the members to promote member's needs.

For the discussion about the general challenges of information-based system development in China, Farmers' Economic Association, as a kind of institutional arrangement, plays active roles in information dissemination under the poor infrastructure conditions, and has the cultural advantage which will help to create new space for using new information and communication technologies, it also has the advantage in feeding farmers' diverse and real needs. However, there is still other macro 'glass ceiling' difficult to be broken just depending on Farmers' Economic Association, such as the gap between information infrastructure building and the challenge in the information supply system. Strategic, long-term and sustainable approach should be taken into account by government and policy decision makers.

Farmers' Economic Association, as 'union is strength' which in the process of rural reform, plays very important roles in information dissemination.

Firstly, individual farmer management institution, with low productivity, and disperse small-scale farmers situation, can hardly be adapted to the market situation as the development of market economy. Agricultural information has emerged as very important linkages between small-scale farms and 'big market'. The development of Farmers' Economic Association plays positive roles in new technology extension, technologic transfer and dissemination of technology and market information.

Secondly, Farmers' Economic Association has the advantage of human factor in information dissemination. According to case studies in some undeveloped rural areas (Miao, 2008), there is a typical feature in information dissemination inner and outside of farmers' economic association, that is, human factors are still the main subject to collect and disseminate information, including market and technology information.

Thirdly, different entities have their own advantages and disadvantages in the establishment of information-based system.

Table 2: Comparison of advantages and disadvantages among different entities

Different entities	Advantages	Disadvantages
Government	Sufficient financial resources; Supply information	Burdened by so many tasks; Inefficient top down policies
Agricultural companies	Guide farmers in planting and breeding areas; Supply information	Interests deviation from farmers when facing risks
Scientific institution and universities	Afford the task of high-tech innovation; Training for people	Distance from farmers; High cost; Practice deviation
Farmers' Economic Associations	Direct interests linkages with farmers; Base for information service; Traditional channel of information dissemination	Potential risk of elite dominant structure; Lack of institution building
Farmer households	Direct interests linkages with farmers; Traditional channel of information dissemination	Lack of resources; Scattered, poor and weak

In the developing process, different developing entities have different interests and objectives. While compared with other entities, farmers' own organizations have the most direct interests' linkages with local individual farmers. This is not to say that the roles played by other entities are not important, from an interest linkage point of view, Farmers' Economic Associations are more direct entry point. Further more, different entities should enhance cooperation with each other, and build cooperative information networks, which will benefit all the stakeholders.

Fourthly, in rural areas, traditional information and communication channels carry out most of the communication among farmers. Communication happens in village meetings, market gathering places, between relatives, from community leaders, etc. The common features of these communications are usually informal, healed behind certain village political and social culture, dependable, reliable, reflect more direct question and answer opportunities. Farmers' associations own the cultural advantages and will keep the daily communication ways which farmers are familiar with. Whether new communication technologies can be employed efficiently depends on whether the technologies can be accepted by local cultures and can be introduced into and combined with traditional channels.

5. RECOMMENDATIONS

In summary, there are some recommendations for a current developing strategy for information-based system establishment, they are mainly on institution building, technology choice and education.

5.1 Institution Building

Generally speaking, an institutional improvement may compensate to some extent for the poor infrastructure conditions in poor areas. While in better infrastructure conditioned areas, it demonstrates farmer's economic cooperatives can play positive roles as the base for further establishment of an information-based system, information covering agricultural skill upgrading, and market demand and supply information. The main advantages will be further explored according to the existing experiences, in terms of responsibility, knowledge and information localization, farmers regional network, partnership, information disseminating, sharing and innovating, cost and risk, etc.

Considering the limitation of resources and rural reality, the role of government should be focused more on facilitating, which helps to establish basic infrastructure, build knowledge capacities, train farming skills and protect farmers' rights and interests. The main facilitating role should be played in combining of resources and institution building, not be confined to infrastructure building.

In terms of policy implications, the following are some recommendations of specific measures:

- In the developing path, farmers' real needs and local diversity should be the main focus. The solution is to let farmers participate in the communication and information dissemination.
- Identify the roles of different factors in the agricultural information development processes. Farmers should be empowered in the developing processes in order to meet the realistic needs while government should act as facilitators.
- Enhance the partnerships and cooperative relationships among different factors. It is helpful to combine various information resources supplied by different factors by enhancing the partnership and cooperative relationships.
- Supply funding to cooperative associations who are dedicated to the farmers' realistic needs. Government and other sponsors have the responsibilities of recognizing and selecting the funding beneficiaries.
- Set up ICTs of supporting centers for the funded farmers' associations. Based on the sufficient finance supports and government assistances, the selected cooperative associations could be facilitating servers and networks for applying communicative online platforms to local farmers, natural villages and other sub-associations.

5.2 Options of technologies

Regarding the current infrastructure circumstances in poor rural areas, it is practical and feasible to adopt simple and adaptive information technologies into agricultural information system, such as mobile technology, computer networks, local television channel services and landline-based services, etc. The case studies in Guanxi province (Miao, 2008) have shown the positive roles played by landline-based services for farmers' associations. There are also some similar cases in Ningxia province. While in some better developed rural areas in the east, more advanced technologies can be considered. At the same time experiences sharing platforms should be bridged between east and west areas.

5.3 Education for farmers

As for challenges of ICTs mentioned before, farmers' skills have been identified as one barrier of the development of ICTs. Under the circumstances, education is the main choice for farmers' capacity building.

This study illustrates that the success of agriculture development requires a well-organized and effectively functioning information-based system, while agriculture information-based system design is an integrated and cooperative process. There is need to embrace human factors more in addition to infrastructure building. There are a lot of challenges in the existing information-based system. The whole process shouldn't be simplified before a timely, reliable, accurate and functional system is established.

NOTES

1.This index conclude five measurement indicators, they are the number of fixed telephone equipment per hundred households, mobile phone number, television number, computer number and radio number.

REFERENCES

Agriculture and Agri-Food Canada. An overview of e-Learning in Canadian Agriculture and Agri-business, 2003.

Cfr. UNESCO. Social Transformation in an Information Society: Rethinking Access to You and the world, 2004.

Daoliang L. Rural Informationize Development Report, China Agricultural Science and Technology Press, 2007.

Kalusopa, T. The challenges of utilizing information communication technologies (ICTs) for small-scaled farmers in Zambia, Library Hi Tech, 2005:414

Mchombu, K. Information Dissemination for Development: An Impact Study, Journal of Information Development, 2003, 19(2):111~126

Miao, J. Cuiling, G. Xiaoye, D. and Leninhan, E. Roles played in agricultural information dissemination by different actors in farmers' economic association in China, working paper for Information System in Agriculture and Forestry 2008.

Obijiofor, L. Africa's Dilemma in the Transition to the New Information and Communication Technologies, Futures, 1998, 30(5):453~462

Richardson, D. The International and Rural & Agricultural Development: Integrated Approach, FAO, Rome, 1997.

Wapakala, W.W. The role of libraries in agriculture, SCECSAL proceedings on libraries and Their Impact on Education, Economics and Culture in Developing Countries, Tanzania Library Association, 1982.

Williams, C. and Williams, S.K.T. Dissemination of (technical) agricultural information, Journal of Administration, 1987, 13(1):87

AN ANALYSIS OF ICT DEVELOPMENT STRATEGY FRAMEWORK IN CHINESE RURAL AREAS

Meiying Duan [1,*], Martyn Warren [2], Yunwen Lang [1], Shaokun Lu [1], Linnan Yang [1]

[1] *School of Information and Engineering, Yunnan Agricultural University, Kunming, Yunnan Province, P. R. China 650201*

[2] *School of Geography, University of Plymouth, Plymouth, Devon PL4 8AA, United Kingdom*

[*] *Corresponding author, Address: School of Information and Engineering, Yunnan Agricultural University, Kunming, Yunnan Province, P. R. China 650201, Tel: +86-871-5823986, Fax: +86-871-5227717, Email: myduanx@yahoo.com*

Abstract: Information and Communication Technology (ICT) development strategy in Chinese rural areas is an indispensable part of national development strategies. This paper reviews the ICT framework in agriculture and rural areas launched by the Department of Agriculture in China. It compares the rural ICT policies and strategies between China and the European Union (EU). The ICT development strategy framework is analyzed based on the situation in Chinese rural area and the experiences of the EU. Some lessons and suggestions are provided.

Keywords: ICT, ICT policy and strategy, ICT initiative, Internet, rural areas

1. INTRODUCTION

China is a big agricultural country with 57.01% of the total population living in rural areas. The toughest challenges related to agriculture, countryside and farmers, which will determine the current society development. The lack of information in rural areas is a dominant factor resulting in the digital divide, which is seriously hampering the rural development. The digital divide is the ever-growing gap between those

---

Please use the following format when citing this chapter:

Duan, M., Warren, M., Lang, Y., Lu, S. and Yang, L., 2009, in IFIP International Federation for Information Processing, Volume 295, *Computer and Computing Technologies in Agriculture II, Volume 3*, eds. D. Li, Z. Chunjiang, (Boston: Springer), pp. 1835–1844.

people and communities who have access to information technology and those who do not. ICT is basically information-handling tool, and particularly refers to the Internet in this paper. The rural ICT policy and strategy plays a key role in promoting an intensive and wide use of ICT, driving economic development and improving quality of life in rural areas.

It is the main aim of China's 11th Five-Year plan (2006-2010) to build a socialist new countryside and to build a socialist harmonious society. Following the Plan, National ICT Development strategy (2006-2010) was published, with an emphasis on speeding up political, economic, cultural and social development through ICT. Based on the strategies, the ICT framework in agriculture and rural areas was developed in 2007. It aims to provide information services for farmers, to narrow the digital divide and help move forward the development of rural economy and society.

2. ICT DEVELOPMENT STRATEGY IN RURAL AREAS

The digital divide is the result of the inability of a large portion of the world's population to access and effectively use ICT and the potential benefits they enable (UNDP, 2001). The rural issue arises from the population in rural areas to be at a disadvantaged position with respect to ICT. This digital disadvantage often results from socio-economic disadvantage, and in turn makes the latter worse: the digital vicious cycle (Warren, 2007). However, "any online process, whether transactional or merely informational, which substitutes for one involving physical movement is likely to be of greater utility to rural than to urban populations, especially given the increasing tendency of governments… to reduce or fail to make provision in areas of dispersed population" (Furuseth, 1998).

ICT development strategy is the plans based on the selection of scenarios and options for applying ICT to national development, ensuring the greatest possible diffusion of ICT, commensurate with national needs, ambitions, specificities and concerns (Labelle, 2005). According to the definition, ICT development strategy in rural areas can be seen as plans for the use of ICT in rural areas and an indispensable part of national development strategies. It needs to be integrated into broad development concerns and mainstreamed into all aspects of society and of development planning, as well as taken into account local, national and international issues. The ICT strategy provides a commentary on the continuing development and use of ICT in rural area.

3. ICT INITIATIVES AND THE LISBON STRATEGY IN EUROPEAN UNION

Lisbon strategy is a wide range of policies launched by the European Union. It aims to make Europe more dynamic and competitive and focuses on growth and jobs in the knowledge-based economy. To achieve the goal of the Lisbon Strategy, eEurope and i2010 were released to build a fully inclusive Information Society across the Europe. They are the political initiatives and actions to bring the benefits of the Information Society to the reach of all Europeans, based on the widespread use of ICT in public services, SMEs and households. In general eEurope had been strong on bringing citizens and businesses online and establishing a framework within which the knowledge economy can grow. i2010 is translating these achievements into tangible economic benefits, higher productivity, improved quality of service, greater social inclusion and non-inflationary growth (European Commission, 2002).

Rural development is also central to the Lisbon process. Lisbon strategy encourages take-up or diffusion of ICT by extension of broadband coverage to help promote new ways of selling and dealing with risk. It assists village ICT initiatives by providing combination of computer equipment, networking and training and facilitates on-farm tourism through ICT use by improving booking systems, promotion and links with recreational activities (European Commission, 2000).

4. ICT IN CHINESE RURAL AREAS

4.1 ICT development

There will be a huge development of ICT in rural areas. The speed of Interne in rural areas has surpassed that in the city, and the annual growth rate reached 127.7%, far higher than that of the urban user, which was 38.2%(CNNIC, 2008). Some students, entrepreneurs, farmers start surfing on the Internet by mobile phone to obtain useful and real time information. More and more farmers and agricultural entrepreneurs are realizing the importance of information. They make an effort to acquire related information to guide their plan, production and sale by using ICT while confront an intensive competition.

The infrastructure is being improved because of the continuous economic growth, which provides a good environment of ICT development in rural areas of China. The average annual rate of growth has exceeded 10% since

2004, and the level of resident income has increased a lot. A large number of capitals with $9.8 million construction funds and $1.3 million maintenance fees were put into the projects of rural ICT infrastructure (CNNI, 2008). For example, the project of Phone for Every Village started in 2004 had made the phone available to 99.5% of administrative villages. A series of policies has been published to support the adoption and use of ICT in rural areas. It was put forward to quicken the construction of agricultural informatization, to enhance the rural ICT infrastructure, and to innovate rural service mode in 2006 for the first time. It is also the first time for government to issue the ICT framework in agriculture and rural areas as a general policy and strategy in 2007.

4.2 Problems and Challenges

There are some problems existing for services to rural areas in spite of improving infrastructure and sound governmental policies. The following analysis is to revolve around the topic of providing services.

A good connectivity is the prerequisite to providing services for farmers. Even the infrastructure of ICT can reach most of villages, but it is not easy to solve the issue of last mile. One factor is related to the income level. The resident income in rural area is $414, only 1/3 of that in urban areas in 2007, which determine the lower rate of owning the hardware device like phone, computer. Another factor affecting the access to the Internet is education level and skill. Lack of knowledge of computer or network keeps people in rural areas away from the Internet. The Internet penetration in rural areas is only 7.1% compared with 27.3% in urban areas (CNNI, 2008). The farmers have used to the traditional way to acquire the information and services, and have no awareness that they can benefit from the Internet.

The lack of useful information and services puts farmers at a disadvantage, which widen the gap between the urban and rural areas. The practical and appropriate content is the most valuable for farmers. However, there are so many irrelative and duplicate resources on agricultural websites. Only 31.2% of official resources on agriculture have established online. The online content obtained by farmer just account for 1.8% of total information (Yang, 2007). Most of rural Internet users only take the Internet as a chatting tool and recreation tool, with shallow Internet application. Respectively 86.4% and 76% of them use the online music and online film and TV functions (CNNI, 2008).

5. ANALYSIS OF ICT DEVELOPMENT STRATEGY FRAMEWORK IN CHINESE RURAL AREAS COMPARED WITH THAT IN THE EU

5.1 ICT strategy comparison between China and EU

There is not a separate rural ICT strategy in EU, but it is covered in eEurope and i2010. China is a developing country, and can learn a lot from the formulation of ICT strategy in EU. With respect to ICT development strategy, the past of that in EU is the present in China, and the present in EU is the future of China.

5.1.1 Aim and objective

Table 1. Aim and objective

	Strategy or initiative	During	Main aim	Objective and action
EU	eEurope 2002	2000-2002	Extending the Internet connectivity	A cheaper, faster and secure Internet Investing in people and skills Stimulate the use of the Internet
	eEurope 2005	2002-2005	Promoting modern online public service and dynamic e-business environment	e-government e-learning e-health services e-business
	i2010	2005-2010	Achieving inclusion and a quality life	A single European information space Innovation and investment in research Inclusion, better public services and quality of life
China	ICT in rural areas	2007-2013	Providing information service and advance rural development	promote modern agriculture advance public service and society management build and improve information organization

In the EU, under the guidance of the Lisbon Strategy, there have been three separate ICT development policies and strategies: eEurope 2002, eEurope 2005 and i2010. The ICT policy and strategy framework in agriculture and rural areas (2006-2015) was published by the Department of Agriculture, which was based on the National ICT development strategy (2006-2020).

From table 1, it can be seen the priorities are different at three stages, but they are related and supportive. eEurope 2002 emphasizes the connectivity of Internet. eEurope 2005 is shifted to providing attractive services and applications, and encourage organizational change. i2010 is the EU policy framework for the information society and media, with purpose of harnessing the potential of ICT to drive innovation and productivity in Europe. The main aim of the Chinese rural ICT strategy is to provide the public service and society management, and meet the need of building the modern agriculture and constructing a new socialist country.

Strategies are achieved using a variety of tools, in which comparative analysis, including studying and comparing international experience and best practices, is a useful tool. The ICT strategies in the EU are continuous and sustainable. The late initiative is built on the previous. Only after reviewing and evaluating the action, can a new strategic framework be proposed. To certain degree the aim of Chinese strategy is similar to the eErurope 2005, but the use of ICT in Chinese rural areas far fall behind that in the EU.

5.1.2 Four elements in ICT strategy

The main components of ICT strategy framework in Chinese rural areas are infrastructure, information resource, human capacity, service and application system, rule system and operation mechanism. The structure can be simplified to four basic elements in ICT strategy, consisting of infrastructure, human capacity, service and environment. They have an impact on economy, politics, culture and society in rural areas.

In a sound environment of legislative framework and mechanism, people use services on Infrastructure to achieve their goals. The following table 2 compares the ICT strategies in the EU and China in terms of four elements.

The strategies regard infrastructure a key part of ICT. eEurope 2002 built a cheaper, faster and secured Internet, and increased the number of citizens and businesses connected to the Internet. It extended the Internet connected to schools, universities, research institution and business. Since 2005, due to transforming to broadband connection and adopting multi-platform technology, the infrastructure had greatly improved. i2010 make the network faster and more security by new digital convergence techniques.

The strategies invest in human capacity. More emphasis was put on human capacity because of the new requirements for information society skills and the problem of skills shortagese. One of objectives of eEurope 2002 was to invest in people and skill. They included three actions: European youth into the digital age, working in the knowledge-based economy and participation for all in the knowledge-based economy. eEurope 2005 and i2010 further to enhance the education and skill by e-learning and e-training.

Table 2. Four elements in ICT strategy

	Infrastructure	Human capacity	Service	Environment
eEurope 2002	Cheaper, faster, secure Internet	Invest in people and skill	Stimulate the use of Internet	Communications network and e-Commerce
eEurope 2005	Broadband Multi-platform	Improve participation and enhance skill	Provide online public services and private services	Private investment
i2010	Digital convergence	Digital skill	offer communications, rich content and digital services	Electronic communication Audio-visual
Rural areas in China 2007	Internet + Broadband	Distant education, Skill training	Agricultural extension of science and technology, e-government	

Service is put in the centre of EU ICT strategies, which is an essential ingredient to drive ICT development. eEurope 2002 stimulated the use of Internet by e-commerce, government online, health online and digital content for global networks. eEurope 2005 offered interactive public services including e-government, e-health and e-learning, and encouraged e-business by private sector. i2010 provides affordable and secure high bandwidth communications, rich and diverse content and better digital services, and quality of life.

The regulatory and legislative framework ensures the development of ICT. In eEurope 2002, legislative framework was put forward to reinforce the competition of market, to reduce the price and stimulate the innovation. eEurope 2005 made policy to encouraged the private sector to invest broadband infrastructure. i2010 established legal framework for audio-visual services.

5.2 Lessons learned from ICT strategy in EU

First, cheap, fast and secure Internet is provided. Even in the EU, the legislative framework is used to lower the cost through the market competition. Low level of income prevents most of Chinese farmers connecting to the Internet. Only the Internet cost is affordable for citizens in rural areas, can the Internet connectivity be extended. Additionally, it is necessary to improve the infrastructure with the new technology. The network was gradually transformed to broadband in the EU since 2005. IP addresses also had transited to Internet Protocol version 6 (IPv6). All the updating makes the Internet more attractive to the user who can use the Internet easily.

Second, people and skill is the most important factor in ICT strategy. The low level of education and skill keep the resident in rural areas away from the Internet. The ICT strategies in the EU pay more attention to the education and training. Education will make a major contribution to developing new skills, and the skills will help people take advantage of the potential of the Internet fully. From the beginning, EU connected the schools to the Internet to promote education, and then persisted in investing in people and skill.

Third, the new content, information and service are the centre of ICT strategy. EU strategies encourage providing the basic and cheaper service to citizens by stimulating the use of the Internet. After that, the new attractive content, service and application are fostered and created with development in multi-platform access, interactive software. People can be access to digital content and media services through PC, digital TV and mobile. It is a priority for the ICT strategy in rural area to offer the practical, cheaper and attractive service.

Fourth, the formulation of ICT strategies is emphasized. The process involves in identifying priority areas, developing action plans, implementing and evaluating the actions. All of ICT strategies of the EU support to realize the Libson aim of growth and jobs. A series of programs and projects are implemented to reach the objectives of the action plan, such as Research Framework Programme and the Competitiveness and Innovation Programme. Mid-term review and annual report are used to analyze and assess the progress in the implementation so that the strategy remains up to date with the rapidly changing ICT environment. In china, the department of making policy is separated from the department of implementation. Therefore evaluation is critical to ensure an achievement of the ICT policy and strategy in rural areas in China.

5.3 Suggestions

ICT policy and strategy in rural areas of China is a general national framework, so the regional context is considered. Different regional strategies should be reformulated and facilitate the exchange of experience, of good practices and demonstration projects, as well as of sharing the lessons from failures. The rural areas in China are divided into three categories which are eastern developed regions, middle medium developed regions and western less developed regions (Mei, 2007). Based on the regional strategy, provincial and county ICT strategies should be launched in different rural areas. They interact and support each other to ensure the ICT strategies of different level are carried out efficiently and achieve their aims.

It is necessary for government to accelerate the setting up of an appropriate legal framework and policy about ICT use in rural areas at national and regional level, which will provide a sound development environment of ICT policy and strategy. The regulatory can encourage the market competition to reduce the cost of access to the Internet and obtain the online public services. The policy of investment in ICT of rural areas will promote more and more rural citizens to surf on the Internet, to get more education, skill training and useful information and services. The legislative framework encourage private sector invest in rural ICT and get more fund support, since only the government investment is not enough to improve the infrastructure, human capacity and service.

The organizations of two-level information service are constructed in county and village. The agricultural information service system is top-down, multilevel and complicated, including too many departments. However, the two kinds of information service points have integrated multi-functions, which can manage the county and village, share the information, converge service, and provide the training and entertainment. For the disadvantage farmers who are illiterate or lack of the IT skill, the local information organization can help and guide them obtaining the useful information and service. The service points manage local content and information, making farmer to participate and benefit from the use of ICT.

6. CONCLUSION

ICT development strategy framework in Chinese rural areas involves in four basic elements of ICT development strategy and identify the services is the priority. It is a universal guidance of regional and local rural ICT strategies. Lessons from the ICT strategy of the EU are helpful to form the rural ICT strategy and plan action. On the one hand, it is very important to provide cheap, fast and secure Internet, invest people and skill, and offer content, information and service. On the other hand, reviewing, evaluating and update the strategy is indispensable. The rural ICT strategy of China should consider regional context, legal framework and policy, and information service organization.

ACKNOWLEDGEMENTS

I would like to dedicate the accomplishment to my supervisor Martyn Warren, the director of Rural Futures Unit, University of Plymouth, who directed and helped me during my staying in the UK. I gratefully acknowledge the support from my colleagues at Yunnan Agricultural University. Yunnan Provincial Department of Education funded me to do research of ICT in the UK.

REFERENCES

China National Network Information Center (CNNIC). Survey Report on Internet Development in Rural China 2008. 2008. http://www.cnnic.cn/uploadfiles/pdf/2008/1/17/104156.pdf

European Commission. 2000. http://www.europe.canterbury.ac.nz/courses/uro227/special_en.pdf

European Commission. 2002. Information Society Benchmarking Report 2002. http://ec.europa.eu/information_society/eeurope/i2010/docs/benchmarking/051222_final_benchmarking_report.pdf

Furuseth, O (1998). Service provision and social deprivation in the geography of rural change. Ilbery, B (ed). Harlow, UK, Pearson: 233-256.

Labelle, R. ICT Policy Formulation and e-Strategy Development. Elsevier, India. 2005. http://www.apdip.net/publications/ict4d/ict4dlabelle.pdf

MEI Fanquan. ICT development strategy and mode in Chinese rural areas. China Rural Science and Technology 2007, (12), 20-21(in Chinese)

UNDP Education Office. Information Communication Technology for Development Essentials: Synthesis of Lessons learned, 2001, http://www.apdip.net/ documents/policy/actionplans/ Essentials05092001.pdf

Warren, M.F. The digital vicious cycle: links between social disadvantage and digital exclusion in rural areas. Telecommunications Policy 2007, 31, (6-7), 374-388.

YANG Bao-zhu. The research of development strategies for agricultural informatics and informatization of agriculture in China. Agricultural Network and Information 2007, (9), 4-8.

FARMERS' INFORMATION USAGE INTENTION IN CHINA BASED ON THE TECHNOLOGY ACCEPTANCE MODEL

Jingjing Zhang [1,2] ,Xiaoshuan Zhang [2] ,Weisong Mu [2] ,Jian Zhang [2] , Zetian Fu [2,*]

[1] *College of Economics & Management,China Agricultural University,Beijing,P.R.China 100083*

[2] *Key laboratory for modern precision agricultural integration Ministry of Education, China Agricultural University, Beijing,P.R.China 100083*

[*] *Corresponding author,Address:Key laboratory for modern precision agricultural integration Ministry of Education, China Agricultural University,17 Tsinghua East Road, Beijin, 100083,P.R.China.Tel:+86-10-62736717,Fax:+86-10-62736717,Email: zhangjingjing1982@cau.edu.cn*

Abstract: Information technology acceptance has received much attention, but little research has been conducted to assess farmers' information adoption. Despite the importance of information, its value will not be realized if farmers are reluctant to accept it. This research aims to study farmers' information adoption in China, in order to provide some decision-making advice for the people and organization who supply the agriculture information. The model of information usage intention has been established based on the Technology Acceptance Model (TAM). A sample of 231 farmers participated in this study. The results show that the factors which influence the usage willingness for information are perceived usefulness, perceived ease of use, learning intention, risk preference and experience in information before. In addition, income and education may also affect the decision.

Keywords: information demand, usage intention, technology acceptance model, perceived usefulness, perceived ease of use, perceived risk

Please use the following format when citing this chapter:

Zhang, J., Zhang, X., Mu, W., Zhang, J. and Fu, Z., 2009, in IFIP International Federation for Information Processing, Volume 295, *Computer and Computing Technologies in Agriculture II, Volume 3*, eds. D. Li, Z. Chunjiang, (Boston: Springer), pp. 1845–1853.

1. INTRODUCTION

Information is a key factor and has been proved to be significant in the agriculture production and lives in recent years. However, there exits a gap between agriculture information supply and acceptance. Previous studies have been mostly focused on users' acceptance of information technology. The issue about whether farmers are willing to accept information has not been widely discussed.

To address this issue, this study seeks to provide insights into farmers' information usage intention in China. In previous studies, the technology acceptance model (TAM) has been widely used in order to gain a better understanding of information technology adoption. In this study, TAM is put forth to the topic and modifications are made to it. Perceived risk is added to the original model according to characters of farmers and agriculture information service. Perceived risk sometimes plays an important part in consumers' decisions. If farmers consider that information will bring them help with little risk, the information usage intention is great. The aim of this paper is to define the key factors impacting farmers' information adoption.

2. THE TAM MODEL

In our research, the TAM model is selected for the main theoretical method.TAM,which originally developed by Davis,has the advantage of being well grounded in established social psychology theory (McKechnie et al.,2006). In the model, two main aspects constitute the model frame: perceived usefulness and perceived ease of use. The former refers to the users' perception of the degree to which using a system will enhance their performance, and the latter means the degree to which users consider the system to be free of effort (Davis et al., 1989).

In the model,perceived usefulness and perceived ease of use affect actual system usage through attitude and intention. Attitude,which is the key determinant of behavioral intention has a direct effect on intention, is determined by both users' perception of usefulness and ease of use, and the actual use is directly determined by behavioral intention. In previous research, it was found that usefulness may impact actual use irrespective of attitude, provided that the use of the system offers direct benefits to the user (Davis et al., 1989; McKechnie et al.,2006). If a system is considered to be easy to master, the perception of the usefulness will be reinforced. Therefore, perceived ease of use may have a direct influence on perceived usefulness. Fig.1 is the TAM model.

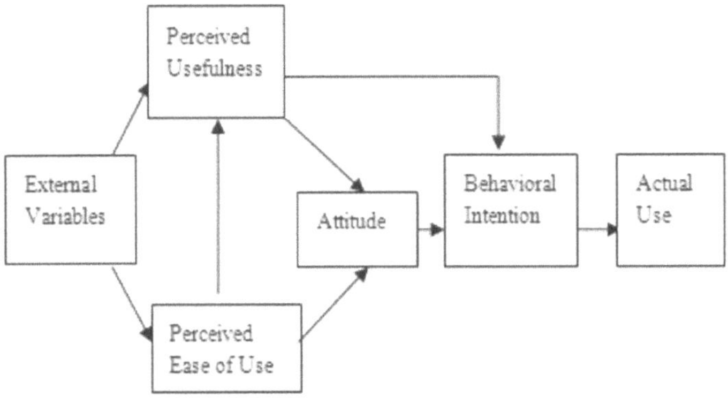

Fig.1: The TAM model

3. METHODOLOGY

In this paper, modifications are made to the original model and several hypotheses are put forward based on the model. Fig.2 is the farmer information usage intention model.

3.1 Factors of the original model

According to the TAM model, perceived usefulness and perceived ease of use are two important determinants.In this study,it is supposed that perceived usefulness and perceived ease of use affect farmers' attitude directly. Furthermore, it is proposed that perceived ease of use not only predicts attitude,but is also an antecedent of perceived usefulness (Davis et al.,1989). Farmers' confidence in information will be increased if it is easy for them to make use of agriculture information.

H1.Farmers' perceived usefulness influences the attitude towards using information positively.

H2.Farmers' perceived ease of use influences the attitude towards using information positively.

H3.Farmers' perceived ease of use influences perceived usefulness positively.

As is the case with the TAM, attitude towards using information may impact information usage intention directly.Similarly,it is assumed that perceived usefulness affect information intention to use without the medium of attitude (Davis et al.,1989).The next two hypotheses are proposed:

H4.Farmers' perceived usefulness affects the information usage intention positively.

H5.Farmers' attitude affects the usage intention positively.

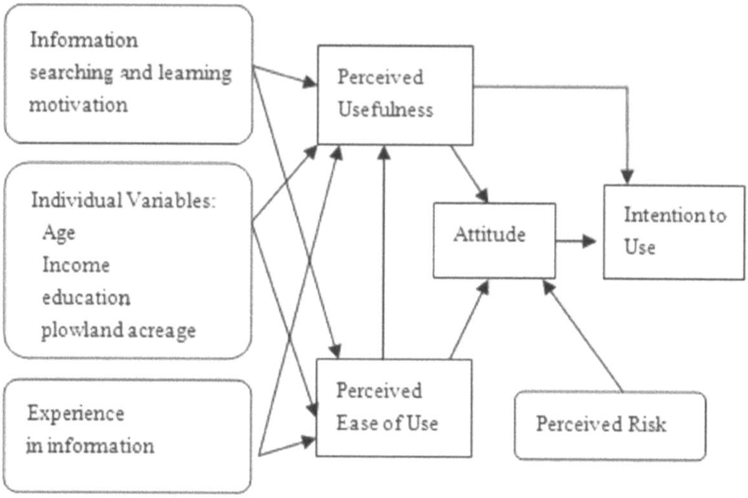

Fig.2: The information usage intention model

3.2 Experience in agriculture information

Prior work has suggested that past experience with the technology was a paramout factor and resulted in the higher degree of the adoption (McKechnie et al.,2006). It is argued that farmers who have experience in using information may have stronger usage intention. There are the hypothesis:

H6.Experience may impact farmers' information usage intention positively with the medium of perceived usefulness.

H7.Experience may impact farmers' information usage intention positively with the medium of perceived ease of use.

3.3 Information searching and learning motivation

Earlier studies suggested that intensive usage motivation would act on the adoption(Lee et al.,2005).We argue that farmers who exhibit higher levels of information searching and learning motivation will be more interested in agriculture information than those with lower levels.

H8.Searching intention is related to perceived usefulness positively.

H9.Learning intention is related to perceived usefulness positively.

H10.Searching intention is related to perceived ease of use positively.

H11.Learning intention is related to perceived ease of use positively.

3.4 Perceived risk

According to consumer behavior theory, the risk that people consider subjectively is not the really exiting one.When deciding whether to adopt information, farmers' perceived risk is crucial. The more likely famrers are to do something venturesome,the more intensive the usage intention is.In this paper, we define farmers' perceived risk as risk preference and the ability to bear risk. It is supposed that perceived risk and risk assuming ability influence intention through information attitude.

H12.Risk preference is related to farmers' information attitude.

H13.Risk assuming ability is related to farmers' information attitude.

3.5 Individual variables

The function of information is recessive and indirect,both of which determine the discrepant outcomes when adopted by kinds of profiles of users.For example,it may be easier to catch on the deeply signification of information for farmers with the higher education. In the same way ,those that have taken on the larger plowland may be more willingly to adopt information.In this paragraph, age,income,education and plowland acreage are included as individual variables.

4. ANALYSIS

4.1 Data collection

Data in the study were collected with a sample of two hundred and thirty-one farmers from thirteen different areas in China. The items were measured on a 5-point Likert-type scale (1=strongly disagree to 5=strongly agree).

4.2 The analysis of the factors

4.2.1 The original factors

Correlation analysis is used in this study. The original TAM model shows a well applicability in the study. Perceived usefulness of information directly affects its intention to use, with a coefficient of 0.501.Perceived ease of use exhibits a positive relationship on perceived usefulness (.581), which means if farmers consider that information technology is easy to master, the usage intention is strong. Besides, there exits a relationship between intention to use and perceived usefulness via attitude, either is perceived ease of use. Attitude influences the usage intention significantly, with a coefficient of .710. Results are shown in Figure 3.

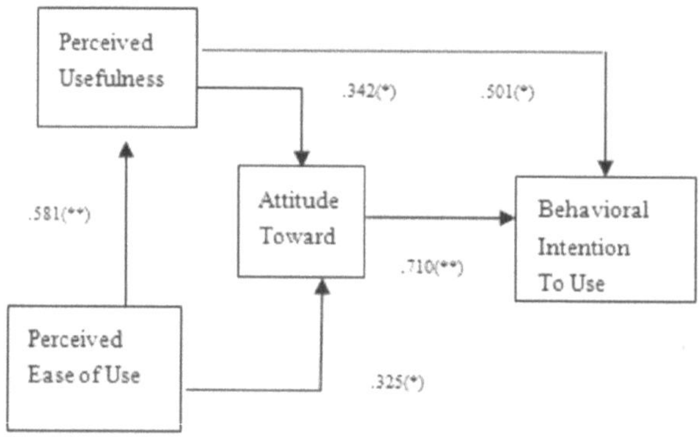

Fig.3: The correlation between factors of the original model

4.2.2 The extended factors

It is indicated that experience is directly associated with perceived usefulness, and more strongly associated with perceived ease of use (.538).Usually, if farmers have contacted with information and received some help, they are willing to consider it useful and the intention to use is stronger. Moreover, experience reduces the difficulty to master information, which in turn enhances farmers' confidence in information usage.

In this paper, information motivation is defined as information searching and learning intention. It is showed that searching intention has impact on perceived usefulness and ease of use positively, but not intensively. This phenomenon could be explained that although farmers are willing to search information voluntarily, they do not know how to do. The correlation both between learning intention and perceived usefulness on one hand and between learning intention and ease of use on the other hand are great, with coefficients of .518 and .631. The relationship is presented in Table 1.

Table 1: The correlation coefficients between the factors

External variable		Perceived usefulness	Perceived ease of use	Attitude
Experience in information		.357(*)	.538(**)	
Information motivation	Searching intention	.133	.271	
	Learning intention	.518(**)	.631(**)	
Perceived risk	Risk preference			.433(*)
	Risk assuming ability			.075

Perceived risk is referred to validate whether it affects farmers' information usage intention via attitude in this study. The result also yields the evidence that perceived risk influences attitude positively, and risk preference has stronger impact on attitude than risk assuming ability.

As for the individual variables, it is showed that education and income affect farmers' usage intention positively, but not so strongly as we have forecast before.

5. DISCUSSION AND CONCLUTIONS

This paper aimed to exam farmers' information usage intention in China based on the technology acceptance model (TAM) added with new variables. TAM model has so far mainly been used to explain the technology adoption topic. The results suggest that the application of the TAM model is helpful in obtaining a better understanding of factors influencing farmers' information usage intention.

It is showed that farmers' attitude and perceived usefulness both directly drive usage intention. The support is found that perceived usefulness and perceived ease of use determine attitude toward agriculture information usage, and perceived ease of use also has impact on perceived usefulness.

The findings highlight that experience in information usage and learning intention are paramount predictors of perceived usefulness and ease of use. Risk preference is positively correlative with attitude. The impact of searching intention is not evident, neither is assuming ability. Our findings also demonstrate that except for education and income, the farmer individual variables such as age and plowland acreage do not have the evident effect.

Based on the conclusions, some suggestions are put forth here:

Varying the categories of information.

To make full use of information, different kinds of farmers could be supplied with various information. Take risk preference for example, on the one hand, farmers who always avoid to do something venturesome may be interested in market information and product information, on the other hand, the ones that are likely to try something new may also concern about the advanced technologies or investment information.

Making the channels of information transmission convenient for farmers.

Even though having known about the importance of agriculture information gradually, most farmers are short of the ability to search for information effectively. To simplify the information transmission methods may lead to information more reachable.

Paying attention to farmers who have experience in information usage. The research shows that farmers are likely to accept agriculture information if they have the similar experience before, moreover, their confidence and behavior will influence others with no experience positively.

6. LIMITATIONS AND FUTURE RESEARCH

Several key findings emerged from this study, and the TAM model has been proved to be an appropriate frame to examine information adoption .However, the number of the sample of farmers is small in the survey, which may have a negative influence on the conclusions.

Various types of agriculture information will appear up in the information market, and some kinds of the information may have the attributes of the private product. Future research could be focused on farmers' willingness to pay for information in order to know about farmers' information demand and adoption deeply.

ACKNOWLEDGEMENTS

This study was supported by the National Science & Technology Support Program (2006BAD10A07-02).The authors are grateful to the farmers that took part in the investigation.

REFERENCES

Fred D.Davis,Richard P.Bagozzi, Paul R.Warshaw.User Acceptance of Computer Technology: A Comparison Of Two Theoretical Models, Management Science,1989,35:982-1003

Fred D.Davis.Perceived Usefulness, Perceived Ease of Use, and User Acceptance of Information Technology, IS Quarterly, 1989, 9:319-340

Hyun-Hwa Lee,Ann Marie Fiore,Jihyun Kim.The role of the technology acceptance model in explaining effects of image interactivity technology on consumer responses, International Journal of Retail & Distribution Management,2006,8:621-644

J.Alberto Castaneda, Francisco Munoz-Leiva, Teodoro Luque.Web Acceptance Model(WAM):Moderating effects of user experience, Information & Management,2007,44:384-396

Juan Carlos Roca,Chao-Min Chiu,Francisco Jose Martinez.Understanding e-learning continuance intention:An extension of the Technology Acceptance Model, Human-Computer Studies,2006,64:683-696

Mattew K.O.Lee,Christy M.K,Cheung,Zhaohui Chen.Acceptance of Internet-based learning medium:the role of extrinsic and intrinsic motivation, Information & Management,2005,42:1095-1104

Sally McKcchnic, Hcidi Winklhofcr, Christine Ennew. Applying the technology acceptance model to the online retailing of financial services, International Journal of Retail & Distribution Management, 2006, 34:388-410

William R.King, Jun He.A meta-analysis of the technology acceptance model, Information & Management, 2006, 43:740-755

Yi-Shun Wang,Hsiu-Yuan Wang,Daniel Y.Shee.Measuring e-learning systems success in an organizational context:Scale development and validation, Computers in Human Behavior,2007,23:1792-1808

STUDY AND ANALYSIS ON TECHNOLOGY AND DEVELOPMENT OF INFORMATION NETWORK OF RURAL POWER GRID

Weiying Li*

College of Information Science and Technology,Agriculture University of HeBeii HeBei,Baoding 071001, China

* *Corresponding author, Address: College of Information Science and Technology, Agriculture University of HeBei HeBei,Baoding 071001, China, Tel: +86-312-7526424, Email: leeweiying@163.com*

Abstract: This paper describes the technology and transferring mode of rural power grid's information network, analyses technology of communication system of electric power grid in rural area, chooses a new develop direction of technique based on the business needs and trend of rural power grids, and gives a route for development.

Keywords: rural power grids, Information network, SDH, MSTP

1. INTRODUCTION

During recent years, the advanced technology and systems are used in electric communication network, the construction of communication and information network for rural power grid is developing at top speed. The electric power enterprises are beginning to construct the information network for rural area to meet the needs of information and communication during the rural power grid's development when compared with the rural power grid upgrade. The information and communication systems of rural power grid are diversified in technical structures and operation patterns based on their own practices and levels of economic development. This paper analyzes and inquiries the mode of planning and developing a network in rural area.

Please use the following format when citing this chapter:

Li, W., 2009, in IFIP International Federation for Information Processing, Volume 295, *Computer and Computing Technologies in Agriculture II, Volume 3*, eds. D. Li, Z. Chunjiang, (Boston: Springer), pp. 1855–1862.

2. SERVICE ANALYSIS FOR RURAL POWER INFORMATION NETWORK

The rural power information network is the platform to provide information and communication service for electric power supply company in the executive grade which is at county level or below.

It serves the domain of production, administration and capital construction for the company. It supports the services of dispatching telephone, administration telephone, and automation of electric network management, relay protection, safety automation, computer linkups and image transmission and so on. If classified by service, they can be labeled to voice traffic service, low speed data service, high speed data service and video service. If classified by specialized field, they can be labeled to dispatching and administration voice information, rural power supply running supervision information, videoconference information, real time video surveillance information and administrative management information.

The description of access and transmission pattern of the services mentioned is listed below:

(1) Dispatching and administration telephone

It is a part of traditional services for voice. As the important platform for dispatching and administration, it must have high reliability and fast connection speed. It is mainly made of PBX or the extend of dispatching exchange and interfaces between exchanges such as FXS、FXO、4W+E/M、E1 and so on. The proportion of voice traffic service continues to drop because of the transform of rural power's communication pattern.

(2) Video surveillance information of substation

Video surveillance is the supplement to the remote control after carrying out the mode of unattended operation in rural substation. It provides the more precise circumstances to operators. And it can help the person in command or on duty quickly determine and make a strategic decision remotely in case of emergency. It adopts the network interface of TCP/IP and 10/100Base-T. (3) Management information system (MIS)

MIS in rural power Supply Company supplies the inquiry function of business enlargement and daily operation, electricity meters, electric power changes, electro-inspections by using control technique, graphics technique, database technology, network technology and multimedia. The computer linkups between various departments and subsidiaries of the company are based on the rural information and communication network of power Supply Company.

(4) The real time data of Automation of Electric Network Management

It is used to remotely monitor and control the equipments and collect real-time data for Control Center. This type of data has high requirement for high reliability and less time delay. It has the forms of voice data, serial data and

TCP/IP data. The TCP/IP data can access the SDH by Network Bridge or direct access the IP port. The RS-232/485 data can access the IP port through the RS-232/485 port of PCM equipment or access the IP port after being converted by terminal server.

(5) Relay protection signal

The relay protection signal has high requirement for high reliability. It can access the SDH through G.703/64kbps port of PCM equipment. The administration information of protective relay unit can be switched in through RS-232/485 port of VPN or PCM equipment.

(6)Video conference service

It is one of the principal services transported by electric communication network. The video conference network system that is based on IP and runs on packet switching network and adopts the H.323 protocol is progressively approved by electric power enterprises along with the perfection of H.323 protocol. The access method of TCP/IP will be adopted more and more hereafter.

(7) others

The subsidiary business of communication supporting networks and monitor message can be carried by the manners of TDM or IP.

In addition, there are services of lightning monitor system, network application, E-mail and document transmission.

The characteristic of the services mentioned above can be concluded in the figure below:

Table1. Traffic analysis of the Power Private Communication Network in rural area

	Transmission rate	Transmission delay	Error rate	Usability	Integrity	Configuration	Protocol
Data traffic							
MIS	4Mbps	classification			high degree	Broadcast/ double action	IP
Multi-media service							
Video conference	2Mbps	≤400ms	$\leq 10^{-5}$	higher	moderate	Broadcast/ double action	TDM/IP
E-mail	256Kbps	classification		moderate	moderate		IP
Network application	2Mbps	second		moderate	moderate		IP
Document transmission	2Mbps	minute		moderate	moderate		IP
Gis	512Kbps	second		higher	higher		IP
Video surveillance	2Mbps	≤400ms	$\leq 10^{-5}$	higher	moderate	Broadcast/ double action	TDM/IP
Telephone service							
Dispatching telephone	64Kbps	≤250ms	$\leq 10^{-5}$	high degree	high degree	double action	TDM/AAL1
Office phone	64Kbps	≤250ms	$\leq 10^{-5}$	high degree	common	double action	TDM/AAL1
IP telephone	16-64Kbps	≤250ms	$\leq 10^{-5}$	high degree	common	double action	IP

Several conclusions flow from this figure: the data communication is replacing speech communication based on TDM by degrees in the information network along with the development of rural power grid; The increasing data traffic based on IP is carried and the various kinds of services have different request of transmission quality; At the same time, the traditional services based on TDM will exist for a long time.

3. TECHNIQUE ANALYSES FOR RURAL POWER INFORMATION NETWORK

With the rural power grid upgrade， a fairly large electric private networks based on SDH are established in many rural areas which support the informationization of the rural power supply. However, restricted by the investor and the works scope, the structure of communication network has weak links because the communication facilities are constructed in the wake of the rural power grid upgrade. The SDH technology that mainly carries the single TDM service can not support the IP data and the view data service efficiently which have increasing proportion, so the bandwidth is wasted. The characteristic that need to be regulated is list below:

(1)The network of chain structure has low reliability.

The topological structure of information network for rural power grid at present depends on the trend of power line which has the star configuration or chain structure. The existent rural power grid has an irregular structure, which causes the low reliability of communication network.

The optical network protection can be classified according to its physical topology as line protection and self-healing ring configuration, the ring pattern can self recovery the network from failure within less than 50ms. But the present situation of rural network only permits line protection and the self-healing ring configuration is unattainable. In some area there is only chain structure and there are even not ADM facilities but only TM facilities, the security of SDH is not fulfilled.

（2）IP service is not supported

The traditional SDH system in rural network mainly transports TDM service such as voice. It holds an advantage on performance monitoring of network, failure recovery and reliability. It is adequate for the time sensitive service. However, SDH is an architecture that has the characteristic of complicated and centralized supply and limited expansibility. It can’t handle the IP service which has the characteristics of burst ness and unbalanced ness. It hasn't got the expansibility and flexibility which is needed in the network that mainly transports the IP service. The Virtual Container can’t match the bandwidth of IP packet. It causes the low utilization rate of bandwidth. The SOH of SDH can’t fit the IP environment. Along with the increasing of data service, SDH can't directly support the 10Mbps, 100Mbps and 1000Mbps ports of IP network.

The SDH system has a large proportion in rural network. At the same time, the solution based on SDH will play an important role in the future because of its maturity, reliability and full cost. So the construction of rural electric communication network should build the architecture that has the characteristics which both faces the IP service in the future and supports the TDM system at present.

4. THE CHOICE OF TECHNICAL STRUCTURES

The demand to rural electric communication network according to the current business is:

(1) Switch in the broadband service efficiently through layers as few as possible.

(2) The proportion of TDM service is brought down but remains high reliability.

(3) Support the video service efficiently.

(4) The dynamic data traffic is carried by dynamic linked network.

(5) Unify the network to build and maintain expediently.

Based on the request for technology mentioned above and to support the increasing IP service preferably, the network operator of rural electric power system are focusing the technology system such as DWDM, RPR and MSTP when building the private network or upgrading the existing SDH network. Furthermore, the pure IP network and the NGN, Soft Switch are receiving more attention. The data traffic is developing rapidly these years, but the TDM service such as dispatching telephone needs being first and foremost considered because of the particularity of electric power system. The pure IP network is not adequate for power grid because it can't meet the need of the reliability of QOS for TDM service.

For the areas where the SDH network already has been established, to protect the investment, it should be upgraded to the communications network that is suitable for IP service from existing equipments and technology system. It means that a plan is needed to transit the static multiplexed mode to dynamic IP network mode.

For the areas that the local rural electric communication network has not been established, a conformable technique structure should be chose to meet the need of specific service. Table2 is the analysis of technical in the power private communication network in rural area.

Table2. Technical analysis of the Power Private Communication Network in rural area

Technology Characteristic	Traditional SDH	MSTP	WDM	RPR
Summarize	Chief component of existing power private communication network	Optimize the data service retaining the advantage of sdh	Transport mass of data	Optimize the sdh network aiming at data service
Emphasis	It is adequate for time sensitive service such as voice	Bring together services on the VC1 level and support varies service interfaces	Transport multi wavelength at high speed	Bring together and transport data service, especially the Ethernet
Technical Description	Fit the mixed service based on the principle of TDM	The integrated function for transmission and traffic grooming; the optional function for data layer gathering	OADM、DWDM、CWDM based on ring network or DWDM、CWDM point-to-point	Data gathering and exchange equipment，the limited ability for voice processing
Advantage	The equipments have been set up in abundance, which have proven technique and high reliability	The existing SDH equipment are compatible, it is rapidly to open any new business and the network equipment is decreased	The capacity is largely increasing. The transparent service based on wavelength can be supported. Any protocol can be transported neatly	Effective data gathering function
Shortcoming	Lower efficiency after the data service becomes the dominant	Packing too more features lead to the difficulty for deployment	Can't gathering the low speed signal to wavelength channel, can't deploy the wavelength dynamically, high cost	Lack of ability for voice processing

A conclusion can be drawn from the table that MSTP has resolved the conflict of transporting TDM service and data service at same time. It is a good option to optimize the network on condition that plenty of SDH systems have been constructed by the rural power supply companies and the voice service is holding important positions.

To better suit the dynamic variation characteristic of data service, MSTP （Multi Service Transport Platform） appends the data processing function to SDH equipment. It integrates the SDH multiplexer, DXC, network switch and PE pouter into one network equipment named MSTP to be jointly controlled. So the MSTP can support both the existent TDM service and the ever-growing IP data service, its technology for building up and protection follows SDH.

MSTP is now in the third generation, its key technology includes 1. Virtual Concatenation: it avoids the bandwidth wasting through providing the more flexible organization of channel capacity to do well meeting the increasing needs of data service.2. Generic Framing Procedure: it is a simple and effective encapsulation method that organizes the data service in denomination of frame. It can transport both fixed-length block data and variable-length data packet in synchronous link.3. Link Capacity Adjustment Scheme: it is control mechanisms to increase and reduce the circuit capacity between the source/sink adapt ion function of virtual concatenation. It can clearly develop the network utilization starting from the premise of guaranteed service quality and in the environment of the burst data service.4. Intelligent Adapter Layer: it is lead in between Ethernet and SDH/SONET. It adapts SDH and Ethernet through adopting LAPS protocol and satisfies the QOS of Ethernet service. So we can say that the MSTP system provides a suitable solution to the construction of rural electric communication network.

5. CONCLUSIONS

The focal point to be considered by the rural electric department is to plan and construct the network with a technical structure that fits the current services.

The current developing trend of the rural electric communication network is to support the IP and Ethernet starting from the premise that the TDM service is transported reliably.

For the rural power supply company who has a mass of SDH network, MSTP can support the data traffic effectively, so it protects the huge investment of intrinsic SDH equipments and guarantees the sustainable development of the rural power grids information network. The MSTP of third generation has lead in many access functions in connection with the different applications. It can optimize the network at low cost through the comprehensive application of multiple functions. The high reliability is preserved at the same time.

In addition, the design practice of ring network should be preferred to give full play to the reliability of it. As much substation as possible should be switched in the ring through breaking the administrative division limit.

The OPGW and ADSS patterns of optical-fiber cable lying should be given priority to because there are the ready resources of electric tower. The communication and information network for rural power grid should form a backbone network for electric information business expansion that adapts the organizational structure and managing mode of electrical grid to provide high-speed and reliable platform for the transmission and switching system in rural electric information network.

REFERENCES

GAO Xin-zhong, The Evolution of Shanxi Electric Power Optical Fiber Communication Network, Shanxi Electric Power 2007.1

Greenville Armitage,MPLS:The Magic Behind the Myths, *IEEE Communications Magazine* 2000,P 124-131

Harry L Bosco ,Evolution of the Wide Area Network , *Bell Laboratories Technical Journal* 2000 , P 46-72

Huston G, Next Steps for the IP QoS Architecture *http://draft -iab- qos- 00. Txt* 2000
Morgan Stanley Research ,The Metro Optical Report: The Return of the RBOCs.
Richard Comerford,Technology 2000 Analysis and Forecast Computing *IEEE Spectrum* 2000 P 4S-49

RESEARCH AND DESIGN OF RURAL INFORMATION SERVICE PLATFORM BASED ON SOA

Jiang Zou, Lu Yang*, Ruizhi Sun

College of Information and Electrical Engineering, China Agricultural University, No. 17, Qinghua Dong Lu, Haidian, Beijing 100083, China

* *Corresponding author, Address: P. O. Box 142, College of Information and Electrical Engineering, China Agricultural University No. 17, Qinghua Dong Lu, Haidian, Beijing 100083, China, Tel: 13311066098, Email: yanglumail@263.net*

Abstract: The paper mentioned using SOA framework and the technology of Web Service, aiming at the problem of "Information Isolated Island", specifically for rural information resources, an agricultural information service platform is constructed based on SOA and Web Service. This platform integrates distributed systems and service platforms and it deals with agricultural information by cooperative processing.

Keywords: web service, SOA, information integration

1. INTRODUCTION

With the development of the global information, there are more and more information resources in Internet. And a lot of information resource systems are developed on the basis of different platform and language, leading to different systems on the heterogeneous platform. With the changing of content, it is difficult to update system once deployed. With more and more duplicated resources on the Internet, these information resources are relatively independent and are not fully utilized, then appears "information isolated islands".

In rural areas in China, problems mentioned above also exist. What is more, agricultural information resources systems are developed by

Please use the following format when citing this chapter:

Zou, J., Yang, L. and Sun, R., 2009, in IFIP International Federation for Information Processing, Volume 295, *Computer and Computing Technologies in Agriculture II, Volume 3*, eds. D. Li, Z. Chunjiang, (Boston: Springer), pp. 1863–1869.

governments at all levels and agricultural institutions and agriculture-related companies. They developed different scale systems, differences in the ability to provide services and the utilization rate of agricultural information resources is not ideal and the sharing of information resources is not enough. To solve above problems, this paper uses Web Service technical and SOA framework for agricultural information system transformation and integration, establishes a unified service platform in rural areas to provide information service. The platform cooperate distributed and different server platforms about agricultural information resources and realizes seamless connectivity on cross-boundary and cross-platform expert system and utilizes information resources more efficiently.

2. SOA AND WEB SERVICE USED TO SOLVE THE PROBLEM OF "INFORMATION ISOLATED ISLANDS"

SOA is the abbreviation of service-oriented architecture. The essence of SOA framework is that all are around business objects or business models, and the "service" is only the form of business models because the uniform form will provide convenient interaction between different suppliers and customers (YUE Kun et al., 2004). SOA is not Web Service, but the Web Service technology applied in the SOA architecture is a practical application of the framework and a very important technology, also the most important and most commonly technology. Web Service can run from all kinds of simple requests to complex business dealings. Once deployed, the other Web Service applications can be found and call the services of other deployments.

Faced with the increasingly serious "information isolated islands", the utilization rate of agricultural information resources is not ideal and many research institutions and universities have made a number of study jobs. Using SOA framework and Web Service technology to solve the problem of "information isolated islands" and realize systems integration such as power, banking, military, hospitals, telecommunications and manufacturing industries.

So far, for solving these problems, SOA framework has been mainly used to provide the whole integration scheme (WANG Yong, 2008; CHEN Si et al., 2007), such as state-owned banks. It focuses on the core business of data management, supporting for banking business innovation and service quality and the improvement of management capacity(XING De-Hai et al., 2007);for the telecommunication industry(CAI Ping et al., 2007) an event-driven enterprise application integration of the telecommunications system structure based on the SOA has been mentioned, and in medical, establishment of medical data based on XML technology has agreed to share the virtual database so that the medical information system to carry out unified information interactive hospital information system integration.

At present in agricultural areas is also carried out some work on the “Information Isolated Island” such as research, has made some progress. This paper proposes a project based on the SOA (YE Jun et al., 2007) and Web Service features and in order to resolve the current problem of "information isolated island"(WU Zhong, 2005; TONG Heng-jian et al., 2007), by using Web Service technology for cross-boundary and cross-platform system of information resources to integrate the information service platform in rural areas.

3. SERVICE PLATFORM ARCHITECTURE

3.1 Platform Overview

The information service platform is based on existing information systems and expert systems of agricultural areas, and other agricultural information resources. To solve the problem that the use of information systems is not ideal, and even the problem of "information isolated island", it provides a unified access to agricultural resources unified platform that provides information integration, coordination, automatically updated data and information evaluation.

3.2 Platform architecture and work principle

The platform contains resources layer, packing layer, information layer, data management layer and performance layer. Platform architecture is shown in Fig.1.

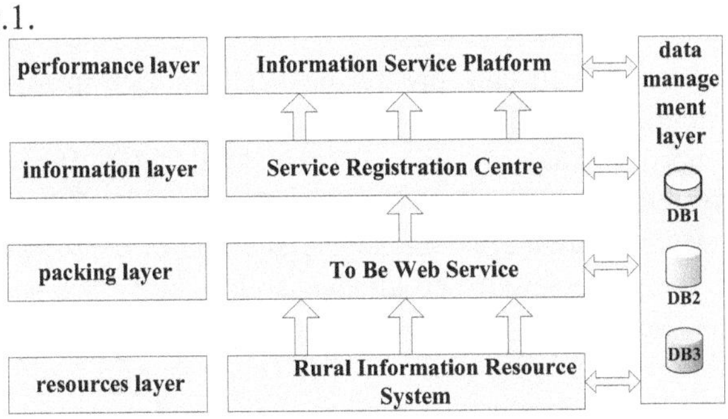

Fig.1: The architecture of service platform

Resources layer of the bottom of the platform includes agricultural information systems and agricultural expert systems resources which implemented by all kinds of technology, which provide relevant information and business, is achieved by the Web Service features and is the main source, is also the platform for all business functions of the sources, as the important support for the packing resources of the system， through the relevant standards on information and Web Service; after all of the processes are packed into the Web Service, all of the Web Service information are registered in the information layer, the Web Service in information layer are used by the performance layer, displayed on the platform, at the same time, relevant information saved in the database. Data management layer services into the data management to better support the operation of the system.

4. INFORMATION RESOURCES INTEGRATION TECHNOLOGY

Resources layer includes many systems which are developed based on different languages and different platforms, how to transform these heterogeneous remote system resources to a certain standard manner and publish, become the key of platform, so the key problem of this platform is how to pack the resources into services in packing layer. Solving this problem, it will provide great convenience for integrating system in resources layer.

Through analyzing the resources of the various information and expert systems, most of these are using the ASP, ASP.NET and JSP, such as technology development, the clustering of these systems, classify the resources into resources which can directly be Web Service and resources which can not directly be Web Service. Resources which can directly be Web Service include MVC model based on the J2EE platform, ASP.NET and other systems belong to resources which can not directly be Web Service. In recent years most of new developed systems are based on the MVC model.

As for resources which can directly be Web Service in the development project, information platforms in the rural areas called development of packing and norms. The system uses axis to develop Web Service. For information systems in J2EE platform, in addition to the JSP pages from the whole packing works as a JAR packages, such as new project fise, packed into fise.jar. Fise project will use all the jar kits and fise.jar tomcat at the same time copy of the directory \webapps\axis\WEB-INF\lib, in the tomcat directory \webapps\axis\WEB-INF, the file server-config.wsdd needs to join in the document issued by the information shown as follows.

<service name="NewsManager" provider="java:RPC">

```
    <parameter name="allowedMethods" value="*"/>
    <parameter name="className" value="com.fise.bl.NewsManager"/>
    <beanMapping languageSpecificType="java:com.fise.News"
    qname="ns:News" xmlns:ns="urn:BeanService"/>
  </service>
```

That is, resources which can directly be Web Service can be published into Web Service. For system developed(LIU Ying-Dan et al., 2003)by ASP.NET, create a new Web Service in the new web applications, and in Web Service in writing to publish into the interface function, as long as in the function added [WebMethod] that this can published function into a Web Service.

As for resources which can not directly be Web Service, because all the web of information resources are based http protocol to communicate and to html files as its content structure, the platform communicate with resources which can not directly be Web Service using the way of simulating http messaging and html documents, through the process of compiling procedure and publishing into Web Service. Specific process is shown in Fig.2.

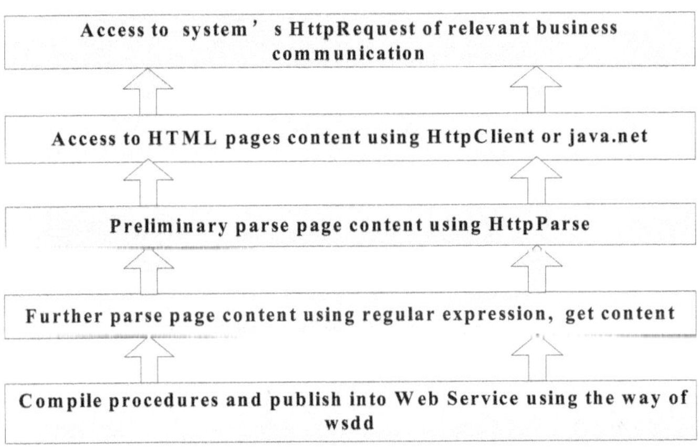

Fig.2: Packing resources which can not directly be Web Service

At this point, we have most contents of the information system packed into Web Service, analyzing the publish way of Web Service, we abandon the way of publication to jws, this way must know the source code of business; adopt a more flexible, dynamic, only need the class document after compiling the source code, which can be published by the wsdd way to publish into Web Service. All of the Web Services are published, registered in register center, using of JUDDI register center, called by information platform.

5. INFORMATION SERVICE PLATFORM FOR PRACTICAL APPLICATIONS

Research in rural information service platform on the basis of key technologies, selects rice planting, cultivation industry in cows, pigs and fish several other business themes, while the selected topic group in the two new agricultural expert systems, seven famous agricultural information systems and expert systems as the resources of platform, according to the research of key technologies referred to in the packing technology, the resources of the packing layer pack and publish into Web Service, which are called by the performance layer, the initial realization of a rural information service platform which can provide business information to the user's query, picture presentation, diseases and insect pests of judging and processing and administration of veterinary drugs. As shown in Fig.3 for rice cultivation of the page.

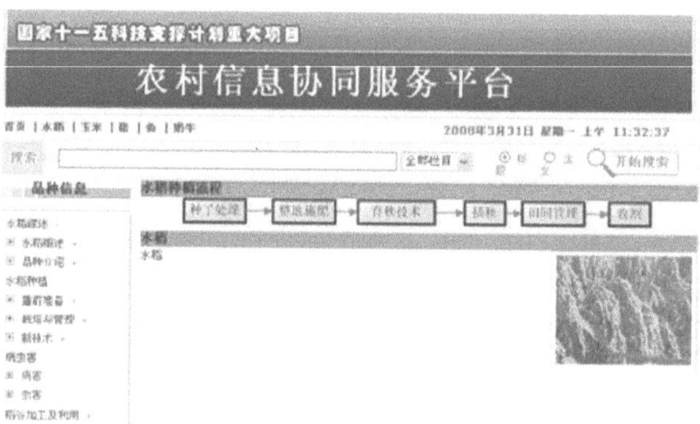

Fig.3: Rice cultivation pages

6. CONCLUSION

Based on SOA and Web Service study, in this paper, we propose a thought which integrates heterogeneous systems and resolving the problem of "information isolated islands" in rural areas, and the initial realization of a rural information service platform to verify the feasibility of this idea. Because this platform uses the SOA framework and Web Service technology, the platform inherited a lot of Web Service application features, such as good encapsulation, loosely coupling, scalability, cross-platform and cross-language ability and fault isolation and so on. To a certain extent, the agricultural information platform solves the problem of "information isolated islands" of agricultural information system and the low utilization rate of agricultural information resources.

ACKNOWLEDGEMENTS

The research presented in this paper was partly carried out in the Institute of Geographic Sciences and Natural Resources Research, CAS and Supported by National Key Technology R&D Program (2006BAD10A0504). The authors are grateful to the experts of Institute of Geographic Sciences and Natural Resources Research, CAS for useful comments during the review process.

REFERENCES

CAI Ping, HUA Qing-yi. Research on China Telecom EAI Based on SOA/ EDA, Computer Technology and development, 2007,17(11):164-170(in Chinese)

CHEN Si, XIA Yang. Research and design of virtual community platform based on web service, 2007,28(10):2446-2449(in Chinese)

LIU Ying-Dan, DONG Chuan-Liang. Implement EAI with Web Service, Computer Applications, 2003,23(7):124-126 (in Chinese)

TONG Heng-jian, SHAO Zhen-feng, PENG Xiu-mei, CUI Le-le. Key technologies of information share and integration based on web service, Computer Engineering and Design,2007,28(22):5400-5401(in Chinese)

WANG Yong. Research and implementation of Web service design time system, Computer Engineering and Applications,2008,44(1):87-90(in Chinese)

WU Zhong. Design of EAI platform in B2B enterprise based on J2EE and web service, Computer Engineering and Design, 2005,26(6):1438-1440(in Chinese)

XIE Gang, ZHANG Wei-Qun. A Framework for Migrating C/S Architecture Legacy Systems to the SOA, Computer Science, 2007,34(11):301-303(in Chinese)

XING De-Hai, QI Er-Shi, DONG Xu-Yuan. Analysis and Design of National Bank Data Centralization Management System Based on SOA , JOURNAL OF BEIJING INSTITUTE OF TECHNOLOGY (SOCIAL SCIENCES EDITION), 2007,9(5):86-89(in Chinese)

YE Jun, LI Zhi-shu, YIN Feng, LI Qin. Research on Enterprise Information System Integration Architecture Using Web Services, Application Research of Computers, 2007,24(6):295-298(in Chinese)

YUE Kun, WANG Xiao-Ling, ZHOU Ao-Ying. Underlying techniques for Web services: A survey, Journal of Software, 2004,15(03):428-442(in Chinese)

STUDY ON RURAL INFORMATION AND FARMERS' INFORMATION QUALITY TRAINING MODELS -- SHIJIAZHUANG CITY AS AN EXAMPLE

Cuiling Wang[1], Yanxia Zheng[2], Xuyang Zhao[3,*]

[1] *Modern Education Technology Center, Shijiazhuang Vocational Technical Institute, Shijiazhuang, Hebei Province, P. R. China 050081*

[2] *Department of Resource & Environment, Shijiazhuang University, Shijiazhuang, Hebei Province, P. R. China 050035*

[3] *Department of Resource & Environment, Shijiazhuang University, Shijiazhuang, Hebei Province, P. R. China 050035*

[*] *Corresponding author, Address: Department of Resource & Environment, Shijiazhuang University, Shijiazhuang 050035, Hebei Province P. R. China, Tel: +86-0311-85960560, 13931981959,Email: log2008@163.com*

Abstract: The information in rural areas of Shijiazhuang survey results showed that there were greater differences between urban and rural areas in the information equipment ownership and consumption of information, the way of farmers understanding information was rather narrow and concentrated. The information gap existed between urban and rural areas. This paper raised the model of cultivating the information-based.

Keywords: Shijiazhuang, rural information, information quality, training model

1. INTRODUCTION

Farmers are the main in building a new socialist countryside, so the information in rural areas is also important to achieve the modernization in rural. In China's rural, information construction is not only information structure construction, but also is to have a skilled user of information and information creators. In order to make farmers adapt to the modern

---

Please use the following format when citing this chapter:

Wang, C., Zheng, Y. and Zhao, X., 2009, in IFIP International Federation for Information Processing, Volume 295, *Computer and Computing Technologies in Agriculture II, Volume 3*, eds. D. Li, Z. Chunjiang, (Boston: Springer), pp. 1871–1879.

information society environment, and cultivate and enhance their information quality level. Therefore, accelerating rural information infrastructure construction, and training farmer's information quality become the top priority of in rural information building.

In recent years, China's rural information construction is quick. In 2005, national agricultural sites were 15, 964(Tan Liang et al, 2006), these sites played a significant role in spreading agricultural policy, technological achievements, and technological information. In addition, China had more than 3,000 species agricultural journals and hundreds of agriculture-related newspapers and professional newspapers, and agricultural programs and columns in radio and television (Yi Ming, 2007). However, Farmer's ability to use information and create information is not attention, so it is necessary to cultivate a higher quality of information users.

In this paper, Shijiazhuang is the background to survey the information situation in rural area. The city's information technology development in rural areas is a better level in the province, but with the modern information society, a far cry from the distance between urban and rural areas, mainly reflects in the information technology infrastructure, and relatively low quality of information farmers. These problems are not conducive to modern agricultural science and technology achievements and the rapid promotion and popularization. With regards to these, the paper raises the cultivating information model about farmers' quality.

2. THE BASIC CONDITIONS ABOUT THE RURAL INFORMATION

2.1 Socio-economic situation

Shijiazhuang city is the center of province's political, economic, technological, financial, and information, as a capital of Hebei Province. There are six districts, twelve counties and five county-level city and a state-level high-tech development zones with a total area of 15,800 square kilometers in Shijiazhuang city. In 2004, the city's resident population is 9.175 million, and which of 5.637 million rural populations, accounting for 61.14 percent. The city's GDP is 16.33 million yuan, and rural per capita net income is 3,799 yuan, per capita consumption expenditure is 2,059 yuan, is the country's major agricultural base.

2.2 Basic conditions of information

Family-owned information equipments are the material basis to get information and dissemination information. This paper investigated the Shijiazhuang family of information equipment at 2004. Fig.1 showed the actual situation. From the survey of 16 counties and county-level city in the 6632 families, until July 2006 with TV subscribers accounted for 106.42 percent, with 80.23%of telephone users, and have accounted for 3.16 % of computer users, Internet users accounted for only 2.66 %. The results showed that every 100 families between urban and rural areas of information equipment conditions existed significant differences. Survey showed that in rural television and fixed telephone had been widespread, mobile phones were still differences between city and rural. The largest differences were computer users and Internet penetration, which proofed the information gap between urban and rural areas.

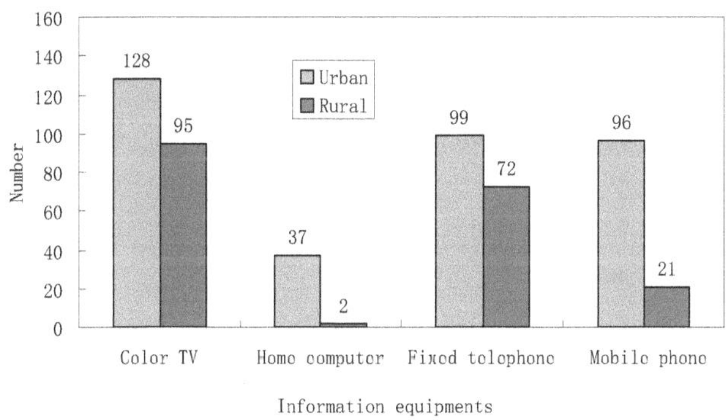

Fig 1 Family-owned information equipments every 100 households between urban and rural

2.3 Differences between information consuming

In 2004, differences between information consuming were 416 yuan in the city and 74 yuan in rural areas. From the survey results showed that the major using of rural fixed telephone and mobile phone were families calling and phone calling. Differences about information equipment also exist at regional space. This difference showed that the exurbs information equipment ownership and consumption level of information were lower and lower from the urban areas to suburbs. For example, the internet users at he suburb of Changan Qu West-Zhaotong village accounted for 7.3%, 6.1% of the computer, 99.6% fixed telephone, mobile phone 103.8 %, 100% TV. But the Internet users and computer users at the suburb of Pingshan county Second-Qingtan village accounted for 0%, 25 percent of fixed telephone, mobile phone 10%, TV 75%.

2.4 The way of farmer's access to information

For farmers access to information, not only by restricting information infrastructure conditions, the impact of another traditional concepts. Today's society, networks, television and newspaper are the primary means of people access to information. According to China Internet Network' survey, people chooses the means of access to information for the respective ratios of 82.6% network, television 64.5%, 57.9% newspapers, magazines, books and radio were the way people access to information, the proportion of choice in 10% to 20%. Therefore, network is the most important means, followed by the popularization of television, then the paper media, and finally the radio. In the paper, the results of the survey showed that TV, go and acquaintances are major market channels to understand information of food, vegetables, fruits, and eggs in their daily life.

2.4.1 The primary means access to career information

Investigation showed that in rural television and radio were the main channels access to career information, followed by newspapers, magazines and introduction from friends and relatives, the Internet accounted for 8 percent.

Table1. The primary means access to career information (%)

Internet	TV and radio	Newspapers and magazines	Introduction from friends and relatives	Go to market
8	38	19	22	13

2.4.2 The primary means of understanding Day-to-day community

Survey results showed that the daily television was the most important channel for the community, accounting for 61.47%, followed by the Go, and accounting for 29.36%. These phenomenon described going to market in the rural areas were not only trading places, but also were important place for the understanding and dissemination information. Newspapers and radio were also a message of the media.

Table2. The primary means understanding Day-to-day community (%)

News Network	Agricultural science and technology and wealth	Agricultural Information	Entertainment	Weather Forecast
30	12	9	24	25

2.4.3 The primary way of understanding food (or vegetables, fruits, eggs, etc.)

According to the survey, TV, go and acquaintances from friends and relatives were the major market channels to understand food, vegetables, fruits and eggs in the daily life. Television accounted for 29%, go for 26%, acquaintances for 21%, and intermediary only for 2%. From that point of view, the way accesses to information of people are blocking and original. The dissemination information of intermediary, newspaper and broadcast may be a delay and distortion. So people are doubt about the authenticity and credibility of information.

Table3 The primary way of understanding agricultural products (%)

Media	TV	Newspapers	Radio	Network	Intermediary	Go to market	Acquaintance	Other
Understanding the market	29	9	7	5	2	26	21	1
Understanding the community	40	12	11	6	\	19	\	12

2.4.4 The main content of watching television

In this survey, the main content of watching TV is news about 30 %, that people concern to the community is improving, weather forecast of 25%, that people are realizing that the weather forecast on the importance of daily life and production. Agricultural information and agricultural technology accounted for 9% and 12%, that agricultural information content is less. The main reasons are not rich, not comprehensive, practical difference, and not satisfy the needs of people.

2.5 Internet situation between urban and rural residents

With the development of the Internet, the networks have become a good helper of lots of people at work and live, but its popularization and application mainly occurred in the city. According to the survey data China's Internet in June 2006 showed that Internet penetration in urban were six times than the rural town, towns of the Internet penetration rate was 18%, 3% in rural areas. The next period of time, the gap between urban and rural may continue to increase.

During the survey none of the online accounted for 54%, with little access to the Internet accounted for 27%, and the regular access to the Internet accounted for only 19%, Internet users accounted for only 2.66%. This descript that the Internet is a blind spot in the rural areas. So there are difficult to eliminate the information gap between urban and rural areas.

3. TRAINING MODEL ON FARMERS INFORMATION QUALITY

3.1 The composition of the information quality

Information quality is the sum of basic quality with information on the identification, analysis, processing, handling, use, innovation, knowledge management, capacity and awareness. Information quality is different types about different learners, information standards and quality of the content. Information quality of new farmers includes the following content:

(1) Information awareness level. Information awareness is the premise and basis of gathering, processing, analysis, synthesis, the use of information. Mainly to realize the importance of the socio-economic information, the relatively strong desire to understand sources of information and information tools.

(2) Information moral level. That is knowledge and understanding of information relating to the economic, legal and social issues, fully aware of information has become a modern morality to comply with the basic norms.

(3) Information knowledge level. That is able to master the basic knowledge and basic skills if information technology, using information technology tools to expand information and communicate information channels. When encountering problems can rely on information to judgment, analysis, use and decision-making.

(4) Information capacity level. Refers to people in their daily life and social practice, with access to information, access to information, the development of information and the ability of self-cultivation.

3.2 Training model on Farmers information quality

At present the masses of peasants are the most eager to the information including three areas: the first is information with own interests related to agricultural and sideline products and labor market. The second is technological information on agricultural science, including the raising, planting, pest control and other disasters practical information. The third is education training about improving the cultural quality and information capability.

New farmers cultivating information quality should focus on training information awareness, information ethics, and information knowledge and information capacity development through the model of education - training - cultivating - nurture (Figure 2). Information education is mainly embodied in the knowledge of farmers to grasp information. Training of farmers is mainly embodied in the comprehensive use of the information capacity, such as the comprehensive analysis and judgments. Cultivating mainly reflected in the information ethics education so that they know and Information relating to the ethical, legal and socio-economic problems. Nurturing is a long-term task, through the process of guidance and application change concepts and increasing capacity and quality of information.

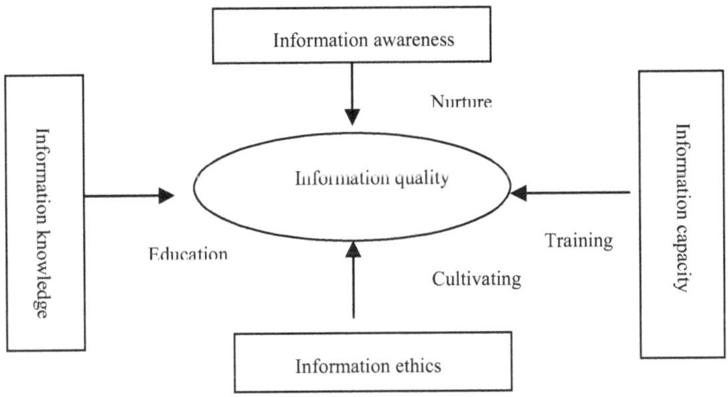

Fig.2. The quality of information and model building

3.2.1 Integration of information quality training and rural vocational skills training

Firstly, teaching concepts and teaching purpose is integrated. Vocational and technical institutions in training the information quality should play to their advantages. Skills training are basic education combined with knowledge and skills. Farmers cultivating information quality are not only to comply with adult characteristics, but also to meet the characteristics of farmers. Training form should diverse because work, family, society, life impacting on farmers learning.

Secondly, the learning patterns are changed. The traditional mode of learning is the main textbook learning resources, students as passive recipients of information. The purpose of the training of farmers is not just to teach them the specific information knowledge and skills, but also is more important to train them information quality with an independent search, processing, processing, and evaluation, with the ability to solve practical problems (Zhang Housheng et al., 2007). "Based on the resource-based" learning is a kind of autonomy and flexibility of learning, such a learning mode in line with the characteristics of the farmers. They can accord their study time, interest and the actual capacity for flexible adjustment. Through the learning resources and materials finding, analysis, judgment and application of information to enhance their cultural skills and learn how to evaluate information and learn to work with others, exchange, the ultimate quality of the ability to access to information.

3.2.2 Training model of information quality through professional associations and Cooperation Organization

In order to meet the agricultural market, the development of the professional associations, cooperative organizations, enterprises and the rural economy and intermediary service organizations and other new organizations, is extremely effective in achieving a market-oriented, with farmers as the basis, leading enterprises as the core, and therefore the dissemination of information on agriculture are more positive, direct and efficient appeal and influence. Each farmer accords its own advantages and conditions to join a professional association or cooperative organizations. In practice, they can place the study in the complex, and vital interests of their closely related to the environment. So they will solve problems in real life and social production by the form of joint cooperation. In the process of solving problems, students receive information quality and ability to cultivate a spirit of cooperation and conducive to cultivating awareness and information pre-empt the sense of lifelong learning. This new mode of education make farmers learners more profound understanding and grasp of information technology than the traditional mode, and building their own quality of the information system and information technology in the dissemination and effective solution to problem that the network technology and television broadcasting not solution.

3.2.3 Accelerating the building of the rural information

First is to establish an information center or information site at each village. Secondly, building information center and the use of information should foster a large number of information talents. Through equipped with computers, touch screen, big screen, and other agricultural services, "autotoll " network works to achieve scientific and technological households, and county, township and village connected to horizontal, vertical linked network of new agricultural technology service system, so that the majority of farmers easier to understand agricultural production and market information. Through the "autotoll " network project is not only able to rapidly increase their use of the capacity and quality of information, more importantly, this is a further extension of the rural economy industry chain and promote peasants to increase income, agricultural efficiency and effective measures. Further promoting rural broadband, telephone, SMS and the universal application of information.

4. CONCLUSION

（1）Shijiazhuang is capital of Hebei Province, the rapidly economic development, the level of information in rural areas have greatly improved in recent years, the owner of mobile phones, fixed phones and TV are relatively high. The wishes of the resident's access to information are relatively strong.

（2）The county economies of Shijiazhuang are less development, mainly are the agricultural economy. The larger differences still exist between urban and rural areas. Especially computer less and the Internet empty become one of the bottlenecks of restricting the further development of the rural areas of information. TV, Go and acquaintances become the way of farmer's access to information. This shows that the means of farmer's access to information is original and concentrated, not satisfying people's needs.

（3）The Government focus on hardware construction on information construction in rural areas, village-development projects faster, but the information provided, publishing, services, management, and other areas are still relatively weak. People's comprehensive, judgments about information should to be strengthened.

According to above findings, this paper raised the model of cultivating the information-based to improve the quality of farmers.

REFERENCES

China'sInternetdevelopmentreport [EB/OL]. [2006-6]

He grams of. Modern educational technology and innovation personnel training. [EB / OL]. Http://www.edu.cn/articledigest2/xian-daihtm. (In Chinese)

Http://www.cnnic.net.cn/images/2006/download/2006011701.pdf (In Chinese)

Shijiazhuang Statistical Yearbook [M]. Beijing: China Statistics Press, 2005. (In Chinese)

Tan Liang. Wang Cheng. Agricultural site evaluation studies [J]. Anhui Agricultural Science, 2007,35 (6). (In Chinese)

Yiming. Agricultural Information: farmers can use is fundamental, 2007. Http://www.sdngx.com. (In Chinese)

Zhang Housheng. Yuanxi Council. Information Literacy [M]. Nanjing: Southeast University Press, 2007. (In Chinese)

STUDY THE EFFECT OF URBAN ECOSYSTEM TO FLOATING POPULATION

Wubin Tu[1,2], Lingxian Zhang[2], Xiaoshuan Zhang[2], Zetian Fu[2,*]
[1] *College of Engineering China Agricultural university, Beijing P.R China 100083*
[2] *Key laboratory for modern precision agriculture integration Ministry of Education, China Agricultural University, Beijing P.R/China 100083*
[*] *Corresponding author, Address: Key laboratory for modern precision agriculture integration Ministry of Education. Address: Key laboratory for modern precision agricultural integration Ministry of Education, China Agricultural University,17 Tsinghua East Road, Beijing 100083, P.R.China. Tel: +86-10-62737617, Fax: +86-10-62737617, Email: tomy_twb@sina.com*

Abstract: The urban ecosystem is a complex system hat are compounded of environment –economic-social ,the factors of urban ecosystem affect the decision of the floating population to chose city .In this paper ,we according to the urban ecosystem to design the questionnaire ,and study the weights of the factors that influence the decision by AHP ,and find that economy is most important which is 0.6806, the society and ecological environment are 0.2014and0.1180.

Keywords: urban ecosystem, floating population, AHP

1. INTRODUCTION

The economic reform in China that started in 1978 has Created a "floating population" Which is one of the most mobile population in the word (Zhigang LU et al ,2006;Yu Zhu,2007), a huge floating population of rural-urban migrants is transforming the urban force (Michael C.Seeborg et al ,2000).There are over 14,735million floating population according to data of the 2005 China 1% Population Sample Survey, of which the inter-province floating population is 47.79 million. Because China has been experiencing a great transition from planned to market economy, and the transition has been accompanied by the increasing mobility of production

Please use the following format when citing this chapter:

Tu, W., Zhang, L., Zhang, X. and Fu, Z., 2009, in IFIP International Federation for Information Processing, Volume 295, *Computer and Computing Technologies in Agriculture II, Volume 3*, eds. D. Li, Z. Chunjiang, (Boston: Springer), pp. 1881–1888.

factors such as: capital, labor ,technology and information (Chaolin Gu et al, 2006). On one hand so huge floating population meet the demand for labor of the rapid economic development of cities, on the other hand they lower the level of the structure of the working population and they become the importing factors that constraint the upgrading of the industrial structure to some extent , meanwhile as so many population move to cities , it glowed the density of urban population and caused many environmental and social issues (Peter Nannestad,2007; Pia M. Orrenius et al ,2007) There are many elements that pulled so great number population moving to cities in China: higher income in cities, better opportunities ,better education ,better quality of life in cities and others (Zhigang Lu et al ,2006;Sandra Poncet ,2006) besides the characteristics of the population themselves ,such as education level ,sex and age(Tracy Simmons,2007) .

The studies on floating population mainly focused on the management of them, the staying time of floating population(Wang guanzhou, et al 2002;Zhu Haiyan,2001) ,and the influence that the floating population to economic of city(Zhou Chang lin et al ,2007) .There are many cities in China, what's the determinants of the city that attract man to move to ? In this paper ,we according to the urban ecosystem theory and analytic hierarchy process(AHP) to study the characteristics of city that attract men to move to.

2. URBAN ECOSYSTEM

The urban ecosystem is a complex system that are compounded of environment –economic-social, characterized by high level, multiple hierarchy and dynamical structure(Forrester JW. 1967,) environmental - economic - social compounds, cities evolve along the life cycle of growth, maturity and stagnation, and behave according to the inner changing mix of the factors involved under the support and constraint of the social and biophysical environment (Button K. 2002) It is shown in Fig. 1 As the center of the system, human is the main activities of social production and life ,and floating population is an important component of the subject.

3. MODEL OF FLOATING POPULATION CHOSE CITY

3.1 Analytic Hierarchy Process (AHP)

The AHP is an efficient approach to make multi-objective decision that employs pair wise comparison to determine the weights and priorities of a variety of factors ,attributes, elements and alternatives. Because Urban ecosystem is a multiple hierarchy, there are many factors ,so we applied AHP to study the degree which affect the moving to city .

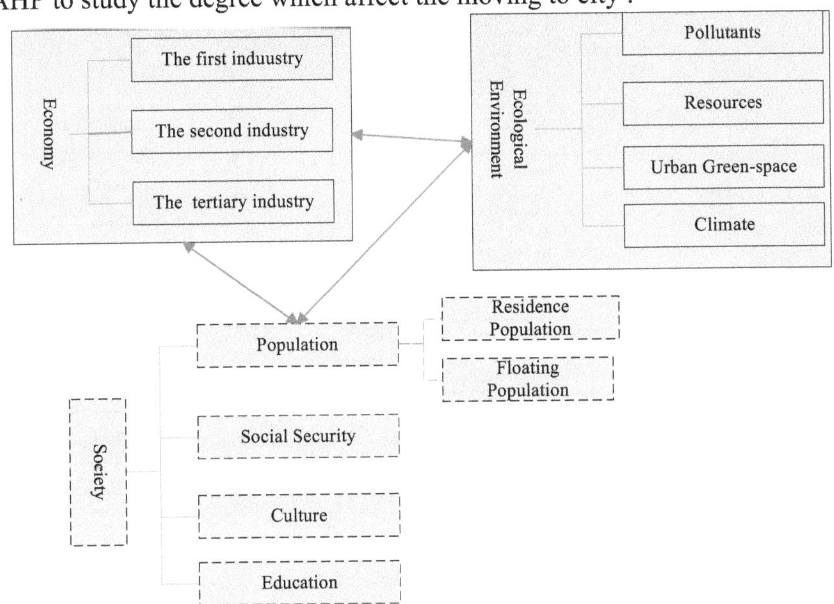

Fig 1. The framework of urban ecosystem

3.2 Hierarchy structural model

To study the determinants of floating population to choose city ,we according urban ecosystem , which is a multiple hierarchy system ,so we adopted AHP to get weights of each factors that influence the decision . The structure of AHP is shown in Fig.2

(1) A—Objective layer: Choose city .First ,the floating population should decide what kind of cities they want to .then we could calculate orderly the power and weight of different factors effect on the general objective in order to know the reason why they chose the city .

(2) C — Criterions layer: This layer includes : society, economy and ecological environment which are the three main compositions . To weigh

and distinguish the estimating standard effecting on the object layer, the criterions layer can be divided into several sub-layers according to the the urban ecosystem theory .

(3) P—Policy alternatives layer: This layer includes 11 factors, they are expressed with expressed with Pi, i=1,2,…,11 respectively. They are the subsystem of society ,economy and ecological environment .

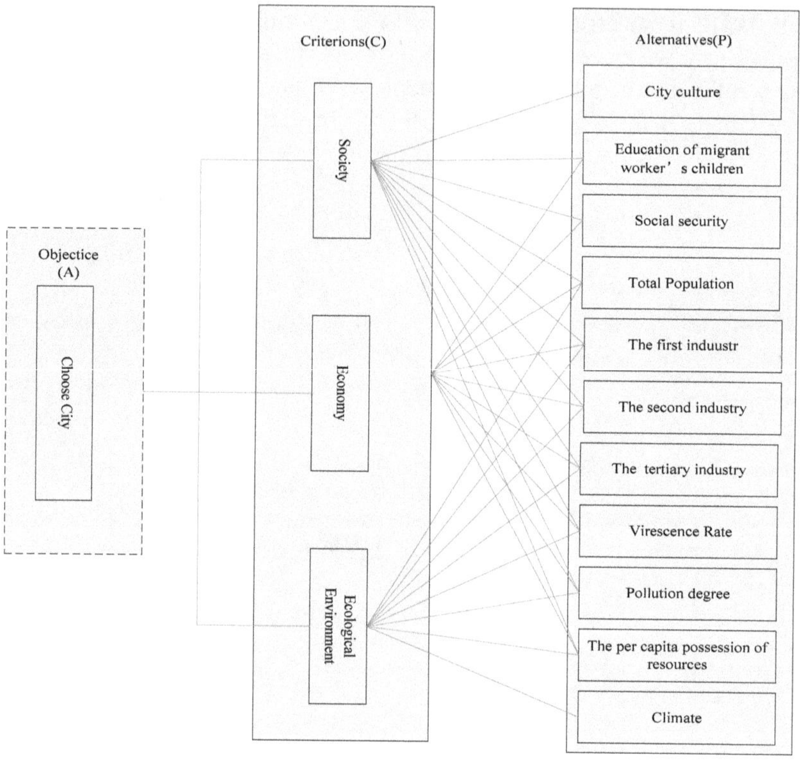

Fig 2. The hierarchy structural model for determinants of floating population chose city

3.3 Assessing policy preferences: analytic hierarchy process

The comparison matrix provides a measure of the weight factor for the decision making, which denotes the relative importance of the element of each layer shown in the hierarchical structural model. Often from top to bottom, the comparison matrix could be established with one elements of above layer pair wise comparing with all the elements of the neighboring below layer according to the relations between the elements or the layers.

Between layer A and C showed in Fig.2, the comparison matrix of A-C layer can be established as Table 1, which elements are evaluated using 9-

point scale shown in Table 2. The pairwise comparison of element i with element j of C layer is placed in the position of c_{ij} of the comparison matrix of A-C layer as shown in Table 1.

The comparison matrix is a square matrix as $A = (c_{ij})_{n\times n}$, and:

$$c_{ij} > 0; \quad c_{ij} = 1/c_{ji}; \quad c_{ii} = c_{jj} = 1 \tag{1}$$

Similarly, the comparison matrix of C-P layers can be obtained in term of the method of forming the comparison matrix of A-C layers. Data on pair-wise comparison matrices were collected from reviewers. Score of c_{ij} were estimated by doing survey in seven provinces in china : Beijing ,Shanghai, Guangzhou , Jiangsu ,Sichuan ,Jiangxi and Hebei then according to the survey ,We choose 300 Valid questionnaires . The participants are all with the local "hukou" ,it includes : farmers and college students from rural areas . we took full account of gender, marital status, academic qualifications and the proportion in the survey .

Table1 The comparison matrix of A-C layer

A	C_1	C_2	C_3	C_4
C_1	C_{11}	C_{12}	C_{13}	C_{14}
C_2	C_{21}	C_{22}	C_{23}	C_{24}
C_3	C_{31}	C_{32}	C_{33}	C_{34}
C_4	C_{41}	C_{42}	C_{43}	C_{44}

4. EVALUATION PROCESS AND RESULTS

4.1 Single ranks of the A-C hierarchical level

We explained the table to the participants then the filled the table according their own judgment about the importance of these factors The results are in Table 3.

Consistency test : λ_{max} =4.0325, CI =(4.0325-4)/3=0.01083, RI=0.89

$$CR=CI/RI=0.01083/0.89=0.01216 <0.1 \text{ (accepted)} \tag{2}$$

Similarly we get the C-P level ranking and *CR* results of which are 0.0583,0.0623 and 0.0303 respectively. They are wholly less than 0.1. It shows that several matrices are very consistent and pass the consistency test (accepted).

Table. 2 The 9-point scale for pair wise comparisons

Score	Description	Comprehension
1	Equal influence	Two elements contribute equally to the objective
3	Moderate influence of one over another	Experience and judgment slightly favours one element over another
5	Essential or strong influence	Experience and judgment strongly favours one element over another
7	Obvious importance	An element is favored very strongly over another, its dominance is demonstrated in practice
9	Absolute importance	The evidence favouring one element over another is affirmed to the highest possible order
2, 4, 6, and 8	Intermediate values between adjacent values	Further subdivision or compromise is needed

*Table 3.*Comparison matrix of A-C layer

A	C_1	C_2	C_3	weight
C_1	1.000	0.2500	2.000	0.2014
C_2	4.000	1.000	5.000	0.6806
C_3	0.500	0.2000	1.000	0.1180

4.2 Overall rank of the hierarchical level

Overall rank of the hierarchical level is the importance rank of the elements of P to the general objective of A. The calculated method is shown as Table 4. The over rank of the hierarchical level is show in Table 5.

Table 4. Calculated method for overall ranks of the hierarchical level

Items	C_1	C_2	C_3	C_4	Overall rank of the hierarchical level
	W_{C_1}	W_{C_2}	W_{C_3}	W_{C_4}	$\sum_{i=1}^{4} W_{C_i} W_{P_j^i}$ $(j=1,2,\ldots, n)$
P_1	$W_{P_1^1}$	$W_{P_1^2}$	$W_{P_1^3}$	$W_{P_1^4}$	$\sum_{i=1}^{4} W_{C_i} W_{P_1^i}$
P_2	$W_{P_2^1}$	$W_{P_2^2}$	$W_{P_2^3}$	$W_{P_2^4}$	$\sum_{i=1}^{4} W_{C_i} W_{P_2^i}$
P_3	$W_{P_3^1}$	$W_{P_3^2}$	$W_{P_3^3}$	$W_{P_3^4}$	$\sum_{i=1}^{4} W_{C_i} W_{P_3^i}$
P_4	$W_{P_4^1}$	$W_{P_4^2}$	$W_{P_4^3}$	$W_{P_4^4}$	$\sum_{i=1}^{4} W_{C_i} W_{P_4^i}$
⋮	⋮	⋮	⋮	⋮	⋮
P_n	$W_{P_n^1}$	$W_{P_n^2}$	$W_{P_n^3}$	$W_{P_n^4}$	$\sum_{i=1}^{4} W_{C_i} W_{P_n^i}$
Σ	1.00	1.00	1.00	1.00	1.00

Notes: In the hierarchy structural model, n represented 11

Table 5. The infection weight of different urban factors that affects the decision of floating population（*CR*=CI/RI）

P	C_1	C_2	C3	Integrated weights
	0.2014	0.6806	0.1180	
P_1	0.0679	0	0	0.0137
P_2	0.1476	0.0913	0	0.0913
P_3	0.1214	0.1192	0	0.1153
P_4	0.0769	0.0806	0.0709	0.0867
P_5	0.0351	0.0462	0.0562	0.0462
P_6	0.2007	0.2057	0.1465	0.1979
P_7	0.2097	0.2237	0.1422	0.2163
P_8	0.0463	0.0758	0.1601	0.0713
P_9	0.0535	0.0964	0.1893	0.0820
P_{10}	0.0409	0.0610	0.1183	0.0664
P_{11}	0	0	0.1159	0.0129
CI	0.095	0.0532	0.0407	0.0818
RI	1.4840	1.4040	1.3410	1.4125

4.3 Analysis

From table3. the weights are: 0.2014, 0.6806, 0.1180 respectively, so in the opinion of floating population that economy is much more important than the others .according Zhigang Lu (2006) rural people migrate to city to seek higher income ,better opportunities ,a better quality of life and a better education for themselves and their children .The most important is higher income and better opportunities in city ,they are all related to economy . The income gap between urban and rural areas is increasing ,it is more than 2.5 now., and there are many Surplus labor in rural areas , so many people in rural areas move into city to earn money, but more of than are in the second and tertiary industry ,so they put them at top . From the table 5, the weight of social security(P3) is 0.1153,is the third important of all，because China's unique household registration (*Hukou*) system，the floating population can not share the same rights and benefits of urban *Hukou* residents, so most of the floating population hope that they can share the security of health, unemployment and old-age insurance Excepted the three ,there are another 8 factors that also effect the decision ,they are culture ,education for children, total population, the first industry, Virescence Rate pollution degree, the per capita possession of resources and climate, and the weights of them are: 0.0137, 0.0913, 0.0867, 0.0462, 0.0713, 0.0820, 00664, 0.0129, 0.0818 respectively.

5. CONCLUSION

The urban ecosystem effects the decision of the floating population's chose of city , the economy is the most important factor they consider , because in city they can get higher income ,but the society and ecological environment also have influence in the decision , the main factors are culture, education for children , social security ,total population ,the first industry, the second industry ,the tertiary , virescence Rate, pollution degree , the per capita possession of resources and climate, and their weights are : 0.0.137, 0.0913, 0.1153,0.0867, 0.0462, 0.1946,0.2163， 0.0713, 0.0820, 00664, 0.0129, 0.0818respectively.

ACKNOWLEDGMENTS

We acknowledge financial support from the National Science & Technology Support Program (2006BAD10A07-02), National 863 Program (2006AA10Z267) from Ministry of Science & Technology of China, and Beijing Municipal Science & Technology Commission (ZZ0521) for financial support.

REFERENCE

Button K. City management and urban environmental indicators. Ecol Econ 2002;40:217 – 33.

Chaolin Gu, Roger C,K.Chan ,Jinyuan Liu ,Christaian Kesteloot .(2006).Beijing's socio-spatial restructuring: Immigration and social transformation in the epoch of national economic reformation ,Progress in Planning 66,249-310.

Forrester JW. Urban dynamics. Cambridge: MIT Press; 1967.

Michael C. Seeborg, Zhenhu Jin, Yiping Zhu. (2000) The new rural-urban labor mobility in China: Causes and implications, The Journal of Socio-Economics, 29, 39-56.

Peter Nanestad .(2007). Immigration and welfare states :A survey of 15 years of research ,European Journal of Political Economy,23,512-532.

Pia M.Orrenius mMadeline Zavodny .(2007). Dose immigration affacr wages? A look at occupation-lebel evidence, Labor Economics， 14,757-773.

Sandra Poncet.(2006) Procincical migration dynamics in China : Borders ,costs and economic motibations ,Regional Science and Urban Economics 36,385-398.

Simmons, T., Sexuality and immigration: UK family reunion policy and the regulationof sexual citizens in the European Union, Political Geography (2007), doi:10.1016/j.polgeo. 2007.10.002.

Wang Guangzhou ,Tong Yu fen ,Lu Jie hua ,(2002) . Staying time of floating population in Beijing and its impact strength,Population &Economic,119,51-54(in Chinese)

Yu Zhu.(2007) .China's floating population and their settlement intention in the cities :Beyond the Hukou reform ,Habit International,31, 65-76.

Zhigang Lu ,Shunfeng Song. (2006). Rural-urban migration and wage determination : The case of Tianjin ,China, China Economic Review ,17,337-345.

Zhu Haiyan .(2001). A study on factors affecting the staying time of in –floating population in beijng,Market and Demographic analysis,1,58-64.(in Chinese)

A STUDY ON THE DOCUMENT INFORMATION SERVICE OF THE NATIONAL AGRICULTURAL LIBRARY FOR AGRICULTURAL SCI-TECH INNOVATION IN CHINA

Qian Xu*, Xianxue Meng

Agricultural Information Institute of Chinese Academy of Agricultural Sciences, Beijing, P. R. China 100081

* *Corresponding author, Address: Agricultural Information Institute of Chinese Academy of Agricultural Sciences, Beijing 100081, P. R. China, Tel: +86-13488769681, Email: xuqian.19840115@163.com*

Abstract: This paper presents the significant function of the Chinese National Agricultural Library (CNAL) in the agricultural sci-tech innovation system in China, analyses the development of collection and service in the CNAL, explores the challenge towards sustain and develop information services for the agricultural sci-tech research and innovation, at last proposes the strategy for sci-tech document information service development.

Key words: document information service, sci-tech innovation , Chinese National Agricultural Library (CNAL)

1. INTRODUCTION

Along with the speed of science and technology increased, sci-tech document information resources have become a necessary and stable basis for the further development of science and technology, and also have become the national strategic resources all over the world. The possessive level of sci-tech document information resources and the efficiency of the information services have become important indicators of a country's

Please use the following format when citing this chapter:

Xu, Q. and Meng, X., 2009, in IFIP International Federation for Information Processing, Volume 295, *Computer and Computing Technologies in Agriculture II, Volume 3*, eds. D. Li, Z. Chunjiang, (Boston: Springer), pp. 1889–1897.

science and technology advancement and the social development level. Sci-tech document information resources construction and utilization have been strongly stressed by all developed countries to support and ensure their sci-tech leading status in the world.

Up to now, there is a large gap between the agricultural technology level and technology innovation capacity in China and the other developed countries. Therefore, China established the Agricultural Sci-tech Innovation System, aiming to promote agricultural knowledge innovation and agricultural technology innovation. Agricultural sci-tech document information resources construction and service have become necessary supporting conditions for agricultural sci-tech innovation, and an essential part of the Agricultural Sci-tech Innovation System.

The Chinese National Agricultural Library (CNAL), initiated in the 1950s by Agricultural Information Institute （AII） of Chinese Academy of Agricultural Sciences（CAAS）to provide document information services for researchers, policy-makers of agricultural research institutes or colleges across the country. Over fifty years, now it has boasted of a unique collection system characterized by document information resources in agriculture and relevant disciplines. The CNAL is the agricultural sciences central library of the National Science and Technology Library (NSTL), which is established in 2000 by the Ministry of Science and Technology, and composed of 7 special libraries from basic sciences, agricultural sciences, medical sciences, and engineering, to form a national reserve library and a supply center for scientific information. Up to now, the CNAL has become the largest agricultural library in Asia, and one of three agricultural libraries in the world.

The main function of the CNAL is to collect, process, storage and delivery Chinese and foreign agricultural sci-tech document information, and to establish an integrated digital scientific information service environment for its users. As the most important agricultural sci-tech document information provision and service system in China, the CNAL plays an important role of supporting national agricultural sci-tech innovation.

Challenges to sustain and develop the CNAL information services lie in three aspects. The first one is the change of science research environment; the second one is the change of operational mechanism of scientific research; the third one is the new needs and information behavior of users in e-science environment.

The objective of this study is to develop the information service strategy for the Chinese National Agricultural Library to improve the quality and capacity of information service for supporting the sci-tech innovation in China.

2. DEVELOPMENT OF DOCUMENT INFORMATION SERVICES

The CNAL integrates the catalog of its physical collection, electronic resources, and special subject research information developed by CNAL or AII into the platform of National Agricultural Information System (NAIS), and provides its information services via this platform. The platform of NAIS is showed in Fig.1.

Fig.1: The platform of National Agricultural Information System

The information service in CNAL can be classified in two types. They are traditional lending service, and network-based information services. As showed in Fig.1, the main functions of information service are listed in the platform of NAIS. The feature of the information services of CNAL can be characterized in two aspects: the resource integration and the service integration.

2.1 The integration of document information resources

The document information resource is the necessary basis for information services. First, document information resources within the library are integrated according to its type via the columns of Catalog Searching, Electronic Resource and Special Subject Research Information in the platform of NAIS. Second, different types of electronic resources are also integrated in the platform, which provides users the convenient services of browsing-online, resource retrieval and documents supply. The integrated retrieval system provides a uniform retrieval platform among different plats databases. In addition, the NAIS realizes the integration of collection within the library and network resources out of the library via the columns of Union

Catalog Searching, Agriculture Website Navigation, Network Resources Recommendation and Link, and Portals for Agricultural Resources, et al.

The information resources integration mentioned above is still an organization of documents and data information at an early stage, and is not able to provide an adequate support for researchers or scientists to filter through mass-scale content to explore unknown knowledge. The real meaning of resource integration is an in-depth integration of document content distributed in different types of resource, constructing the relationship among content points in one database or among different databases.

2.2 The integration of services

In terms of the function of service, the platform of NAIS provides integrated one-stop information services. It developed the interlibrary loan service to delivery required articles to user's e-mail box. Opened digital reference service to enable users to receive librarians' help via the internet and provided training sessions for individual agricultural research institutes within and outside CAAS.

In addition, agriculture resources navigation service introduces and offers important sci-tech websites from all over the world, including that of important agricultural scientific and technology institutions and information institutions, aiming at guiding the users' choosing their useful resources.

What's more, the platform provides the users who register the Green Channel service with authentication and authorization services which enable the users to request NAIS's services no matter where they are.

3. CHALLENGES TO SUSTAIN AND DEVELOP DOCUMENT INFORMATION SERVICES

3.1 Changes of the scientific research environment

Scientific research environment is or will be a ubiquitous digital environment, which can be summarized in two aspects. First, not only the formal scholar information sources, such as journals, books, technical reports, conference papers, theses and dissertations, patents, etc., are almost totally digitally accessible, but an ever increasing volume of informal information published as research results by individuals or institutions is also available online (Kayvan and Mike, 2008). The digital information environment and the quick and easy access it provides to vast amounts of digital information from a wide range of sources have greatly affected the way in which users seek information and make use of the resource. Secondly, it becomes a normal way for most scientists to depend upon internet tools, such like e-mail, web-site, discussion forums, and file transfer, to communicate with each other, conduct information exchange.

In addition, the environment of scientific and scholar information communications is also undergoing a dramatic change. Access to information is no longer inter-mediated by and channeled through a library when forces like Google Scholar/Print, open access materials, and institutional repositories are creating a new information supply chain. Publishers, database producers, commercial search engine producers, citation index service providers and other content brokers are all going into the information service market, not only providing information services more directly and effectively to users by an open or commercial system, but also becoming increasingly active in knowledge-oriented and high added-value services such as personalized services, and reference services. This calls for a new definition of information resources, information organization, and information service integration (Zhang Xiaolin, 2005).

3.2 Science is increasingly operating in a digital fashion

The operational mechanisms of scientific research is even increasingly become digital. E-science or cyber-infrastructure creates a networked infrastructure rich in data grids, computing grids, digital libraries, and collaboration. All kinds of scientific objects, including people, programs, facilities, data and documents, procedures and workflows, and even policies and strategies, can be and will be digitally represented, accessed, interconnected, and invoked through networked interaction. Scientists are able to share observation and experiment resources, share data and information recourses, share computing and analysis capabilities cross time, space, and administrative boundaries (Zhang Xiaolin, 2005). Information becomes the target of research analysis and processing for data mining and knowledge discovery. Networked collaboration, multi-system cooperation, and process management of virtual research organization constitute the basic supporting mechanism for large research undertakings.

3.3 Changes of users' information needs

Due to the ubiquitous digital information environment, the bottleneck of the users' information activities has shifted from document-based information retrieval to relationship-based knowledge discovery. Traditional library service usually covers collecting, processing, retrieval and delivery of information resources, with less analyzing the inter-relationship of the content or the concepts. It can not be an adequate support to scientists' core missions. Effective services are required to help users to mine knowledge and knowledge relations among mass of various scientific objects. The information activities and information services should also diffuse into and interact with the totality of users' problem-solving processes and users' information environment.

4. STRATEGIES FOR DEVELOPING THE DOCUMENT INFORMATION SERVICE

The CNAL will take those challenges as an opportunity to develop its document information services for a better support to agricultural sci-tech innovation in China. In this part, some strategies for improving information services are proposed.

4.1 In-depth integration of the service resources

Central to scientific research, users put high value on the systematization and specialization of information or knowledge. Information resource integration not only requires a systematic collaborative system consisting of printed collections, electronic information databases, and third-party resources, but also expects to re-organize those information objects and construct a systematic knowledge organization for the further creation of knowledge by conducting knowledge mining, which reveals the hidden relationship between concepts. What is more, based on the knowledge-based organization, inter-related information or knowledge resources concerned with one study or a special subject could be put together into a desired pattern and structure. They can be organized into horizontal information networks in different study fields, or linked together to form vertical information chains according to demands of the users.

In addition, facing with the increasingly urgent needs for varied information resource, together with the situation of the decentralized distribution of documents, CNAL should pay more attention to the sic-tech information resource sharing with diversities libraries and agricultural sci-tech document information service institutions in China by establishing an information resources processing and service system on the basis of digitalizing technology and networking technology. This is not only meaningful for the actually better exploit and utilizing the current agricultural sci-tech document information resources, but also brings significant impacts on the communication and collaboration between Chinese and international information network systems.

4.2 Enhancing the information analysis service

Information analysis mentioned goes beyond regular catalog compilation or literature review, but knowledge mining, knowledge discovery and knowledge organization concentrating on problem-solving, which supports strategy planning, discipline trend analysis, sci-tech policy analysis, research evaluation, and deals more with unknown or potential problems, with concept and relationship analysis, and knowledge creation. Information analysis not only supports the current sci-tech research, but also could stimulate new research. Therefore, information analysis enables the library the better service for decision-making and scientific research.

To strengthen the capacity of information analysis, on one hand, CNAL needs to develop technical tools, network-based collaboration platform to identify meaningful, unknown relationships between concepts, enable broader inquiry of the scientific literature (Bernard and Jeffrey, 2006), and depict content points and relationship hiding behind. On the other hand, while knowledge discovery techniques can uncover hidden relationships in the data and documents, it depends on the user's expertise to give those relationships meaning (Bernard and Jeffrey, 2006). Therefore, knowledge discovery techniques should be designed into the scholar process for discipline analysis, opportunity discovery, development evaluation, trend forecast, strategy planning, and should be integrated into user's own information system for the sake of the personalized needs.

4.3 Developing the knowledge service mechanism

The core mission of information service in the current situation is to support users' information needs and resolve their problem directly and effectively. Since the collection-oriented workflows and library-centered service are not adequate, the new user-centered and problem-solving-oriented service mechanism of library is called upon, and services diffusing into and interacting with users' research processes are essential for success in such a situation.

Traditional libraries often center their information services on the "circulation of collections", and put themselves out of the research workflow, leading to a less active enrollment of the librarians to the science environment, whose ability to follow research needs closely and sufficiently is severely limited.

Aiming to develop user-centered digital library services, which diffuse into research environment, collaborative mechanisms should be introduced to libraries for librarians to work together with researchers in terms of knowledge process and to provide further personalized services.

To sum up, user-centered service is the active and interactive personalized service. It is conducted by libraries in accordance with users' needs in their research process, with the aim of helping users to resolve their problem and accomplish the research task. This service patter poses emphasize on forming a partnership with researchers, and operating closely into and together with research or knowledge process.

4.4 Adopting CRM to enhance the study on user's information needs

Customer relationship management (CRM), evolving from business processes, emphasizes not only a comprehensive strategy, but also a process of acquiring, retaining and partnering with selective customers to create superior value for the company and the customer (Parvatiyar and Sheth, 2000).

The concept of CRM aims at facilitating the realization for more effective customer personalization management. Although more and more organizations and companies have "begun to use the Internet to obtain customer information in their database marketing processes to enhance customer relationship management" (O'Leary et al., 2004), few libraries have employed sophisticated CRM information systems. When it comes to digital library, CRM could be interpreted as a method of management on customer information, which is an important resource for libraries. Taking advantage of deep analysis on the customer information, libraries could have a better know about the clients' information needs and behavior. Accordingly, it could improve their information services. Central to the conception of developing customer relationships, libraries put high value on the discovery of potential customers, and also on developing potential users to real ones, so as to keep the expansion of the scope of information service.

Therefore, the CNAL could think about adopting more CRM concept and applications as one of the possible solutions to secure users' satisfaction while facing the challenges brought by proliferated information service channels, information explosion, and users' higher expectations.

5. CONCLUSION

As the most important agricultural sic-tech document information provision and service system in China, the National Agricultural Library plays an increasingly important role in supporting national agricultural sci-tech innovation system. Facing with the limitation of development and the challenges proposed by environment outside, strategies for developing the document information service are proposed in terms of the service resource development, the service content, the service mechanism and the service management.

There are still various factors or aspects needed to be considered and studied to develop a systematic strategy for improving the document information service in CNAL with regarding to serving the sci-tech innovation. But what important is the vision and the approaches for CNAL development. It is the author's believe that the viability and sustainability of CNAL lie in its in-depth integration of document information services and in its supporting for the scientific decision-making. The strategies outlined are the experimental steps toward this goal.

REFERENCES

A. Parvatiyar, J. N. Sheth. The domain and conceptual foundations of relationship marketing, in J. N. Sheth, A. Parvatiyar (Eds), Handbook of Relationship Marketing, Sage Publications, Thousand Oaks, CA, 2000: 3-38

C. O'Leary, S. Rao, C. Perry. Improving customer relationship management through database/internet marketing: a theory-building action research project, European Journal of Marketing, 2004, 38(3/4): 338-354

D. Bernard, D. Jeffrey. Knowledge Discovery in the Digital Library: access tools for mining science, Information Services and Use, 2006, 26(1): 39-44

http://www.ifla.org/IV/ifla71/Programme.htm

K. Kayvan, T. Mike. Sources of Google Scholar Citations outside the Science Citation Index: A comparison between four science disciplines, Scientometrics, 2008, 74(2): 273-294

Zhang Xiaolin. From Digital Library to E-Knowledge Development Strategy, Journal of Library Science in China, 2005, 31(158): 273-294(in Chinese)

Zhang Xiaolin. Sustainable Digital Library Development for Scientific Communities. In The Proceedings of World Library and Information Congress. 71th IFLA General Conference and Council, August 14th - 18th, 2005, Oslo, Norway.

RESEARCH ON ASSESSMENT METHOD FOR RURAL INFORMATIZATION LEVEL BASED ON AHP

Du Jing[1,3], Daoliang Li[2], Hongwen Li[1,*], Yanjun Zhang[2]

[1] *College of Engineering, China Agricultural University, Beijing, P. R. China 100083;*

[2] *College of Information and Electrical Engineering, China Agricultural University, Beijing, P. R. China 100083;*

[3] *China Agricultural University Library, Beijing, P. R.China, 100083*

[*] *Corresponding author, Address: P. O. Box 46, College of Engineering, China Agricultural University, 17 Tsinghua East Road, Beijing, 100083, P. R. China, Tel:+86-10-62737631, Fax:+86-10-62737300, Email: lhwen@cau.edu.cn*

Abstract: Based on rural informatization connotation and five essential elements that affect rural informatization assessment, which are development environment, information infrastructure, information resource, information service system and application of information technology in rural areas, This paper designs an indicator system for rural informatization level assessment. Through AHP method, it sets up the hierarchical construction model of rural informatization assessment and weight of each indicator is calculated. Thus the evaluation method for assessment on rural informatization level is proposed in this paper. It combines subjective evaluation with objective appraisal and will help direct rural informatization management departments with jobs and promotes rural informatization development.

Keywords: Rural informatization; Assessment, Analytic Hierarchy Process(AHP); Indicator system

1. INTRODUCTION

Agriculture and rural informatization is an important component of the national economy informatization. Rural informatization is the inevitable requirment to realize countryside modernization and an important means of

Please use the following format when citing this chapter:

Jing, D., Li, D., Li, H. and Zhang, Y., 2009, in IFIP International Federation for Information Processing, Volume 295, *Computer and Computing Technologies in Agriculture II, Volume 3*, eds. D. Li, Z. Chunjiang, (Boston: Springer), pp. 1899–1907 resolving the problems facing our agriculture, rural areas and farmers (Zang Chunrong et al., 2004). By strengthening rural broadcast television networks, telecommunication and computer networks and other information infrastructure construction, fully developing and using each kind of information resource, constructing comprehensive information platform and service system, promoting the exchange of information and knowledge sharing, Rural informatization will promote the application of advanced information technology in countryside production and management, public services, government affairs management, life expense and so on each aspect(Tan Guoliang et al., 2007).

Rural informatization assessment is to evaluate the level and effect of rural informatization construction by adopting mathematical statistics, operation research principle and specific indicator system (Liu Shihong et al., 2007).To do that, It should compare the unified standard, compute quantitativly and qualitatively and make contrastive analysis in accordance with certain procedures. By evaluating rural informatization level of our country through quantitative analysis method, the rural informatization conditiongs of different area and rural informatization degree of the same area can be revealed quantitatively (Sheng Qifeng et al., 2005). Evaluating rural informatization level comprehensivly and scientificly, measuring and grasping the information development level of each rural area accurately will help to find problems and shortcomings in the process of rural informatization, thus to direct every area to get information technology work forward and provide basis for formulating long-term rural informatization development planning and information-based strategy (Yi Yaxin et al., 2004).

2. INDICATOR SYSTEM OF RURAL INFORMATIZATION ASSESSMENT

Rural informatization assessment is a huge system engineering that synthesizes technological innovation and institutional innovations. Its implementation involves all aspects of rural business and information techonology (Qin Xiangyang et al., 2008). The comprehensive assessment of rural informatization belongs to information system appraise category and involves assessment on many aspects of rural informatization such as exterior development environment, information infrastructure, rural information resource, information organizations and agencies engaged in the countryside information service work and application of information technology in rural areas.

2.1 Selection of indicator

The core content of rural informatization assessment includes five mutually and closely contacted planes: development environment, information infrastructure, information resource, information service system and application of information technology in rural areas. Rural informatization assessment bases on the construction of scientific and rational evaluation indicator system. In order to objectively evaluate the level, indicators that affect rural informatization level should be selected according to results of expert consultation and questionnaire survey. Factor choice should obey the following principles(Liu Shihong et al., 2007; Lu Lina et al., 2007):

1) Comprehension: The factor must reflect each side of the appraised object and study its connotation and extension deeply in the process of construction, making every effort to fullly and actually manifest it.

2) Operability: Weight of each indicator shoule be determined based on the actual development of rural informatization. The evaluation method stresses on availability of indicator data.

3) Guidance： Prospective and tendency factors should be introduced to guide the direction of rural informatization development.

4) Connection: The indicator system should link up with national and regional informatization assessment index system and reflect the development characteristic of rural informatization downward.

5) Comparability: International and domestic general targets should be introduced in. Each target should be fairly comparable corresponding to every appraised object.

2.2 Construction of indicator system

According to the concept of rural informatization and above principles, This paper chooses twenty indicators from the above five planes of rural informatization. Its schematic diagram is shown in Fig.1.

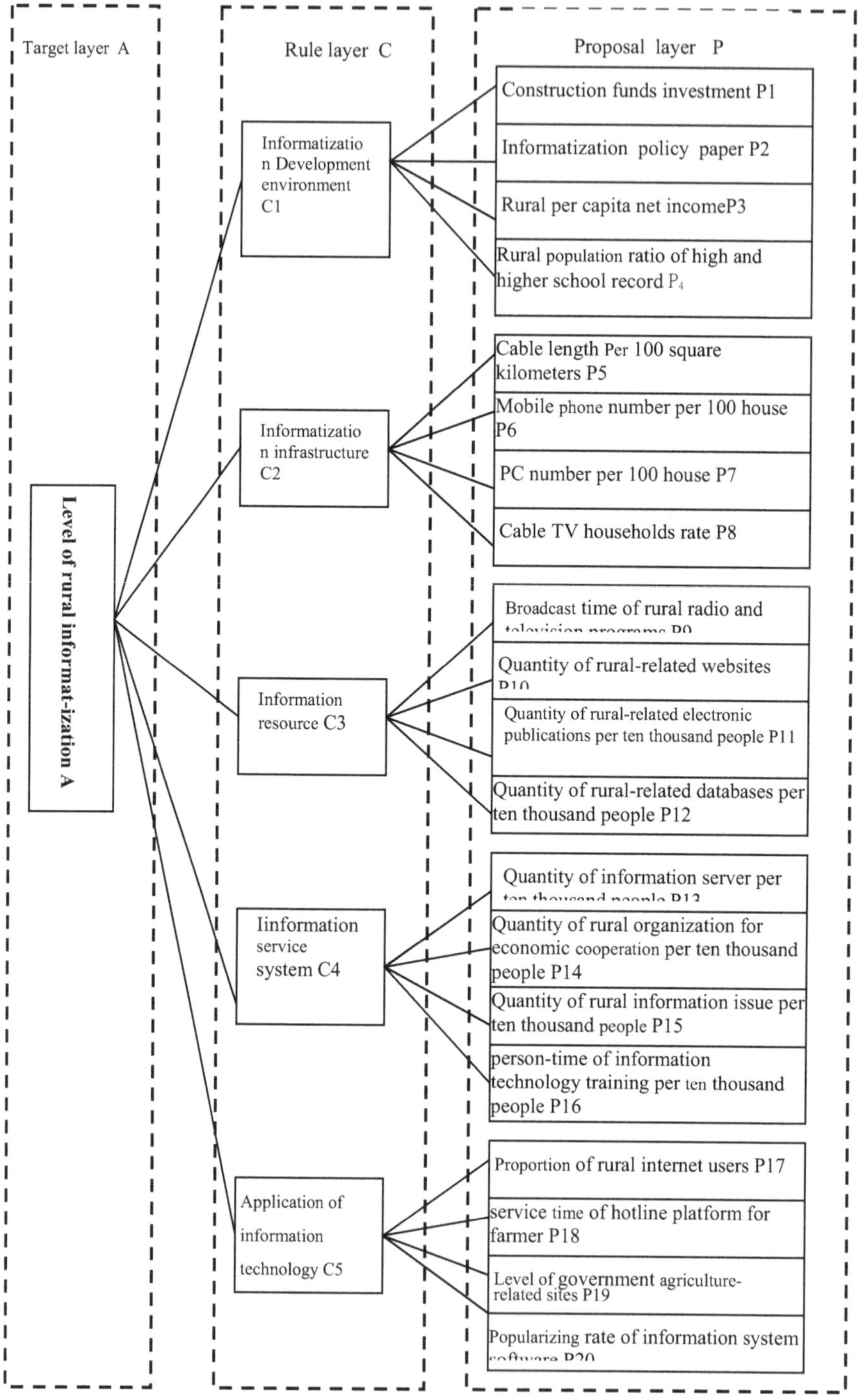

Fig.1: Hierarchical construction of rural informatization assessment

3. MODEL OF RURAL INFORMATIZATION ASSESSMENT

3.1 Form assessment indicator set

Based on the indicator system above, the assessment factor set X could be formed and expressed as $X = \{X_1, X_2, X_3, X_4, X_5\}$,And $X_i = \{X_{i1}, X_{i2}, \cdots, X_{ij}\}$, $(i = 1, 2, 3, 4, 5; j = 1, 2, 3, 4)$.

3.2 Establish stepped hierarchical construction

The graded hierarchical construction model of rural informatization assessment includes three layers, they are: the level of rural informatization development, the main content of rural informatization and the rural informatization assessment factors. Its structure drawing is shown in Fig.1.

3.3 Ascertain the weight of the indices

The weight coefficients of indicators at all levels are fixed by Analytic Hierarchy Process algorithm (AHP) developed by Saaty. (An Shaoshan et al. 2003). AHP algorithm is a multi-objective decision analytical method that combines qualitative with quantitative analysis and also a kind of optimization techniques. AHP algorithm combines qualitative analysis with quantitative analysis organically during programmed decision-making, so it can be used to calculate the weight of each indicator(Cai Kangying et al., 2008; Chen Ruyun et al., 2007)

1) Construct comparison judgment matrix:

By pairwise comparing elements of the same layer in accordance with certain upper factor, each element of the judgment matrix is able to be defined.The relative importance of each element followes 1-9 proportion evaluation scale. Several judgment matrixes can be eatablished based on the scores that experts provided.The judgment matrixes of rural informatization assessement are shown in Table1-6.

Table 1. A-C_i Comparison judgment matrix

A	C_1	C_2	C_3	C_4	C_5
C_1	1.000	0.7639	1.0234	0.6806	0.7556
C_2	1.4263	1.000	1.6226	0.9250	1.0694
C_3	1.1592	0.9528	1.000	0.8222	0.8250
C_4	1.5929	1.1194	1.7782	1.000	1.0167
C_5	1.4925	1.1639	1.5048	1.0333	1.000

Table 2. C_1-P Comparison judgment matrix

C1	P1	P2	P3	P4
P1	1.000	1.819	1.522	1.825
P2	0.625	1.000	0.995	1.119
P3	0.739	1.414	1.000	1.2361
P4	0.626	1.097	0.862	1.000

Table 3. C_2-P Comparison judgment matrix

C2	P5	P6	P7	P8
P5	1.000	0.944	0.814	0.962
P6	1.083	1.000	0.869	1.045
P7	1.264	1.181	1.000	1.191
P8	1.139	1.083	0.907	1.000

Table 4. C_3-P Comparison judgment matrix

C3	P9	P10	P11	P12
P9	1.000	1.097	1.200	1.053
P10	1.002	1.000	1.193	1.058
P11	0.867	0.953	1.000	0.851
P12	1.093	1.213	1.233	1.000

Table 5. C_4-P Comparison judgment matrix

C4	P13	P14	P15	P16
P13	1.000	1.197	1.170	1.035
P14	0.910	1.000	1.014	1.008
P15	0.881	1.083	1.000	0.957
P16	0.970	1.137	1.013	1.000

Table 6. C_5-P Comparison judgment matrix

C5	P17	P18	P19	P20
P17	1.000	1.138	1.268	0.885
P18	0.940	1.000	1.155	0.822
P19	0.810	0.890	1.000	0.696
P20	1.217	1.352	1.486	1.000

2) Monolayer weights order

The importance that each element of No.(k+1) layer relative to No.k layer is orderd according to the judgment matrix. The order is expressed by the relative value such as follows:

$$W_C = \begin{bmatrix} W_{C_1} \\ W_{C_2} \\ W_{C_3} \\ W_{C_4} \\ W_{C_5} \end{bmatrix} \qquad \sum_{i=1}^{5} W_{C_i} = 1 \tag{1}$$

Where: W_{Ci} is the importance value of the ith C layer relative to A layer.

W_{C_i} is calculated by adopting geometric average algorithm.The steps are as follows:

a) Calculate the product of elements of each row:

$$M_{C_i} = \prod_{j=1}^{5} C_{ij} \qquad (i = 1,2,3,4,5) \tag{2}$$

b) Calculate the 5th root of M_{C_i}:

$$\overline{W}_{C_i} = \sqrt[5]{M_{C_i}} \tag{3}$$

c) Do standardization with $\overline{W}_{C_i}$:

$$W_{C_i} = \frac{\overline{W}_{C_i}}{\sum_{i=1}^{5} \overline{W}_{C_i}} \tag{4}$$

3) **Consistency test**

To ensure the monolayer weights order are correct, the judgment matrix should be dealed with consistency test. They are judged by the random consistency ratio CR. If CR<0.1, the results can meet the requirements. Consistency should be tested as follows:

a) Computing consistency index *CI:*

$$CI = (\lambda_{\max} - n)/(n-1) \tag{5}$$

Where: λ_{max} is the eigenvalue of maximum, n is the number of the order of judgment matrix.

b) Checking the value of the average random consistency index *RI* from Table7.

Table 7. Values of *RI*

n	1	2	3	4	5	6	7	8
RI	0.00	0.00	0.58	0.89	1.12	1.24	1.36	1.41
n	9	10	11	12	13	14	15	
RI	1.46	1.49	1.52	1.54	1.56	1.58	1.59	

a) Calculating consistency ratio CR:

$$CR = CI/RI \tag{6}$$

If $CR < 0.1$, the consistency of the judgment matrix can be satisfied and so the order weights accepted.

By means mentioned above, all of the monolayer weights order can be get: W_1 = [0.344, 0.210, 0.245, 0.201], W_2 = [0.226, 0.243, 0.281, 0.251],W_3 = [0.259, 0.253, 0.218, 0.270]; W_4 = [0.269, 0.241, 0.239, 0.252]; W_5 = [0.262, 0.239, 0.207, 0.308]; W = [0.153, 0.217, 0.174,0.232,0.224],

The values of λ_{max} of the the judgment matrix are: 4.262, 4.094, 4.184, 4.084, 4.082, 5.389.

The values of CR are:0.098, 0.036, 0.069, 0.031, 0.031,0.087. they are all less than 0.1, so can pass the consistency test.

3.4 Value of Comprehensive assessment

Based on the assessment factor matrix X and the monolayer weights matrix W_{Ci}, the value of comprehensive assessment of rural informatization level can be determined, the formula is described as follows:

$$V = \sum_{i=1}^{5} \overline{X_i}\, W_{Ci} \qquad (i = 1,2,3,4,5) \tag{7}$$

Where: V is the level of comprehensive assessment on rural informatization development, $\overline{X_i}$ is the assessment factor matrix that made by standard operation.

4. CONCLUSION

From five main contents of rural informatization which are: development environment, information infrastructure, information resource, information service system and application of information technology in rural areas, This paper establishes an indicator system for rural informatization assessment and make a study of evaluation method furthermore. By use of AHP algorithm, the paper constructs the hierarchical structure model of rural informatization assessment and calculates the weight of each factor. The comprehensive assessment model we design has preferable operability and practicability. It combines subjective evaluation with objective appraisal, qualitative with quantitative study, and fully considers each assessment factor. Therefore, it is scientific and reasonable and suit to be used to evaluate the level of rural informatization development.

ACKNOWLEDGEMENTS

This study is part of the work of the Project:Index system for rural informatization assessment of JiangSu province and supported by JiangSu Information Industry Department.

REFERENCES

An Shaoshan, Chang Qingrui. Study on the characteristic properties and genesis of land desertication cation in the agriculture and husbandry interlace zone, ACTA Ecologica Sinica, 23(2003): 106-111.

Cai Kangying, Ding Zhuo, Zhou Yannian. Research on evaluation method for government functions based on AHP model, Computer Engineering and Design, 2008,29(7):1834-1837

Chen Ruyun, Fu Baolong. Application of AHP in college library information management level evaluation, Science Technology and Engineering, 2007,7(17):4425-4428

Liu Shihong. Study on the indicator system for measuring the rural area informatization level in China, Library And Information Service, 2007,51(9):33-35

Lu Lina. Construction of index system for IT application in agriculture, Journal of Library and Information Sciences in Agriculture, 2007,19(4):178-183

Qin Xiangyang, Zhang Xicai, Zhang Xingxiao. Study on assessment index system for rural informatization in Beijing, Journal of Beijing Agricultural Vocation College, 2008,22(1):42-46

Sheng Qifeng. Study on agricultural informazation construction and assessment. Master degree dissertation, Anhui Agricultural University, 2005

Tan Guoliang, He Wei. A study on contents, obstacles and countermeasures of rural informatization in China, Journal of Jiangxi Agricultural University, 2007,6(2):86-88

Yi YaXin. Assessment about the agricultural informatization development in Guangzhou, Master degree dissertation, South China Agricultural University, 2004

Zang Chunrong, Chen Xiao, Liu Shanwen, et al. Present situation analysis of China rural informatization development, Journal of Hunan Agricultural University(Social Sciences), 2004,4(5):10-13(in Chinese)

THE DEVELOPMENT AND PRELIMINARY APPLICATION OF PLANT QUARANTINE REMOTE TEACHING SYSTEM IN CHINA

Zhigang Wu [1] , Zhihong Li [1,*] , Ding Yang [1] , Guozhen Zhang [1]

[1] *China Agricultural University, Beijing, 100193, China*

[*] *Corresponding author, Address: Department of Plant Protection and Quarantine, College of Agronomy and Biotechnology, China Agricultural University, Beijing 100193, China, Tel: 010-62731299, Fax: +86-10-62733404, Email: lizh@cau.edu.cn*

Abstract: With the development of modern information technology, the traditional teaching mode becomes more deficient for the requirement of modern education. Plant Quarantine has been accepted as the common course for the universities of agriculture in China after the entry of WTO. But the teaching resources of this course are not enough especially for most universities with lack base. The characteristic of e-learning is regarded as one way to solve the problem of short teaching resource. PQRTS (Plant Quarantine Remote Teaching System) was designed and developed with JSP (Java Sever Pages), MySQL and Tomcat in this study. The system included many kinds of plant quarantine teaching resources, such as international glossary, regulations and standards, multimedia information of quarantine process and pests, ppt files of teaching, and training exercise. The system prototype implemented the functions of remote learning, querying, management, examination and remote discussion. It could be a tool for teaching, teaching assistance and learning on-line.

Keywords: plant quarantine, remote teaching, system development and evaluation, e-learning

1. INTRODUCTION

China need more professional workers at plant quarantine field given that 2008 Olympic Games will be hold in August 2008, Beijing and increasing

---

Please use the following format when citing this chapter:

Wu, Z., Li, Z., Yang, D. and Zhang, G., 2009, in IFIP International Federation for Information Processing, Volume 295, *Computer and Computing Technologies in Agriculture II, Volume 3*, eds. D. Li, Z. Chunjiang, (Boston: Springer), pp. 1909–1917.

quality safety concerns. At present, most of universities apply TE to plant quarantine education, i.e., classroom teaching (CT). With the development and popularization of computer and multimedia technology, many teachers also use such as video and power point for teaching, but comparing with e-learning, CT have many shortage.

- The information of book education (BE) is limited. It is not well if we only read book. In order to satisfy the need of training more persons with ability, learner should get great number of information so that they can success in huge competition with others.
- Some experiment video and power points which are applied to teaching reformation make courses active. However, the arrangement is only considered by the teacher, not learners. Thus, we can not adjust measures to different learners' conditions.
- For some graduated learners who attend in a advance studies lesson, their study time is hard to meet. Then they can't go to class completely as normal learners.

As the result, the insufficient of professional workers will never be overcome within a short period of time under the context of the traditional teaching mode. With the development and popularization of computer and internet technology, remote teaching, a new education mode, appeared and is developing quickly. And on-line learning (i.e., e-learning) is attracting more and more attention. E-learning refers to the use of electronic devices for learning, including the delivery of content via electronic media such as Internet/Intranet/Extranet, audio or video tape, satellite broadcast, interactive TV, CD-ROM, and so on (Kaplan-Leiserson, 2000).E-learning applications may appear with different forms of designation such as web-based learning, virtual classrooms, and digital collaboration (Kaplan-Leiserson, 2000; Khalifa et al., 1999). This type of e-learning places a greater emphasis on the enabling or facilitating role technology plays in data search and transmission, interactivity, and personalization (Piccoli et al., 2001).

E-learning offers new possibilities in learning. It share education resource, individual study condition, and hasten study socialization. Many research showed that e-learning is better than traditional education (TE), and it may replace TE at some terms （Hertz-Lazarowitz et al., 2002; Buchanan, 2002）. Thus, a learner can get immediate feedback on problem, learning paths can be individualized, etc. At present, a lot of e-learning tools with varying functionality and purposes exist (Aroyo et al., 2002; Murray et al., 2003). E-learning is an alternative concept to the traditional tutoring system. The course tutor in a software tutoring system controls learners relatively weaker than (s)he does in the traditional one, where the (human) tutor is in charge of the contents and sequence of instructions. Therefore, in order to obtain better tutoring outcomes, a software tutoring system should emphasize engaging learners in the learning process and be adaptive to each individual learner (Anatoly et al., 2008).

It is showed from some data that 72% of American universities adopted e-learning in 1999. And the proportion increased sharply to 99% in 2001 (Li Rongsheng, 2005; Yuan Qiu, 2002). Some universities, middle schools and elementary schools also adopted some e-learning courses in China. There are many plant quarantine information system, but most of them are decision assistance system, alarm analysis system, authentication assistance system etc (Sun Guanying et al., 2003; Zhou Weichuan et al., 2004; Li Zhihong et al., 2003; Li Zhihong et al., 2005). However there have few of report about plant quarantine learning and assistance system.

This paper discusses our experience in developing and evaluating an e-learning system for plant quarantine. Section 2 describes the questionnaire to acquire user' need and part of domain knowledge. Section 3 shows the system architecture and discusses the system implementation. Then the last section discusses and draws some conclusions.

2. USER NEED ANALYSIS

When applying a learning tool or system for learners, it is necessary to investigate both teachers' and learners' attitudes toward that tool or system. Essentially, understanding their perceptions toward learning environments is a crucial issue for enhancing teaching performance and learning effects. The results of Liaw's study confirm that instructors are willing to use e-learning environments to aid their teaching activities. Learners also respond favorably to e-learning environments for complementing to their learning activities (Liaw et al., 2007).

The potential users of plant quarantine e-learning system are undergraduates and graduate students and staffs who work at plant quarantine. To better meet their needs, a questionnaire was designed to investigate their problems and needs (Fig.1). The questionnaires were sent to some Chinese universities, the plant quarantine departments and plant protection stations of Beijing, Tianjin and Guangdong. The result gave us much good suggestion and idea.

- The system must has a own user database. Thus, not only it can run by itself, but also can be added to other systems.
- For different users in different level, the interface should be very friendly and simple.
- The modules of testing by themselves and chatting on-line are necessary.
- The modules of Learning note, statistic and analysis, email awoke and study plan are also suggested.

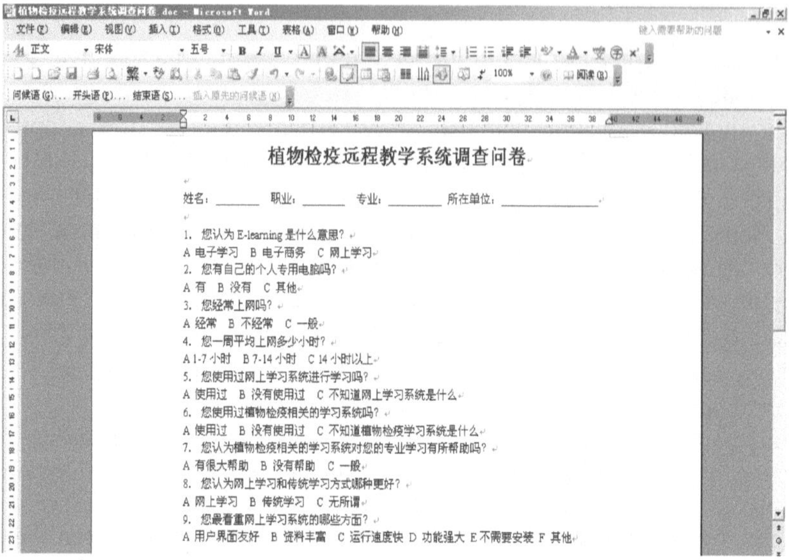

植物检疫远程教学系统调查问卷

姓名：______ 职业：______ 专业：______ 所在单位：______

1. 您认为 E-learning 是什么意思？
A 电子学习 B 电子商务 C 网上学习
2. 您有自己的个人专用电脑吗？
A 有 B 没有 C 其他
3. 您经常上网吗？
A 经常 B 不经常 C 一般
4. 您一周平均上网多少小时？
A 1-7 小时 B 7-14 小时 C 14 小时以上
5. 您使用过网上学习系统进行学习吗？
A 使用过 B 没有使用过 C 不知道网上学习系统是什么
6. 您使用过植物检疫相关的学习系统吗？
A 使用过 B 没有使用过 C 不知道植物检疫学习系统是什么
7. 您认为植物检疫相关的学习系统对您的专业学习有所帮助吗？
A 有很大帮助 B 没有帮助 C 一般
8. 您认为网上学习和传统学习方式哪种更好？
A 网上学习 B 传统学习 C 无所谓
9. 您最看重网上学习系统的哪些方面？
A 用户界面友好 B 资料丰富 C 运行速度快 D 功能强大 E 不需要安装 F 其他

Fig.1: part of questionnaire

3. SYSTEM DESIGN AND DEVELOPMENT

To explore the assistant effect of e-learning system on plant quarantine teaching, we designed and developed Plant Quarantine Remote Teaching System (PQRTS). The system included lots of plant quarantine teaching material, such as quarantine pest, international glossary, regulations and standards, multimedia information of quarantine process and pests, ppt files of teaching, training exercise and so on. The system prototype implements some function, such as remote learning, querying, management, examination and remote discussion. It could be a tool for teaching, teaching assistance and learning on-line.

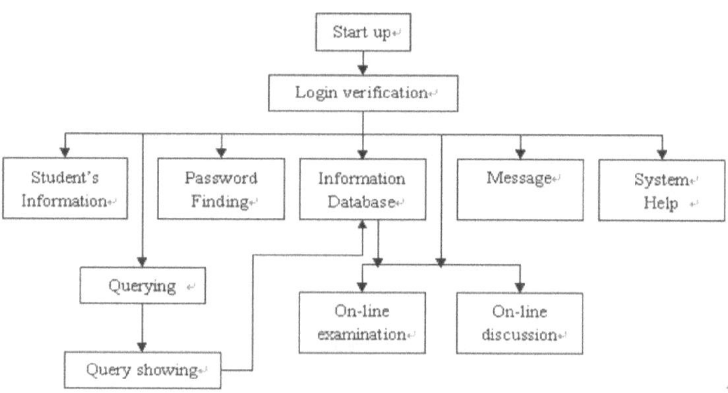

Fig.2: the functional diagram of the system

3.1 Database

The database takes a very important part and is responsible for storing all the information in the whole system. In view of transplant and money, the database was designed with MySQL. The databases include 8 teaching materials databases and 3 system function databases. Fig.3 shows the logic relationship among the databases.

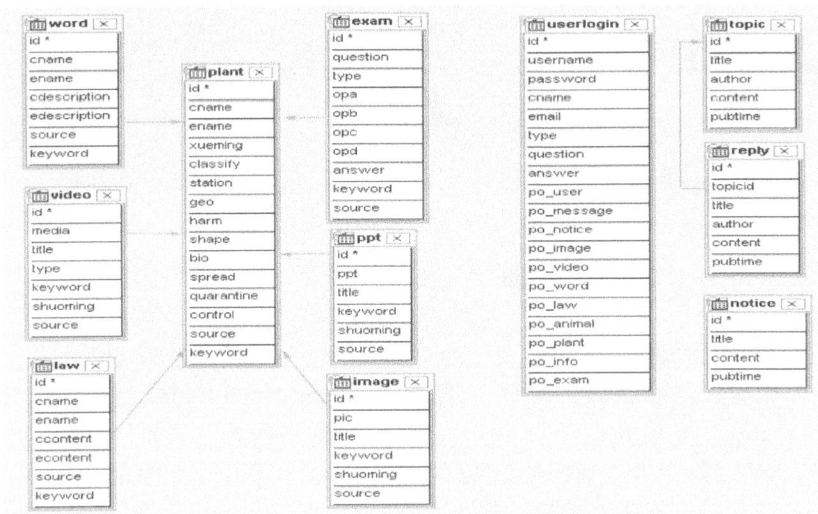

Fig.3: the logic relationship among the databases

3.2 User interface

User interface is the direct media between users and computers. To best meet the different users' need, designers should insist on friendly principle. In another word, they should consider all users' knowledge level and ensure the interface complete, compact and easy. We made the system interface with considering the characteristic of plant quarantine teaching as much as possible. For instance, the index page includes 6 areas, such as system navigation, system function, system introduction, fast querying, user login, user register and system notice (Fig.4).

Users can use the function of examination on-line to test themselves easily at any time. After they choose the number of test subject, the system will select randomly from exercises database and form test paper. When users click the submit button, the result of the examination will be showed in front of them（Fig.5）.

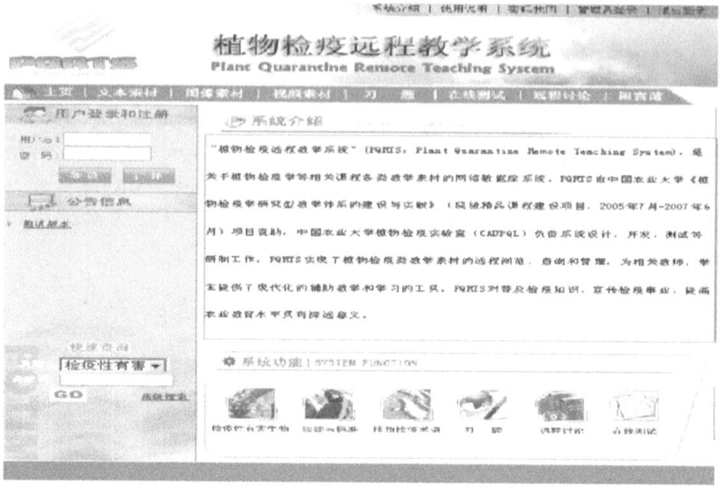

Fig.4: the index page of the system

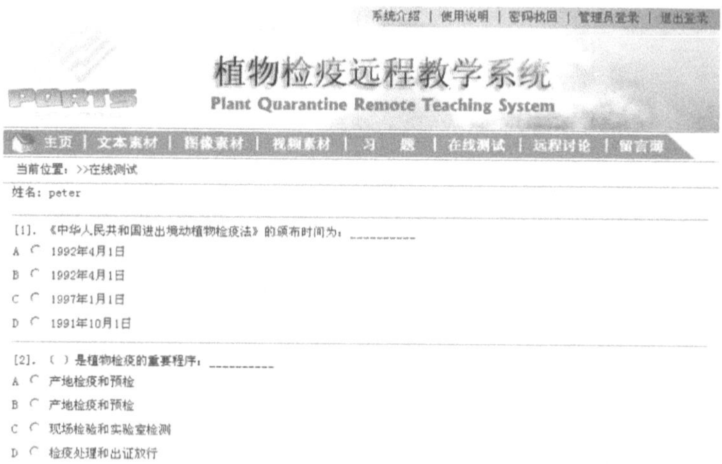

Fig.5: the user interface on on-line examination

3.3 Development language and environment

In view of easy development, easy transplant, little money, the system took B/S structure. Users only need use Brower such as Internet Explorer, firefox and so on to visit the system website via internet or LAN, not install any other software to their PC. There are many solutions for developing the web-based systems. We adopted the solution by Sun, i.e. JSP (Java Sever Pages) + MySQL + Tomcat due that it is prone to master by non-professional system developer. Among the development plan,

- JSP is serving as the main develop language to bridge user interface and web server and database server.

- MySQL4.0, a GPL (free) SQL relational database management system and data warehouse development tool, is serving as the back-end of the workstation to facilitate data storage and retrieval.
- Tomcat 5.0 is serving as a web server to provide information service.
- Microsoft Windows Server 2003 is serving as an operational system.
- Serv-U 6.0 is serving as an FTP system.
- In addition, JavaBean technology is adopted to access the database (Fig.6). It make the system more safe.

```
.......
public class DBconn
{
String DBDriver="org.gjt.mm.mysql.Driver";
String ConnStr="jdbc:mysql://localhost/pqrts?useUnicode=true&characterEncoding=GBK";
String MyUser="sa";
String MyPassword="";
Connection conn = null;
ResultSet rs = null;
int i=0;
public DBconn()
{
try
{
      Class.forName(DBDriver);
}
.......
public ResultSet executeQuery(String sql)
{
rs = null;
try {
      conn = DriverManager.getConnection(ConnStr,MyUser,MyPassword);
      Statement stmt = conn.createStatement();
      rs = stmt.executeQuery(sql);
}
.......
```

Fig.6: part of JavaBean code for access to database

4. DISCUSSION

The system were applied to the plant quarantine course of China Agricultural University (CAU) in 2006. During the last 2 years, the system ran well and got good appraise by users. It played a good role in plant quarantine teaching assistance. We think that the system could be introduced to other agricultural universities and plant quarantine departments as a method for teaching reform. By using such e-learning system, students could get more acquaintance and promote their interesting in plant quarantine. It may also provide some material for Chinese remote teaching.

- The system has over 100 pieces of international glossary, 20 pieces of regulations and standards, 200 pieces of quarantine pest, 20 ppt files of teaching, 500 images, 20 Flash files, 20 videos and so on.

- After surveying and ensuring of the user group, we designed the system in term of different users' need. It's interface was friendly and can be operated easily.
- The system took B/S structure. Users only need use Brower to visit the system website via internet or LAN.
- The system not only can run by itself, but also can be added to other systems.
- JSP is serving as the main develop language, so the system can run in many platforms.
- Besides some shortage mentioned above, there are still some items should be addressed, such as:
- Most of the materials come from Chinese data. We should add some materials from foreign language data.
- The examination on-line module should add the multinomial question into exercise database.
- The system only practice in CAU because of time limit. In order to better finish the study, we should put it into practice in other universities and departments.
- Due to time limit and money limit, we only developed an system antetype. The modules of Learning note, statistic and analysis, email awoke and study plan should be developed in future.

Remote teaching is a new teaching mode, not the copy of TE, broadcast education or TV education. We argue that the aim of remote teaching is that we can implement share and individuated teaching. Then a environment of teaching and learning will be built by modern information technology（Lim, 2001）. The remote teaching system based computer technology is modern tool for modern education. It provide a platform which be used to explore new method and new mode of modern education.

ACKNOWLEDGMENTS

This program is supported by the graduate student teaching reform project of China Agricultural University. The authors would like to thank Zhang Baofeng, Huang Guansheng, Lin Wei, Liang Guangqin, Wu Jiajiao, Deng Yuliang, and all other officers of China Inspection and Quarantine to support to the study. This study received financial support from teaching reform projects of China Agricultural University.

REFERENCES

B M Balchunas, C E Hucks, Management tools at your fingertips tapping the power of the web. American Water Works Association, 2005, 97(5): 94~100

C P Lim, The dialogic dimensions of using a hypermedia learning package. Computers and Education, 2001, 36(2):133-150

E Kaplan-Leiserson, e-Learning glossary. Available from http://www.learningcircuits.org/glossary.html.,2000

F Moti, R Nurit, H Keith, Respecting the human needs of students in the development of e-learning. Computers and Education, 2003, 40: 57–70

G Piccoli, R Ahmad, B Ives, Web-based virtual learning environments: a research framework and a preliminary assessment of effectiveness in basic IT skills training. MIS Quarterly, 2001, 25(4): 401–426.

G Y Sun, X X Chen, J A Cheng. The web-based entry-exit plant quarantine information management & decision assistance system. Journal of Zhejiang University (Agric.& Life Sci.), 2003，29(4): 407-412 (in Chinese)

H Y Chen, K Y Liu, Web-based synchronized multimedia lecture system design for teaching/learning Chinese as second language. Computers and Education, 2008, 50: 693–702

L Aroyo, D Dicheva, Courseware authoring tasks ontology, In Proceedings of the international conference on computers in education. IEEE Computer Society, 2002

M Khalifa, R C-W Kwok, Remote learning technologies: effectiveness of hypertext and GSS. Decision Support Systems, 1999, 26(3): 195–207

R Hertz-Lazarowitz, I Bar-Natan, Writing Development of Arab and Jewish Students Using Cooperative Learning (CL) and Computer-mediated Communication (CMC). Computers and Education, 2002, 39(1): 19-36.

R S Li, If E-learning can arouse the revolution of enterprise training. Training In China, 2001, (8): 29-31

S S Liaw, H M Huang, G D Chen, Surveying instructor and learner attitudes toward e-learning. Computers and Education, 2007, 49: 1066–1080

T Buchanan, The Efficacy of a World Wide Web. Computing Research, 2002, 23(2): 203-216.

T Murray, S Blessing, S Ainsworth, Authoring tools for advanced technology learning environments: Towards cost-effective adaptive, interactive, and intelligent educational software. 2003

W C Zhou, X F Wu. The development of Remote Service System of Quarantine Pests based on Multimedia database. Plant Quarantine, 2004, 18(5): 277-278 (in Chinese)

Yuan Qiu, The Spring of e-Learning is coming. Software Engineer, 2002, (2): 34-37 (in Chinese)

Z H Li, B F Zhang, Z R Shen, et al. The development of Quarantine Pests Information and Identification System in China. Plant Quarantine, 2003, 17(5): 273-276 (in Chinese)

Z H Li, G Q Liang, W X Li, et al. The development of Fruit Fly Monitoring Web System in Guangdong province. Plant Quarantine, 2005, 19(4): 214-216 (in Chinese)

SOA BASED INTEGRATION INFORMATION SERVICE PLATFORM STRATEGY IN RURAL INFORMATIZATION

Baoqing Dong [1,*], Lei Chen [2], Yuefeng Zhou [3]

[1] *School of Geoscience and Space Science, Peking University, Beijing, 100871 P. R. China*

[2] *Network & Information Center of Liaoning University, Shenyang, 110036, P.R.CHina*

[3] *College of Economics and Management, Dalian University, P.R.China*

[*] *Corresponding author, Address: School of Geoscience and Space Science, Peking University, Beijing 100871, P. R. China, email:qingbd@263.net*

Abstract: This paper compares the hierarchical structure between SOA and integration information service platform in rural informatization, analyse the main problems in the macro level of rural information and gives some strategy advices in service-oriented integration information service platform on rural informatization. Through the implementation of pilot projects, it is proved that the SOA architecture has good effect in the rural information resources integration.

Key words: SOA, Rural informatization, related agriculture information platform, integration

1. INTRODUTION

With the development of information technology and government function reform, the demand in rural informatization is increasing. It involves multiple sectors and industries. Each sector builds their own business syetem in accordance with their needs. However, These business systems base on different technical standards, different software and hardware platforms, and these heterogeneous systems are developed by different software providers. The status of information island has been formed

---

Please use the following format when citing this chapter:

Dong, B., Chen, L. and Zhou, Y., 2009, in IFIP International Federation for Information Processing, Volume 295, *Computer and Computing Technologies in Agriculture II, Volume 3*, eds. D. Li, Z. Chunjiang, (Boston: Springer), pp. 1919–1927.

In order to achieve high efficiency, low cost of promoting the rural information and intensive construction and resource integration has become an important initiative. To this end, government departments must be resolved between the integration of information systems integration. Web-based services of a service-oriented architecture, to achieve heterogeneous platform for information integration, inter-governmental support within and across sectors and between the end-to-end integrated services through the newly-formed new services, agriculture and the rural information Integrated agriculture-related information service platform to consider the top-level design provide a new opportunity.

2. SOA ARCHITECTURE AND THE KEY TECHNOLOGY

2.1 SOA architecture

SOA is a coarse-grained, loosely coupled, platform-independent and programming language, standards-based software architecture model, as shown in Figure 1.

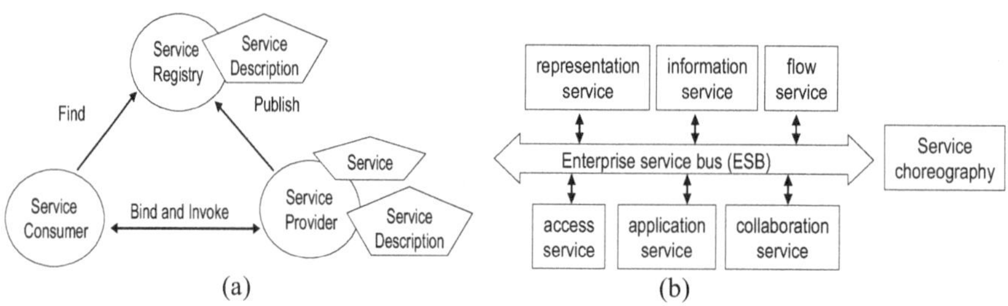

Fig.1: SOA basic structure

(1)Service provider:

An agent that can be addressed through a network, Its acceptance and implementation of the request from the users. It will own services and interface services contract issued to the registration centers, so that users can find and access the service.

(2)Service consumer:

An application or a software module or the need for a service of another service, which launched the registration centre in the services available through the transmission bundled services, and implementation of services in accordance with the user interface services contract to perform the services.

(3)Service registry:

Services find that supporters. It includes a repository of available services, and allows users to find services of interest to service providers interface information.

Service-oriented architecture of each entity is playing a service provider, the user registration centre and three in a certain role in one or more.

2.2 Functional properties based on SOA

SOA is a style of architecture, it features the application of the different units - services, through service between the definition of a good interface and contract linked. Interface defined by a neutral, independent of the concrete realization of the hardware platforms, operating systems and programming languages, making construction of the system of services can use the unified and standard approach to communications. Such a neutral definition of the characteristics of the interface between loosely coupled services called.

In the SOA model, all the business logic are in the form of packaging services, they can be shared, reuse and configuration. An application is composed by any number of services component, which is through the Enterprise Service Bus (ESB) for communication, interaction and presentation management. Each service through the standard interface can be invoked by other services. Each application by the service providers, consumers and service agent component, which is in a standard form of interactive services complete business functions. Its interactive relationship is shown in Figure 1(a).

SOA model of thinking to achieve service-oriented application to unity and a common way of interactive service, can be easily achieved between the business applications of data integration, application integration, business integration.

2.3 Operation of SOA

SOA realization of the entire service is the core. SOA is an essential element of services, SOA designated a group of entities (service providers, consumers, service registry, the terms of service, service agents and service contract), these entities a detailed description of how to provide and receive services. These services are interoperable, independent, modular, location specific, the loosely coupled through the Internet and can find their address.

Service-oriented architecture in the operation include the following three points.

Publish: In order to make services accessible, the need to issue service description to enable service users can find it.

Find: Service to those requesting it location-based services, it is for the registry to find services to fit their standards of service.

Bind and invoke: To describe the search service, the service users continue to service description information to call services.

3. SOA BASED INTEGRATION DESIGN IN RURAL INFORMATION SERVICE

3.1 The overall framework of integration information service platform in rural informatization

The rural information integrated information service platform for the general overall structure of the ‘five horizontal two longitudinal.

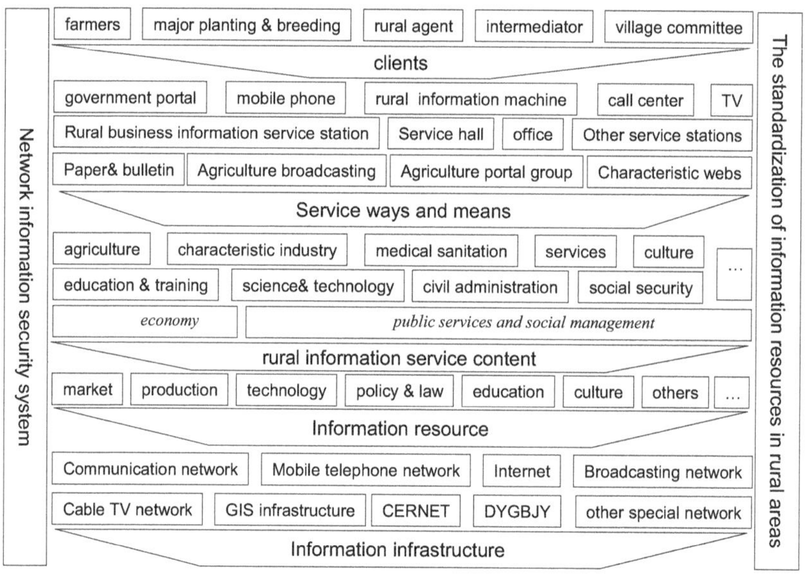

Fig 2: The overall framework of related agriculture information platform

‘Five horizontal’ refers to the information infrastructure, information resources, services, channels and service establishments, service targets. ‘Two vertical’ refers to the rural areas of standardization system of information resources and information network security system.

The platform in the field of information technology in generally refers to all kinds of resources and services to the exchange, integration and operation of a hardware and software environment. Agriculture-related information service platform is the exact meaning of the entire information technology work in rural areas involved in a variety of technical means, a variety of methods of work, organizational structure and sharing of data transmission, such as the form of a standardized definition, as the internal integration of business support, The total external environment to provide services in this environment, including all the services.

From the rural areas of the information building to the actual situation analysis, integrated information service platform external interface is a classic example of the portal, agriculture-related portal sites and call centers, including on the agriculture-related portal in Internet, the Government of the agriculture-related portal in special network, also include support them operation, the network, computer hardware and software system, security measures, information resources and service teams, service establishments (access points), services content, and so on.

3.2 The demand of Agriculture-related information service platform

1) Basic requirement of integration

Platform moves up, service moves down. It is better to promote the deployment of agriculture-related information service platform in high-level administrative region, and intensive construction in province or city or county. It isn't necessary to establish agriculture-related information service platform in the region of below county. And the agriculture-related information service platform can be directly established in province if there is no condition in city level.

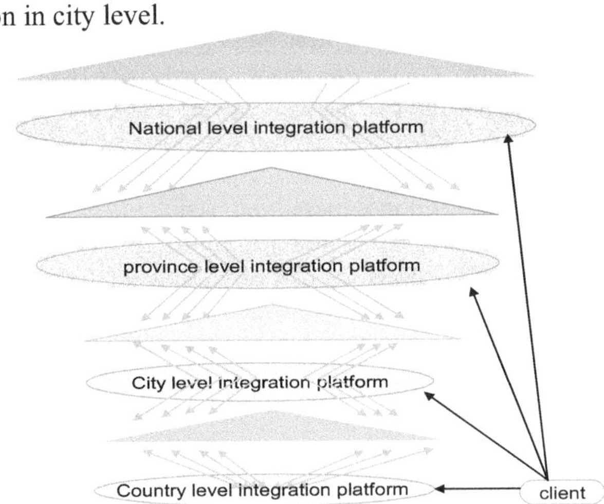

Fig 3: The integration of agriculture-related information service platform

2) One network carries variety of business

The net resources share in rural region should be promoted suitably. And the department of telecommunication, broadcast and television should be encouraged to display their respective advantages by government, which can promote a variety of terminal services business and business integration. At the juncture of speeding up agricultural and rural information construction, removing institutional obstacles and exploring the rural areas "Sanwangronghe" mechanism should be promoted. Regardless of which kind of network formed in the past, under quality assurance, it is necessary to develop a variety of services by way of IPTV and cable Internet

3) Many purposes based on one terminal

Relying on fixed communications, mobile communication, Internet portals, radio, television and newspapers, books, columns and other publicity channels of information dissemination, information services and applications can be promoted into villages and peasant households.

3.3 The technologies design of platform integration

Compared with the traditional integration method, SOA-based integration has more advantages, such as reused, platform-independent, standards-based, coarse-grained, loosely coupled, and change according to business needs, and so on. In addition, SOA structure itself has a good scalability, flexibility and adaptability, and it is easy to achieve interoperability and information share among applications, and to avoid “isolated island” among the applications of information systems. Therefore, it is a better choice to use SOA framework to establish rural synthetical information service platform. The structure is shown in Figure 4.

In this structure, unified application platform is composed of supporting software, infrastructure, application service framework. The difference from other application supporting platform is that this platform encapsulates every functional component by forms of service, which form into enterprise service bus (ESB) to carry out standard service of interoperation and information exchange. Services based on standards become the knot of application. Through the ESB the interoperability between different services can be achieved, and the scheduling and combination of services increase the flexibility, reusability and integration, which provide flexible and convenient method for application and business process to access into.

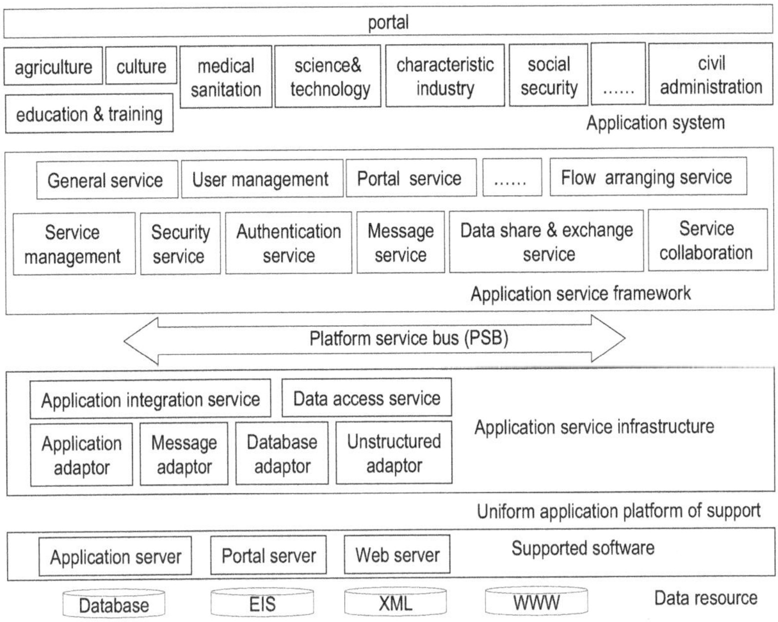

Fig 4: The technology structure of rural synthetical information service platform based on SOA

1) Basic infrastructure of application service.

It uses middleware to achieve interoperable, reusable infrastructure and services and to provide a service-oriented and application-oriented basic environment, and can shield difference of software platforms, all of which make the service platform to be cross-platform, cross-integration technology. Application service infrastructure is composed of enterprise service bus, service delivery channels and interactive environment.

2)Framework of application service

Framework of application service is the service set and service interface, which is acquired by abstracting common functions in specific field of agricultural-related synthetical information service platform, such as portals and user management, identity and some common service. SOA-based agricultural-related synthetical information service platform can quickly reflect to business change and provide new service application, because the application deployed on the SOA application supporting platform can easily reuse from application service platform and recompose basic service and common service, also can add new function and service.

3) Typical basic services. Include the following aspects.

a) Data access services.

b) Data share and exchange services.

c) Application and integration services.

d) Collaborative business services.
e) Portal services.
f) Unified security services.
g) Unified management services.

4. EXPERIMENT AND APPLICATION

According to the design mentioned above, the Ningxia Hui Autonomous Region has been selected as a trial area to implement the fully integrated agriculture-related information resource. The SOA-based agriculture-related information service platform was built in Ningxia to integrate rural information resources, and has achieved better results, which are listed in the following.

1) The utilization of existing resources

Service-oriented structure can be developed based on existing investment system, without the need of a completely new system. By ways of appropriate SOA framework, business service can be constructed into component set, which can be used only if you know the name of interface, the internal details of service. The complexity of data transmission among components of the service is hidden from outside, which makes existing resources be used and can be packed by ways of service to be provide to new system or others to access.

2) Commercialisation of infrastructure

Utilization of SOA framework will make the development and deployment uniform between different application programs of government web. The components set, which are combined with existing components, new components and components bought from manufacturers in a defined framework, will be deployed on agricultural-related information service platform as service.

3) Costs cut

As the development of business demand and introduction of new demand, costs of strengthening present and new application programs and establishing new service can be reduced greatly by way of the SOA framework and services library.

5. CONCLUSION

Practice shows that there are more problems existing in agriculture, for example wide field, a variety of information source, information resources of agricultural and agricultural-related departments. And there is no uniform management system, which makes agricultural information to be owned by departments itself without share. And also there is no uniform standard for technologies to develop agricultural system leading to the lack of effective information integration tools. The utilization of loosely coupled SOA architecture in the integration of agriculture-related information systems resources can solve these problems mentioned above, which will achieve narrow the urban-rural digital gap by spending relatively less time and financial resources.

REFERENCES

CHAPPELL D. 2004. Enterprise service bus[M]. Sebastopol, CA: O'Reilly Media, 2004.

Dreyfus P. 2006. "What's the best software to implement as a service if you're just starting SOA?" http://www-128.ibm.com/developerworks/library/ar-itio4/.

Gartner P.M. 2006. "Service-Oriented Architecture Craves Governance." http://www.gartner.com/DisplayDocument?id=488180.

Gronmo R., Skogan D., Solheim I. 2004. Model-driven Web services development. In: Proc of the 2004 IEEE Int Conf on e-Technology, e-Commerce and e-Service (IEEE'04). Taipei.

IBM. 2005. Service2oriented architecture and Web services [EB /OL]. http: / /www. ibm.com/services/us/imc/html/soa.html.

Li Daoliang. 2007. Report of China's rural information development.

Newcomer E. Greg L. 2005. Understanding SOA with Web Services. Boston, MA: Addison-Wesley.

PapazoglouM P. 2003. Service-oriented computing: concepts, characteristics and directions. In: Proc. Int Conf on Web Information Systems Engineering (WISE03). Los Alamitos. 2003: 3-12

Yefim V. N., Roy W. S. 2006. "Advanced SOA for Advanced Enterprise Projects." http://mediaproducts.gartner.com/reprints/oracle/141940.html.

STUDY ON THE DEVELOPMENTAL STRATEGY OF COUNTY REGION ECONOMY INFORMATIONIZATION IN CHINA

Guizhen Sun[1], Huijun Wang[2,*]

[1] *Humanities College, Hebei Agricultural University, Baoding, Hebei Province, P. R. China 071001*

[2] *Hebei Academy of Agricultural and Forestry Sciences, Shi jiazhuang, Hebei Province, P. R. China 050051.*

[*] *Corresponding author, Address: Hebei Academy of Agricultural and Forestry Sciences, Shi jiazhuang 050051, Hebei Province, P. R. China, Tel: +86-311-87652003, Fax: +86-311-87066140, Email: nkywanghj@yahoo.com.cn*

Abstract: The concept of county region economy and its informationization were limited. The characteristics and functions of county region economy informationization were pointed out. On the basis of analyzing the present situation and the existent problems, the developmental strategy of county region economy informationization in China were put forward from guiding ideology, basic principles, system frame and the way of implementation.

Key words: county region economy, informationization, strategy

1. INTRODUCTION

As one of important contents of region informationization, the county region economy informationization is an efficient path to reduce "the digital gap" between urban and rural areas. At present, Japan, the US, Singapore, Canada, France, Germany and so on are in leading position in region economy informationization in the world. The informationization accelerates the development of traditional industries and distinctive industries, improves the management level and the grade of service through information technology, and provides fair and convenient information services for the

---

Please use the following format when citing this chapter:

Sun, G. and Wang, H., 2009, in IFIP International Federation for Information Processing, Volume 295, *Computer and Computing Technologies in Agriculture II, Volume 3*, eds. D. Li, Z. Chunjiang, (Boston: Springer), pp. 1929–1938.

resident through sharing the information in these countries. The informationization expanded the open degree and propelled the regional economy development (Tian Ye et al., 2001; Liu Guicai et al., 2006; Zong Yu et al., 2006). On the basis of the successful experiences of other countries and the reality of our country, it is the inevitable choice to realize county region economy informationization in China.

2. OUTLINE OF COUNTY REGION ECONOMY INFORMATIONIZATION

2.1 Concept of county region economy informationization

County region economy, the most basic regional economy unit of the national economy, is a kind of regional economy with bright regional features and complete functions. It is divided into the geography space by county level administrative area, takes market as guidance and takes county level political power as the important impetus. It is the intersection of economy not only urban and rural areas but also industry and agriculture.

County region economy informationization is an entire process of using information technology in industry and agriculture, the government affairs and the village affairs and public services and so on. Through gathering, transmitting and processing the information, developing and applicating the information resources in breadth and depth, it can reduce the consumption and the pollution to enhance the efficiency of industry and agriculture, the government affairs and the village affairs and public services unceasingly, and make them intercoordination and fusion to enhance the competitive power of county region economy.

2.2 Features of county region economy informationization

2.2.1 Multiplication of main body

County region economy informationization involves many main body, such as each kind of enterprise, the individual engaged in commerce, the family dealing with industry or agriculture, the township government and the related department in government affairs or village affairs and all kinds of intermediary organizations in public services and so on.

2.2.2 Diversification of content and form

County region economy informationization involves many operational systems in industry and agriculture, government affairs and village affairs or public services, multiple message transmission networks (eg, broadcasting network, telecom network, internet, etc) and multiple message terminals (eg, computer, telephone, television, newspaper, etc).

2.2.3 Integration of each kind of service

County region economy informationization is a process to promote intercoordination and fusion among the production or operation management of industry and agriculture, government affairs and village affairs and public services. The integration enables the society to become an organic whole.

2.3 Highly county region economy informationization's form

The informationization makes information resources as an important resources and power of social development and brings the following changes.

2.3.1 Agricultural production of precision

Farmers can promptly gain information about agricultural science and technology, technical advisory work, meteorology, agricultural product demand and so on. They applicate technological production in agriculture to raise the land productivity, the yield rate and the availability of resources, and realize the sustainable development of agriculture.

2.3.2 Industrial production of specialization

Information technology was widely applicated in industrial enterprise. It causes the digitization and network in product design, manufacturing, operation and management. With the infinite information network and the strong message resources, the low consumption of energy and pollution, the specialized and characteristic production in industry will have stronger competitive power and higher benefit in the world.

2.3.3 Service process of digitization

Based on the information network and the strong message resources, the omni-directional and the three-dimensional modern service structure will be formed. It includes the foundation service industry such as communication services and information services, production and market service industry such as e-procurement, online sale and e-logistics, online individual consumption service industry such as education, healthcare, accommodation, catering, cultural entertainment and traveling and public service industry sunch as government public administration and public health and so on.

2.3.4 Service function of government

The flattening structure will become the basic construction of county township government and village committee. The interactive and open administration according to the law will be the basic way of government operation. The comprehension services will be the core responsibility of county township government or village committee.

3. SITUATION AND PROBLEMS OF COUNTY REGION ECONOMY INFORMATIONIZATION

In order to do well county region economy informationization, the State Council Informationization Office of China selected six state-level county regions as the experiment site, which are respectively Anping county in Hebei province, Fangshan district in Beijing, Cixi city in Zhejiang province, Wuyuan county in Inner Mongolia, Ningguo and Shucheng county in Anhui province. County region informationization has made great progress.

3.1 Present situation

3.1.1 Facilities of information infrastruction

In China, during the 10th Five-Year-Plan period (2001-2005) the synthesis communication ability obviously strengthened in county region and county-wide information communication network was formed basically. Each county laid down the fiber optics. The communication network has covered all villages or small towns in plain area. "The wide band from village to village" was basically realized in developed areas. The user of cable TV and computer increased. Many technologies such as fiber optic communication, data communication, satellite communication, mobile communication and multimedia correspondence were widely used in communication network.

3.1.2 Development and application of information resources

In China, the county region has mostly established special agricultural message center, e-government local area network and government website. In view of the local distinctive industry, some characteristic data systems such as industrial policy, market supply and demand and suitable technology of animal husbandry have been established. The media of network, publication, magazine, broadcast and television has unified mutually. Based on these, intermediary organizations and leading enterprises were built and became the first to share the information resources. Many kinds of service modes and agricultural information added farmers' receiving and promoted the agricultural industrialization.

3.1.3 Popularization and application of information technology

The popularization and application situation of information technology is different for the differences of counties. In enterprise informationization, the difference is very big. Generally speaking, the county with the characteristic industry developments quickly, otherwise is relatively slow. For example, Anping county with famous silk screen industry is a clear example of the aspect. In e-government aspect, the development in various samples is very uneven. For example, more than 80% institution at Ningguo county has implemented government online project. In the slow pace of development of the counties, e-government is building and developing.

3.2 Main problems

3.2.1 The weak information infrastructure

For the lower computer penetration rate, the bigger investment of network infrastructure and the higher price or cost of online in rural, the ability to gain information by network is insufficient and the channel to disseminate information is not unimpeded. In the existent agricultural information service system of rural areas, the backward means of gathering, processing, distributing and issuing information, non-uniform standard and non-standard method of information and so on cause the obsolete or the distortive information and the serious information asymmetry.

3.2.2 The low exploitation and application of information resources

The information resources construction in different departments of county government is relative backwardness. The technical methods and operation patterns of information dissemination exists flaw and the technology appears the insufficiency in the depth and breadth to await improvement in the usability, serviceability, visualization and dynamic. In agricultural information, many are macro, outdated and inapplicable so as not to truly instruct the farmer's decision-making. At present the information resources in different departments cannot be shared.

3.2.3 The poor qualities of information recipients

The informationization requests the information personnel or recipients have high qualities. However, the existent information service personnel with uneven qualities, mostly without specialized training, are not familiar with the information service business. Most of them engaged in several jobs at the same time. These have affected the quality of information service.

The farmers, as the recipients of information, have a lower quality. According to the census results, among the rural labor forces at Wuyuan county, there is 87.91% person with junior middle school education, 10.07% with high school education, 1.63% with specialized middle school education, and only 0.39% with college education. The reality has directly limited the farmers to learn information technology and network knowledge and lead to a lower level of information application.

3.2.4 Unsound policy environment of informationization

Informationization policy environment is weak and unsound, especially in informationization management, investing system, operation mechanism, informationization standard and the legal environment etc. The overall frame of informationization does not yet constitute.

3.2.5 Expanding information difference between urban and rural areas

Since the 1990s, the establishment and deepening of marketability led to the essential factors of production such as rural labor force, fund, land, etc flowing out from the rural massively. Therefore, the difference between urban and rural areas unceasingly expanded. Taking the computer penetration rate at Anping county as an example, 59.09% families from urban residents have the computer, but only 2.68% farmer families. The low income of the farmer directly influences the obtaining information and rural informationization.

4. DEVELOPMENTAL STRATEGY OF COUNTY REGION ECONOMY INFORMATIONIZATION

4.1 The guiding ideology

Under the strategy and policy of the national informationization, county region economy informationization should closely regarde economy and society development as the central task, take the promotion of education, agriculture and government from informationization as the goal, take the exploitation and application of information resources as the core, take the use of information technology as the method, and display the booster function and the multiplier effect of informationization.

4.2 The basic principles

4.2.1 Overall conforming information resources

County region economy informationization must take the scientific development concept as a command, use the existent each kind of channel and information resources fully, eliminate the redundant building project and the unshared information and realize the sharing and the optimization disposition of information resources.

4.2.2 Prominent application of information

According to local condition and the demand among enterprises, farmers, middle organizations and county township government departments, county region economy informationization must stress on the application of informationization to solve the key questions related the overall situation and manifest the multiplier effect of informationization.

4.2.3 Foothold in system innovations

County region economy informationization must unify informationization and institutional innovations, consummate gradually policies and regulations, the management and the coordination mechanism of informationization, study the low cost and the high benefit development model and be sure to seek the actual application effect of informationization.

4.2.4 General services of informationization

County region economy informationization must insist on the principle of general services to advance the informationization application in different groups of people or industries, reduce "the digital gap" gradually, and make full use of informationization to enhance county region economy.

4.3 The system frame

In view of the present situation and problems, the system frame of county region economy informationization should involve the following six aspects.

4.3.1 The main body of services

In county region economy informationization, the public, enterprises, middle organizations, county township government and its departments are all the main body and the object. It is noteworthy that the public include farmers, enterprises include the rural enterprises, middle organizations mainly refer to the agricultural association and the rural service station of science and technology.

4.3.2 Using platform

In generally, the using platform of informationization, including the using platform of government affairs information, the enterprise information and the public services, provides the application services for the county township government, the village committee, each kind of enterprise, the public, and the agricultural association and so on.

4.3.3 The application and the service system

Without the universal available criterion, the application and the service system should be according to the local actual conditon to operate and share the resources of different provinces or cities to avoid the redundant informationization construction.

4.3.4 Information resources

The information of enterprises, government affairs, geography, population, agriculture, etc are all the information resources of informationization. The development and application of it is important content of informationization. Without it, the main body cannot obtain the information to make decisions.

4.3.5 Information infrastructure

Information network infrastructure is the foundation of informationization. In informationization, the network operator's enthusiasm should be aroused. The resources of hardware and software should be conformited, and the principle of "the resource sharing, mutually for uses" should be insisted on.

4.3.6 Regulations and standards of management

As a key to standard and impetus the development of county region economy informationization, the policies and the laws and regulations are the safeguard of the informationization. Various county (city) areas must profit from the valuable experiences which accumulates in the policies and regulations aspect of the experimental sites to serve informationization.

4.4 The implementating way

The informationization must take the significant demonstration projects informationization as a breakthrough point, take the leading strategy of mainstay industry as an engine and the promoting strategy of the environment and personnel as a safeguard. Thus lead to the key and related industrial informationization. The way is shown in Figure 1.

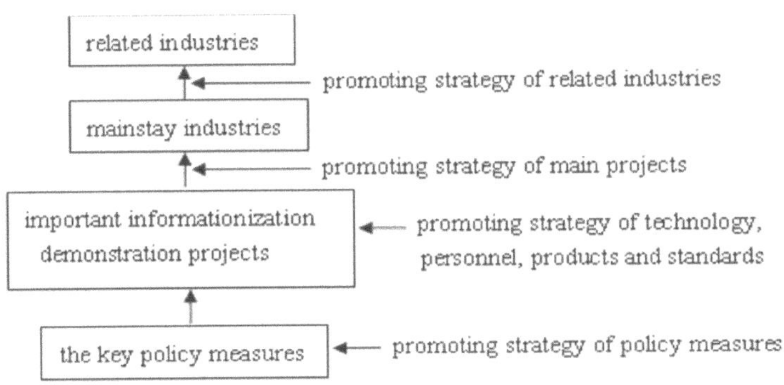

Figure 4. The implementating way of county region economy informationization

5. CONCLUSION

The county region economy informationization, an important component of the region informationization in China, is an efficient path to reduce "the digital gap" between urban and rual areas. The article has limited the concept of county economy informationization and analyzed its characteristics. On the basis of the present situation and main problems of informationization, the developmental strategy of county region economy informationization in China was discussed from the guiding odeology, the basic principle, the system frame and the implementating way.

ACKNOWLEDGEMENTS

Funding for this research was provided by Informationization Office of the State Council of China. The authors are grateful to the Hebei Agricultural University and the Hebei Academy of Agricultural and Forestry Sciences for providing them with help in the research.

REFERENCES

Liu Guicai. The Agricultural Informational Work in American, Journal of Agricultural Outlook, 2006, (1):32-33 (in Chinese)

Tian Ye. Agricultural Informationization and Its Inspiration inJapan, Journal of Global Technology Economy Lookout, 2001, (1):47-48(in Chinese)

Zong Yu, Xu Honglin. An Analysis on Rural Information Construction in China, Journal of Rural Economy, 2006, (12):96-98(in Chinese)

DESIGN OF SYSTEM SCHEME AND OPERATION MECHANISM ON AGRICULTURAL SCIENCE & TECHNOLOGY INFORMATION SERVICE SYSTEM '110'

Yongchang Wu[1,*], Zhiquan Hu[2], Bilin Xiao[1], Quanxin Li[1]

[1] *Institute of Agricultural Resources and Regional Planning(IARRP), Chinese Academy of Agricultural Sciences(CAAS) ,Beijing 100081,P.R China*

[2] *Institute of Agricultural Economics and Development, Chinese Academy of Agricultural Sciences (CAAS), Beijing 100081,P.R China*

[*] *Corresponding author. Current Address: Institute of Agricultural Resources and Regional Planning, Chinese Academy of Agricultural Sciences, No.12 Zhongguancun South Street, Haidian District, Beijing, 100081,P.R China.Tel: +86 10 82109633-807; Fax: +86 10 82109633-807; Email address: wuyc@mail.caas.net.cn*

Abstract: Abstract: Agricultural science & technology information service system '110' (ASTISS-110), connected through unitary telephone hotline as well as multipurpose service of the network, television and video etc, is one of the most characteristic content of the Chinese rural informatization. ASTISS-110 is a low cost and high efficiency way to make the agricultural science & technology achievements extension and achieve the combination of science & technology with farmers in the rural area. This paper would primary focus on the ASTISS-110 foundation and system principle. On basis of its main functions and system objectives, we put forward the combination of the 'Sky-Land-People' technical solution, and analyze the management operation mechanism from commonweal service, enterprise management and commercialization operation.

Keywords: ASTISS-110, system, technical scheme, operation mechanism

---

Please use the following format when citing this chapter:

Wu, Y., Hu, Z., Xiao, B. and Li, Q., 2009, in IFIP International Federation for Information Processing, Volume 295, *Computer and Computing Technologies in Agriculture II, Volume 3*, eds. D. Li, Z. Chunjiang, (Boston: Springer), pp. 1939–1947.

1. INTRODUCTION

During the transformation period from the conventional agricultural world to an industry and informatization society in China, issues concerning Agriculture, Countryside and Farmers are the biggest challenge (Mei Fangquan, 1997; Xue Liang, 1999,). Nowadays technology extension has made great achievement, but main extension modes in most of regions still concentrate on on-site instruction and training by technician at countryside (Shen Guiyin et al., 2003). However, the communication limited by the factors involving time, location and people etc. has caused the extension method unitary. The original agricultural extension system is difficult to meet with epoch requirements on account of the fund shortage, job hopping and vulnerable for expanding the use of agricultural technology. The only way to achieve the agricultural modernization and informatization and make the rural economy flourish is to spread and exploit the information technology which makes farmers fast master the science & technology and information resources in time, convenient(Jia Shangang,2000; Bai shuo,2003;Chen liangyu,2005).

The farmers' demand for science & technology information is limited by the space-time and economy. China has already established the agricultural information service network coverage at provincial, city and county level. However, when the information spreads to the countryside, village and rural household, the price of the computer is almost half of the farmers' annual revenue. Inconvenient transportation, poor education condition, traditional production mode and limited ability to get information resources seriously effect the construction of the agricultural information service system. 'Last one-mile' and 'Digital Gap' have become the most popular issues in the construction of the agricultural information service system. Spot check results indicate that about 0.5% of online users are usual farmers and about 1.3% of the farmers have experience to harvest the science & technology and information online(Hu Zhiquan et al.,2005).

Considering the situation of the rural area in our country, farmers can fast, conveniently and accurately get help by choosing the cheapest telephone to communicate with the science & technology extension people, agricultural experts or technical support and information service, computer system and networks. The number of the telephones has exceeded 640 million, customers of fixed telephone and mobile phone separately amount to 310 million and 330 million. The rural fixed telephone customers have reached 102 million accounting for 32.5% of the total (Wu Yongchang, 2006). Since the concept of agricultural science & technology service system '110'(* Note: '110' is an emergency call number in China, it was referred to quick response in the Agricultural science and technology Information service system.)(ASTISS-110) has emerged by Quzhou City, Zhejiang province,

ASTISS-110 has made speedy development in this period. Farmers could get the technological advice and information service by mean of receiving signal by terminal modes of remote satellite, internet, video and telephone etc. that forms some new thought and mechanism. In relative to traditional rural agricultural science & technology transmission modes, ASTISS-110 mode has overcome the influences by space-time limitation and weak interactivity, and has satisfied farmers' growing needs on agricultural science & technology service. It provides a fast channel to make agricultural and rural science & technology information spread in fast, practical, effective and real-time way so that could demonstrate the rich connotation and tremendous vitality (Wu Yongchang,2007).

2. SYSTEM PRINCIPLE OF ASTISS-110

Agricultural science and technology service system '110'(abbreviation ASTISS-110) takes uniform service hotline as the ligament, telecommunication and broadcast network as the foundation, as well as information resources database as the core. By means of connecting with the internet, using the comm. & networking and digital technology, adopting the multiple teminals, perfecting the rural information network and integrating the rural service power, farmers can get the information cheaper, faster and more effective. Information has functions for farmers in making rich in a rapid reaction and zero distance service.

2.1 System structure of ASTISS-110

ASTISS-110 is composed of three networks involving three sub-systems of the sky, the land and the people. The system structure can be seen in details from Fig 1.

The 'sky' network composed of internet network at state, region and county levels is used for information integration and resources optimization. Nowadays 100% of provinces, cities and municipality, 80% of prefecture cities and about 50% counties are equipped the information network facilities. The 'land' network comprises with receiving terminals and call centers hotline network distributed in various cities. Hotline number 12396 funded by Chinese National Spark Program （CNSP） of the Ministry of Science and Technology (MOST) has already put into operation, 20 provinces or cities have constructed 188 ASTISS-110 service sub-centers which are the combining cites of network resources and telecommunications service. It fulfills the information resources distributing to grassroots level and feedback function. The 'people' network made up of grassroots level service organizations and messengers spread all over the country including experts, special correspondents, schools, leaders, leading enterprises, transformation centers, intermediaries, 7000 agricultural materials companies. 110 thousand agricultural information messengers from the farmer broker, cropping and poultry managers, farmers' specialized cooperative economic organization and intermediate agencies people are the key to convert information resources to real productive power and implement zero-distance information service.

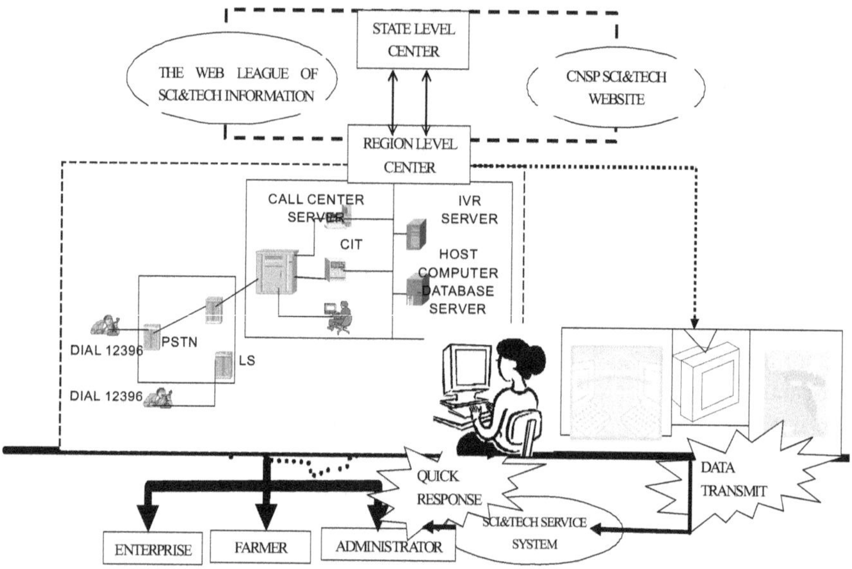

Figure 1: Diagram of 'Sky-Land-People' network structure

2.2 System functions of ASTISS-110

ASTISS-110 system functions consist of information resources integration, sharing, service and feedback.

Integrating and sharing. To gathering, arrange and process the dispersed scientific and technological outputs from different departments, internal departments, districts, fields and enterprises. Through the 'Sky' and 'Land' network, it can form the uniform incessant database which can be queried and searched by the information resources providers and users.

Serving and feeding-back. ASTISS-110 institution, the information and control center, could organize the rural science & technology service staff and capital goods dispatching enterprise to put the science & technology and capital goods to rural labors. It could accelerate the outputs transformed and applied by means of the 'Land' and 'People' network.

2.3 Information sharing system and service mechanism

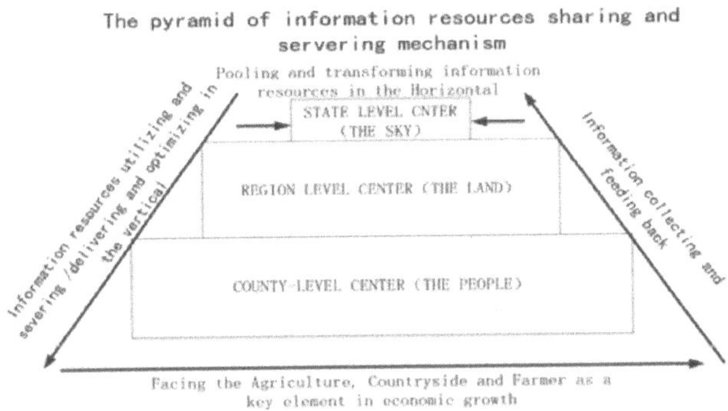

Figure 2: Diagram of information resources sharing and service mechanism

According to the ASTISS-110 operating principle shown in Fig 2, ASTISS-110 has formed sharing mechanism of pyramidal data, service and technology. It is relationship of resources distribution and services from the top to low levels that transfers gradually from outputs information-intensive to scientific and technological service-intensive. It is an information feedback system which forms a smooth information upload channel from low level to the top. The 'Sky' network is based on the principle of resource integration, the 'Land' network mainly is based on network resources and telecommunications service and the 'People' network focus on service gathering and rapid reaction. According to the characteristics of different districts, it uses the uniform hotline 12396 to put forward the demand, adopts various of service modes of the internet, telephone or mobile-phone, television, satellite, broadcast to get the information resources. The core is to combine the technology with agro-material supply and information service in operation.

3. ASTISS-110 TECHNOLOGICAL SCHEME

According to the scheme, the main body of ASTISS-110 is composed of database, network, call center, internet service websites and users' receiving system.

3.1 Database system

Under the uniform classification and standard, database as the resources foundation of the whole platform, at least include two types of supply and need , 9 databases including agricultural science & technology production database, agricultural organization database, agricultural experts database , agricultural policy and rule database, market information database, agricultural training courseware and teaching material database, directory database, demand information database and information feedback database, has the function on inquiry, search, update and maintenance.

3.2 Scientific information sharing and service network system

The Web Services platform, which relies on the website of Spark Program of China, is the important carrier and channels of the whole platform, with which function as rural science & technology outputs exhibition, transaction of the rural science & technology outputs and information, science & technology decision supporting, rural science & technology training. It provides the channel of users' information demand and feedback, and it also has the ability of network platform system management.

3.3 Call center service system

Call Center system, which relies on local junction center, is the hinge in the whole platform with the function of information consultant, information service, information gathering, arrange and feedback, help information automatic classification system, Major Animal Epidemics, plague emergency and disaster emergency report to the leaders. It provides functions on the service mode of automatic alternate voice, manual work, messages, science & technology outputs Information customization and transaction etc.

3.4 ASTISS-110 service websites

ASTISS-110 service websites are the basic service units in the whole platform which provide the direct and fast service for issues concerning Agriculture, Countryside and Farmer. By means of constructing front-line science & technology service platform with multiple technical services that include department of agriculture, forestry, aquatic products, livestock, water resources, weather and science & technology extension units resources which involve colleges, science &technology special correspondent, schools. There are still including social intermediary agent, agricultural technological entrepreneurship, guilds, rural organizations for economic cooperation etc.

3.5 Users receiving system

Users' receiving platform is multiple terminals platform that is the target group of agricultural science & technology and diversifying receive mode. The receptive system could be information receiving terminal such as telephone, cell-phone, television, wire broadcasting or wireless broadcasting, internet, satellite, publication, blackboard newspaper etc.

4. OPERATION AND MANAGEMENT MECHANISM

4.1 Scheme I: Commonweal service

ASTISS-110 is a commonweal in a long run except multi-channel financing. Supporting and training the disadvantage group farmers are all base on the principle of government and public supporting around the world. We suggest that Ministry of Science and Technology (MOST), All China Women's Federation (ACWF), Central Committee of Communist Young League, Ministry of Information Industry, Ministry of Agriculture, Ministry of Education work together to set up non-profit organization Chinese ASTISS-110 Information Service Fund which could accept donation by the caste, enterprise, individuals and government branch.

This scheme has strong advantage in building up the ASTISS-110's framework at the beginning, but lack of flexible and long-term inefficiency operation and management mechanism could be the government burden and adverse to sustainable development for ASTISS-110.

4.2 Scheme II: Enterprise management

Enterprise invest in ASTISS-110 with comprehensive enterprise management that could develop enormous commercial opportunity, Guangdong province has already tried to adopt the operation mechanism. The advantage of this scheme is so flexible that beneficial to develop ASTISS-110 and its convenient operation and less burden for the government. However, it is difficult to form the service network at the beginning. When conflict of interests between the enterprise and farmers has emerged, farmers always were at a disadvantage for enterprise pursuing the maximal profit.

4.3 Scheme III: Commercialization Operation

The implement of ASTISS-110 is tight combined with the extension mode of agricultural science & technology expert yard and special correspondent. It can achieve the goal of government administration; enterprises dominant, farmers benefiting to make the government guidance combined with market-oriented, non-profits service combined with productive service.

It should depend on the local government to construct the ASTISS-110 service platform and with the given necessary polices and financial support to make full play to serve issue concerning Agriculture, Countryside and Farmers. The departments of science & technology, agriculture, forestry, telecommunications should work together to organize the colleges, scientific research institutes, corporate champion and various service organization to carry out technical training, develop expert database and service standards and also supervise and check the process to promote the healthy development.

To enable ASTISS-110 sustainable development, the ideas of market, corporation management and capital efficiency play a key role in pushing such mentioned organizations to undertake the ASTISS-110's construction and management to form a famous brand of public interest social system. By means of the ASTISS-110's service, the profits could be brought out from market service to choose quality agricultural products, virtue seed and non-gratuitous technical guidance and service from the agricultural experts and special correspondents.

5. CONCLUSIONS

To construct the rural information service network ASTISS-110. On the one hand, it could meet the farmers' growing demand for agricultural science & technology informatization, and it is an effective path in resolving the problem 'last one-mile' in rural informatization. On the other hand, it could integrate the technology, talents and outputs resources from colleges, universities, and institutes and accelerate the commercialization of research finding and technology extension. It is not only propitious farmers to shake off poverty and set out on a road to prosperity, but also play an important role in constructing the new socialist countryside and helping farmers, agriculture and countryside to achieve leaps in economic development.

ACKNOWLEDGEMENTS

Funding for this research was provided by Chinese National Spark Program （CNSP） of the Ministry of Science and Technology (MOST) (P. R. China) and Scientific Research Fund for National Non-profits Academy & Institute of Ministry of Finance (MOF).

REFERENCES

Bai Shuo. Analyzing Agricultural Informatization and Stimulating Farmers' Income. Rural Economy 2003, 6, 57-59(in Chinese)

Chen Liangyu, Chen Ai Feng. Present situations and development orientations of China's Rural Inform ation Development. Review of China Agricultural Science and Technology, 2005, 7, 2 ,67-70(in Chinese)

Fu Hongxun. Chinese Agricultural Informatization Development. Problem of Agricultural Economy. 2002, 11,44-47(in Chinese)

Hu Zhiquan, Wu Yongchang, Cheng Guangyan. Informatization and issues concerning Agriculture, Countryside and Farmers. cChina Information Times, 2005, 16,16-18 (in Chinese)

Jia Shangang. Golden Agricultural Project and Agricultural Informatization. Agricultural Information Study. .2000, 1, 5-10(in Chinese)

Lu Xiaoyan. Agricultural Modernization and informatization. Chinese Agricultural Science & Technology Press. 2002, Beijing (in Chinese)

Mei Fangquan. Prospects of China Informatization Construction. Computer and Agriculture. 1997, 3, 1-3(in Chinese)

Mei Fangquan. Strategy Analysis of Agricultural Informatization for Agricultural Modernization. Chinese Rural Economy, 2001, 12 (in Chinese)

Shen Guiying, Gu Huang Zhang. Trends of Agricultural Informatization Development in Foreign Countries. World Agriculture, 2003, 8, 4-6(in Chinese)

Wu Yongchang. ASTISS-110's current situation and system assumption. Proceedings of ASTISS-110 high level forum. Hainan, 2007 (in Chinese)

Wu Yongchang. Chinese Spark ASTISS-110 system principle and design scheme. Proceedings of 20th annual Spark program high level forum Hainan , 2006(in Chinese)

Xue Liang, Fang Yu. Agriculture Informatization. Jinghua Press, 1999, Beijing. (in Chinese)

Zhao Ping. Theory of Information Resources Development in 21th Century Agricultural Informatization. Economy Science Press. 2000, Beijing(in Chinese)

INFORMATION SERVICE SYSTEM FOR SMALL FOREST OWNERS

Shaochen Zhang, Yun Li [*]
School of Information Science & Technology, Beijing Forestry University, Beijing, Beijing, P. R. China
[*] *Corresponding author, Address: School of Information Science & Technology, Beijing Forestry University, Beijing 100083, P. R. China, Tel: +86-10-62338246-426, Email: liyun@bjfu.edu.cn*

Abstract: Individual owned forests have boomed in the last decade in China. Hundreds of millions of private forest owners have emerged since years of afforestation practice and collective forest ownership reform. Most of those private forest owners are former peasants living in afforestation areas. They thirst for forestry information, such as technique knowledge, forestry policies, finance, marketing, etc. Unfortunately the ways they could get certain information are very limit. Before internet time, Local governments are the main channel they search helps for useful information and technique supports. State and local governments have paid much attention to provide necessary forestry technique supports to those small forest owners and provided varies training projects, issued official forestry information through their websites. While, as state government expands household contract system in the management of collective forestry land, the number of individual forest owners is bumping up in future 5 years. There is still a gap between supplying ability and requirement of forestry information. To construct an effective forestry information service system in next 3-5 year can bridge the gap. This paper discusses the framework of such an information service system.

Key words: small forest owners, information service system, forestry information

1. INTRODUCTION

The total area of plantation forests in china has exceeded 53.6 million hectares. It ranks the first in the world (SFA, 2008a). Most of these

---

Please use the following format when citing this chapter:

Zhang, S. and Li, Y., 2009, in IFIP International Federation for Information Processing, Volume 295, *Computer and Computing Technologies in Agriculture II, Volume 3*, eds. D. Li, Z. Chunjiang, (Boston: Springer), pp. 1949–1955.

plantations own by collectives and individuals. Since year 2003, China has implemented a household contract system in the management of collective forests in Fujian, Jiangxi, Liaoning and Zhejiang provinces. Millions of forestry lands have contracted to farmers. This system will be expanded nationwide from this year. It was recognized as another milestone in the country's transformation of rural relationship of production after it adopted the system for collective farmland thirty years ago. Through 70-years contracts, farmers would be put more effort into planting and growing trees. They expect to get more income from the forest land they leased.

China has a mature forestry service system for state and collective ownerships of the forestry that provides investments, forest plans, technique expansions and trainings etc. Local governmental forestry agencies response to technique supports for seedling, tree plant, forest pests and diseases prevention. Local forestry bureaus made forest management plans. Individuals needn't consider much to forestry information. Now, the situation is different. Forest lands have been divided into small pieces. Their owners must think about how to manage the land and timbers to get more productivity and earn more money. They need varies forestry information which support them to achieve their objectives. Where they can seek help and where they can find useful information? Government, industry, research institute and university are places they could ask for help. While, only few of those private forest owners know where the right direction is they could find the proper information. Even they know where the source of information they looked for is, they might not know to whom they could contact and get the information. A former research (Chen Yongfu et al., 1997) showed that the extensive management, poor commercial awareness and lack of science and technology of forest owners significantly hinder the forest production and management.

As the development of information technology, using internet, mobile, together with tradition channels, establishing the information service system for small forest owners can help them get enough forestry information.

2. THE CURRENT INFORMATION SERVICE FOR SMALL FOREST OWNERS

At present, government plays a main role in providing information service system for small forest owners in China. Local forestry bureaus public newsletters, give seminars and training courses to small forest owners periodically. Many expansion projects put forward every year. Technicians and specialists give on the spot instructions aperiodically. They are welcomed by private forest owners, while the number of technicians and specialists are far from enough. Most individual forest owners have no chance to meet a specialist who could answer all the questions about their tree planting and timber management. The first place they seek help is local government. We once run into a farmer when we did survey in Wenan County, Hebei province in last November. He went to local forest bureau to ask where he could find an agency that could help him analyze his soil and the water he irrigated his timbers. He hoped the forest bureau had such a department which did test and gave him a report. He said he suspected the water quality because his trees were dying. Wenan forestry bureau has no such a department. The officer who received him suggested him to try his fortune in Hebei Agriculture University. When that farmer left, the officer told us forest owners came to forestry bureau asked any kind questions about their trees every day, while only partly of them could get satisfied answers. Because many questions were out of forestry categories and officers could not always give them suggestion.

Another example was a farmer who planted 4 mu (0.27 hectare) poplar 3 years ago. His trees grew much slower than his neighbor. We met him in his field when we went on a tour of selecting trail spots last spring. Fortunately, he met us, a deputy direct general of local forestry bureau and a specialist from Hebei Forestry Science Institute accompanied with us. After talking with him, they suggested him to do heave thinning in the fall and promised to help him sell timbers he thinned. The deputy direct general told us since those farmers lacked of necessary knowledge and got no effective information, their plantations were hardly managed. Local forestry bureau had no enough force to deal with so many questions.

Besides tree cultivating knowledge and technologies, those small forest owners also need finance, market and insurance information. No forest owner's associations in China at present. Some individuals cooperate with paper mills or panel mills. Their timbers are parts of raw material forests for certain mills. The timbers owners need to know when the right time they cut woods and get satisfied revenues is. For the insurance issue, few private forest owners considered about buying insurance for their woods before the snow disaster in south China in early 2008. It destroyed 20.86 million hectares of forests and direct cost 57.3 billion yuan (about 8 billion U.S. dollars) according to SFA deputy director Zhu Lieke (Xinhua, 2008) said in a news conference. Over 2.6 million forestry population has been affected by the disaster. Now they know the important of insurance. While, what kind of insurance should they buy and where and how much to buy are still questions they need solve. Is there a portal they could find proper information timely? An information service system could meet the demand.

3. THE IDEA AND PLAN OF CONSTRUCTION AN INFORMATION SERVICE SYSTEM FOR SMALL FOREST OWNERS

The construction of the information service system for small forest owners should be planned from an overall perspective and carried out gradually. Government gets the leadership of providing services for small forest owners from the very beginning. A technology service action for forest ownership reform has put forward from this year. Small forest owners should establish their associations and provide services for individual members. China forestry industry association had come into existence since August 2007. Its website provides information surrounding to the industry. At present few small forest owners joined into this association. Perhaps most of them have not known about such an organization. Forestry education institutes, public libraries, media and other information services providers are all having opportunities to involve in this system and providing varies kind of forestry information resources and services. Building such a system will spend 3-5 years and need lots of investment. It also need participants focus on information demands and integrate varies relevant resources. Specifically speaking it contains two stages as follows:

3.1 Stage one—the government plays the leading role in the course of the construction of the information service system and provides public welfare services

In the first stage, the government plays the leading role and constructs the information service system frame from three aspects, the organization system, the infrastructure and the resources and the service mode. In this stage public welfare services are provided.

(1) The construction of organization network

The organization network is the service system's support. Service stations can be set up at the forestry bureaus, such as the forestry administration and the forestry station to form a third-level organization system——"district service center, township service station and village information point". The district service center as the district-wide information gathering and exchanging center is responsible for collecting, reorganizing and processing the information and also issuing the information to the forest owners through the township service station and the village information point. As a middle level, the township service station takes charge of information classification and releases it to the village information point according to the different demands of villages. It is also responsible for giving information from the village information point in to the district service center. The village information point is responsible for collecting information and demands from forest owners directly and gives that in to the township service station.

(2) The construction of the infrastructure and the resources

The construction of the infrastructure and the resources is the service system's foundation. Rural area informationization gives many successful experiences to cite. Such as village service points and agriculture websites, call centers, local training TV programs are all good information service forms. SFA has already made plans to provide service for newly emerged small forest owners. Government, forestry research and education institutes, technology expansion groups are all welcome joining in this action. Many projects, such as research results demonstration, technical training, special websites construction, education textbooks, special manuals, brochures and CDs publishing, etc. will start up in next 3 years (SFA, 2008b).

Human resources are very important in building information service system. A human resource development plan has made by SFA. An action plans to train ten thousand householders to master one specialized forestry technique respectively (SFA, 2008c). Specialists going to mountain areas to give instructions and seminars to forest owners, evening school and training courses are included in this plan.

(3) Service mode

Forest owners' characteristics must be taken fully consideration when the service mode is selected. Three kind of service mode is feasible.

(a) Setting up the network service platform

Set up a comprehensive service platform based on a forestry website which provides information for the forest owners to meet their daily life needs, such as the forestry policy, the forestry science and technology, market information, loan and insurance, the recruiting of workers information, the weather and the transportation and so on. Furthermore, voice mode is also supported. Forest owners can contact the Technicians through the 24 hours hot line. Daytime the forestry bureau technicians are responsible for the hot line, at night the hot line answers the call automatically and records the request. The second day the technicians will answer the request according the record. SMS and mobile application software are promise service resources, as mobile is popular in mountain area. It might be a better solution of 'last one kilometer' problem.

(b) Establishing floating technical service team

The floating technical service team mainly has two responsibilities: first, it is in charge of giving the unsolved questions through the hot line a face-to-face solution. Second, transform the passive consultation into active service. It can give on the spot instructions and exchange information with the forest owners regularly in persons and get feedback directly.

(c) Establishing forestry science and technology center

Establish the multifunction forestry science and technology center in the district service level and provide "one-stop" service. This center supposes to be five areas as follows.

■ Information inquiry and release area. This area provides free internet surf service for forest owners. Furthermore, both desktop and touch-screen computers are equipped.

■ Expert consulting area. Forest owners can come here for free consultation from experts. Telephone consultation is also supported.

■ Product demonstrating area. This area exhibits materials and tools for forest owners, such as good quality seeds, tree measurement tools etc.

■ Reading area. Forest owners can read and order books and periodicals in this area.

■ Immediate news releasing area. Release the market quotation information, supply and demand information and science and technology news.

3.2 Stage two—various service providers participate under the government's guidance to form ASP operation mode

During the first stage, the government is the main service provider who can not only provide the powerful policy safeguard but also create a favorable environment in all respects for building the information service system. However, that the government is the main service provider exists several pitfalls. First, the services provided are non-profit. As a consequence, the related service group's initiative could not be fully aroused. Second, the government is responsible for both the administration and affairs which is supposed to be duty of the public institutions. This leads to the information services initiated to a great extent by the government's arrangement not by the forest owners' needs (Yu Liangzhi et al, 2007). Therefore, in the second stage, the operation mode should be changed and let more service providers participate under the government's guidance.

In the second stage, ASP mode is a good choice. ASP (Application Service Provider) is a new outsourcing mode which appears along with development in technology and trends in the business environment. ASP offers individuals or enterprises access over the Internet to ASP software and related services that would otherwise have to be located in their own personal or enterprise computers and ASP is also responsible for the maintaining and updating (Zhou Nande, 2004). All the individuals and enterprises have to do is to pay for the software and services. In this stage, the government can rent the software and related services from the ASP, hand over the responsibility of providing information service to ASP and concentrate on administration.

ASP possesses several benefits. First, service providers such as software providers, equipment manufacturers and information providers, can take advantage of their professional skills to provide professional services in all aspects, while the government can concentrate on the administration and provide fund and powerful policy safeguard. Second, ASP can provide the forest owners the ability to access a broader range of applications and professional technical support. Along with the development, the ASP alliance can be formed. More service providers such as the physical distribution dealers and the forest dealers can join the ASP alliance to form an industrial chain and provide a better service for the forest owners.

4. CONCLUSION

With the implementation of forest ownership reform, thousands of millions individual small forest owners come forth. They need varies kind of information service to help them manage their woodlands and increase land productivity and get rich. Government has acted to provide essential services for them, while, it could not meet all demands. To build an information service system and involve non-government service providers can solve the problem. This paper analyses the present situation of the information service for forest owners, and gives the specific ideas for the construction of an information service system. Much further work should be focus on. For example, forest owners' objectives and information demands, their information behaviors are issues should be considered of.

REFERENCES

Chen Yongfu et al. Problems and Countermeasures in Development of Rural Forest Farms[J]. Journal of Zhejiang Forestry College, 1997, 14(2): 199-203. (In Chinese)

SFA . 'Hundred Counties, thousand villages and ten thousand householders' Science and technology demonstration action implement plan [D], State Forestry Administration, 2008c, 7pp. (In Chinese)

SFA. National forest resource 2006 report[OL], State Forestry Administration, 2008a, http://www.forestry.gov.cn/data/lytj/2006q/tj_01.htm.(In Chinese)

SFA. Science and technology service action for forest ownership reform[D], State Forestry Administration, 2008b, 4pp. (In Chinese)

Xinhua. Winter storms cost China's forestry sector $8 bln[OL],2008.2.19, http://english.people.com.cn/90001/90776/90884/6356519.html

Yu Liangzhi et al. Establishing a Rural Information Service System for New Farmers: a Case Study in Tianjin[J]. The Journal of the Library Science in China, 2007, 33(6): 33-35. (In Chinese)

Zhou Nande. Study on ASP-based Information Technology with Multiple Tier[J]. Application Research of Computers, 2004, 21(4): 55-57. (In Chinese)

RESEARCH ON DECISION-MAKING SUPPORT OF CHINESE RURAL LAND TENURE INFORMATION SYSTEM

Jun Tan[1], Hongyou Su[1,*]

[1] *Dept. of Land & Real Estate Management, Renmin University of China, Beijing 100872, China*

[*] *Corresponding author, Address: Dept. of Land & Real Estate Management, Renmin University of China, Beijing100872, China. Tel: +86-10-62517572, Cell Phone: +86-13581684146, Email: hustshy@163.com*

Abstract: Since 1949, the information of land tenure has a positive effect on defining the scope of collective land and state-owned land, implementing the system of cultivated land protection and land use control, designing general land use planning, etc. But as the economic and social development, the existing land tenure information is not appropriate anymore and results in many problems. The emphasis in the near future should be placed on establishing rural land tenure information system including cadastral management system, the uniform property registration system and cadastral management information system, defining the scope and content of various collective land ownership, securing peasants' land tenure rights, shortening the gap between urban and rural areas, all of which will guarantee the effective use of information of land tenure for the government's decision-making.

Keywords: Land Tenure Information, Cadastral Management, Tenure Rights Investigating, Land Registration

1. INTRODUCTION

The Chinese current land right system in force composes of land ownership, land use right and the other rights on land.

Until the establishment of Property Right Law last year, the long history of taking the registration of real estates as an administration measure had

---

Please use the following format when citing this chapter:

Tan, J. and Su, H., 2009, in IFIP International Federation for Information Processing, Volume 295, *Computer and Computing Technologies in Agriculture II, Volume 3*, eds. D. Li, Z. Chunjiang, (Boston: Springer), pp. 1957–1966.

resulted in a mess in the regulations of real estate registration that can be seen in diversified independent civil laws or administrative regulations or judicial interpretations, e.g. Real Estate Administration Law, Mortgage Law, Land Administration Law, Agriculture Law, Contractual Management Law, Grassland Law, Forest Law, Land Registration Regulations, Forest and Forest Land Tenure Registration Administration Method, Sea Area Use Right Registration Method, etc.. And since a Real Estate Registration Law has never been existed so far in China, there is still no uniformed provision governing the registration of real estates. In addition, adversely affected by the impact of the society conversions, our legislation bodies have obviously placed too much importance on their administration while pay less attention to the function of guaranteeing transaction safety（Tan Jun,2005）.

In late 1980s, the land and resources administration authorities have set out the registration of the use right of state-owned land, the ownership of rural collective's land, the use right of collective's land and other rights on land. With great efforts for decades, the registration for state-owned land in urban areas has been finished almost completely. Now, the registration of the ownership of rural collective's land has been progressing all over the country. However, the registration of the use right of collective's land has been carried out only in a small part of the vast rural areas (Evelyn Chia, 2006).

2. CONSTRUCTION OF COLLECTIVE-OWNED LAND TENURE INFORMATION SYSTEM IS AN IMPORTANT FACTOR AFFECTING THE STABILITY OF RURAL AREAS

According to analysis of Ministry of Land and Resources on letters of complaint from the people in recent years, the visits due to condemnation of rural collective-owned land, land compensation, and vague land ownership took up considerable proportion of the letters of complaints from the people.

2.1 Upper rural collective organizations infringe the rights and interests of lower rural collective organizations.

Major embodiments: The township (town) allots land of administrative villages at will or the administrative village freely allots the land of village teams; the township or administrative village freely allocates land compensation belonging to villages or teams. According to initial analysis of department for letters from and calls by the people of Ministry of Land and

Resources, visits due to above disputes and cases of teams accusing villages and villages accusing townships have been growing in recent years.

2.2 Land users owe land compensation for long.

It occurs from time to time that land users owe land compensation. Certain chemical plant in Panjin City of Liaoning Province owed land compensation totaling 1,600,000 Yuan to certain villager team for 12 years, peasants of the team urged payment and appealed to higher authority several times in vain, and they even blocked the plant gate several times. Hubei Zhongxiang City Bureau of Public Security, Nianpanshan Water Resources Development Company, etc. owed land compensation totaling 3,000,000 Yuan to Gongbeiqiao Village successively since 1993, which seriously affected resettlement of peasants and economic development of the village, and caused great dissatisfaction of peasants (Wang Shuzhen,2004; Tan Jun,2005).

2.3 Villages and teams in the part between urban and rural areas resisted land condemnation due to low condemnation price and risk of unemployment.

Such problems become more striking in economically developed regions. For instance, there are nearly 200,000 peasants in Panyu City of Guangdong Province who have no land or lack land and are much annoyed against land condemnation. And land condemnation is rarely paid enough according to existing standard. For land condemnation by Linying County Development Zone of Henan Province, the price should be 20,000~25,000 Yuan per MU according to legal standard and yet the government fixes the price only at 3500~7000 Yuan per Mu. No. 201 National Road condemns 700Mu of cultivated land in Wanli Town, Dongsheng City, Yi League, Inner Mongolia, land compensation should be 4000~5000 Yuan per Mu according to existing laws, yet only 1500 Yuan is offered per Mu, and so the peasants resist condemnation (Yang Sui,2005).

2.4 Adjust land by taking advantage of second round of land contracting.

Some localities freely break the boundary of villager teams to out-contract land during second round of land contracting, which leads to land disputes; some local villages and townships adjust some land for intensive management by taking advantage of second round of contracting, which leads to disputes; some localities out-contract, within the scope of the whole village, the land, which the village has been developing and running for dozens of years and which once belonged to villager teams, but the villager teams ask for return of the land ownership, which results in land dispute. These disputes exist to different degrees in all localities, and some of them affect spring farming and have to postpone second round of contracting after autumn (James Kai-sing Kung,2002).

Above disputes are striking problems in current development and stability of rural areas and more complicated, many of them are related to construction of collective-owned land title system. The reasons are exhibited in two aspects: First, collective-owned land has vague ownership and boundaries; second, the contents of ownership of collective-owned land in rural areas are unclear. If not handled properly, the disputes will affect the normal operation of local governments, demand much energy of all levels of cadres, and be unfavorable for protection of the rights and interests and stability of rural areas, unfavorable for arousing the enthusiasm of peasants to remain and input investment in cultivated land, and unfavorable for the planning, management, protection, and reasonable land use.

3. CURRENT PROGRESS OF OWNERSHIP INFROMATION SYSTEM CONSTRUCTION OF COLLECTIVE-OWNED LAND IN RURAL AREAS OF CHINA

Chinese government began to establish the rural land tenure information system by promoting ownership registration and certificate issue of collective-owned land in rural areas of China. The trial project of adjudication investigation of collective-owned land in rural areas of China started formally in 1999. By the end of 2005, Ministry of Land and Resources had set up trial places for rural collective-owned land adjudication investigation in 24 counties (cities) of 11 provinces in China. The special cost for investigation of land and resources totaling 7,300,000 Yuan is input in the trial places, all trial districts of the project have fund support, and the ratio of national financial cost over local match cost is about 1:4. Meanwhile, most districts have started such operation on the basis of self-trial successively.

According to comprehensive statistics of Ministry of Land and Resources on cadastral management, contrast to 2006, the initial use right registration and certificate granting of state-owned land increased 2,494,000 in the year of 2007, while the initial ownership registration and certificate granting of collective-owned land increased 75,000 and the initial use right registration and certificate granting of collective-owned land increased 3,510,000. China had accomplished 53% ownership registration of collective-owned land and 73% use right registration of collective-owned construction land.

Ownership registration and certificate issue in China falls into three categories according to the degree of completion: 1) provinces (regions) of rapid progress, mainly include Shandong, Xinjiang, Heilongjiang, Fujian, Anhui, and Gansu, etc. In these regions, adjudication investigation and certificate issue of rural collective-owned land started early, and some of them are trial regions of the state. 2) Provinces (regions) of ordinary progress, mainly include Hebei, Henan, Guizhou, Jiangsu, Hainan, Hubei, Shanxi, etc. These regions started registration and certificate issue in 2000~2002, and the work went on smoothly since the government paid high attention and took effective measures. 3) Provinces (regions) of slow progress mainly include Sichuan, Guangdong, etc. Registration and certificate issue was restricted by all kinds of factors.

4. ANALYSIS OF THE EFFECT OF OWNERSHIP REGISTRATION AND CERTIFICATE ISSUE OF COLLECTIVE-OWNED LAND IN RURAL AREAS OF CHINA

4.1 Strengthen land ownership construction to maintain the rights and interests of collective-owned and state-owned land.

Land registration and issue of land ownership, use right, and other right certificate to land obliges is not only an administrative act but also a very solemn legal act. For a considerably long period to come, title construction and improvement of collective-owned land in rural areas of China will still be the most important problem among "three-agriculture" problems. Current adjudication survey, registration, and certificate issue of rural collective-owned land makes the ownership of collective-owned land in counties (cities, districts), townships (towns), and villages more definite, and confirm them legally, which will play an active role in improving land ownership system, developing land use right and other right systems, establishing land title registration system, clarifying the legal effect of land registration, strengthening rural contract responsibility system, establishing the modern title system that renders clear ownership, definite rights and liabilities, strict protection, and smooth transfer, strengthening mediation and treatment of land disputes, and maintaining the rights and interests of collective-owned and state-owned land.

4.2 Improve cultivated land protection mechanism and strictly implement cultivated land conservation system.

It has been the goal of all levels of departments of land and resources to improve cropland protection mechanism, and implement most strict cropland conservation system. Title investigation, registration, and certificate issue of collective-owned land in rural areas reflect the quantity, type, distribution, and ownership of counties (cities, districts), townships (towns), and villages in data table and also put such information on geographical maps to provide scientific basis for achieving dynamic balance of cultivated land acreage, establishing long-term mechanism for protection of basic farmland, strengthening and improving cropland protection target responsibility system, cultivated land occupation and supplement responsibility system, and land price appraisal system, and lay sound foundation for implementing the most strict cropland conservation system, discovering problems in use and protection of cultivated land, and investigating and punishing illegal acts involving land.

4.3 Enhance the technical level of land management.

The major results of ownership investigation of rural collective-owned land and certificate issue may be used to update the land use database at county (city, area) level and establish rural cadastral database, which will greatly promote network management of land market information, cadastral information, land use information, and land market information, as well as standard management of rural cadastre. Not only speeds up information of land and resources and provides the service platform for sharing land information, but also promotes transfer of land management to modern method and creates favorable conditions for enhancing land management level and social services.

4.4 Solve land ownership disputes to maintain social stability.

The existence and occurrence of land ownership disputes seriously affect the normal production and life of people in the disputed areas, and bring about disastrous consequences to agriculture, forestry, and husbandry; in some districts, land ownership disputes lead to fight and personal casualty, affect national unity, and endanger social stability. In the process of ownership registration and certificate issue of rural collective-owned land, a large group of land ownership disputes are investigated and settled, which enhances social stability. Take Hainan Province for instance, since land ownership starts to be fixed, all levels of governments and relevant departments concentrate to solve a group of land disputes influential in the society, investigate and settle 2031 cases totaling 588,200Mu of land, and maintain social stability. Since the ownership of land starts to be fixed, the appeals and complaints of the masses have declined rapidly; the ratio in total cases declined from 64% in the first half of 2004 to below 40% at the end of 2004, and the effect promoting social stability was gradually exhibited.

5. EFFECTIVE APPROACH TO ESTABLISH LAND TENURE INFORMATION SYSTEM IN RURAL AREAS OF CHINA

5.1 Existing problems

The land in rural areas of China has gone through several big adjustments since founding of People's Republic of China, vague ownership exists in large quantity of land between collectives, collective and the state, collective and military forces, enterprises, schools, etc., and most land ownership disputes have complicated reasons, involve broad aspects, last long period of time, and are much difficult for settlement. For 50 years, due to improper settlement of land ownership disputes by some local governments, new contradictions have appeared and influenced society's stability.

Determination of land ownership is governmental act, yet the infringement of peasants' land ownership is usually exhibited in governmental examination and approval of land use in violation of laws and regulations, and the protection of peasants' interest usually depends on the investigation and punishment of local government and land management departments over illegal act, so it is not practical to solve infringement of peasants' interests only through land registration. But we can start from enhancing the consciousness of local governments and land management departments to protect peasants' interests, and strengthening investigation and punishment of infringing peasants' interests in violation of laws and regulations (He Kangning, 2005).

Land registration should be a legal act and the oblige should make application actively. Current ownership registration of rural collective-owned land is actually administrative work. Usually village cadres and peasant representatives believe that such work is arranged by the county government and land department. They do not recognize how much benefit the results of ownership registration can bring about to the peasants. Since the ownership subject of rural collective-owned land is not definite, the peasants pay more attention to the contracting right and use right of land.

Some leaders of local governments and land resources departments have not sufficient understanding of ownership registration and certificate issue of collective-owned land, and can not relate such work to solving three-agriculture problems, protecting the legal rights and interests of rural land, maintaining social stability in rural areas. They even believe falsely that to start ownership registration and certificate issue of collective-owned land, fix the ownership of collective-owned land, and clarify the rights of peasants over land will be unfavorable for local condemnation of collective-owned land (Loren Brandt et al.,2004).

At present, ownership registration and certificate issue of collective-owned land mainly makes use of materials on current land use survey. Yet this information poorly reflects the practicality, and new maps are formulated slowly through numerous procedures and can not meet the actual need of adjudication survey. Such situation is fairly striking in economically backward regions, such as Inner Mongolia, Guangxi, Qinghai, etc.

5.2 Solution channels

First, unify regulations, clarify the scope and contents of all kinds of collective-owned land ownership, and solve all kinds of ownership disputes effectively.

Second, considering reduction of peasant burdens and maintenance of rural stability, in view of the welfare nature of ownership registration and certificate issue of collective-owned land, it is proposed to list the costs for such work into fiscal budget to be shared by all levels of finance or to allow such costs listed from paid use of land, the central authority offers certain sum, and the provinces, cities, and counties provides some amount to meet the costs.

Third, fully display the function of "Collective-owned Land Ownership Certificate" in using agricultural land for other purpose, examination and approval of built-up and associated land, and establishment of land development and reclamation arrangement, take "Collective-owned Land Ownership Certificate" as the key examination document in system and implement it strictly, Combine construction of land use database and land use updating survey, cadastral management etc. to undertake ownership registration and certificate issue of collective-owned land.

Forth, to avoid the problems due to data difference between the result of ownership registration and certificate issue of collective-owned land and the result of field survey for land condemnation effectively, town planning area and planning area of townships, farms, and towns having completed 1/500 cadastral survey, investigate the land ownership, register and issue certificate according to the survey results; town planning area and planning area of townships, farms, and towns having not completed 1/500 cadastral survey, and ownership of other rural collective-owned land within counties, only investigate land ownership, and both sides of ownership sign "Collective-owned Land Ownership Investigation Table" for confirmation. Next, divide the results acquired through field survey into parcels for filing. "Collective-owned Land Ownership Certificate" for the part of rural collective-owned land shall be registered and issued after completion of 1/500 town cadastral survey.

Fifth, considering China has vast rural areas and economic level varies a great deal with different regions, the localities may adopt different technical methods during investigation and certificate issue .In economically well-off regions, make use of orthographical image or aerial picture superimposed with current land use drawing to divide and determine land ownership boundary. In regions with ordinary economic conditions, make use of field investigation and plotting of current land use drawing to make supplement survey of changing land class and divide ownership boundary. In regions with poor economic conditions, if there is no dispute about ownership boundary, current land use drawing may be based to fix ownership boundary in reference to definite ground object (Tao Yinghu, 2005).

Sixth, ownership registration and certificate issue of collective-owned land is highly technical and of strong policy nature, and concerns land ownership survey, land ownership dispute mediation and treatment, cadastral mapping, land registration, cadastral information system construction, etc., so that the work staff should not only be familiar with laws, regulations, and polices on land management but also grasp technologies of cadastral management comprehensively. It is, therefore, necessary to strengthen training, stabilize and construct a highly qualified team, especially township land management team to meet the requirements of modern cadastral management.

6. CONCLUSION

The existing land tenure information system of Chinese rural area can't follow the economic and social developments and leads to many problems. To make use of information of land tenure for the government's decision-making, we should pay more attention on defining the scope and content of various collective land ownership, promoting the ownership registration and certificate issue, applying high technical investigating methods, securing peasants' land tenure rights and using all these measures to establish a new rural land tenure information system.

ACKNOWLEDGEMENTS

Some dates and conclusions come from the study on the law system and theoretic by Ministry of Justice L.R.C in year 2006: the land's property right security and the registration system (06SFB5027).

REFERENCES

Evelyn Chia. Institutions in Transition: Land Ownership, Property Rights and Social Conflict in China. The China Journal, 2006, 55(1):157-158

He Kangning. Explore the rural property right reformation. Theory and Contemporary, 2005(9):20-24(in Chinese)

James Kai-sing Kung. Choice of land tenure in China: The case of a county with quasi-private property rights. Economic Development and Cultural Change, 2002, Vol. 50, Iss. 4:793-815

Li Yuan. Persist on system innovation and push the reformation of rural collective land used for building system. National Land & Resources Information, 2003(6):34-37(in Chinese)

Loren Brandt, Scott Rozelle, Matthew A Turner. Local Government Behavior and Property Right Formation in Rural China. Journal of Institutional and Theoretical Economics, 2004, Vol. 160, Iss. 4:627

Tan Jun. Research of real property right registration system. Beijing: China Literary History Press, 2005

Tao Yinghu. Establish modern rural property right system adapted to market economy. Rural Economy, 2005(11):22-24

Wang Shuzhen. Rural property right system and protection of ploughed land. Theory Front, 2004(18):37-38(in Chinese)

Yang Sui. The basic character of rural property right system in China. Rural Economy, 2005(6):42-43

DEVELOPMENT STRATEGY FOR MOBILE COMMUNICATIONS MARKET IN CHINESE RURAL AREA

Liwei Zhang[1], Yanjun Zhang[1], Liying Xu[1], Daoliang Li[1,*]

[1] *College of Information and Electrical Engineering, China agricultural University, Beijing, P. R. China, 100083*

[*] *Corresponding author, Address: P. O. Box 121, College of Information and Electrical Engineering, China Agricultural University, 17 Tsinghua East Road, Beijing, 100083, P. R. China, Tel:+86-10-72737741, Fax:+86-10-62737679, Email: li_daoliang@yaohoo.com*

Abstract: Based on full analysis of rural mobile communication market, in order to explore mobile operators in rural areas of information services for sustainable development model, this paper presents three different aspects, including rural mobile communications market demand, the rural market for mobile communications business model and development strategies for rural mobile communications market research business. It supplies some valuable references for operators to develop rural users rapidly, develop the rural market effectively and to get access to develop a broad space.

Keywords: Rural, Mobile Communications, Development Strategy

1. INTRODUCTION

China has a rural population of 737 million, accounting for 56% of the total Chinese population; More than 210,000,000 households, which accounted for 67.6 percent of total family, therefore, it is a huge potential consumer market. In other words, any communication product families as long as the rate increased by 1 percentage point, then it will increase more than 2,000,000 demands. If communication fee grows one yuan in each family, it will generate nearly 2 billion's communication needs. Evidently, compared to facing an increasingly saturated market of communication

---

Please use the following format when citing this chapter:

Zhang, L., Zhang, Y., Xu, L. and Li, D., 2009, in IFIP International Federation for Information Processing, Volume 295, *Computer and Computing Technologies in Agriculture II, Volume 3*, eds. D. Li, Z. Chunjiang, (Boston: Springer), pp. 1967–1975.

difficulties in cities, the rural area is a very attractive large cake for telecom operators.(Jingyu Shi & Jianqiu Zeng, 2007)

However, since farmers have weak awareness of informationization and low-level purchase, in order to bear the fruits of information technology on barren land, it needs a huge investment and long-term market foster. In that case, how will China's telecom operators to balance their own benefits and to open up rural information technology market?

According to analysis of the rural communications market, it is indicated that the primary reason of farmers using the communication tools is to facilitate the hope that rapid access to get useful information, be able to make the best of social and dynamic market for agricultural production and management services. Therefore in the progress of developing rural market, it is important for mobile operators to hold a tight grasp of the characteristics of rural communications to carry out the broad masses of the information service for farmers as a breakthrough point, so that farmers can get the communications tangible benefits, which spurs on mobile operators to occupy the rural market.(Jierong Guo,2006)

To explore sustainable development model for the rapid development of the rural operator customers, effectively in developing the rural market and get access to a broad the development of space for reference, this paper will analyze market analysis of the rural mobile telecommunications from the following three aspects: demand analysis, operation mode and development strategy.(Qifeng Gu & Chengjie Yan, 2007)

2. RURAL MOBILE COMMUNICATIONS MARKET DEMAND ANALYSIS

Rural mobile user market can be divided into agriculture-related government work, agriculture-related enterprises, township cadres in village, township agricultural extension workers, agricultural marketing major, major processing of agricultural products breeding, farmers brokers, Common farmers, migrant workers and so on .(Qifeng Gu & Chengjie Yan, 2007)

To develop the rural mobile telecommunications market in rural areas, it is first necessary to maximize the excavation of the main information needs, and then analyze the needs of the main requirements of the information needs of different groups of people have great differences.

(1) Major agricultural marketing, agricultural planting and processing large brokers

They are large brokers and business links with the outside world needs to understand the outside world, the mobile communications market in the rural areas of focus groups. The big brokers belonging to high-income groups in rural areas, communications and information services have a strong demand for highly value the development of the market. For such groups, whose main demand is to call, in general the number of calls, call for a long time and overhead charges greater demand for information are relatively large, charges and price adjustment on the role of the market segments are more obvious, they are the most potential users of mobile communications. These users belong to the rural market "customer" , mobile operators can refer to the city's large customer preferences and give appropriate incentives for rural areas, at the same time, doing SMS advertising business specific to some of the bosses , enabling them to become backbone of mobile communication services in rural income growth.

(2) Township and village cadres in the village of the township agricultural extension agents

This type of users have a relatively high education and relatively stable income, in the rural market, they are opinion leaders, seizing some of this kind of people as a development target, it means that the core of the rural market has been obtained. Such users are extremely sensitive to price, because of their higher level of education, relatively good price, the price, the price comparison is understood that all types of information are the main spreaders of the farmers have a greater influence. As a result, seizing this group in rural area is the key point to develop the rural market.

(3) Agriculture-related government officials and agriculture-related business staff

This type of users accounts for an absolute small population, but they are the largest source of customers for mobile communications company. In general leaders in government enterprises and institutions have a handset subsidy costs, the use of mobile phones is out of practical demand for work, they would not consider whether the cost of calls per month expenditure is much higher than average or not. Government civil servants in general have stable revenue, the expenditure on phone calls is also higher than average. In addition, due to the workers of the institutions have relatively high level of education, the value-added services and other services can be quickly accepted.

(4) Common farmers

They are the farmers specializing in agricultural production activities. These staff is characterized by low income, small mobility, less contact with the outside world; but the groups are in large scale, high-status at home, with the power to decide on the need to buy products, it is necessary to tap the market potential. The group stayed in rural areas (the family members of people working out), they are mainly women, children and elders having no direct revenue. However, compared to full-time farmers they have more stable consumer spending and stronger demand for communication.

(5) Migrant workers

It is primarily about local workers or other non-agricultural labor force work staff. They are township enterprises workers who are mainly young people with characteristic of following fashion, liking SMS new business volume, being more sensitive to price, these people belong to low consumer groups, but these large population has a relatively stable income, it is the major forces of rural cell phone numbers growth. It should be noted that: different working groups in the province of its location are mainly working and living in cities, they no longer belong to the rural market as the target group.

3. RURAL MOBILE TELECOMMUNICATIONS MARKET BUSINESS MODEL

The development process of the mobile communications market is mainly divided into three stages including communication services, information services and application services, corresponding to these three stages of market development, the business model in rural areas are basic mode of communication, information and services model and transaction service model.(Yan Wang & Yufei Gao,2006)

3.1 Basic mode of communication

The basic mode of communication is consistent with business model in mobile operator's primary strategic form, the mode of the main mobile operators is located in mobile communications service provider, which is for users to provide access to information transmission, the provision of business service and more single to voice, point-to-point messaging service. It is shown in Figure1.

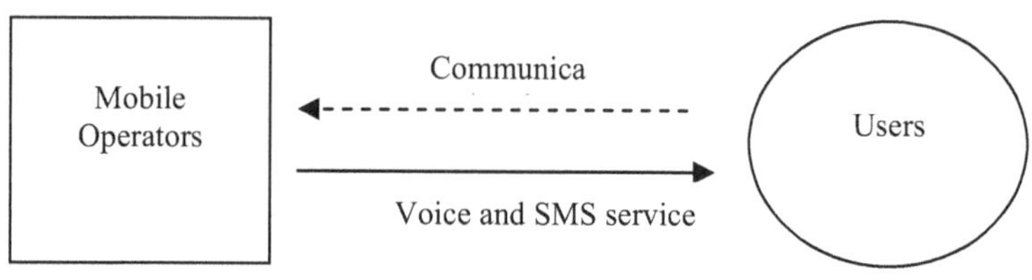

Figure1 : Communication Mode

Mobile voice communication is a basic mode of communication of the most important business, the basic needs of mobile users, the first mobile communications and most common services. SMS is a telecommunications business accompanied with digital mobile communication system. In the early development of mobile communication, it is mainly used to achieve the communication between inter-network text messaging. Through the basic point-to-point voice and text messaging between mobile users, it can meet the communication needs of real-time information, point-to-point transmission.

In the rural areas, voice service is still the largest source of income, secondary to the voice messaging business is SMS, though in far less than the amount of voice service but it has a higher rate and will continue growing. The reasons for SMS business with low levels of certain major cities are mainly because in most parts of the users having low level of education and more lower-age young people working outside.

3.2 Information service model

Information service model is in the process of transacting from the mobile operators in the mobile communications service provider to the mobile information service provider, the formation of a new business model. Now thanks to the content and service providers' growing, the mobile operators are able to provide more rich mobile communications services, which are integrated by the content provider, transmit a large number of information through mobile operators to the hands of mobile users. It is shown in Figure2.

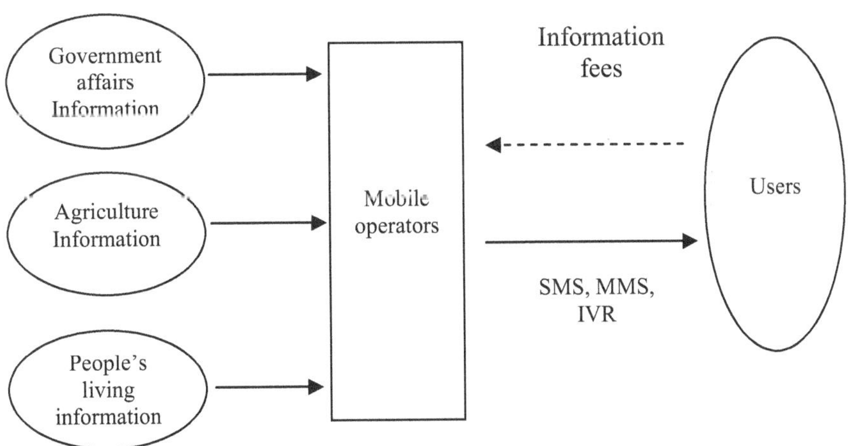

Figure2. Information service model

Custom SMS, MMS and IVR are the main modes of information services business. SMS Custom business refers to mobile phone users to customize Web site or on-demand by mobile operators to provide various types of short message service; MMS custom business, refers to mobile phone users, including customized web site or on-demand video and audio information, pictures , the contents of the text, such as multimedia messaging, text messaging and custom letter is relatively rich and varied content, the amount of information, but it need support MMS mobile phone services; IVR voice service means users dialing the designated number to obtain the necessary information or to participate in interactive services.

Information service model are widely used in the rural market, in the early period China Mobile launched the ICT, primarily by the way of text messages sent to the farmers, such as markets information and technology information, the following introduction of MMS products have been widely welcomed by users. Currently, IVR application in the rural market is in good condition, the voice of agricultural information service hotline is now available for users with automatic voice service, voice messaging services, voice services and artificial voice services.

3.3 Transaction service model

Referring to the main mode of transaction services means to complete transactions online and offline combination of services for customers, such as: trade intermediaries, agents, electronic payment, credit management and logistics services, which are named as depth of services. Mobile operators and application service providers cooperated with achieving customer service for both sides through mobile phon.It is shown in Figure3.

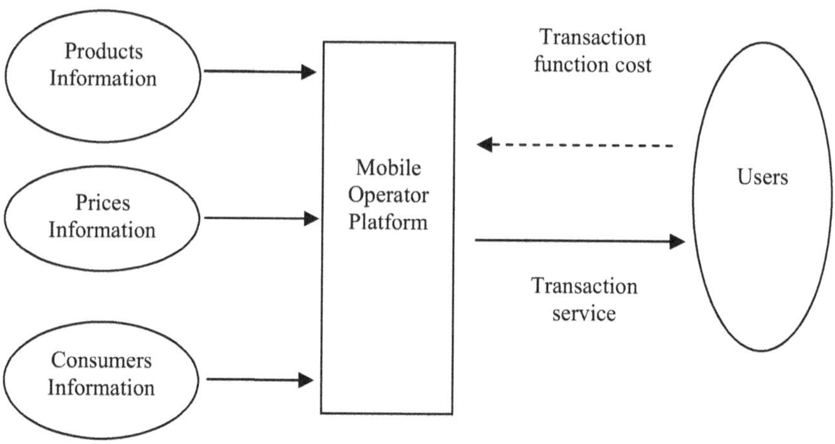

Figure3. Transaction service model

In view of the current situation in rural areas, it could classify information which are uploaded by sale customer by regions, variety, quantity and time of the transaction information, accurate information to trade matching, to solve the problem of difficulty in selling hard to buy; can also be set up agriculture-related supply and demand information exchange center The Center is currently available in addition to providing the functions of the

Internet, the information provided by the transaction, you can enter text messages a central database, the information will first be sent to the user.

4. RURAL MOBILE TELECOMMUNICATIONS MARKET DEVELOPMENT STRATEGY

4.1 Take carrying out rural information service as a breakthrough point to occupy the rural communications market rapidly

After several years' rapid development, China's mobile market has exceeded 400,000,000 users, which become the world's largest market of users. In the growing number of the users, it is worth noting that, from the second half of 2003 to so far, the new low-end prepaid subscribers increased significantly, which means that China's mobile communication market is gradually moving the market to mature, high-end users The growth of stable, low-end users is the reason why the operator open up new users of the main objectives of the user base. With the mobile users in large and medium-sized cities gradually saturated market, it is foreseeable that the next few years, with a large number of low-end users of the 34 cities, towns and rural economic development, such as the extent of the region's poor, will be for new mobile users in the next major battlefield.

Due to China's rural communications market and the cities having their own characteristics including the pursuit of urban consumers personalized, fashion-oriented products and a variety of value-added services, rural consumers pay more attention to send and receive communications with voice and text messages and other basic functions, and practical information services, hoping that the purpose of communication is to facilitate rapid access to get useful information, which can be able to make the best social and dynamic market for agricultural production and management services, In order to develop the rural market mobile communications operators should grasp closely to the rural communications, which is a unique feature for the broad masses of farmers to carry out a service as a breakthrough point, so that farmers can get the communications tangible benefits, which will help them to occupy the rural market quickly.

4.2 For farmers' different spending power in different parts to implement different services strategy

Because of different level of economic development, education and awareness of information in China's rural areas, there is a sense of a significant gap, resulting in different parts of the farmers having different purchasing power of information services, mobile communications operators should carry out information services according to different areas of information consumption level, which should be the implementation of differentiated service strategy.

1. For coastal and eastern suburbs of big cities with more developed economy, farmers have more spending power on information in these region, information services should rely mainly on market mechanisms, mobile operators should adopt the paid services strategy and improve the quality of information services and enhance information services benefits to attract customers at the core gradually open up the market.

2. For the areas with low level in economic development, farmers with limited spending power of information in the central grain-producing areas, information services should be free of charge lower fees and the road through the government sector, farm produce processing enterprises, agricultural businesses, such as the provision of information transmission platform for recovery of the agricultural information service The cost and the current focus on increasing awareness will expand services and improve their awareness of consumer information as well as the nurturing of rural communications markets.

3. For western region with underdeveloped economics, due to the low income of the farmers, who do not have a strong sense of the information, enabling them to spend money buying the information is not practical, to carry out the information services the mobile operators should provide free service, service costs as much as possible So that the supply of agricultural materials and agricultural products processing enterprises and develop the western region as an opportunity to focus the Government on seeking financial and policy support.

4.3 Accurately grasp the profit, to intervene various types of information services in an orderly pattern

For China's rural areas there is a wide range of information services model, from perfect platforms services, improve service efficiency point of view, mobile operators having the space to intervene in China's agricultural market information has not been fully formed, the Government Information Service in rural areas is still in a dominant position, mobile operators must accurately grasp each mode of information services which may link the earnings and profits in accordance with its ease which are involved in the development of the order of priority.

At present, agriculture-related government departments hold a wealth of agricultural information resources, the grass-roots level in rural China is the leading force in the information services, mobile operators should take advantage of its strengths information transmission, the priority to participate in government-led agricultural information services, for the country And local government funding and policy support for the Government to make full use of information resources, organizational systems and influence, and gradually expand its information services business, and through local governments for the provision of appropriate information service application platform to achieve profitability.

Secondly, mobile operators can be involved in agricultural products processing enterprises, agricultural enterprises and farm produce wholesale markets, such as the provision of information services for farmers among the models, to provide them with more convenient and efficient transmission platform to reduce the cost of information services to improve The effects of information through the supply and demand for the release of information companies must charge the cost of profits.

As the farmers association service model is peasant farmers self-service, and our farmers generally don't have high spending power on information, if mobile operators want to involve in the short-term model, which will be difficult to generate considerable economic benefits, if mobile operators can associate with peasant farmers to lead them to participate in the service mode, , for all aspects of operating profit, it is also necessary to seek other kinds of cooperation in a wide range of agriculture-related departments.

REFERENCES

Jierong Guo. 2007. Strategies of how to develop market in rural informatization by telecom companies. *Communication World:* 10: 24-25.

Jingyu Shi, Jianqiu Zeng. 2007. A discussion on the thought of mobile operators in rural informatization. *Mobile Communications,*9: 91-94.

Qifeng Gu, Chengjie Yan. 2007. The opportunities of mobile operators in the rural informatization: to find a market in rural areas. *Management of Communication Enterprise*, 2: 48-49.

Yan Wang, Yufei Gao. 2006. Research on the Value Chain and Business Models for Mobile Commerce. *Logistics Technology*, 9: 63-66.

DEVELOPMENT AND POLICY PROPOSALS OF RURAL INFORMATIZATION IN CHINA

Yanjun Zhang[1], Liwei Zhang[1], Liying Xu[1], Daoliang Li[1,*], Jing Du[1,2]

[1] *College of Information and Electrical Engineering, China agricultural University, Beijing, P. R. China, 100083*

[2] *China Agricultural University Library, Beijing, P. R. China, 100083*

[*] *Corresponding author, Address: P. O. Box 121, College of Information and Electrical Engineering, China Agricultural University, 17 Tsinghua East Road, Beijing, 100083, P. R. China, Tel:+86-10-72737741, Fax:+86-10-62737679, Email: li_daoliang@yaohoo.com*

Abstract: Based on the full analysis of the current development situation of rural informatization in China, combined with the development practices and trends in recent years, this paper presents some policies proposals for promoting the development of rural informatization mainly from five aspects, including infrastructure construction, integration of the resources, system of information service, application of information technology and policy environment.

Keywords: Policy Proposals, Rural Development, Rural Informatization

1. INTRODUCTION

The People's Republic of China is the fourth largest country in terms of area and the largest in terms of population. It has a population of 1320 million, with rural population of 730 million, accounting for 55.1% of its total population (National Statistic Bureau, Feb., 2008). 30 years' reform and opening up has brought about tremendous changes to China's economic and social conditions. The Gross Domestic Product (GDP) of 2007 reached 24,661.9 billion yuan, which is at the fourth place all over the world; meanwhile, the income of the farmers keeps rising. However, there are still a number of problems and difficulties to encounter in rural development.

Please use the following format when citing this chapter:

Zhang, Y., Zhang, L., Xu, L., Li, D. and Du, J., 2009, in IFIP International Federation for Information Processing, Volume 295, *Computer and Computing Technologies in Agriculture II, Volume 3*, eds. D. Li, Z. Chunjiang, (Boston: Springer), pp. 1977–1988.

Economic development is regionally uneven, resulting in considerable differences among areas. The gap of economic development between western parts and eastern coastal areas is significant. Inhabited by 40% of the national population, the coastal areas nevertheless took up 60% of GDP, with per-capita GDP 1.5 times higher than the national average level. In rural areas, the situation is even graver. The income gap between urban and rural areas is getting wider and wider, on account of dual structure in urban and rural economy. In 2007, the per-capita net income of farmers was 4,140 yuan, which was 3.33 times lower than that of urban residents, that is, the absolute gap reached 9646 yuan (The State Council. Aug., 2008). There really has a long distance to go in raising farmers' income. With the improvement of agricultural productivity, a great many surplus labor force has emerged in rural areas. The labor force primarily transferred to urban areas and relatively developed areas, forming a new social class with 200 million migrant workers from rural areas, resulting in a series of new social problems (Research Office of the State Council. Apr., 2006).

All in all, China has higher density of population when compared with geographic areas, and its average natural resources are rather low. Regardless of the fact that social services, like education, health care, culture, have made some progress in rural areas, there remain a number of problems and difficulties in rural development. Uneven economic and social development continues to exist between regions, and between urban and rural areas. Moreover, problems about how to deal with the shrinking of arable land, how to increase farmers' income, how to transfer labor force, are crying out for solutions.

Rural informatization is a situation and process to fully develop and employ the information resources and establish an information service system through the infrastructure construction of rural broadcasting and TV network, telecom network, computer network, etc, with a purpose of promoting information exchange and sharing of knowledge so as to realize the application and popularity of modern information technologies in the fields of rural production and operations, public service, administrations and life consumptions, etc. Rural informatization is a necessity of speeding up the rural modernization at the information era when the rural economy, farmers' income and cultural qualities have reached a certain level and the corresponding informatization environment including systems and standards have been gradually established (Daoliang Li, Oct., 2008). The development of rural informatization will contribute to solve a range of problems and encounter difficulties.

2. DEVELOPMENT OF RURAL INFORMATIZATION

2.1 Infrastructure Construction

2.1.1 Radio and television networks in rural areas

There are 378 million families in China, while 240 million families are scattered in rural areas; and nonetheless only 5490 rural families have access to cable television networks, compared to 140 million in the whole country. More than 70% of rural people mainly receive radio and television programs through wireless way. The Ministry of Finance gave full support to wireless networks, with total investment of 3,000 million yuan (500 million yuan in 2006 and 2,500 million yuan in 2007). After joint efforts, radio and television programs could be brought to 95.04% and 96.23% of the population at the end of 2006. Currently, there are 66,000 wireless television transmitting and relay stations throughout the country. The large-scale and potent radio and television networks have been built up, covering both rural and urban areas in the country. People can listen to or watch programs more easily and clearly in those areas.

2.1.2 Popularity of fixed and mobile phone

Although China Telecom ranks first in the world in terms of scale and the number of users, troubles just came out in the extension of communications in rural areas, especially in those old, small, remote, mountainous and impoverished areas. The Ministry of Industry and Information Technology has initiated the Village Coverage Project to popularize rural communications since 2004. By the end of 2007, China Telecom and other 5 operators have invested more than 30 billion yuan. As a result, 99.5% of administrative villages in the country have got the access to telephone networks, and telephone calls can be made in all the administrative villages of 29 provinces (The Ministry of Industry and Information Technology, Jan., 2008). On the other hand, China maintains a good momentum in the popularity of mobile phone. Mobile phone subscribers have increased to 5,230 million by the end of September of 2007. Nearly half of the new subscribers are from rural market (The Ministry of Industry and Information Technology, Nov., 2007). With the popularity of fixed and mobile phone, communications in rural areas have been greatly improved.

2.1.3 Access to internet in rural areas

China's net-surfers have amounted to 253 million by the end of June in 2008, ranking first in the world (CNNIC, Jul., 2008). However, the gap of internet development between urban and rural areas stays wide. In June of 2007, rural net-surfers were 37.41 million, accounting for 5.1% of rural population. While the number of rural net-surfers was only 29.9% of that of urban ones, with rural internet popularity rate less than one fourth of the urban. The infrastructure of internet is weak. In 2006, the computers owned by rural families were 2.7 every hundred households, well below urban areas with 47.2 computers every hundred households. Among rural net-surfers, males were comparably more than females, with a proportion of 61.7%; youngsters were the largest part, net-surfers under the age of 25 accounting for 69.1%. There are two reasons for why farmers seldom use internet: incapability of using it and lack of computers, separately resulting in 28.3% and 39.5% of non-net-surfers. Accordingly, they become the biggest obstacles in the expansion of net-surfers (CNNIC, Aug., 2007).

2.2 Information services system

2.2.1 Establishment of information service stations

Based on the information technology like computers, internet, intranet, telephones, rural information service stations, which are leaded by the government, supported by network operators and partly run by social organizations, adopt a approach of paid operation and free services. They can provide various information services to farmers, such as searching, skimming, collecting and releasing information, communicating with others and entertaining themselves. At the beginning of 2006, 97% of cities and 80% of agricultural sectors at the county level have launched information management and service organizations, and simultaneously 41% of information service stations in villages and townships possesses computers and are able to surf on the internet (Yu Fang. 2006).

2.2.2 Foundation and development of farmers' organizations

It is considered as an important part of modern agriculture and new village construction to develop and improve farmers' organizations. Economic cooperation organizations are the most universal type that mainly refers to organizations such as special cooperative society, community cooperative society, special society or specialized technical associations, various united economic entities and associations established by co-operatives. Currently, there were more than 1,400 thousand special economic cooperation organizations, of which 150,000 run in a standardized manner, with 23,630,000 members, accounting for 9.8% of famers in the whole country. This kind of organization concerns all sorts of industries, ranging from fruit and vegetable industry, animal husbandry, aquaculture, forestry, to agricultural machinery services, transportation, water conservancy construction, resource development, production of handicrafts and so on.

2.2.3 Information tellers in rural areas

With the continuous development of market economy in rural areas, farmers yearn for more specialized and updated information about the market and agricultural technology. Information tellers could transmit all sorts of agriculture-related information to farmers in different forms, making great contribution to the development of rural economy. At present, an agricultural information service system has been set up from the central to local governments; more than 200 thousand information tellers around the country are playing their roles in information transmission through varied channel, such as computers, telephones, television.

2.3 Application of information technology

With a late start, the application of information technology to agriculture however enjoys rapid development. Assorted agriculture-related networks, databases and application systems have been widely employed. Meanwhile, the research and extension of information technology has yielded fruit.

Agricultural information network platform has taken initial shape. Centering on Agriculture Information Network of China, assembling over 20 professional networks into a whole, a national gateway is now basically in place. The gateway has been linked with over 3000 websites, with daily page view of 2400 thousand. The system for information broadcast on the national agricultural network has launched a relatively perfect indicator system for information collection. With unified data standard and communal module, information can be issued through a website but shared in the whole system, which has promoted the development and sharing of information resources in an all-round way (Yu Fang. 2004).

In coastal areas and eastern developed areas, a large quality of information technology has been widely applied, such as information technology in agricultural production and management (like agricultural expert systems, control technology in agricultural facilities, agricultural remote-diagnosis technology), technology in acquiring and processing agricultural information (like "3S", namely, GIS, GPS, RS), technology in rural information terminals(like call centers, set-top boxes, SMS platforms, hotlines), information technology in management of rural communities (like the integrity office for government affairs of villages and townships, digital libraries, systems of serious epidemic and disaster forecasting and alarm, systems about population, social security and medical care).

3. POLICIES FOR THE DEVELOPMENT OF RURAL INFORMATIZATION

3.1 Strengthening Infrastructure Construction and Accelerating the Route to Success by the Magnitude Project

The construction of agricultural and rural informatization is progressing smoothly under the guidance of macro policies. The State Council Information Office started informatization experiments in counties; based on the project of "gold farmers", the Ministry of Agriculture advanced the construction of the system for agricultural informatization and agricultural information network platform, together with project of "nine major activities"; the Ministry of Industry and Information Technology initiated a range of projects, like five projects for agriculture, rural areas and farmers, projects of "village to village" and post chain distribution networks; the Commerce Department launched projects for "building new rural commercial information system" and "new rural business network"; the Ministry of Science and Technology started activities to make science and technology enter villages; the Publicity Department, the Spiritual Civilization Office of CPC and the State Administration of Radio, Film and Television launched the project of "television in thousands of families"; the Central Committee of CPC, the Education Department and the Ministry of Culture separately set up long-distance education for rural Party members and cadres, long-distance education in primary and middle schools, project of sharing cultural resources around the country.

According to preliminary estimation, more than 30 ministries and commissions have made policies about rural informatization, involving 40-odd projects, which key projects powerfully promoted infrastructure construction and the popularity and application of informatization. These important projects provide the greatest stimulus to the development of rural informatization in the longer term. It should be the current focus to implement these major projects in depth, consolidate the existing results, shorten the distance between farmers and informatization, improve economic and social development in rural areas and consequently increase farmers' income.

3.2 Strengthening the integration of all information resources to find solutions to "The first mile"

The ultimate goal of integrating rural information resources is to enhance the comprehensive effectiveness of rural informatization and service quality, which can be achieved by reallocating the relevant resources to focus the intensity of resource allocation and to find the best match between demands and resource utilization.

3.2.1 To strengthen the integration of information resources and information platforms

Currently, it is very important to strengthen the integration of information resources and platforms in rural informatization. Firstly, in view of database construction, we should develop the data exchange interface, establish the resource directory system, create data format technical standards for information exchange, set up data sharing platform to provide environment and technology for information resources congregating, exchanging and sharing. Secondly, the information platform shall be an integration of multi technical and working methods by connecting all kinds of agricultural service process, hardware, application software and standards, enabling the integrating of the inner business and providing service to the outside. Thirdly, the construction of system should depend on the improvement of the communication and coordination of all agricultural-related sectors and branches. The improved information exchange system should avoid repeating collection of information, the divided possession of information, the monopoly use of information and low effectiveness of information system, developing, but accelerate the open and share of information among agricultural sectors and branches in all levels, agricultural enterprises and scientific research institutions instead.

3.2.2 To improve the integration of information service and information dissemination channels

The integration of information service places must be based on the principle of sharing, under which, the extended information service stations and resources from different agricultural sectors and branches shall be integrated, while information blockage and monopoly shall be abandoned. Information service stations, shall play multi-role(business, society service and information etc.), provide reliable information and service, , what's more, run at lowest cost. The integration of information dissemination channels is mainly encouraged agricultural enterprises, organizations and institutions to build information dissemination channels in the areas where rural informatization infrastructure is incomplete. Based on the principle of "Acts as circumstances permit, to adapt measures according to time", the strategy to develop dissemination channels should stick to multi channels presence with the Internet as the main channel. In which case, it means that modern means and traditional medias including Internet, fixed communication, mobile communications, radio and televisions, newspapers, books, blackboards etc., are all carriers of information promoting information services and applications available to every village and rural householder.

3.2.3 To strengthen the integration of information services and information service personnel

The mechanism to integrate of information service should be guided by the government with cooperation of all agencies under the principal of "combination of public welfare bodies and non-public welfare bodies". Government plays important roles in integrating information service provision and promoting the development of information services. Through government purchasing service or contracting with the service providers, service providers, communication operators, IT enterprises and other relevant business enterprises are eager to participate in the activities of rural informatization adoption.

3.3 To strengthen both the construction of rural informatization service systems and innovation in organizing and delivering information services in rural areas

3.3.1 Make the best use of the joint forces involved in rural informatization, underlining the leading role of government, demonstrating local strengths especially the functions of associations

People involved are the most important force in rural informatization, which tells us that agricultural informatization should start at the grass roots. First of all, it is government who will guide the whole process of informatization. At the same time, it is equally important to bring the local forces into active playing, including the dragon enterprises, wholesale markets, intermediary organizations, farmers' brokers, large scale farmers and breeders. The special position, knowledge and demand of agriculture information make the above mentioned folk forces outstanding in informatization provision and adoption. As to the farmers, it is very important to help farmers realize the benefits of agriculture informatization and to help them initiate their demand for agriculture informatization by training, demonstration and experience of the intermediaries who have already adopted the informatization.

3.3.2 The most practical information service mechanism is to provide farmers with free information services and charge for the application of the technology carriers and logistics

Under the free information service principle, use the profit from the technology carriers and logistics to compensate the shortage of funds, so leads the dissemination of countryside science and technology information to march into the benign development track and make the information flow and logistics supplement each other during the process.

3.4 To strengthen the application of information technology in agriculture and achieve the informatization in rural affairs management

The integration of information technology and traditional agriculture is increasing. And information technology has penetrated widely into all fields of agricultural production and management, which becomes an important feature of modern agriculture. Therefore, we should focus on the following aspects of the application.

i. To strengthen the availability and transparency of information about agricultural products and inputs including seeds, plants, pesticides, fertilizers etc., and to improve information availability about the release of agricultural products and market information, to make it the convenient for farmers to purchase production materials, verify the product information and complain when needed.

ii. To strengthen the building of Monitoring, Forecasting, and Early Warning Systems for national staple agricultural products market and to provide accurate production-oriented information.

iii. To speed up the utilization of information technology for better controlling and cost-efficiency of agricultural production.

iv. To increase information technology support capability of resisting heavy natural disasters and animal and plant epidemic situation and to enhance the capacity of agricultural production system to cope with disaster prevention, disaster relief and disaster reduction.

v. Actively support informatization provision and adoption in the agricultural products processing, preserving and packaging sectors.

vi. To enhance the match between of agricultural products supply and demand information. To promote the construction of e-business and brand-based web sites and marketing for specific agricultural products.

vii. Focus on the transformation of information technology in distribution channels, to the develop e-commerce, to support the development of ordering agriculture in rural areas, chain operating, logistics and distribution, and to promote the integration of modern agricultural logistics flow and information flow.

viii. To implement the whole china tracing management from agricultural products from the supply of raw materials, processing, packaging, marketing with support of information technology.

ix. To improve the reclamation agriculture system's functions such as environment protection, tourism and leisure, cultural heritage via information technology. To exploit the specific agricultural resources, the leisure and tourist information service network should be developed and completed.

The informatization in rural affairs management is still in its infancy, especially in the less developed western regions. Therefore, we should give priority to develop the following aspects of application.

i. To speed up the construction and integration of Rural Government Affairs Management System to assist the decision-making of the rural grass-root units.

ii. To set up the Digital community Management System to promote the process of digital countryside construction.

iii. To stress the construction of rural areas culture informatization to promote the process of rural informatization.

iv. To stress the construction of rural areas' security system to ensure the security of rural areas.

v. To build and complete the labor security and society help information system to increase government management and service ability.

vi. To build a rural distance-learning education platform and to make use of the platform to promote the sharing of excellent education resources, consequently the education quality in rural areas could be increased.

vii. To speed up the construction of new rural cooperatives medical service information system to increase the public welfare medical information service.

3.5 To create a favorable environment for the development of rural informatization provision, and promote all-round development of rural informatization adoption

The success of rural informatization can not be achieved without regulations and laws. So it is important to make and perfect supportive policies, to issue relevant laws and regulations, to establish and improve the management system along all the agricultural products supply chain, and to promote the healthy development of an agricultural information service system. On the other hand, it is not enough to just depend on government; all forces should participate actively by offering funds with the demand of market. By funding input with all kinds of society forces, a diversified rural informatization investment pattern is achieved, which is guided by government and invested mainly by society.

The capacity of rural informatization staff that provides information services should be enhanced by regular training and inspection. The training activities shall help update information teller's notion, improve their computer and internet application skills, enhance computer and network applications, strengthen their accepting and analyzing ability of knowledge and information. At the same time, rural informatization is nothing if farmers have no intention to accept or adopt it. The understanding of information and information services is a new concept to farmers. Therefore, it is critical to increase farmers' awareness and acceptance of informatization by all kinds of means like training and guiding.

4. CONCLUSIONS

With the rapid development in rural areas, it remains lots of problems and difficulties to encounter. One of the effective ways to solve rural problems is to promote informatization and to enhance the popularity of information technology in rural areas. This paper has given the status and policies for promoting the development of rural informatization from five aspects like below: construction of infrastructural facilities, integration of information resources and system of information service, application of information technology and environmental of policy. It will help staff in agriculture and rural to grasp the realities of rural informatization and to promote the development of rural informatization better.

ACKNOWLEDGEMENTS

This work is subsidized by the EU-China Information Society Project (ASIE/2004/003-265), and it is part of the work of the Project: Study on ICT for Rural Development in China and EU. Our special thanks should also go to Mr. Fang Yu at Advisory Committee for State Informatization (China) for his support and guidance.

REFERENCES

CNNIC (China Internet Network Information Center). Aug., 2007. *China Internet Survey Report in Rural Areas.*

CNNIC (China Internet Network Information Center). Jul., 2008. *Article 22 of the China Internet Development Report.*

Daoliang Li. Oct., 2008. *Annual Report on ICT for China Rural Development 2008*: 19~20

National Statistic Bureau. Feb., 2008. *Statistics Bulletin of National Economy and Social Development in 2007*

Research Office of the State Council. Apr., 2006. *An Investigation Report on Rural Migrant Workers of China.*

The Ministry of Industry and Information Technology, Jan., 2008. *A Suggestion on Promoting Village Coverage Project and Developing Rural Information Service in 2008.*

The Ministry of Industry and Information Technology, Nov., 2007. China Telecom ranks first in the world in terms of scale, *Industry Information*, 338: 14~15

The State Council. Aug., 2008. *A report on Promoting Higher Income for Rural Workers.*

Yu Fang. 2004. Let Information Services more Substantial. *Chinese Computer Users,* 21: 40

Yu Fang. 2006. Reality and Prospects: Informatization under the "three rural" issue. *China Information Times,* 24: 16~18

CONTROL OF CO_2 INSIDE AUSTRALIAN GRAINS FREE AIR CARBON DIOXIDE ENRICHMENT (AGFACE) FACILITY

Mahabubur Mollah [1,*], Robert Norton [2], Jeff Huzzey [3]

[1] *Department of Primary Industries, 110 Natimuk Road, Private Bag 260, Horsham, Victoria, Australia,*

[2] *The University of Melbourne, 110 Natimuk Road, Private Bag 260, Horsham, Victoria, Australia*

[3] *Land Technology Pty Ltd, 445 Smiths Road, Laharum 3401, Victoria, Australia*

[*] *Corresponding author, Address: 110 Natimuk Road, Private Bag 260, Horsham, Victoria, Australia, Tel: +61-3-53622111, Fax: +61-3-53622187, Email: Mahabubur. Mollah @ dpi.vic.gov.au*

Abstract The Australian Grains Free-Air Carbon dioxide Enrichment (AGFACE) facility was established at Horsham (36°45′07″S latitude, 142°06′52″E longitude), Victoria, Australia in 2007 to conduct research under Australian conditions with the aim to generate field response data in broad-acre wheat to elevated levels of CO2. Carbon dioxide concentrations within the rings in the AGFACE facility are computer controlled, using inputs of sensed carbon dioxide concentration, wind speed and direction at the centre of each ring. The control system uses a number of pneumatic valves and associated fittings driven by a control algorithm. On average, the control system at the Horsham site has enabled the AGFACE rings to maintain the concentration of CO2 at the centre of the ring at ≥90% of the target (ie. ≥495 ppm) for 97% of the operating time. The typical mean CO2 concentration at the centre of an AGFACE ring was 550 ± 28 ppm and this meets the generally accepted standards for defining FACE systems operating limits.

Keywords: spatial variation, climate change, CO2 meter, CO2 sampler, CO2 analyser

---

Please use the following format when citing this chapter:

Mollah, M., Norton, R. and Huzzey, J., 2009, in IFIP International Federation for Information Processing, Volume 295, *Computer and Computing Technologies in Agriculture II, Volume 3*, eds. D. Li, Z. Chunjiang, (Boston: Springer), pp. 1989–1998.

1. INTRODUCTION

A collaborative project between the Victorian Department of Primary Industries (DPI) and the University of Melbourne commenced in 2007 to establish the Australian Grains Free-Air Carbon dioxide Enrichment (AGFACE) facility. The main aim of the research is to generate field response data of field grown wheat crops to elevated levels of CO_2 (eCO_2) across a range of temperature and water supply conditions.

The AGFACE uses direct injection of CO_2 in the atmosphere and is based on the Brookhaven National Laboratory design (Lewin et al. 1994) similar to other systems such as Rice FACE (Okada et al. 2001) and POPFACE (Miglietta et al. 2001). In these systems, CO_2 is injected into the atmosphere on the upwind sides of an octagon around the target through 0.3 mm holes. In the process, CO_2 gas jets reach sonic speeds and create shock-waves at the jet outlets which enhance the mixing of CO_2 gas with air. This mixed air-CO_2 is then quickly transported across the ring by the prevailing wind. According to Hendrey et al. (1993) control of CO_2 concentrations within the FACE system should be within a criterion of ±20% for set point of 1-min averages at least during 80% of the experimental period. This is generally the accepted design criterion for FACE experiments (Okada et al. 2001; Hovenden et al. 2006).

This paper describes the major components used to maintain and control the target CO_2 concentration of 550 ppm within the protocol tolerance at the centre of an eCO_2 AGFACE ring and presents basic performance results for the control system used.

2. MATERIALS AND METHODS

The AGFACE facility consists of 1) a supply of CO_2 gas to the field from a bulk container, 2) power supply to each ring, 3) central control system for individual ring, 4) AGFACE rings, 5) CO_2 gas analyser, 6) wind sensors and 7) data monitoring and acquisition system (Fig.1).

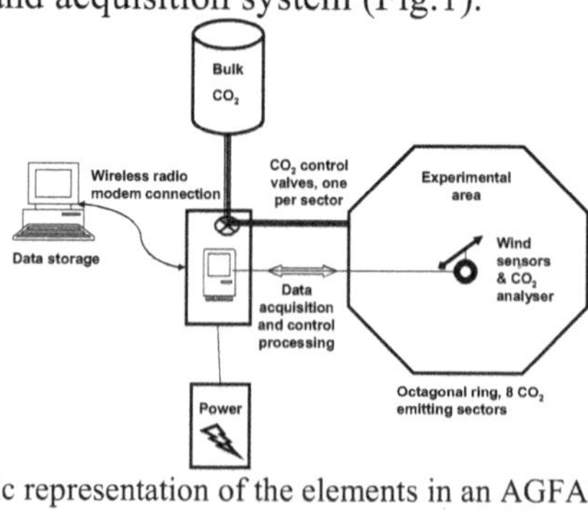

Fig. 1. Diagrammatic representation of the elements in an AGFACE ring in the field.

The experimental system at Horsham consists of a 7.5 ha site with eight eCO_2 rings and eight control ambient CO_2 (aCO_2) experimental areas.

2.1 Valve assembly for CO_2 control

A number of pneumatic valves and associated fittings (SMC Pneumatics, Australia, Pty Ltd, VIC) are assembled in a particular order and then connected to AGFACE ring in the field to control CO_2 gas supply (Fig.2). The function of each component is described below.

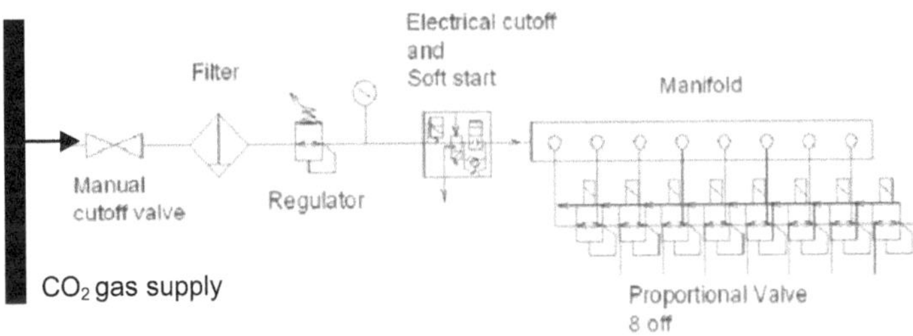

Fig.2. The order of connecting pneumatic valves for CO_2 control at each ring (adapted from Huzzey and Pyke 2007).

Manual cut-off valve: The manual cut-off valve (Fig.2) allows the CO_2 gas supply to each ring to be turned off independently for maintenance.

Filter: A filter is next in the flow line (Fig.2) to trap moisture or small particles, so subsequent components do not malfunction.

Regulator: The regulator (Fig.2) helps to stabilise the subsequent manifold pressure irrespective of the supply line pressure and enables a different pressure to be used at the manifold than the supply line, if required.

Electrically *controlled cut-off / soft-start valve:* From the regulator, the gas flows through an electrically controlled cut-off valve / soft start valve (Fig.2). This valve enables the gas flow to be turned on and off by the microcontroller.

Manifold: The gas flows from the ‘soft start’ to an eight port manifold (Fig.2). Eight proportional valves, one for each side of the octagon (“ring”) (Fig.1), are connected to the manifold.

Proportional valves: The proportional valves (Fig.2) control the rate of flow of gas to the fumigation tubes. These valves can be controlled either using a voltage or current input signal. AGFACE proportional valves use a 4 to 20 mA current signal, because unlike voltage-control, current-control allows the valves to be located a significant distance from the control system, which can be very useful in the design of the FACE array. The voltage drop that occurs over a length of cable can cause unreliable operations.

2.2 CO_2 delivery lines and fittings

Sufficiently flexible low density polyethylene tubes (Brotec Services, VIC, Australia) with 12 mm OD and 800 kPa pressure rating has been used to deliver CO_2 gas to each fumigation tube (a sector of an octagonal ring) through a proportional valve. Push-in fittings (SMC Pneumatics (Australia) Pty Ltd, VIC) have been used throughout to connect the delivery tubes to the valves and the stainless steel fumigation lines. The gas is emitted through 0.3 mm laser drilled holes spaced 50 mm apart. The precise laser drilling of the orifices means flow rates are very similar between each hole.

2.3 CO_2 sampler and analyzer

AGFACE uses ten Infra Red Gas Analysers (IRGA) manufactured by PP Systems (McVan Instruments Pty Ltd, VIC, Australia) to determine CO_2 levels in each eCO_2 rings and in two ambient (aCO_2) rings. AGFACE uses the base SBA-4 model which is an Original Equipment Manufacturer (OEM) board that comes as a complete package, enabling a significant cost saving. All IRGAs used are single frequency design and use an absorber column to obtain a zero reference.

The SBA-4 OEM board required the addition of an absorber column, an air pump capable of 0.15 to 1 L/min flow, an adjustable flow indicator, a calibration potentiometer and a filter. All these components are housed in a waterproof and dustproof enclosure (Rittal Pty Ltd, VIC, Australia). The analysers used to measure the aCO_2 in the two control rings are of the same design except that a wireless modem is housed in the same enclosure to send data directly back to the computer in the site office.

2.4 Wind sensors

The Gill MicroVane and 3-Cup anemometer (Campbell Scientific Australia, QLD) are used in AGFACE for measuring wind direction and speed at a height of 2.0 m above the ground. This sensor is more accurate at low wind speeds and available with a frequency wind speed output which is expected to be more reliable and require less maintenance than other models.

2.5 Power supply

All components in the control system of AGFACE use 12 V DC and or 24 V DC supply. Employing the two different DC voltages gives more flexibility and reliability reducing the risk of interference between components.

2.6 Control system and control algorithm

The 'Z-World' (now Rabbit Semiconductor) SR9000 system as used at SoyFACE (University of Illinois 2007) was chosen for the AGFACE controller. The control algorithm is written in 'Dynamic C', providing the programming flexibility of 'C' and is of industrial quality, design and manufacture (Huzzey and Pyke 2007). Also, the modular design of the algorithm allows for flexibility in the inputs and outputs used. This type of controller is widely used in direct CO_2-injection FACE systems because of its cost-effectiveness. It has a low power requirement.

The core proportional integral derivative (PID) control section of the algorithm was developed by Miglietta *et al.* in Italy (Huzzey and Pyke 2007). The controller software used at SoyFACE was made available to AGFACE and permission was given to modify the algorithm to suit AGFACE requirements.

The control system is programmed to sense the CO_2 concentration, wind speed and direction at the centre of the ring. This is to determine the location of upwind sectors from which the necessary volume of CO_2 need to be released to maintain the target concentrations (550 ppm) at ring-centres. The control system is also programmed to turn the CO_2 off if the average wind speed reaches or exceeds 10 m/s (maximum limit) for 45 seconds and release relatively smaller volumes from all sectors when wind speeds are below 0.3 m/s for 45 seconds.

Carbon dioxide is vented only during the daylight hours (photoperiod) from emergence until maturity of the annual crops studied.

2.7 Data monitoring and data acquisition

A wireless Ethernet local area network (LAN) is installed to monitor and collect data from the individual AGFACE rings. A wireless modem is located within each controller enclosure and each controller has its own Internet Protocol (IP) address. The Elpro 905U-E Ethernet wireless modem (EDAC Electronics, VIC, Australia) was selected for data monitoring and data acquisition for AGFACE as the same unit can be setup to either receive data in the site office or send data from a controller or from the aCO_2 analyser.

The CO_2 concentration, wind speed and wind direction, voltage to proportional valves (Vvalue), and the identification number of the main sector (one of the eight fumigation tubes) releasing CO2 are recorded by the controller every second and these data are averaged over a minute. Then wireless modems transmit one second data every fourth second and one minute data every minute to a computer in the site office. Every fourth second's data are logged to a daily second data files (eg "Ring 9_SecData_011007.csv") for each ring. One minute average data are also logged to a daily file and appended to a cumulative one minute data file for each ring. The time recorded in the cumulative minute data file is the time at the end of the relevant minute.

If a ring stops releasing CO_2 the Vvalues are recorded as zeros. These values are >0 when a ring is releasing CO_2, however the controllers override these values to zeros to turn the CO_2 off if the average wind speed stays ≥10 m/s (maximum limit) for 45 seconds.

A Visual Basic monitoring and data logging program was obtained from SoyFACE and then modified extensively to suit the local AGFACE array (Huzzey and Pyke 2007). This program currently displays one second data, updated every four seconds. The data includes CO_2 concentration (ppm) at the centre of the ring, voltage values for the proportional valves, wind speed (m/s) and direction (degrees). Current-input signals are used to operate these valves but proportional-voltage values are displayed avoiding further modification of logging program. The program has the ability to plot the CO_2 concentration data at the centre of any ring.

A wireless NextG Modem (Telstra, Australia) and LogMeIn software (LogMeIn Inc., USA) are being used for remote access and downloading of data.

2.8 Temporal variation of CO_2

The sensed CO_2 concentration from each ring is recorded every second and averaged over a minute. The data for every fourth second and minute-averages are logged to a computer in the site office. These data are available to evaluate the effect of wind speed on the stability of CO_2 concentration in the middle of the ring.

2.9 Spatial variation of CO_2

A compact, lightweight, field portable CO_2 analyser (EGM-4 Environmental Gas Monitoring for CO_2, McVan Instruments Pty Ltd, VIC, Australia) was used to measure the variability of CO_2 concentrations in a horizontal plane inside and neighbouring outside of each AGFACE ring on four occasions (18 and 19 September 2007, and 24 and 25 October 2007) during the growing seasons. During this period four readings were taken at each of 36 sampling locations at the height of the CO_2 sampler (top of crop canopy) at the centre of each ring.

3. RESULTS AND DISCUSSION

3.1 Temporal variation

Fig.3 shows a trace of the wind speed and 1 minute averaged CO_2 concentrations for one ring on one day. On this day, the wind speed ranged from less than 1 m/s to more than 5 m/s, however the CO_2 concentrations was maintained near 550 ppm irrespective of wind speed (Fig.4).

When taken over a longer period, on average the control system maintained the concentration of CO_2 at the centre of the ring at ≥90% of the target (ie. ≥495 ppm) for 97% of the time (Table1). While operational (during the daylight hours), the typical mean CO_2 concentration inside the ring was 550 ppm with the standard deviation of 28 ppm.

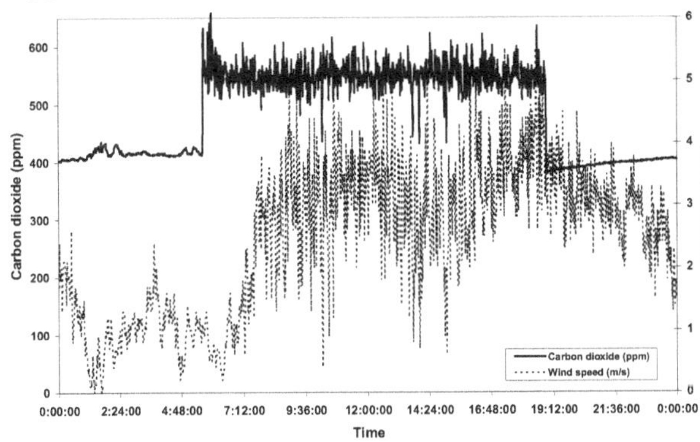

Fig.3. CO_2 and wind speed data logged on 24 October 2007 from an AGFACE ring.

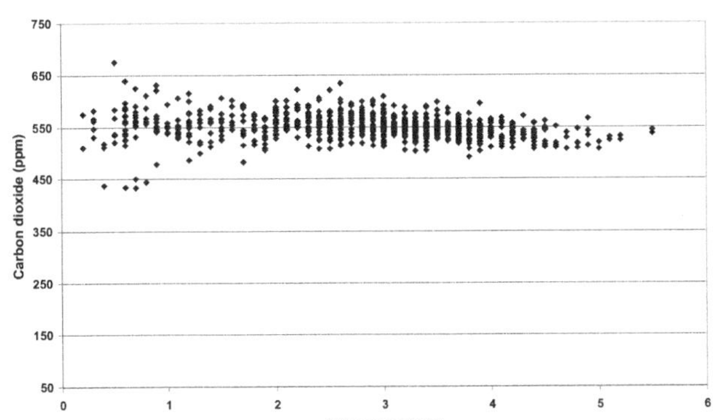

Fig.4 CO2 concentration at the centre of an AGFACE ring graphed against wind speed from data logged on 24 October 2007 (Fig.3).

Table1. Average performance of 8 AGFACE rings in terms of controlling CO_2 concentration at the centre of the ring (the range is given in square brackets).

Total CO_2 injection (hr)	Average CO_2 concentration inside the ring while operational (during daylight only)				% of time CO_2 ≥ 90% of the target.
	Target (ppm)	Median (ppm)	2.5 centile (ppm)	97.5 centile (ppm)	
786	550	548 [547-550]	484 [405-505]	625 [609-701]	97 [92-98]

3.2 Spatial variation of CO_2

The data collected within the AGFACE rings using the portable CO_2 analyser showed that in general, the rings complied with the protocol requirement, that at least 80% of the ring area received ≥90% of the target CO_2 concentration. For example, on 24 October 2007, 98% of the area inside a ring received ≥90% of the target CO_2 concentration (*ie.* ≥495 ppm) (Fig.5).

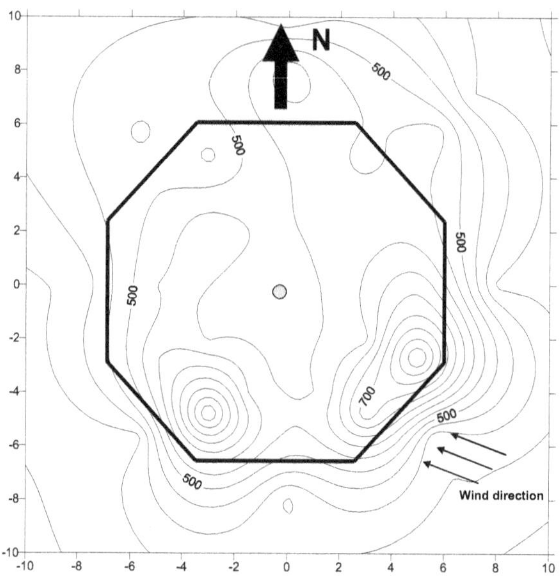

Fig.5. An example of CO_2 distribution in and around a ring on 24 October 2007.

Fig.5 was derived using data collected when the wind was predominantly blowing from the south-east. Consequently, three upwind sectors from south-eastern corner released CO_2 upwind at concentrations >550 ppm, and then the prevailing wind blew CO_2 over the ring. As expected the concentration was gradually diluted as CO_2 travelled to the opposite end of the ring. It is important that the crop grown inside an AGFACE ring should receive uniform concentrations of CO_2 during a growing season to maximise the area for core agronomical experiments.

Wind direction, and hence the locations of CO_2 releasing sectors change very rapidly as well as vary between days and seasons. These variations result in an even distribution of CO_2 inside the ring in the long-term providing the direction of the wind is reasonably even.

Long-term and frequent data collection is planned for 2008 to understand the system performance in detail and fine tune the control system, if needed.

4. CONCLUSION

Careful choice of locally available industry standard components and use of local expertise have enabled significant cost savings in AGFACE construction. Operations in 2007 have shown that the control system for AGFACE facility in Horsham, Victoria, Australia has met the general design standards for FACE experimentation. More data will be collected in 2008 to better understand system performance and fine tune the system.

ACKNOWLEDGEMENTS

The project (UM00027) is funded by Department of Climate Change, Grains Research and Development Corporation, Victorian Department of Primary Industries (DPI), The University of Melbourne and BOC Limited. The authors acknowledge the invaluable technical support from Russel Argall, Janine Fitzpatrick and Dennis Pye (DPI) and Peter Howie from The University of Melbourne. Authors acknowledge the assistance with data analysis from Dr Jenny Smith (DPI). Authors also acknowledge the assistance of Will Pyke, Land Technology Pty Ltd with the construction of AGFACE rings.

REFERENCES

Hendrey GR, KF Lewin, J Nagy (1993). Free-air carbon dioxide enrichment: Development, progress and results. Vegetatio, 104, 17-32.

Hovenden MJ, Miglietta F, Zaldei A, Vander Schoor JK, Wills KE, Newton PCD (2006). The TasFACE climate-change impacts experiment: design and performance of combined elevated CO2 and temperature enhancement in a native Tasmanian grassland. Australian Journal of Botany, 54, 1-10.

Huzzey EH, Pyke WJ (2007) National grains-FACE project. Part2, design specifications for FACE arrays. Internal report to University of Melbourne and Department of Primary Industries, Victoria. Land Technology, Laharum, Victoria, Australia.

Lewin KF, GR Hendrey, J Nagy, RL LaMorte (1994) Design and application of a free-air carbon dioxide enrichment facility, Agricultural and Forest Meteorology, 70, 15-29.

Miglietta F, Peressotti A, Primo Vacari F, Zaldei A, De Angelis P, Scarscia Mugnozza G (2001) Free-air CO2 enrichment of a poplar plantation: the POPFACE fumigation system. New Phytologist, 150, 465-476.

Okada M, Lieffering M, Nakamura H, Yoshimoto M, Kim HY, Kobayashi K (2001) Free-air enrichment (FACE) using pure CO2 injection: system description. New Phytologist, 150, 251-260.

University of Illinois (2007). SoyFACE. http://www.soyface.uiuc.edu/index.htm. Accessed on 1.11.07.

A TWO-PHASE METAHEURISTIC FOR FARM WORK SCHEDULING

Senlin Guan[1,*], Morikazu Nakamura[1], Takeshi Shikanai[2], Takeo Okazaki[1]

[1] *Faculty of Engineering, University of the Ryukyus, 1 Senbaru Nishihara, Okinawa, Japan*

[2] *Faculty of Agriculture, University of the Ryukyus, 1 Senbaru Nishihara, Okinawa, Japan*

[*] *Corresponding author, Address: Faculty of Engineering, University of the Ryukyus, 1 Senbaru Nishihara, Okinawa, Japan 903-0213, Email: guansid@gmail.com*

Abstract: This paper proposes a two-phase metaheuristic approach to planning daily farm work for agriculture production corporations. The two-phase metaheuristic contains the optimization of resources assignment and searching schedule based on Genetic Algorithm and hybrid Petri nets model. In the experiment, the effect on optimizing the resource assignment and priority list, initializing population of GA with sorted chromosomes by waiting time, inheriting priority list from tasks in the previous resources assignment enhanced the evolutionary speed and solution quality. The computational experiment revealed high effectiveness for constructing farm work schedule with high ratio of resource utilization. The proposed approach also contributes a referential scheme for combining metaheuristic to solve scheduling problem under constraints.

Keywords: farm work planning, scheduling, metaheuristic simulated annealing, genetic algorithm, hybrid Petri nets, modeling, sugarcane

1. INTRODUCTION

In Japan, there are over 190,000 management units serving as farmers/agricultural production corporations (The Ministry of Agriculture, Japan, 2006), which lease and manage the geographically scattered farmlands with full mechanization in order to accomplish the low-cost and high-efficient cultivation in the corporation style. For these agricultural production corporations, an adequate farm work planning system may guide

---

Please use the following format when citing this chapter:

Guan, S., Nakamura, M., Shikanai, T. and Okazaki, T., 2009, in IFIP International Federation for Information Processing, Volume 295, *Computer and Computing Technologies in Agriculture II, Volume 3*, eds. D. Li, Z. Chunjiang, (Boston: Springer), pp. 1999–2009.

them to carry out farm work in an organized and planned manner for efficient management. However, constructing an optimum farm work plan is difficult for farmers in these corporations, because that making a rational schedule requires not only traditional experiences but also mathematical and statistical approaches.

In agriculture, some methods for farm work planning presented valuable references (Arjona et al., 2001; Daikoku, 2005; Haffar and Khoury, 1992; Nanseki, 1998; Tsai et al., 1987; Santiago et al., 2005; Suliman, 2000) have been developed. Astika et al. (1999) also proposed a stochastic farm work scheduling algorithm based on short range weather variation. These researches are usually designed for a specific farm work problem, especially unsuitable for the daily schedule of entire growth cycle for assigning necessary resources to field operations for geographically dispersed farms.

An integration farm work management has been developed for sugarcane production in Okinawa, south of Japan. It comprises farm data recording system (Guan et al., 2006), web-based management system, modeling (Guan et al., 2008) and scheduling system. In which farm data recording system serves indicating daily work schedule, recording daily farm work progress, changes of cancellation of work, machine breakdown and so on. Web-based management system provides a platform for maintenance of database and outputting analysis report. Modeling system models farm work flow and uncertainty into mathematical data, and acts a simulation tool for mastering overall status of farm work progress and resources. Farm work scheduling system plays most important role of providing workers with appropriate daily work schedule.

In this study, we propose a two-phase metaheuristic for farm work scheduling with uncertain resources and constraints. At the first phase, the resource assignment was optimized using simulated annealing (SA) algorithm; and at the second phase, the optimization is based on genetic algorithm (GA), which searches the optimal schedule according to firing rules of hybrid Petri nets (David and Alla, 2001). The study is applicable to address the problem of production scheduling with arbitrary idle time, processing time, supplement or cancellation of work, cooperative work, moving time of machinery, machinery breakdown.

2. PROBLEM DEFINITION AND FORMULATION

2.1 Constraints in practical farm work

For each farmland in an agricultural production corporation, there are a series of works from work of planting to harvesting in a crop growth cycle. Many farm machineries and workers are available for corresponding work. In order to more theoretically describe conditions in practical farm work, we designate N_F, N_W, N_R as the total number of farmlands, works in a crop growth cycle, and resources, respectively. Other notations and their description are listed in Table 1. Notice that m_{ij} represents amount of scheduled work W_j in F_i. W_j will be scheduled if $m_{ij} > 0$, otherwise, this work will be not performed in F_i.

Table 1. Definition of variables

Notation	Definition
F_i	Farmland $i, i \in \{1,\ldots,N_F\}$
W_j	Work $j, j \in \{1,\ldots,N_W\}$
R_k	Resource $k, k \in \{1,\ldots,N_R\}$
A_i	Area of F_i
m_{ij}	Amount of scheduled work W_j in F_i, $m_{ij} \in [0, A_i]$
S_{jk}	$S_{jk} \in \{0,1\}$, 1: R_k is available to perform W_j, otherwise, 0
S'_{jk}	$S'_{jk} \in \{0,1\}$, 1: R_k is scheduled to perform W_j, otherwise, 0
v_k	Working speed of R_k
W_{ij}	Waiting time between end time of W_{j-1} and start time of W_j in F_i
$P_j(s), P_j(e)$	Pre-defined work period [start time $P_j(s)$, end time $P_j(e)$] for W_j
T_{ijk}	Task performed in F_i by R_k for W_j
$t_{ij}^{R_k}(s), t_{ij}^{R_k}(e)$	Start (end) working time of work j in F_i by R_k
v'_k	Moving speed of R_k
D_{ab}	Distance between F_a and F_b, $a,b \in \{1,\ldots,N_F\}$

A resource assignment must guarantee that at least one resource is assigned to perform W_k, and the total number of assigned resources is less than $\sum_k S_{jk}$, that is the total number of resources available to perform W_k. "A resource" is defined as a set of the minimum necessary machinery and labor for the farm work, not as an individual resource. If more than two resources are assigned to a same work k ($\sum_k S'_{jk} > 1$), it is possible to perform cooperative work. The cooperative farming work is defined as a process where multiple machines perform the same work, and the entry time of a resource to perform cooperative farming work is arbitrary.

$$1 \le \sum_k S'_{jk} \le \sum_k S_{jk} \quad (1)$$

The amount of scheduled work m_{ij} are completed by certain resources R_k during $t_{ij}^{R_k}(s)$ and $t_{ij}^{R_k}(e)$ on working speed v_k. For any resource allocation scheme, the following equations exist:

$$\begin{bmatrix} t_{1j}^{R_1}(e)-t_{1j}^{R_1}(s) & t_{1j}^{R_2}(e)-t_{1j}^{R_2}(s) & \cdots & t_{1j}^{R_k}(e)-t_{1j}^{R_k}(s) \\ t_{2j}^{R_1}(e)-t_{2j}^{R_1}(s) & t_{2j}^{R_2}(e)-t_{2j}^{R_2}(s) & \cdots & t_{2j}^{R_k}(e)-t_{1j}^{R_k}(s) \\ & \vdots & & \\ t_{ij}^{R_1}(e)-t_{ij}^{R_1}(s) & t_{ij}^{R_2}(e)-t_{ij}^{R_2}(s) & \cdots & t_{ij}^{R_k}(e)-t_{ij}^{R_k}(s) \end{bmatrix} \begin{bmatrix} v_1 \\ v_2 \\ \vdots \\ v_k \end{bmatrix} = \begin{bmatrix} m_{1j} \\ m_{2j} \\ \vdots \\ m_{ij} \end{bmatrix} \quad (2)$$

To avoid the superposition of durations $\left[t_{ij}^{R_k}(s), t_{ij}^{R_k}(e)\right]$ and $\left[t_{pq}^{R_k}(s), t_{pq}^{R_k}(e)\right]$ $(p \in \{1,\ldots,N_F\},\ q \in \{1,\ldots,N_W\})$, we have the following conditions:

$$\forall i,j,p,q,k, t_{ij}^{R_k}(s) < t_{ij}^{R_k}(e)$$
$$t_{ij}^{R_k}(e) < t_{pq}^{R_k}(s) \cdots \text{if } t_{ij}^{R_k}(s) < t_{pq}^{R_k}(s) \quad (3)$$

For the purpose of timeliness work, start working time $t_{ij}^{R_k}(s)$ and end working time $t_{ij}^{R_k}(e)$ suffer from additional conditions stated in Equation (4), where $k' \in \{1,\ldots,N_R\}$.

$$\forall i,j,k,k', t_{ij}^{R_k}(s) \geq max(P_j(s), t_{i(j-1)}^{R_{k'}}(e) + W_{ij})$$
$$t_{ij}^{R_k}(e) \leq max(P_j(e), t_{i(j-1)}^{R_{k'}}(e) + W_{ij}) \quad (4)$$

The works for farmland i are subject to *precedence constrained* relation [4]; in other words, a latter work W_j can only start after the completion of a former one W_{j-1}. This condition is defined by Equation (5).

$$\forall i,j,k,k', t_{ij}^{R_k}(s) > t_{i(j-1)}^{R_{k'}}(e) \quad (5)$$

Considering the moving time between farmlands, the start time of the next work should be the sum of completion time of the previous work and the moving time (Equation (6)).

$$\forall a,b,j,k, t_{aj}^{R_k}(s) \geq t_{bj}^{R_k}(e) + D_{ab}/v'_k \quad (6)$$

Based on the above conditions, the objective of farm work scheduling is formulated as the following equation:

$$\min(\sum_{a,b,j,k}[t_{bj}^{R_k}(s)-t_{aj}^{R_k}(e)])$$

(7)

where task $t_{aj}^{R_k}$ is a latter task of $t_{bj}^{R_k}$ $(t_{bj}^{R_k}(e) \geq t_{aj}^{R_k}(s))$. Minimizing idle time between works conduces to high ratio of utilization of machinery. The objective is same with minimizing make-span in common scheduling problem.

2.2 Formulating farm work scheduling on hybrid Petri nets

Hybrid Petri nets informally contain a discrete part and a continuous part of Petri net. Figure 1 illustrates a hybrid Petri nets modeling for scheduled farm work. The discrete part of Petri net is comprised by discrete places drawn as single-line circles, and discrete transitions drawn as bars. The state of resource R_k is represented by token, which is a black dot within place. The continuous part contains continuous places P_{ij} drawn as double-line circles, and continuous transitions drawn as box. Continuous transition, whose naming is same as task T_{ijk}, denotes performing the task in farmland F_i by R_k, for work j. Each continuous transition, place is associated with predefined work duration $P_j(s) \to P_j(e)$ and waiting time W_{ij}, respectively.

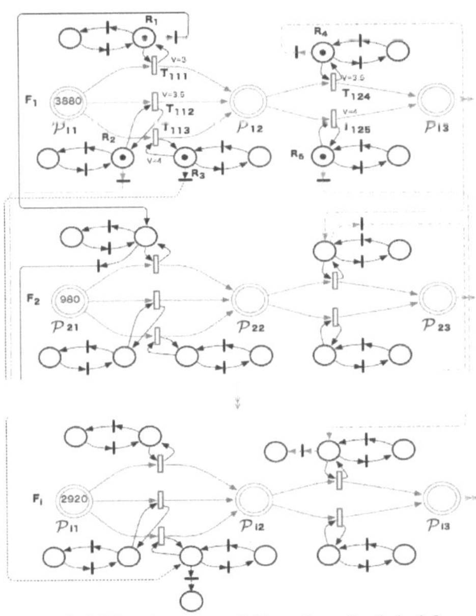

Fig. 1. Hybrid Petri nets modeling for scheduled farm work

This hybrid Petri nets not only acts on modeling the farm work process, but also scheduling farm works. Using hybrid Petri nets, a substantial reduction in the complexity of problem formulation is achieved. A detailed description on hybrid Petri nets modeling for farm work flow can be found in (Guan et al., 2008).

3. EXPERIMENTAL RESULTS

A farm work schedule includes assigning resources and arranging work sequence. At the first phase, a scheme of assigning resources is determined and optimized. At the second phase, the work sequence is designated as a priority list in which works are arranged according a specific priority. The priority list is optimized for minimizing the idle time between tasks, according to firing rules of hybrid Petri nets.

3.1 Optimizing resource assignment and priority list

Figure 2 displays the contrastive effect on optimizing resour-ce assignment and priority list correspon-ding to different gen-eration size in GA. The curves are drawn by current best fitness against execution time. “gen-100" represents the evolution process for high frequency of optimizing resource assignment but short computation time for optimizing the priority list. Comparing with “gen-100", “gen-1000" emphasizes optimizing priority list but results in reduction of frequency for optimizing resource assignment at same computation time.

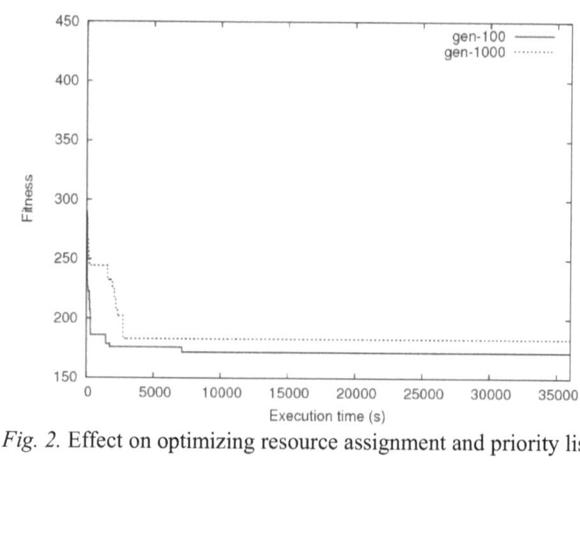

Fig. 2. Effect on optimizing resource assignment and priority list

In the figure, not only speedy evolution but also good solution quality appears in "gen-100", especially at early evolution stage. It reveals that increasing the frequency of optimizing resource assignment conduces to speedy evolution and convergence in computation. It is considered that resource assignment is an important factor for producing an efficient schedule.

3.2 Strategies of initializing population of GA

Generally, the waiting time between works (W_{ij}) has considerable influence on evolution of solution quality. In practical, however, the waiting time between works is most same because of uniform farm works in all farmlands. In the case of most same waiting time between works, the effect on solution quality under sorting works by W_{ij} in the initial population is shown in Figure 3.

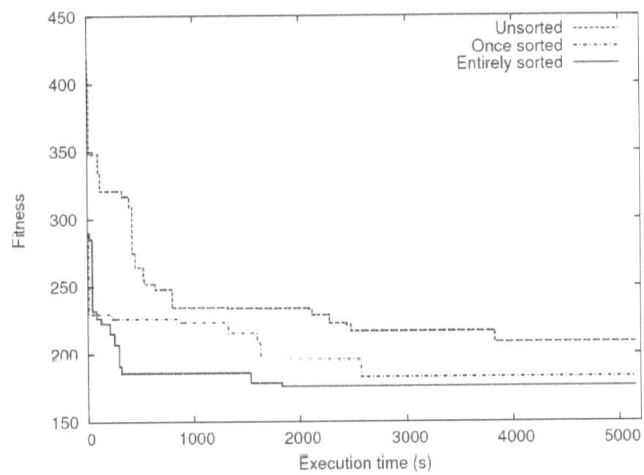

Fig. 3. Effect on initializing population by sorted chromosomes

The curves show evolution process star-ted from three initialized populations with raw chromosomes (un-sorted), one sorted chromosome and entirely sorted chromosomes. It is obvious that evolution speed of the curve titled "Unsorted" is slowest in comparison with that of other two curves. Both high evolution speed and solution quality are obtained in case of initializing population by entirely sorted chromosomes. For the curve representing one sorted chromosome, the fitness will suffer from other constraints like moving time, so that both evol-utionary speed and sol-ution quality are wea-ker than the curve titled "entirely sorted". Beca-use of the chromoso-mes sorted by most same waiting time, the-se chromosomes may have further variations. Therefore, the populat-ion comprised of entir-ely sorted chromoso-mes by W_{ij} may enhance the possibility of approaching best sequence. These three curves distinctly illustrate that sorting tasks by waiting time between tasks contributes to speedy evolution.

Inheriting best priority list for the previous resource assignment starts at initializing the population for the second resource assignment. From on computation for the second resource assignment, the priority list is optimized, and present best work sequence for each resource is ascertained. Inheriting the present best work sequence for chromosomes in the initial population can reserve and further improve solution quality. We investigated the effect on inheriting present best priority list at different inheriting rate, and the comparison result is shown in Figure 4. In order to avoid the relevant influence on sorting chromosomes by waiting time between tasks, we initialized the popula-tion with unsorted chr-omosomes by waiting time between tasks.

In Figure 4, “cpr-0%" represents that the chromosomes in initial population are entirely randomly generated; “cpr-10%" means that 10% chromosomes inh-erit from the best pri-ority list from the pre-vious resource assign-ment, and the remain-ing chromosomes are randomly generated. Although several curves cross at the beginning of evolution, the final best fitness is arranged by descending order of inheriting rate. The comparison result firstly validates that partial inheriting operation may improve evolution speed and solution quality. And secondly, inheriting operation for all chromosomes (“cpr-100%") exhibits high evolution speed and also solution quality among all cases. In conventional consideration, inheriting operation for all chromosomes in the initial population may bring disadvantage because of same chromosomes without variety. Nevertheless, in the experiment, the chromosomes generated by inheriting operation still own their varieties because that resource assignment is renewed and the chromosomes are generated randomly before inheriting operation.

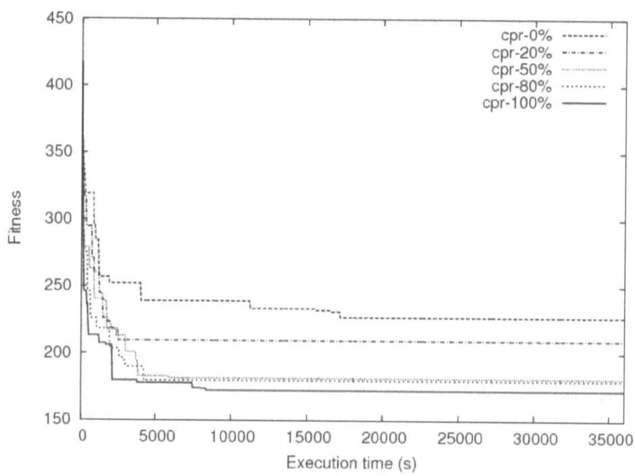

Fig. 4. Effect on inheriting the present best priority list

3.3 Scheduling result

The information on generated schedule with best fitness is listed in Table 1. Resource $R_1 \rightarrow R_2$, $R_7 \rightarrow R_9$ are available to perform W_1 and W_6, respectively. In the schedule, ten tasks will be performed cooperatively. The average rate of utilization for each resource reaches 93.9%, in which the moving time is not counted. The total amount of work, that is the amount of all works in all farmlands, is less than the product of total area of farmlands and number of works. This is because that some farm works have no necessity to schedule, or the amount of scheduled work is less then area of the farmland. The schedule length here represents the time period between the start of the first task and the completion of the last task. The schedule length is applicable to farm work in a growth cycle, because that the sugarcane producing corporation usually requires time to carry out extra farm works entrusted by individual farmers. These extra farm works may increase income for the corporation.

Table 1. Information on generated schedule

Resource	R_1	R_2	R_3	R_4	R_5	R_6	R_7	R_8	R_9
Moving time (h)	20.2	16.3	15.5	25.2	24.0	25.2	14.7	16.5	11.7
Idle time (h)	1.0	0 .0	0.0	0.0	0.0	0.0	0.0	1.83	0.33
Number of tasks	52	52	47	76	71	74	46	53	46
Work Duration (h)	297.7	298.7	308.0	436.5	406.2	463.2	210.0	208.0	209.7
Rate of utilization	0.932	0.945	0.950	0.942	0.941	0.946	0.930	0.921	0.944
Times of performing work cooperatively	10								
Total number of farmlands	76								
Total area of farmlands (hectare)	9.36								
Total amount of work (hectare)	49.2								
Schedule length (h)	2306.0								

3.4 Discussion

In this study, two-phase optimization is adopted for solving the farm work scheduling problem. In single phase optimization, the conflict of resource utilization has to be inspected for deadlock removal. In contrast, assigning resource first at two-phase optimization may reduce infeasible solution greatly, and deadlock caused by resource conflict may not exist. Moreover, inheriting operation at the second phase avoids resuming search from an unknown origination; so that the searching efficiency is improved. On hybrid Petri nets model, the breaks caused by uncertainties during the farming process and online status of resources were well handled. The computation results validated the rationality of the proposed approach.

The computing consumed many hours because of the big problem size for practical farm work planning. The computing time is most consumed by iteration of GA and simulation on hybrid Petri nets. Although the computation time is within allowable scope, the proposed algorithms should be further improved. The effective method for speedy convergence such as subtour exchange crossover, Partially Mapped Crossover (PMX) and Edge Recombination Crossover (EX) are expected to better than the one-point order crossover adopted in this research. In addition, parallel computing is attractive strategy for speeding up the computation.

For agricultural production corporation, some extra farm works entrusted by individual farmers should be also taken into account. These extra farm works are possibly treated as stochastic object if referring to historical data. In addition, the weather may also be stochastic variable derived from weather forecast. These constraints with probability distribution lead us to consider stochastic scheduling for farm works.

As discussed above, our extension works will focus on reduction of computation time, online scheduling and stochastic scheduling for uncertainty.

4. CONCLUSION

In this study, a two-phase metaheuristic was developed for farm work scheduling under constraints. The experiment results on solution evolutionary reveal that speedy evolution and good solution quality were obtained by emphasizing resource assignment optimization, initializing priority lists sorted by waiting time between works, and initializing priority lists inherited from present best task sequence in the previous resource assignment. The generated schedule with high ratio of resource utilization was applicable for practical farm work plan in some agricultural corporations, when considering conventional activities such as cooperative work, moving time of machinery, waiting time between works.

REFERENCES

Arjona E., Bueno G., Salazar L., (2001). An activity simulation model for the analysis of the harvesting and transportation systems of a sugarcane plantation, Computers and Electronics in Agriculture, 32, 247-264

Astika I W., Sasao A., Sakai K., Shibusawa S. (1999). Stochastic farm work scheduling algorithm based on short range weather variation, Journal of the Japanese Society of Agricultural Machinery, 61(2), 157 164, 61(3), 83 94, 61(4), 141-150

Bassett M. H., Pekny J. F. and Reklaitis G. V. (1997). Using detailed scheduling to obtain realistic operating policies for a batch processing facility, Industrial Engineering and Chemical Research, 36, 1717-1726

Chekuri C., Motwani R. (1999). Precedence constrained scheduling to minimize weighted completion time on a single machine, Discrete Applied Mathematics, 98, 29-39

Daikoku M. (2005). Development of computer-assisted system for planning work schedule of paddling and transplanting in distributed fields, Japanese Journal of Farm Work Research, Vol. 40(4), 210-214 (In Japanese)

David R., Alla H. (2001). On hybrid Petri nets, Discrete Event Dynamic Systems, 11, 9-40

Guan S., Nakamura M., Shikanai T.,Okazaki T. (2008). Hybrid Petri nets modeling for farm work flow, Computers and Electronics in Agriculture, 62(2), 149-158

Guan S., Shikanai T., Minami T., Nakamura M., Ueno M. (2006). Development of a system for recording farming data by using a cellular phone equipped with GPS, Agricultural Information Research, 15, 241 254

Haffar I., Khoury R. (1992). A computer model for field machinery selection under multiple cropping, Computers and Electronics in Agriculture, 7, 219-229

Murata T. (1989). Petri nets: properties, analysis and applications, Proceedings of the IEEE, 77 (4), 541-580

Nanseki T. (1998). Operations of FAPS97: a decision support system for evaluating agricultural technology and farm planning, Misc. Publ. Tohoku Natl. Agric. Exp. Stn. 21, 1 119 (In Japanese)

Santiago-Mozos R., Salcedo-Sanz S., DePrado-Cumpli M. and Bousono-Calzon C. (2005). A two-phase heuristic evolutionary algorithm for personalizing course timetables: a case study in a Spanish university, Computers and Operations Research 32(7), 1761-1776

Suliman S.M.A. (2000). A two-phase heuristic approach to the permutation flow-shop scheduling problem, International Journal of Production Economics 64, 143-152

The Ministry of agriculture, forestry and fisheries of Japan, (2006). Japan's post-war agricultural land reform and subsequent agricultural land system, National Report for the International Conference on Agrarian Reform and Rural Development.

Till J., Sand G., Urselmann M. and Engell S. (2007). A hybrid evolutionary algorithm for solving two-stage stochastic integer programs in chemical batch scheduling, Computers and Chemical Engineering, 31, 630-647

Tsai Y. J., Jones J. W., Mishoe J. W. (1987). Optimizing multiple cropping systems: a systems approach, Transaction of the ASAE, 30(6), 1554-1561

DEVELOPMENT OF A GENERAL PRINCIPLE SOLUTION FOR ISOAGRINET COMPLIANT NETWORKING SYSTEM COMPONENTS IN ANIMAL HUSBANDRY

Arne Kuhlmann [1,*], Daniel Herd [1], Benjamin Rößler [1], Eva Gallmann [1], Thomas Jungbluth [1]

[1] *University of Hohenheim, Institute of Agricultural Engineering (440b), Garbenstr. 9, 70599 Stuttgart, Germany*

[*] *Corresponding author, Contact: Tel.: +49 (0)711 459 22506; Email: arne.kuhlmann@uni-hohenheim.de*

Abstract: In pig production software and electronic systems are widely used for process control and management. Unfortunately most devices on farms are proprietary solutions and autonomically working. To unify data communication of devices in agricultural husbandry, the international standard ISOagriNET (ISO 17532:2007) was developed. It defines data formats and exchange protocols, to link up devices like climate controls, feeding systems and sensors, but also management software.

The aim of the research project, "Information and Data Collection in Livestock Systems" is to develop an ISOagriNET compliant IT system, a so called Farming Cell. It integrates all electronic components to acquire the available data and information for pig fattening. That way, an additional benefit to humans, animals and the environment regarding process control and documentation, can be generated.

Developing the Farming Cell is very complex; in detail it is very difficult and long-winded to integrate hardware and software by various vendors into an ISOagriNET compliant IT system. This ISOagriNET prototype shows as a test environment the potential of this new standard.

Key words: Data exchange, farm network, database, ISOagriNET, Farming Cell, IT FoodTrace, TINI, NSLU2

Please use the following format when citing this chapter:

Kuhlmann, A., Herd, D., Rößler, B., Gallmann, E. and Jungbluth, T., 2009, in IFIP International Federation for Information Processing, Volume 295, *Computer and Computing Technologies in Agriculture II, Volume 3*, eds. D. Li, Z. Chunjiang, (Boston: Springer), pp. 2011–2019.

1. INTRODUCTION

Electronic systems and sensors are widely used in livestock farming. Especially in pig housing systems they are often applied to control and monitor processes. Unfortunately, most manufacturer specific solutions are working proprietary and in isolation, as they are not developed for network tasks. As a consequence, advantages like documentation and process control, arising from communication between components, remain unused.

In order to standardize data communication within farm networks, the international standard ISOagriNET (ISO 17532:2007) had been developed. Its aim is to enable the uniform communication between devices in the stable just as with external partners (Paulsen et al. 2007). On the one hand ISOagriNET defines the data formats to use. On the other hand it specifies the procedures for network based data exchange, whereby the complete process for data transmission is covered.

The essential goals of this standard are: enhancing data transmission and communication between local devices and through the internet, implementation of demand-oriented services in farm networks (Paulsen et al. 2007).

Only few products are available on the market which link up hard- and software products e.g. feeding, ventilation and management system in pig housing systems. In practice the PIG NET integrates some of these devices by using the ISOagriNET standard (Quinckhardt 2006).

2. OBJECTIVES

Main target of this project is to build an ISOagriNET compliant on farm network for livestock systems, called Farming Cell, which integrates existing electronic devices and software products. It serves for linking up components and collects, transmits and consolidates data for different purposes e.g. traceability of food and feed, animal welfare or environmental protection.

The Farming Cell is developed and implemented at the research station for fattening pigs, Unterer Lindenhof, at the Universität Hohenheim. The barn there consists of two separate but identical compartments, each holding 56 fattening pigs. Standard equipment is installed to operate the barn. These devices like the liquid feeding system, the ventilation system or the electronic scale are controlled by separate terminals or alternatively by software applications running on a personal computer. Beyond this standard equipment, additional sensors are installed, measuring emissions, water or energy consumption as well as activity for research purposes.

The central task is to adapt these currently used hardware and software solutions to the international standard ISOagriNET. To harmonize non-ISOagriNET compliant hardware is the main challenge. The devices need to be linked to the Ethernet and their functionality must be extended by newly developed software providing the necessary ISOagriNET functionality.

For a successful implementation companies have to be cooperative. Their software products often have to be modified or separate software interfaces for hardware components need to be programmed. The development of the Farming Cell is complex and only promising with the involvement of experts from different disciplines. For this reason the development is done in close cooperation with the manufacturers of agricultural equipment and software products.

3. BASIC APPROACH

The technical challenge of adapting commonly used hardware and software solutions to the ISOagriNET standard consists of two major tasks.

First of all, the present systems and sensors have to be connected to the Ethernet, as it is the transport medium defined in the standard. Secondly the features of every single component have to be evaluated and individual concepts for further development have to be elaborated.

On the one hand there are system units that are accessible via software and providing complex functionality (system units: climate and feeding control, scale with electronic ID). On the other hand, there are simple sensors just outputting a voltage or a counting impulse (sensors: water, electricity, NH3, CO2, etc.). These two types differ by functionality as well as hardware connectivity. While system units have standard hardware interfaces, sensors do not have these. The following chapter "Hardware Components" will go into some hardware details of the different devices.

The subsequent chapter "Software Components" deals with the second task mentioned above. In this chapter the general software concept of the Farming Cell and the differing software solutions to contact system units or sensors are described.

3.1 Hardware Components

This chapter explains the different device properties regarding the hardware level, which means the possibility to attach each device to the Ethernet.

There are three system units located in the research barn.

Table 2. Hardware properties of the system units

System unit	Hardware interface	Ethernet connectivity
climate control	LON-Bus	- No - The device is controlled by a computer connected via LON-Bus to RS232 adapter unit
scale with included electronic animal identification (RFID)	RS232	- Yes - RS232 to Ethernet adapter unit
feeding control	RS232	- Yes - RS232 to Ethernet adapter unit

Table 1 shows the system units with their type of hardware interface and the way they are connected to the Ethernet.

Currently the scale and the feeding control are physically available in the Ethernet, but not accessible through software layers. Therefore they have to be invented in a next step. Further steps with regard to the climate control are unsettled so far.

Table 2. Hardware properties of sensors

Sensor	quantity	signal
water meter	16	impulse
electric meter	9	impulse
heat meter	2	impulse
differential pressure	1	voltage
NH3	1	voltage
CO2	1	voltage
Temperature	1	voltage
Brightness	1	voltage
Humidity	1	voltage

All present types of simple sensors, their quantity and the signal type they output are listed in table 2.

In contrast to the system units, simple sensors do not provide a hardware interface which is directly convertible to Ethernet. In order to link these sensors to the Ethernet, additional hardware components are necessary. Since there are no ISOagriNET compliant systems available for sensor linking, they have to be developed.

To solve this problem, the programmable and universal customizable TINI platform from Maxim, shown in figure 1, was chosen.

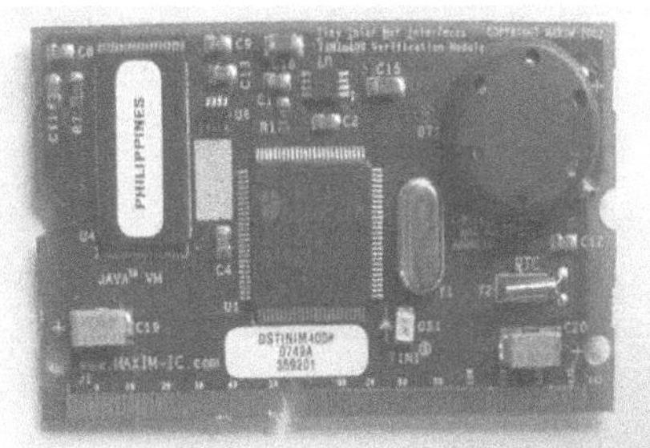

Figure 5. TINI platform

TINI (Tiny INternet Interface) is a microprocessor based platform, offering different hardware interfaces like Ethernet and One-Wire, a fully implemented TCP/IP network stack and a Java Virtual Machine 1.1.

The TINI serves as a basis for the developed One-Wire analogue/digital converter and counter modules (see Figure 2. below).

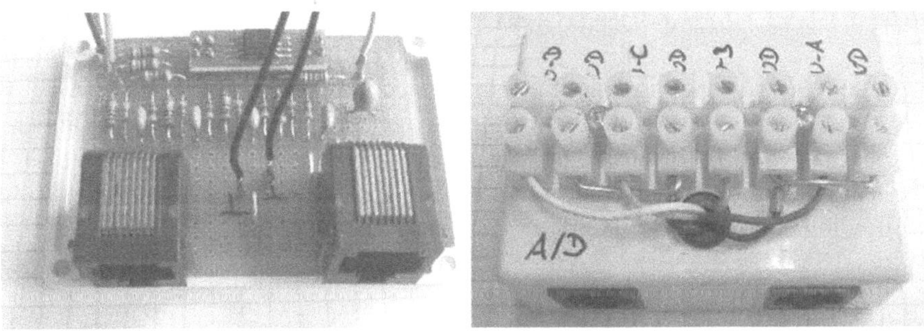

Figure 2. One-Wire A/D converter

That way both kinds of sensors, those outputting voltages and those outputting counter impulses, can be linked to the Ethernet via TINI. Beside that, each sensor connected to the TINI is configurable. The configuration of signal type, conversion formulas and other features can be performed via a web service hosted on the TINI.

The ability to connect system units and sensors to the Ethernet in the described ways leads to a complex IT-system, producing high amounts of data. This data has to be stored in a permanently available database. As a low cost and low power database server, the Network Storage Link for USB 2.0 (NSLU2) by Linksys is used (see Figure 3. below).

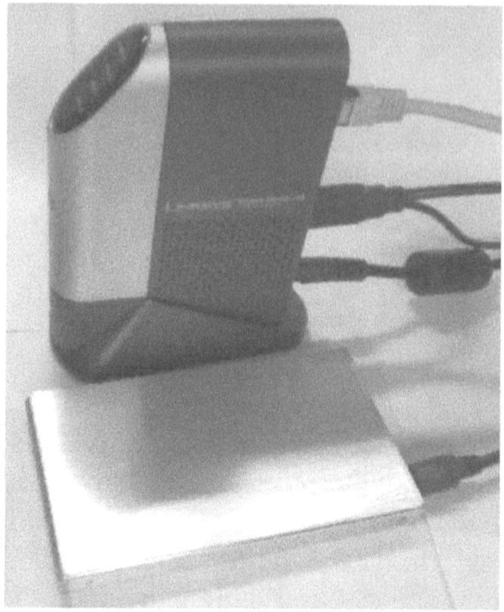

Figure 3. NSLU2 with attached 2,5″ hard disk drive

The NSLU2 was originally built to connect hard disc drives as Network Attached Storage (NAS) to the local network. Due to its popularity a Debian Linux distribution for the NSLU2 is available opening up many possibilities of usage. In this case MySQL 5.0 database server was installed (compare following chapter).

3.2 Software Components

The previously described hardware is linked to the Ethernet, but the ISOagriNET functionality has not been implemented by software yet. Therefore individually programmed software, providing this functionality is needed. This application is written in Java and runs on the TINI platform. Sensors and system units are connected to the Ethernet via several TINI platforms hosting individual software for each type of component. The microcontrollers collect data, convert it to an ISOagriNET compliant data format and send it to a multicast group or to the originator of an enquiry.

Figure 4 shows the software concept and the way data is transmitted and collected. Within this model the Management Computer is hosting the management software for pigs which is a commercial product by agrocom GmbH. A database is integrated in this product and stores the animal data. In addition the Management Computer will host a web server offering a web service making basic data of the farm and the pigs accessible.

In addition to the MySQL database an ISOagriNET Server was developed and installed on the NSLU2. It has three basic functions.

1st it collects data send by sensors and devices.
2nd it holds the ISOagriNET data dictionary to convert this encoded data.
3rd it stores collected data in the database for sensor data.

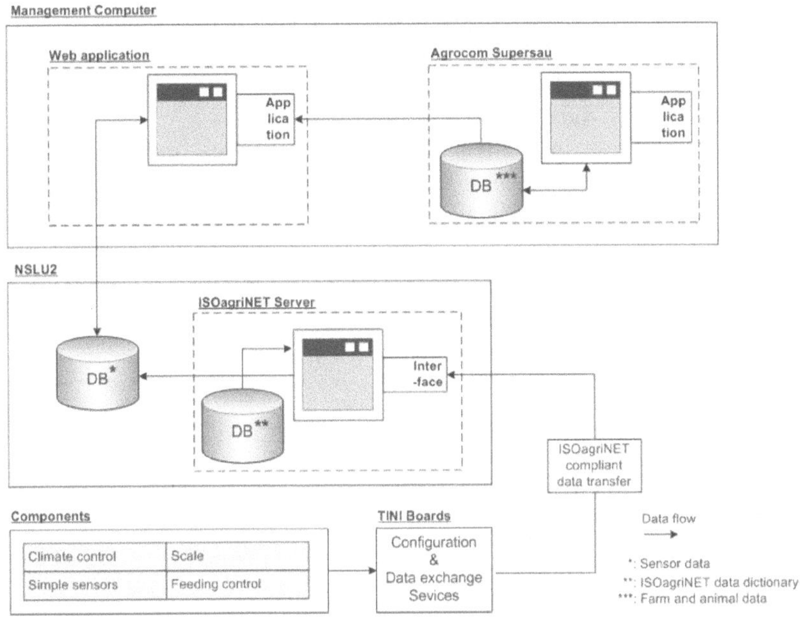

Figure4. Software Concept

The web application on the management computer to control system units has not been implemented yet.

4. CHALLENGES

During the implementation process there were many constraints to struggle with. The companies that are producing technical equipment for pig housing systems are using different kinds of hardware interfaces. Therefore adapters and converters to transform these interfaces to Ethernet standard are used. Software drivers were installed to transform different protocols and signals. Moreover, sensors could not be linked to the Ethernet directly, but a specific solution, the TINI board, had to be used.

The integration of many devices into the IT-system, the installation of software and in particular the development of new software has to be done individually. Therefore trained and experienced persons with very good IT-skills are needed.

5. EVALUATION

Many sensors and system units are under development to meet the ISOagriNET standard. Thus, they are not yet available on the market and are restraining an automated integration of devices and software into the farm network. To get around this restriction, gateways for data exchange and communication have to be created.

By knowing these potentials but also challenges there are several advantages and disadvantages for the implementation of a Farming Cell.

Advantages of the Farming Cell:

- redundant data input is avoided
- automation of process documentation is simplified
- availability of electronically stored data is enhanced
- data analyses for different purposes, like traceability is easy to realize
- work load and actual costs can be reduced

Disadvantages of the Farming Cell

- implementation and operation needs experts
- only few products which are ready for the market exist
- data security has to be guaranteed
- investment for hard- and software can increase

6. CONCLUSION AND OUTLOOK

The implementation of an over all IT-system by linking up individual hard- and software components produced by different manufacturers is very complex, mainly due to incompatibility. There is still a lot industrial effort needed to standardize on farm equipment. New open source gateways or standards are on the way. Companies and administrative bodies are supporting the development.

Process documentation in pig housing systems has not to be done electronically. But there are advantages for this, e.g. improved speed and effectiveness for data analysis. Thus, central database or management systems will become more important. On the other side, the implementation and operation of these complex IT-systems need experts with hard- and software skills.

This project with its ISOagriNET prototype will help to evaluate the benefits and drawbacks of the standard. Further it contributes to the implementation of standards within a heterogeneous hard- and software world.

ACKNOWLEDGEMENTS

The research project „Information and Data Collection in Livestock Systems“ is part of the "Federal Ministry of Education and Research" project „IT FoodTrace“ (maturity June 2006 till May 2009), www.itfoodtrace.de.

REFERENCES

ISO 17532 – INTERNATIONAL ORGANIZATION FOR STANDARDIZATION: ISO 17532:2007 - Stationary equipment for agriculture - Data communications network for livestock farming, CH Geneva.

Paulsen, C.; Martini, D.; Kunisch, M., 2007: Austausch von Daten aus der Tierhaltung mit agroXML – Konzeption der Zusammenarbeit mit ISOagriNET: p. 97 – 104. In: KTBL-Schrift 454: agroXML – Informationstechnik für die zukunftsorientierte Landwirtschaft. KTBL - Vortragstagung vom 17. bis 18. April 2007 in München, Germany. Darmstadt, Germany.

Quinckhardt, M., 2006: Vernetzte Managementhilfen zur Steuerung und Dokumentation von Prozessabläufen in der Tierhaltung: p. 119 – 124. In: VDI-Bericht Nr. 1935, 2006: Tagung Tier Technik 2006, Lebensmittel aus dem Stall. Stuttgart, Düsseldorf, Germany

APPLICATIONS OF ZIGBEE WIRELESS TECHNOLOGY TO MEASUREMENT SYSTEM IN GRAIN STORAGE

Huiling Zhou[1,*], Xuechuan Chen[1], Xiangdong Liu[2], Jun Yang[1]

[1] *Automation School, Beijing University of Posts and Telecommunication, Beijing, China, 100876*

[2] *College of Engineering, China Agricultural University, Beijing, China, 100083*

[*] *Corresponding author, Address: P. O. Box 107, Automation School, Beijing University of Posts and Telecommunication, Beijing, 100876, P.R.China, Tel: +86-10-62282927, Fax:+86-10-62282629, Email: huiling@bupt.edu.cn*

Abstract: Wireless sensor network (WSN) has attracted more and more attentions in the computer science and communication research community. It has appeared a great potential in numerous industrial applications. In this paper a wireless sensor network based on ZigBee protocol for the measurement system in grain storage was developed and tested as an alternative to the cable system, which showed better efficiencies of the cost and time in the installation and maintenance than a cable system. Based on above study a new ZigBee sensor-node was built to measure the grain temperature and the temperature, moisture and pressure drop of the air through a bulk grain system. The wireless data transmission and the power consumption were tested. This paper proposed a practical solution for the ZigBee protocol.

Keywords: wireless sensor network, ZigBee, grain storage, date transmission, power consume

1. INTRODUCTION

The scientific researches have already shown that the grain quality in the barn was strongly related to the grain storage condition, the environment and the storage time (Li Tiepan et al., 2004). An effective monitoring of temperature and humidity is fundamental for ensuring the grain quality

Please use the following format when citing this chapter:

Zhou, H., Chen, X., Liu, X. and Yang, J., 2009, in IFIP International Federation for Information Processing, Volume 295, *Computer and Computing Technologies in Agriculture II, Volume 3*, eds. D. Li, Z. Chunjiang, (Boston: Springer), pp. 2021–2029.

during the storage. According to the real time information of grain temperature and humidity, the aeration, as well as optimal aeration time, are necessary to inhibit from microbial growth, sprouting and germination and insect population growth.

Wireless Sensor Network (WSN) is a system comprised of radio frequency (RF) transceivers, sensors, microcontroller and power source. This system with self-organizing, self-configuring, self-diagnosing and self-healing capability has been developed to solve problems or to enable application that traditional technologies could not address. WSN originates from military and environmental monitoring, moves towards machine to machine communication and is eventually applied to all aspects of our life (Wang, N. et al., 2006). Compared with the fieldbus system, WSN for monitoring has the following advantages: simple structure, less problems than connecting wires; feasibility and low cost of installation; and reliable and stable communication (Maxwell, D. et al., 2002).At the present, ZigBee within ISM band of 2.4GHz based on provided license-free operations, huge spectrum allocation and worldwide compatibility, has been specifically developed to address the demands of wireless sensor networks, including the need to handle a large number of nodes (more than 65,000). ZigBee built upon the physical (PHY) and medium access control (MAC) layers defined by the IEEE 802.15.4 standard (IEEE, 2003) enables the creation of a higher network flexibility, such as star, cluster tree or mesh network, and is very suitable for industrial, agricultural, vehicular, residential and medical environments. The intent of IEEE 802.15.4 is to provide an ultra low power consumption and very short wake-up time capabilities at very low cost to devices operating in a Low Rate (250 kbps) Wireless Personal Area Network (LR-WPAN). The IEEE 802.15.4 assumes that the data transmitted are short and that transmissions occur at a low-duty cycle (active/sleep times ratio), reducing the overall power needs and enabling the application of battery–powered embedded systems (Raul Morais et al., 2008).

Wireless sensor technology based on ZigBee protocol is still at its early development stage. Applications in grain storage are very limited. This paper is to give a kind of ZigBee node (Tech1) for monitoring the grain temperature, the temperature-humidity and pressure of the air through the bulk grains during storage. The node's hardware architecture and the experiments to get the data of the temperature, temperature-humidity, transmission distance and the battery life are major issues studied in this work.

2. HARDWARE OVERVIEW OF ZIGBEE NODE

ZigBee node Tech1 is integrated by a microcontroller board (ZigBee board) and a sensor board with DC-DC transducer (3V-3V, 3V-5V). Node Tech 1 architecture prototype is shown in Fig.1.

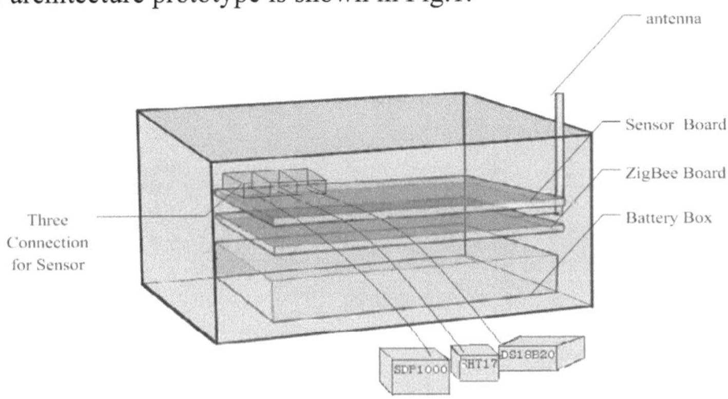

Fig.1. ZigBee node Tech1 architecture prototype

The ZigBee board hosts a TI CC2430 that is a true SoC solution tailored for IEEE802.15.4 and ZigBee applications. An enhanced 8051 microcontroller core, a 14bitA/D, a 128MB flash and a 8kB RAM are embedded in the CC2430. The CC2430 also combines CC2420 RF transceiver, and combined with the industry leading ZigBee protocol stack (Datasheet for SmartRF® CC2430, 2005). It is specifically suited for the monitoring system of grain storage for which ultra low power consumption is required.

Sensor board consists of the sensors and DC-DC model. And three types of precise digital sensors are used: a maxim DB18B20 for monitoring the grain temperature, a Sensirion SHT71 for measuring the temperature and moisture of the air through the bulk grains, and a Sensirion SDP1000 for collecting the pressure information through the bulk grains during the aeration. These wireless sensors have signal conditioning and processing units and transmit signals in the digital form, therefore the noise pick-up becomes a less significant problem. The sensor block diagrams are given in Fig.2 (Datasheet for DS18B20, Datasheet for SHTxx(v2.04) and SDP1000).

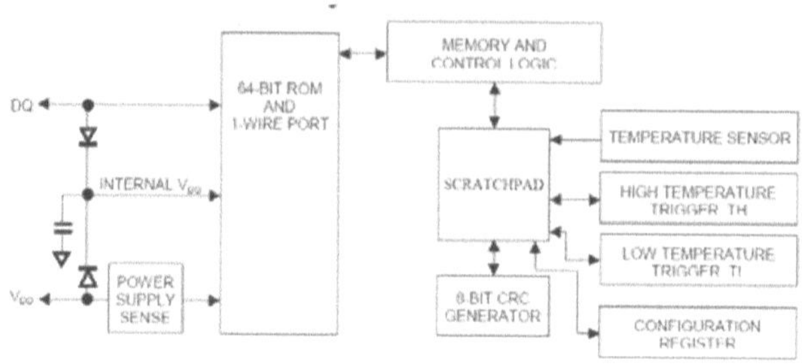

(a) DB18B20 block diagram

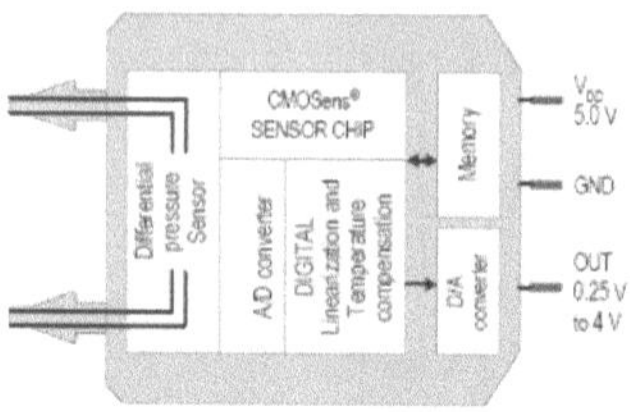

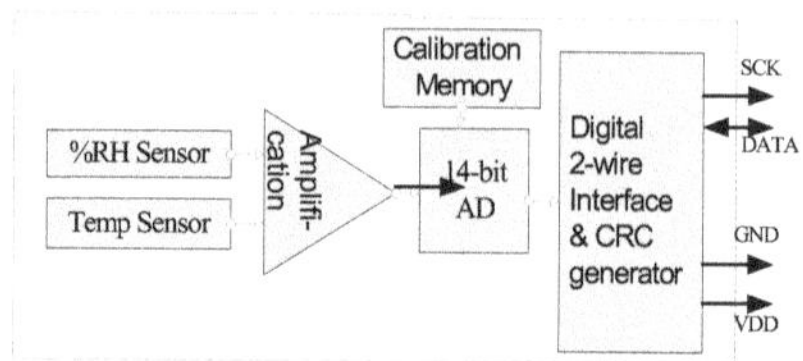

(b) SDP1000 block diagram (c) SHT71 bolck diagram

Fig.2. Sensor block diagrams

The node power comes from two 1.5V AAA batteries. Because the ZigBee board and sensors need different voltages to provide power, the compact and high efficient DC-DC converter (Maxim Max1677) was designed. The input voltage range is +0.7-+5.5V, the output voltage is 3V to ZigBee board and sensor SHT71, 5V to sensors DB18B20 and SDP1000. The photograph of Tech1 node is shown in Fig.3.

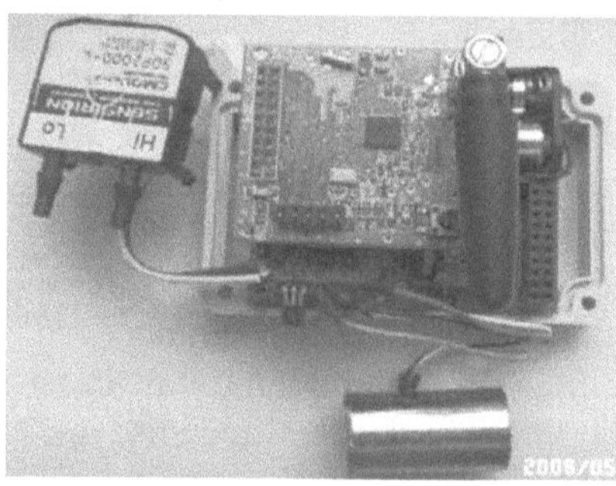

Fig.3. Sensor block diagrams

3. EXPERIMENT OF LABORATORY AND RESULTS

In order to test the node work reliability and the transmission distance through grain, a ZigBee star network with 4 Tech1 nodes is constructed in the laboratory (see Fig.4): one is the full function device(FFD) as a coordinator, three are reduced function devices (RFD) as end devices. This categorization makes power saving possible, because RFD devices need to be connected to the network only to check network messages and transmit sensor data to coordinator. A monitoring system runs in PC, which communicates with the coordinator by RS232 interface, in order to display the information of the temperature, humidity, pressure, and RSSI data from each node timely, as well as battery voltage. As test material, the steel grain bin is filled with millet.

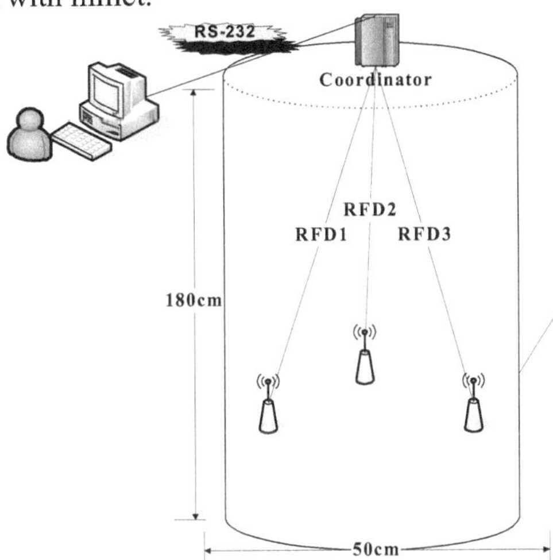

Fig.4: End devises and coordinator placement within grain

3.1 Distance estimation

In order to estimate the transmission distance of Tech1 within the grain, Received Signal Strength Indication (RSSI) technology is used. RSSI measures the attenuation in radio signal strength between the coordinator and the end devise. Although RSSI technology for distance estimation is not very inaccurate(J. Hightower et al., 2000), this data is mainly used to get approximate maximum grain depth that the node can communicate and, as well as to estimate the total nodes of the real ZigBee wireless monitoring network within grain in the future.

Distance estimation based on RSSI needn't use any extra hardware support. CC2430 has a built-in RSSI giving a digital value that can be read from the 8 bit, signed 2's complement RSSIL.RSSI_VAL register bits. The RSSI value is always averaged over 8 symbol periods (128 μs), in accordance with IEEE Standards (IEEE std. 802.15.4, 2003). The RSSI register value RSSI.RSSI_VAL can be referred to the power P at the RF pins by using the following equation

$$P = RSSI\_VAL + RSSI\_OFFSET[dBM] \quad (1)$$

Where: RSSI_OFFSET is approximately –45. E.g. if reading a value of –20 from the RSSI register, the RF input power is approximately –65 dBm. A typical plot of the RSSI_VAL reading as function of input power is shown in Fig.5. It can be seen from the figure that the RSSI reading from CC2430 is very linear and has a dynamic range of about 100 dB(Datasheet for SmartRF® CC2430, 2005).

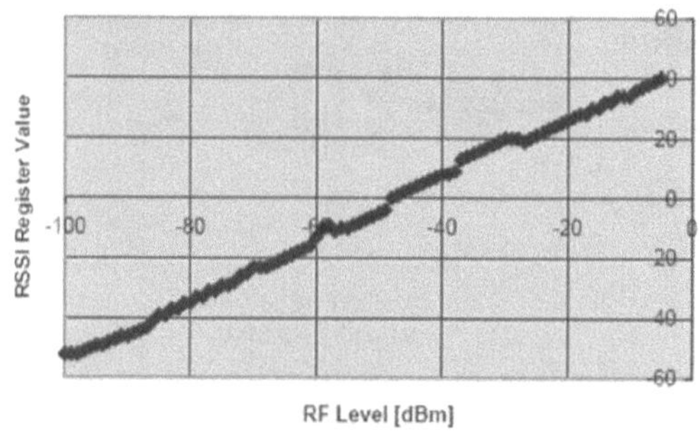

Fig.5. Typical RSSI value vs. P

In the tests, three nodes and the coordinator were first placed above the grain, and then the nodes were pushed slowly into grain, the depth and RSSI_VAL values were recorded. The average signal attenuation is shown in Fig.6. These data indicate that the node communication distance is approaching 115cm. Beyond this range the power of the radio signal falls off so rapidly and can't connect with network. But this result doesn't accord well with the equation given by Nelson(Nelson, S.O., 1996), with the steel grain bin the signals' high reflectivity and the relation with the direction of the antenna maybe are the reasons.

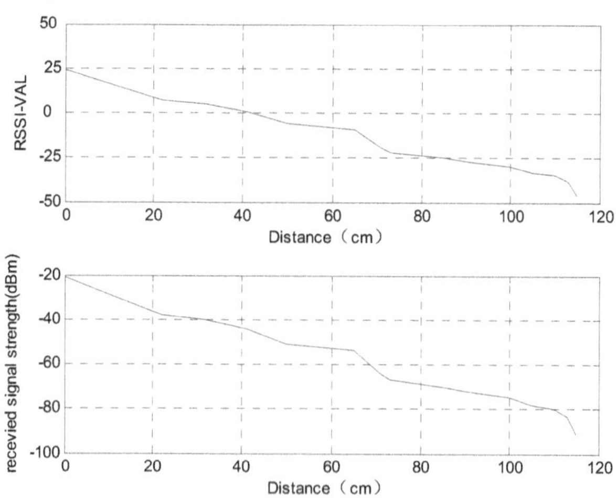

Fig.6. Signal attenuation for distance (ambient air temperature 22.9°C,%RH 81.5%)

3.2 Temperature, temperature-humidity, pressure data collection and transmission

In order to test the node's reliability many experiments were made in the laboratory. The plots in Fig.7 are a part of data in one experiment. These data are shown that Tech1 can reliable work within grain. These results are very useful for the next research which will focus on the control of grain storage. As the plots shown that during the aeration the humidity changed very rapidly, the temperature decreased at first, and then increased gradually.

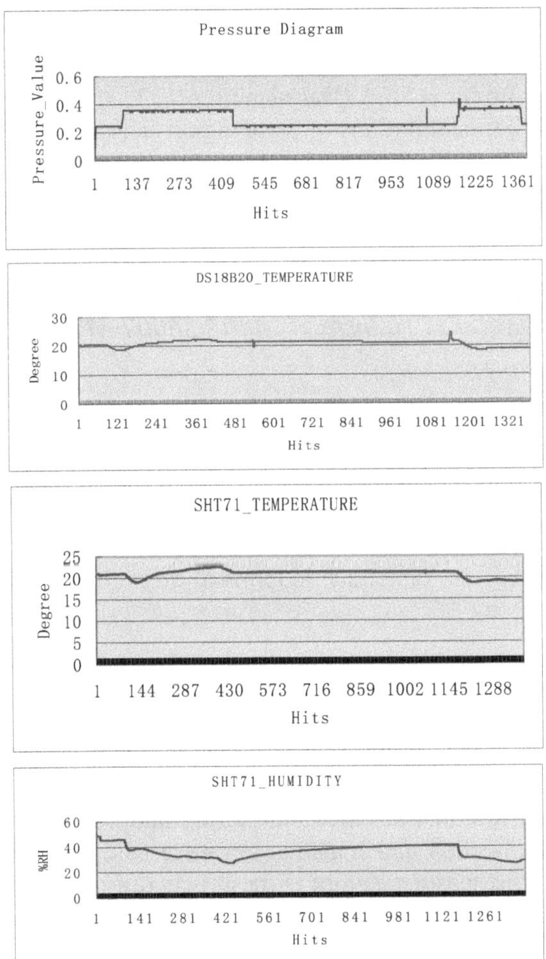

Fig. 7. Temperature, temp-humidity and pressure data from Tech1(Test: sending data pre 2 second, 15:23:21-16:46:09,2008/5/10,Beijing)

3.3 Node lifetime estimation

The node lifetime decides whether a grain storage monitoring network based ZigBee protocol can give an application-level solution. In order to estimate Tech1 lifetime, our test strategy is:

- One end devise with 2 new 1.5V AAA alkalescent batteries
- Battery capacity about 2200mAh(Mainwaring A et al., 2002)
- CC2430 MCU still active and TX mode, 0dBm
- sensors data transmission Per 2 second

In theory, the current consumption of CC2430 MCU active and TX mode is about 24.7mA, ADC current consumption is 0.9mA, Timer is 19uA (Datasheet for SmartRF® CC2430, 2005), DB18B20 work current is 1mA, SHT71 work current is 550uA (voltage source 5V,Temp.25°C) and SDP1000 work current is 5mA (Datasheet for SHTxx(v2.04) and SDP1000). Because the work time durations of these digital sensors are all less than 1ms, and then transfer into sleep state that power consumption is very low. The time of data sending one time of the end devise is also less than 1ms, the node total power consumption per one second is more lower than 31.169mA. As SHT71 work voltage range is 5V- 2.4V, the total runs needed for the end devise sending data are at least

$$\begin{aligned} 440mAh/31.169mAs &= 440*3600/31.62 \\ &\approx 50818(times) \end{aligned} \quad (2)$$

In according to this work mode, the result of our node lifetime estimation experiment is: 56206 times. The test result proved the theory calculation validity.

The end devises in grain operate normally at really low duty cycles. If the network nodes wake up, perform their tasks and then revert back to sleep state, this would be very beneficial in terms of extending node lifetime.

4. FUTURE WORK

The ZigBee Protocol is the only international standard wireless sensor network protocol in existence, and is suitable for the special needs of low-power, low-cost, low maintenance monitoring and control systems. In this paper the data collection and transmission were based on a simple ZigBee star network. If the sensor network is mesh network or cluster tree network topologies, routers will need to be capable of running in lower power sleep modes, waking up in time to perform any of their assigned tasks and then going back to sleep mode. Thus how to create a routing optimization mechanism to increase the network lifetime will be the future work. Another challenge work is to find the relationship between the grain temperature and the temperature humidity of the air through the bulk grains during the aeration.

REFERENCES

Datasheet for DS18B20. http://www.maxim-ic.com.

Datasheet for SHTxx(v2.04) and SDP1000， 2004. http://www.sensirion.com.

Datasheet for SmartRF ® CC2430, 2005. http://www.maxim-ic.com.

IEEE std. 802.15.4, 2003: Wireless Medium Access Control (MAC) and Physical Layer(PHY) specifications for Low Rate Wireless Personal Area Networks (LR-WPANs).

J. Hightower, R. Want, and G. Borriello. "SpotON: An Indoor 3D Location Sensing Technology Based on RF Signal Strength" Technical Report, UW CSE 00-02-02, University of Washington, Seattle, WA, February 2000.

Li Tiepan, Song Youlin. Distribution function of temperature in grain storehouse and its application, Journal of Zhengzhou Institute of Technology, 2004， 25(3):77-79(in Chinese).

Mainwaring A, Polastre J, Szewczyk R, Culler D, Anderson J. Wireless sensor networks for habitat monitoring, In: Proc of ACM WSNA'02,2002.

Maxwell, D., Williamson, R., 2002. Wireless temperature monitoring in remote systems, sensors, Sensorsmag, October, 2002.

Nelson, S.O., 1996. Review and assessment of radio-frequency and microwave energy for stored-grain insect control. Transactions of the ASAE. 39(4): 1475-1484.

Raul Morais, Miguel A. Fernandes. Computers and electronics in agriculture 6 2 (2 0 0 8) 94–106, A ZigBee multi-powered wireless acquisition device for remote sensing applications in precision viticulture.

Wang, N., Zhang, N., Wang, M., 2006. Wireless sensors in agriculture and food industry – recent development and future perspective. Computer and Electronics in Agriculture 50, 1–14.

THE DESIGN AND REALIZATION OF EMBEDDED WIRELESS DATA COLLECTION SYSTEM

Junfeng Zhang[1], Feng Yu[1], Jichun Zhao[1,*], Sufen Sun[1]
[1] *Institute of Information on Science and Technology of Agriculture, Beijing Academy of agriculture and forestry Sciences, Beijing, China, 100097*
[*] *Corresponding author, Address: Institute of Information on Science and Technology of Agriculture, Beijing Academy of agriculture and forestry Sciences, Beijing, China, 100097, P. R. China, Tel: 13552902579, Email: zhaojichun123@tom.com*

Abstract: The paper gives a method that greenhouse data collection and supervisory system based on GPRS (General Packet Radio Service) network is designed and implemented. It discusses chiefly a scheme, which uses ARM microcontroller to control GPRS module and make use of SMS(Short Messaging Service) to complete greenhouse data collection and supervision, and supplies a design program of the hardware and software, the communication way has better timeliness ,without room limitation.

Keywords: GPRS, SMS, ARM, Data Collection

1. INTRODUCTION

The data collection and supervision in the greenhouse profession still is limited to the local area network with the 485 communication protocol. Although this way can meet the data supervision and processing, but the cover scope is small, the line maintenance quantity is big, the noise pollution in the electric power carrier will cause the correspondence not to be unreliable, moreover, intelligence supervisory system of greenhouse based on GPRS uses GSM/GPRS wireless network as the communication way, not only can solve the above problem, but also can widely be applied in other professions.

Please use the following format when citing this chapter:

Zhang, J., Yu, F., Zhao, J. and Sun, S., 2009, in IFIP International Federation for Information Processing, Volume 295, *Computer and Computing Technologies in Agriculture II, Volume 3*, eds. D. Li, Z. Chunjiang, (Boston: Springer), pp. 2031–2036.

2. SYSTEM PROFILES

Intelligence supervisory system of greenhouse is a set of comprehensive computer system used in the upscale greenhouse, each greenhouse installs one, supervises the greenhouse warm target and so on humidity through sensors and carries on the automatic control to the greenhouse corresponding equipment to realize the climate parameter value which the greenhouse needs. Realizing one to multi-spot communication through the GPRS network, according to certain communication protocol, it may realize data gathering, the user forces to intervene the active state of the microcontroller and the equipment failure long distance reports to the terminal.

3. SYSTEM REQUEST AND FUNCTIONAL DESIGN

The system construction needs to meet and complete some basic long range data gathering and the transmission request, simultaneously, considering the demand of satisfying the market development and the economical dispatchment, having the dominative ability for input or output of switch quantity and prohibiting disturbance ability. According to the system demand, the function which the system needs to provide is as follows:

(1) data collection function: voltage or electric current standard signal which converted through the senor and bridge circuit turns the digital data after A/D transformation, These data may carry on the GPRS long-distance transmission through the monolithic integrated circuit by the short news way.

(2) data submission function on SMS: the control center may transmit the communication order to carry on the data collection, phone terminal number revision and to force to control the running state of switch quantity.

(3) control ability: the controller can dominate pad_pump、evaginable window、fan and so on through supervising the temperature and humidity in the greenhouse.

4. HARDWARE DESIGN

4.1 Hardware structure

The remote data collection terminal is consisted of the sensor, the monolithic integrated circuit, the GPRS module and so on, it chiefly completes the environment data gathering and communicate with phone terminal device through certain protocol, its realization principle of hardware is shown in figure1.

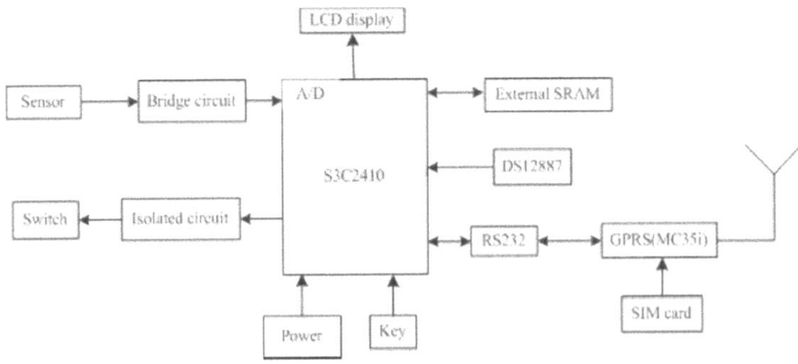

Figure 1

4.2 Hardware explanation

(1) CPU: The S3C2410 offers outstanding features with its CPU core, a 16/32-bit ARM920T RISC processor designed by Advanced RISC Machines. The ARM920T implements MMU, AMBA BUS, and Harvard cache architecture with separate 16KB instruction and 16KB data caches, each with an 8-word line length.

(2) Data collection: the Pt100 of temperature sensor can output 4～20mA standard electric current through bridge circuit,the humidity sensor SA801 applies standard current output so that it does not carry on the non-standard electric quantity transformation.

(3)Store unit: data storage can deposit the data outside which gathered, it selects 32kb static RAM, because of the battery in its interior, so, after the system power failure, it can preserve each gathering data and the user setting values.

(4) Clock part: it selects the DS12887 of American DALLAS corporation which arrives precisely the second and reads the current time for CPU.

(5) Serial port part: because GPRS module brings the RS232 level by itself, but MCU serial port is the TTL level, so, it should carry on the level transformation through the MAX232 chip.

(6) Man-machine contact surface: the data display part uses the LCD12864 liquid crystal, the keyboard response part uses the THINK82C79 control.

(7) Control output: in order to reduce the electromagnetic interference, the output unit uses the photoelectricity isolation technology, thus, selecting light pair MOC3061 and the silicon-controlled rectifier Z4090 chip.

5. SOFTWARE DESIGN

5.1 AT instruction

For realizing SMS transmission, we must conform to AT instruction of the GSM communication, AT instruction which chiefly used in the article is shown in table1.

Table 1

AT instruction	Function
AT+CMGF	Select SMS message format (0=TEXT, 1=PDU)
AT+CMGS	Send SMS message
AT+CMGR	Read SMS message
AT+CMGD	Delete SMS message
AT+CSQ	GPRS network exist?

For example, the article send short message by text model, the procedure transmission data as follows:

AT+CSQ<cr>

+CSQ: 21,99 //network exist

AT+CMGF=1<cr> // set text model

AT+CGMS="phone number"<cr>send message <ctrl/z>

condition. The SMS communication makes use of serial port interrupt characteristic of the MCU, it can reach the data collection and compulsion operation to the terminal device, its SMS procedure process is shown in figure2.

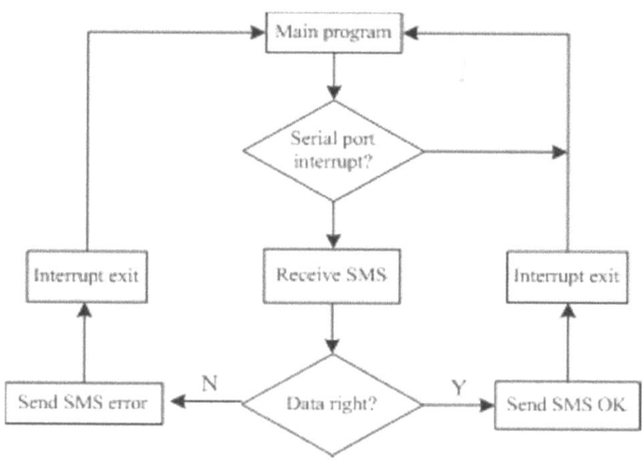

Figure 2

5.2 SMS subprogram

(1) the subprogram that MCU receives AT instruction is follow:

```
RECEIVE:LD   AX,#BDH;storage unit
            LDB CL,#06H;datacount
RECEIVE_1:DI
RECEIVE_2:LDB BL,SBUF
            JBC SP_STAT,6,
RECEIVE_2 STB BL, [AX]+ DJNZ CL, RECEIVE_1 RET
```

(2) the subprogram that MCU receives AT instruction is follow:

```
SEND:LDB  CL,#30H;data  count// sen  message to certain phone
number+CMGS:199//sendback  OK
```

5.3 Program design

The main program unceasingly compare with setting values of the user through gathering external environment condition, it will carry on the corresponding control command when satisfying the LD AX,#00H ;data base address

```
SEND1:LDB   BL,#00H
      LDB   BL,DATA4[AX]
      LDB   SBUF,BL
      JBC   SP_STAT,5,$
      INC   AX
      DJNZ  CL,SEND1
      RET
```

6. EXPERIMENT RESULT

The user can edit the corresponding communication instruction which sent to the GPRS(TC35i) module via AT command orders, in order to manipulate the controller, for instance, one group operating results are as follows:

(1) access to the current temperature and humidity:

the back result which is shown on the cell phone by sending communication order LY1 is that: the temperature: 16.6℃and humidity: 26.6%.

(2) open and off equipment: order LY2 which is sent denote that the equipment is ran ,on the contrary, it will close the one by the order LY3, the back result is: success, if the result do not return, it may the SIM card debt or equipment failures.

(3) host set: by sending order LY4, in order to a cell phone always can controll single or multiple device, the back result : host lock.

(4) alarm on and off set: the back result on the cell phone is that the alarm setting is ran by sending order LY5, when it reach the alarm condition, it will send SMS to the host phone, on the contrary, it will close the alarm setting program by sending order LY6.

(5) targeted set: LY6+temperature number+humidity number of the order format is sent, for example, the order LY6+27+29 denote that current goal temperature: 27℃, humidity: 29%, the precision only can reach integer here, the back result: targeted set success.

7. CONCLUSION

Along with the constant development of industrial modernization and new technologies evolvement, data collection and remote supervision is entering a phase of rapid development. Because of adopting GPRS netmwork, the system is a more advanced technology, it is easy to replace a lot of cable-based communications systems similar, whether for automatic meter reading system, or temperature measurements, environmental supervision, marine oil fields, have very good prospects for development.

REFERENCES

Aljadhai, A.; Znati, T.F.;"Predictive mobility support for QoS provisioning in mobile wireless environments", Selected Areas in Communications, IEEE Journal on , Volume: 19 Issue: 10 , Oct. 2001, pp. 1915 — 1930.

Application of Image Gather Based on the S3C2410 Platform and Embedded Linux [M].Micro computer Information, 2006, 22

C. F. Grecas, I. S. Venieris, and D. -D. Koutsouris; "Uninterrupted Data Transmission over GSM and Application to Health.

Camargo J.R. An evaporative and desiccant cooling system for the air conditioning in humid climates. Journal of the Brazilian society of Mechanical Science and Engineering[J], 2005,27(3):243-247.

Egginton. Evaporative cooling saves energy costs. Engineering Technology[J],2005,8:39-408(6):1667-1675.

Engineering and Manufacturing Co. Muskofgee, Oklahoma, 2003, 5 (7) :727~741.

R. Pandya, "Mobile and Personal Communication System and Services", IEEE Press , ISBN 0-7803-4708-0,2000.

APPLICATION OF SUPPORT VECTOR MACHINE BASED ON TIME SERIES FOR SOIL MOISTURE AND NITRATE NITROGEN CONTENT PREDICTION

Shaoe Yang [1,2] , Yuanfang Huang [1,*]

[1] *Key laboratory of plant-soil interactions, MOE; Key laboratory of soil and water, MOA; College of Resources and Environment, China Agriculture University, Beijing, P. R. China 100193*

[2] *Institute of remote sensing applications Chinese academy of sciences, Beijing, P. R. China 100101*

[*] *Corresponding author, Address: College of Resources and Environment, China Agriculture University, Beijing, 100193, P. R. China, Tel: +86-010-62732963, E-mail: yfhuang @china.com*

Abstract: Support Vector Machine based on Time Series (SVM-TS) was applied to predict soil moisture and nitrate nitrogen (NO3--N) content. For the prediction of soil moisture, the statistical result (t-test) indicate that there is no obvious difference between predicted and observed values in 0-20cm and 20-60cm soil layers, and that SVM-TS is capable for soil moisture prediction. For the prediction of NO3--N content, there is no obvious difference between predicted and observed values in 0-30cm soil layer, although the difference in 30-60cm soil layer is obviously, compared to the variability of observed value, a large predicted error is acceptable, SVM-TS is applicable for NO3--N content prediction.

Keywords: support vector machine, time series, soil moisture, nitrate nitrogen

1. INTRODUCTION

Prediction of soil moisture and nitrate nitrogen (NO_3^--N) content can supervise irrigation, fertilization and reduce pollution. However, alteration of

Please use the following format when citing this chapter:

Yang, S. and Huang, Y., 2009, in IFIP International Federation for Information Processing, Volume 295, *Computer and Computing Technologies in Agriculture II, Volume 3*, eds. D. Li, Z. Chunjiang, (Boston: Springer), pp. 2037–2045.

soil moisture and NO_3^--N content is uncertain, nonlinear and complicated. Process-based predictive models rely on accurate, reliable physics theories, but their reliability can be diminished by the paucity of data, assumptions, and limited information (Zhang Shirong. 2002). To overcome these limitations, some approximation methods were utilized as surrogate for process-based models.

Many attempts have been made to simulate soil moisture in large scope (Peng Shizhang et al. 2004, Shen Huijuan et al. 2003). In specifically region, alterations of soil moisture and NO_3^--N content close to time series principle cause by the well-regulated precipitation and traditional cultivate manners; it makes the simple time series predictive models practicable (Yang Shaohui et al. 2006, Liu Hongbin et al. 2004). Since fluctuant soil moisture and NO_3^--N content affected by many environmental efforts, relatively factors should be considered in time series models for more accurate estimate. Artificial neural network (ANN) is a common tool for modeling complex "input-output" dependencies; it is another feasible method for soil moisture prediction (Zhou Liangchen et al. 2005, Liu Hongbin et al. 2003, Fan Aiwu et al. 2002, Jin Long et al.1998). The advantage of ANN is that it does not require knowledge of the mathematical form of the relationship between the inputs and corresponding outputs, but it is limited by the potential to fail to generalize well when trained with limited data (Khalil A. et al. 2005).

Support Vector Machine (SVM) was proposed by Vapnik which rely on the statistical learning theory (SLT) (Vapnik V. N.1995,1998a), it can effectively avoid overfitting and multidimensional problem. Good performances have been reported in many SVM applications (Khalil A. et al. 2005, Wang Jinglei et al. 2003, Li Qingguo et al. 2005, Shie-Yui liong et al. 2002).

The objective of this work is to combine the advantages of time series model and SVM algorithm, explore the applicability of Time Series Based Support Vector Machine (SVM-TS) model for prediction of soil moisture and NO_3^--N content.

2. MATERIALS AND METHODS

2.1 Support vector machine

SVM's essential idea is the linearly separability, or hyperplane classifier. Assume a training set like (x_1, y_1) , (x_2, y_2) ,..., (x_l, y_l) with the input $x_i \in R^n$ and the output $y_i \in R$, the goal of linear SVM is to construct a function $y = f(x) = \omega \cdot x + b$ With ω , $x \in R^n$, $b \in R$. According to statistical learning theory and Lagrangian theory (Dongil Kim et al. 2008), linear SVM regression function can be educed to formulate as $f(x)=\sum_{i=1}^{l}(a_i^*-a_i)(x_i\cdot x)+b$. For nonlinear problems, SVM employ a kernel function like $K(x,y)=\varphi(x)\cdot\varphi(y)$ which follows Mercer's theory, it makes SVM regression function can be transformed to $f(x)=\sum_{i=1}^{l}(a_i^*-a_i)K(x_i,x)+b$, each feature vector is projected into a higher dimensional feature space; then SVM be able to utilized a hypothesis spacc of lincar functions in a high dimensional feature space. The popular kernels contain linear kernel, polynomial kernel $K(x,y)=[s(x\cdot y)+c]^d$ with $d\in N$, radial basis function (RBF) $K(x,y)=\exp(-s\|x-y\|^2)$ and sigmoid kernel $K(x,y)=\tanh[s(x\cdot y)+c]$.

In this study, SVMdark which is based on SVMlight is chose as the SVM learning software (Alex Smola et al.). SVMdark provides four kernel functions (linear, polynomial, RBF, sigmoid) for both classification and regression. It can reduce the number of potential models by "zooming in" on those which perform well on the validation set according to the training mean squared error (MSE), and select the parameters that performed optimally on the validation set.

2.2 Support vcctor machine based on time series

A time series is a sequence of data consisting of continuous values changing with time. Any time series can to be split to trend component, periodical component and noisy component. Preparatory analysis showed that the fluctuation of soil moisture and NO_3^--N content is a nonstationary time series process with the annual periodicity.

Assume $x_i\in R^n$ are the variables for soil moisture and NO_3^--N content prediction, y_i is the predictive value of soil moisture and NO_3^--N content, the purpose of SVM is to establish the relationship between x_i and y_i as follow:

$$f\colon R^n\rightarrow R \qquad (1)$$

$$y_i=f(x_i)\ (i=1,\dots,k) \qquad (2)$$

In this work, R^n is the trend component and periodical component of the time series in soil moisture and NO_3^--N content fluctuation, R is the predictive value of soil moisture and NO_3^--N content. Trend component ($T_{j,k}$) and periodical component (S_k) was abstracted as follow:

$$T_{j,k} = \frac{1}{p}\sum_{k=1}^{p} x_{j,k} \quad (1 \le j \le q) \tag{3}$$

$$S_k = \frac{1}{q}\sum_{j=1}^{q} \left(x_{j,k} - T_{j,k}\right) \quad (1 \le k \le p) \tag{4}$$

Where j is the number of years, k is the number of observed values, $x_{j,k}$ is the observed value, p is the number of observed value in one period, q is the number of periods.

2.3 Data measurement

Experiment was performed in Quzhou experimental station of China Agricultural University in Chinese Hebei province. Experimental area is one hectare and there are winter wheat-summer maize planting rotations. Three soil sections were set and sign as section-A, section-B and section-C. In these three sections, moisture was measured in 0-20cm and 20-60cm layers, NO_3^--N content was measured in 0-30cm and 30-60cm layers. Soil moisture was measured by neutron probe and NO_3^--N content was measured in Analytical Conssle-TRAACS2000 about once a week. There was one more measurement 24 hours later after fertilize, irrigate or rain. The experiment was last from 1999 to 2002 and March to October in each year.

For soil moisture study, 16 observed values were selected in each year in layer 0-20cm, 14 observed values were selected in each year in layer 20-60cm. For NO_3^--N content study, 15 observed values were selected in each year in both layer 0-30cm and 30-60cm. Observed values extension of soil moisture and NO_3^--N content is showed in table 1; and the difference between the three sections is estimated by relative error is showed in table 2. Relative error is calculated as follow:

$$\frac{\sum_{i=1}^{n}\left(\frac{\left|x_{p,i} - x_{q,i}\right|}{\left(x_{p,i} + x_{q,i}\right)/2}\right)}{n} \tag{5}$$

Where p, q presents different section, i is the number of observed value, n is the total number of observed values.

Table 1. Observed value extension of soil moisture and NO_3^--N content

	Layer(cm)	Extension	Section-A	Section-B	Section-C
Moisture	0-20	Maximum(cm^3/cm^3)	0.419	0.443	0.487
		Minimum(cm^3/cm^3)	0.113	0.084	0.084
	20-60	Maximum(cm^3/cm^3)	0.424	0.402	0.394
		Minimum(cm^3/cm^3)	0.148	0.236	0.144
NO_3^--N	0-30	Maximum(mg/l)	218.37	78.83	156.02
		Minimum(mg/l)	2.05	0.21	0.35
	30-60	Maximum(mg/l)	142.39	44.15	180.42
		Minimum(mg/l)	0.11	0.05	0.24

Table 2. Relative errors of soil moisture and NO_3^--N content observed values between soil sections

	Layer(cm)	Section-A to section-B		Section-B to section-C		Section-C to section-A	
		Mean relative error(%)	Max relative error(%)	Mean relative error(%)	Max relative error(%)	Mean relative error(%)	Max relative error(%)
Moisture	0-20	21.30	69.76	25.64	75.68	27.72	108.60
	20-60	10.39	69.58	22.86	52.37	18.53	58.60
NO_3^--N	0-30	122.64	194.51	93.89	191.16	83.14	176.15
	30-60	131.74	204.45	105.09	206.66	91.22	197.47

2.4 Dataset setting

In this work, trend component and periodical component of soil moisture and NO_3^--N content were set as the inputs of SVM, soil moisture and NO_3^--N content are the outputs. For soil moisture study, in layer 0-20cm, 48 data groups between year 1999 and 2001 were setup as training set, 8 data groups between March and June in 2002 were setup as validation set, 8 data groups between July and October in 2002 were setup as test set; In layer 20-60cm, 42 data groups between year 1999 and 2001 were setup as training set, 7 data groups between March and June in 2002 were setup as validation set, 7 data groups between July and October in 2002 were setup as test set. For NO_3^--N content study, 45 data groups between year 1999 and 2001 were setup as training set, 8 data groups between March and June in 2002 were setup as validation set and other 7 data groups between July and October in 2002 were setup as test set, in both 0-30cm layer and 30-60cm layer.

2.5 WHNCS model

For the purpose of compared with physical model, here present the WHNCS model (Zhang Shirong. 2002). WHNCS is a component object model for the joint simulation of one dimensional soil water, heat and nitrogen behaviors, it is capable for soil moisture and NO3--N content prediction in different soil layers. WHNCS is based on the physical models including crop evaporation model, soil water solution model, soil NO3--N transport model and soil temperature estimate model (Zhang Shirong. 2002).

3. RESULTS AND DISCUSSION

Trainings were based on the training sets and validation sets; and selected the kernels and parameters according to the training mean squared errors. Selected kernels and relevant parameters for soil moisture and NO3--N content prediction are showed in table 3.

Table 3. Selected kernels and parameters for soil moisture and NO_3^--N content prediction

	Section	Layer(cm)	Kernel	*C*	ε	*d*	*s*	*c*
Moisture	A	0-20	polynomial	830256.1	0.105	4	2.130	3.813
		20-60	linear	195745.7	0.008	—	—	—
	B	0-20	linear	910702.8	0.016	—	—	—
		20-60	linear	540665.9	0.038	—	—	—
	C	0-20	polynomial	612353.9	0.060	3	4.215	3.486
		20-60	linear	497573.8	0.032	—	—	—
NO_3^--N	A	0-30	polynomial	489455.9	0.184	1	0.823	0.866
		30-60	polynomial	620502.3	0.408	3	3.904	4.871
	B	0-30	polynomial	688345.0	0.411	1	0.651	2.784
		30-60	linear	424298.8	0.045	—	—	—
	C	0-30	linear	149357.6	0.057	—	—	—
		30-60	polynomial	305703.9	0.174	4	3.093	0.759

Apply the selected kernels and parameters to predict soil moisture and NO3--N content for the test sets. Predicted results are showed in table 4. For soil moisture prediction, all statistical results (t-test) present $|t| < t0.05$, it indicates that there is no obvious difference between predicted values and observed values, SVM-TS is capable for soil moisture prediction in 0-20cm and 20-60cm layers. For NO3--N content prediction, $|t| < t0.05$ in 0-30cm layer, but $|t| > t0.05$ in 30-60cm layer in section-A and section-B, there is no obvious difference between predicted values and observed values in 0-30cm layer, but the difference is obvious in 30-60cm layer. It supposes that SVM-TS is capable for NO3--N content prediction in surface soil layer but not capable in deep layer. That may cause by the more complicated transformation in deep soil layer than in the surface soil.

Table 4. SVM-TS predicted results for soil moisture and NO_3^--N content

	Section	Layer (cm)	Mean relative error (%)	Max relative error (%)	*t* value
Moisture	A	0-20	18.41	26.26	-0.527
		20-60	26.59	77.82	0.056
	B	0-20	15.36	34.83	-0.318
		20-60	6.82	13.61	-1.551
	C	0-20	17.86	29.97	-0.491
		20-60	15.57	26.49	-0.679
NO_3^--N	A	0-30	84.55	291.75	1.126
		30-60	31.38	44.07	-3.485**
	B	0-30	37.01	114.16	-0.363
		30-60	48.72	89.87	-3.117**
	C	0-30	50.80	75.15	-1.440
		30-60	69.14	92.01	-1.105

Note: in 0-20cm layer sample number is 8, $t_{0.05}$=2.145, $t_{0.01}$=2.977; else sample number is 7, $t_{0.05}$=2.179, $t_{0.01}$=3.055.

Although most of the statistical results show that there is no obvious difference between predicted values and observed values, but the mean relative errors are quite large. Compared with the relative errors of observed values between different sections in table 2, the maximum of observed relative errors of soil moisture is 27.72%, and 131.74% of NO3--N content, predicted values' relative errors is smaller with the maximum of soil moisture is 26.59% and 84.55% of NO3--N content. It indicates that the large predicted mean relative errors are acceptable; SVM-TS can be another capable new method for soil moisture and NO3--N content prediction.

The data is also applied to WHNCS model to predict soil moisture and NO3--N content. Because of the absence of model parameters of soil water, NO3--N content and soil temperature, we only predict soil moisture and NO3--N content between June and December in 1999 in section-C; and compared the predicted mean relative error with SVM-TS in Fig.1 and Fig.2 They show that the SVM-TS predicted results are better than WHNCS model for both soil moisture and NO3--N content prediction.

WHNCS model is base on reliable physics theories, considering water, heat and solution conditions, but numbers of parameters are required and much strict limits for model parameters lead to the limit of accuracy. Reversely, SVM-TS model is simple but practical, superior for soil moisture and NO3--N content prediction.

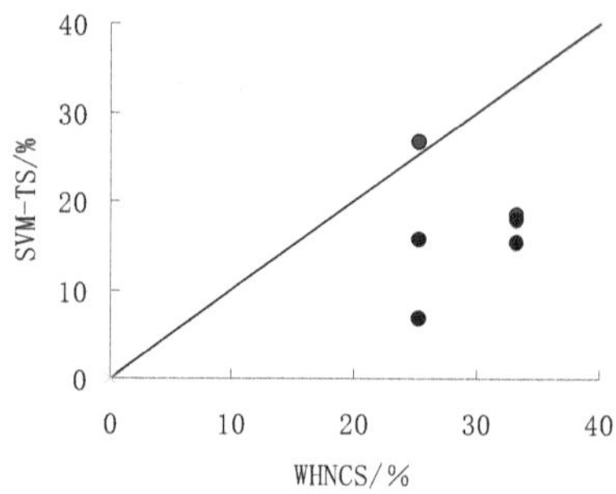

Fig. 1. Predicted mean relative error comparison between SVM-TS and WHNCS for soil moisture

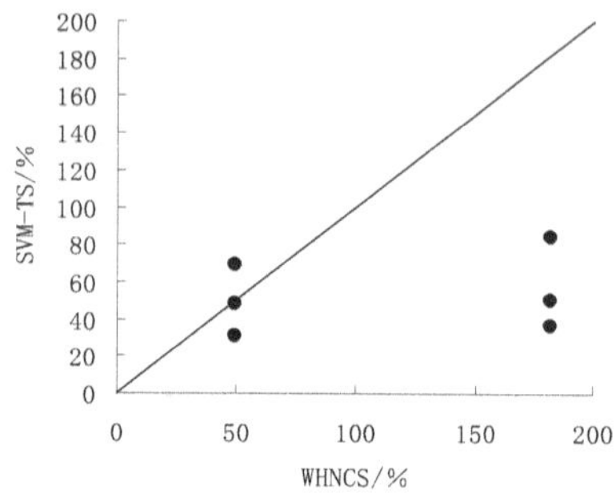

Fig. 2. Predicted mean relative error comparison between SVM-TS and WHNCS for NO_3^--N content

4. SUMMARY AND CONCLUSIONS

Support vector machine base on time series (SVM-TS) was applied to predict soil moisture and NO3--N content in this work. Statistical results indicated that SVM-TS is capable for soil moisture prediction; and capable for NO3--N content prediction in surface layer but not capable in deep soil layer. Although the predicted mean relative errors are quite large, but it is acceptable compared with the relative errors of observed values between different sections. Meanwhile, predicted results indicate that the overall performance of SVM-TS was better than of the WHNCS physical model.

ACKNOWLEDGEMENTS

The authors acknowledge the financial support provided by the National Natural Science Foundation of China (No. 40671084), The 863 Program (2008AA10Z216, 2006AA10Z224), Beijing Natural Science Foundation (6072017), the National Key Technologies R&D Program (2006BDA10A01), and by the Program for New Century Excellent Talents in University (NCET-06-0107).

REFERENCES

Alex Smola, Bernhard Schölkopf. http://www.kernel-machines.org/index.html

Dongil Kim, Hyoung-joo Lee, and Sungzoon Cho. Response Modeling with Support Vector Regression, Expert Systems with Applications, 2008 (34), 1102–1108

Fan Aiwu, Liu Wei, Long Yan. Study on the Prediction of Soil Water Content Based on BP Network, Journal of Huazhong University of Science and Technology (Natural Science Edition), 2002, 30(5):85-87

Jin Long, Luo Ying, Miao Qilong, et al. Forecast Model of Farmland Soil Moisture by Artificial Neural Network, Acta Pedologica Sinica, 1998, 35(1):25-32

Khalil A, Almasri M. N, McKee M, et al. Applicability of statistical learning algorithms in groundwater quality modeling. Water Resources Research, 2005, 41:5, W05010

Li Qingguo, Chen Shouyu. A SVM regress forecasting method based on the fuzzy recognition theory, Advances in Water Science, 2005, 16(5):741-746

Liu Hongbin, Wu Wei, Wei Chaofu, et al. Soil Water Dynamics Simulation by Autoregression Models, Journal of Mountain Science, 2004, 22(1):121-125

Liu Hongbin, Wu Wei, Wei Chaofu. Study of Soil Water Forecast with Neural Network, Journal of Soil and Water Conservation, 2003, 17(5):59-62

Peng Shizhang, Hu Ling, Zhang Lixin. Research Status and Dynamic Analysis of Irrigation Forecast in Domestic and Abroad, Journal of Irrigation and Drainage, 2004, 23(4):6-10

Shen Huijuan, Yan Changrong, Dai Yaping. Progress and Application of Soil Moisture Monitoring and Forecasting Models, Ecologic Science, 2003, 22(4):366-370

Shie-Yui liong, Chandrasekaran Sivapragasam. Flood stage forecasting with support vector machines, Journal of the American Water Resources Association, 2002, 38(1):173-186

Vapnik V. N. Statistical learning theory. John Wiley and Sons Inc, New York, 1998a

Vapnik V. N. The nature of statistical learning theory. Springer, New York, 1995

Wang Jinglei, Wu Jingshe, Sun Jingsheng, et al. Application of Support Vector Machine Method in Prediction of Ground Water Level, Journal of Hydraulic Engineering, 2003, 5:122-128

Yang Shaohui, Wang Yiming, Guo Zhengqin, et al. Research on Soil Moisture Forecast Based on ARIMA Model, Agricultural Research in the Arid Areas, 2006, 24(2):114-118

Zhang Shirong. GIS-Based Simulation of Regional Soil Water and Nitrogen Behavior and Analysis of Agricultural Management, PhD candidate paper of China Agriculture University, Beijing, 2002

Zhou Liangchen, Kang Shaozhong, Jia Yunmao. Application of BP Artificial Neural Network on Prediction of Soil Water Content, Agricultural Research in the Arid Areas, 2005, 23(5):98-102

APPLICATION OF PROJECTION PURSUIT MODEL IN SOIL EVALUATION OF CONSERVATION TILLAGE

Junjing Yuan, Hongwen Li*

College of Engineering, China Agricultural University, Beijing, China, 100083;

* *Corresponding author, Address: P. O. Box 46, College of Engineering, China Agricultural University, 17 Tsinghua East Road, Beijing, 100083, P. R. China, Tel:+86-10-62737631, Fax:+86-10-62737300, Email: lhwen@cau.edu.cn*

Abstract: This paper established a conservation tillage evaluation model based on projection pursuit to evaluate the soil composite achievement of conservation tillage. After optimizing the project direction, the multi-dimension data of the seven evaluation indices are synthesized to one dimension, and the author could evaluate each item with the projection data easily, which avoided the jamming of weight matrix. The results of the evaluation mode in Linfen, Shouyang and Linghai are accordant with the production, which indicated that the model was available and provided a new method or thought to evaluate the composite achievement of conservation tillage.

Key words: conservation tillage, projection pursuit, evaluation index

1. INTRODUCTION

In 1974, Friedman and Tukey put forward the term projection pursuit and proposed the first algorithm, and the projection pursuit method is well known since its first formulation. They considered "projection pursuit will automatically tend to avoid projections involving those measurement variables that do not contribute to the data structure" (Friedman et al., 1974). Many researchers (Huber, 1990; Eslava et al., 1999; Simone et al., 2002; Choulakian, 2006; Daniela, 2007) have studied the projection pursuit models and introduced them to overcome the problem of an excessive number of

Please use the following format when citing this chapter:

Yuan, J. and Li, H., 2009, in IFIP International Federation for Information Processing, Volume 295, *Computer and Computing Technologies in Agriculture II, Volume 3*, eds. D. Li, Z. Chunjiang, (Boston: Springer), pp. 2047–2055.

parameters in multivariate analysis. The application of PPC model in irrigation schemes obtained better results and provided a new thought for the evaluation of irrigation schemes (Jiang et al., 2006). Projection pursuit learning networks (PPLNs) were used in image processing and were successful at developing an inverse blur filter to enhance blurry images (Lesa et al., 2000). Compared with the time-consuming and expensive traditional approaches, it is quicker, easier and lower-cost to apply the projection pursuit regression (PPR) method in the study of shower cooling tower (Qi et al., 2008).

Conservation Tillage began in the 60's and the systematic study began in the 90's in the 20th Century in China. The contents of study consist of yield of crops (Peng et al., 2006; Zhao et al., 2006; Li et al., 2000), production cost (Wang et al.,2007; Li et al., 2000), soil properties and texture (Kay, 1990), water utilization (Liu et al., 2000; Zhang et al., 2002), greenhouse gas emission (Liu et al., 2006; Li et al., 2003; Lar et al., 2004) and etc.. And the methods of study consist of pairing comparison of captive test (Peng et al., 2006; Wang et al.,2007; Li et al., 2000), cost-benefit method (Zhao et al., 2006; Li et al., 2000), Delphi (Wang et al.2004), AHP (Wang et al., 2004; Tian et al., 2006) and so on. Pairing comparison of captive test is suitable for one single index or fewer indices when comparing the achievement between Conservation Tillage and traditional tillage. Generally, the cost-benefit method utilized fixed price to calculate the cost, benefit and static pay back period. Although the method is easy and simple, it is difficult to compare the achievement among different areas and different farming models. The methods of Delphi and AHP took into account the experts' experiences and intuition sufficiently, on the one hand, they solve the quantification of qualitative indices, on the other hand, they make use of the original information insufficiently and much jamming is unavoidable when put up weights (Wu et al. 2004).

It is from these considerations that this paper tries to apply projection pursuit to evaluate the soil achievement of Conservation Tillage in different areas.

2. PROJECTION PURSUIT MODEL

Projection pursuit is a recent statistical method and a new technique for nonparametric multivariate regression which can overcome the curse of dimensionality. The projection pursuit method is not affected by the presence of noisy variables and it is aimed at solving the difficult problem of identifying structure in high dimensional data. It does this by projecting the high-dimensional data onto low-dimensional subspace and clustered the observations into homogeneous groups (Angela, 2001). Compared with the methods of Delphi, AHP and Eigenvector, the projection pursuit method is an external evaluation method based on the characteristics and structure of original data (Yang et al., 2005).

The projection pursuit method does not establish direct relationship between the independent variables and the dependent variables, but conduct projections on high- dimensional independent variables; after a low dimensional projection variable is obtained, the relationship between the projection variable and independent variable can be determined (Qi, et al., 2008).

The steps of projection pursuit model for soil evaluation of conservation tillage are as follows (Zhang, 2007; Jin et al., 2005).

Step 1. Establishment of evaluation function. Given the matrix $X^*_{n\times p}$ present the index values of samples, where p is the number of observed variables, n is the number of units and x^*_{ij} denote the value of the index j of unit i.

Step 2. Method of dimensionless parameter. In order to make the dimension of all values coincident, dimensionless calculation for original values of evaluation indices is necessary. All the evaluation indices are divided into three kinds:

Evaluation indices are normalized to be maximizing ones, and after normalization, matrix $X^*_{n\times p}$ could be replaced by matrix $X_{n\times p}$.

$$x_{\mathrm{ij}} = \frac{x^*_{\mathrm{ij}} - \min(x_{\mathrm{j}})}{\max(x_{\mathrm{j}}) - \min(x_{\mathrm{j}})} \tag{1}$$

Evaluation indices are normalized to be minimizing ones,

$$x_{\mathrm{ij}} = \frac{\max(x_{\mathrm{j}}) - x^*_{\mathrm{ij}}}{\max(x_{\mathrm{j}}) - \min(x_{\mathrm{j}})} \tag{2}$$

Evaluation indices are normalized to be minimizing ones,

$$x_{\mathrm{ij}} = \frac{\left|x^*_{\mathrm{ij}} - mid(x_{\mathrm{j}})\right|}{mid(x_{\mathrm{j}})} \tag{3}$$

In the above-mentioned three formulas, $\max(x_{\mathrm{j}})$ means the maximum of the index j, $\min(x_{\mathrm{j}})$ means the minimum of the index j, and $mid(x_{\mathrm{j}})$ means the intermediatevalue of the index j.

Step 3. Analysis of linear projection. The aim of projection is to decrease the dimension and avoid the 'curse of dimensionality'. In this paper, the high-dimension values are projected to one-dimension. Assume unit vector a is the optimum projection direction and z_i is the projection eigenvalue of matrix X, the formulas are as follows.

$$a=\{a_1, a_2, \cdots a_p\}, (j=1,2,3\ldots,p) \tag{4}$$

$$z_i=\sum_{j=1}^{p} a_j \cdot x_{ij}, (i=1,2,3\ldots,n; j=1,2,3\ldots,p) \tag{5}$$

Step 4. Establishment of objective function. According to the classification principles, the distribution of projection values should possess the following characteristics: (a) the dots of projection in the graphics should be as dense as possible in the local, namely it is better that most dots mass together; (b) the points mass in the graphics should be as sparse as possible in the whole. Therefore, it is better that the distances between classes and the densities within classes are the maximum at the same time. And the objective function could be denoted as product of the distances between classes and the density within classes.

$$\begin{aligned} Q(a) &= S_Z \cdot D_Z \\ S_z &= \sqrt{\frac{\sum_{i=1}^{n}(z_i - E(z))^2}{n-1}} \\ D_z &= \sum_{i=1}^{n}\sum_{k=1}^{n}(R - r_{ik})) \cdot f(R - r_{ik}) \end{aligned} \tag{6}$$

In the above functions, S_z means the Standard Deviation of z_i, and D_z means the partial density of z_i.

Step 5. Optimization of objective function. For the given indices values of the samples, the value of $Q(a)$ may change when changes take place in the projection direction a, and the optimal projection direction reflected certain structures of high-dimension furthest. Thus, the representation of optimal function is as follows.

$$\begin{aligned} &MaxQ(a) = S_z \cdot D_z \\ &sub.to: \sum_{j=1}^{p} a^2(j) = 1 \end{aligned} \tag{7}$$

Due to $a=(a_1, a_2, \cdots a_p)$ is as a variable for the complex nonlinear optimization function, it is difficult for the traditional optimization methods to deal with the problem, and real coding based accelerating genetic algorithm (RAGA) is effectual.

Step 6. Classification and evaluation. Suppose a^* is the optimal projection direction and $z^*(i)$ is the eigenvalue of the optimal projection direction, we can obtain a^* according to $MaxQ(a) = S_z \cdot D_z$, and also we can obtain $z^*(i)$ according to a^* and function (5). Comparing $z_1^*, z_2^*, \ldots, z_n^*$, the maximum means its corresponding sample is the best and the minimum means its corresponding sample is the worst and we also could classify the samples in terms of the approximate degree of $z_1^*, z_2^*, \ldots, z_n^*$. Comparing $a_1^*, a_2^*, \ldots, a_p^*$, the result magnitude means its contribution degree or sensitivity degree of the corresponding index to the system.

3. APPLICATION

Conservation Tillage, a kind of technology of dry farming, is different from traditional agriculture, and the main contents include No Tillage, Reduced Tillage, Straw Mulching, and so on. Table 1 is the standard grade of soil confirmed by the experts, and there are seven evaluation indices and six samples (grades).

Table 1. soil standard grade

Grade	I	II	III	IV	V	VI
Ph	6.5~7.5	6.0~6.5 7.5~7.8	5.5~6.0 7.8~8	5.0~5.5 8~8.5	4.5~5 8.5~9	≤4.5 >9
Bulk density g/c m3	1.25~1.35	1.2~1.25 1.35~1.40	1.15~1.20 1.40~1.45	1.10~1.15 1.45~1.50	1.00~1.10 1.50~1.60	>1.6 <1
Water storage %	14~16	16~18 12~14	18~20 10~12	20~22 8~10	22~24 6~8	<6 >24
Organic matter %	>4	3~4	2~3	1~2	0.6~1	<0.6
Total N %	>0.2000	0.15~0.20	0.10~0.15	0.075~0.10	0.05~0.075	<0.050
Available P mg/kg	>40	20~40	10~20	5~10	3~5	<3
Available K mg/kg	>200	150~200	100~150	50~100	30~50	<30

Adopt RAGA method to optimize the projection direction and adopt MATLAB7.0 to program, suppose the parameters are as follows, initial population size n=400, crossover probability p_c=0.8, mutation probability p_m=0.03, number of superior individualsis 20, α=0.05, times of acceleration are 11, and after running software we can obtain the optimum projection direction and projection eigenvalue, namely,

$$a_s^* = (0.1897, 0.3416, 0.4419, 0.4723, 0.3997, 0.4987, 0.4235)$$

$$z_s^* = (2.6138, 1.7216, 1.2366, 0.9895, 0.6585, 0.5165)$$

Fig. 1 is about the relationship between projection values and evaluation grades, consider y*represent the grade and the evaluation model of soil achievement based on projection pursuit is as follows.

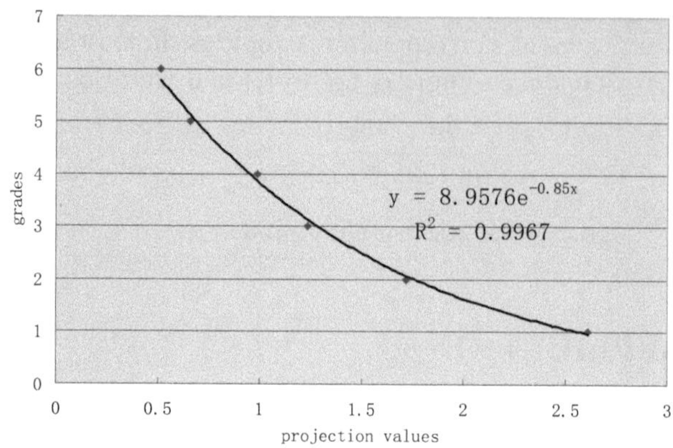

Fig. 1 scatter diagram of grade and projection value

The error analysis in table 2 indicate that the precision of evaluation model possesses a much higher precision and reliability

Table 2. Error Analysis

Empirical value	Projection value	calculated value	absolute error	Relative error
1	2.6138	0.9712.	-0.0288	-2.8805
2	1.7216	2.0733	0.0733	3.6651
3	1.2366	3.1311	0.1311	4.3708
4	0.9895	3.8629	-0.1371	-3.4266
5	0.6585	5.1181	0.1181	2.3614
6	0.5165	5.7746	-0.2254	-3.7559
Average			0.1190	3.41

Test data in Table 3 root in the conservation tillage research center of Agricultural Ministry.

Table 3. Test Data

	pH	bulk density g/c m3	water storage %	organic matter %	Total N /%	available P mg/kg	available K mg/kg
A1	8.08	1.30	13.51	1.45	0.0816	48.91	268.63
A2	8.02	1.41	14.43	1.72	0.0877	36.77	362.16
A3	7.98	1.39	14.49	1.91	0.1287	39.42	350.46
B1	7.97	1.26	12.51	1.46	0.0702	21.93	190.69
B2	8.03	1.45	14.21	2.28	0.0736	38.63	405.02
B3	8.04	1.37	15.04	2.60	0.1001	46.04	412.82
C1	7.02	1. 27	19.9	2.47	0.0765	21.05	152.69
C2	7.01	1.31	21.9	2.50	0.0772	20.46	234.53
C3	7.02	1.32	23.4	2.51	0.0782	19.95	227.71

There are seven indices and nine samples in this system, and tests of class A, B and C are processing respectively in Linfen, Shouyang and Linghai. And the detail significations of the samples are in table 4.

Table 4. Significations of the sample

Sample	Signification
A1, B1, C1	Traditional tillage
A2,B2	Shallow tillage and straw cover
A3, B3	No tillage and straw cover
C2	No tillage and standing stubble
C3	No tillage and break straw cover

With the same method to deal with the test data, the projection eigenvalue and the soil grade with different farming models are as table 5.

Table 5. Soil grade of the sample

Sample	Projection Values	Grades
A1	0.8592	4.3154
A2	1.2408	3.1200
A3	1.4512	2.6090
B1	0.8369	4.3979
B2	1.2378	3.1279
B3	1.4758	2.5551
C1	1.1701	3.3132
C2	1.2311	3.1458
C3	1.2384	3.1263

It is obviously that these nine samples are classified three clusters, A1 and B1 belong to the grade Ⅳ; A2, B2, C1, C2 and C3 belong to t the grade Ⅲ; A3 and B3 belong to the grade Ⅱ.

4. RESULTS AND DISCUSSION

The principal contributions of this article are the following:

(1) The model based on projection pursuit avoided the jamming of weight matrix, and take full advantage of the inter-structure of original data.

(2) In Linghai, conservation tillage has been applied 2 years, and the soil grades of C2, C3 are promoted a little compared to C1. It is showing that the effect of short-term conservation tillage is not significant in improving soil properties.

(3) In Linfen and Shouyang, conservation tillage has been applied 15 years, and the grade of soil promoted from 4 to 3 and 2. Therefore, the effect of long-term conservation tillage is significant, which is accordance with productions.

REFERENCES

Angela Montanari, Laura Lizzani. A projection pursuit approach to variable selection. Computational Statistics & Data Analysis 35 (2001) 463~473

Daniela G. Calò. Gaussian mixture model classification: A projection pursuit approach. Computational Statistics & Data Analysis 52 (2007) 471 – 482

Friedman, J.H., Tukey, J.W., 1974. A projection pursuit algorithm for exploratory data analysis. IEEE Trans. Comput C 23, 881~890.

G. Eslava, F.H.C. Marriott. Projection indices and multimodality for mixtures of normal distributions. Statistics & Probability Letters 43 (1999) 289~297

Huber, P.J., 1990. Data analysis and projection pursuit. Technical Report PJH-90-1. MIT, Dept. of Mathematics.

Jiang Guoyong, Cao Hualiang. Application of PPC model for comprehensive evaluation of irrigation schemes. Research of soil and water conservation, 2006, vol.13, No.5: 87~89

Jin Juliang, Hong Tianqiu, LI Ruzhong. Objective weighted model for evaluating dynamic eco-environmental comprehensive quality of Chaohu lake basin. Journal of Hydroelectric Engineering, 2005,24(5):99~103

Kay, .B. D. Rates of change of soil structure under different cropping systems. Advance in Soil Science. 1990,12:1～2

Lal R, Griffin M, Apt J. Managing soil carbon [J]. Science, 2004,304(4):393

Lesa M. Kennedy, Mitra Basu. Application of projection pursuit learning to boundary detection and deblurring in images. Pattern Recognition 33 (2000) 2019~2031

Li Changsheng, Xiao Xiangming, Frolking S,Moore B,etc. GREENHOUSE GAS EMISSIONS FROM CROPLANDS OF CHINA. Quaternary Sciences, 2003, 23(5):493-503.

Li Hong-wen, Gao Huan-wen, Zhou Xing-xiang, Mao Ning. Economic analysis of maize production in dryland with conservation tillage. Agricultural research in the arid areas, 2000,9,18(3):44~4

Liu Jianmin1, Hu Lifeng, Zhang Aijun. Research Progress in Greenhouse Effect Caused by Conservation Tillage. Chinese Agricultural Science Bulletin, 2006,8,22(8):246～249

Liu Li-jing, Gao Huan-wen, Li Hong-wen, Jin Shuang-yi. Experimental study on influencing factors of evaporation under conservation tillage. Journal of China Agricultural University 2006,11(4): 78～82

Peng Wen-ying- , Zhang Ya-bin. Review of impacts of no-tillage on crop yield and economic benefit. Agricultural Research in the Arid Areas, 2006,7,24(4):113～118

Simone Borra, Agostino Di Ciaccio. Improving nonparametric regression methods by bagging and boosting. Computational Statistics & Data Analysis 38 (2002) 407 – 420

Tian Shu-min, Liu Rui-han, HOU Fu-qiang. Study on Evaluation Indicator System of Conservation Tillage Technology Implementary Benefit in Beijing Rural Areas. Journal of Beijing Agricultural College, 2006,10,21(4):50～53

V. Choulakian. L1-norm projection pursuit principal component Analysis. Computational Statistics & Data Analysis 50 (2006) 1441~1451

WANG Xu-mei, ZHANG Ying, AN Xiao-guang. Evaluating indexes system and weight about agricultural eco-system. Journal of Northeast Agricultural University , 2004,35(3):386～388

Wang Yu-zhen, Discussion of the economic benefit and ecological benefit of conservation tillage. Agricultural economy,2007,5:30

Wu Dianting, Li Dongfang. SHORTCOMINGS OF ANALYTICAL HIERARCHY PROCESS AND THE PATH TO IMPROVE THE METHOD. Journal of Beijing Normal University(Natural Science, 2004, 40(2):264～268

Xiaoni Qi , Zhenyan Liu, Dandan Li. Prediction of the performance of a shower cooling tower based on projection pursuit regression. Applied Thermal Engineering 28 (2008) 1031~1038

Yang Xiao-hua, YANG Zhi-feng, SHEN Zhen-yao, LI Jian-qiang. A Multi-Objective Decision-Making Ideal Interval Method for Integrated Assessment of Urban Environmental Quality. Systems Engineering-theory & Practice, 2005,(8):119~123

Zhang Bin. Application of PPC Model to Grasses of Sloping Field for Soil and Water Conservation. Research of Soil and Water Conservation, 2007,14(2):299~301

Zhang Hailin , Chen Fu , Qin Yaodong , Zhu Wenshan. Water Consumption Characteristics for Summer Corn Under Under No-Tillage With Mulch. Transactions of The Chinese Society of Agricultural Engineering, 2002,18(2):36-40

Zhao Hai-tao, Shi Shuai-bing. Benefit analysis on the technology of mechanized conservation tillage. Journal of Agricultural Mechanization Research, 2006, 8:74~76

THE ARCHITECTURE OF INFORMATION FUSION SYSTEM IN GREENHOUSE WIRELESS SENSOR NETWORK BASED ON MULTI-AGENT

Wenting Zhu [1], Ming Chen [1,*]

[1] *Department of Information Technology, Shanghai Ocean University, Shanghai, P. R. China 200090*

[*] *Corresponding author, Address: Department of Information Technology, Shanghai Ocean University, Shanghai 200090, P. R. China, Tel: +86-13371935523, Fax: +86-21-65710862, Email: mchen@shou.edu.cn*

Abstract: In view of current unprogressive situation of factory breeding in aquaculture, this article designed a standardized, informationized and intelligentized aquaculture system, proposed a information fusion architecture based on multi-agent in greenhouse wireless sensor network (GWSN), and researched mainly the structural characteristic of the four-classed information fusion based on distributed multi-agent and the method to construct the structure inside of every agent.

Keywords: GWSN, multi agent, information fusion

1. INTRODUCTION

For a long time aquaculture in china mostly uses traditional ways that consume lots of resources, especially shrimp breeding uses the mode of production of high-density, high devotion, high yield and a high rate of water change, at the same time the lower level of management mainly depends on the experiences. It is Inconsistent with principles and policies in china that vigorously develop digital fishery and accurate fishery. With accelerating of information course in the whole society, networking and digital information environment forms in accelerating. To achieve

---

Please use the following format when citing this chapter:

Zhu, W. and Chen, M., 2009, in IFIP International Federation for Information Processing, Volume 295, *Computer and Computing Technologies in Agriculture II, Volume 3*, eds. D. Li, Z. Chunjiang, (Boston: Springer), pp. 2057–2066.

comprehensive information system of fisheries and sustainable development has become a great challenge of 21st century. Therefore, standardized production and operation, industrialized and intelligentized greenhouse breeding methods will become one of the directions of the future development in the aquaculture industry.

Therefore, breeding environment has great significance on High-density factory aquaculture. Timely monitoring and controlling the changes in water quality factors and improving the condition of breeding water quality can effectively enhance the output and efficiency of breeding. based on the establishment of a multi-sensor nodes wireless sensor network (Akyildiz et al., 2002) of real-time monitoring system, this paper researches the technology of data fusion (Cheng Hongbing et al., 2006) on deferent sampling frequency of sensor layer under different types of Integrated sensor nodes such as temperature, pH, DO, ORP, light, etc, Combines testing data of laboratory, real-time data that collected by the sensor network, sensor data, historical data and growth of biological data collected through offline, data of feeding conditions and Expert knowledge, fuses data from data center layer, establishes a waters micro-ecology environment and biological growth model. At the same time, the multi-agent technology will be introduced in all levels of data fusion (Fan Bo et al., 2003). Using the characteristics of autonomy, coordination, self-learning and reasoning of multi-agent technology (Michael et al., 1995), the system can be reasoning, integration and learning by capture data information and state from environment, and it makes a great improve on the Intelligence of data fusion technology. This paper constructs a decision-making model based on multi-agent by information exchange at all levels of data fusion to intelligently control and breed micro-ecological environment and growth process of the whole factory breeding. The model will optimize the breeding technology, reduce production expanse and laboring intensity, increase productivity, achieve the modernization and intelligence of aquaculture, and accelerate the commercialization of aquaculture.

This paper is organized as follows: the 2nd section introduces related works; the 3rd section proposes the architecture of multi-agent information fusion of GWSN; the 4th section brings forward the function and internal structures of each agent; the last section will give a conclusion to the whole system.

2. THE ARCHITECTURE OF MULTI-AGENT INFORMATION FUSION SYSTEM IN GWSN

2.1 GWSN

The purpose which we established GWSN is to regulate the water areas and environment of greenhouse, reduce the production-costs and labor intensity, improve the efficiency of production through collect the key factors of relative water areas and greenhouse's environment in real-time, and combine the regulation models of establishing water areas and environment such as chemical analysis laboratory factors, biological growth factors and so on.

An intensive breeding factory is composed by several greenhouses (workshops), every greenhouse workshop is composed by many standardized ponds, and the factory regulates the water quality and the greenhouse environment through the uniform control system. It used a cluster structure when constructed the monitoring network (Wang Xue et al., 2007). It designed a cluster which is managed by a clustering node in a greenhouse workshop; a greenhouse workshop equipped with a number of wireless multi-sensor nodes which connects five sensors: PH, DO, light, dissolved oxygen, temperature sensor and a wireless control node which controls the equipment such as water pump, aerobics machine, lighting, curtain and so on. Several clusters are connected with central monitoring room. The features of GWSN which are different from other wireless sensor networks because:

(1)Cluster Structure. The wireless sensor nodes only transmits data to cluster node and send control order to the control node in the same cluster, and the sensor nodes do not transmit the messages to each other; there is only a control node in a cluster; the control node can receive the order of sensor nodes in this cluster, the control order of cluster nodes and the control order of central node passed through cluster node.

(2)It has good power supply, and the node which is easy to maintain have the small possibility to failure; but it greatly influences the stability of the sensors' signal and the control equipment.

(3)A wireless sensor node integrated a number of sensors; a control node integrated a number of control equipment.

2.2 Construction of multi-level information fusion system

According to the features of the wireless sensor network system in factory breeding, this article researches multi-source information fusion technology (Hu Xuejun et al., 2002) such as a variety of sensors information, chemical analysis information, production information and the metabolism of biological growth under the high-density breeding and closed environment, and establishes multi-level information fusion system, shown as Fig.1.

(1)First-class fusion is information fusion of sensor and information processing. Every sensor collects information on single parameter, and mainly solves how to keep the collected information according with the actual situation. The reasons which mainly influence the data different from the fact are: (i) Environmental noise influence greatly on sensor's signal. (ii) The long-term use of sensor will lead to the surface passivation, reduce the signal sensitivity and possibly appear the drift phenomenon. In the first-class fusion, it mainly deals with the data-different situation.

(2)Second-class fusion is information fusion of several sensor nodes. The wireless multi-sensors node include the water quality sensor such as PH, salinity, DO, ORP and greenhouse environment sensor such as lighting, temperature and so on. Through long-term experiments and practical breeding productions, it founds that different kinds of water quality parameters have the different relativity under different situation. We focus on analyzing the mutual-change relation models of every sensor's parameter when the parameter of water quality is under the external condition such as feed delivery, medication, control water quality and change temperature, illumination, then establishing the rules of changing water quality's parameter and judging whether the collected data every time is same as the current actual water quality and greenhouse environment in real-time according to the rule-based learning. Once the data relativity disaccords, the system should establish the alarm model of sensor nodes and send the order to the control nodes to regulate the water environment.

(3)Third-class fusion is information fusion of greenhouse workshop (cluster node). The focus of third-class fusion is that: (i) It preprocesses the information of the several wireless sensor nodes, then sends the preprocess results to the central control node in order to accomplish the variance analysis and the weighted average analysis; (ii) It combines relation models of sensors' parameter, researches the relation between the change of ammoniac nitrogen, nitroso nitrogen, COD in the water and PH, salinity, Do; (iii) according to the environmental conditions of the whole greenhouse, it establishes the control model under the exceptional condition and send the control order to the control nodes.

(4)Fourth-class fusion is information fusion of data center (central node). In the information fusion of this class, it should deal with three problems: (i) Filter the mainly attribute. First we should find the mainly relative attributes from a mass of attributes in order to find out the mainly influencing attribute in every kind of parameters; then we consider to design a statistical learning way via setting up confidence interval in order to regulate the weight of attributes step by step and achieve the propose of filtering mainly attributes; (ii) Multi-model analysis. Every attribute has complex relationships, some have the statistics regularity such as water quality's parameter and the delivery number of different feeds, some can bring the specific rules such as biological metabolism rule, some regularity hides in a mass of information such as the influence of PH and ammoniac nitrogen in production of micro-algae and the component of breeding biologic metabolites. We use the method of data mining such as fuzzy neural network model, multivariate statistics and rule-based learning according to the different conditions to establish the different relation model. (iii) Decision analysis. We make every kind of regulated parameters of greenhouse factory breeding as the target of decision, and establish the decision analysis model of multi-target according to every attribute relation models to make decision on every kind of target parameter, then regulate the water quality via regulation of relative parameter to establish the environment of greenhouse and water quality which is suitable for biological growth.

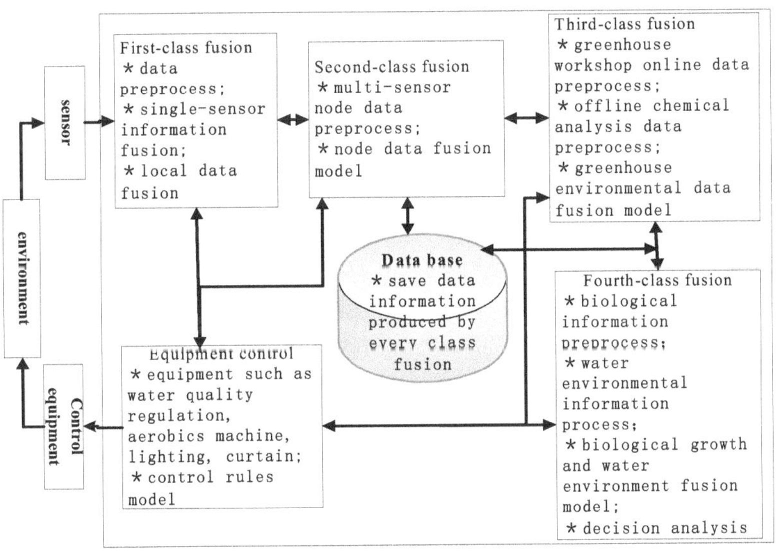

Fig.1: structure of multi-level information fusion in greenhouse monitoring network

2.3 Design of multi-agent information fusion system in GWSN

According to the features of the information fusion system in GWSN, it designs a multi-agent system (Wang Jun et al., 2008) to reflect the autonomy, adaptability and coordination of information fusion in different level, specific structure shown as Fig. 2.

In the whole multi-agent system, it emphasizes the independent work of every local agent to solve the own problem and the functional requirements; the adaptability of agent is an ability of adapting local requirement and self-regulation. At the same time, according to the requirement of every class information fusion, all agents communicate and coordinate each other under the coordination of routing coordination agent. The mainly function and mutual relation of every agent in GWSN as follows:

- Single Sensor Information Process Agent: solve the information drift problem of sensors; collect real-time information then send the information and equipment failure information to local multi-sensor information fusion agent.
- Local Multi-Sensor Information Fusion Agent: establish the relation of multi-sensor's real-time parameter to form the environment of biological growth and the required area of water quality in local pond area, then send the fusional environmental information to area information fusion agent and send the equipment control order to equipment control agent.
- Area Information Fusion Agent: combine the real-time information of different multi-sensor information fusion and water environment regulation model which is formed by ammoniac nitrogen, nitroso nitrogen, COD analyzed by lab termly in the range of local greenhouse, then send the fusional greenhouse environmental information to central decision agent and send the control order to control agent according to the requirement.
- Central Decision Agent: combine the environmental information of area information fusion agent, biological growth information and production output information, establish the environmental optimization model to form the decision analysis system of production output, biological growth and environment control, and send the control order to equipment control agent according to the requirement of environmental optimization.
- Routing Coordination Agent: according to the communication protocol and the features of GWSN, it can create the routing rules between different agents and send decision according to the requirement of different agent packet.
- Equipment Control Agent: generally speaking, the control order of multi-sensor nodes is instant control orders that the one or several parameters of water quality or environment is in the critical range (alarm range) of biological growth. The control order of area information fusion agent is based on the every kind of real-time information of greenhouse water environment and lab information. The control order of central decision agent is based on the optimal environmental requirement of biological growth and requirement of production decision, so the control order of every class has the different priority. Equipment control agent should create the rule priority of different equipment control and form the rule-base, when the equipment does not work, equipment control agent will send the alarm of fail information to the relative node.

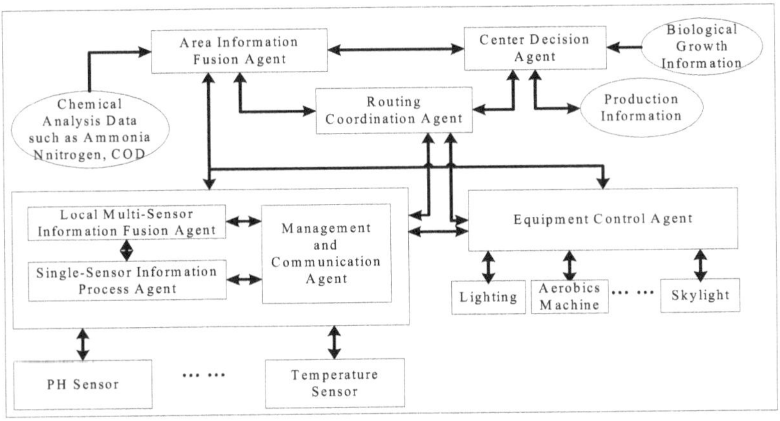

Fig.2: the architecture of multi-agent information fusion of GWSN

3. RESEARCH ON THE STRUCTURE OF AGENTS IN MULTI-AGENT SYSTEM

3.1 The internal structure of agent

Based on the features of greenhouse multi-agent information fusion system, the independent ability and coordination relation of every kind of agent (Kraus, 1997), it considers that the inner structures of agent includes data cache, triggered events which trigger agent, knowledge base(KB) or rule base(RB) of agent which process the information fusion intelligently, processing module which process the inside of agent according to knowledge rules and triggered events, coordination module which coordinate the inner components and data exchange according to relative strategies, outer communication interface and so on. The structure is shown as Fig.3.

And with all the modules, different Agents have different inside functions, their internal structures of every module are as follows:

- <Agent> : : =<data cache><event trigger><RB, KB><processing module><coordination processing><outer communication interface>
- <data cache>: : = [local data][outer data]
- <event trigger>: : =<real-time information trigger of sensors><failure information trigger><trigger control order>
- <RB, KB> : : = [data structure][routing rules] [fusion arithmetic][control trigger rules]
- <processing module> : : =<information collection><information

transmits><equipment control signal><routing and packet processing>

- <coordination module>: : =< coordination protocol>< coordination strategy>< coordination process>
- <outer communication interface> : : =<information collection communication><failure alarm communication><communication with other agents>

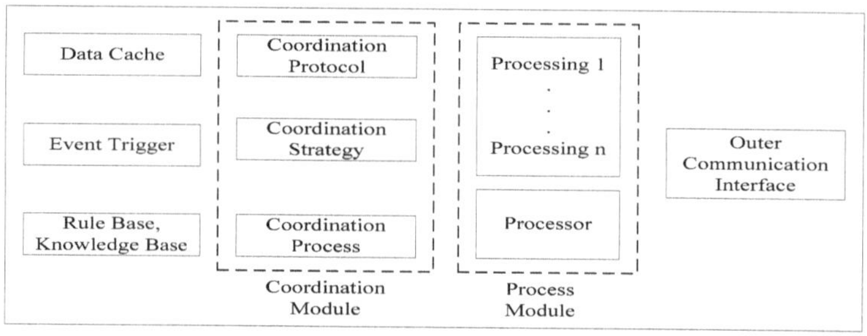

Fig.3: structure mode of agent

3.2 Establishing RB of Agent

It is important for each agent to build a RB in information fusion, because different agent corresponds to different level of information fusion and has different function; however, the fusion algorithm is different too. In this section we mainly research on establishing RB of Local Multi-sensor Information Fusion Agent which corresponds to the second-class information fusion. We used mainly rule based reasoning to fuse factors of water quality in Local Multi-sensor Information Fusion Agent. Every factor such as salinity, PH value, and water temperature was transformed into rules according to the standard range which is suitable for biological growth and saved as two-dimensional form. The basic form of rules is:

IF P THEN Q

Thereinto, P represent the condition, Q represent the result. It means if P is reached, then does the action Q defined.

When Local Multi-sensor Information Fusion Agent received data, the system calls the rule of knowledge base and does the fusion processing to data. The pseudo codes of part knowledge are as follows:

If Type of Prawn = JUVENILE

If Water Temperature>23 or Water Temperature<20 Then Call Alarm Module and Call Control Module

Else Submit Fused Information to Area Information Fusion Agent

Else If Type of Prawn = ADULT

If Water Temperature>30 or Water Temperature<25 Then Call Alarm Module and Call Control Module

Else Submit Fused Information to Area Information Fusion Agent

If Type of Prawn = JUVENILE

If Salinity>6 or Salinity<3 Then Call Alarm Module and Call Control Module
Else Submit Fused Information to Area Information Fusion Agent
Else If Type of Prawn = ADULT
If Salinity>3 Then Call Alarm Module and Call Control Module
Else Submit Fused Information to Area Information Fusion Agent
……

Due to the lots of factors affecting water quality, we just listed the rules of water temperature and salinity. After establishment of rule-base, we used RETE algorithm to do the rule based reasoning. It's a high-efficient pattern-matching algorithm. When the Local Multi-sensor Information Fusion Agent receives water quality data, the processor module will call the RB and do some rule based reasoning to the data. It is worth noting that the processor handles incidents at two levels. Equipment failure is the first process level. When equipment failure, the processor will trigger alarm module for a failure signal. Data fusion is the second process level. The processor calls RETE algorithm to fuse several kinds of water quality factors. If the factor of any kinds is beyond the standard range, the processor will trigger alarm module immediately and send the regulating signal to Equipment Control Agent to regulate the relevant factors of water quality to standard range; if all factors of water quality are in the standard range, the processor will send these water quality data to the superior information fusion. Now I will take out 4 sets of data for analysis from a local multi-sensor node (cluster node) of an adult prawn breeding pond to prove that the second-class information fusion algorithm is reliable and efficient. The data is shown in Table1.

Table 1. Analysis of water quality factors fusion

	Temperature	PH	DO	Result
1	31.2℃	7.84	4.56 (mg/l)	Alarm
2	29.2℃	8.07	6.95 (mg/l)	Submit
3	28.2℃	6.94	2.57 (mg/l)	Alarm
4	26.6℃	7.65	4.48 (mg/l)	Alarm

We selected the three factors of temperature, PH and DO for fusion, and the numerical value beyond the standard range was marked by yellow in table 1. The first set of data in Table 1 shows that the temperature exceeds the standard range, because the temperature range of adult prawn is 25℃-30℃. In the second-class information fusion, when the Local Multi-sensor Information Fusion Agent receives the first set of data, the processor will trigger the alarm module after fusion for an alarm signal and conduct the Equipment Control Agent to open the cooling device. So the column of "Result" in table 1 shows "Alarm". The second set of data is in the standard range after fusion, and the processor will send these data directly to the superior information fusion, so the column of "Result" in table 1 shows "Submit". Facts have proved that this approach of information fusion in Local Multi-sensor Information Fusion Agent can process data effectively and regulate the aquaculture environment of prawn to an optimal condition.

4. CONCLUSION

This article researches the modern, intelligentized, factory aquaculture system deeply, and proposes a multi-agent information fusion system structure in GWSN. It focuses on every class information fusion technology and the information fusion structure model based on multi-agent. Due to the characteristics of agent such as autonomy, distribution and coordination, the function of the multi-sensor information fusion system become more powerful and the overall capability of the system is farther improved.

ACKNOWLEDGEMENTS

This research is supported by National 863 Plan, No. 2006AA10Z253, 2006AA10Z311, 2007AA10Z238 (P. R. China).

REFERENCES

Akyildiz L F, Su WeiLian, Sankarasubramaniam Y, Cayirci E. A Survey on Sensor Networks. IEEE Communications Magazine, 2002, 40(8):102-114

Cheng Hongbing, Huang Xiao, Yang Geng. Research of Data Fusion System Architecture in Wireless Sensor Network. Computer Science, 2006, 33(7):60-62(in Chinese)

Fan Bo Etc. The Technology of Multi-Agent Introduced to Information Fusion System. Computer Engineering and Applications, 2003(22):100-102(in Chinese)

Hu Xuejun, Luo Zhongliang. Method of multi-sensor information fusion based on statistics theory. Journal of Transducer Technology, 2002(08):38-43(in Chinese)

Kraus S. Negotiation and cooperation in multi-agent environments. Artificial Intelligence, 1997, 94(1):79-97

Michael W, Nicholas R J. Intelligent agents: theory and practice [J]. Knowledge Engineering Review, 1995, 10 (2):115-152

Wang Jun, Wang Yuanyun. Multi-agent and the Application in the Information Fusion. Computer and Digital Engineering, 2008, 36(3):128-130(in Chinese)

Wang Xue, Wang Sheng, Jiang Aiguo. Cluster-based data fusion decision method in wireless sensor. Control and Decision, 2007, 22(11):1208-1212(in Chinese)

STUDY ON AOTF-BASED NEAR-INFRARED SPECTROSCOPY ANALYSIS SYSTEM OF FARM PRODUCE QUALITY

Xiaochao Zhang*, Xiaoan Hu, Yinqiao Zhang, Hui Wang, Hui Zhang

[1] *Institute of Mechatronics Technology and Application, Chinese Academy of Agricultural Mechanization Sciences, Beijing, P. R. China100083*

* *Corresponding author, Address: Box25, No.1 Beishatan, Dewai St, Beijing, 100083, China, Tel: +86-10–64882584, Fax: +86-10-64882652, Email: zxc@caams.org.cn*

Abstract: The technology of near-infrared spectroscopy analysis is one of the important ways to test agricultural products inside quality quantitatively. The near-infrared spectral analysis system which used the technology of sound and light tunable filter (AOTF) can be applied to on-line measurement. The system constructed by AOTF, analysis optimization wavelength together with genetic algorithm, execute data filtering with wavelet analysis technology, process data of signal with second differential and Mahalanobis-distance, establish Partial Least-Squares model, tested the wheat and powdered milk. Reflects the development of the near-infrared spectral analysis system can be applied to the quality of agricultural products preferably.

Keywords: Near-infrared, AOTF, Milk Powder, Quality Analyses, Wavelet Analysis, Support Vector Machine

1. INTRODUCTION

Information collection, processing and monitoring technology are the key problem of digital agriculture. Near-infrared spectral analysis method is a better way to measure food and milk power quality rapidly (Hao, 2003; Geladi, 1985).

Acousto-optic tunable filter (AOTF) (Pozhar V E et al, 2002) is a new technology appeared in the early 1990s. Using birefringent crystal, by changing the radio frequency to change the wavelength, AOTF made homochromatic ray.

Please use the following format when citing this chapter:

Zhang, X., Hu, X., Zhang, Y., Wang, H. and Zhang, H., 2009, in IFIP International Federation for Information Processing, Volume 295, *Computer and Computing Technologies in Agriculture II, Volume 3*, eds. D. Li, Z. Chunjiang, (Boston: Springer), pp. 2067–2076.

The device has no mobile components, with a high scanning speed and better stability (R. TSENKOVA, 2000), so, it is suitable for on-line analysis particularly, is one of the development directions of near-infrared on-line measurement technology (Xu Kexin, 2002). This is a more mature technology in west countries but at the initial stage in china. Since this method can be widely used in quality measurement and control during the process of agricultural production, it will bring new vitality for agricultural equipment industry, and promote the development of quality of agricultural products on-line analysis rapidly with the character of low-cost, non-contact and real-time (Guyon I, 2002).

2. HARDWARE DESIGN OF AOTF SPECTROMETRY SYSTEM

2.1 Structure of the system

The system is composed of seven departments: rotating sample pool and control unit, high stability light source, AOTF optical system, temperature control unit of detector, detection signal acquisition system, DDS frequency selection control system and computer signal processing system, which is shown in Fig. 1.

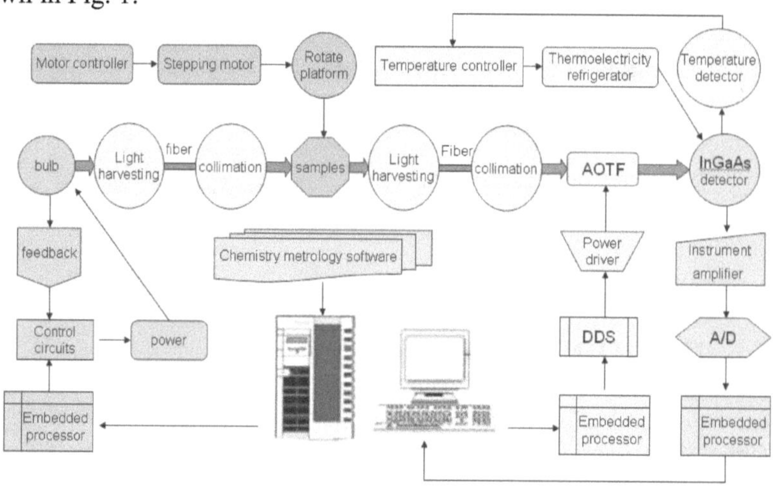

Fig.1: AOTF based spectroscopy analysis system

Use the revolving sample pool to test samples adequately; current-feedback-based light source control system provide 50 to 150 watts of light; AOTF spectrometry system uses diffuse reflectance measurement method; temperature control unit provide stable environment with an error of ± 0.2 degree for detectors; the computer issue 20 MHz ~ 70MHz inspirit signal through the DDS frequency selective circuit; and 16 bit A/D converter collect the corresponding spectrum signals to the computer in real-time.

2.2 Principle of AOTF

Sound-light tunable filter is an electro-optical modulator device. Its main principle is to use the Prague dissemination of sound wave to the light insinuated into medium when the sound wave spreading in anisotropic medium. Sound-light tunable filter was composed of single crystal birefringence (common material is TeO_2), piezoelectric transducers that glued to one side of the single crystal, and high-frequency signal source that used on the piezoelectric transducer. When certain frequency radio signals were imported, AOTF will diffract multicolor light, and choose homochromy light with the wavelength of λ. The homochromy light wavelength λ and radio frequency f has corresponding relations. Choose the second refraction crystal tellurium dioxide designed AOTF, which has the characters as follows:

Wavelength:	800nm ~1700nm;
Aperture:	Φ4mm;
Frequency:	20~70MHz;
Spectral resolution:	2nm at 800nm;
9nm at 1700nm	
Diffraction efficiency:	40% at 800nm;
	20% at 1700nm;
Drive power:	1W.

The examination showed that the centre-correlation coefficient between the radio driver frequency and the number of diffraction light waves center.

2.3 Design of DDS radio frequency driver

According to the frequency range of 20 MHz ~ 70MHz that AOTF RF driver needed, DDS (Direct Digital Synthesizer) clip AD9854 was selected. The chip has a 300 MHz internal clock, 4 to 20 times internal programmable frequency multiplier, which generated an external input clock frequency range of 15 ~ 75 MHz. In addition, the clip has a 100 MHz parallel interface bus, and an inside orthogonal dual-channel DAC output, working with a variety of programming fashion, and can produce complex signals such as linear and nonlinear FM signals and so on.

The working process is as follows:

(1) Translate digital waveform stored in several forms to Analog waveform by D/A.

(2) Change the frequency of the output signal by changing the step of Addressing. The step is the incremental phase of digital waveform. The accumulator value is search address; accumulate according to the phase

accumulator.

(3) Output low-sound analog waveform by low-pass or band-pass filter.

The core department of DDS System is phase accumulator, which is composed of an adder and a phase-based register. Suppose the bit width of the adder is 2^N, the high P-bit accumulator is to search Sin function table, and the frequency of clock is f_c, then the output frequency is:

$$f_{out} = \frac{f_c}{2^N}$$

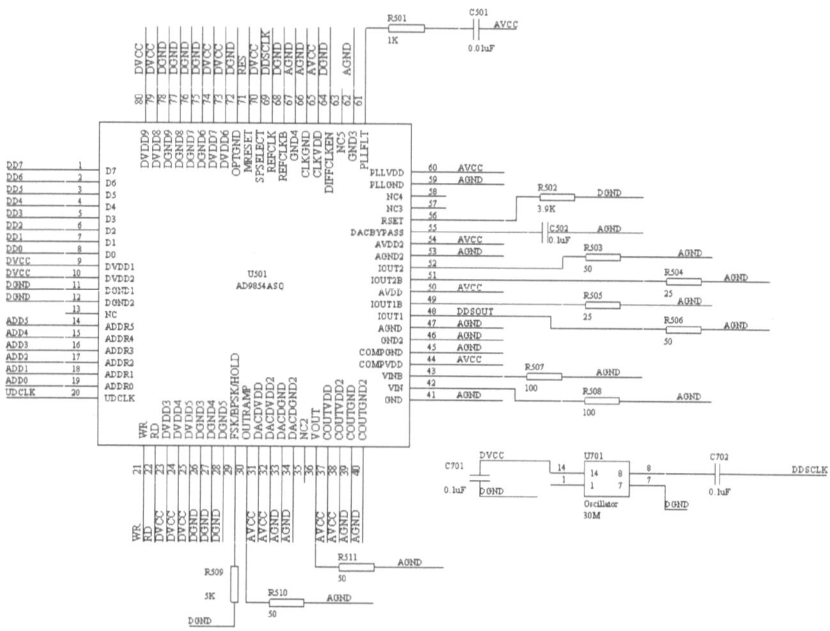

Fig.2: circuits of DDS

2.4 Design of RF low pass filter

The requisition of designing an elliptic function low pass filter is: the decay from 0 to 70 MHz should less than 1 dB and greater than 50 decibels if the frequency bigger than 75 MHz, terminal match resistance is 50 ohm. The design steps are as follows:

(1) Elliptic function obtained should have an order of seven.

(2) Search for components value from the normalization table according to the value of A_P, A_S, ω_C, ω_S and n.

(3) Calculate components value with integrated calibration formula and design circuit.

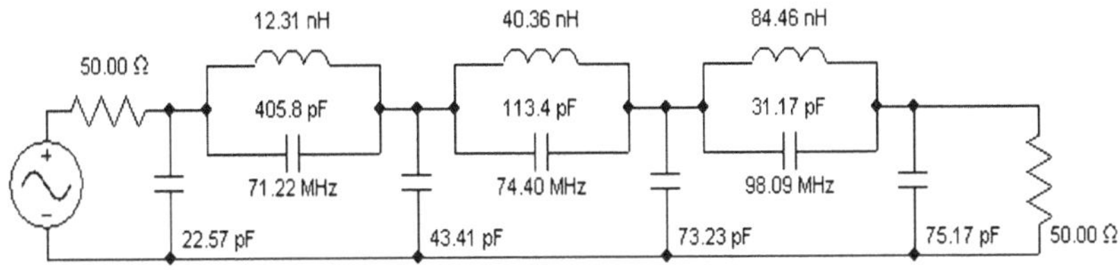

Fig.3: design of filter

The transfer function is as follows:

$$\frac{5.179e{+}07{*}S^6 + 4.136e{+}25{*}S^4 + 1.05e{+}43{*}S^2 + 8.607e{+}59}{S^7 + 4.013e{+}08{*}S^6 + 5.49e{+}17{*}S^5 + 1.744e{+}26{*}S^4 + 9.871e{+}34{*}S^3 + 2.315e{+}43{*}S^2 + 5.793e{+}51{*}S + 8.607e{+}59} \quad (1)$$

3. SOFTWARE DESIGN OF AOTF SPECTROMETRY SYSTEM

To avoid the effect of equipment measuring noise, temperature drifting, light fluctuations and density variety, the following technology were adopt:

(1) Multiplicative scatter correction, standard normal variety, de-trending algorithm and vector normalize.

(2) One-level and two-level window difference technology.

(3) Digital signal processing technology based on wavelet analysis and orthogonal signal correction filter.

(4) Best wavelength automatically search technology based on genetic algorithm.

(5) Principal component based mahalanobis-distance together with spectrum error forecast technology.

(6) Build model whit the technology of support vector machine.

3.1 Signal processing technology

Solve the noise interference problems whit the combination of wavelet analysis and orthogonal signal correction. The technology can fully retain effective information when signal filtering processing, and extract objectives related signal. With the characteristics of multi-resolution of wavelet analysis, filter all layers of signal.

Ultimate of orthogonal signal correcting: before establish correction quantitative models; remove the spectrum signal that has nothing to do with concentration matrix with the method of orthogonal mathematical method, and then conduct multiple corrections, so as to simplify the model and improve the ability forecast ability.

Ultimate of wavelet filter: in wavelet transform, the signal is divided into the cluster overlay of a series wavelet functions, decompose the signal to be analyzed to different scales with the method of multi-scale analysis, and achieve the purpose of signal processing through the hierarchical signal processing and remodeling. Wavelet packet analysis further decomposed not only low frequency but also high frequency part of coefficient; so that it enhanced the time-frequency differentiate.

The formula of fast discrete wavelet packet is:

$$\left.\begin{aligned} d^k_{j+1,2n} &= \sum_l d^l_{j,n} h_{l-2k} \\ d^k_{j+1,2n+1} &= \sum_l d^l_{j,n} g_{l-2k} \end{aligned}\right\} \quad (2)$$

The formula of discrete wavelet packet reconstruction is:

$$d^k_{j,n} = \sum_l d^l_{j+1,2n} \bar{h}_{k-2l} + \sum_l d^l_{j+1,2n+1} \bar{g}$$

With the method of compulsory noise cancellation, set zero to some high-frequency and low frequency coefficient of wavelet packet, and reconstruct the multi-signal.

3.2 Model construction technology

Construct model with the method of Support Vector Regression. The use of characteristics of minimum rise, get more stable and precise calibration model.

Suppose the sample set is:

$$(y_1, x_1), \cdots, (y_l, x_l), x \in R^{n,} y \in R \quad (3)$$

And regression function was expressed as the following linear equations:

$$f(x) = \langle w, x \rangle + b \quad (4)$$

The best regression function is the minimum value to the following function:

$$\Phi(w, \xi^*, \xi) = \frac{1}{2}\|w\|^2 + C\left(\sum_{i=1}^{l} \xi_i + \sum_{i=1}^{l} \xi_i^*\right)$$

Where C is set penalties values, ξ and ξ^* are upper and lower limits of the slack variable.

Use the following insensitive loss function:

$$L_\varepsilon(y) = \begin{cases} 0 & \text{for} \quad |f(x) \\ \left||f(x) - y\right| - \varepsilon & \text{otherwise} \end{cases}$$

And the optimize function:

$$\max_{\alpha,\alpha^*} W(\alpha,\alpha^*) = \max_{\alpha,\alpha^*} \left\{ \begin{array}{l} -\frac{1}{2}\sum_{i=1}^{l}\sum_{j=1}^{l}(\alpha_i-\alpha_i^*)(\alpha_j-\alpha_j^*)(\\ +\sum_{i=1}^{l}\alpha_i(y_i-\varepsilon)-\alpha_i^*(y_i+ \end{array} \right.$$

Under the construction of:

$$0 \le \alpha_i \le C, \quad i = 1,\dots,l$$
$$0 \le \alpha_i^* \le C, \quad i = 1,\dots,l$$
$$\sum_{i=1}^{l}(\alpha_i - \alpha_i^*) = 0$$

Calculate the formula:

$$\bar{\alpha},\bar{\alpha}^* = \arg\min_{\bar{\alpha},\bar{\alpha}} \left\{ \begin{array}{l} \frac{1}{2}\sum_{i=1}^{l}\sum_{j=1}^{l}(\alpha_i-\alpha_i^*)(\alpha_j-\alpha_j^*)(x_i^T x_j, \\ -\sum_{i=1}^{l}(\alpha_i-\alpha_i^*)y_i+\sum_{i=1}^{l}(\alpha_i+\alpha_i^*)\varepsilon \end{array} \right.$$

We get the coefficient α_i and α_i^* of Lagrange equation, and then regression coefficient and constant as follows:

$$\begin{aligned} \bar{w} &= \sum_{i=1}^{l}(\alpha_i-\alpha_i^*)x_i \\ \bar{b} &= -\frac{1}{2}\langle \bar{w},(x_r+x_s)\rangle \end{aligned} \tag{5}$$

3.3 Practical technologies for products

Practical technologies adopted in this thesis include:

Designed Low-power and high-precision portable light source, designed core components of temperature control process; dynamic adjust technology of double-calibration of standard light and background; completed signal and model standardization; Rotating dynamic correction technology and density light automatically tracking measurement method.

4. APPLICATIONS IN FARM PRODUCE CHARACTER DETECTION

Examination was done to test the quality of grain and milk powder with developed apparatus. Take dairy products as an example, choose 50 samples of milk powder, and set the measure range of wavelength of 800～1700nm, points of wavelength is 400, establish model calibration for pretreatment spectral data of water, ash, acidity and protein respectively. We found that milk ingredients in different wavelength absorption strength were very different, for example, water had a strong absorption peak in the wavelength of 960 nm and 1440nm (Fig. 4).

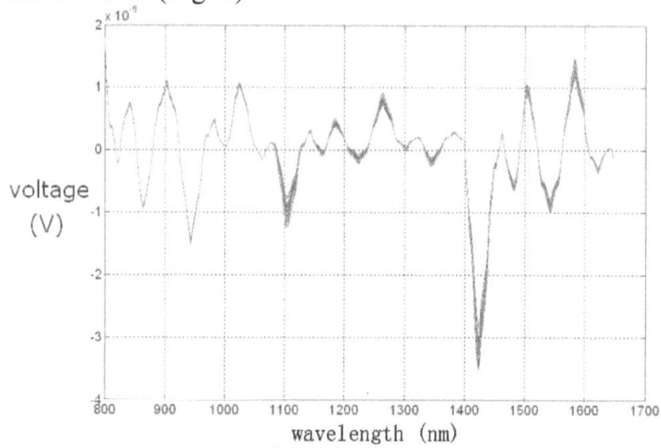

Fig .4: absorption peak of water

The best main components are determined by the square of forecast error of correction aggregate and the predictive residual error sum of squares.

Considering the original baseline drift of the absorption spectrum, a first derivative and second derivative processing technology were adopted. Fig. 5, Fig. 6 and Fig.7 are separately the model correlation of water, ash and protein.

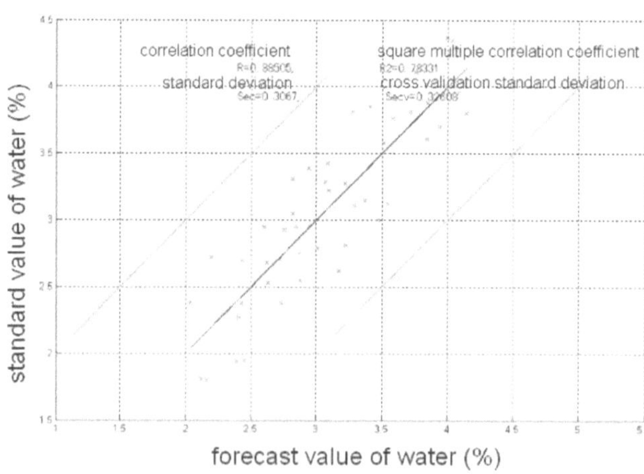

Fig.5: model correlation of water

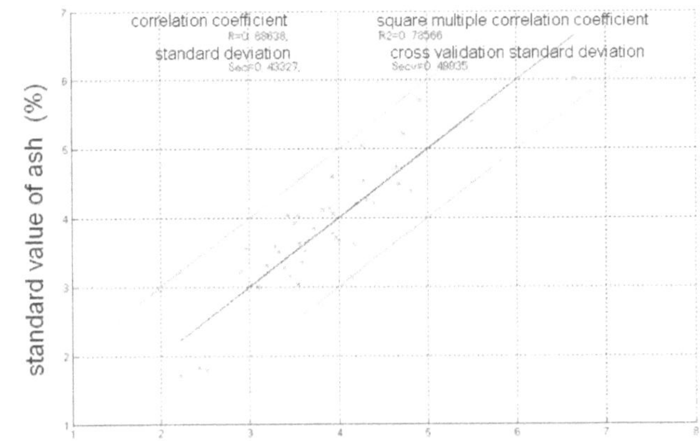

Fig.6: model correlation of ash

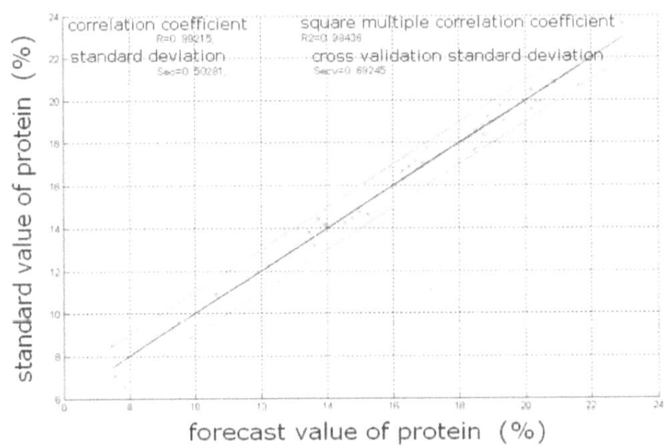

Fig.7: model correlation of protein

Obligate 20% samples as forecast ones, four regression models was forecasted separately. Predicted results of protein were shown in table 1 and measurement repeatability test results in table 2.

Table.1: analysis data of protein

type	Model or not?	number	R	SEC
Revise aggregate	YES	40	0.99215	0.5
Forecast aggregate	NO	10	0.9958	0.47

Table.2: errors and repetition of each component

Component	Water (%)	Ash (%)	Acidity (T)	Protein (%)
Error	0.4	0.6	1.2	0.7
Repetition	0.1	0.2	0.5	0.3

5. CONCLUSIONS

The examination shown that it was feasible to detect the component of milk powder with the AOTF based principle of near-infrared spectroscopy system. But there were still disparity in errors and repetition with national standard. In order to ensure the accuracy and stability, we will continue to study the following aspects:

(1) Improve the capability of AOTF apartments,

(2) Improve the stability of the frequency signal drive, and,

(3) Enhance the stability and sensitivity of detectors.

The diffuse reflection we used in the spectrum system can detect grain liquid and powder agricultural products directly.

The system is especially suitable for the rapid analysis of large number of repetitive samples, which can detect all the elements in 40 seconds, and the method is simple and repeatable.

REFERENCES

B Sholkopf ,K Sung, CJ C Burges,et al.Comparing support vector machine with Gaussian kernels to radial basis function classifiers. IEEE Trans. Signal Processing,1997,45:2758-2765

Guyon I, Weston J, Barnhill S, et al. Gene selection for cancer classfication using support vector machines. Machine Learning,2002, 46(1): 389-422.

Hao Li,Tong Liu,Changyun Wen. All-fiber acousto-optical tunable filter with loop structure. Optical Engineering.2003,42(12):26-31

Johan A.Westerhuis,Sijmen de Jong,Age K.Smilde.Direct orthogonal sighal correction. Chemom.Intell.Lab.Syst.,2001,56:13-25.

O. Svensson,T. Kourti,J. F.MacGregor.An investigation of orthogonal signal correction algorithms and their characterics. J. chemometrics,2002,16:176-188

P Geladi,D MacDougall,H Martens. Linearization and scatter-correction for near-infrared reflectance spectra of meat. Applied Spectrotroscopy,1985,39(03),491-500.

Pozhar V E, Pustovoit V I. Long-pathoptical spectral AOTF-based gas analyzer[J]. SPIE, 2002, 4574:174-178.

R J Barnes, M S Dhanoa,S J Lister.Standard normal variate transformation and de-trending of near-infrared diffuse reflectance spectra. Applied Spectroscopy,1989,43(05):772-777

R. TSENKOVA, Y. OZAKI, et al. Near Infrared Specerosoopy for Biomonhoring: Cow Milk Composition Measurement in a Spectral Region from 1100 to 2400 Namometers[J].J. Anim, Sci., 2000, 78:515 — 521

Xu kexin, Ni Yong. The optical characteristics of the acousto-optic tunable filter with the equivalent point[J]. Chinese Journal of Lasers, 2002, B11(2):133-138.

THE NEW FIBER-OPTIC TEMPERATURE SENSOR FOR GREENHOUSE

Xueguang Wang*, Zenghuan Liu, Xiaowei Du

College of Information and Electrical Engineering, Hebei University of Engineering, Handan, Hebei province, 056038, P.R.. China

* *Corresponding author, Address: College of Information and Electrical Engineering, Hebei University of Engineering, Handan, GuangMing South Street 199, Hebei province, 056038, P.R.. China, Tel: 0310-8578738, Fax:0310-8578746, Email: wxg1317@126.com*

Abstract: In this paper, we introduced a new fiber-optic temperature sensor with white light interferometric principle based on Fabry-Perot, which is capable of providing measurement for the temperature of greenhouse. The signal processing of system is that the optical signal wavelength is modulated in the Fabry-Perot cavity and is demodulated during the optical-electricity transform. Compared with a common optical fiber strain sensor, it has more advantages, such as low cost, high stability, and high anti-interference. The resolution of the fiber optical temperature sensor is up to 0.01℃ within the test range being -40℃～100℃.The fiber-optic temperature sensor system made up of this fiber-optic temperature sensor is put into use in agriculture fields for monitoring in real time and absolute temperature measurement.

Keywords: optic fiber, temperature, interferometric, Fabry-Perot, greenhouse

1. INTRODUCTION

Along with national economic quick growth, the agricultural research and the application technology take more and more recognition, especially the hothouse has become the important part of the high-efficiency agriculture. It may not only provide the off-season fresh agricultural products, but also can create good economic efficiency. The growth vigor, quality and output of the crops in the hothouse are closely related with the temperature, humidity and soil moisture of the hothouse. Therefore the temperature measurement to the

Please use the following format when citing this chapter:

Wang, X., Liu, Z. and Du, X., 2009, in IFIP International Federation for Information Processing, Volume 295, *Computer and Computing Technologies in Agriculture II, Volume 3*, eds. D. Li, Z. Chunjiang, (Boston: Springer), pp. 2077–2082.

hothouse is a very important link. The traditional temperature measurement usually adopts temperature elements, thermocouples, thermistor as well as integrated temperature sensor and so on. But this kind of elements is easy to be limited by the fields measured by fields measured and the environment. And they should be examined and replaced regularly for the capability can be declined while high temperature or long-term use. Therefore the practical application is greatly inconvenient. Optical Fiber Temperature Sensor can be used in every kind of condition and environment (P.M Nellen, 1999, V.K.Varadan, 1999, Yang YH, 2007). In this paper, we introduced a new fiber-optic temperature sensor with white light interferometric principle based on Fabry-Perot, which is capable of providing measurement for the environment temperature. White light interferometry is a low coherent interferometer that is used to avoid error and noise caused by bending of the fiber and fluctuation of the light source (S.Vered, 2002). Compared with a common optical fiber temperature sensor, it has more advantage, such as low cost, high stability and absolute measurement.

2. SENSOR CONFIGURATION

The proposed temperature sensor structure is shown in Fig.1, the fiber sensor system is based on white-light interferometric principle, which signal processing is that the optical signal wavelength is modulated in the Fabry-Perot cavity with fiber sensor system, and is demodulated during the optical-electricity transform. For the sensor system, the Light emitting diodes (LED) is used as a white-light source, and the detector system is mainly made up of CCD photoelectric transform circuit. The designed fiber sensor is of fine linearity strain characteristic, highly precision and highly sensibility.

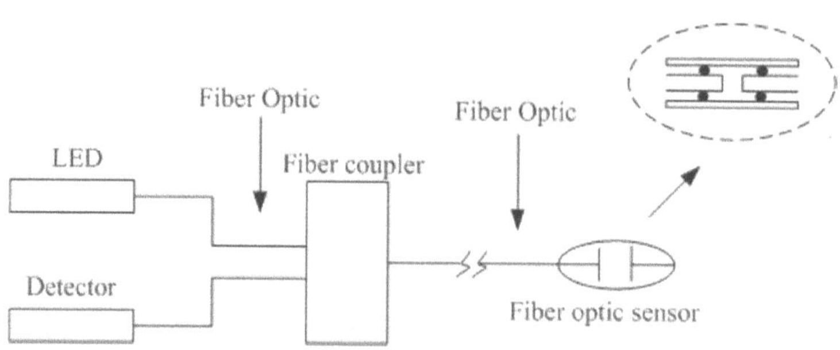

Fig.1. The fiber temperature sensor system

The optical fiber temperature sensor designed in this paper, is that the measured temperature is transformed into optical fiber Fabry-Perot the interference cavity length L change. Fig.2 shows the structure of fiber temperature sensor; the reflective fiber (RF) and the incident fiber (IF) are both infixed the hollow fiber which inner diameter (D_{in}) is micron magnitude and are cemented at the hollow fiber ending. M_1 and M_2 are the coating end faces for avoiding the loss. The distance between two fixed points is named gauge length (L_g), which influences the measurement range and accuracy of fiber strain sensor.

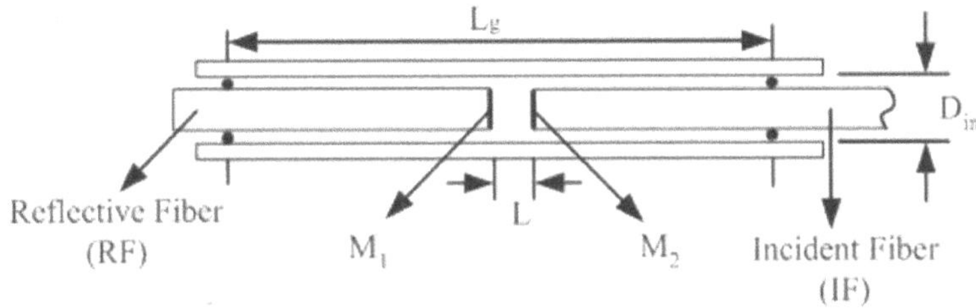

Fig.2. The fiber temperature sensor structure

The external strain changes modified the length of Fabry-Perot cavity, so the strain change $\Delta\varepsilon$ is proportional to the change of the length of Fabry-Perot cavity ΔL in the following way:

$$\Delta\varepsilon = \frac{\Delta L}{L_g} \tag{1}$$

According to stress-strain and temperature relations, in Cartesian coordinate, the component of strain caused temperature is:

$$\varepsilon_{xx} = \varepsilon_{yy} = \varepsilon_{zz} = kT \tag{2}$$

$$\varepsilon_{xy} = \varepsilon_{yz} = \varepsilon_{zx} = 0 \tag{3}$$

In the formula, k is object thermal-expansion coefficient; ε_{ij} $(i, j=x, y, z)$ is measured body's in i, j direction component of strain. If only considers the optical fiber temperature sensor's axial strain $\varepsilon_{zz} = \Delta L / L_g$, the result in by the formula (2):

$$\Delta L / L_g = kT \tag{4}$$

When the reflective coefficient of surface $R << 1$, the multi-beam interference in Fabry-Perot cavity can be approximately considered as two-beam interference, the reflective light intensity I_R is (Bi Weihong, 1999):

$$I_R = 2I_0R(1 + con\phi) \tag{5}$$

Where I_0 is the incident intensity, ϕ is the phase difference of two beams. Eq. (5) shows that the extreme point of reflective intensity appears with the follow equation meeting,

$$\phi = 2k\pi \qquad (k = 1,2,3\cdots) \tag{6}$$

The phase difference ϕ can be expressed:

$$\phi = \frac{4\pi}{\lambda} L \tag{7}$$

Where λ is the optical wavelength, L is the length of Fabry-Perot cavity. So the length of Fabry-Perot cavity can be showed:

$$L = \frac{\phi}{4\pi}\lambda = \frac{2k\pi}{4\pi}\lambda = \frac{k}{2}\lambda \qquad (k = 1,2,3\cdots) \tag{8}$$

Eq. (8) shows that the reflective intensity maximizing when the length of Fabry-Perot cavity is integer multiple of the optical half-wavelength. We design the optical strain sensor based on white light interference, which a short coherence length white source is used, and the white light source is LED that produces broadband emissions. In a Fabry-Perot light interferometer, the incident polychromatic wavelength is $\lambda_1, \lambda_2, \cdots \lambda_n$, so Eq. (8) can be written as:

$$L = \frac{k}{2}\lambda_i \qquad (i = 1,2,3\cdots n) \tag{9}$$

Then projected onto a CCD camera, the fringes can be captured and analyzed on an accompanying microcomputer. The advantages of the system include the absolute measurement capability and the potential for multiprocessing.

3. EXPERIMENTAL RESULTS AND DISCUSSION

Because the temperature is decided to the optical fiber sensor's influence by the optical fiber temperature sensor carrier nature, the carrier volume, the carrier and between the optical fiber temperature sensor's cementation and uses the rubber with the rubber thickness. Regarding metals and so on carrier like copper, iron along with the temperature change are linear, simultaneously when the seal is not too thick with the rubber, optical fiber's volume to be opposite is very small in the carrier, the temperature and the strain relations are approximate are decided by the carrier temperature characteristic. The optical fiber temperature sensor mentioned in this paper, is housed in the diameter is in the 2.4mm copper pipe by rubber, which the thermal-expansion coefficient of copper is *k*=14.8×10-6/℃。

We analyze the experiments of temperature characteristic of fiber temperature sensor, and we measured the temperature characteristic of designed fiber strain sensor and the results presented in Fig.3, which black block expression for actual survey data, straight line expression fitting curve.

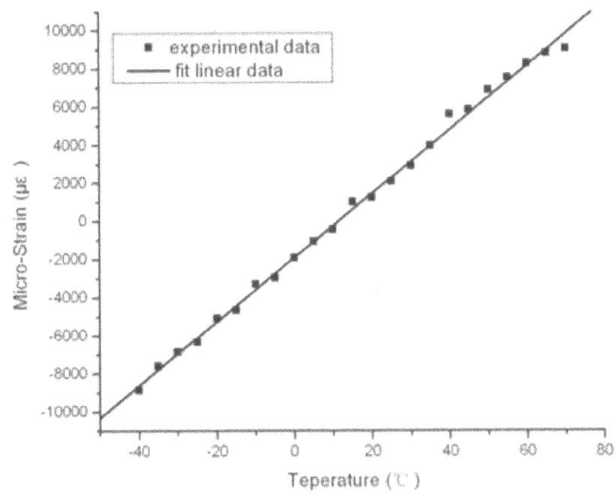

Fig.3. The temperature characteristic of fiber strain sensor

The results of temperature characteristic showed in Fig.3 were measured by the attemperator from –40℃to 70℃, the resolution achieves 0.01℃ and they presented goodness linearity strain characteristic, which linear fitting is 0.9943. But when the external environment temperature surpasses 90℃, the empirical datum and the theory fitting have the deviation, the reason possibly is that the temperature hysteresis quality of rubber influence measurement result which appears under the high temperature, causes the optical fiber temperature sensor's temperature change and the metal carrier temperature change is inconsistent.

Theoretically, the designed optical fiber temperature sensor unit cavity long change value is: $\Delta L / L_g \Delta T = 14.8 \times 10^{-6} / °C$; But the cavity long change value which obtains by the empirical datum is: $\Delta L / L_g \Delta T = 15.06 \times 10^{-6} / °C$. Because the experiment error in reading and the seal create with the rubber to the optical fiber temperature sensor's influence, the theoretical value and the experimental results are not completely same. This kind of error may carry on the compensation through the algorithm.

4. CONCLUSION

A new fiber-optic temperature sensor with white light interferometric principle based on Fabry-Perot was presented and investigated theoretically and experimentally, which the resolution achieves 0.01℃ from –40℃to 70℃ and it adopt to use in the greenhouse big awning the temperature survey, specially to temperature request high crops greenhouse big awning.

Compared with a common optical fiber strain sensor, it has more advantages, such as low cost, high stability, high anti-interference and absolute measurement. So it will certainly to have the widespread application prospect, further consummates its production process, definitely may achieve practical the request.

REFERENCES

Bi Weihong, Mathematical models for Fiber-Optical Fabry-Perot interferometic cavity. ACTA PHOTONICA SINICA, 1999, 28(8): 744-747

P. M. Nellen, P Mauron. Mechanical and Optical Reliability of Fiber Bragg Grating Strain and Temperature Sensors at High Temperature. Proceeding of SPIE, 1999, (20): 131-135

S.Vered, Katzir, Abraham, Four-band Fiber Optic Measurement During an Exothermal. Optical Radiometry for True Temperature Engineering SPIE, 2002, 41(7): 1052-1056

V. K. Varadan, V.V. Varadan. Wireless Electro-optic Switching Network for optical Fiber Sensor Array Using MEMS-IDT Devices. Proceedings of SPIE, 1999, (22): 131-135

Yang YH, Xia HY, Jin W, Practical polarization maintaining optical fibre temperature sensor for harsh environment application, MEASUREMENT SCIENCE & TECHNOLOGY, 2007, 18(10): 3235-3240

MODELING AND DESIGNING OF A NONLINEAR TEMPERATURE-HUMIDITY CONTROLLER USING IN MUSHROOM-DRYING MACHINE

Xiuhua Wu*, Haiyan Luo, Minhui Shi
Information and Electrical Engineering Institute, Shenyang Agricultural University, Shenyang 110161, China

Abstract: Drying-process of many kinds of farm produce in a close room, such as mushroom-drying machine, is generally a complicated nonlinear and time-delay cause, in which the temperature and the humidity are the main controlled elements. The accurate controlling of the temperature and humidity is always an interesting problem. It's difficult and very important to make a more accurate mathematical model about the varying of the two. A math model was put forward after considering many aspects and analyzing the actual working circumstance in this paper. Form the model it can be seen that the changes of temperature and humidity in drying machine are not simple linear but an affine nonlinear process. Controlling the process exactly is the key that influences the quality of the dried mushroom. In this paper, the differential geometry theories and methods are used to analyze and solve the model of these small-environment elements. And at last a kind of nonlinear controller which satisfied the optimal quadratic performance index is designed. It can be proved more feasible and practical than the conventional controlling.

Key words: nonlinear controller, mathematical model, temperature-humidity control, mushroom-drying machine

1. INTRODUCTION

In recent years, the yield of mushroom has been increased year by year. It makes the demand of high quality drying-technology urgent. Drying

Please use the following format when citing this chapter:

Wu, X., Luo, H. and Shi, M., 2009, in IFIP International Federation for Information Processing, Volume 295, *Computer and Computing Technologies in Agriculture II, Volume 3*, eds. D. Li, Z. Chunjiang, (Boston: Springer), pp. 2083–2090.

mushroom needs to gain an ideal goal in a limited period or as quickly as it can. In order to reduce the cost of drying process, proper drying methods must be selected. The requirements can be satisfied by the drying equipments which used heat energy associated with solar energy. They not only make best use of the regenerative solar energy but also can ensure the drying time with heat energy when solar energy is not sufficient. It is a kind of combined and efficient drying method.

The environment of the drying machine likes a small green house. The main indexes are illumination, temperature, humidity, air-flowing speed and the quantity of heat. The temperature can't be too high in a long time during the process of mushroom-drying, otherwise the shape, color and the nutrition of the dried mushroom are damaged. The drying time can not be very long. The check of the products depends on the humidity of dried mushroom. In drying process, changes of temperature and humidity are not simple linear but complicated nonlinear. The traditional linear control can not content the needs to exactly control and get high quality products. A comparatively strict and exact control method is needed to get high quality and grades of the products. So, to make an ideal mathematical model about the varying of temperature and humidity in the drying machine is the important thing.

1. MODEL

Consideration of the cost, glass wall is used to make full use of solar in day. It's paved at night. The ventilation-holes is used to ventilate naturally. Heat energy associated with solar energy control the temperature in machine. So the temperature control dynamic model satisfies the following heat balance equation (Daskalov,1997):

$$V\rho C_p \frac{dT_i}{dt} = Q_{rad} + Q_{heat} + Q_{crad} - Q_{tran} - Q_{vent} - Q_{cac} \quad (1)$$

Where V is the volume of the machine (m^3) , ρ the air density $(1200g/m^3)$, C_p the specific heat of air $(1.006J/g\cdot K)$, T_i inner air temperature , Q_{rad} the solar radiant energy, Q_{heat} the heat energy by burning fuels, Q_{crad} the radiant energy of long waves, Q_{tran} the heat losses for transpiration , Q_{vent} the heat losses for air flowing , Q_{cac} the heat losses due to transmitting, and as the value very low compared with the others, it can be ignored. The right parts of the Eq. (1) as follows:

$$Q_{rad} = A_s \cdot I_a \cdot \tau_a \quad (2)$$

Where A_s is the areas of the sunshine（m2）, τ_a transmitance of the glass(0.89), I_a the solar radiant intensity（W/m^2）. I_a is a variable changed with the time. For simplification, the average radiant intensity of a day in summer or autumn is used. It's approximately 70 W/m2.

$$Q_{heat} = h_p(T_r - T_i), \tag{3}$$

Where T_r is the desired inside temperature (K), T_i the inside air temperature (K), h_p the heat of burning fuels per temperature unit (J/K).

According to Stefan-Boltzmann law, the radiant energy of long waves as follow:

$$Q_{crad} = \varepsilon_{12} A \sigma T_i^4 \tag{4}$$

Where $\varepsilon_{12} = (\varepsilon_1 + \varepsilon_2 - 1)^{-1}$ is emissivity between interfaces, $\varepsilon 1, \varepsilon 2$ are the emissivities of glass and air respectively, $\varepsilon 1 = 0.9, \varepsilon 2 = 0.3$. σ coefficient of Stefan-Boltzmann, $\sigma = 5.67 \times 10^{-8}\, W/m^2 \cdot K^4$, $K = -273^\circ C$.

$$Q_{tran} = \lambda \cdot E_m A \tag{5}$$

Where Q_{tran} is the absorbing energy of transpiration of the products(J/h), A the area of the products(m^2), $\lambda = 4.1868(597 - 0.57t)$ the parameter of the transpiration heat(J/g), t the temperature of the products in transpiration, in this case it used the T_r , E_m the intensity of transpiration($g/m^2 \cdot h$), E_m can be denoted as:

$$E_m = 1.2k_e[3.9 \times 10^{-3} \exp(\frac{T_i}{15.2}) - q_i] \tag{6}$$

Where k_e is the coefficient of the evaporation which has relation to the air speed, $k_e = 0.223 + 0.149 \times f(v_{air})$ (g/h) (Daskalov,2006) ,for comprehensive consideration, it uses 0.5(Govriachev, 1968).

Substitute Eq.(6) into Eq.(5), it gains:

$$\begin{aligned} Q_{tran} &= 4.1868(597 - 0.57T_r) \times 1.2 \times 0.5 \times (3.9 \times 10^{-3} \exp(T_i/15.2) - q_i)A \\ &= 5.85A\exp(T_i/15.2) - 1500Aq_i - 0.005T_rA\exp(T_i/15.2) + 1.432AT_rq_i \\ &= (5.85A - 0.005T_rA)\exp(T_i/15.2) + (1.432AT_r - 1500A)q_i \end{aligned} \tag{7}$$

$$Q_{vent} = \rho C_p v_{na}(T_r - T_i) \tag{8}$$

Where Q_{vent} is the heat losses of the air flowing, v_{na} the natural ventilation ratio ($m^3/\min$), it uses the average value 2.5 $m^3/\min$.

Substitute Eqs.(2)-(4)and (7)-(8) into Eq.(1), it can get:

$$\begin{aligned} V\rho C_p \frac{dT_i}{dt} &= A_s I_a \tau_a + h_p(T_r - T_i) + (\varepsilon_1 + \varepsilon_2 - 1)^{-1} A\sigma T_i^4 - (5.85A \\ &- 0.005T_rA)\exp(T_i/15.2) - (1.432AT_r - 1500A)q_i - \rho C_p v_{na}(T_r - T_i) \end{aligned} \tag{9}$$

The humidity dynamic model of the drying machine is as follow:

$$\frac{dq_i(t)}{dt} = \frac{AE_m}{V} - \frac{v_{na}}{V}[q_o - q_i] \tag{10}$$

Where q_o the outside absolute humidity (g/m^3), q_i the inside controlled absolute humidity(g/m^3).

Substitute the Eq.(6) into Eq.(10),it can get :

$$\frac{dq_i}{dt} = \frac{A}{V}\times 0.6\times[3.9\times 10^{-3}\exp(T_i/15.2) - q_i] + \frac{v_{na}}{V}q_i - \frac{v_{na}}{V}q_0$$
$$= 0.002\frac{A}{V}\exp(T_i/15.2) + \frac{v_{na} - 0.6A}{V}q_i - \frac{v_{na}}{V}q_o \tag{11}$$

In addition, as the temperature and humidity inside are higher than those outside, the temperature and humidity outside have little effect on those of inside. So T_r, q_o, v_{na} can be regarded as constants.

We take place all the constants in Eqs.(9) and (11) with $a_0,\ldots,a_5, b_0,\ldots,b_2$, then we can observe:

$$\begin{cases} \dfrac{dT_i}{dt} = a_0T_i^4 + a_1T_i + a_2e^{kT_i} + a_5 + a_3q_i + (a_4 - T_i)h_p \\ \dfrac{dq_i}{dt} = b_0e^{kT_i} + b_1q_i + b_2 \end{cases} \tag{12}$$

Where $X = [T_i q_i]^T, X \in R^n$ is the state vector, $\dot{X} = [\dot{T}_i \dot{q}_i]^T$ the differential of the state variables, $u = h_p$, $u \in R$ the control variable, $y(t) = q_i$, $y \in R$ the output variable.

Apparently, the differential Equation set contents the normal form:

$$\begin{cases} \dot{X}(t) = F(X) + g(X)u \\ y(t) = h(X) \end{cases} \tag{13}$$

So it is a SISO multivariable affine nonlinear control system. Where

$$F(X) = \begin{bmatrix} a_0T_i^4 + a_1T_i + a_2e^{kT_i} + a_3q_i + a_5 \\ b_0e^{kT_i} + b_1q_i + b_2 \end{bmatrix} \tag{14}$$

$$g\ (X) = \begin{bmatrix} a_4 - T_i \\ 0 \end{bmatrix} \tag{15}$$

$$y(X) = h(X) = q_i \tag{16}$$

The control purpose is to find an optimal control law u which can make the state variables reach the idealist values and the energy consumed is the least. At the same time, it's to find a coordination transform $\varphi(\mathrm{Z})$, changing the nonlinear system into a linear system as equation set (17) , namely exactly linearization.

$$\begin{cases}\dot{\mathrm{X}} = \mathrm{AX} + \mathrm{Bu} \\ \mathrm{Y} = \mathrm{CX} + \mathrm{Du}\end{cases} \tag{17}$$

2. METHODS AND RESULTS

According to the theory of differential geometry about exact linearization, the affine nonlinear system must have the relative degree r equal the order of the state variables n. so we first calculate the following Lie derivative (Lu Q.,1996).

$$L_g L_f{}^0 h(X) = L_g h(X) = \frac{\partial h(X)}{\partial X} g(X) = [0\,1]\begin{bmatrix} a_4 - T_i \\ 0 \end{bmatrix} = 0 \tag{18}$$

$$L_f h(X) = \frac{\partial h(X)}{\partial X} F(X) = [0\,1]\begin{bmatrix} a_0 T_i^4 + a_1 T_i + a_2 e^{kT_i} + a_3 q_i + a_5 \\ b_0 e^{kT_i} + b_1 q_i + b_2 \end{bmatrix} \tag{19}$$

$$= b_0 e^{kT_i} + b_1 q_i + b_2$$

$$L_g L_f h(X) = L_g (b_0 e^{kT_i} + b_1 q_i + b_2) = [b_0 k e^{kT_i} b_1]\begin{bmatrix} a_4 - T_i \\ 0 \end{bmatrix} \tag{20}$$

$$= (a_4 - T_i) b_0 k e^{kT_i} \neq 0 (a_4 \neq T_i, also when T_r \neq T_i)$$

Where $L_f{}^n h(X)$ denotes the rth order Lie derivative of $h(X)$ along f. $L_g L_f\ h(X)$ the mix Lie derivative of $h(X)$ along f and g.

So the relative degree of this nonlinear system r-1=1, r=2=n .

Now looking for a coordination mapping $\varphi(\mathrm{Z})$:

$$\mathrm{Z}_1 = h(X) = q_i \tag{21}$$

$$\dot{\mathrm{Z}}_1 = \mathrm{Z}_2 = \frac{\partial h(X)}{\partial X} \dot{X} = \frac{\partial h(X)}{\partial X}(F(X) + g(X)u) \tag{22}$$

$$= L_f h(X) + L_g L_f{}^0 h(X) u = L_f h(X) = b_0 e^{kT_i} + b_1 q_i + b_2 (T_r \neq T_i)$$

$$\dot{\mathrm{Z}}_2 = \mathrm{Z}_3 = L_f{}^2 h(X) + L_g L_f{}^1 h(X) = \frac{\partial (L_f h(X))}{\partial X} F(X) + L_g L_f h(X) u \tag{23}$$

$$L_g L_f h(X)u = (a_4 - T_i)b_0 k e^{kT_i} u \tag{24}$$

$$L_f{}^2 h(X) = [b_0 k e^{kT_i} b_1]\begin{bmatrix} a_0 T_i^4 + a_1 T_i + a_2 e^{kT_i} + a_3 q_i + a_5 \\ b_0 e^{kT_i} + b_1 q_i + b_2 \end{bmatrix}$$

$$= b_0 k e^{kT_i}(a_0 T_i^4 + a_1 T_i + a_2 e^{kT_i} + a_3 q_i + a_5) + b_1(b_0 e^{kT_i} + b_1 q_i + b_2) \tag{25}$$

$$= a_0 b_0 k e^{kT_i} T_i^4 + a_1 b_0 k e^{kT_i} T_i + a_2 b_0 k e^{2kT_i} + (a_5 b_0 k + b_1 b_0) e^{kT_i} +$$

$$(a_3 b_0 k e^{kT_i} + b_1^2) q_i + b_1 b_2$$

$$\dot{Z}_2 = L_f{}^2 h(X) + L_g L_f{}^1 h(X)u \tag{26}$$

$$= \alpha(X) + \beta(X)u = V$$

So, we get the Brunovsky normal form:

$$\begin{cases} \dot{Z}_1 = AZ_2 \\ \dot{Z}_2 = BV \\ y = h(X) = q_i \end{cases} \tag{27}$$

Compared with the Eq. set (17), we can get:

$$\mathrm{A} = \begin{bmatrix} 0 & 1 \\ 0 & 0 \end{bmatrix}, B = \begin{bmatrix} 0 \\ 1 \end{bmatrix}, D = 0, y = h(X) = q_i \cdot$$

While $\begin{cases} V^* = \alpha(X) + \beta(X)u^*, \\ u^* = -\dfrac{\alpha(X)}{\beta(X)} + \dfrac{V^*}{\beta(X)}, (\beta(X) \neq 0) \end{cases}$ and

$\begin{cases} \alpha(\mathrm{X}) = L_f{}^2 h(X), \beta(X) = L_g L_f h(X) \\ u^* = \dfrac{-L_f{}^2 h(X) + V^*}{L_g L_f h(X)} \end{cases}$ is the control law.

Then we will determine the input variable V which is the control input of the linear system of the Brunovsky normal form. To the most reasonable way is using the linear optimal control design method with the quadratic performance index (LQR method) to produce the V*.

It can be proved that the system is a controllable system as the rank of the matrix $[B \mid AB \mid A^2 B \mid ....]$ is n.

Selected the linear quadratic performance index (LQR) as follow:

$$J = \frac{1}{2}\int_0^\infty (Z^T QZ + V^T RV)dt \tag{28}$$

Then we can get the control vector V which makes the performance index functional J reach its extremum.

$$V^* = -R^{-1} B^T P^* Z(t) = -K^* z(t) \tag{29}$$

Where V^* denotes the optimal control vector, K^* is the optimal feedback gain matrix:

$$K^* = R^{-1} B^T P^* \tag{30}$$

P^* is the solution of the Riccati algebraic equation, Q, R are semi-positive and positive definite weighting matrices, respectively.

$$A^T P + PA - PBR^{-1} B^T P + Q = 0 \tag{31}$$

Generally, Q is a diagonal matrix. In this question, $Q = \begin{bmatrix} 1 & 0 \\ 0 & 1 \end{bmatrix}, R = 1$, we can deduce $\begin{bmatrix} 0 & 0 \\ 1 & 0 \end{bmatrix} \mathrm{P} + \mathrm{P} \begin{bmatrix} 0 & 1 \\ 0 & 0 \end{bmatrix} - \mathrm{P} \begin{bmatrix} 0 \\ 1 \end{bmatrix} \begin{bmatrix} 0 & 1 \end{bmatrix} \mathrm{P} + \begin{bmatrix} 1 & 0 \\ 0 & 1 \end{bmatrix} = 0$.

Here we suppose $\mathrm{P} = \begin{bmatrix} x_1 & x_2 \\ x_3 & x_4 \end{bmatrix}$, substitute it into the upper equation. We get:

$$\begin{bmatrix} 0 & 0 \\ 1 & 0 \end{bmatrix} \begin{bmatrix} x_1 & x_2 \\ x_3 & x_4 \end{bmatrix} + \begin{bmatrix} x_1 & x_2 \\ x_3 & x_4 \end{bmatrix} \begin{bmatrix} 0 & 1 \\ 0 & 0 \end{bmatrix} - \begin{bmatrix} x_1 & x_2 \\ x_3 & x_4 \end{bmatrix} \begin{bmatrix} 0 \\ 1 \end{bmatrix} \begin{bmatrix} 0 & 1 \end{bmatrix} \begin{bmatrix} x_1 & x_2 \\ x_3 & x_4 \end{bmatrix} + \begin{bmatrix} 1 & 0 \\ 0 & 1 \end{bmatrix} = 0 \tag{32}$$

Then,

$$\begin{bmatrix} 0 & 0 \\ x_1 & x_2 \end{bmatrix} + \begin{bmatrix} 0 & x_1 \\ 0 & x_3 \end{bmatrix} - \begin{bmatrix} x_2 x_3 & x_2 x_4 \\ x_3 x_4 & x_4^2 \end{bmatrix} + \begin{bmatrix} 1 & 0 \\ 0 & 1 \end{bmatrix} = 0 \tag{33}$$

Solving the equation, we gain: $\mathrm{P}^* - \begin{bmatrix} \sqrt{3} & 1 \\ 1 & \sqrt{3} \end{bmatrix}$,which is a positive definite matrix. Substitute it into the expression (30), we can get

$$K^* = R^{-1} B^T P^* = \begin{bmatrix} 1 & \sqrt{3} \end{bmatrix} \tag{34}$$

Then,

$$\begin{aligned} V^* &= -K^* Z(t) = -Z_1(t) - \sqrt{3} Z_2(t) = -h(X) - \sqrt{3} L_f h(X) \\ &= -q_i - \sqrt{3}(b_0 e^{KTi} + b_1 q_i + b_2) \end{aligned} \tag{35}$$

$$\begin{aligned} u^* &= \frac{-\alpha(X) + V^*}{\beta(X)} = \frac{-L_f{}^2 h(X) + V^*}{L_g L_f h(X)} \\ &= -\frac{a_0 b_0 k e^{kT_i} T_i^4 + a_1 b_0 k e^{kT_i} T_i + a_2 b_0 k e^{2kT_i} + (a_5 b_0 k + b_1 b_0) e^{kT_i}}{a_4 b_0 K e^{KT_i} - b_0 K T_i e^{KT_i}} \\ &+ \frac{(a_3 b_0 k e^{kT_i} + b_1^2) q_i + b_1 b_2}{a_4 b_0 K e^{KT_i} - b_0 K T_i e^{KT_i}} - \frac{q_i + \sqrt{3}(b_0 e^{KT_i} + b_1 q_i + b_2)}{a_4 b_0 K e^{KT_i} - b_0 K T_i e^{KT_i}} \end{aligned} \tag{36}$$

It is the optimal control law of the original affine nonlinear control system.

3. CONCLUSIONS

By analyzing the real working environment of the drying-machine and all kinds of effects, dynamic mathematical model of the complicated drying-process was made. It is an affine nonlinear control model. An optimal control law was obtained by exactly linearizing the model with differential geometry theories and solving the Riccati equation. The final control law is satisfied the optimal quadratic performance index. It can make loss of the energy least and the control object ideal.

REFERENCES

Cai Jiabing. Estimatingreference evapotranspiration with the FAO Penman–Monteith equation using daily weather forecast messages. Agricultural and Forest Meteorology . 2007. 145, pp.22–35

Chua K.J, etc. A comparative study of different control strategies for indoor air humidity. Energy and Buildings,2007.39, pp.:537-545

Daskalov P.I,etc.Non-linear adaptive temperature and humidity control in animal buildings .Biosystems Engineering ,2006.93(1),pp.1-24

Garcia, Magali el al. Dynamics of reference evapotranspiration in the Bolivian highlands (Altiplano). Agricultural and Forest Meteorology. 2004.125, pp.67-82

Lu, Q etc. Decentralized nonlinear optimal excitation control. IEEE Transactions on power systems, 1996.11(4) ,pp.1957-1962

Lu, Qiang el al. Nonlinear control systems and power system dynamics. Kluwer Academic Publishers.Boston.2001

Mesquita L.C.S.,etc. Modeling of heat and mass transfer in parallel plate liquid-desiccant dehumidifiers. Solar Energy .2006. 80,pp. 1475–1482

Ríos-Moreno G.J.,etc. Modeling temperature in intelligent buildings by means of autoregressive models. Automation in Construction.2007. 16,pp.713–722

Soldatos A.G..etc. Nonlinear robust temperature humidity control in livestock buildings. Computers and Electronics in Agriculture.2005.49, pp.357-376

Stefan Heinrich,etc. Modeling of the batch treatment of wet granular solids with superheated steam in fluidized beds, Chemical Engineering and Processing, 1999. 38,pp.131–142

Yan Zhengyong,etc. Mathematical modeling of the kinetic of quality deterioration of intermediate moisture content banana during storage. Journal of Food Engineering.2008. 84,pp.359–367

Zhang yun,etc. Cold–humid ecological effects of the Sanjiang Plain. Ecology and environment,2004.13, pp.37-39

NUMERICAL SIMULATION OF NATURAL VENTILATION IN TYPICAL PIGGERY OF SOUTH-EAST CHINA

Mingwei Shen[1], Yong He[*], Feilin Hao[2]

[1] *Institute of Biosystems Engneering and Food Science, Zhejiang University, Hangzhou, Zhejing Province, P.R. China 310029, Email:shenhao@zju.edu.cn*

[2] *Dept. of Environment and public utility, China National Air Separation Plant Corporation, Hangzhou, Zhejing Province, P.R. China 310004*

[*] *Corresponding Author. Address: Institute of Biosystems Engineering and Food Science, Zheiang University, Hangzhou 310029, Zheijiang province, P.R.China, Tel:+86-0571-86971143, Fax:+86-27-86971143, Email:yhe@zju.edu.cn*

Abstract: In order to understand the airflow pattern of natural ventilation in typical piggery in South-east China, a three dimensional ventilation simulation study was carried out based on the Navier-Stocks Equation by the aid of the fluid dynamics CFX® software. The inlets are mounted in windward sidewall in equidistance along the ridge of the piggery, the airflow guiding passages are set along the roof, on which installed a lambdoidal coping. The sty was incorporated into the simulation model as porous medium, and the factor of outer wind velocity varying with the height was considered. The simulation results shows: there is a significant difference in mass flow of different location of inlets, the central ones is bigger than the side ones. For the guiding openings, the central, right and left account for 64.76%, 29.53% and 5.71% respectively. The airflow inside the piggery is not even, some anticlockwise and clockwise circumfluence are formed in different direction. The airflow on the vertical plane through the inlets are much intense than that in between. Inside the stys, there displays unbalanced and lower airflow velocity, the average airflow velocity in the right sty is higher than that of the left.

Key words: airflow simulation, natural ventilation, piggery

Please use the following format when citing this chapter:

Shen, M., He, Y. and Hao, F., 2009, in IFIP International Federation for Information Processing, Volume 295, *Computer and Computing Technologies in Agriculture II, Volume 3*, eds. D. Li, Z. Chunjiang, (Boston: Springer), pp. 2091 2099.

1. INTRODUCTION

The living environment has an important role in affecting health of pig that live inside, and further lead to the economical efficiency of pig production. So the environment control is one of the priorities in piggery management, good-functioned ventilation design could make air be mixed evenly, and help creating a comfortable environment through modulating humidity and air movement. At the same time, the ventilation design should avoid inadequate air velocity and turbulence, which would increase suspended granule, pathogen and accelerate diffusion of ammonia, sulfureted hydrogen, carbon dioxide, etc. and pollute and deteriorate the inside and outside environment.

The industrialized piggery usually equipped with natural and mechanical ventilation system. Forced ventilation mostly works in situation of that natural ventilation could not meet the environment demand such as extreme hot or humid case. In most time of a year, the natural ventilation plays a primary role with the consideration of cost and increased energy stress, so, how to exert the efficiency of the natural ventilation as much as possible deserves main concern.

There are many factors to be considered in natural ventilation, and some factors are highly inter-connected, which makes the theoretical research rather complex. The traditional research concentrates on experiments which compared the featured structural design of natural ventilation systems (M.P. Foster, et al., 1987; B.L. Brockett, et al., 1987; K.J. Andonov, et al., 1989 and 1994; M.T. Karimipanah, et al., 1994 and 1996; K. Andonov, et al., 2003,), they did experiments of characteristic piggery and special design natural ventilation systems, studied the method of ventilation control to the effect of the airflow pattern, etc. With the development of computing technology, the computational fluid dynamics (CFD) provide a powerful tool in simulation airflow movement for greenhouse and barn.

In order to have a better understanding of the typical ventilation arrangement on the effect of airflow in South East China region, based on Navier-Stocks Equation and by the aid of fluid dynamics CFX® software, three dimensional steady simulation were undertaken to analyze and evaluate the air movement inside the piggery. The understanding of the detailed airflow movement inside could help enhancing the environment management and make better living condition for the pig growth.

2. VENTILATION THEORY

The ventilation theory is based on Navier-Stocks Equation (Mingwei.Shen et al., 2004; CFX Handbook Ver. 5.6. 2002.), after setting up parameters of initial value and boundary conditions, the software CFX® is used in resolving the relations.

The existence of sty would undoubtedly affect the surrounding airflow intensity. The method that uses to describe porous medium "Darcy-Forcheimer" Equation is applied in this case to approximate this effect (Mingwei.Shen et al., 2004):

$$-\frac{\partial p}{\partial x_i}=\frac{\mu}{K}U_i+K_{loss}\rho\left|\vec{U}\right|U_i \tag{1}$$

$-\frac{\partial p}{\partial x_i}$ is pressure gradient; ρ is fluid density; μ is fluid dynamic viscosity; U_i is airflow velocity; K is medium penetration ability; K_{loss} is non-linear momentum lose parameter.

The piggery that constructs comparatively high to achieve better natural ventilation effect, to achieve a more precise simulation result, the outer wind velocity change along different height are also taken into consideration (A. Shklyar et al., 2004) in this case.

$$U_{wind}=\frac{u_{fric}}{K}\ln\left(\frac{z+z_0}{z_0}\right), \qquad u_{fric}=\frac{Ku_h}{\ln\left(\left(h+z_0\right)/z_0\right)} \tag{2}$$

U_{wind} is outside wind velocity; u_{fric} is frictional wind velocity; K is Von Karman constant (= 0.42), z, z_0 is the height in y direction, in which z_0 denotes ground surface coarseness , which takes the value of 0.01m (Bartzanas et al., 2004) with referring to agrarian surface; u_h denotes the reference wind velocity, which takes the average value of summer time in Hangzhou City; h is the height of reference wind velocity.

3. AIRFLOW SIMULATION

The simulation is taken in a widely used and typical piggery in South East China, which has span length is 8.0m, eave height 2.5m, ridge height 3.72m, roof gradient 4:12, transom window height 4.36m, piggery length 32m. The sty is assembled with the height of 1.0m, length of 3.35m, the width could be adjusted according to pig growth requirement. Fig.1 shows the structural view of the piggery, there lay equably 10 rectangular inlets in windward sidewalls, with the size of 1.6m×0.4m each. Outlets lay along the transom window, with the width of 0.4m and have lambdoidal protection cap above, airflow guide board are laid along the roof. The simulation is extended to a wide scope of 48.0m(Length)×40.0m(Width)×6.4m(Height) to achieve a more accurate estimation of the surrounding airflow movement.

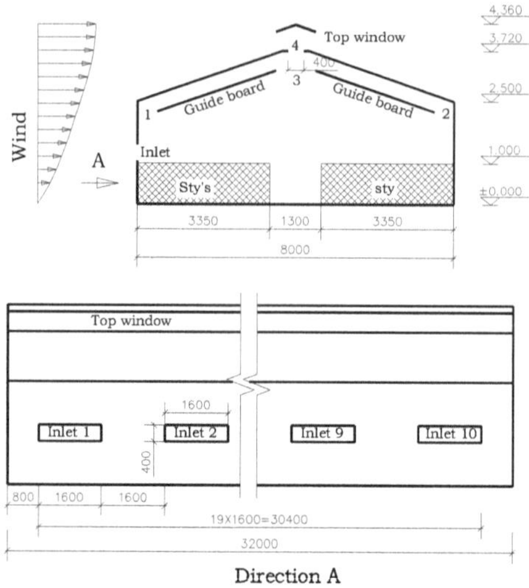

Fig.1: Schematic view of pig house

Limited cell gridding is the basic unit for CFD resolution, so in the first place, the space inside and outside of the piggery should be partitioned to form cells. Fig.2 shows the mesh in overall and partial region, it could be seen that the area within and surrounding ventilation windows (inlets and outlets) lie higher density of cells, which is in an effort to simulate the local airflow fluctuation better. In consideration of friction and cut effect, the cell are partitioned into thinner 5 layers in all walls, airflow guide board, sty boundaries, ground etc. to meet the non-slippage model.

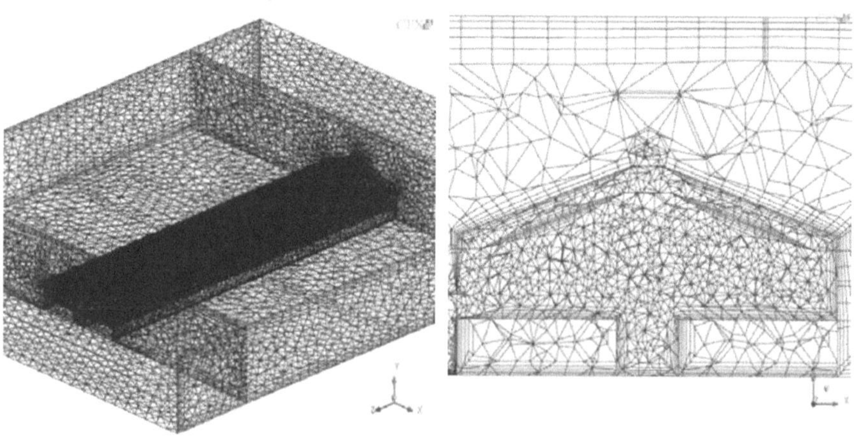

(a) Overall meshing arrangement (b) Local meshing arrangement

Fig.2: Schematic view of CFD mesh for pig house

In consideration of structural characteristic and environment condition, some basic parameters and intuition condition is set in table 1., the simulation result of different sections is shown in Fig.3 to 5.

Table 1. Some basic parameters and initial condition of the model

Variable	Value	Variable	Value
Initial value:		Computational mesh:	
Reference Pressure	101325 Pa	Mesh type	Tetrahedron mesh
Reference temperature	298.15 K	Maximum mesh size	1000mm(surrounding space of the piggery), 500mm(within the piggery), 100mm(within area of ventilation windows)
Turbulent degree	5%, medium	Mesh expanding control: control suface/layer/proportion factor	Walls including both inside and outside of the piggery, sty boundaries/5/1.3
u_h	1.85 m/s (average wind velocity of Hangzhou city)	Others:	
h	10.0 m (the height of anemoscope)	Turbulent model	K- ε
Boundary condition:		N-S Equation discrete format	Self-adaption two factorial windward format
Wall surface(fixed boundary) condition	Lubricity, non-slippage model	N-S Equation resolution method	Overall: AMG acceleration ILU iteration; velocity and pressure: SIMPLEC method
Inlet	Hexahedron windward surface of piggery periphery	N-S Equation resolution precision(average radical sign error)	RMS<0.0001
Openings(inlet&outlet)	Hexahedron leeward surface of piggery periphery		

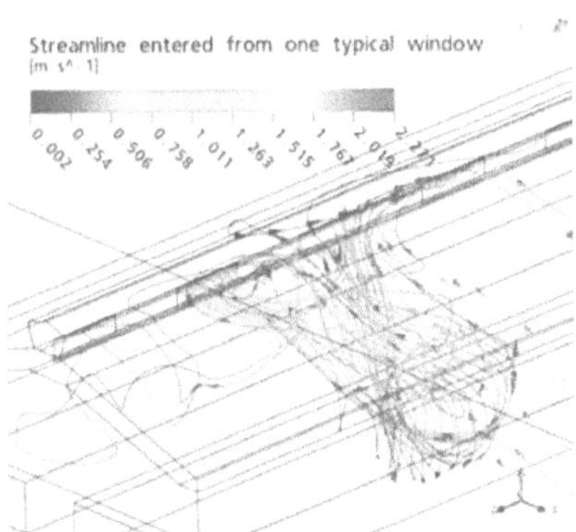

(a) Air entered from one typical window

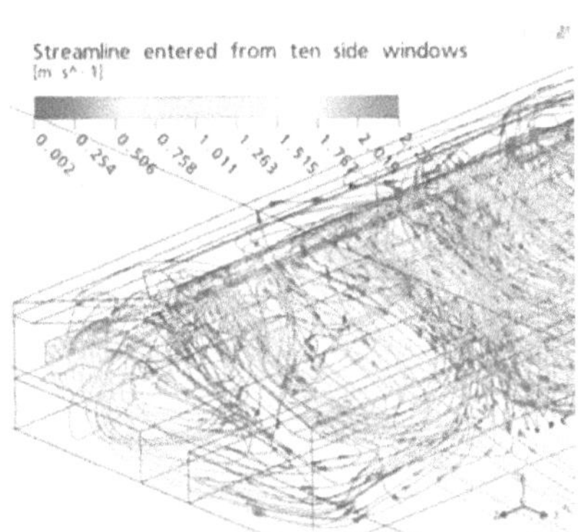

(b) Air entered from ten side windows

Fig.3: Simulated airflow streamlines entered from inlets

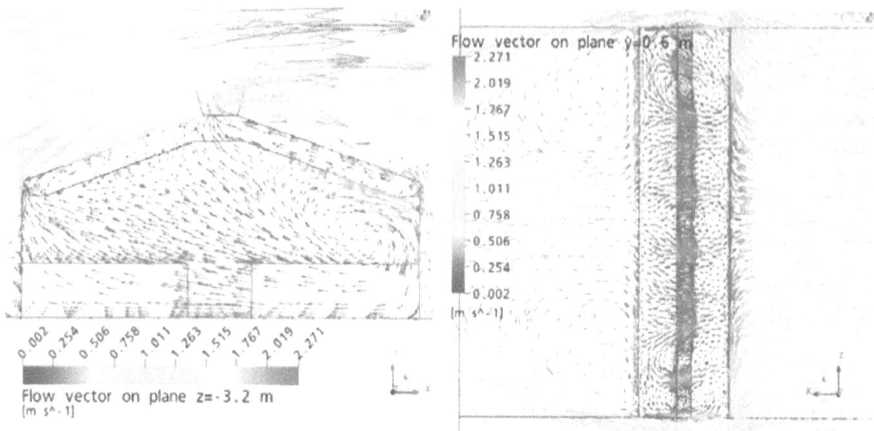

Fig.4: Simulated airflow vector on some typical sections

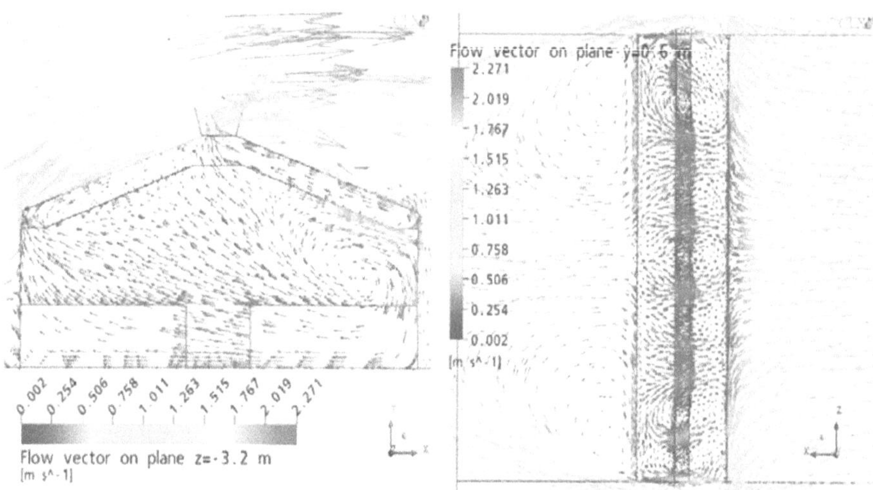

(b) Section through no windows (b) Horizontal section through pig columns

Fig.5: Simulated airflow vector on typical sections

Table 2. shows some statistical value of several typical regions, it could be seen that the different position inlets along the sidewall have different air flux, the central location has relatively higher and the outskirts lower, the maximum difference is 2.15% to total inlet flux, the average mass flux is 0.00682kg/s, The airflow displays symmetrical basically along the sidewall. The difference is obvious in three guiding opening above, in which 64.76% of total outflow is through the central one (flow guiding opening 3), 29.53% of total is through right one (flow guiding opening 2), only 5.71% is through left one (flow guiding opening 1). There is difference in airflow velocity inside the piggery, the area within the sty area is relatively slower and outside quicker, the average airflow velocity for outside area of the sty is 0.25m/s. The difference could also be observed in left and right side sty, the average airflow velocity is 0.21m/s and 0.16m/s for right and left sty respectively, it could still satisfy the requirement of pig growth in terms of airflow velocity (Derang Wu, 1994).

Table 2. Statistic data in some typical regions of the pig house

Ventilation window/region	Average mass flux/velocity (kg/s /m/s)	Ventilation window/region	Average mass flux/velocity (Kg/s /m/s)
Inlet 1	0.00589 kg/s	Inlet 10	0.00592 kg/s
Inlet 2	0.00672 kg/s	Flow guiding opening 4	0.23955 kg/s
Inlet 3	0.00699 kg/s	Flow guiding opening 1	0.00436 kg/s
Inlet 4	0.00723 kg/s	Flow guiding opening 2	0.02252 kg/s
Inlet 5	0.00722 kg/s	Flow guiding opening 3	0.04938 kg/s
Inlet 6	0.00736 kg/s	Upper guiding aisle of piggery	0.159 m/s
Inlet 7	0.00722 kg/s	Inside space of piggery	0.251 m/s
Inlet 8	0.00700 kg/s	Left side of sty	0.166 m/s
Inlet 9	0.00666 kg/s	Right side of sty	0.214 m/s

Fig.3(a) shows the airflow movement coming through the inlet 2 (in order to be more clearly displayed, the airflow from the other inlet are hidden in this case). Fig.3(b) is the overall airflow movement inside the piggery. It could be seen the overall airflow is basically symmetrical along the ridge, while the airflow coming through inlet 1 and 10, the inlet 2 and 9, and the remaining inlets are different. The airflow from the inlet 1 or 10 is divided into two branches, of which one remains at the height of inlet and a bit of leaning to the nearer frontispiece, flow in the direction to the opposite sidewall, it decreases to about 1/2~1/3 when it arrives the right sty, then turns back, deviates to the nearer frontispiece and flow upward, finally the majority are flow out through central guiding opening , the remain is out through the left guiding opening; The other that flow downward also lean a bit to the nearer frontispiece, flow into the left sty in about half length of left sty, continue flow forward to the right sty, the airflow velocity decreases gradually to about 1/10 of the inlet, turns back and deviates to flow upward , finally the majority of airflow move out through central guiding opening and the remain through the left one. There presents two obvious circumfluence which permeate most space of the piggery, the airflow in the left sty is less than the right one, the distribution is not even.

Fig.3 and Fig.4(b) shows the airflow movement coming from inlet 2 and 9, it could be seen that the incoming air flows slowly and slightly to the neighboring frontispiece, on the whole, it does not flow into the left sty, the part that airflow influx into right sty are distributed in partial above space.

When the airflow arrive at the opposite sidewall, the majority turn back and upward to flow out of the piggery through right guiding opening, the remaining which has velocity less than 0.1m/s form large and mild turbulence, finally flow out the central and right guiding opening.

The airflow movement inside the piggery coming from inlet 3 to 8 is approximately similar. The incoming airflow deviates slightly in horizontal plane (XY plane) and flows in a direction of upward. These airflow flow through mostly above the stys, when arriving the opposite sidewall, some flow forward and out from right guiding opening, much more airflow turn downward to flow into right sty, in which some turn back to flow forward, some flow through sty, finally flow out from piggery through central and left guiding openings. The intensity of airflow movement in left sty is less than that of the right one, but much more even.

The Fig.5 shows the airflow movement on vertical plane between two inlets and horizontal plane 0.6m above the ground, the one of 0.6m is selected to better represent the pig growth living area. The overall airflow in vertical plane through the inlet is more intense than that of in between one, which is more so in two side of the piggery.

4. RESULT AND DISCUSSION

From the simulation result, the airflow movement is basically symmetrical along the ridge direction. There is difference in mass flow of different location of inlets, the central ones is bigger than the side ones, the maximum difference is 2.15% to total inlet flux.

The three guiding opening is functioning differently, in which the central one accounts for more than half of the total airflow flux, the right one lays second and that is less than the half of the central one, the left one the is least and that is about 1/11 of the central one.

There shows three different fluxions after air flow into the piggery, the air coming into from inlet 1 and 10 have the characteristics of downward and inclining to nearer frontispiece, the air that coming into from inlet 2 and 9 would flow downward, some small part would flow in the direction to the nearer frontispiece, the anticlockwise circumfluence are form in opposite Z direction. The air coming from inlet 3 to 8 see less deviation from the horizontal plane (XY plane), move slightly upward and forms clockwise circumfluence in the direction of opposite Z direction.

The airflow inside the piggery are not even, the airflow on the vertical plane through the inlets are much intense than that in between, the area of the middle aisle is the most intensive. In consideration of the two side stys, the average airflow velocity in the right side is higher than that in the left, all could meet the airflow requirement of pig growth.

From the overall ventilation status, the airflow movement is much higher and even in non-living surroundings of the piggery, and shows lower velocity and unbalanced inside the stys, which could arose dirt, accelerate production and diffusion of poisonous gas, and pollute the environment both inside the piggery and outside. There still exists some place that need to be improved in terms of ventilation design.

ACKNOWLEDGEMENTS

Funding for this research was provided by the Natural Science Foundation of Zhejiang Province, China (Project NO: Y307166). Education Office of Zhejiang Province, China (Project NO: 20070128).

REFERENCES

A. Shklyar, A. Arbel. Numerical model of the three-dimensional isothermal flow patterns and mass fluxes in a pitched-roof greenhouse. Biosystems Engineering, 2004, 88(4):479~940

B.L. Brockett, T.D. Albright. Natural ventilation in single airspace buildings. Journal of Agricultural Engineering Research, 1987, 37:141~154

CFX. 2002. CFX Handbook Ver. 5.6. U.K. CFX Company.

Derang Wu. Agricultural Architectonics[M], Agricultural Publishing House, Beijing, 1994

K. Andonov, P. Daskalov, K. Martev. A new approach to controlled natural ventilation of livestock buildings. Biosystems Engineering, 2003, 84(1): 91~100

K.J. Andonov, P.I. Daskalov, K.V. Martev. Microclimate system for livestock building with controlled narural ventilation. Agricultural Engineering, 1989, 2:55~62

K.J. Andonov. An aeration configuration in a stock breeding building. Agricultural mechanization, 1994, 5:16~18

M.P. Foster, M.I. Down. Ventilation of livestock buildings by natural convection. Journal of Agricultural Engineering Research, 1987, 37:1~13

M.T. Karimipanah, M. Sandberg. Deflection and influence of room-size of a two-dimensional wall jet in a ventilated room. Proceeding of ROOMVENT'94, Fourth International Conference on Air Distribution in Rooms, 15~17, June, 1994, Krakow, Poland

M.T. Karimipanah, M. Sandberg. Maximum velocity of return flow close to the floor in a ventilated room-experimental and numeric results. Proceedings of the 17th AIVC Conference, 17~20, September, 1996, Gotenburg, Sweden

Mingwei.Shen, Feilin.Hao. Simulation of natural ventilation of two wind direction in tunnel greenhouse. Journal of Agricultural Engineering, 2004, 20(6): 227~232

tation.

THE DESIGN OF RURAL POWER NETWORK POWER QUALITY MONITORING AND ANALYSIS PLATFORM ON LABVIEW

Chunling Chen*, Xiaofeng Wang, Tongyu Xu, Yong Yang
College of Information and Electrical Engineering, Shenyang Agricultural University, Liaoning province, P.R.China 110161
* *Corresponding author, Address: College of Information and Electrical Engineering, Shenyang Agricultural University, shenyang 110161,Liaoning province, P.R.China , Tel: +86-024-88760090, Fax: +86-024-88487122, Email:snccl@163.com*

Abstract: A design of power quality monitoring and analysis platform based on the virtual instrument technology is proposed which aims at the key issues affecting quality rural power grids. This article discusses the typical algorithm of harmonic analysis, voltage deviation calculation and unbalance in three-phase calculation. Completing software design of power quality monitoring and analysis platform based on LabVIEW, which including visual design algorithm of the monitoring and analysis, graphical display of the calculation and analysis of the results, and the results of the export data. Simulation results show that the platform test results are accurate, friendly interface, stable performance and practical.

Keywords: Power quality; LabVIEW; Harmonic analysis; Voltage deviation; Unbalance in three-phase

In recent years, with the development of rural economy and the country's industrial structure adjustment, township enterprises have continuously increased and a large number of industrial and mining enterprises transferred from the city to the rural areas(Chen et al., 2007). The non-linear, non-symmetry and volatility in the rural power network has become increasingly serious and caused many incidents such as electrical burning, electric energy measurement error, overheated transformers, capacitors can not work in normal operation, relay protection and automatic device malfunction tripping, and causing serious harm to the rural power network, users' electrical equipments and all kinds of appliances. In order to ensure the rural power

Please use the following format when citing this chapter:

Chen, C., Wang, X., Xu, T. and Yang, Y., 2009, in IFIP International Federation for Information Processing, Volume 295, *Computer and Computing Technologies in Agriculture II, Volume 3*, eds. D. Li, Z. Chunjiang, (Boston: Springer), pp. 2101–2109 supply system and the majority of users of electrical equipment and all kinds of appliances are running safely and economically, it is necessary to monitor and analysis each index of the power quality, therefore it can provide accurate historical data and basic data to project construction and accident analysis.

1. DESIGN OF HARMONIC MONITORING PLATFORM

Power quality monitoring system of rural power network is designed achievement by using virtual instrument technology, with the least hardware supporting, and uses software programming instead of hardware circuit to complete the functions of data processing, calculation, analysis and results display, so it changes the idea from the hardware to the software(Ding et al., 2004). The framework chart of this system include software and hardware system structure. The software part has power quality analysis and display functions. The hardware part mainly includes transducer, signal adjustment and data acquisition card, the system architecture in Fig.1.

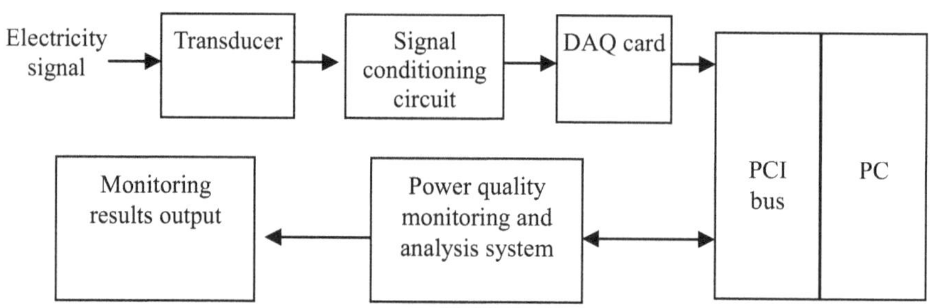

Fig. 1. The power quality monitoring platform topology of rural power network

1.1 Hardware circuit

The detailed processes of PC-based data Acquisition system and interpolating board as follows: External signal input → Signal connect terminals → Signal sensors → Signal conditioning (Signal amplification → Isolation) → Acquisition.

In order to obtain good results and effective measurement accuracy, transducer and transmitter output signals must be adjusted before it access to the data acquisition card. Signal conditioning typically include signal amplification, filtering, electrical isolation and multi-channel technology. Passband filter is used to prevent the noise and interference from the aliasing, overcome the spectrum-aliasing which may lead to the distortion. Data acquisition card (DAQ) is the key to hardware devices, and the accuracy and speed of the conversion have a major impact on system real-time and accuracy. The system chooses LingHua DAQ2010 interpolating data acquisition card, which based on PCI, bus structure, with the advantages of speed, technical characteristics: 14-bit A/D resolution, 4 channels simultaneous analog input, the maximum sampling frequency is up to 2MHz. According to the sampling theorem, the sampling rate should be at least equal to or greater than twice of the highest frequency, the data acquisition card is sufficient to meet the requirements of 50th harmonic analysis.

1.2 Design of system software

This system uses LabVIEW, a graphics-based programming language development environment from National Instrument. LabVIEW uses flow chart programming, the style of programming is simple and effective, and very direct-viewing, LabVIEW have abundant powerful functions of numerical analysis, function and signal processing.

Using virtual instrument technology as the core, and PC-BASED mode to construct power quality monitoring platform, it can take full advantage of the PC resources. Especially when system requirements change, or when the monitoring system used in different environments, we only need to carry out some necessary structural adjustments, amendments and supplements, without change hardware, thus it can save development time and the costs(Hu et al., 1997; Hu et al., 2007).

Power quality monitoring and analysis software application modules are shown in Fig.2. The software modules include: parameters setting module, real-time data acquisition and display, software analysis module. Software parameters setting module includes sensor variable ratio, data acquisition card sampling frequency, data storage path and power quality monitoring relevant parameters, etc. Software analysis module is the core of this software, including harmonic analysis, voltage deviation calculation, unbalance in three-phase and exporting the harmonic analysis data.

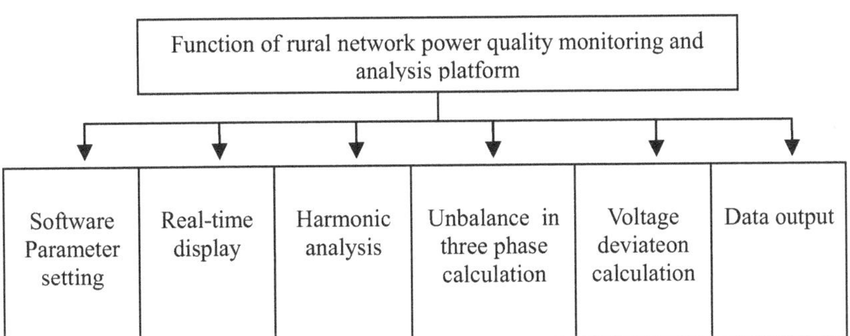

Fig. 2. Function diagram of software

2. THE MAIN ALGORITHMS OF POWER QUALITY MONITORING PLATFORM IN LABVIEW

2.1 Voltage deviation(Luo et al., 2006; Pang et al., 2003)

Sampling according to the sample rate and calculating voltage deviation ΔU per minute.

$$\{\Delta U\}_{(\%)} = \frac{V - V_N}{V_N} \times 100 \quad (1)$$

The system will cumulate monitoring voltage time, calculate the percent of pass everyday, and obtain the maximum positive and negative from statistics.

2.2 Harmonic analysis(Luo et al., 2006; Pang et al., 2003)

After the signal being pretreatment and A/D convert acquisition, it uses Discrete Fourier Transform (DFT) to Fast Fourier Transform (FFT) basic theory to calculate the voltage and current harmonics. The continuous-time signals *f(t)*, divide a sampling period *T* into *N* equal points, and sampling period is $T_s=T/N$, it will obtain the discrete time signal, the discrete time point is $t = k\,T_s$, (k=0,1,2,···,N-1), and the sampling value is f_k, so

$$\dot{F}_h = \frac{1}{N}\sum_{k=0}^{N-1} f_k W^{kh}, h = 0,1,2,\cdots, \mathrm{N}-1 \quad (2)$$

In the formula (2), F_h is the complex value of each harmonics, fundamental waves and DC components. So we can deduce the occupancy of the *h*^th-harmonic of voltage and current:

$$HRU_h = \frac{V_h}{V_1} \times 100, (\%)$$
$$HRI_h = \frac{I_h}{I_1} \times 100, (\%) \quad (3)$$

THD of voltage and current:

$$THD_u = \frac{\sqrt{\sum_{k=2}^{N/2} V_h^2}}{V_1} \times 100, (\%)$$
$$THD_i = \frac{\sqrt{\sum_{k=2}^{N/2} I_h^2}}{I_1} \times 100, (\%) \tag{4}$$

Because FFT algorithm have barrier effect and spectrum leak in the harmonic analysis module, this problem affects the harmonic measurement accuracy. So it should compensate the original signal with adding window function, it can reduce the leak effect. LabVIEW provides us 12 window functions such as Hanning and Hamming window.

2.3 Unbalance in three-phase

This device use component of symmetry to calculate unbalance in three-phase, and divide the unbalance voltage into positive-sequence, negative-sequence and zero-sequence, because the calculation of the harmonic sequence component is very complicated, so here only considering the fundamental-wave signal sequence component. Using FFT to obtain three-phase fundamental-wave voltage, the following formulas are to calculate each voltages and unbalance in three-phase.

$$\begin{vmatrix} V_0 \\ V_1 \\ V_2 \end{vmatrix} = \frac{1}{3} \begin{vmatrix} 1 & 1 & 1 \\ 1 & e^{-j120^0} & e^{-j240^0} \\ 1 & e^{-j240^0} & e^{-j120^0} \end{vmatrix} \begin{vmatrix} V_a \\ V_b \\ V_c \end{vmatrix} \tag{6}$$

$$\varepsilon = \frac{|V_2|}{|V_1|} \times 100\% \tag{7}$$

3. CONSTRUCTION OF THE POWER QUALITY MONITORING PLATFORM

In LabVIEW platform, using virtual instrument to construct the power quality monitoring platform, Fig.3 shows a part of Block-Diagram of the monitoring platform—Harmonic analysis module

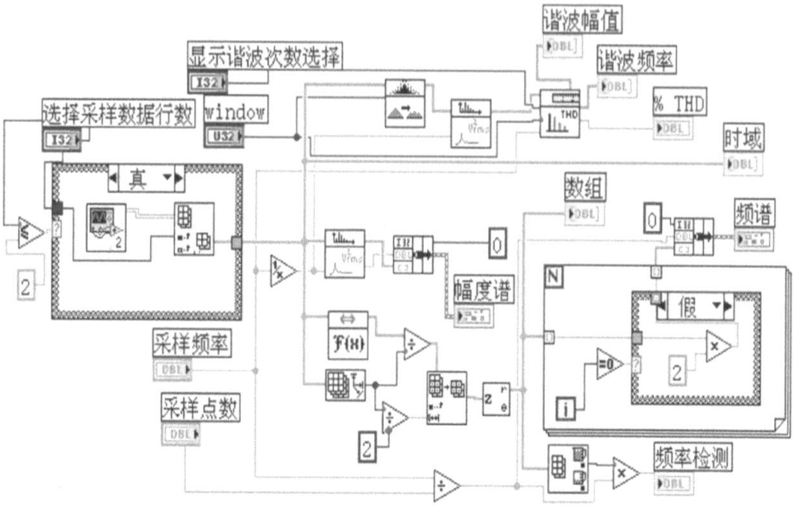

Fig. 3. Diagram of rural power network power quality analysis

4. SIMULATION RESULT ANALYSIS

Real-time waveform of the three-phase voltage and current, is shown in Fig.4.

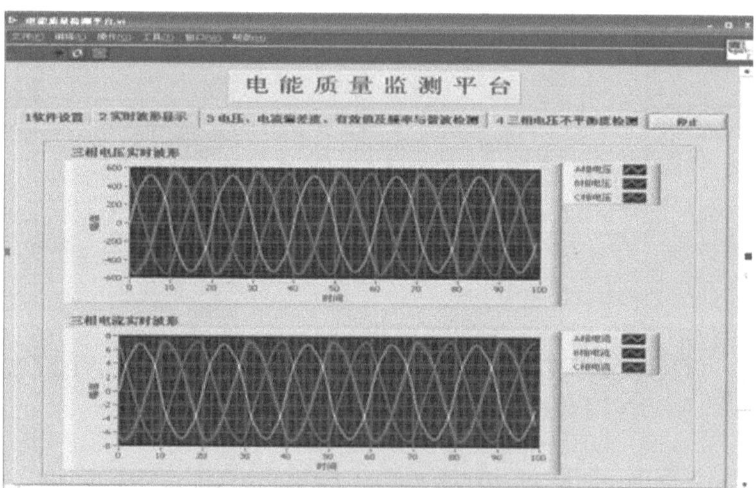

Fig. 4. Interface of real-time waveform

Here it measuring and analyzing 1-phase voltage, as the result in Fig.5. The voltage U1 has obvious distortion, and its corresponding FFT spectrum display the 3rd, 5th, 7th and 9th-harmonic, the THD reaches 13%. It is also can choose any times harmonic frequency or amplitude, and with the graphics and data showing the voltage frequency, effective value and degree of voltage deviation. With the powerful LabVIEW, we can export spectrum of harmonic analysis with data file, display harmonic amplitude. The specific data is shown in Table1.

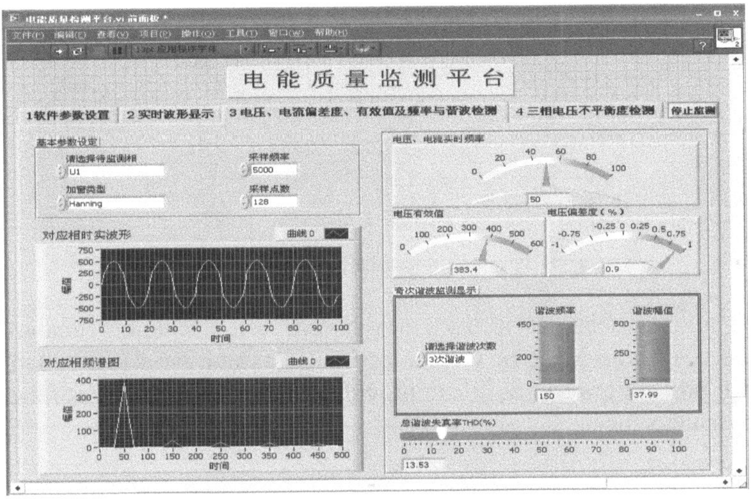

Fig. 5. Interface of harmonics analysis

Table 1. Harmonic amplitude

Harmonic times	Harmonic frequency（Hz）	Harmonic amplitude（V）
1	50	383.40
3	150	37.99
5	250	26.60
7	350	18.99
9	450	11.39

Running the platform with disconnecting one of the load, and measuring voltage unbalance in three-phase, the result shown in Fig.6. The three-phase voltage effective value, vector graph and THD show the voltages are complete balance; while there is no current B, so the THD reaches up to 53%.

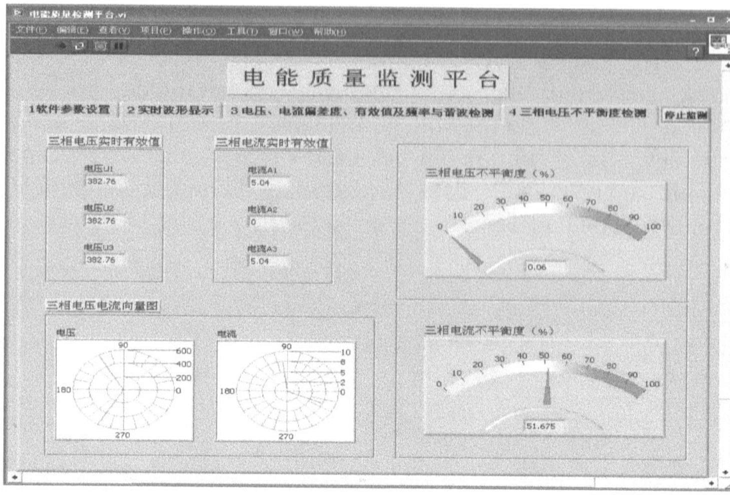

Fig. 6. Interface of unbalance in three-phase analysis

Simulation results show that this system can accurately analyze the unbalance in three-phase of voltage and current, and the actual situation of the rural power network. So it can provide reliable scientific basis for power quality management.

5. CONCLUSION

The system uses "Data acquisition card + PC + LabVIEW" to construct the power quality monitoring and analysis platform of rural power network, is flexible and open. In terms of hardware, it needs configure different accuracy and speed of the DAQ card according to the requirements; In the terms of software, it can take full advantage of LabVIEW's powerful signal processing and analysis, variety of integrated measures of power quality measure and analysis; we also can construct a user-friendly, feature-rich power quality analysis and management platform base on PC resources. This platform has a broad prospect to against the rural power network which power quality increasingly complex.

REFERENCES

Chen Xihui, Zhang Yinhong. LabVIEW 8.20 Programming[M]. Beijing:Tsinghua University Press, 2007

Ding Yifeng, Cheng Haozhong, Zhan Yong, Sun Yibin, Yan Jiangyong. Present status and development in power quality monitoring. Electric Power, 2004, 37(7):16-19

HU Guangshu. Digital Signal Processing[M]. Beijing:Tsinghua University Press, 1997

HU Qian, TANG Zhen-zhou. Electric harmonic test system based on virtual instrument. Electronic Measurement Technology,2007,30(2):90-92

Lei Zhenshan. LabVIEW 7 Express Technique Tutorial[M]. Beijing: China Railway Department Press, 2004

Luo An. Control of power network and reactive compensate technical and equipment[M]. Beijing: Electric Power, 2006

Pan Wen, QIAN Yutao, ZHOU E. Power Harmonics Measurement Based on Windows and Interpolated FFT(I) Transactions of China Electrotechnical Society,1994,9(1):50-54.

Pang Hao, LI Dong-xia,ZU Yun-xiao. An Improved Algorithm for Harmonic Analysis of Power System Using FFT Technique[J].Proceedings of the CSEE,2003,23(6):50-54.

PRC National Standard. GB/T14529-93, Power Quality-Public power network harmonic, Beijing, National Bureau of Technical Supervision, 1993

Wu Jingchang. Harmonic of Power System[M]. Beijing: China Electric Power Press, 1998.

Xiao Xiangning, XU Yongmei. Problems and Comprehensive Control of Power System Harmonics[J].Electric Power,1998,31(4):59-61

Zhan Yong, Cheng Haozhong, Ding Yifeng. Design and development of power quality monitoring analysis system. HuaDong Power, 2004, 32(10): 10-14

Zhao Chengyong, GAO Benfeng, JIA Xiufang. Comprehensive power quality detecting system based on LabVIEW. Journal of North China Electric Power University, 2006, 33(2):63-67

Zhao Wenchun ,MA Wei-ming, HU An.FFT Algorithm with High Accuracy for Harmonic in the Electric Machine[J]. Proceedings of the CSEE, 2001,21(12):83-87

Zhou Zhiyu, LI Yuneng, GUO Songmei. Design and Implementation of On-line Harmonics Monitoring System Based on Technology of Virtual Instrument. Process Automation Instrumentation）, 2005, 26(12):34-36

DEVELOPMENT OF AN OPTICAL DEVICE TO INVESTIGATE CHLOROPHYLL CONTENT OF TOMATO LEAVES

Di Cui, Minzan Li*, Xiuhua Li
Key Laboratory of MOE on Modern Precision Agriculture System Integration Research, China Agricultural University, Beijing 100083, China
* *Corresponding Author- Tel: +86-10-62737924, Email: limz@cau.edu.cn*

Abstract: Chlorophyll content is an important indication for evaluating crop growth status and predicting crop yield. The NDVI (Normalized Difference Vegetation Index) is commonly used as an indicator in practical crop healthy monitoring. Hence, a spectroscopy-based device for indirectly measuring crop growth conditions in terms of NDVI is developed. This device consists of four channels: two are designed to measure the intensity of the sunlight and the other two are used to measure the reflected light from the crop canopy at the same time. An electronic control unit was designed to control the sensing and data recording processes, as well as to calculate the NDVI based on the sensed data. The measurable two wavelengths are 610 nm and 1220 nm. A series validation tests, comparing the measurement result against spectroradiometer readings, are conducted to evaluate the performance of the device. Leaf samples are collected to measure chlorophyll contents in laboratory. The correlation coefficient between the NDVI readings from the developed device and the chlorophyll content data measured by the UV-VIS Spectrophotometer reaches 0.81, which shows that the device can be used in practical crop management.

Keywords: NDVI, crop growth condition, chlorophyll content, sensor

1. INTRODUCTION

Crop chlorophyll content is an important index used to evaluate the growth status and to predict the yield and the quality for many crops.

Please use the following format when citing this chapter:

Cui, D., Li, M. and Li, X., 2009, in IFIP International Federation for Information Processing, Volume 295, *Computer and Computing Technologies in Agriculture II, Volume 3*, eds. D. Li, Z. Chunjiang, (Boston: Springer), pp. 2111–2118.

Previous studies revealed that chlorophyll had a great effect on the spectral reflectance of crop leaves (Cartelat et al., 2005). When the nitrogen content is deficient in crop plants, the chlorophyll level in those plants will go down. It will cause a reduction in leaf spectral reflectance at NIR band and an increase at visible band (Li et al., 2006). Such a phenomenon makes it possible to estimate crop nitrogen stress in terms of the measurable leave spectral reflectance of crop canopy.

Since 1970s, scientists have conducted many basic researches to evaluate crop growth condition by using spectral reflectance of leaf or canopy. And researches on developing ground-based sensing systems for detecting plants nitrogen content have been reported. Stone et al. (1996) reported the development of optical sensors for detection of nitrogen availability in winter wheat plants. They reported that that device could provide a close correlation between the plant nitrogen contents and a plant nitrogen spectral index (PNSI) calculated as $|(NIR+red)/(NIR-red)|$. Kim et al. (2001) investigated applicability of using a multispectral camera as a ground-based sensor for assessing corn plant nitrogen stress, also found it could provide a reasonable correlation between NDVI and leaf chlorophyll content. Sui et al. (2005) developed a device for detecting nitrogen status in cotton by measuring the spectral reflectance of cotton canopy in four wavebands (blue, green, red, and NIR) with modulated illumination to the canopy under the Sunlight. Min et al. (2006) investigated the use of a visible and near infrared spectroscopy to estimate nitrogen concentrations of Chinese cabbage. Campbell et al. (2007) investigated the use of reflectance (R) and fluorescence (F) information to monitor the vegetation stress responses. Heege and Thiessen (2002) developed an on-the-go control system for site-specific nitrogen top dressing in terms of measured crop reflectance. Noh et al. (2005) developed an equipment-mount multispectral image sensor to detect corn plant nitrogen deficiency in real-time for supporting sensor-based variable-rate nitrogen side dressing.

On the other hand, it is necessary to develop a portable detector for small scale farm in China. In order to create a handy device for in-field use, the overall objective of this study was to design a spectra-based hand-hold device to measure spectral reflectance of crop canopy and the sunlight at Red and NIR wavebands, and then calculate NDVI to indirectly detect the chlorophyll content in crop plants.

2. METHODOLOGY AND PROCEDURE

2.1 Theoretical background for design

Several vegetation indices have been developed for revealing crop growing condition information. A widely used index is the NDVI (Normalized Difference Vegetation Index). NDVI is calculated as a ratio differences between measured canopy reflectance in the red and near infrared bands, respectively. These two spectral bands are chosen because they are mostly affected by the absorption of chlorophyll in leafy green vegetation. The equation calculating NDVI is as follow (Tucker,1979).

$$\mathrm{NDVI} = \frac{R_{\mathrm{IR}} - R_{\mathrm{R}}}{R_{\mathrm{IR}} + R_{\mathrm{R}}} \quad (1)$$

where, R_{R} and R_{IR} are canopy reflectance data in the red and near infrared bands, respectively.

2.2 Sensing unit design

In order to obtain R_{R} and R_{IR} data shown in Equation (1), an optical sensing device was designed, It consisted of four channels: two were designed to measure the intensity of the sunlight and the other two measured the reflected light from the crop canopy at the same time. An electronic control unit was designed to control the sensing and data recording processes, as well as to calculate the NDVI based on the sensed data. Figure 1 illustrates the system schematics of the designed device.

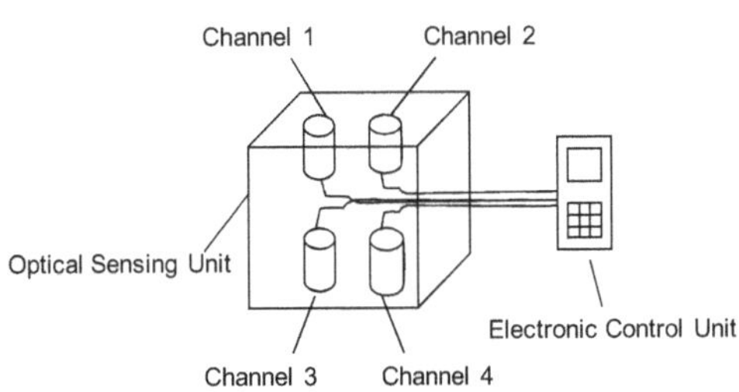

Figure1. The system schematics of designed device

2.2.1 Design of optical sensing unit

As introduced above, the designed optical sensing device used four channels for acquiring light reflectance data (Figure 1). Those sensing channels were designed identically, each consisting of an optical window for collecting the light, a filter for transmitting only the light of specific wavelength, a lens for making the light get together, and a transducer for converting the optical signal into electrical signal. The structure of each channel is shown in Figure 2.

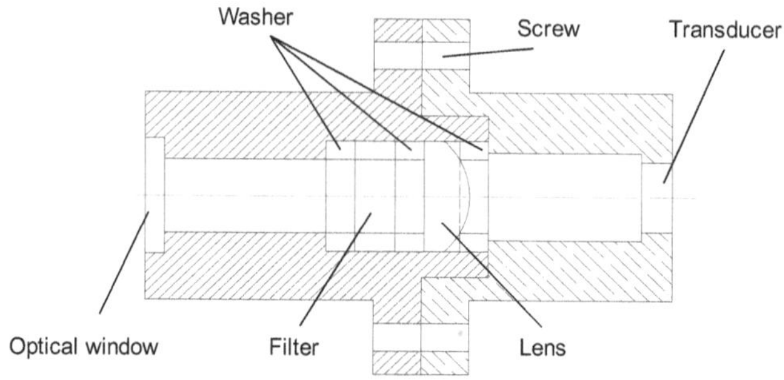

Figure2. The structure of each channel

The optical windows of two channels were configured upwards to measure the intensity of the sunlight, and the other two windows pointed downwards to sense the intensity of the reflected light from plant canopy. Each pair of the opposite arranged channels equipped the same filters and transducers. Since the 610nm-centered red band and 1220nm-centered NIR band were reported being sensitive to chlorophyll content variation of crop plants (Xue et al., 2004), these two bands were chosen for constructing the optical sensing unit. The type of the filters in channel 1 and 3 (NIR band) was 1220±6 nm center wavelength of 20 mm diameter, and the type of transducers was InGaAs photodiode with a light active area of 1.5×1.5 mm^2 and the spectral response range of 1000-1650 nm. Similarly, in channel 2 and 4 (red band) the type of filters was 610±6 nm center wavelength of 20 mm diameter, and the type of transducers was Silicon photocell transducers with a light active area of 5×5 mm^2 and a spectral response range of 400-1100 nm. To reduce the influence of solar elevation angle on transducer outputs, a milky diffuse glass was used as the optical window for two up-view channels.

2.2.2 Design of electronic control unit

The electronic control unit was designed to control the sensing process, to convert the light induced electric currents in optical sensing elements to voltage signals, and to calculate crop NDVI in terms of the recorded voltage signals. Figure 3 shows the system schematics of the unit.

The converter changed the electric current into the voltage and the amplifier changed the amplitude of voltage into 0-5V. The A/D transducer converted this analog signal into the digital one, so that the digital signal could be measured by the microprocessor. The microprocessor measured the voltages of the four channels, calculated the NDVI using these four values and sent the results to the LCD.

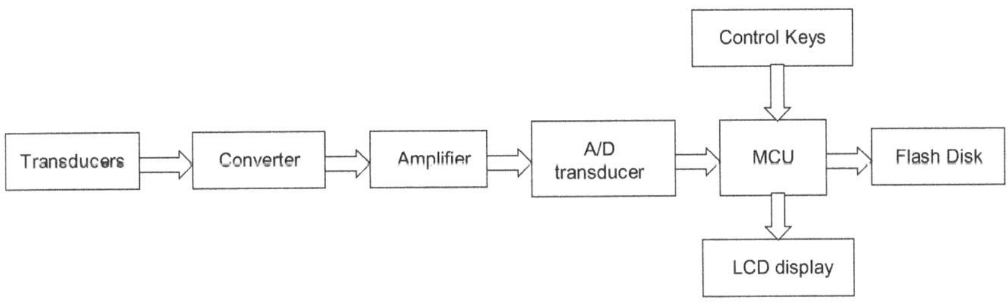

Figure 3. The system schematics of the electronic control unit

3. TESTS AND RESULTS ANALYSIS

3.1 Repeatability analysis

Before using the developed device in detecting crop plant reflectance to indirectly measure crop nitrogen stress, a performance test was conducted to check the repeatability of the device in measuring the known reflectance. According to the design requirement, the standard deviations (σ) of the measured data should be less than 0.1. In this performance test, device data were collected in a 20 minutes interval between 10:00 AM and 15:40 PM. Each measurement was repeated three times and the average was taken as the measured datum. A white panel of 100% reflectance and a grey panel of 35% reflectance were put on the ground out of the shadow as the known reference. The device was perpendicularly positioned to the panel at 10 cm above the panels.

As shown in Figure 4, the measured reflectancedata at the NIR and red bands are closely distributed within a small range around 1.0 between 10:00 AM and 14:40 PM, but those data experienced a dramatic increase after 5:00 PM. Through visual observation, the intensity of sunlight was noticeably decreased after 15:00 PM. In addition, since a milky diffuse glass was used as the optical window for two up-view channels, when intensity of sunlight decreased, the light entered the up-view channels would much more decreased than that in down-view channels. This result revealed that without any additional compensation the developed device could provide a reasonable repeatability when the sunlight intensity is above a certain threshold value.

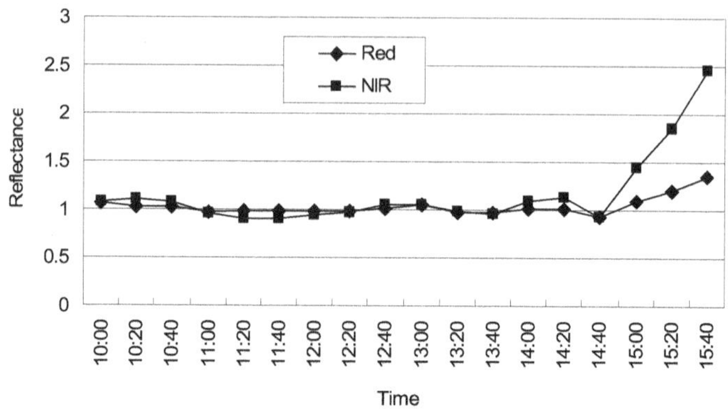

Figure 4. Measured reflectance of the white panel at different time of the day

As expected, the measured reflectance of the gray panel was found constantly lower than that of the white panel (Figure 5). However, a similar pattern to white panel result was found in this test that the measured reflectance data at NIR and red bands were around 0.35 between 10:00 AM and 14:40 PM, and significantly increased after15:00 PM. The standard deviation (σ) of measured reflectance was 0.027 if excluding the last three measurements. Such result showed that the device satisfied the design requirement when the sunlight intensity was above a certain threshold value.

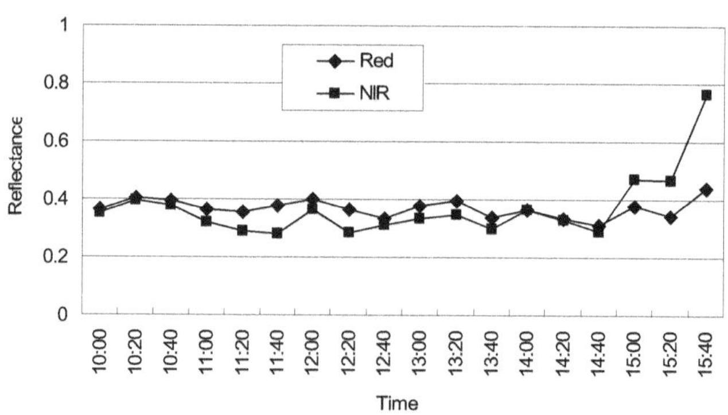

Figure 5. Measured reflectance of the gray panel at different time of the day

3.2 Field test results analysis

A series of field tests were performed to evaluate performance of the developed device in a greenhouse environment. The instrument was placed about 10 cm above the crop plants. Total 24 samples were measured in this evaluation test. After measurement, leaf samples from the testing plants were collected for measuring the chlorophyll contents in a laboratory procedure. In this laboratory test, 0.4 g leaf sample was filled in test tubes with 25ml solution mixed acetone and ethanol in a 2:1 ratio. After place those tubes in dark room for 24 hours to dissolve the chlorophyll in the solution, the chlorophyll content of those processed samples were measured using an UV-VIS Spectrophotometer.

The correlation coefficient between the NDVI readings from the developed device and the chlorophyll content measured by the UV-VIS Spectrophotometer is shown in Figure 6 (r=0.81). The result shows that the NDVI readings from the developed device had a positive linear correlation with the chlorophyll content of tomato leaves, and the developed device could be used in practical measurement because of high correlation coefficient.

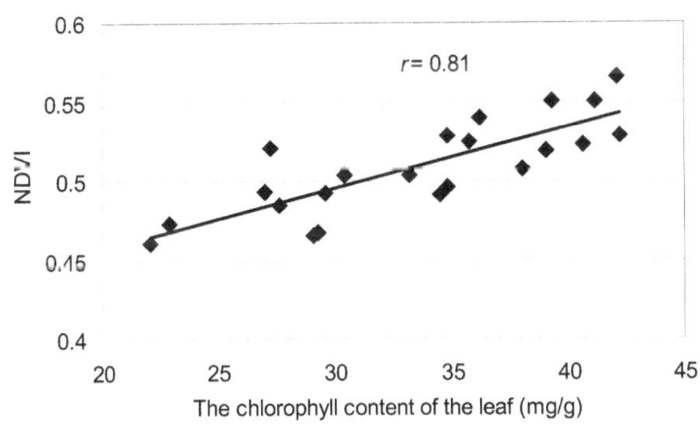

Figure 6. Correlation between the chlorophyll contents measured in laboratory procedure and NDVI readings from the developed device

4. CONCLUSION

A hand-held multispectral sensing device of crop chlorophyll content was developed. The initial test results indicated that the device could satisfied the design requirement (σ<0.1) in field data collection. And the test results obtained in both laboratory and field environment showed that it could be used to detect the chlorophyll content level of tomato plants. Through calculating, the correlation coefficient between the NDVI readings from the developed device and the chlorophyll content data measured by the UV-VIS Spectrophotometer was r=0.81, which was high enough to be used in practical use. More intensive field tests is needed to verify and improve the developed device for its capability and limitation for actual field data collection.

ACKNOWLEDGEMENTS

Funding for this study was provided by National High Technology Research and Development Program of China (863 Program): Digital Agriculture (2007AA10Z207) and (2008AA10Z225).

REFERENCES

Campbell P, Middleton E M, McMurtrey J E, et al. Assessment of vegetation stress using reflectance or fluorescence measurements. Journal of Environmental Quality, 2007, 36(3): 832-845.

Cartelat A, Cerovic Z G, Goulas Y, et al. Optically assessed contents of leaf polyphenolics and chlorophyll as indicators of nitrogen deficiency in wheat (Triticum aestivum L.). Field Crops Research, 2005, 9(1): 35-49.

Heege H J, Thiessen E. On−the−go sensing for site−specific nitrogen top dressing. 2002 ASAE Paper No. 021113.

Kim Y, Reid J F, Hansen A, et al. Ambent Illumination effect on a spectral image sensor for detecting crop nitrogen stress. 2001 ASAE paper No. 01-1178.

Li M, Han D, Wang X. Spectral Analyzing technique and Applications. Science Press, 2006, Beijing, China.

Min M, Won S L, Yong H K, et al. Nondestructive detection of nitrogen in Chinese cabbage leaves using VIS-NIR spectroscopy. HortScience, 2006, 41(1): 162-166.

Noh H, Zhang Q, Han S, et al. Dynamic calibration and image segmentation methods for multispectral imaging crop nitrogen deficiency sensors. Trans. ASAE, 2005, 48(1): 393-401.

Stone M L, Soile J B, Raun W R, et al. Use of spectral radiance for correcting in-season fertilizer nitrogen deficiencies in winter wheat. Trans. ASAE. 1996, 39(5): 1623-1631.

Sui R, Wilkerson J B, Hart W E, et al. Multi−spectral sensor for detection of nitrogen status in cotton. Applied Engineering in Agriculture, 2005, 21(2): 167-172.

Tucker C J. Red and photographic infrared linear combinations for monitoring vegetation. Remote Sensing of Environment, 1979, 8(2): 127-150.

Xue L, Cao W, Luo W, et al. Correlation between leaf nitrogen status and canopy spectral characteristics in wheat. Acta Phytoecologica Sinica, 2004, 28(2): 172-177.

CONFOCAL FLUORESCENCE MICROSCOPY OF MUNG BEAN LEAVES

Zhiwei Chen [1,2,*], Dongwu Liu [1,2]
[1] *Analysis and Testing Center, Shandong University of Technology, Zibo, Shandong Province, P. R. China 255049*
[2] *School of Life Sciences, Shandong University of Technology, Zibo, Shandong Province, P. R. China 255049*
[*] *Corresponding author, Address: Analysis and Testing Center, Shandong University of Technology, Zibo 255049, Shandong Province, P. R. China, Tel: +86-0533-2786781, Fax: +86-0533-2786781, Email: chen@sdut.edu.cn*

Abstract: Recently, confocal microscope has become a routine technique and indispensable tool for cell biological studies and molecular investigations. The light emitted from the point out-of-focus is blocked by the pinhole and can not reach the detector, which is one of the critical features of the confocal microscope. In present studies, the probes acridine orange (AO) and rhodamine-123 were used to research stoma and mitochondria of mung bean leaves, respectively. The results indicated that the stomatal guard cells and mitochondria were clearly seen in epidermic tissue of mung bean leaves. Taken together, it is a good method to research plant cells with confocal microscope and fluorescence probes.

Keywords: confocal microscope, mung bean leaves, chloroplast, mitochondria, guard cell

1. INTRODUCTION

Confocal microscope is one of the important advances in the development history of optical microscopes. At present, confocal microscope has become a routine technique and indispensable tool for cell biological studies and molecular investigations (Lichtman, 1994). Confocal microscope works by exciting fluorescence with a highly focused beam of laser light. The light emitted from the point out-of-focus is blocked by the pinhole and can not

Please use the following format when citing this chapter:

Chen, Z. and Liu, D., 2009, in IFIP International Federation for Information Processing, Volume 295, *Computer and Computing Technologies in Agriculture II, Volume 3*, eds. D. Li, Z. Chunjiang, (Boston: Springer), pp. 2119–2124.

reach the detector, which is one of the critical features of the confocal microscope. Thus only an image of the fluorescence from the focal plane is observed. In addition, with the laser scanning over the sample from point to point, a single two-dimensional image of the optical section is obtained (Lichtman, 1994). Moreover, a series of optical sections at different focal planes can be acquired, and a three-dimensional image of the sample reconstructed (Webb, 1999). Since sectioning is performed using optics, living cells can also be analyzed, which is another important feature of the confocal microscope (Blancaflor and Gilroy, 2000).

With confocal microscope, the researchers could obtain the images of the cells much more clearly than conventional microscope. There have been numerous scientific papers employing confocal microscope in plant biology since it was first used in plant science research (Hepler and Gunning, 1998; Taira et al., 2004; Meckel et al., 2007). Confocal microscope has been applied to determine heterogeneity of plant mitochondrial responses (Armstrong et al., 2006), chloroplast structure and arrangement of chlorophyll-protein complexes (Garstka et al., 2007), oxidative stress in plant cells (Li et al., 2007), microtubule arrays and arabidopsis stomatal development (Lucas et al., 2006), and mitochondrial localization (Yang et al., 2006). However, previous reports mainly concern the cultured plant cells, and there are few reports about analyzing the whole plant leaves. Since confocal microscope is an appropriate method for qualitative analysis of plant cells, the aims of the present studies are to analyze the mitochondria and stoma in mung bean leaves with confocal microscope.

2. MATERIALS AND METHODS

2.1 Plant material and growth conditions

Seeds of mung bean were germinated on moist filter paper at room temperature. Seedlings were then transferred to 4-inch pots containing clay soil and were watered three times per week. Plants were grown in extraventricular chambers. Mature leaves from 25-day-old plants were used for analyses.

2.2 Mitochondria visualization

Leaf disks were incubated in the presence of 10 μM Rhodamine-123 (Molecular Probes) at room temperature for 1 h. At the end of the incubation period, the mung bean leaves were rinsed briefly with deionized water to remove excess dye from the surface of the leaves, wiped to remove excess water. Samples treated with water alone served as controls.

2.3 Stoma visualization

Leaf disks were incubated in the presence of 0.01% acridine orange (AO) (Molecular Probes) at room temperature for 10 min. Excess dye was eliminated by washing the disks three times in PBS buffer.

2.4 Laser-scanning confocal microscopy

For microscopy, 3×5 mm leaf cuttings were put under an inverted microscope (Leica DM IRB, Germany). A laser-scanning confocal microscope (Leica TCS SP2, Germany) with an air-cooled, argon-ion laser as the excitation source at 488 nm was used to view the sites of stoma, and 514 nm was used to view the sites of mitochondria. Moreover, He-Ne laser as the excitation source at 633 nm was used to view the sites of chloroplast. Images were obtained with a HCX PLAN Apo 20.0×0.7 PH2 objective lens. Mitochondria and stoma were respectively detected in the yellow and green channel. The channel settings of pinhole, detector gain, and amplification offset and gain were adjusted to provide an optimal balance of fluorescent intensity of the targeted cells and background. Data were collected by a computer attached to the instrument, stored on the hard drive, processed with a Leica TCS Image Browser, and transferred to Adobe Photoshop CS2 for preparation of figures.

3. RESULTS AND DISCUSSION

Exchange of water and CO_2 between leaves and the ambient air is one of the important plant processes. In this plant process, heat is dissipated and a primary substrate for photosynthesis is taken up. Stoma plays a key role in the control of water efflux and CO_2 influx between the inside leaf and the ambient air. In the epidermis of plant leaves, stoma is surrounded by two guard cells. Recently, guard cells have become a highly developed model system for characterizing early signal transduction mechanism in plants (Garcia-Mata and Lamattina, 2007; Garcia-Mata et al., 2007; MacRobbie and Kurup, 2007). It has been found that acridine orange (AO) is the most popular fluorochrome for studies on whole blood, reticulocyte counting, and identification of nucleic acids (Anderson, 1957; Armstrong and Niven, 1957; Armstrong, 1956). In present studies, stomatal guard cells stained with AO were imaged with confocal microscope, and guard cells and chloroplasts in guard cells were clearly seen (Figs 1B, C, D). The results indicate that AO is easily detectable using the confocal microscope, and AO is an appropriate probe for living guard cells studies.

Rhodamine 123, which is a potentiometric sensor and nontoxic dye, can accumulate in mitochondria for the mitochondrial membrane potential. Moreover, rhodamine-123 is easily detectable using an epifluorescent or confocal microscope, which highlights heterogeneity of mitochondrial shape, size, position, and dynamic within living plant cells. Mitochondrial morphology and dynamics have been studied extensively in eukaryotic cells. The results of Logan and Leaver indicate that the shape of mitochondria is varied depending on the cell type (Logan and Leaver, 2000). As early as 1914, the mitochondrial shape was studied with simple bright field microscope (Lewis and Lewis, 1914). Since then there have been numerous studies investigating mitochondrial polymorphism and physiology. Ultrastructural mitochondrial shapes and sizes within fixed cells are usually studied using the electron microscope. However, it is not possible to dismiss the possibility that any heterogeneity observed resulted from preparation or fixation artifacts (Logan and Leaver, 2000). Here we used rhodamine 123 as fluorescence probe, and mitochondria was clearly seen in epidermic tissue of mung bean leaves (Fig. 2B, C).

Taken together, the developments in technology and reagents that brought fluorescence-based methods to microscopy have become prevalent in plant cell biology. Our results indicate that it is a good method and simple way to research mung bean leaves with confocal microscopy and fluorescence probes AO and rhodamine-123.

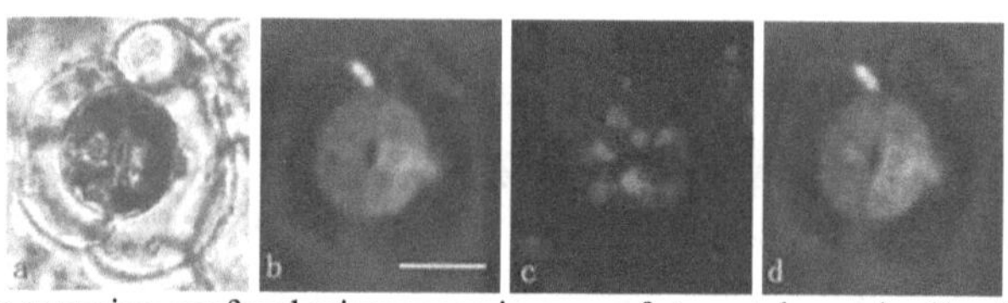

Fig.1: Laser-scanning confocal microscope images of stomatal guard cells of mung bean leaves. Stomatal guard cells were labeled with AO. (A) transmission channel, showing phase contrast; (B) green channel, showing DNA and RNA fluorescence; (C) Red channel, showing chlorophyll fluorescence in chloroplasts of epidermic cells; (D) B and C combined. Scale bars, 12.58 μm.

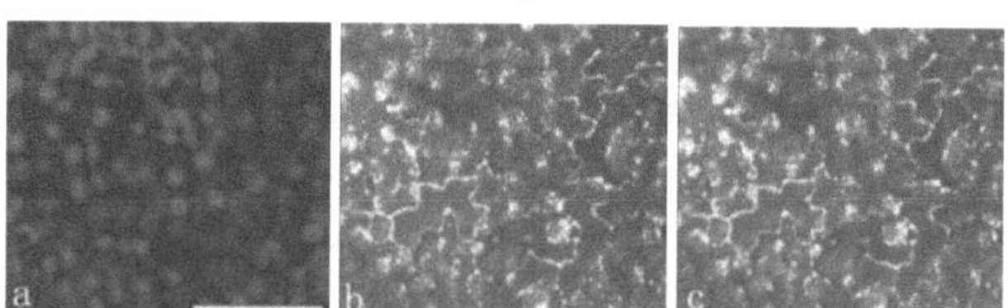

Fig.2: Intracellular localization of mitochondria. Laser-scanning confocal microscope images of epidermic tissue of mung bean leaves. Epidermic cells were labeled with Rhodamine-123. (A) Blue channel, showing chlorophyll fluorescence in chloroplasts of epidermic cells; (B) yellow channel, showing mitochondria fluorescence; (C) A and B combined. Scale bars, 75μm.

4. CONCLUSION

Taken together, the developments in technology and reagents that brought fluorescence-based methods to microscopy have become prevalent in plant cell biology. Our results indicate that it is a good method and simple way to research the mitochondria and stoma with confocal microscope and fluorescence probes AO and rhodamine-123.

ACKNOWLEDGEMENTS

This work was supported by the Natural Science Foundation for Outstanding Young Scholars of Shandong Province, China (Grant No. 2007BS06021).

REFERENCES

A. F. Armstrong, D. C. Logan, A. K. Tobin, et al. Heterogeneity of plant mitochondrial responses underpinning respiratory acclimation to the cold in Arabidopsis thaliana leaves, Plant Cell Environ, 2006, 29: 940-949

B. Li, D. Xing, L. Zhang. Involvement of NADPH oxidase in sulfur dioxide-induced oxidative stress in plant cells, Photochem Photobiol Sci, 2007, 6: 628-634

C. Garcia-Mata, L. Lamattina. Abscisic acid (ABA) inhibits light-induced stomatal opening through calcium- and nitric oxide-mediated signaling pathways, Nitric Oxide, 2007, 17: 143-151

D. C. Logan, C. J. Leaver. Mitochondria-targeted GFP highlights the heterogeneity of mitochondrial shape, size and movement within living plant cells, Journal of Experimental Botany, 2000, 51: 865-871

E. A. MacRobbie, S. Kurup. Signalling mechanisms in the regulation of vacuolar ion release in guard cells, New Phytol, 2007, 175: 630-640

E. B. Blancaflor, S. Gilroy. Plant cell biology in the new millennium: new tools and new insights, Am J Bot, 2000, 87: 1547-1560

E. S. Anderson. Visual observation of deoxyribonucleic acid changes in bacteria during growth of bacteriophage, Nature, 1957, 180: 1336-1338

J. A. Armstrong, J. S. F. Niven. Fluorscence microscopy in the study of nucleic acids, Nature, 1957, 180: 1335-1336

J. A. Armstrong. Histochemical differentiation of nucleic acids by means of induced fluorescence, Exp Cell Res, 1956, 11: 640-643

J. R. Lucas, J. A. Nadeau, F. D. Sack. Microtubule arrays and Arabidopsis stomatal development, J Exp Bot, 2006, 57: 71-79

J. W. Lichtman. Confocal microscopy, Scientific American, 1994, 271: 40-53

M. A. Dong, J. L. Bufford, Y. Oono, et al. Heat suppresses activation of an auxin-responsive promoter in cultured guard cell protoplasts of tree tobacco, Plant Physiol, 2007, 145: 367-377

M. Garstka, J. H. Venema, I. Rumak, et al. Contrasting effect of dark-chilling on chloroplast structure and arrangement of chlorophyll-protein complexes in pea and tomato: plants with a different susceptibility to non-freezing temperature, Planta, 2007, 226: 1165-1181

M. R. Lewis, W. H. Lewis. Mitochondria (and other cytoplasmic structures) in tissue cultures, American Journal of Anatomy, 1914, 17: 339-401

M. Taira, U. Valtersson, B. Burkhardt, et al. Arabidopsis thaliana GLN2-encoded glutamine synthetase is dual targeted to leaf mitochondria and chloroplasts, Plant Cell, 2004, 16: 2048-2058

P. K. Hepler, B. E. S. Gunning. Confocal fluorescence microscopy of living cells, Protoplasma, 1998, 201: 121-157

R. H. Webb. Theoretical basis of confocal microscopy, Methods in Enzymology, 1999, 307: 3-20

T. Meckel, L. Gall, S. Semrau, et al. Guard cells elongate: relationship of volume and surface area during stomatal movement, Biophys J, 2007, 92: 1072-1080

X. D. Yang, C. J. Dong, J. Y. Liu. A plant mitochondrial phospholipid hydroperoxide glutathione peroxidase: its precise localization and higher enzymatic activity, Plant Mol Biol, 2006, 62: 951-962

DETECTION OF VOLATILE OIL CONTENT OF SINGLE-GRAIN ZANTHOXYLUM SEED BASED ON NIR

Yun Xu [1], Yiming Wang [1,*], Jingzhu Wu [2], Shiping Zhu [3]
[1] *College of Information and Electrical Engineering, China Agricultural University, Beijing, China 100083*
[2] *College of Information Engineering,. Beijing Technology and Business University, Beijing, China, 100037*
[3] *College of Engineering and Technology, SouthWest university, Chongqing, China, 400716*
[*] *Corresponding author, Address: P. O. Box 1631, College of Information Engineering, China Agricultural University, 17 Tsinghua East Road, Beijing, 100083, P. R. China, Tel:+86-62737591, Email: ym_wang@263.net*

Abstract: A NIR model was established to predict the volatile oil content of single particle red Zanthoxylum seed in this paper. With the characteristic of irregular surface, A single Zanthoxylum seed will reflect the great difference in response to spectrum signals the entire spectrum detection and exceptional sample rejection method were employed before model optimization. As a result, the NIR model for predicting the content of volatile oil were built up by 74 red Zanthoxylum seed and results indicated: the NIR model of the single grain Zanthoxylum seed had good stability and predictability (RSD<10%,RSD>3). Results of this paper suggested that NIR could be used as a quick and convenient method for predicting the volatile oil content of Zanthoxylum seed, which is useful for breeding and the quality evaluation of it.

Key words: NIR, Zanthoxylum seed, Model Optimization

1. INTRODUCTION

Zanthoxylum is economic cultivars in many countries of Asia in people's daily life. It is not only an important spice but also has great medicinal value. The volatile oil content of Zanthoxylum seed is one of important purchase

Please use the following format when citing this chapter:

Xu, Y., Wang, Y., Wu, J. and Zhu, S., 2009, in IFIP International Federation for Information Processing, Volume 295, *Computer and Computing Technologies in Agriculture II, Volume 3*, eds. D. Li, Z. Chunjiang, (Boston: Springer), pp. 2125–2132.

standards, as well as deep processing, sale and breeding. At present, the detection method of volatile oil is very time-consuming (at least five to six hours per sample) based on the existing GB "Zanthoxylum" (SB/T10040-92) and "Zanthoxylum grade quality" (LYT1652-2005). It is difficult to detect the quality of Zanthoxylum for each sample need to do pre-processing and consumption of chemical. It also has brought great inconvenience in market circulation links, breeding and other requirements for nondestructive testing. So how fast, and how to make non-destructive testing on Zanthoxylum quality is an urgent problem to solve.

NIR spectroscopy technology has been well developed recently in China because this technology is quick, no chemical-consuming and nondestructive. NIR is gradually applied in the field of food and agriculture. The technology has good performance especially for on-line detection and locale detection. Duan S. Q. (Duan et al, 1996) used infrared spectroscopy, i.e. atomic absorption spectrometry and elemental analysis methods to identify the organic component and inorganic elements of Zanthoxylum. Lin P. Y. (Lin et al, 2003) used ATR-Fourier Transform Spectrometry, with cluster analysis method to identify Zanthoxylum herbs. However, there is no report in the literature that the Zanthoxylum seed volatile oil was detected rapidly and non-destructive with NIR.

Establishing a fast and non-destructive detective method has important significance to promote the Zanthoxylum breeding, deeply products and quality detection. It also has practical value to establish an effective system of quality assessment and quality standards. NIR model is established to predict the volatile oil content of single particle red Zanthoxylum seed in this paper. The authors discuss the practicality and feasibility of this method as a new, fast, and non-destructive green detection.

2. MATERIALS AND METHODS

2.1 Experimental design

The sample set is composed of single particle red Zanthoxylum seeds of 74 groups. They are collected from farmer's trade wholesale market of Chongqing, CaiYuanBa and BeiPei, some of which are purchased from several big supermarkets of Chongqing. The Zanthoxylum seed volatile oil density scope is 1.1937-3.4304(ml/100mg). The volatile oil content is obtained by food scientific institute of SouthWest University according to GB.

2.2 Instruments and spectral data

Near infrared spectra were collected on the Zanthoxylum seed sample set by diffuse reflectance from 12500 cm^{-1} to 4000cm^{-1}, Resolution is 4cm^{-1} , scanning 64 times each samples by the Fourier near-infrared spectroscope MATRIX-I spectrometer (the diffuse reflection integral ball appendix; the PbS detector) (BRUKE Corp Germany). Figure 1 shows original NIR spectrum curves.

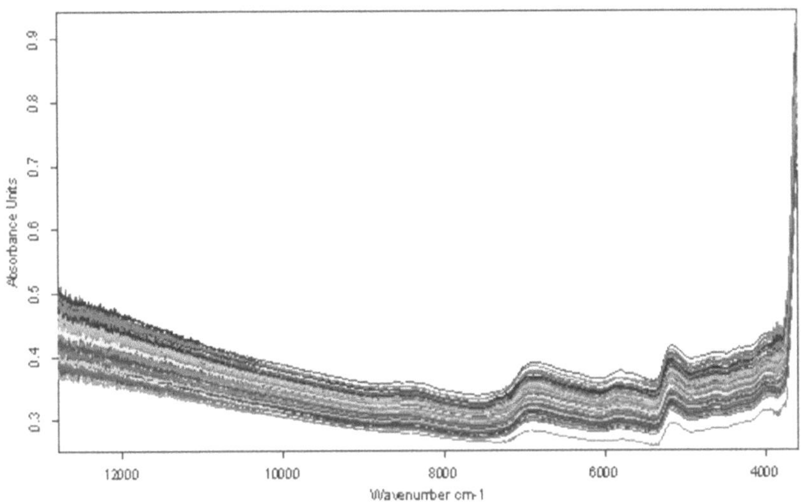

Fig1:74 unit original NIR spectrum of single red Zanthoxylum seed

2.3 Evaluation parameters of Near-Infrared model

To check the model if forecast ability and the usability, two evaluation parameters were used: one is relative standard deviation RSD(Lu et al,2000)and another relative analysis error RPD(Lu et al,2000;Yan et al,2005). RSD indicates the prediction ability of the model,and if RSD<10% ,it means that the model can be applicable in actual detection. While RPD index shows that if the model is stable and predictive. If RPD>3, the model has the high steady capability predictive ability (D.Cozzoline et al, 2004).

3. RESULTS AND DISCUSSION

3.1 Selection of sample

The spectrum of a single Zanthoxylum seed is different from one seed to another in one sample. In this experiment, each scanned spectrum curve as a sample was processed. Figure 2 shows original NIR spectrum curves of three particles from the same sample. And the total number of primitive spectrum is 222. In the experiment, the effect of the spectrum difference on the model influence is bigger than the density difference. Here first primitive sample centralism outlier samples 22 are elimated based on the F statistics significance examination method. We selected 140 samples as the training set. The testing set is made up of the other 60 samples. The classifier is based on Kennard-Stone (R W Kennard et al,1969;Lu et al,2004) method .

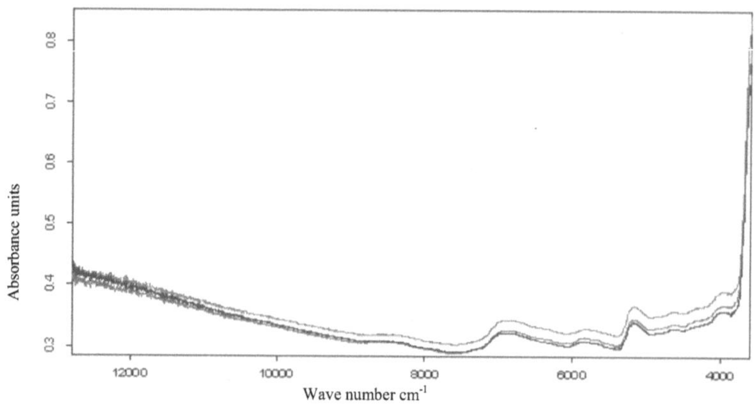

Fig 2: spectrum picked up three particles from the same sample

3.2 Elimation of outlier samples

The sample is red single particle Zanthoxylum seed. Its irregular surface can make the scanning spectrum inaccurate. Therefore, in order to establish a good NIR model of single particle Zanthoxylum seed, it is necessary to reject the unusual sample and model optimization.

The so-called unusual sample refers to the sample with big error of density value or the spectrum data, the common solvent used in rejection unusual sample: method based on the forecast density residual error, restructuring spectrum residual error, principal components score cluster analysis, as well as the release lever value and the student T-test and so on (Wang H W,1999;Wang D J et al,2004;Xu et al,2000).

In this paper we elimate the unusual samples through software OPUS6.0 (BRUKER Corp.). After unusual samples rejection, the training sample set is 116 integers, and the test sample set is 37.

3.3 Model calibration

Conventional spectrum denoising methods and their combination are used in this experimental, and PLS is the most popular linear calibration method in NIR quantitative analysis. We use software OPUS6.0 to pretreatment the spectrum and optimize the model. Figure 3 shows the spectrum after prctrcatment.

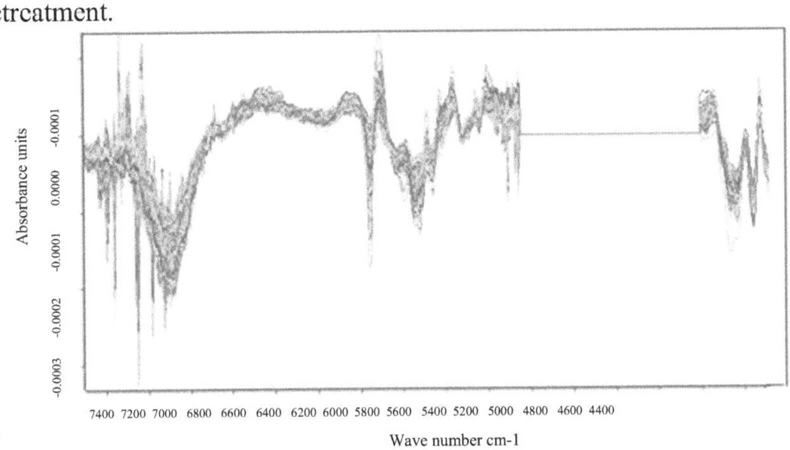

Fig 3 first derivative + MSC spectrum after wave pretreatment

pretreatment of spectrum the first derivative and MSC ,methods were applied to make pretreatment of spectrum , and the scope of spectra is adopted from 4601.6 cm^{-1} to 4248.7 cm^{-1}, and from 7500.2 cm^{-1} to 5448.2 cm^{-1}; the number of principal component is 7; Root Mean Square Error of Estimation is 0.209; Coefficient of determination is 0.9049. The relationship of target value and the prediction value of training set are shown in Figure 4. It shows that a good linear relationship was obtained between the chemical analysis curve and the NIR predictive curve.

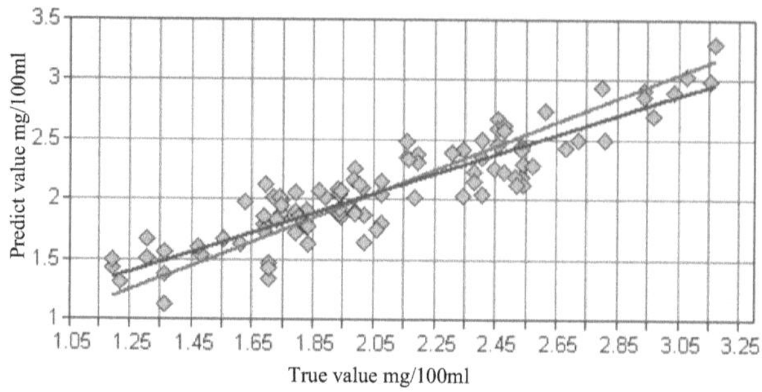

Fig 4:model Prediction analysis on target value of training Set

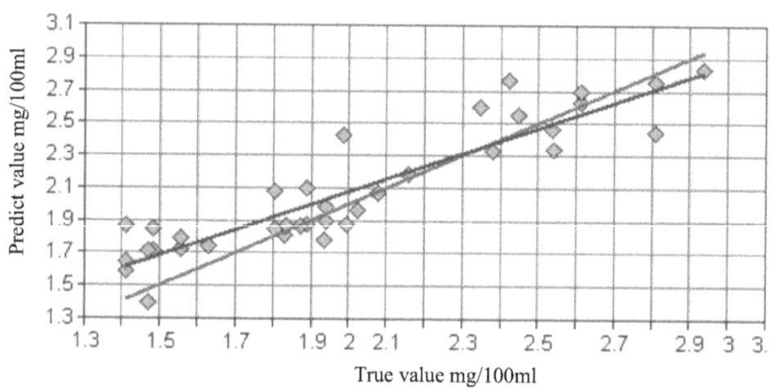

Fig 5: model Prediction analysis on target value of testing set

Use the above model to predict the test samples. The decision coefficient is 0.9136 and root mean square error of estimation is 0.197. The relationship of true value and the predict value of test set are shown in Figure 5. It shows that there is a good linear relationship between the chemical analysis and the NIR predictive value.

In order to verify that the predictability and practicality of the model which predict the volatile oil content of the single red particle Zanthoxylum seed, the relative standard deviation (RSD)and the relative error (RPD) of the training sets and test sets are calculated and shown in table 1.

Table 1, Verification result of NIR mode*l*

Sample set	Sample number	R	RMSEE	RSD	RPD
Train set	116	0.9049	0.209	8.86%	3.17
Test set	47	0.9136	0.197	6.87%	3.21

From the table above, RSD of the training set and test sets is less than 10% and RPD is more than 3. It has a good calibration result, so the NIR model for forecast the volatile oil content of single particle red Zanthoxylum seed can be used for actual testing.

The experiment shows that the spectral data is not accurate because of the irregular surface of Zanthoxylum seed, so a large number of abnormal spectral data will appear. Therefore, if we do some pre-processing work before scanning spectral curve, such as changing Zanthoxylum seed into powder with Special equipment, or using a big sample cup while scanning, it can reduce the spectrum of differences, therefore reducing the abnormal sample. However, there are some drawbacks in pre-treatment work: on the one hand, the powder will make the samples destructed; on the other hand the powder will increase the detection time. It needs a certain number of samples by using big sample cup. Therefore, the suitable actual demand model can be established only according to the different application situation.

4. CONCLUSION

NIR models were established to predict the volatile oil content of single-particle red Zanthoxylum seed in this paper in order to enhance the breeding and the detection technology of the Zanthoxylum seed. As a result, the NIR model for predicting the content of volatile oil were built up by 74 red Zanthoxylum seed which buy from ChongQing and results indicated: the NIR model of the single grain Zanthoxylum seed had good stability and predictability ($RSD<10\%$,$RSD>3$). Results of this paper suggested that NIR could be used as a quick and convenient method for predicting the volatile oil content of Zanthoxylum seed, which is useful for breeding and the quality evaluation of it.

ACKNOWLEDGEMENTS

Funding for this research was provided by State Natural Sciences Foundation, P. R. China [Study on mechanism and the model optimization of fast detection for Zanthoxylum quality based on the near-infrared spectrum] (project No. 30671198).

REFERENCES

D. Cozzolino, MJ Kwiatkowski, M. Parker, et al, Prediction of phenolic compounds in red wine fermentations by wisible and near infrared spectroscopy. Analytica Chimica Acta 513, 2004:73 ~ 80

Duan S Q, Gong M C Research on Zanthoxylum ingredient, Sichuan Chemical,1996, 4:34 - 36

Lin P Y, Tao C Y, Identification of genus Zanthoxlum by Fourier transform infrared spectrometry " Academic Journal of Second Military Medical University " 2003.9 -1029 to 1031

Lu W Z, Yuan H F, Xu G T, "Modern near-infrared spectral analysis technology". Beijing: China Light Industry Press, 2000

Lu Y J, Qu Y L ,Piao R G, The Algorithm of Eliminating the Similar Samples in the Process of Calibration and Prediction. Spectroscopy and spectral analysis,2004，24（2）：158~161

People's Republic of China forestry industry-standard "Zanthoxylum grade quality" (LYT1652-2005)

People's Republic of industry-standard "Zanthoxylum" (SB/T10040-92)

R W Kennard,L A Stone. Computer Aided Design of Experiments. Technometrics 11 (1969)：137~148

Wang D J, Application of Near Infrared Spectroscopy Detection Technology to Agriculture and Food Analysis .Spectroscopy and spectral analysis, 2004 (04): 447 - 450

Wang H W, Partial least-squares regression method and its application. Beijing: National Defense Industry Press, 1999

Wu J Z Study on Quality Detection of Agricultural Products Based on Near Infrared Spectroscopy Technology, Doctoral dissertation 2006.6

Xu Guangtong, Hong-Fu Yuan, Wan-Zhen Lu. Development of Modern Near Infrared Spectroscopic Techniques and Its Applications. Spectroscopy and spectral analysis, 2000,20(02) :134 - 142

Yan Y L, Zhao L L Han D H, "Near-infrared spectral analysis foundation and application [M]. Beijing: China Light Industry Press, 2005

KEY TECHNOLOGIES IN WSN-BASED COTTON FIELD SOIL MOISTURE MONITORING SYSTEM

Ruirui Zhang [1] , Liping Chen [1,*] , Gang Xu [1] , Yanji Wang [1]
[1] *National Engineering Research Center for Information Technology in Agriculture, Beijing, P. R. China 100089*
* *Corresponding author Address: Room A517, Beijing Nongke Building, No.11, Shuguanghuayuan Mid-road, Haidian District, Beijing,P.R.China,100089 Tel: +86-10-51503425, Fax: +86-01-51503425, Email: chenlp@nercita.org.cn*

Abstract: To level up the application of wireless sensor network in cotton field soil monitoring, wireless sensor network technology has been studied deeply. Some difficulties, including the design of soil moisture sensor, covering problem in wireless sensor network and diagnosis algorithm for wireless sensor networks and node, are been proposed and discussed. Combined with cotton field soil moisture monitoring application, coverage and deployment based not only on plane space but also on time are been discussed and a fault diagnosis algorithm based on entropy is been put forward in the article.

Key words: wireless sensor network, field soil moisture monitor, coverage algorithm, fault diagnosis algorithms

1. INTRODUCTION

Precision agriculture (PA) is an agricultural concept relying on the existence of in-field variability. It's about doing the right thing, in the right place, in the right way, at the right time. It's connotation is regulate the crop inputs according to the current soil properties, which will reach the goal of optimizing formula, scientific management, activating the soil production so that we can get equal or high production by the same input. PA is the

---

Please use the following format when citing this chapter:

Zhang, R., Chen, L., Xu, G. and Wang, Y., 2009, in IFIP International Federation for Information Processing, Volume 295, *Computer and Computing Technologies in Agriculture II, Volume 3*, eds. D. Li, Z. Chunjiang, (Boston: Springer), pp. 2133–2140.

direction of future agriculture and can get the result of improving environment. Acquirement of the soil parameter is an important part in PA.

Soil moisture is an important factor to cotton plant, which will influence its quality and quantity. Insufficiency of water leads to thirsty of cotton plant, and overabundance of water results in not only the waste of water but also shallow root and weak ability for anti-lodging. Getting the current cotton-field soil moisture content precisely and irrigating the field quantitatively and distinctively become an urgent problem to solve.

Research in existing technology of soil moisture monitoring reveals that precise and prompt result can not be got properly. There are mainly two ways to get soil moisture content. The first is getting soil sample manually, then after firing and weighing, to get the field soil moisture result. This method requires a lot of manpower and lack real-time property, which can not satisfy the demand of PA obviously. The second popular way is via the WLAN (Wireless local area network, WLAN) technology. But there are also some shortcomings of the way. High power consumption of WLAN module and monotonous topology of network restrict the geographical distribution of sample point, which can lead to inaccuracy result. In light of the above situation, system of WSN-based cotton field moisture monitoring is proposed.

2. WIRELESS SENSOR NETWORK

Wireless sensor network(WSN) considered as one of the most affected technology to people's daily life in 21st century, is a wireless network consisting of spatially distributed autonomous devices using sensors to monitor physical or environmental conditions cooperatively at different locations. (Jing Qi et al., 2007) Comparing to traditional technique, the WSN-based monitoring system has following advantages:

a) Low impact on field environment. For one thing, each sensor node is very small, so being placed in field hardly have affect to work of farming machinery. For another, nodes are powered by AA alkaline battery and communicate with each other by radio, so that nodes can be placed according to our need and free from cable. Up to some extent, WSN-based farmland moisture monitoring system has the least impact on field local environment.

b) High dense sampling point. As the nodes are tiny and can relay message each other, they can be placed at high density. By this, more accurate soil moisture content can be got.

c) Low cost comparing with manual sampling and WLAN- based way.

3. SYSTEM FRAME AND KEY TECHNICAL DIFFICULTIES

3.1 System framework

Work principle of the monitoring system: sensor nodes, equipped with RF communication module and distributed in a cotton field, sense the cotton-field soil moisture content and send data in relay through the radio wave from a node to another. At the end, all the sensor data is collected and swarms into a WSN-GPRS gateway which processes the data package, and then retransmits the data to data management center through the GPRS network. Data management center provides decision reference for irrigation after analyzing the data. People can get access to the decision reference message by an Internet-connected PC or a PDA. Figure 1 shows the system concrete construction. (Chen et al., 2004)

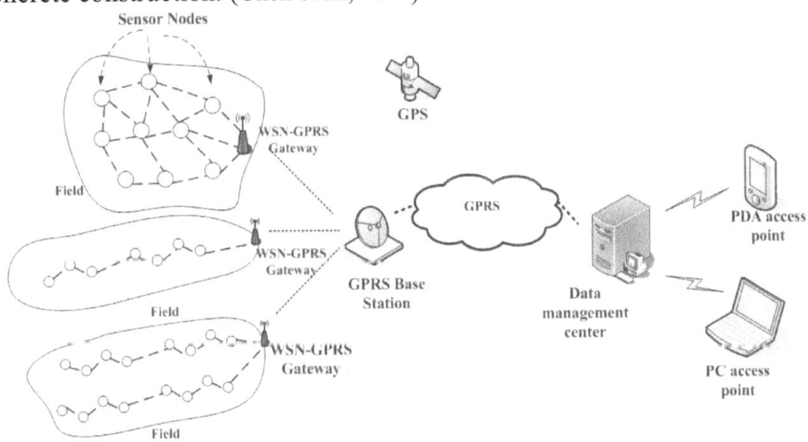

Figure 1 Scheme of the WSN structure

In the application of cotton field, mainly three aspects of technology need to study: soil moisture sensor, coverage algorithm and fault diagnosis algorithm.

3.2 Soil moisture sensor suitable for WSN

Soil moisture sensor, an important part of the wireless sensor network node, which has the following constrains.

a) Power energy is limited. The nodes are very small which just equipped with alkaline battery as energy source and replacing batteries in time is infeasible, which requires that the system must work in low power dissipation and so as the sensor. According to the application requirement, continuous working current of sensor must under 0.5mA in our system. (Blumberg et al., 2001; Li et al., 1999)

b)Finite volume. In general, micro size is a greater advantage of WSN-based technology and bigger size would destroy the environment of the field, so the size of sensor is expected the smaller the better. Ideal size above ground is smaller than 200cm3. (Ma et al., 1993; Wang et al., 1991)

c)Finite computing and storage capability of WSN node result into that the sensor output of result must short. Too complex result certainly will cost the microprocessor more time and power. An 8 bits binary result or 0~ 3V analog signals are proper.

d)The sensor must have flexible work mode and highly measuring speed. The node works alternatively with sleep mode and mode awake, most time being in sleep. Sensor requires the ability of adjust the working mode and output the sampling result immediately, so as that the WSN nodes can go into sleep as quickly as possible to save as much energy as possible.

Based on above conditions, ECH2O soil sensor of Decagon Devices Company is selected in the system (Figure. 2).

The ECH2O probe gets the volumetric water content by measuring the dielectric constant of the soil. Since the dielectric constant of water is much higher than others, that of water is 80, soil minerals 4, air 1, the dielectric constant of the soil is a sensitive measure of water content. The ECH2O probe has a very low power requirement and high resolution which gives us the ability to make as many measurements as we want over a long period of time, with minimal battery usage. (Hu et al., 2004; Luo et al., 1990)

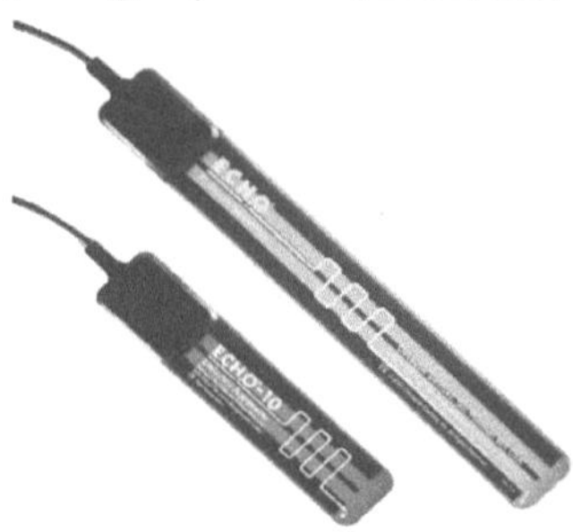

Figure 2 Photo of ECH2O

3.3 Covering algorithm based on time and space concept

Coverage is an important part of the WSN technology. As application oriented, definitions of WSN coverage is different from each other. Aiming at cotton field soil moisture monitoring, single coverage is defined as that for a certain region in the field, its soil moisture can be monitored by at least one node (Mao et al., 2007; Ren et al., 2006) .So does it for k-times repetitive coverage, which is that soil moisture can be monitored by at least k nodes for certain region in the field. Considering the features of soil moisture monitoring and reducing the energy consumption, single coverage is concerned in the following only. (Zhang et al., 2005; Wang et al., 2007)

Coverage in plane space has many mature algorithms and models, such as Boolean sensing model, grid deployment model and so on. Best deployment in certain time can be got through one model. Although the growing of cotton plant has influence on the communication distance between nodes, researching covering algorithm more deeply and finding an algorithm which can utilize the energy furthest but not just by adding redundant nodes become an difficult problem.(Xu et al 2007)

Besides, the following are also what should be studied in this part:

a) Algorithm of initial deployment used in cotton field, to find an easy and rapid method for implementation;

b)How to guarantee the quality of data from the important area. K-coverage can be considered if needed.

3.4 Election of cluster head/ cluster building and fault diagnosis algorithms

3.4.1 Election of cluster head/ cluster building

Because that the topology is based on plane space and time, the cluster head and cluster change as time elapsing and cotton growing. Whether election of cluster head and cluster building is reasonable or not can directly affect the quality of service of network. A reasonable routing table and a rapid election algorithm is the key for solving the upper problem.

3.4.2 Health diagnosis algorithms

WSN-based soil moisture monitoring field network has the feature of large mount of nodes, so it is difficult and impossible to maintain it manually. It requires that the node has nice noise tolerance and equips with nice fault diagnosis algorithm in order that the node can adjust itself as timely as possible.

The feature for the diagnosis algorithm is its statistical calculation, or based on entropy (Wang et al., 2004; Zhang et al 2004). Because nodes are exposed in open air and suffered from corrosion, high-low temperature and restricted energy, that the node gets out of order accidentally is very normal. Diagnosis algorithm must have the ability to distinguish accidental fault from frequent fault. One way to solve this is that diagnosis algorithm judges the accident and counts its occurring time and then the diagnosis algorithm estimates the state of the node by using statistics. What needs to further study is that how to distinguish the fault before the statistic model is established, how to identify the wrong data and how to reduce the work time of diagnosis algorithm to adjust the node as quickly as possible.

Further more, stub-network is also one part of diagnosis. Paths by which sensor node data is transferred to sink node are expected to distribute in several different ways statistically, for prevention that part nodes would die in advance. If data from one node or one cluster is delayed in only way, the cluster are called stub-network. Stub-network is not allowed to appear in the network. Building a diagnosis for the sub-work and locate its position are need.

4. EXPERIMENTAL STUDY

The studying team preliminary structures soil moisture monitoring platform in National Precision Agriculture Demonstration Bases in Beijing for the long-term experiments. In order to facilitate the analysis of statistics, server software was developed. The server can receive sensor nodes data, and playback of each sensor node dynamic data transmission path.

Figure2 shows distribution of sensor network. Experiment was carried on in a cotton field with 5000 meter long and 2000 meter width.

The diagnosis algorithm and coverage algorithm experiment are further developed on the platform, and the exhausted nodes are found and alarm. The soil moisture monitoring system is improved.

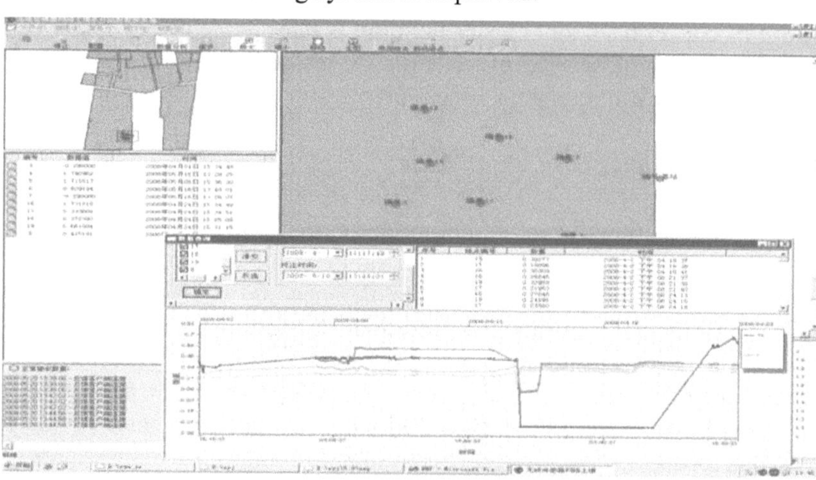

Figure 3 Photo of PC analysis software

5. CONCLUSION

Wireless sensor networks has unparalleled advantage to traditional technologies in soil moisture monitoring. The article shows the framework of WSN application in cotton soil moisture monitor system. Combining with experiments, thinking on how to solve the technical difficulties is also been shown in the article. Which worth to be emphasized is that the system can also be used in other application by just a little alteration, such as temperature and humidity monitor in field, monitoring in wheat field and so on.

ACKNOWLEDGEMENTS

This work was supported by a grant from Key Technology R&D Program of China (No. 2007BAD44B06).

REFERENCES

Blumberg DG, Freilikher V (2001), "Soil water-content and surface roughness retrieval using ERS-2 SAR data in the Negev Desert", Israel. J. Arid Environ. 49:449–464;

Chen Z, Zhou Q, Tang H (2004), "The future agricultural applications demanding on space-born remote sensing in China", In: H Tang (ed) Research on the Agricultural Resource Utilization and Regional Sustainable Development. China Outlook Press, Beijing, pp 236–250;

Hu Jiandong, Duan Tiecheng, He Sailing, "Study of a Capacitance-Based Self Model-Establishing Sensor for Measuring Water Content of Soil", Journal of Transcluction Technology, 1st, 2004;

Jing Qi, Chen Zhoufeng, Guan Zhi, Wang Zhao, "Design and deployment of wireless sensor networks", Computer Engineering and Applications, 21st, 2007;

Li J, Islam S (1999), "On the estimation of soil moisture profile and surface fluxes partitioning from sequential assimilation of surface layer soil moisture". J. Hydrol. 220:86–103;

Luo Rufang, Lou Guichang, LuYuping, Yian Ronhua, "Study of the Si-MOS Capacitive Type Micro Humidity Sensor", Journal of Transcluction Technology, 3rd, 1990;

Ma Xiaoyi, Xiong Yunzhang, Sun Mingqin, "The Development of Capacity Soil Moisture Sensor", Journal of Transducer Technology, 1st, 1993;

Mao Yingchi, Feng Guofu, Chen Lijun; Chen Daoxu, "A Location-Independent Connected Coverage Protocol for Wireless Sensor Networks", Journal of Software, 7th, 2007;

Ren Yan, Zhang Sidong, Zhang Hongke, "Theories and Algorithms of Coverage Control for Wireless Sensor Networks", Journal of Software, 3rd, 2006;

Wang Chao, Wang Hailing, Shi Xiangyong, Gong Xu, "Routing Algorithm for WSN Based on Swarm Intelligence Optimization", Journal of Shanghai University (Natural Science Edition), 4th, 2007;

Wang Gang, Zhao Hai, Wei Shouzhi, "Decision-Level Fusion Model Based on Threat Game Theory", Journal of Northeastern University (Natural Science), 1st, 2004;

Wang wei, Qi Changyong, "Soil-moisture Sensors", Journal of Transducer Technology, 2nd, 1991;

Xu Congfu, Li Shijian, Rao Dazhan, Pan Yunhe, "Triangle-Arranged Method for Coverage with Connectivity in Sensor Networks", Chinese Journal of Sensors and Actuators, 3rd 2007;

Zhang Qing, Xie Zhipeng, Ling Bo, Sun Weiwei, Shi Baile, "A Maximum Lifetime Data Gathering Algorithm for Wireless Sensor Networks", Journal of Software, 11th, 2005;

Zhang Xiaodan, Zhao Hai, Wang Gang, Wei Shouzhi, "Fusion Algorithm for Uncertain Information by Fuzzy Decision Tree", Journal of Northeastern University (Natural Science), 7th , 2004;

DESIGN OF BEE PRODUCTS QUALITY MONITORING INFORMATION SERVICE PLATFORM

Yeping Zhu[1], Jing Zhao[2], Shengping Liu[1], Yue E[1]
[1] *Institute of Agricultural Information, Chinese Academy of Agricultural Sciences, Beijing, China, 100081;*
[2] *Bee Research Institute, Chinese Academy of Agricultural Sciences, Beijing, China, 100093*

Abstract: The bee products quality monitoring and tracing information service platform was researched and developed. This paper describes the design concept and critical technologies for the construction of the bee products quality monitoring and tracing information service platform. The system has functions of remote information collection, production-purchase-processing automatic coding, bar code generation and identification, product quality tracking and tracing, information release, origin evaluation, market prediction, and analysis based on Geographic Information System (GIS), providing information technology(IT) tools for bee products quality safety control and quality tracing in China.

Keywords: bee products, quality monitoring, information, traceability

1. INTRODUCTION

Today, the modern management of apiculture has been clearly seen in many countries. For example, modern computer management technology has been adopted for bee products production and processing in Brazil, Colombia and Australia; GIS has been implemented in Brazil's leading apiculture.(K. Simoes, et al, 2005, F Foret, et al, 2005, K H Mcllvride,2005, B D Mahaman, et al, 2002, Qi Yamei,2007). Therefore, the apiculture will witness the trend of information technology applications, and modernization.

Please use the following format when citing this chapter:

Zhu, Y., Zhao, J., Liu, S. and Yue, E., 2009, in IFIP International Federation for Information Processing, Volume 295, *Computer and Computing Technologies in Agriculture II, Volume 3*, eds. D. Li, Z. Chunjiang, (Boston: Springer), pp. 2141–2149.

In China, the construction of information system platform from fork to farm and quality tracing for agricultural products has become an important project, including in develop agricultural products quality inspection system, market analysis and prediction system, digital agricultural products trading system, and distribution system; establish integrated agricultural products production and circulation information platform for safe production and efficient circulation of high-quality agricultural products; and create space information management analysis platform and agricultural products safety quality tracing system platform to realize holistic information management(Zhao Chunming, 2007, Liu Haifeng, etc,2007, Liu Dongmei,2006, Zhang Ting,ect,2007, Zhang Bing,ecty,2007, Shen Guanglei,ect,2007).

China's apiculture still features looseness, small scale, mobility of apiarists, it is difficult for monitoring and controlling of product quality(Zhao Jing, 2006,2002), so it is very necessary to track and control product quality through information technology.

2. DESIGN OF BEE PRODUCTS QUALITY MONITORING INFORMATION SERVICE PLATFORM

The investigation and analysis show that the production and processing of bee products have the following features: apiarists are isolated and dispersed though each of them has many beehives; apiarists often migrate and move their apiaries according to the florescence and season; apiarists are relatively undereducated and their production facilities are very simple. At present, organizations (like apiary association) and governments help apiarists improve production conditions, raise the awareness of bee products quality safety, and boost the development of apiculture. IT is indeed an effective approach for information communications, quality control and tracking.

The effective control, monitoring and tracing of bee products quality shall rely on the modern measures and IT. Therefore, we conduct the technical research on the bee products quality monitoring information service platform, and design relative technical framework and functions. According to the status quo of the apiculture(Chen Lihong, 2003), technical roadmap and standards should be first designed and determined. Let us take honey as example: the apiarist puts honey into the honey bucket, and the merchant attaches a bar code to indicate the origin and source after buying it; if the factory plans to mix the honey, another bar code is generated to indicate components before the processing; the final bar code and label of honey product will be generated before the sales. The system is designed to use PDA(personal digital assistant) as the data collection tool, so it should develop PDA-based bee products data collection system. To facilitate the field data collection and use, it should simplify the information input, and allow the merchant to generate bar code in PDA, print and attach bar code through portable printer, enter information of apiarists and honey sources in computer system and submit it to the processing factory. However, it is a challenge for the extensive apiculture to track product quality, so the construction of bee products quality tracking platform should satisfy the features of apiculture (esp. the migration of apiary) from the perspectives of technology and operation.

By using existing communications network platform and technologies such as Java EE, GIS, Agent and Web Service, we design the framework of bee products circulation tacking information management platform, realize the digital bee products management, visualize the tracking and management on the whole bee products process from production to sales, including environmental quality assessment, production prediction, logistics and distribution management, circulation tracking and monitoring, and market analysis and prediction.

The platform can provide governments and administrative departments with bee products circulation information, serve as a tool for supervision management and quality inspection, and support the development of policy, regulation and decision. Furthermore, the platform can provide information and consulting services for enterprises, farmers and consumers, and facilitate the information communications for bee products trading and circulation. In fact, the platform will significantly enhance the holistic quality tracking and management of production-processing-circulation of bee products, and offer important IT measures for addressing issues of bee products quality control and export. The infrastructure of the platform is shown in Figure 1.

3. KEY TECHNOLOGY

3.1 Information collection

In view of scattered bee products producers, information collection device should contain PDA or handheld PC, portable barcode printer, and WindowCE OS for the system development. In view of mobility and swiftness, the collected simple data such as apiarist code, origin, products species, quantity and batch are used to generate bar code and conduct onsite printing.

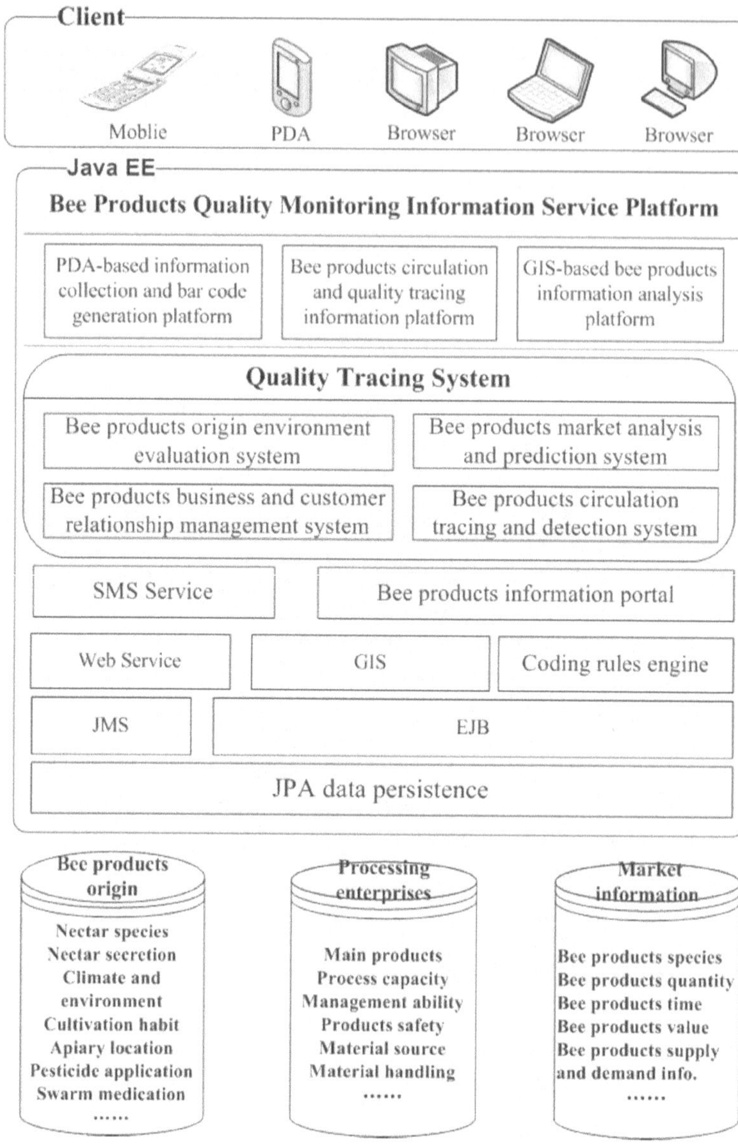

Fig.1: Architecture diagram

3.2 Database design

Besides bee products database and database management system, bee products data warehouse should also be established. The databases contain:

■ Bee products origin information database: includes data of space distribution, environment, apiarist, distribution of main crop and nectar plants.

■ Bee products processing enterprises database: include data of address, products, processing, management, product safety, materials source and handling of enterprises.

■ Bee products market information database: include data of species, quantity, time, quality, price, origin, source, supply and demand information, and trading information of bee products traded in China and other countries.

3.3 Tracing coding proposal

As global standard system EAN/UCC-128 international numbering system has commonality, universality, maintainability, expandability, low price and is widely used in food safety traceability system. In this research we employ EAN/UCC-128 as bee product mark. A mark unit includes honey which was encapsulated in the same pail and bar code is made of apiary code, product number, honey category and production date. Apiary code is made of seven-digit, which includes four postal codes and three apiary codes. Product number is made of four-digit, which includes 2 digits for year and 2 digits for product batch. Honey category is made of two-digit to show raw material. Production date is made of six-digit, such as 080326 shows that production date is march 26, 2008.

(7030)1566115001(10)0821(251)01(17)080326

Fig. 2: Traceability label of bee product quality

It is bee product's chain and basis that every production link has respective product feature code and trace code, so that inputting correct information and protect information safety appear especially important. We can easily read bar code information through optical scanner. When some link completes and transmits another link system would automatically produce and print new bar code. The normal bar code can only contain 13 digits and not describe product. Information traceability and transmitting mainly depend on database support so that quality training of apiculture stuff and system manager in charge of system client data inputting becomes very important. If we have enough money we should use 2-dimensional bar code which can save a great deal of information.

3.4 Tracing key link control

Mainly links of bee product quality safety tracing include apiary, purchaser, professor and retailer as four product quality reference point.

Apiary control mainly record and manage bee product original product information, gather data which includes place environment, sources of honey, water source situation, swarm condition and implements of production. Because swarm have distribution and movement characteristic we need gather information and inspect honey farmers' production conditions and supervisor mode through purchaser and beekeepers' association.

Purchasers and professors mainly realize bee product transit monitoring. They need input essential information in system client which includes bee product origin, test method, qualified situation, typeofproduction and packing information. Tracing code is used to control production information transformation. Retailer not only control basis information but also is responsible for consumer information inquiry. Consumer can inquire bee product interrelated producer and geographical origin information through product tracing label, mobile message, telephone and web site.

4. DESIGN OF SYSTEM FUNCTION

4.1 Formulation of digital coding rules and barcode of bee products

The formation of bee products digital coding rules allows bee products to be coded during the whole production and circulation process, giving a full picture of production through bar code at each step. In addition, the engine of coding rules is created to provide coding service for practical application components in upper layer.

4.2 Evaluation of bee products origin environment

Based on bee products origin information database, an origin environment quality evaluation system is established to evaluate the origin environment and predict quality change. Through association analysis on environment data and bee products information data, it can use historic data to predict products quality and quantity.

4.3 Bee products circulation tracing and detection

The quality safety issues of bee products include: residual pesticide in nectar plants, residual veterinary drug during ill treatment, residual bio-toxic substance caused by contamination of heavy metal (such as iron, lead, etc) during production or by improper storage, and misuse of additive, adulteration, and fake; internal quality issues such as immature production, mechanical impurities in production, etc.

To address those issues, it shall track and manage bee products circulation through product coding; query information about apiarist and enterprises as well as circulation route of bee products, and sum up circulation amount; establish SMS-based bee products inquiry system, and get the detail information from origin to current stage by inputting bee products codes to server via mobile terminals like mobile phone; create GIS-based bee products information management platform, and realize the spatial information management, query, analysis and visualization of nectar source, bee migration and product production; realize quality safety monitoring and control from material procurement, acceptance to product production; realize quality safety monitoring of package, storage, transport, processing and sales of bee products, and quality safety control of various analysis, inspection and test. As the quality safety monitoring and control covers the entire production process, it can make timely response, warning, modification, adjustment, and recovery when there is a fault in a step. Meanwhile, the qualify issue can be traced back in accordance to the false steps in the whole production process.

4.4 Market analysis and prediction of bee products

The analysis of bee products circulation tracing, detection, and market supply and demand can get the law of circulation of bee products on the market. Furthermore, it should analyze circulation progress, establish an analysis system and forecast the law of circulation of bee products. Compared with the true result, it can further predict the market trend of bee products and instruct the production according to the analysis.

4.5 Bee products e-Commerce platform

The functions of e-settlement, price collection, logistics control, inventory management, products auction, and online marketing are established on the basis of bee product information database and digital platform, integrating the inventory management, e-settlement, capital allocation, info collection, price release, etc.

4.6 Information management and publication

The establishment of bee products quality safety management information system and the release of latest quality safety standard will highlight the importance of safety among apiarists. The timely release and update of apiculture news will guide the correct production for apiarists. The establishment of the bee products quality safety management information system can enhance the apiarist' awareness of quality safety, improve the bee product monitoring and management system, promote the apiculture productivity and manageability, raise the bee products inspection level and realize the industrialization to ensure the quality safety of bee products.

Meanwhile, the system can publish the latest development, vision, approach and consulting of technology information in the apiculture, thus to publicize technology knowledge of bee products, accelerate the transformation from technology to productivity, combine technology demand with technology supply, promote the application of technology in production, and consequently improve the quality safety of bee products.

5. CONCLUSIONS

This paper deals with the key technologies used in the production, circulation and quality tracing of bee products. The research and design of relative network application platform and framework realize the feasible technological scheme and design thought for the resources sharing and joint decision making. The research of standard, specification and coding rules of bee products provides an information processing criterion for the standard production of bee products. The database of origin environment, quality monitoring, product inspection and circulation provides a tool for tracing the bee product quality. The portal offers an information exchange platform for the digitized control of apiculture pattern, production, sales, circulation and quality tracing, thus transferring the bee products from field to table.

ACKNOWLEDGEMENTS

This research was supported by the National Department (Agriculture) Public Benefit Research Foundation of China(No. nyhyzx07-041)

REFERENCES

B D Mahaman, P Harizanis, I Filis, et al. A diagnostic expert system for honeybee pests[J]. Computers and Electronics in Agriculture, 2002,36(1), P17-31.

Chen Lihong, Standard Production Technique of Bee Products [M], China Agricultural University Press（in Chinese）

F Foret, et al, New Software for beekeping administration on Tropical Countries [J]，2005，P.60.

K H Mcllvride, "Bkeeping" A Computer Program [J], 39th Apimondia International Apicultural Congress, 2005, P.63.
K. Simoes, C.R.Sluter, L.Simoes, Geographic Information System For Apiaries Management [J], 39th Apimondia International Apicultural Congress,2005，P.60.
Liu Dongmei, Construction of the Information System for Agricultural Product Quality and Safety [J], Journal of Anhui Agricultural Sciences, 2006.34(21), P5752-5755（in Chinese）
Liu Haifeng, Meng Xiangbao, etc. Design and Development of Agricultural Products Quality Safety Monitoring Information Management System [J], Modern Agricultural Equipments, 2007.6, P47-51（in Chinese）
Qi Yamei, EU Food Safety Alarm System [J], China Animal Husbandry Bulletin, 2007.7, P72-74（in Chinese）
Shen Guanglei, Zan Linsen, etc. Implementation of Beef Quality and Safety Traceability System via Internet [J], Transactions of the Society of Agricultural Engineering, 2007.7 (23)，P170-173（in Chinese）
Zhang Bing, Huang Zhaoyu, etc. Design and Implement of Vegetable Quality and Safety Traceability System [J], Food Science, 2007. 8(28), P 573-577（in Chinese）
Zhao Chunming, A Study On Techniques Systems Of Agricultural Products Quality Control [J], Agriculture & Technology, 2007. 27 (4)，P13-17（in Chinese）
Zhao Jing, A Study on the Quality and Safety Technology of Bee Products[J], Apiculture of China, 2006.12(57) , P30-31（in Chinese）
Zhao Jing, Quality Safety of China Bee Products [J], Apiculture of China, 2002. 53 (1) , P29-30（in Chinese）
Zhang Ting, Han Jianzhong, The Application and Development of the Traceability System in Food Safety Control and Management [J], Food Research and Development, 2007.8 (28) , P171-175（in Chinese）

INTERNET USAGE IN THE FRESH PRODUCE SUPPLY CHAIN IN CHINA

Xiaoxiao Xu[1], Yanqing Duan[1,*], Zetian Fu[2], Xue Liu[2]
[1] *University of Bedfordshire Business School, UK*
[2] *China Agricultural University, PR China*
[*] *Corresponding author: Prof Yanqing Duan, University of Bedfordshire Business School, Park Square, Luton, LU1 3JU, UK, Yanqing.duan@beds.ac.uk.*

Abstract: Although effective implementation of the Internet technologies has a great potential for improving efficiency and reducing wastage within the fresh produce supply chain. the situation of the Internet usage by SMEs (small and medium sized enterprises) in the fresh produce supply chain is still unclear in China. As the main players, SMEs haven't been given enough attention from both academics and governments. Therefore, this research attempts to address this issue by, first, investigating the current usage of the Internet and related software by Chinese SMEs in the fresh produce supply chain, and then, by identifying enablers and barriers faced by SMEs to call government's attention. As a part of an EU Asia IT&C funded project, a survey was carried out with SMEs in this industry from five major cities in China. The results reveal that in the relatively developed areas of China, SMEs in the fresh produce supply chain are rapidly adopting the Internet and software packages, but the level of adoption varies greatly and there is a significant lack of integration among the supply chain partners. Chinese SMEs are keen to embrace emerging technologies and have acted to adopt new software and tools. Given that cost of implementation is not a barrier, their concern over legal protection and online security must be addressed for further development.

Key words: Internet technologies, fresh produce supply chain, supply chain software and tools, small and medium sized enterprises (SMEs), China.

Please use the following format when citing this chapter:

Xu, X., Duan, Y., Fu, Z. and Liu, X., 2009, in IFIP International Federation for Information Processing, Volume 295, *Computer and Computing Technologies in Agriculture II, Volume 3*, eds. D. Li, Z. Chunjiang, (Boston: Springer), pp. 2151–2160.

1. INTRODUCTION

China is the biggest agricultural country in Asia and its entry to the WTO has further increased its influence. As the fastest developing country, China is currently experiencing dramatic changes in almost all industry areas, and it is facing the most challenging changes in agriculture, especially in the fresh produce sector. Having been traded in the street market and wholesale market for hundred years in China (Datamonitor, 2006), fresh produces are converting into the sales channels owned by supermarket, influenced by international multi-retailers, such as Carrefour and Walmart. There is also a call from the government to facilitate this change, as customers can benefit from food safety traceability, guaranteed produce quality, and a relatively cleaner shopping environment (Ministry of Agriculture of the People's Republic of China, 2006).

The Internet and information technologies play a critical role in helping SMEs (small and medium sized enterprises) to embrace the technological challenges and improve the business competitiveness in a global market. However, several issues, such as concerns over the online security, low level of users' skills, cost of implementation, lack of legal protection of online transaction and lack of awareness of using technologies, discourage SMEs to adopt and use the Internet and related software. Funding from the Europe Asia IT&C programme by European Commission was granted to a project called VEGNET: 'Enhancing Vegetable Supply Chain Management with Internet Technologies' (http://vegnet.beds.ac.uk/vegnet). The aim of this project was to transfer managerial and technological know-how to Chinese SMEs from Europe and to introduce supply chain management concepts and the use of Internet technologies to support supply chain management (SCM) in Chinese agri-businesses.

The research presented in this paper aims to analyse the current Internet usage in the fresh produce supply chain in China, to benchmark the supply chain management status for Chinese fresh produce companies, and to evaluate the factors influencing the Internet usage in the fresh produce industry. The results provide a contemporary overview of the current development, pinpoint the issues in the Internet adoption and usage in Chinese agribusiness, and provide suggestions for future research and practice.

2. IMPORTANCE OF SMES IN THE FRESH PRODUCE SUPPLY CHAIN

Chinese government has put tremendous efforts on so-called "informatisation" to enable people in agriculture obtain in-time information from the market, especially for those in remote rural areas. The development of agricultural websites grew rapidly since 1998, which include government websites, educational websites, and commercial websites (Xiang et al., 2006). From all agricultural websites, commercial websites owned by companies are 82.6% in 2003 (Yang et al., 2003), and progressed to 84% in 2006 (Xiang et al., 2006). However, those data did not differentiate the contributions made by SMEs.

Chinese publications indicate that most agribusiness companies are SMEs, and recognize the importance of SMEs in a macro-economic point of view (Li, 2000). However, there is no clear indication of the percentage of agribusiness SMEs in the industry. A scholar, Mr. Zhou (Zhang, 2003) complained about inconsistence in statistics of SMEs among different departments of the central government. More ironically, China SMEs Online, the official SMEs department of China, did not cover the SMEs in agriculture in its national "Survey Report of Informatisation in Chinese SMEs (SME online, 2006). Those facts demonstrate that agribusiness SMEs are ignored by previous research in terms of "informatisation", although they are believed to be important for the agriculture industry.

Previous research is mostly based on anecdotal evidence when analyzing the role of the Internet in agriculture (Wang and Liu, 2005), benefits that the Internet can bring to farmers (Wan, 2007), difficulties of farmers to accept it, and suggestions to promote Internet usage by farmers (Tian and Chen, 2006). Those literatures may serve as a good foundation to propose ideas, but no empirical investigations have been conducted to find out the situation of SMEs acceptance of the Internet, what factors enable them to use it, and what factors discourage their usage. This research aims to fill this gap by conducting a survey study with SMEs in the fresh produce supply chain and attempts to reflect the current situation on the adoption of Internet technologies. Findings shed light on the current situation on the use and further development of Internet technologies along the Chinese fresh produce supply chain and provide valuable implications for Chinese government and policy makers.

3. METHODOLOGY

3.1 Survey design

The questionnaire contained three sections. Section one collected information about the company and respondent profile, including the role of the company in the fresh produce industry, company size, business scope, managers' understanding and skills. Section two was about the technology issues of the company, such as the software used in the supply chain, Internet functions, the identification of the development level, and level of Internet-based supply chain development etc. Section three dealt with the internal and external factors influencing the level of Internet usage. A five-point scale was used to measure the degree of the influence factors.

3.2 Survey administration

The survey was targeted at SMEs in the Chinese fresh produce supply chain, including growers/ farmers, food processors/ manufacturers, wholesalers, importers/exporters, retailers and transporters. Because the definition of SMEs in China varies by industries (not including agriculture) (China SME online, 2003), it is difficult to follow. To be comparable with the data in the UK survey, we define a SME as a company with 10 to 250 employees, by adopting the definition by the UK Department for Business, Enterprise and Regulatory Reform (BERR, formerly DTI), which is also a widely accepted definition in the EU.

The survey was conducted by using a data collection service in China, who carried out a Computer Assist Telephone Interview (CATI) in 5 representative cities in China, covering five major representative cities, Beijing (North), Shanghai (East), Guangzhou (South), Chengdu (West), and Wuhan (Central), with 30 samples from each city. From the data provided by the service, overall 2,995 companies were contacted. Every effort was made to reach a diverse but valid sample group. 151 valid responses were received, which gave a 5% overall response rate. By listening to the interview recordings, the authors gained tremendous information about the viewpoint of agribusiness managers in China. In this paper, quotes are provided wherever possible for a further explanation.

3.3 Profile of respondents

The responses covered almost all of the players in the fresh produce industry in China. A majority of them come from wholesaler (68%) grower/ farmer (55%), and retailer (52%), and the rest of them come from food processor / manufacturer (27%), transporter (17%) and importer/ exporter (17%). Most SMEs play a multiple role in the supply chain. For example, some of the SMEs have their internal vertical integration by participating in growing, manufacturing, wholesaling and retailing. However, most growers and wholesalers outsourced transportation service to a third party.

The questionnaire was required to be completed by the company owners and managing directors, or by other senior managers who had a broad knowledge of the fresh produce supply chain operation and information systems presently used. From the responses, 66% of the replies were from company owners, CEOs, or managing directors, while 16% were completed by logistics/supply chain/operation directors and 18% were answered by IT managers or others.

According to BERR's definition, we included businesses with a current headcount of up to 249 staff employed directly by the business. This clarification is critical, since in the fresh produce industry, most companies employed temporary/ casual/ seasonal staff. About 40% of the respondents belong to medium sized companies (50-249 employees), while there were 60% of the surveyed companies are small companies (10-49 employees).

4. SURVEY RESULTS

4.1 Current use of Internet technologies

The situation of the use of Internet technologies is covered by mainly two parts, Internet and website application, and related software packages.

As shown in Table 1, only 7% of the companies have no Internet applications, while email systems and online communication tools are very common for the surveyed SMEs. The respondents believe that these tools are much easier and cheaper than traditional ways of communication, and they can also obtain more product and price information by contacting prospective customers through the online communication tools such as QQ, MSN (with overseas partners) or other tools developed by online marketplaces such as Alibaba. The usage of Intranet and Extranet are limited compared with other usages.

Table 1 Internet Applications in China's Fresh Produce SMEs

Internet Applications		Website Functions	
Email system	83%	Company information	97%
Online communication tools	77%	Products information	96%
Company website	59%	Online ordering	53%
Intranet	36%	Customer account management	33%
Extranet	30%	Online payment	30%
No Internet	7%	Supplier account management	28%

More than half of the companies have company websites, and most of the websites contain company information and product information. Over half of the respondent companies provide online ordering functions. Some of them also manage their online payment and customer and supplier accounts on the Internet. From their own explanations, they usually have an account in the online marketplace such as Alibaba, which hosts a homepage of their company. Orders are redirected from their website to the online marketplace to be processed. However, most deals are made after face-to-face meeting or phone calls. They express their concern over the authenticity of online information.

There are some companies, which had bad experiences in online trading. When asked if they have online communication tools or a company website, here are some examples of responses: "I don't trust any information online." "It is too easy for people to cheat online. Nobody would know that other people are telling lies". "You will never see a real price online."

Regarding the use of the supply chain management software (Table 2), results show that the technologies are becoming accepted by a limited group of SMEs. Of the software packages currently used, Barcode systems (31%) and WMS (warehouse management system) (30%) are the most popular, while GTS (geo-code track system) (11%) and RFID (radio frequency identification) (9%) are least popular by fresh produce companies.

Results also show that Chinese agribusiness managers have a limited knowledge of supply chain software packages, for example they argued that "we are just growing/selling vegetable/ fruits, and those high technologies are not possible to use." They haven't realized the possible impact that those emerging technologies could bring, and still believe that agribusiness ought to be not technology driven. For many of them, it was their first time to hear those software names or explanations.

Table 2 Software Used in SMEs

	Barcode systems	WMS	CRM	EDI	AQC	TMS	ERP	GTS	RFID
Currently using	31%	30%	27%	23%	19%	15%	12%	11%	9%
Plan to use in 2 years	19%	13%	22%	18%	19%	16%	16%	12%	7%
Plan to use over 2 years	8%	14%	8%	10%	9%	10%	19%	11%	13%
No plan to use	23%	19%	25%	26%	27%	31%	29%	29%	42%
No need to use	15%	18%	14%	14%	15%	25%	15%	31%	17%
Don't know	5%	6%	4%	8%	11%	3%	10%	6%	12%

Please note that a full explanation of the software list can be found in Duan et al. (2007).

Looking at the future, companies have little intention to use RFID, as 42% of them had no plan to use the system, 17% of them felt no need to use it. CRM (customer relationship management) is more likely to be used in two years. In the longer term, ERP (enterprise recourse planning) is more likely to be used. It is also quite common that many respondents felt hard to clarify their future plans, though they were at the top management of the companies.

"Whether to use any software could not be decided by me. I have to see the trend of the industry."

"It is my first time to hear those strange words. I have to decide whether to use them according to the development of my company as well. Our company is still small, not like the big companies."

"I usually contact our trade partners by telephone, and seldom use the Internet. I don't trust anything online."

"Fresh produce business is quite different from others. You have to see the real products. Every orange is different, even though they grow in the same plant. You have to see them by your own eyes. It is impossible to have online transaction for this type of business. The Internet just works as a marketing tool."

4.2 Factors affecting Internet adoption

The results presented in Table 3 and 4 show the perceived motivators and barriers of using the Internet by Chinese agribusiness managers.

Managers believe that "Pressure from market competition" is the biggest driver for them to use the Internet. Some managers commented that "the real benefit that the Internet could bring to us is really limited, but we really have to be present online as a marketing strategy." "Geographic scope of the company" is ranked in the second place, because companies feel that the large scope of their business within China or Asia enables them to use the Internet with more prospective clients. However, companies feel less "Pressure from customers and suppliers", because it is unlikely to lose a customer or supplier only because of not using the Internet. They explained that they are very flexible in using the technologies, no matter what tools their clients use.

Table 3 Motivators of Internet Usage

Motivators	N	Mean	S.D
Pressure from market competition	145	3.84	1.27
The geographic scope of your company	144	3.39	1.21
The nature of the fresh produce industry	147	3.26	1.41
Government provided incentives	145	3.17	1.38
Perceived benefits of IT implementation	147	3.16	1.21
Pressure from top management	146	3.00	1.36
Bank requirements for transactions	142	2.95	1.41
Pressure from customers for integration	145	2.90	1.34
High complexity of business operation	143	2.81	1.28
Pressure from suppliers for integration	143	2.66	1.31

Regarding barriers to the Internet adoption, a very strong concern of legal protection and security issues are demonstrated by Chinese SMEs. Many respondents expressed their concern over security. Most managers reflected their dissatisfactions with those out-of-date and poor qualities of the information online. They also regarded the lack of know-how and training as significant barriers. For example, in one of the cases in Wuhan, the local government helped farmers to launch a website for them to sell fresh produce, but the manager complained that the government people only teach them basic IT skills, without any commercial information of the trend of the market, and they have no idea where to find this information.

Table 4 Barriers of Internet Usage

Barriers	N	Mean	S. D
Inadequate legal protection for online trade	147	3.24	1.33
Concern over Internet security	149	3.19	1.42
Lack of know-how	148	2.99	1.35
Lack of training	147	2.87	1.32
Resistance to changes of adopting the Internet in the supply chain	146	2.77	1.37
Lack of collaborative partners	149	2.68	1.28
Lack of trust in trading partners	147	2.65	1.40
Poor IT infrastructure	144	2.65	1.36
Reluctance of sharing information	143	2.50	1.37
Cost of implementation	145	2.30	1.28

Surprisingly, "Cost of implementation" and "Poor IT infrastructure" are ranked as the least barriers to Internet adoption. Their comments not only reflect their limited understanding of the cost of Internet implementation, but also reinforce the fact that the concern over Internet security discourages their usage of the Internet. The managers commented that "Computers and Internet access are not a problem for us. Paying money to online marketplace is also not a big deal. But all we are worried is how much genuine information we will be able to obtain? We don't want to spend money for rubbish information."

5. DISCUSSION AND CONCLUSION

This paper represents a key part of a research effort in understanding the current situation and promoting a better adoption of Internet technologies in the Chinese fresh produce supply chain.

Literature reviews reveal a lack of investigation of Internet technology usage by SMEs in the fresh produce industry. By using a survey approach, this paper provides both qualitative and quantitative information on the current development in the use of ICTs in the fresh produce supply chain in China. The paper also helps to identify the key motivators and barriers of the Internet usage in Chinese SMEs.

The findings from the survey interviews show that fresh produce SMEs are using the Internet to communicate via online tools. SMEs perform well in using the online marketplaces for information searching. There are more SMEs doing online business transactions compared with those in the UK (Duan et al., 2007). In terms of supply chain software, a small group of SMEs are using online software, but most SMEs do not have any intention to use software or emerging technologies. They haven't realized that supply chain software packages play a key role in improving website functionalities and internal supply chain efficiency.

Market competition is the most significant driver for SMEs. To survive in the very competitive business environment, companies have to have an online presence to attract more potential customers. Pressures from their trade partners seem not to be a major driver for the Internet adoption.

Cost of implementation and poor IT infrastructure of the Internet have been considered as barriers by previous research in Western countries (Peng, 2005, Wang and Liu, 2005, Chen, 2005), but this research reveals that among all the barriers, cost of implementation is considered as the least important one, while poor IT infrastructure is considered as the third least one. In contrary, lack of legal protection and concern over Internet security are considered as the biggest barriers by SMEs. Several Chinese publications also mentioned that the government should develop appropriate laws to protect online transaction (Peng, 2005, Feng, 2007). The findings also suggest that Chinese SMEs are keen to embrace emerging technologies and have acted to adopt new software and tools. Given that cost of implementation is not a big barrier, their concern over legal protection and online security must be addressed by policy makers and the government.

It should be noted that the above findings are based on the survey in 5 large cities in China, and may not represent SMEs in other small cities and less developed remote regions. However, the research findings have a significant contribution to a better understanding of Internet applications in agriculture sector and new knowledge on factors affecting the technology adoption and further development. Findings also offer practical guidelines and managerial implications to agribusiness mangers and government policy makers.

ACKNOWLEDGEMENTS

This research is partially funded by European Commission's Euro Asia ITC program (CN/ASIA-IT&C/005 (89099). The authors would like to acknowledge the financial support of EC and all the project partners for their collaboration and contribution.

REFERENCES

Chen, X. (2005) Exploitation and Utilization of Agricultural Information Resource under Network Environment- Focus on Network Exploitation and Construction of models Used by Farmers. Journal of Library and Information Sciences in Agriculture, 17, pp. 56-58.

China SME online (2003) Temporary Definition of SME, http://www.sme.gov.cn/web/assembly/action/browsePage.do?channelID=1172982990367 &contentID=1172982990524,SME Department of National Development and Reform Commission of P.R. China.

Datamonitor (2006) Food Retail in China. Datamonitor.

Duan, Y., Xu, X., Liu, X., Fu, Z., Zografos, K. & Bemeleit, B. (2007) Accelerating Internet Adoption in China's Fresh Produce Supply Chain: a VEGNET Approach. New Zealand Journal of Agricultural Research, 50, pp. 1299–1305

Feng, W. (2007) Research on E-business in Agriculture. Wenzhou Agriculture Technology, 1, pp. 20-21 and 45.

Li, Z. (2000) How to Transfer Surplus Labour in Rural Area in China, http://www.ncer.tsinghua.edu.cn/research/trend/papers/27.htm, accessed by Jun 1st, 2008

Ministry of Agriculture of the People's Republic of China (2006) Design and Plan for Agri-product Market System for the "11th Five Year" Period, http://www.agri.gov.cn/xztz/t20061212_738312.htm, Department of Market and Economic information

Peng, J. (2005) A Brief Discussion on How to Develop E-commerce in China's Agricultural Corporations. Research in Agricultural Mechanisation 4, pp. 54-56.

SME online (2006) Survey report of Informatisation of Chinese SMEs, http://www.sme.gov.cn/web/assembly/action/browsePage.do?channelID=1154936200894 &contentID=1154936201212, SME Department of National Development and Reform Commission of P.R. China.

Tian, M. & Chen, H. (2006) Problems and Solutions of the Development of E-business in Agriculture in China. Jiangxi Agriculture Academic, 18, pp. 203-204.

Wan, Q. (2007) Research on the Electronic Commerce of Agriculture. Agriculture Network Information, 9, pp. 113-114.

Wang, X. & Liu, Y. (2005) Using E-business to Improve the Development of Agri-business. Economist, 7, pp.185 and 187.

Xiang, J., Xiao, L., Meng, X., Cui, L., Xiao, N. & Jin, C. (2006) The Situation and Evaluation of china Agriculture Information Website Development. Agriculture Network Information, 2.

Yang, X., Jia, S. & Zhao, Y. (2003) Current Situation and Evaluation of Agricultural Websites in China. Computer and Agriculture, 9, 19.

Zhang, L. (2003) How Many SMEs are There in China? http://www.china.com.cn/chinese/OP-c/421148.htm, accessed by May 31st, 2008

PRACTICAL USE OF IT IN TRACEABILITY IN FOOD VALUE CHAINS

Jon Ratcliff[1], Michael Boddington[2]

[1] *F.A.C.S. Ltd, The Old Stores, Binton, Stratford upon Avon, Warks, CV37 9TN, United Kingdom, Email:* jon@facs.uk.com

[2] Asian Agribusiness Consulting, Suite C2201, Bai Jia Zhuang Commercial Centre, San Li Tun South Rd, Chaoyang District, Beijing 100020, Email: michael@*boddingtonconsulting*.com

Abstract: Traceability is today considered an essential requirement for the food value chain due to the need to provide consumers with accurate information in the event of food safety recalls, to provide assurance with regard the source and production systems for food products and in certain countries to comply with government legislation. Within an individual business traceability can be quite simple to implement, however, in a global trading market, traceability of the entire supply chain, including logistics is extremely complex. For this reason IT solutions such as TraceTracker have been developed which not only provide electronic solutions for complete traceability but also allow products to be tracked at any point in the supply chain.

Keywords: traceability, global food value chain, tracking, IT solutions, TraceTracker

1. INTRODUCTION

Today's consumers are informed and aware and they demand much more information about the food they buy than their parents did. Beyond simply wanting to know how much it costs, today's shoppers also ask: Is it safe? Is it healthy? Does it contain any allergens? Where was it produced? Was it grown organically? Is the farming and harvesting done in a sustainable manner? Is the packaging recyclable?

As a result, brand and private label owners need to communicate more information about their products than ever before. To respond to the modern

---

Please use the following format when citing this chapter:

Ratcliff, J. and Boddington, M., 2009, in IFIP International Federation for Information Processing, Volume 295, *Computer and Computing Technologies in Agriculture II, Volume 3*, eds. D. Li, Z. Chunjiang, (Boston: Springer), pp. 2161–2175.

consumer's need for food they can trust, they need to be able to obtain accurate information from within their own business processes, from their supply networks and from their suppliers' suppliers and intermediaries. Even for a simple food chain example, the amount of data required and the time scale for response to provide complete traceability is one reason that IT solutions can be an invaluable tool to provide the information required within the food value chain. For the purpose of this paper I will refer to the TraceTracker software solution package.

2. BUSINESS DRIVERS FOR IMPLEMENTING TRACEABILITY

Food safety is probably the primary basis for the current drive for traceability in the food industry. Food scares are usually based on solid facts of unsafe foods, and it then takes solid facts to keep the public from becoming scared of all products in the same group. Traceability and traceability systems provide such facts. In cases of unsafe foods, it is vital to be able to differentiate the safe from the unsafe. Without traceability, this may be impossible, and scenarios of very costly product recalls may result. Some products and companies may take major blows from recall scenarios, and may never recover their market shares or even go out of business. There are many examples of such incidents, and there will undoubtedly be more to come. Traceability is an important form of insurance from such situations, and traceability systems are the tools for using the traceability information registered. Companies that implement traceability systems may actually get lower insurance premiums, both for their recall insurance and their general liability insurance.

Governments all over the world have recently started to focus more and more on food safety and traceability. Politicians do not want to be held responsible for food incidents, and in this case, they are doing their job right, pushing businesses to implement traceability by defining new laws and directives.

In today's markets, competition is often fierce, and differentiating higher quality products from the general commodities can be a difficult task without hard facts to prove marketing claims. Traceability can supply the needed proofs, facilitating high-end markets for superior products. For example, in many countries, eggs from free range chickens are preferred due to both welfare and quality issues and they are willing to pay a higher price for such eggs. However, consumers do not like to be cheated, and there have been a number of incidents in Europe whereby cage eggs have been sold as free range. It is imperative for the honest farmers to have good traceability practises to be able to restore consumer confidence in their brands.

3. DEFINITIONS OF TRACEABILITY

The most widely used definition of traceability, is probably the ISO9001 definition from 2000. It defines traceability as:

«The ability to trace the history, application or location of an item or activity by means of recorded identification. When considering product (3.4.2.), traceability can relate to the origin of materials and parts, the processing history, and the distribution and location of the product after delivery».

This definition of traceability is considered by a number of experts as being too broad in scope for use in the specific context of the food value chain. The EU Commission[1] (defines traceability within food value chains as:

«the possibility to find and follow the trace, throughout all the stages of production, processing and distribution of a food stuff, feed stuff, an animal destined for food production or a substance destined to be incorporated in food stuff or feed stuff or with a probability of being used as such»

The EU definition is a little bit more specific, but still vague, for instance in what it means by follow the trace.

It is important to stress that traceability really involves more than only the minimum of what is required to provide the one-up, one-down type of information required by different legislations which tend to look at traceability as something internal to each company in the chain. However, it is important not to miss the complete chain view on traceability. As an example, consider the following figure:

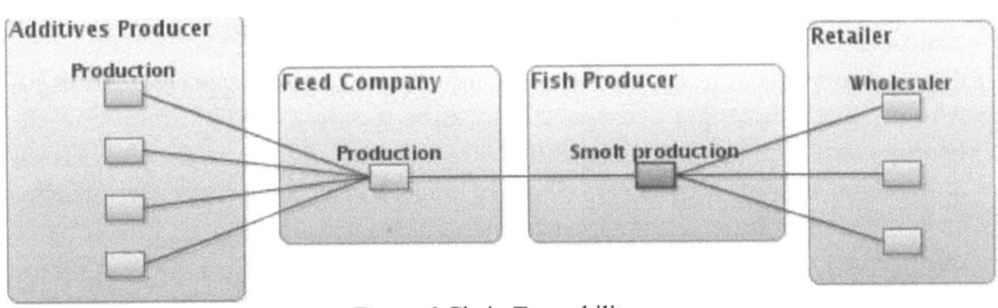

Figure 1 Chain Traceability

In this scenario, all four companies in the chain are able to trace and track their raw materials, their internal production batches and their outgoing trade units, not only one step up and one down, but across the whole chain.

TraceTracker offers the following definition:

«Chain Traceability means the ability to transparently trace and track the production and manufacturing history and content of any food item (including ingredients) exchanged between companies in the chain. This requires each company to

1. register all their incoming raw-materials in a globally unique way at the batch/lot level, and who the supplier is
2. link the raw material lots to the internal production steps they are entering
3. link together all internal production steps up to and including the final step of production where the «trade units» are created
4. for each such traded unit, register who the traded units are being sold to»

Two words which are frequently used, are the words tracing and tracking. Some tend to use these interchangeably, but there are different semantics associated with them, which let us make more precise definitions. Tracing is normally defined as following a path backwards towards its origin, also called upstream, in the meaning towards where the stream starts. In this case, the stream is the food value chain. Tracking is normally defined as following a path forwards towards its end, also called downstream.

4. BUSINESS EXAMPLES

In a small business, such as a single fishing boat, traceability would consist of logging when and where a catch (in traceability terminology called a lot or a batch of fish was caught and when it was sold and delivered to a specific wholesaler or other bulk customer. The log could be electronic or on paper, this does not matter for traceability, except that electronic documentation may make it much simpler and faster to communicate the traceability-information to the customer or other interested parties. When a customer might ask: Where did the fish I bought come from? The fish person would consult their log and have a ready answer. If the government asked who bought fish caught at a specific location within a certain timeframe, the fisherperson could again consult their log and after a little work would come up with a list of customers. This is traceability. Due to smart internal processes, the little fishing business can be said to have full internal traceability.

Out there in the real world, there are businesses of greater complexity than the simple business described above. Some of these may be vertically integrated, meaning that they have subsidiaries that produce the parts or raw materials they use (backward integration) or they may have subsidiaries that use or distribute their products (forward integration), or both. Vertical integration means extending the value chain of the enterprise to a higher number of traceable links. For example, a fish farming company may decide to buy both a feed producer, a fillet plant and a distribution company, thus covering the complete value chain from raw materials to supermarket chain Traceability for such a company would require detailed logs of many manufacturing steps to determine where the fish products came from, but all the steps could be local to a single enterprise. Such traceability would be hard to handle on paper, but can certainly be well handled electronically, given good traceability routines and reasonably standardized IT systems across the whole enterprise chain.

When different businesses anywhere in the world trade products with each other, global value chains are a fact. For instance, a food additives producer in Germany may sell chemicals to a fish feed producer in Norway, which provide raw materials (feed) to a fish farmer in Chile, while this company may sell slaughtered fish to the big Japanese retail market. This would be a truly global value chain, and traceability in this chain could be hard to imagine. However, with today's telecommunication infrastructures and the Internet in particular, such distances are not a problem. The companies mentioned should, however, have good traceability processes and internal traceability systems implemented, and of course some effective tools for interconnecting their traceability data. Such tools happen to be just what TraceTracker provides.

5. LEVELS OF TRACEABILITY

It is sometimes useful to distinguish between different levels of traceability, as both business use-cases and the required capabilities of the (traceability) software systems involved will differ slightly among these levels. The lowest level of traceability is called internal traceability. This is about traceability inside and internal to one single company. This is sometimes also called one-up-one-down traceability, which means keeping records of the sources of everything received and of the destinations of everything sent out. But note that one should actually differentiate between one-up-one-down and full internal traceability. The latter is more extensive than one-up-one-down. One-up-one-down traceability only require that a business keeps records of who their supplies come from and to whom products are sold. There may not be records inside the company of which supplies are used for which products. Full internal traceability on the other hand, also requires records of which raw materials are used for which products, and this is clearly more extensive than on-up-one-down. Internal traceability can for simple internal chains sometimes be handled by general or special purpose Enterprise Resource Planning (ERP) systems or other types of enterprise software. The reason is that the business scope of such software is per definition a company internal scope, hence the name Enterprise Resource Planning. A reasonably modern ERP system will therefore most often be able to handle one-up, one-down traceability, given that the required data is fed into the system. The same may be true when it comes to full internal traceability, but unless good traceability practises are implemented in the company, one often see that the required data is not logged, even if the system was capable of modelling full internal traceability. For more complex internal value chains however, ERP systems tend to come short in modelling the required dynamics, and special purpose traceability software should be installed on top of the ERP systems.

Enterprise traceability is just full internal traceability extended across two or more locations or companies within a larger enterprise, for instance within a vertically integrated company. Requirements and challenges are therefore the same as for internal traceability, but complicating factors may be that more than one ERP system are involved, and with probably missing links between the different systems.

To accomplish Chain Traceability, it is necessary to be able to trace and track between different companies in a given value chain, as stated in the definition given at the beginning of this document. Clearly, this requires some kind of interconnected value chain, and standardized ways of communicating with each other. In addition, each company needs to have full internal traceability in place, so that it is possible to trace and track through the internal chain of each company. TraceTracker provides a software solution called the Global Traceability Network (TT GTNet) which is able to connect different internal traceability systems from different companies.

6. PREREQUISITES TO TRACEABILITY

Traceability is not only about having good software systems to administer all the needed information, it is also about having internal production processes that support traceability and not break it. In fact, this is the single most important thing to focus on before doing any installation of traceability software.

All internal processes supporting the internal production, from the reception of raw materials to the shipment of finished products, must support traceability, not break it.

Examples of processes which do not support traceability, can be

1. feeding silos which are hardly ever emptied completely or cleaned between each refill
2. water tanks similar to feeding silos
3. recycling ingredients in production processes

1. and 2. are sometimes described as the never-ending-batches problem, both often seen within all kinds of industrial animal production. It is not always obvious whether existing processes need to be re-engineered to

guarantee a satisfactory level of internal traceability, so it is recommend that companies conduct a traceability survey done by traceability professionals.

7. UNIQUE IDENTIFICATION SCHEMES

One of the most important things a traceability survey will address, is whether the traceable entities (batches and trade units in our model) are being identified properly. By properly, we mean in a unique way within the scope of such a traceable entity. The scope of a batch is per definition an internal scope, and internal to a company or probably an enterprise. Supporting software used in the internal production must then be able to generate enterprise wide unique identifications for all batches in the enterprise value chain. If the numbers generated for some reason are recycled at some time in the future, this must not happen until the old batches with the same numbers are completely out of scope. Trade units on the other hand, are exchanged between companies, so the scope of a trade unit is really a global scope. This means that trade units must be identified in a globally unique way. As it happens, there is an organization which has been working with global identification schemes for a long time, namely the GS1 – formerly EAN.UCC organization.

EAN.UCC, recently renamed to GS1, is a globally active industry organization with members in 103 countries. The organization handles standardization of numbering systems for many purposes. The mission statement for GS1 is as follows:

By creating Open, Global, Multisectoral Standards based on Best Business Practices, and by driving their implementation, play a leading role in Supply & Demand Chain Management improvement worldwide.

The organization is probably best known for their standardization of bar code systems world wide, but they have recently broadened their business scope, for example by looking also into RFID based technologies. When talking about traceability, it is really the numbering systems GS1 offer which are interesting, not the technology used to carry the numbers, i.e. bar codes, RFID tags, EDI or XML messages and so on. GS1 provides unique numbering systems for a wide range of applications, and also suggests how several such numbers can be concatenated to form globally unique traceability keys.

7.1 The Global Trade Item Number – GTIN

The GTIN is one of the most important of the GS1 numbering systems. GTINs are allocated to GS1 member companies according to criteria decided by GS1. They consist of two main parts:

1. A company prefix
2. A product number

In addition there is an extra GTIN type digit, and a check digit. The GTIN is easily encoded in different bar code schemes or alternatively in an RFID tag following a product. GTINs are used by the companies to identify specific products. GS1 suggests allocating the numbers sequentially, so two consecutive numbers may specify two totally different products, but this is really up to the company. The numbers may be reused after 4 years if a product is discontinued. Each company that has a company prefix can freely allocate their own GTIN product numbers, but quite detailed rules for how to do so are given by GS1 in order to have uniform practises around the world. An example of a GTIN is given below, here encoded as an EAN-13 bar code.

5 412345 678908 >

GTINs are perfect for identifying a manufacturer of a product and identifying a product's exact type. However, there may be millions of identical products all carrying the same GTIN number. If there is no way to differentiate these products from each other, tracing the product to the correct production batch on the correct day may be impossible. Therefore, a batch or lot number that is unique to this batch of products is needed. This information must be carried along with the product to its point-of-sale (POS) if traceability from consumer to ingredient production shall be possible. Note that a company that manufactures a specific product is uniquely specified by the company prefix of the GTIN number, so the company only needs to allocate its batch/lot numbers unique internally within the company for the combination of GTIN and batch/lot number to be globally unique.

7.2 Bar Codes

The bar codes you normally find on all supermarket items are normally used to carry GTIN information. They are read electronically with a bar code reader, but the number information is also written at the bottom of the bar code, so it can be read by humans and manually entered into a point-of-sale computer if necessary. There are many types of bar codes, and some can carry only a few digits of information while others can carry more. GS1 has a variety of standardized bar codes, such as EAN-13, UPC-A, ITF-14 and UCC/EAN-128. These can carry different amounts of data, and can be used to carry for example a batch or lot number and more in addition to the GTIN.

We will look at the EAN-128 in some more detail, as this is the most modern and flexible bar code scheme for carrying information between trading partners.

EAN-128 defines a concept called Application Identifiers (AI), which is used to encode different fields in the bar code in a flexible manner. When the scanner equipment reads the bar code, it will identify the AIs. Each AI defines the meaning of the next N (the number N depends on the given AI) digits in the code, and they can occur in any order. Two trading partners can agree on what to encode in the bar code, based on the requirements of the receiver. With many customers, the manufacturer should of course try to standardize what is encoded in the bar code. The screenshot below shows a fictive EAN-128 bar code, where the numbers below the bars represent the encoded information in a human readable way. Application identifiers are often visualized by putting parentheses around them. In this example, we see the AI 01 which represents the GTIN, indicating that the next 13 digits represent the GTIN number.

(0 1) 0 5 4 1 2 3 4 5 6 7 8 9 0 8

GS1's globally unique numbering systems are perfect for constructing traceability keys that are globally unique, and that's just what is needed for identifying trade units. So, how can a company, through their ERP or production systems, create a unique traceability key for each trade unit they ship out? GS1 has a suggestion for what the key should look like:

Figure 2 The GSI suggestion for traceability keys on trade units

The suggested fields are shown above. The GLN is optional, but should be used when there is a risk that the same batch numbers may be generated across different production plants of the same company, for the same GTIN. The serial number is a way to serialise the production batch, and this is in fact necessary to be able to end up with a globally unique traceability key, which uniquely distinguishes between two trade units coming from the same production batch, the same GTIN and of course the same company. The serialisation is up to each company to implement, and the number can be anything as long as it gives rise to a unique traceability key. Sometimes companies use a datetimestamp to accomplish this.

7.3 RFIDs

RFID technology has been around for a while, its origins lie in the Identify Friend or Foe (IFF) systems developed for military aircraft in the second World War. The first trials of commercial applications began in the 1970s, but it was only in the late 1990s that RFID started to become viable for widespread business use. The technology has over the years steadily been improved, and today RFID devices are used for many practical purposes within a multitude of functions that include purchasing, production, storage, shipping and sales. It is for this reason that RFID is now common within many retail chains as a means to automate restocking, track products in transit and also track consumer purchase behaviour. The specific characteristics and functionalities of the software components of an RFID system vary greatly depending on the applications and other needed requirements. The components can be categorised as follows:

- **RFID system software:** This is needed for the collection of applicable software functions necessary to effectuate the basic interaction between a tag and reader. The functions are: read/write, Anti-Collision, Error detection/correction, encryption, authorization and authentication (security).
- **RFID middleware:** This consists of software components that acts as a bridge between the RFID system components (tags and readers) and the host application software. It performs two primary functions:
  - Monitors the device (reader) health and status
  - Manages RFID-specific (tag & reader) infrastructure and data flow.
- **Host application:** This receives processed and normalized data sent from the tag, via the reader and the RFID middleware software.

8. TRACETRACKER'S TRACEABILITY MODEL

It is useful to start this description with an illustration which depicts the chain view of the model, i.e. Companies in the chain trade with each other, and one single company is typically both a buyer and a seller. In Figure 3, this is shown.

The dashed rectangle in the figure symbolizes the internal production chain of a company. At the start of this chain, the company receives raw materials which enter the internal production chain. At the end of the production chain, some kind of final product is created and then sold. Both the received raw materials and sold products may be assembled in logistic units such as pallets, but this is a modelling option. The model defines two types of traceable entities – batches and trade units. They are described in the following:

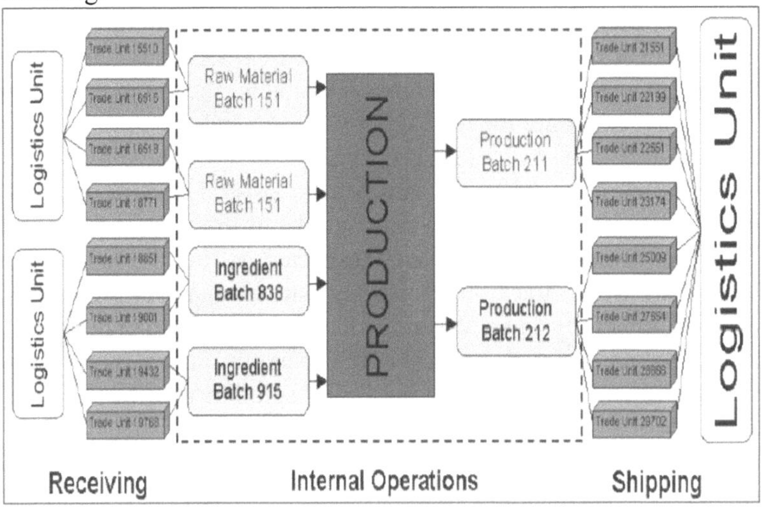

Figure 3 TraceTrackers traceability model

Batches are one of the basic units of TraceTracker's traceability vocabulary and represent an object of production internal to a company. They will seldom need to be exposed to the outside world, and are mainly of interest to quality managers and staff working with the internal production of a company. However, batches carry with them the history a final product has gone through, from the mix of raw materials through intermediate internal batches up to a final product. This history is modelled through two other modelling concepts described below – transformations and batch properties.

A typical batch in for example a bakery would be the last morning's production of a certain type of bread. These loaves of bread may all be expected to share the same characteristics since they are made from the same raw materials. If they were made from two different batches (for example two shipments from different mills) of flour, the baker might have separated them into two batches of bread for traceability reasons, but chances are, if something was wrong, he wouldn't know which loaves of bread came from which flour. (Unless flour is known to vary much in quality, in which case an expert might be able to trace batches based on quality differences). A Production Batch is a normal batch which is marked as being a production batch, which should be taken to mean that it is the last batch of the internal production chain. The reason for giving these batches special attention, is that these batches are the primary targets when doing recalls and withdrawals.

Trade Units are another kind of traceable entities, but they are used to model trade between companies and from retailers to consumers. The distinction between batches and trade units is especially useful when we look at the different levels of traceability described earlier. Internal traceability systems and processes have the **batches** in focus for the tracing and tracking, while the **trade units** are the primary focus in global traceability scenarios. In reality, both batches and trade units need focus when there is a food crisis where some contaminated traded goods have crossed several organizations in the value chain. A typical scenario would be to first locate the production batch of the company which were suspected to be the cause of the contamination, then to track this batch downstream again, as well as to trace it backwards internally to be able to isolate the cause of the contamination. To be able to handle such scenarios transparently and fast, both internal traceability systems and a global traceability solution which can connect all the internal traceability systems are needed.

Both batches and trade units can appear in clusters, and they will then be named as cluster batches or cluster trade units. The most typical use of this modeling concept, is to use it for modeling of logistic units, for instance pallets. Going back to the previous example – in the bread business, one loaf of bread would be a single trade unit at the consumer retail level, but the bakery might not operate with less than a dozen loaves as the smallest quantity of sale. This could be modeled as a trade unit cluster. Any batch or trade unit can have any number of so called properties associated with it, but all batches of the same type have the same set of properties. This must not be confused with the actual values of the properties. As an example, consider a model of a value chain for salmon farming. Let us assume that we had defined a batch type in this chain called Fry, and with two mandatory properties called Genetic strain and Number of fish. Two batches of this type would automatically have the two properties defined, but with possibly different values.

9. STATIONS AND TRANSFORMATIONS

Station is a useful modeling concept used to describe some kind of influence that batches or trade units are exposed to without being physically transformed. That is, they are kept intact throughout the period of influence. The traceable entities are said to enter and leave the station, and these two timestamps should be recorded. The station may be associated with measurable physical environment parameters, such as a log of temperature, humidity, pressure etc. A real life example of a station could be a truck. Modern trucking companies have many sophisticated means of recording data during a transport, for instance temperature loggers. If such a log was electronically available from the transporter, it could easily be associated with the station (the truck) and the batches or trade units which had been in the truck within a certain time frame.

Transformations model the changes raw material batches or intermediate batches go through as they are mixed and split and handled through a production process. Both batches and incoming trade units may take part in transformations. Transformations can also be thought of as the glue which links one traceable entity in the chain to the next step, which always is another batch or trade unit, or a station.

10. SOFTWARE SOLUTIONS

To be able to deal with traceability issues other than only the most trivial ones, efficient software solutions are needed. As previously mentioned, ERP systems or other types of internal enterprise software may be of some help, but primarily as tools for solving internal traceability issues. Even then, modifications or add-ons are often needed. Such software has not been written with traceabilty in mind from the beginning, and it may turn out to be more costly to build traceability functionality into such software than installing a native traceability application on top of the enterprise software. Some big vendors of ERP software have come up with dedicated modules targeting traceability in food value chains, examples could be SAP and Navision (Microsoft). Industry specific solutions also exist, but they tend to be very narrow and specific to the industries. They are not necessarily well suited for global traceability scenarios, at least not without add-ons on top of the solutions.

10.1 TraceTracker Enterprise Solution TT ES

TraceTracker Enterprise Solution is an Enterprise wide traceability system which can be used as a standalone system for internal and enterprise wide traceability. In addition, it is an enabling platform for being able to handle global traceability challenges, where the company or enterprise needs to be able to exchange traceability information with trading partners. TT ES integrates smoothly with any reasonably modern ERP system or other enterprise software, as long as the underlying systems are capable of modeling internal traceability according to the definitions given earlier in this document.

10.2 The Global Traceability Network – TT GTNet

TT GTNet is short for the TT Global Traceability Network. GTNet is created by TraceTracker and offered as a subscription service (Software as a Service) to members of any supply chain. By subscribing to the network and contributing own data to the network, the subscriber gets basic and transparent access to upstream and downstream trading partners in the same supply chain (not only one step up and one step down) they are members of. However, all access to data is controlled by the company who contributed the data in the first place. A robust web application is offered for easy access to all GTNet services. Public APIs are available for both data capture and data lookup. These APIs can be used to create custom applications. GTNet also provides rich and full blown services for internal traceability, and a subscriber may choose to use only this functionality (not sharing data with trading partners). The Global Traceability Network delivers traceability right across the businesses involved, allowing customers, governments and other interested parties to trace products from company to company. Figure 4 illustrates a real global example where a fillet of smoked salmon found in a supermarket in France may be traced through many links upstream in the value chain, through trucking, transport and fish farming to end up at a manufacturer of chemical additives used in the fish feed. As a result, users of the system always known the answer to such questions as;

Where are my products located?
Where, when, and in what quantities were they produced?
Which ingredients and processes were used to make them?
What quality assurance data is available for each of them?
Where, when, and in what quantities were they shipped to/from?

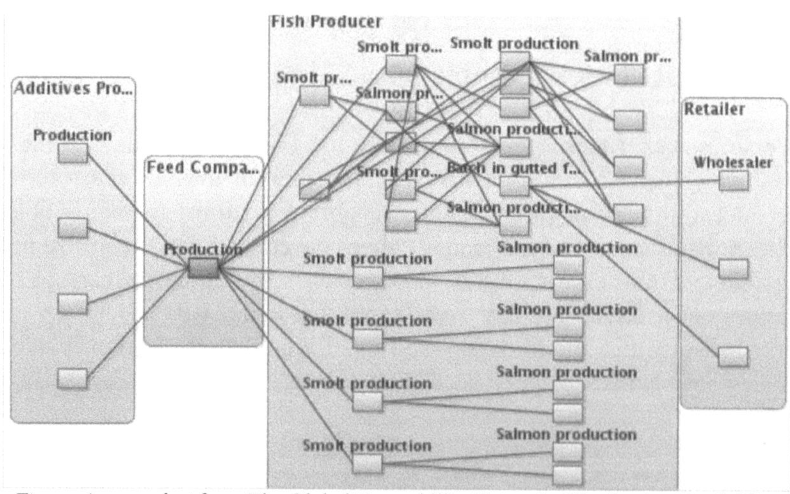

Figure 4 screenshot from The Global Traceability Network, "Production Batch View"

11. CONCLUSION

Traceability has been driven by issues of food safety and product recalls and fraudulent claims for higher value food products such as free range and organic. Traceability is ultimately about trust. Consumer trust for the products and brands that they purchase, and supplier trust within a global food value chain. It is possible to manage a system for a simple internal one step up and one step down traceability, but for more complex supply chain situations and the ability to provide real-time tracking IT software solutions are required. IT solutions are able to utilise internationally recognised identity code schemes to provide web based access for both traceability and tracking and provide real time answers that ultimately can enhance a brand or product or save cost through efficient management of the supply chain.

ACKNOWLEDGEMENTS

The authors are indebted to Dr. Ulrich Heindl for his help in the preparation of this paper and providing detail about TraceTracker

REFERENCES

Art. 3. Regulation (EU) No 178/ 2002 of the European Parliament and the Council of 28 January 2002.

FRACTAL FEATURE ANALYSIS OF BEEF MARBLING PATTERNS

Kunjie Chen [1,*], Chunfang Qin [1]

[1] *College of Engineering, Nanjing Agricultural University, Nanjing, Jiangsu Province, P. R. China 210031*

[*] *Corresponding author, Address: College of Engineering, Nanjing Agricultural University, Nanjing 210031, Jiangsu Province, P. R. China, Tel: +86-25-58606513, Fax: +86-25-58606699, Email: kunjiechen@njau.edu.cn*

Abstract: The purpose of this study is to investigate fractal behavior of beef marbling patterns and to explore relationships between fractal dimensions and marbling scores. Authors firstly extracted marbling images from beef rib-eye cross-section images using computer image processing technologies and then implemented the fractal analysis on these marbling images based on the pixel covering method. Finally box-counting fractal dimension (BFD) and informational fractal dimension (IFD) of one hundred and thirty-five beef marbling images were calculated and plotted against the beef marbling scores. The results showed that all beef marbling images exhibit fractal behavior over the limited range of scales accessible to analysis. Furthermore, their BFD and IFD are closely related to the score of beef marbling, suggesting that fractal analyses can provide us a potential tool to calibrate the score of beef marbling.

Key words: beef marbling, fractal, fractal dimension, image processing

1. INTRODUCTION

The beef marbling score, a dominant indicator in deciding beef quality in most current beef quality grading systems (Chen and Ji 2006), is a measurement of the abundance levels of intramuscular fat on the cross-section of the longissimus dorsi muscle (rib-eye) between the 12th and 13th or 6th and 7th ribs. In order to determine the beef marbling scores, the content and distribution of marbling flecks in the rib-eye region have to be

Please use the following format when citing this chapter:

Chen, K. and Qin, C., 2009, in IFIP International Federation for Information Processing, Volume 295, *Computer and Computing Technologies in Agriculture II, Volume 3*, eds. D. Li, Z. Chunjiang, (Boston: Springer), pp. 2177–2186.

taken into account. However, most studies on automatic grading techniques of beef marbling only pay attention to determining the grade of beef marbling in terms of variables such as the ratio of the area of fat regions over the entire beef rib-eye cross-section (abbreviated as fat ratio), the circumference of the fat regions, the number of fat particles and so on (Kuchita and others 1993, Shiraniata and others 1996, Jeyamkondan and others 2000). These feature quantities, in fact, are only associated with the overall fat content; not the distribution of fat flecks. Unfortunately, the differences in eating quality of beef can be attributed to not only the total fat content but also the fat distribution within the muscle (Albrecht and others 1996). Moreover, as previously indicated in beef studies (Kuchida and others 1992, Kurosawa and Nakanishi 1995, Shiraniata and others 1996), a considerable error would occur in grading beef marbling if the scores were evaluated only in terms of the above variables. A few of studies have focused on the distribution of intramuscular fat flecks. Yoshikawa and others (2000) proposed to use run length processing as a tool which takes into account the spatial information on the distribution of fat particles. Shiranita and others (2000) employed the amount of scatter of the distribution of marbling to take into consideration the geometry and location of fat flecks in the rib-eye region. As a further improvement of this approach, it is necessary to pay much attention to characterization of the geometric structure of the beef marbling so as to find out a more useful and appropriate feature quantity by which the overall distribution of fat particles over the cross-sectional rib-eye region can be identified correctly.

The concept of fractal geometry, which was created by Mandelbrot (1983), provides an excellent explanation of erratic natural patterns and many other natural phenomena. Its application to many fields in science and engineering has made a great progress in recent years. For example, many investigators employed fractal concept in the graphical simulation of natural phenomena like trees, clouds and so on (Huan and others 1997, Zhang and others 2000). In addition, other authors found that geometric characteristics of natural patterns can be described based on the fractal; and the recognition of human face, the letters, leaf and fingerprint etc., complex, irregular natural patterns, could be carried out by analysing fractal features (Zhang and others 1995, Borkowski 1999, Chen and Qi 2000, Yuan and others 2002).

The fractal dimension is a primary tool describing the fractal features, which can be employed to quantitatively measure the irregular degree, complexity and fragmentation of a set or pattern. Tian and others (2003) pointed out that the box-counting dimension values can be used to effectively characterize the spatial occupation degree of the crevices in grassland. Lian (2001) proposed that the box-counting dimension can reflect the spatial occupation degree of the Carpinus pubescence population, while the information dimension can quantify the variation degree of pattern intensity and the unevenness of individual distribution. Therefore, supposing that the beef marbling pattern is fractal, it is possible to use the fractal dimension to quantitatively describe the geometric features of distribution and amount of fat flecks.

Many methods including varying megascopic degree, interrelation function, distribution function and spectrum function etc. have been proposed for estimating the fractal dimension and analysis of the fractal features (Chen 1999; Sun 2004). From amongst these methods, the box-counting approach based on varying megascopic degree, although not quite exact, is the simplest to undertake using a personal computer, and it appears to be more suitable for the estimation of fractal dimensions (Tanaka and others 1999). Therefore, most researchers employ this method for the fractal analysis of a fractal pattern.

This study attempts to demonstrate that beef marbling exhibits fractal behaviour and to relate the fractal dimensions with beef marbling scores. Three aspects of this problem have to be addressed. The first question involves in the extraction of the beef marbling from sample images by image processing technologies. The second problem relates to the application of the method of varying megascopic degree to measure BFD and IFD of beef marbling images. Finally, the third aspect of this study discusses relationships between the score of beef marbling and fractal dimensions

2. MATERIALS AND METHODS

2.1 Hardware

The computer used in this study was a 1.8GHz PC equipped with a 40 GB hard drive, 256 MB of RAM. The actual images of beef rib-eye cross-section were selected by a digital camera, with a resolution 2048 by 1536 pixels and output in RGB format.

2.2 Image samples

One hundred and thirty-five images of beef rib-eye cross-section captured in a factory in Shandong Province, China, were used for the fractal analysis and discussion of the relationship between fractal dimensions and marbling scores. A three-graders panel was established to assign marbling scores based on four official China standard pictures with marbling scores ranging from 1 to 4 (Showed as Fig. 1) . All images were digitized to 528×256 (W×H) pixels.

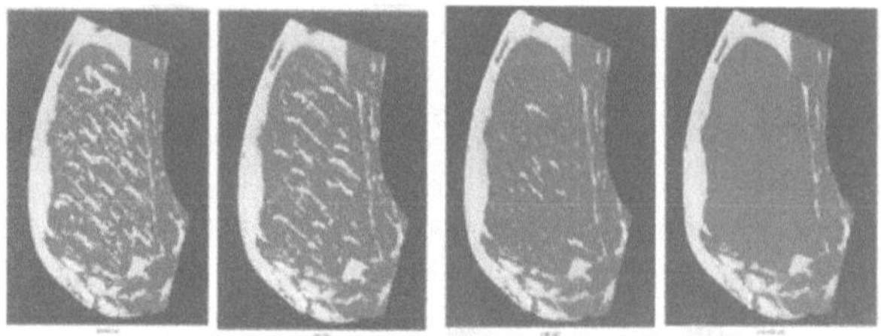

Fig. 1: BMS images from China

2.3 Pre-processing of images

The image pre-processing refers to the segmentation of marbling, involving removal of the background, segmentation of the rib-eye area and extraction of marbling. These operations were presented in detail by Chen and others (2007).

2.4 Examination of the fractal dimensions

2.4.1 Box-counting fractal dimension

The pixel covering (Feng and Zhou 2001) was applied to the measurement of box-counting fractal dimension based on the changing megascopic degree. Its primary principle is that a big rectangle is firstly used to cover the given pattern, a pre-processed image of marbling (In fact, it is expressed as a pixel matrix in computer), then the rectangle is divided into many small grids with a given side length r ($r \leq$ W and $r \leq$ H), which, in general, is named as box (sub-matrixes) and r refers to the length scale. Finally, the number of boxes which covers any part of the given pattern is recorded by counting the sub-matrixes which do not equal to zero and note it with $N(r)$. By changing the length scale r and counting the "un-empty" boxes, two data arrays can be derived. According to the box-counting method, if the data arrays fit the following relationship:

$$N(r) \propto (1/r)^{D_B} \tag{1}$$

Then the given pattern is designated a fractal object and D_B is defined the box-counting fractal dimension. Usually, the above relationship is transformed into a logarithm equation:

$$\ln[N(r)] \propto D_B \ln(1/r) \tag{2}$$

Thus, the BFD can be obtained by fitting the values of r and $N(r)$ to Equation 2 by the regression analysis.

2.4.2 Information fractal dimension

When estimating the box-counting fractal dimension, we only pay attention to the fact that the box is empty or not. According to the definition of information fractal dimension, how much information is included in a box has to be taken into account for the measurement of IFD. After being binarized, a beef marbling image is saved in computer by means of data matrix consisting of *1* and *0*, which represents a pixel of marbling and a pixel of lean meat, respectively. Further *1* also represents one information unit of marbling. Therefore, the fat information amount included in one box can be determined by simply counting the number of *1s* in one sub-matrix. According to Zhu and others (2005), to estimate IFD, the given pattern was firstly covered with a big rectangle, and divided into many small grid boxes of size *r*. Then the number of *1*s in the matrix, *N,* and the number of *1*s in a sub-matrix N_i were also recorded. According to the definition of IFD, the distribution density of information and information amount in a small grid or box are, respectively, defined as:

$$Pi = \frac{N_i}{N} \qquad (3)$$

$$I_i = -P_i \ln P_i \qquad (4)$$

The total information of marbling is given by the following equation:

$$I(r) = \sum_{i=1}^{N(r)} P_i \ln P_i \qquad (5)$$

Where Pi — distribution density of information

I_i — information amount in a small rectangle or box

$I(r)$ — total information of marbling under a given length scale, r

$N(r)$ — the number of sub-matrix which dose not equal to zero.

By varying the length scale *r*, different $I(r)$ values can be obtained; and if the relationship between *r* and $I(r)$ follows:

$$I(r) \propto (1/r)^{D_i} \qquad (6)$$

Then the given pattern may be regarded as a fractal object with D_i representing the IFD. The values of *r* and $I(r)$ were fitted to the linearised form of Equation (6) to deduce D_i:

$$ln[I(r)] = k - D_i ln(r) \qquad (7)$$

Where *k* is a constant.

3. RESULTS AND DISCUSSION

3.1 Box-counting fractal dimension

To begin with, an arbitrary picture taken randomly from one hundred and thirty-five samples was chosen as an example for executing a fractal analysis. Then similar analyses have been made for the remaining samples but they are not shown here for simplicity. Fig. 2 shows this image and its pre-processed image obtained by the method described above is shown in Fig. 3 (For visualization clarity, here shows its grey level image).

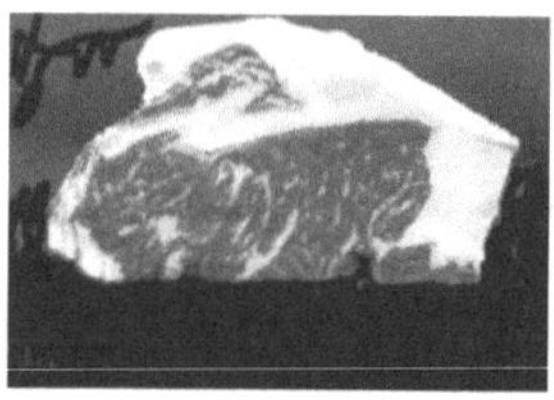

Fig. 2: An original image of sample

Fig. 3: The pre-processed image

Fig. 4 shows the graph of ln $N(r)$ versus ln(r), for the pre-processed image. The plot can be divide into three sections based on observation: the plots can be considered to be linear in the length scale intervals covering [1,11] and [12,116] pixels, whereas it is curved when r>116 pixels. The regression analyses also show that ln $N(r)$ is highly significantly linearly correlated with ln(r) in the two intervals: [1, 11] and [12, 116] pixels (Significant level α=0.01). Thus, the totall number of "un-empty" boxes as a function of length scale of the fractal analysis satisfies a power law behaviour shown in Equation. (1), indicating that this image of beef marbling is scale-free and can be characterized by multi fractal in the interval r = [1, 116] pixels. The corresponding BFDs obtained from slopes gave values 1.3463 in interval r = [1, 11] pixels and1.5226 in r = [12, 116] pixels. Afterward, the other one hundred and thirty-four samples were subjected with the above fractal analysis and also demonstrated the same multi fractal feature in the interval r = [1, 11] and r = [12,116] pixels. Their BFDs range from 1.0888 to 1.8181 in the box size [12, 116].

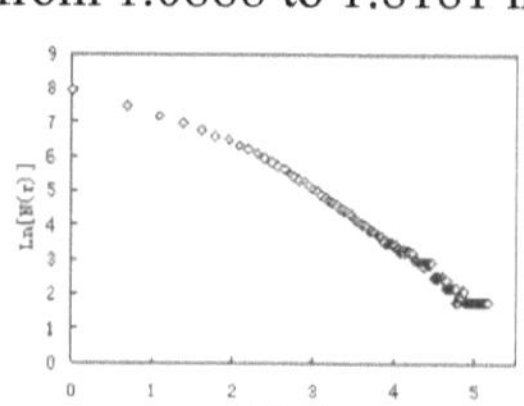

Fig. 4: Plot of $\ln[N(r)]$ versus $\ln(r)$ for the image shown in Fig. 3

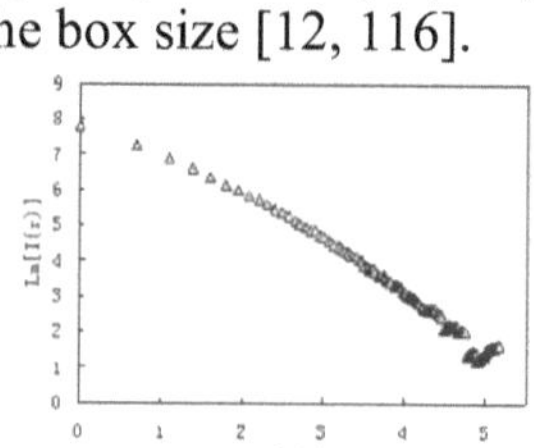

Fig. 5: Plot of $\ln[I(r)]$ versus $\ln(r)$ for the image shown in Fig. 3

Especially, four parameters: ratio of fat area to l.d. muscle area (RFA), total number of fat flecks (TNFF), number of big fat flecks (NBFF, refers to these that their areas determined by pixel count are bigger than the average area of all fat flecks) and number of small fat flecks (NSFF, whose areas determined by pixel count are smaller than the average area of all fat flecks) were calculated for every image. After making a comparison, we found that the images which have the same total amount of marbling have different BFDs due to difference in the distribution of marbling. For example, for two images which have fat area ratio, 0.1340 and 0.1339, respectively, their BFDs are 1.6067 and 1.7643 respectively. The former has the TNFF 249, NBFF 80 and NSFF 169 while the latter has TNFF 293, NBFF 46 and NSFF 247. Obviously, the former's TNFF is less than the latter and its NBFF is more than the latter's, suggesting that marbling in the image with lower dimension is more concentrative than that with higher dimension.

3.2 Information fractal dimension

The image shown in Fig. 3 was also subjected to the determination of IFD and fractal analysis. The decrease in total information of marbling [$I(r)$] as the length scale *(r)* increases, is plotted in Fig. 5 in order to correlate the ln $I(r)$ with ln (r). It is clear from Fig. 5 that ln $I(r)$ — ln (r) dependence also consists of two lines and one curve. Further regression analyses also justify the two significant linear relationships between the ln $I(r)$ and *ln* (r) in intervals [1, 17] and [18, 116] pixels (α = 0.01), indicating that the beef marbling pattern is multi fractal in a given interval r = [1, 116] pixels with slopes 1.5123 and 1.5609 in intervals [1, 17] and [18, 116] pixels, respectively. Similarly, the fractal analysis of all other samples show that beef marbling is a fractal object over two specific length scale intervals: [1, 17] and [18,116] pixels, with IFDs ranging from 1.0843 to 1.956 in the interval [18, 116].conference is agriculture-oriented and targeted participants are from government, institutes and large companies, industrial districts, clusters and regional development agencies and consultants all over the world.

3.3 Relationships between fractal dimensions and beef marbling scores

Descriptive analysis for BFD and IFD data of all beef images was carried out. The effect of marbling score on the BFD and IFD was examined by means of an ANOVA. Pearson's correlation coefficients were calculated in order to evaluate the relationship between the BFD and IFD of beef marbling and its score to study weather the score of beef marbling could be predicted from the BFD or IFD value.

BFDs of all beef marbling images in intervals [1,11] and [12,116] pixels were plotted against marbling score in Fig. 6 and Fig. 7, respectively. BFDs for different marbling scores were significant distinct from each other as observed from Fig. 6 and Fig. 7 although there are questions of a little overlap between the BFD values of adjacent score of beef marbling. There was a visible decrease in thc BDF values with the increase in marbling score from 1 to 4. The beef marbling having higher BFD value showed lower score. This fact suggested that the more abundant the beef marbling is, the higher its BFD value is. A correlation analysis was applied to analyze the relationship between beef marbling scores and its BFD values. Our results indicated those BFD values of beef marbling in intervals [1,11] and [12,116] pixels were significant correlated to beef marbling score at the 0.01 level, respectively.

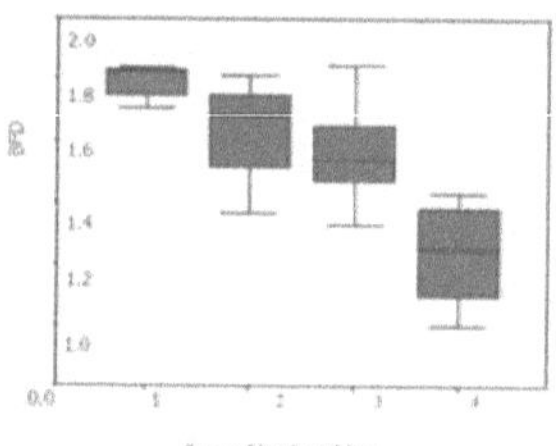

Fig. 6: Variation in BFD in interval [1,11] pixels with Score of beef marbling

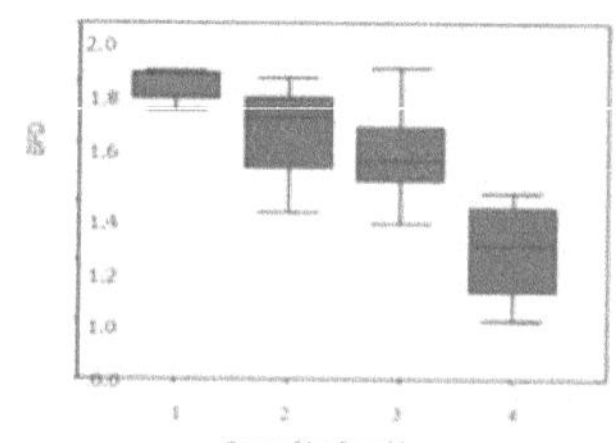

Fig. 7: Variation in BFD in interval [12,116] pixels with Score of beef marbling

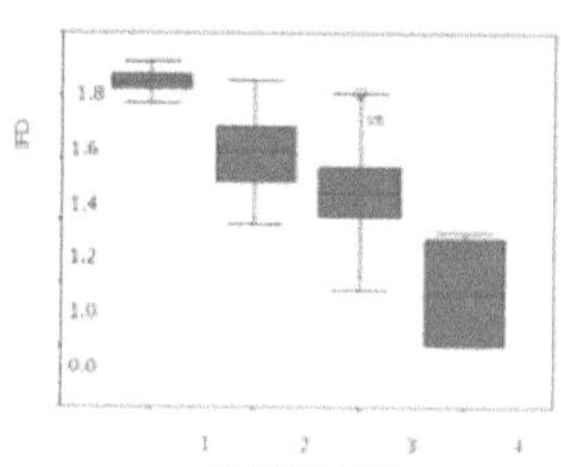

Fig. 8: Variation in IFD in interval [1,17] pixels with Score of beef marbling

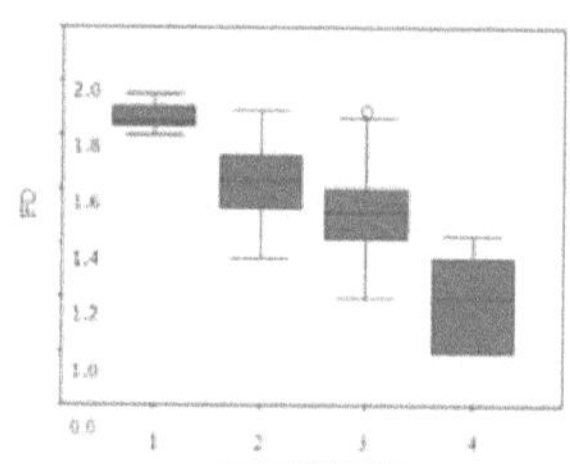

Fig. 9: Variation in IFD in interval [18,116] pixels with Score of beef marbling

Similar results could be obtained from Fig. 8 and Fig. 9 illustrating the variations of IFD values in intervals [1,17] and [18,116] pixels as beef marbling scores change, respectively. IFD also showed significant correlation with Beef marbling scores at the 0.01 level.

For comparison, marbling features including RFA, TNFF, NBFF and NSFF were calculated. Note that all the features were normalized with respect to l.d. muscle area. The mean, SD and simple correlation coefficients of each predictor with beef marbling score were listed in Table 1. Correlation coefficient is a measure of the linear association between two variables. All feature parameters of beef marbling significantly correlated with beef marbling score ($P < 0.01$). RFA showed the highest correlated with marbling score (r=0.661). The BFD has slightly higher correlation with marbling score in smaller length scales [1,11] pixels (r = 0.628) than in more one [12, 116] (r = 0.560). The IFD has also slightly higher correlation with marbling score in smaller length scales [1,17] pixels (r = 0.644) than in more one [18,116] (r = 0.592). although the correlation coefficients between BFD and IFD and marbling score are lower than that between RFA and marbling score, both BFD and IFD in smaller length scales have higher correlation with marbling score compared with TNFF (r = 0.608), NBFF (r = 0.582) and NSFF (r = 0.574), which are useful in predicting beef marbling score. This suggested that BFD and IFD of beef marbling image are potential indicators which could be used in predicting marbling scores.

Tab. 1 Means, Standard deviations and correlation coefficients of BFD, IFD, RFA, TNFF, NBFF and NSFF with beef marbling score

Variables	Means and Standard deviations				Correlation coefficients
	No. 1	No. 2	No. 3	No. 4	
BFD*	1.7792±0.0042	1.6327±0.1077	1.5300±0.1138	1.1859±0.1749	0.628**
BFD**	1.7893±0.0054	1.6340±0.1287	1.5548±0.1125	1.2333±0.1809	0.560**
IFD*	1.8415±0.0046	1.6184±0.1182	1.4843±0.1480	1.1716±0.1698	0.644**
IFD**	1.1813±0.0058	1.6312±0.1439	1.5147±0.1541	1.1938±0.1879	0.592**
RFA	0.1662±0.0034	0.1458±0.0039	0.0085±0.0042	0.0025±0.0012	0.661**
TNFF	360±82	316±87	213±84	67±40	0.608**
NBFF	57±12	57±14	36±17	13±5	0.582**
NSFF	304±91	259±83	174±75	54±35	0.574**

BFD* in intervals [1,11]; BFD** in intervals [12,116]; IFD* in intervals [1,17]; IFD** in intervals [18,116]

**P<0.01

4. CONCLUSIONS

The beef marbling is significantly characteristic of the fractal in given intervals. The multi fractal characterization of beef marbling image implies the existence of more complex distributions of marbling within the image, indicating there is marked variation in local density of marbling.

Both BFD and IFD of beef marbling images are significantly correlated with beef marbling score. BFD and IFD in smaller length scales showed higher correlation with marbling score than TNFF, NBFF and NSFF. Fractal dimensions over a specified interval are potential parameters for reflecting the abundance degree of beef marbling.

REFERENCES

Albrecht E, Wegner J, Ender K. 1996. A new technique for objective evaluation of marbling in beef. Fleischwirtschaft 76:1145-1148.

Borkowski W. 1999. Fractal Dimension based Features Are Useful Descriptors of Leaf Complexity and Shape. Canada Journal of Forest Research 29:1301-1310.

Chen G, Qi F. 2000. Face Recognition Based on Fractal and Genetic Algorithms. Journal of Infrared Millim Waves 19 (5):371-376.

Chen J. 1999. Definition and measurement of fractal dimension. Science and Technology of electron 1: 44-46.

Chen K 2007 Segmentation of beef marbling image based on image operations. Transactions of the Chinese Society of Agricultural Machinery 38(5):154-157.

Chen K, Ji Ch. 2006. Research on techniques for automated beef steak grading. Transactions of the Chinese Society of Agricultural Machinery 37(3):152-156.

Features of the Spatial Structure of Crevices in Grasslands. Arid Zone Research 20(1):53-59

Feng Z, Zhou H 2001. Computing Method of Fractal Dimension of Image and Its Application. Journal of Jiangsu University of Science and Technology (Natural Science) 22 (6):146-151.

Huan Z, Ji K, Cao W, Song Q 1997. The Simulation and Formation of Fractal Images. Publications of Yunnan Observator 1:1-9.

Jeyamkondan S, Ray N, Kranzler GA, Biju N 2000. Beef Quality Grading Using Machine Vision. Conference on Biological Quality and Precision Agriculture II (pp.91~101), Boston, USA.

Kuchita K, Yamagi K, Yamagishi T 1993. Meat quality evaluation method by image analysis and its application. Chikusan-Kenkyu 47:71-73.

Kuchita K, Yamaki K, Mizuma Y 1992. Evaluation of meat quality in Japanese beef cattle by computer image analysis. Animal Feed Science and Technology 63(2):121-128.

Kurosawa M, Nakanishi S 1995. A beef grading system by fuzzy inference and neural networks. T IEE Janan 115-C (12):1490-1498

Liang Sh 2001. Fractal Characteristics of Distribution Pattern of Carpinus pubescensPopulation. Journal of Wuhan Botanical Research 19(4):263-268

Mandelbrot B B. 1983. The fractal geometry of mature. New York: Freeman. 468 p.

Shiraniata K, Hayashi K, Otsubo A, Miyajima T, Takiyama R 2000. Grading Meat Quality by Image Processing. Pattern Recognition 33 (1):97-104

Shiraniata K, Miyajima T, Takiyama R 1996. Grate meat quality by image processing and neural network techniques. Proceedings of the international Symposium on Nonlinear Theory and its Application (NOLTA'96):437-440

Sun B 2004. Fractal Dimension and Its Metrical Method. Journal of northeast forestry university 32 (3):139-143

Tanaka M, Kayama A, Kato R, Ito Y 1999. Estimation of the Fractal Dimension of Fracture Surface Patterns by Box-counting Method. Fractals 7(3):335-340

Tian H, Zhou D, Li Zh, Hou Y, Zhong Y, Liu X, Xu Y, Lu Z 2003. Study on the Fractal

Yoshikawa F, Toraichi K, Wada K, Ostu N, Nakai H, Mitsumoto M, Katagishi K 2000. On A Grading System for Beef Marbling. Pattern Recognition Letters 21(12):1037-1050

Yuan T, Yu T, Ernest CM 2002. New method for feature extraction based on fractal behavior. Pattern Recognition 35:1071-1081

Zhang Y, Li X, Wang, Zh, Wang Y 2000. Fractal Simulation of Natural Tree's Appearance. Computer Engineering 26(9):112-114

Zhang Zh, Wang J, Wang J 1995. Fractal Matrix Model for Fingerprint Pattern Classification. Journal of Tianjin University 28 (5):628-686

Zhu Y, Li Zh, Lu K, Cui L 2005. Relationship between fractal dimensions of watershed topography characteristics and grid cell size. Shuili Xuebao 36(3):146-150

NARROW BAND RATIO VEGETATION INDICES AND ITS RELATIONSHIPS WITH RICE AGRONOMIC VARIABLES

Fumin Wang , Jingfeng Huang*

Institute of Agricultural Remote Sensing & Information Application Huajiachi Campus, Zhejiang University Hangzhou 310029, P. R. China

* *Corresponding author, Address: College of Environmental and Resource Sciences, Zhejiang University, Hangzhou 310029, Zhejiang Province, P. R. China, Tel: +86-571-86971830, Fax: +86-571-86971831, Email: Email:hjf@zju.edu.cn*

Abstract: The present study aims to determine spectral bands that are best suited for characterizing rice agronomic variables. The data for this study came from ground-level hyperspectral reflectance measurements of rice at different stage. Reflectance was measured in discrete narrow bands between 350 and 2500 nm. Observed rice agronomic variables included leaf area index (LAI), wet biomass (WBM including aboveground wet biomass-AGWBM, leaf wet biomass-LWBM, stem wet biomass-SWBM), and dry biomass(DBM: including aboveground dry biomass-AGDBM, leaf dry biomass-LDBM, stem dry biomass.) Firstly, narrow band ratio vegetation index (NBRVI) involving all possible two bands combinations of discrete channels were tested. The second part of the paper describes a rigorous search procedure to identify the best NBRVI predictors of rice agronomic variables. Special narrow band lambda (λ1) versus lambda (λ2) plots of R2 values illustrates the most effective wavelength combinations (λ1 and λ2) and band-widths (Δλ1 and Δλ2) for predicting rice agronomic variables at different development stages. The best of the NBRVI models explained 58% to 83% variability rice agronomic variables at different development stage. A strong relationship with rice agronomic variables is located in red-edge, 700 nm to 750 nm, the longer portion of red (650nm to 700nm), the shorter portion of green (500nm to 550nm), a particular portion of NIR (800nm to 850nm). They are followed by moisture-sensitive NIR(1150nm to 1200nm), and two portions of SWIR (1600nm to 1650nm).

Keywords: hyperspectral remote sensing, narrow band ratio vegetation indices, rice agronomic variables

---

Please use the following format when citing this chapter:

Wang, F. and Huang, J., 2009, in IFIP International Federation for Information Processing, Volume 295, *Computer and Computing Technologies in Agriculture II, Volume 3*, eds. D. Li, Z. Chunjiang, (Boston: Springer), pp. 2187–2196.

1. INTRODUCTION

Recent literature has shown that the narrow bands may be crucial for providing additional information with significant improvements over broad bands in quantifying biophysical and biochemical variables of agricultural crop. These hyperspectral studies were conducted for rice yield (Shibayama and Akiyama, 1991), chlorophyll content of plants (Blackburn, 1998), coniferous forest LAI (Gong et al., 1995), pinyon pine canopy LAI (Elvidge and Chen, 1995), and photosynthesis and stomatal conductance in pine canopies (Carter, 1998). The indices discussed in this paper involve two bands. According to Lawrence and Ripple (1998) and Gong et al.(1995), the use of two-band vegetation indices unnecessarily constrains the regression analysis. They argue that if one requires knowledge of an ecological variable (e.g., above-ground biomass or LAI), the researchers must ultimately analyze the relationship between the spectral index used and the ecological variable through a regression analysis. As the crop conditions vary due to factors such as management conditions, soil characteristics, climatic conditions, and cultural practices, different band combinations can be used (Shibayama and Akiyama, 1991; Shibayama et al., 1993). In the same time, theoretical models concerning the indices have been developed to describe the reflectance from single leaves to complete canopies. Many investigations demonstrated that reflection of red, green, and near-infrared radiation contains considerable information about crop biomass owing to the contrast between soil and vegetation. Thenkabail et al(1999) describe a rigorous search procedure to identify the best narrow band NDVI predictors of crop biophysical variables.

During the past several years, estimates of biomass, ground cover and leaf area indices (LAI) as a function of spectral reflectance measurements have shown promising results (Leblon et al.,1991; Clevers,1989).However, paddy rice being grown in continuously flooded fields, is a special crop which deserves particular consideration, as the layer of water on the soil surface modifies the spectral reflectance of the soil-plant system (Patel et al.,1985).

Based on the above background, the main goal of this research is to evaluate the performance of NBRVI in characterizing rice agronomic variables. The final goal is to determine and recommend an optimal number of hyperspectral bands, their centers and widths, in the visible, NIR portion and short-wave infrared portion of the spectrum (350–2,500 nm) required to best study rice, thus reducing the redundancy in hyperspectral data. A rigorous and exhaustive approach is adopted in computing and evaluating narrow band ratio vegetation index. Rice variables used are leaf area index (LAI), wet biomass (WBM: including aboveground wet biomass—AGWBM, leaf wet biomass –LWBM, stem wet biomass—SWBM, and root wet biomass—RWBM), dry biomass (DBM: including aboveground dry biomass—AGDBM, leaf dry biomass –LDBM, stem dry biomass—SDBM, and root dry biomass—RDBM), which are the best indicators of rice growth and yield (Bartlett et al., 1988; Wiegand and Richardson, 1990; Shibayama et al., 1993; Thenkabail et al., 1995; Fassnacht et al., 1997).

2. EXPERIMENT SITE

The study area is at Zhejiang University Experiment Farm, Hanghou, Zhejiang province, China, located 30° 14′ N, 120° 10′ E. The data were obtained from three nitrogen fertilizer levels and five species treatments of rice in 2002 growing seasons. Rice cv. 'Xiushui 110(S1),Xieyou 9308(S2), Jiayu 293(S3), Jiazao 312(s4) and Z00324(S5)' were selected for the investigation. Xieyou 9308, Jiayu 293, Jiazao 312 and Z00324 are indica rice, and Xiushui japomica rice. Xieyou 9308 is a hybrid rice, others common rice. The three nitrogen fertilizer treatments(N0,N1,N2) of rice are fertilized 0,140, 240kg/ha pure nitrogen respectively. Each treatment has four repeats. A completely randomized design consisting of 60 plots of 4.81m×4.68m was used, and 12 plots for each variety. Each plot was planted in east-west. The experiment area is characterized by monsoon climate with a hot summer and a cool winter, marked seasonal variations in precipitation. The average annual rainfall is 1374.7 mm from 1961 to 1990. The sandy loam paddy soil had the following properties: pH 5.7, organic matter with 16.5g/kg and total N with 1.02g/kg.

3. METHODOLOGY

Spectral and agronomic variables data were gathered during the growing season of rice from June to October. A spectroradiometer with a range from 350 nm to 2500 nm, manufactured by Analytical Spectral DevicesTM (FieldSPEC, 2002), was used to gather spectral data of rice at different stages. Due to severe noise in the water absorption regions, only the data gathered in 350 nm through 1330 nm and in 1480 through 1780 nm and 1990 nm through 2400 nm were used. The spectroradiometer unit consisted of a main spectrometer, a personal computer, fiber optic cable, a pistol grip, and different field of view (FOV) cones. Inside the spectrometer instrument, light is projected from the fiber optics onto a holographic diffraction grating where wavelength components are separated and reflected for independent collection by the detector(s) (FieldSPEC, 2002).

Rice Spectra and agronomic variables were observed seven times totally at their distinctive growth stages. These growth phases were: tillering, elongation, booting, heading, flowering, milk ripe, ripe, and harvest. At each plot, 10 reflectance measurements were consistently taken, with a nadir view from a height of 1.0 m above canopy, using a 25° FOV. This resulted viewing an area of 6,939 cm2 at the canopy level. Spectroradiometer data of rice were analyzed using Data Processing System (DPS) version 6.12 (Tang, and Feng, 2002)

At the field of rice, a representative area was chosen for all measurements. The same rice sample was taken for laboratory analysis. In the laboratory, rice samples were analyzed for leaf area, wet weight, and dry weight. Rice samples were weighed on a simple weighing machine. Leave areas measurements were also made for all the rice samples, and then LAI were acquired.

Narrow band ratio vegetation indices(NBRVI) were computed using 1690 spectral channels of percentage reflectance data from the spectroradiometer. The performance of these indices in establishing rice agronomic variables were then evaluated and compared with the well-known reference Landsat TM broadband indices.

$$NBRVI=R_i/R_j$$

where i,j=1,N is reflectance(percent) of narrow bands with N=narrow bands=1690 R=reflectance of narrow bands

4. RESULTS AND DISCUSSION

4.1 Narrow band RVI relationships with rice agronomic variables

As many literatures depicted, RVI can give good prediction to LAI and WBM. Compared with NDVI, RVI is a simple, practical index. The availability of hyperspectral data in 1690 (N) discrete narrow bands allowed the computation of N×N=2856100 NBRVIs for any rice agronomic variable. Regression coefficients R^2 between all possible two-band NBRVIs and rice agronomic variables were determined. Therefore, the matrixes were obtained. The results of this comprehensive analysis are illustrated in contour plots of R^2 values for all the pairs of wavelength. We determined a contour interval of 0.05 for R^2 values. Based on the criterion, contour plots were made for all the agronomic variables of rice (Fig. 1). Since the R^2 values are not symmetrical, the whole diagonal was occupied to display the matrix. Then band centers and band widths that combine to form the best nine indices were determined for rice biophysical variables of each observation date. After the results for various rice variables were evaluated, a strong relationships were showed, which centered in red-edge portion (700nm to 750nm), the longer portion of red (650nm to 700nm), the shorter portion of green (500nm to 550nm), a particular portion of NIR (800nm to 850nm). They are followed by moisture-sensitive NIR, and two portions of SWIR (1150nm to 1200nm, 1600nm to 1650nm). For the precise R^2 values, visually scan through 1690(N) by 1690(N) matrix of R^2 values. Identify the λ1 and λ2 band centers having highest R^2 value. This is the band center for the index, which has highest correlation. Observe the R^2 values in the immediate vicinity of these band centers (λ1 and λ2) .Often the R^2 values remain the same or nearly the same for a few narrow bands in immediate vicinity of λ1 and λ2. This range of similar or near-similar spectral response to rice variable will constitute the bandwidth. Using the method, nine best he band centers (λ1 andλ2) and band widths (Δλ1and Δλ2) determined, only part of them are presented here (Table 1).

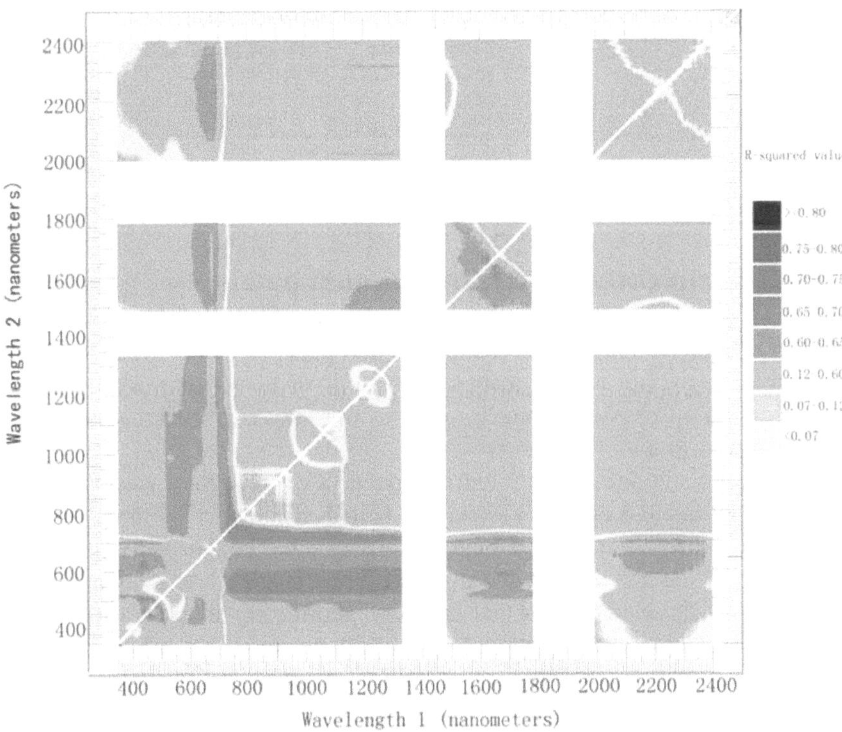

Fig. 1. Contour plot showing the correlation (R^2) between AGWBM and narrow band RVI values on 31st ,August calculated for 1690 narrow bands spread across λ1 (350 nm to 2500 nm) and λ2 (350 nm to 2500 nm).

Table 1. Nine best NBRVIs band centers (λ1 and λ2) and band widths (Δλ1 and Δλ2) in nm for the linear NBRVI models for the different rice variables

		Band Centers(λ1 and λ2) and Band Widths((Δλ1 and Δλ2) for NBNDVI							
date		LAI				LWBM			
		λ1	Δλ1	λ2	Δλ2	λ1	Δλ1	λ2	Δλ2
23th, Jul.	Index 1	726	27	990	107	730	12	984	60
	index 2	722	26	1233	191	728	8	1236	181
	index 3	716	20	1630	297	720	16	1629	294
	index 4	708	10	2189	321	712	7	2274	29
	index 5	523	14	613	54	523	7	589	34
	index 6	574	111	2003	24	889	107	731	23
	index 7	901	76	726	13	1071	63	726	19
	index 8	1074	5	723	8	755	26	727	14
	index 9	598	46	525	9	591	23	521	6
30th Jul.	Index 1	689	10	1515	67	698	5	1190	41
	index 2	617	93	1515	67	699	9	1318	18
	index 3	519	27	1529	96	567	82	1319	17
	index 4	514	18	2366	16	580	122	1630	298
	index 5	593	202	1629	298	573	23	1173	12
	index 6	517	13	1320	17	610	78	2019	6
	index 7	691	11	979	45	694	12	2313	37
	index 8	578	123	721	25	605	22	2312	34
	index 9	721	20	565	88	774	51	547	27

4.2 Frequently occurring optimum bands

To gain a broad view of the results in this paper, it is useful to consider bands that frequently appear in optimum indices. It can be said that these bands contain an overwhelming fraction of the total rice information in the full spectrum.

The results obtained using narrow band RVI models was used for determining the band centers and widths (Fig 1. Table 1). The procedure of obtaining the band centers and band widths was explained previously. The narrow bands that appear in two-band RVI type models were evaluated. The percentage of these narrow bands in every 50nm interval were established (Fig. 2)and their clusters plotted for the 350 nm to 1330 nm, 1480nm to 1780nm, and 1990nm to 2400nm. In NBRVI model type, an overwhelming proportion of crop information was in red-edge (700 nm to 750 nm), the intermediate portion of SWIR (1600 nm to 1700 nm), and a portion of green (500 nm to 550 nm). These were followed by the moisture-sensitive NIR (950 nm to 1000 nm), a particular portion of NIR (1100 nm to 1150 nm), and the longer portion of red (650 nm to 700 nm). A remarkable about 25% of all the bands that occur in RVI type model were clustered in 700 nm to 750 nm region of the spectrum (Fig. 1, Table 1).

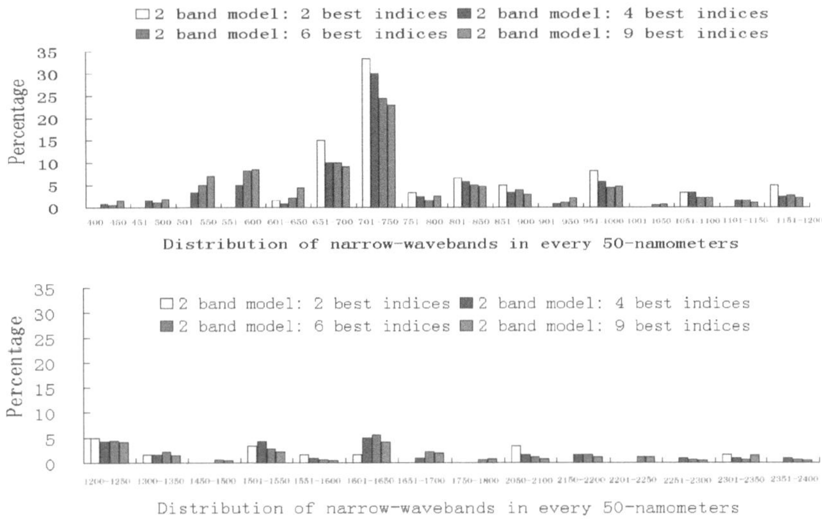

Fig. 2. Percentage of occurrences of hyperspectral narrow bands in the nine best RVI models

The band widths ($\Delta\lambda 1$, $\Delta\lambda 2$) were determined from the $\lambda 1$ versus$\lambda 2$ plots such as the ones illustrated in Fig. 1. The $\Delta\lambda 1$ and $\Delta\lambda 2$ of the nine best narrow bands RVI models were used to determine the frequency of occurrence of various band widths as follows:

(1) very narrow bands (1 nm to 15 nm), 49 occurrences,18.1%

(2) narrow bands (16 nm to 30 nm), 85 occurrences, 31.5%

(3) intermediate bands (31 nm to 45 nm), 25 occurrences, 9.3%

(4) broad bands (greater than 45 nm), 111 occurrences, 41.1%

It is clear from these results that nearly 50% of the bands have narrow (16 nm to 30 nm) or very narrow band (1 nm to 15 nm) widths. If we consider them as a whole, the percentage of them is 49.6% for RVI. This indicates that narrow bands are very useful to predict rice LAI and LWBM.

The red-edge portion around 720 nm which represents the maximum band of the first order derivative is the single most frequently occurring band across the variables in NBRVI type of models. This in not surprising since this wavelength has been affected by growth stage and nitrogen levels. The band widths ($\Delta\lambda 1$ or $\Delta\lambda 2$) for the red-edge narrow bands typically range between 5 nm to 28 nm. The best suggested band widths for any red-edge narrow band is 15 nm. The other two red narrow bands, $\lambda 4$(710 nm) and $\lambda 6$(730 nm), are of importance depending on type of rice, growth stage, and growing conditions, including cultural practices, wherein there is a likelihood of red-edge shifting to these bands from the most commonly occurring red-edge band (720 nm). Indeed red-edge varies significantly depending on host of these variables.

The λ2 (550 nm) is the green narrow band at the reflectance peak in the visible range. The band λ1(530 nm) is the point of greatest positive change in slope per unit change in wavelength in visible spectrum, and λ3(580 nm) is the greatest negative change in slope per unit change in wavelength in visible spectrum for most dates. Similarly, λ6 (720 nm) represents the point of greatest positive change in slope per unit change in wavelength in red-edge portion of the spectrum. The λ5 (710 nm) and λ7 (735 nm) are the possible greatest derivative values. The red absorption maximum occurs around 680 nm (λ4), which is another frequently occurring band. The NIR shoulder can be represented by a broad band or a narrow band centered at 845 nm (λ8). The λ9 =920 nm and λ11 =1080 nm represent the possible peak for maximum reflectance region of the NIR spectrum. Which band the peak reflectance will occur is dependent on a lot of conditions. The λ10 = 982 nm is the center of the moisture-sensitive "trough" portion of NIR, which is strongly related to change in rice moisture and biomass (Penuelas et al., 1993). The size of "trough" in 940 nm to 1040 nm region increases with an increase in growth stages and is strongly related to biomass. The λ12 = 1125 nm represents the sudden decrease after the second possible peak NIR reflectance. The λ13=1190 is the second center of the moisture-sensitive "trough" portion of NIR. The λ14 =1630 nm is the premaxima reflectance of intermediate SWIR. With respect to λ16 =1680 nm, it is the most particular band of high occurrence frequency in the paper. The fifteen bands were selected based on their frequency of occurrence in various models rice variables as discussed throughout this paper. The distribution of the rice information clusters in different waveband portions also shows an overwhelming proportion of quantitative rice information in the band centers of these 15 bands. Many of the optimum bands that were selected make physical sense given what we know about plant chemistry, canopy structure, and plant spectra (Elvidge,1990).

5. SUMMARY AND CONCLUSIONS

This research established the optimum number of hyperspectral bands, centers, and widths (Table 1) in the visible, near-infrared and short-wave infrared spectrum for establishing relationships with rice biophysical characteristics. This recommendation is based on the study of eight dates of observation (representing eight growth stages), and wide ranging rice characteristics. A remarkably strong relationship with rice variables is in red-edge (700 nm to 750 nm), the longer portion of red (650nm to 700nm), the shorter portion of green (500nm to 550nm), a particular portion of NIR(800nm to 850nm).They are followed by moisture-sensitive NIR, and two portions of SWIR(1150nm to 1120nm, 1600nm to 1650nm).An overwhelming proportion of these channels had band widths that were classified as: (1) very narrow (1 nm to 15 nm wide) or (2) narrow (16 nm to 30 nm wide). Most were on the order of 10 nm to 20 nm wide (narrow or very narrow). A rigorous procedure adopted here to identify the best narrow band RVI type models showed that the best two-band combinations often involve: (1) a very narrow or a narrow red-edge band and a narrow or a broad NIR band, or (2) a very narrow red-edge band and a narrow or a broad SWIR band, or (3) a broad NIR band and a narrow red band or (4) a narrow green band and a broad NIR and SWIR band.

The study recommends 15 specific narrow bands, centers, and widths, which provide optimal crop information in the whole spectra. These narrow bands include 710 nm, 720 nm, 735 nm, 680 nm, 530 nm,550 nm, 580 nm, 845 nm, 920 nm, 982 nm,1080 nm, 1125 nm , 1190 nm, 1630 nm and 1680 nm. Most of them make physical sense.

ACKNOWLEDGEMENTS

This work was supported by the National Natural Science Foundation of China (40571115) and the Hi-Tech Research and Development Program (863) of China (2006AA120101)

REFERENCES

B. Leblon, M. Guerif, F. Baret. The use of remotely sensed data in estimation of PAR use efficiency and biomass production of flooded rice. Remote Sening of Environment, 1991, 38: 147-158

C. D. Elvidge, and Z. Chen. Comparison of broad-band and narrow-band red and near-infrared vegetation indices. Remote Sening of Environment, 1995: 54, 38–48

C. D. Elvidge. Visible and near infrared reflectance characteristics of dry plant materials. International Journal of Remote Sensing, 1990, 11(10):1775–1795

C. L. Wiegand, and A. J. Richardson. Use of spectral vegetation indices to infer leaf area, evapotranspiration, and yield: I. Rationale. Agronomy Journal, 1990, 86: 623–629

D. S. Bartlett, M. A. Hardisky, R. W. Johnson, et al.. Continental scale variability in vegetation reflectance and its relationship to canopy morphology. International Journal of Remote Sensing, 1988, 9: 1223–1241

G. A. Blackburn. Quantifying chlorophylls and caroteniods at leaf and canopy scales: An evaluation of some hyperspectral approaches. Remote Sening of Environment, 1998, 66: 273–285

G. A. Carter. Reflectance bands and indices for remote estimation of photosynthesis and stomatal conductance in pine canopies. Remote Sening of Environment, 1998, 63: 61–72

J. P. G. W. Clevers. The application of a weighted infrared-red vegetation index for estimating leaf area index by correcting soil moisture. Remote Sening of Environment, 1989, 29: 25-37

K. S.,Fassnacht S. T. Gower, M. D. MacKenzie,et al. Estimating the leaf area index of north central Wisconsin forests using the Landsat Thematic Mapper. Remote Sening of Environment, 1997, 61: 229–245

M. Shibayama, and T. Akiyama. Estimating grain yield by remote sensing of crop of rice canopies using high spectral resolution reflectance measurements. Remote Sening of Environment, 1991, 36: 45–53

M. Shibayama, W. Takahashi, S. Morinaga, and T. Akiyama. Canopy water deficit detection in paddy rice using high resolution field spectroradiometer. Remote Sening of Environment, 1993, 45: 117–126

N. K. Patel, T. P. Singh, B. Sahai, et al. Spectral response of rice crop and its relation to yield and yield attributes. International Journal of Remote Sensing, 1985, 6: 657-664

P. Gong, R. Pu, J. R. Miller. Coniferous forest leaf area index estimation along the Oregon transact using compact airborne spectrographic imager data. Photogrammetric Engineering and Remote Sensing, 1995, 61: 1107–1117

P. S. Thenkabail, A. D. Ward, and J.G. Lyon. Landsat-5 Thematic Mapper models of soybean and corn crop characteristics. International Journal of Remote Sensing, 1995, 15: 49–61

P. S. Thenkabail, B.S. Ronald, and D.P. Eddy Hyperspectral Vegetation Indices and Their Relationships with Agricultural . Remote Sening of Environment, 1999, 71: 158–182

Q. Y. Tang, and M. G. Feng. DPS© Data processing system for practical statistics, Chinese Science Press. 2002, 304-311

R. L. Lawrence, W. J. Ripple. Comparisons among vegetation indices and bandwise regression in a highly disturbed, heterogeneous landscape: Mount St. Helens, Washington. Remote Sening of Environment, 1998, 64: 91–102

A MODEL TO PREDICT SHELF-LIFE LOSS OF HORTICULTURAL PRODUCE DURING DISTRIBUTION WITH FLUCTUATED TEMPERATURE AND VEHICLE VIBRATION

Weiwei Gong [1], Daoliang Li [2], Xue Liu [2], Jun Yue [2,3], Zetian Fu [1,*]

[1] *College of Economics and Management, China Agriculture University, Beijing, P. R. China 100083*

[2] *College of Information and Electrical Engineering, China Agriculture University, Beijing, P. R. China 100083*

[3] *College of Management, Ludong University, Yantai, Shandong Province, P. R. China 264025*

[*] *Corresponding author, Address: College of Economics and Management, China Agriculture University, Beijing 100083, P. R. China, Tel: +86-010-62736717, Fax: +86-010-62736717, Email: gongweiwei040625@yahoo.com.cn*

Abstract: Fresh fruits and vegetables has become a public concern from the food security aspect. And the prediction of shelf-life loss under the fluctuated temperature becomes one of the key problems in food supply chain operation. So this paper identifies the impact aspects of produce decaying during distribution. For the key temperature factor, the process is divided into three phases: sorting, traveling and door-opening. Based on time-temperature function, a model of shelf-life loss of horticultural produce during distribution is developed by evaluating respiration rate of vegetables and fruits considering both the environment fluctuated temperature and vehicle vibration during traveling. Taking eggplant as an example, the numerical experiment result demonstrates that the average cost for ambient distribution is 2.8 times of the insulation way.

Keywords: shelf-life, horticultural produce, distribution, respiration rate

Please use the following format when citing this chapter:

Gong, W., Li, D., Liu, X., Yue, J. and Fu, Z., 2009, in IFIP International Federation for Information Processing, Volume 295, *Computer and Computing Technologies in Agriculture II, Volume 3*, eds. D. Li, Z. Chunjiang, (Boston: Springer), pp. 2197–2206.

1. INTRODUCTION

Food is vital to human being which plays an important and fundamental part in sustaining life. Nowadays, consumers' increasing awareness of food safety and quality has become the major internal drivers for the government to play an efficient and effective security role of produce safety. "Special provisions on strengthening food safety supervision and management by State Council","2008 Beijing Olympic Food Safety Action Program" have been promulgated which indicate the Chinese Government has paid high attention to the national physical fitness. On another side, food business today is operating in a complex and changeable commercial environment. If companies want to survive, they must remain ahead of food security surveillance to keep perishable food fresh. So, the prediction of shelf-life loss naturally becomes one of the key problems in food logistics operation.

The unique characteristics of fresh food which differs form other processed items is the high rate of spoilage. The product moves along the "from farm to table" supply chain network going with sharp chemical and biochemical changes any time. In the last two decades many papers have been devoted to research on application and modeling for the food quality. (Johanne Rønnow Olsen, 2008) proposed that quality is a main determinant of consumer food choice and they investigate managerial quality goals and how these may be linked to product development competences. For animal produce such as meat, poultry and fish, Lynette(M. Johnston, 2006) pointed out that the loss of produce quality is the result of bacterial growth. (Fanbin Kong, 2007, Tomas Skoglund and Petr Dejmek, 2008) explained the relationship between food quality transformation by predictive microbiology and kinctics. For thc horticultural producc, Uchino, (Nei,2004) develop a respiration model for time and temperature dependence to simulate the respiration rate for fresh produce during transport, retailing and storage. Roy, Nei et al. (2008) had done the life cycle inventory analysis of fresh tomato distribution systems in Japan considering the quality aspect. The result of (Li Guo,2008) demonstrated that the speed of green bean quality loss is influence by storage temperature and strength of CO_2 with the characters of respiration rate, weight loss, change in titrable acidity and sugar. (G. Urrutia-Benet etc 2007) showed at laboratory and pilot scale that potatoes' decaying speed was not increased after freezing and thawing processes when pressure was applied, being even slightly reduced in the metastable region. In the distribution process, the main quality affecting factors are environment temperature and vehicle-shake. Bogataj, (Bogataj,2005) forecasted the temperature evolution in cold logistic chain, and analysis the sensitivity of its fluctuation. (Van Zeebroeck, Tijskens ,2006) and (Van Zeebroeck,2007) used discrete element method to simulate the apples' damage during transportation. However, during the distribution process, the temperature is not constant but changeable and the vibration has to be considered at the same time. But there are few models on it up to now. That is the reason why we present this paper.

The rest of paper is organized as follows. First, impact factors of produce decaying during distribution are identified. And based on time-temperature function, a model of shelf-life loss of horticultural produce during distribution is developed by evaluating respiration rate. And computational results and discussion are presented later. Finally conclusions are drawn.

2. IMPACT FACTORS

Perishability arises mainly in agri-product such as fruits and vegetables, meat and poultry, fish and shellfish, milk etc. All of them deteriorate at a high speed as well as their semi-processed foods. In accordance with the composition of the food itself, they can be divided into two groups: animal produce and horticultural one. Raw materials can be strictly divided into these two categories, fruits and vegetables for the horticultural produce, meat and poultry for the animal produce of which decaying rate is easy to measure. But if the food is processed by both types of raw produce, take quick-frozen dumpling and bento as an example, the prediction of quality decreasing rate is much more complex.

Food deteriorates and spoils in a very complicated way with many chemical and biochemical reactions taking place at the same time. However, the main activities are microbial growth of meat and enzymes activity of plant. Fig.1 describes the primary factors affecting the perishable food during delivery. For animal food, the life is end without energy absorption from the environment to maintain its basic metabolism. The food corruption causes mainly by different types of microbial growth whose concentration impact by temperature greatly. Moreover the wind speed, solar radiation, air temperature and the times of door opening will cause the change of temperature. The temperature of perishable food distribution by cold chain in the urban area is different from the long-distance transportation's. For transportation, the period between one door opening and another is longer and less frequently with light temperature change. On the contrary, the travel time is shorter between two delivering customers and the door has to be opened more often for distribution. This is why the distribution temperature is fluctuated dramatically. For the plant food, the produce is still alive after harvest. They are breathing all the time in the supply chain. As they can not absorb nutrition from the nature, the respiration decomposes the food body itself then get rotten. That is the reason why the total amount of exhaled carbon dioxide is used to measure the waste quality of vegetables and fruits. In this paper we ignore the influence to respiration rate by water activity and gas composition but only temperature and vehicle vibration. The road condition, vehicle acceleration and position on the compartment will affect the intensity of vibration. Generally, items at the rear suffer greater vibration than ones on the front and the intensity at the bottom is less then the top.

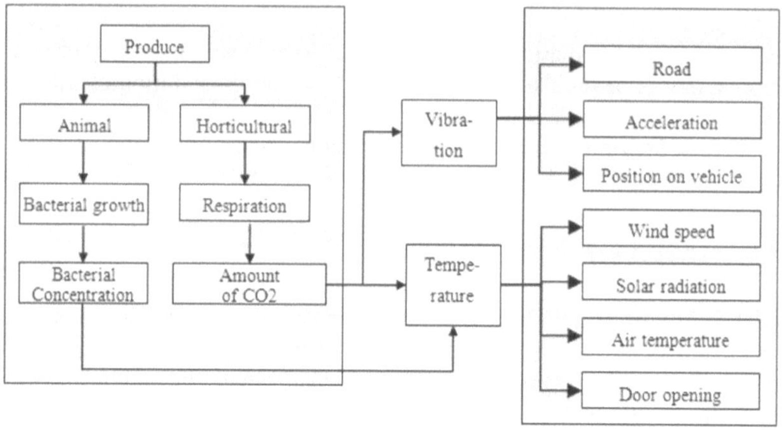

Fig.1: Primary factors affecting the perishable food during delivery

3. TIME-TEMPERATURE MODEL

Temperature plays a key role in the food quality decision in distribution the same as warehousing and transportation. So cold chain research today focuses on temperature control. Although the International Institute of Refrigeration recommended the fresh produce transport temperature, each country has its own regulation. For example, the temperature can not exceed 7℃ for meat product distribution in Australia. However, because of the high frequency switching doors and oxygen contact, the outside world impacts the food dramatically. Even with the refrigerator, the temperature inside will keep on fluctuating. According to it, the temperature variety is divided into three stages stage:

Sorting: Goods are sorted based on the food type and customer requirement artificially or by equipment and taken from the shelf to the handling area. Then they are ready for distribution. According to the plan all produce will be loaded into the trucks. In this process, the fresh suddenly get exposed in the air making the food hot with heavy breathing.

Traveling: The quality is influenced by the vehicle type and acceleration, outside temperature, wind speed, solar radiation and road condition with temperature increases.

Door opening: the door switching is an inevitable operation in discharging loads. After the door is opened, the air outside will flush into the compartment which results in the heat exchange leading to the food temperature increases.

The time-temperature function is described by a sub-linear relationship:

$$T(t) = \begin{cases} a_s t + b_s &, \quad t \in [s_{tart}, s_{end}] \\ a_c(i)t + b_c(i) &, \quad t \in [c_{start}(i), tc_{end}(i)] \\ a_t(i)t + b_t(i) &, \quad t \in [t_{start}(i), t_{end}(i)] \\ a_o(i)t + b_o(i) &, \quad t \in [o_{start}(i), o_{end}(i)] \end{cases} \qquad (1)$$

Where: a(i) is the temperature evolution rate of period i, b(i) denotes the temperature adjustment of period i, s denotes sorting period, c cooling period while traveling, t is only traveling period, o is door opening period, T is temperature, t is time, i denotes the period i.

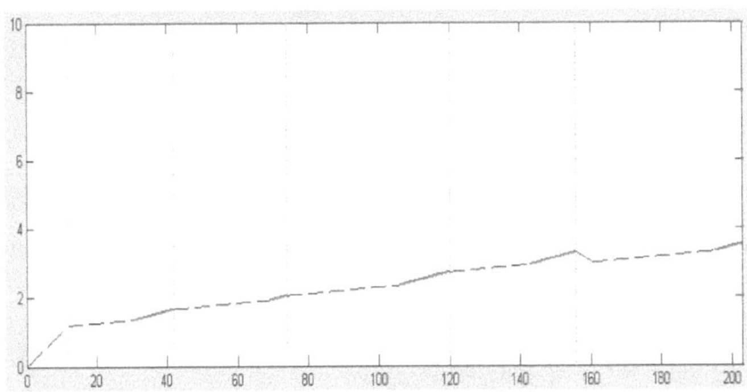

Fig.2: Time-temperature evolution

The vertical dashed separate time-temperature evolution into six periods that is to say the vehicle serves six customers (see fig.2). The first period is for sorting. Form the second to the fifth the dashed line is for traveling and the real line is for door opening which has a stronger temperature shift. In the last one, there are three sub-lines, first one is for cooling with decreased temperature and then the other is for traveling and door-opening.

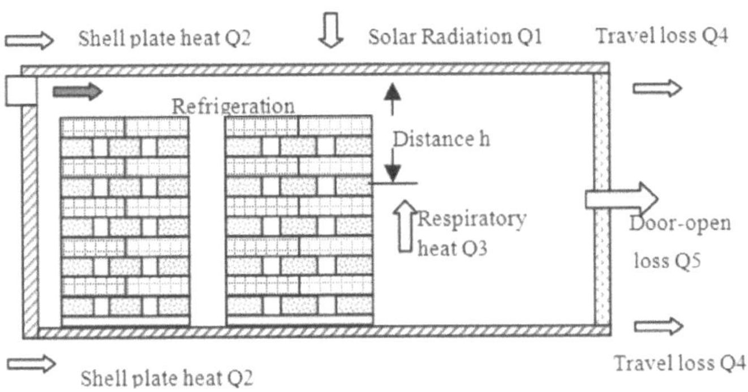

Fig.4: Cold consumption id the delivery process

In real life, there are many factors change the temperature inside the delivery vehicle, food's breath will produce heat and conduct the thermal by both conductivity and radiation with the environment which is a very complicated process. To simplify the calculation, only the main factors are considered. As shown in Fig. 4, the refrigeration vents is usually installed in the upper front of the compartment with refrigeration Q0; the impact from air temperature and solar radiation is Q1; cold consumption from the shell is Q2; respiration hot is Q3; cold loss because of internal and external pressure is travel loss Q4; the loss by opening the door and loading cargo is Q5. The total consumption of cold Q to influence the temperature is:

$$Q=Q1+Q2+Q3+Q4+Q5-Q0 \tag{2}$$

4. SELF-LIFE LOSS PREDICTION MODEL

Horticultural produce is different from meat because life still continues. The respiration of fruits and vegetables is the symbol of continuity of living. In the cultivation process, the individual can obtain nutrients form the natural world. However, in the circulation process, it can not, so it decomposes to maintain the life of its normal activities. There is a correlation between the corruption velocity and respiration rate with temperature as a major factor. Therefore, the self-life loss is evaluated by the amount of CO_2 caused by reparation which is affected by fluctuated temperature and truck vibration.

Toshitaka Uchino(2004) simulate the respiration rate of fresh produce with a model based on the enzyme kinetics. They respiration rate is presented as follows:

$$R_c = A \cdot \exp\left(\frac{-Ea}{RT}\right)\left\{\frac{k_s}{k_d} - \left(\frac{k_s}{k_d} - [E]_0\right)\cdot \exp(-k_d \cdot t)\right\} \tag{3}$$

Where: R_c is the respiration rate (mmolkg^{-1}h^{-1}), A is pre-exponential factor(h^{-1}), Ea shows the apparent activation energy(J mol^{-1}), R is ideal gas constant(J mol^{-1}K^{-1}) , T absolute temperature(K), k_s denotes rate constant of synthesis of enzyme(mmolkg^{-1}h^{-1}) and k_d is rate constant of decomposition(h^{-1}),[E] is the enzyme concentration (mmolkg^{-1}), $[E]_0$ is the initial concentration, t is time(h).

Based on Toshitaka's Eq.(3), substitute time-temperature Eq.(1) into Eq.(3) transforming the respiration rate Rc(T , t) to Rc(t) with only one variable time t:

$$R_c = A \cdot \exp\left(\frac{-Ea}{R(a\cdot t+b)}\right)\left\{\frac{k_s}{k_d} - \left(\frac{k_s}{k_d} - [E]_0\right)\cdot \exp(-k_d \cdot t)\right\} \tag{4}$$

As the distribution food has to be transported to overcome space distance, the produce will get a certain degree of injury which will lead to the respiration increase during the sorting and delivering operations. Japanese scholar proposed a quantitative model to evaluate the traffic injury according to Palmgren-Miner linear cumulative rule (Heye Chengfu, 1999). The model described as follows:

First, suppose the injury is D caused by the harmonic vibration while the acceleration amplitude vibration Gi for the frequency n_i ($n_i = 1,2,\ldots$), then:

$$D = \frac{1}{\beta}\sum n_i G_i^{\alpha} \tag{5}$$

α、β denote constant of food characteristics. The injury caused by random vibration:

$$D = \frac{f_0 t}{\beta}(\sqrt{2\sigma})^3 \Gamma(1 + \frac{\alpha}{2}) \tag{6}$$

f_0 is the average frequency of vibration, σ is transient stress probability density standard deviation, Γ represents the gamma function(Heye Chengfu, 1999).

Suppose there is a positive correlation between the degree of injury and respiration rate then:

$$R_c = m \cdot D \cdot A \cdot \exp(\frac{-Ea}{RT})\left\{\frac{k_s}{k_d} - (\frac{k_s}{k_d} - [E]_0) \cdot \exp(-k_d \cdot t)\right\} \tag{7}$$

m denotes the injury adjustment factor. Integral the respiration rate to compute the amount of CO2 (TCO2):

$$TCO_2(t) = \int_{t_0}^{t} R_c(t)dt \tag{8}$$

Then, substitute respiration rate Eq.(7) into Eq.(8)

$$TCO_2(t) = \int_{0}^{t} m \cdot D \cdot A \cdot \exp(\frac{-Ea}{R(a \cdot t + b)})\left\{\frac{k_s}{k_d} - (\frac{k_s}{k_d} - [E]_0) \cdot \exp(-k_d \cdot t)\right\} \tag{9}$$

Let $K_1 = \frac{K_s}{K_d} A$, $K_2 = \frac{K_d}{K_s}[E]_0 - 1$, then

$$TCO^2(t) = \int_{0}^{t} m \cdot D \cdot K_1 \cdot \exp(\frac{-Ea}{R(a \cdot t + b)})\{1 + K_2 \cdot \exp(-k_d \cdot t)\} \tag{10}$$

Don W. Brash(1995) discovered that a similar strong negative correlation obtained between shelf-life and a accumulated heat units and CO2 is linearly related to accumulated heat units. So the shelf-life of customer (SLP) can be presented as:

$$SLP = b_{co2} - a_{co2} \times TCO_2 \tag{11}$$

b_{co2} and a_{co2} is shelf-life constant.

Next we have to calculate the $Load_i$ which represents the truck load from customer i to customer i+1 (see fig.5):

$$Load_i = \sum_{k=i+1}^{M} g_k, \quad i = 0,1,\cdots M-1 \tag{12}$$

M is the number of customer, g_i denotes the request of customer i (i=1,2,⋯,M), i=0 is distribution center. The total shelf-life loss of plant food is described as:

$$TSLP = \sum_{k=0}^{i-1} load_i \times (SLP_i - SLP_{i+1}) \tag{13}$$

C_{PSLP} is the value for each unit. Then, we evaluate the total value loss C_{TSLP}:

$$C_{TSLP} = C_{PSLP} \times TSLP \tag{14}$$

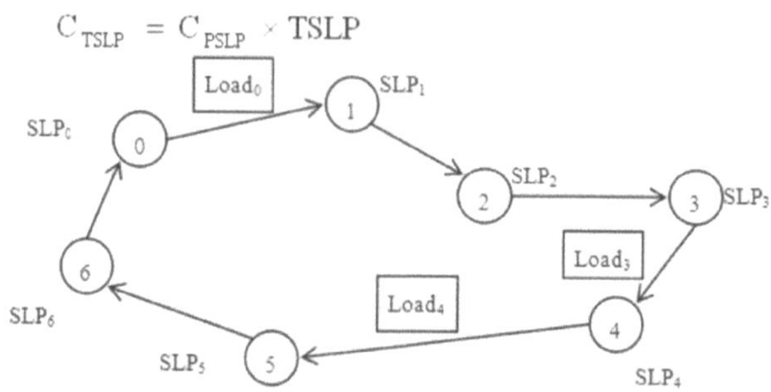

Fig.5: Shelf-life loss parameters during distribution

5. CASE STUDY

The model is coded in Matlab 7.0 on a computer with AMD Sempron(tm) Processor LE-1100 1.91GHz , 512M RAM , Windows XP. The temperature evolution rate, sorting time, traveling time from one customer to another, door opening time, loading time and customer requirement are generated by random number. The vehicle travel input parameters are in Table 1. The traveltimes, opentimes and loads are array generated by the size of amount of customers.

Table 3. Travel input data

Parameter	Meaning	Range	Unit
a_s	Evolution rate of sorting	[0.08,0.12]	℃/m
a_c	Evolution rate of cooling	[-0.03,-0.07]	℃/m
a_t	Evolution rate of traveling	[0.006,0.01]	℃/m
a_o	Evolution rate of door opening	[0.01,0.03]	℃/m
picktime	Sorting time	[10,30]	m
traveltimes	Traveling time	[15,45]	m
opentimes	Door opening time	[5,20]	m
loads	Customer requirements	[50,100]	kg
size	Amount of customers	[5,10]	-

The numerical experiment takes eggplant as an example and the respiration related input parameters are in Table 2.

Table2. Eggplant respiration data

Parameter	Value	Parameter	Value	Parameter	Value
m	1.00	Ea	193.00	α	3.00
K_1	2760.00	R	0.83	β	2.27×105
K_2	0.79	f_0	2.80	σ	0.12
k_d	0.29	a_{co2}	0.002	C_{PSLP}	0.1

Because the absolute value of result can not fully reflect the quality loss of agricultural products, in this paper we simulate two types of distribution respectively: ambient distribution of which the temperature evolution is the above curve in figure 6 and insulation distribution of which the temperature evolution is the below curve. The initial temperature of ambient distribution is a random number between [37K, 42K] and the insulation distribution is a constant 32K.

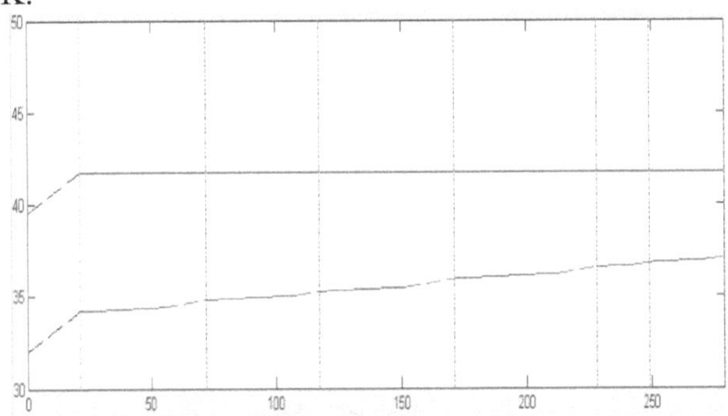

Fig.6: Time temperature evolution curves

According to the input parameters in Table 1 and 2, operate the Matlab procedure for 30 times. The average load is 415.82kg, the average cost for insulation distribution is 141.68 RMB, the average cost for ambient distribution is 397.45 RMB which is 2.8 times of insulation way.

6. CONCLUSION

Temperature and vibration are considered to be the most important factors in evaluating the shelf-life loss of horticultural produce during distribution. So, a model of the loss is developed by evaluating the respiration rate. In order to simulate the distribution process, the mathematical tool Matlab is used taking eggplant as an example. The numerical experiment result demonstrates that the average cost for ambient distribution is 2.8 times of the insulation way. It is concluded that this approach to predicting shelf-life loss of horticultural during distribution has considerable potential in real time food safety surveillance in food logistics.

ACKNOWLEDGEMENTS

This study was supported by Beijing Municipal Science and Technology Commission doctoral thesis fund special funds (No.ZZ0742) and China's 863 project (Contract No. 20060110Z2005) by Chinese government.

REFERENCES

Bogataj, M., L. Bogataj, et al. Stability of perishable goods in cold logistic chains, International Journal of Production Economics, 2005, 93-94:345-356

Don W. Brash, Chris M. Charles, Sandy Wright and Bruce L. Bycroft, Shelf-life of stored asparagus is strongly related to postharvest respiratory activity, Postharvest Biology and Technology, 1995, 5(1-2):77-81.

Fanbin Kong, Juming Tang, Barbara Rasco and Chuck Crapo. Kinetics of salmon quality changes during thermal processing, Journal of Food Engineering, 2007, 83(4):510-520

G. Urrutia-Benet, T. Balogh, J. Schneider and D. Knorr. Metastable phases during high-pressure–low-temperature processing of potatoes and their impact on quality-related parameters, Journal of Food Engineering, 2007, 78(2): 375-389

Heye Chengfu, Yanyuan Mufu, Song Baofeng. Simulation Technology of Transportation Package, Technology Journal of Zhuzhou Institute, 1999(1):8-12

Johanne Rønnow Olsen, Hanne Harmsen and Alan Friis. Linking quality goals and product development competences, Food Quality and Preference, 2008, 19(1):33-42

Li Guo, Ying Ma, Da-Wen Sun and Peng Wang. Effects of controlled freezing-point storage at 0 °C on quality of green bean as compared with cold and room-temperature storages, Journal of Food Engineering, 2008, 86(1):25-29

Lynette M. Johnston, Lee-Ann Jaykus, Deborah Moll, Juan Anciso, Brenda Mora and Christine L. Moe. A field study of the microbiological quality of fresh produce of domestic and Mexican origin, International Journal of Food Microbiology, 2006, 112: 83-95

Roy, P., D. Nei, et al. Life cycle inventory analysis of fresh tomato distribution systems in Japan considering the quality aspect, Journal of Food Engineering, 2008,86(2): 225-233

Tomas Skoglund and Petr Dejmek. A dynamic object-oriented model for efficient simulation of microbial reduction in dispersed turbulent flow, Journal of Food Engineering, 2008,86(3):358-369

Toshitaka Uchino, Daisuke Nei, Wenzhong Hu and Hussain Sorour, Development of a mathematical model for dependence of respiration rate of fresh produce on temperature and time, Postharvest Biology and Technology, 2004, 34(3):285-293.

Uchino, T., D. Nei, et al. Development of a mathematical model for dependence of respiration rate of fresh produce on temperature and time, Postharvest Biology and Technology, 2004,34(3): 285-293

Van Zeebroeck, M., E. Tijskens, et al. The discrete element method (DEM) to simulate fruit impact damage during transport and handling: Case study of vibration damage during apple bulk transport, Postharvest Biology and Technology, 2006, 41(1):92-100

Van Zeebroeck, M., V. Van linden, et al. Impact damage of apples during transport and handling, Postharvest Biology and Technology, 2007, 45(2): 157-167

STUDY ON IDENTIFICATION SYSTEM OF MAIZE SEEDS VARIETIES BASED ON MACHINE VISION

Xianxi Liu[1,*], Yuliang Wang[1], Qingtang Su[2], Zhaona Wang[1]

[1] *Mechanical & Electrical Engineering College Shandong Agriculture University, Taian Shandong 271018*

[2] *Modern Education Technology Department Lu Dong University, Yantai Shandong 264025*

[*] *Corresponding author, Address: .Mechanical & Electrical Engineering College Shandong Agriculture University, Taian 271018, Shandong Province, P. R. China, Tel: +86-538-8243669, Fax: +86-538-8249752, Email: bullish@sdau.edu.cn*

Abstract: This paper systematically studies maize seeds varieties identification technology and algorithms using advanced machine vision technology. A multi-object contour extraction algorithm adapting to maize seeds varieties identification was proposed. Geometric features and color features parameters of maize seeds were defined and analyzed, and a multi-object geometric features and color features extraction algorithm is realized. Maize seeds image processing strategies and varieties identification algorithms, which is based on the machine vision, is optimized. The precision and speed of maize seeds varieties identification is improved. Through maize seeds varieties identification test on four species including Nongda 108, Ludan 981 and Zhengdan 958, identification accuracy is more than 95%.

Key words: machine vision, multi-object, feature extraction, maize seeds, variety identification

1. INTRODUCTION

Artificial identification has great limitations because it is affected by various factors. Variety identification and quality inspection of corn and other cereal seeds using machine vision instead of the human vision is an inevitable trend of achieving detecting agricultural products automatically

Please use the following format when citing this chapter:

Liu, X., Wang, Y., Su, Q. and Wang, Z., 2009, in IFIP International Federation for Information Processing, Volume 295, *Computer and Computing Technologies in Agriculture II, Volume 3*, eds. D. Li, Z. Chunjiang, (Boston: Springer), pp. 2207–2216.

and intelligently because it has the merits such as real-time, efficiency, objective, accuracy and without injury (Cheng et al.,2001; Wang et al.,2003). It can guarantee the seed quality, increase the economic benefits of planting and promote the development of corn planting with high-quality, high yielding and stable yielding. It said regarding the China this kind of large agricultural nation that, has extremely important significance.

In the field of corn kernel detection and identification, Liao K .etc.(Liao K et al.,1994) applied real-time feature extraction algorithms to online testing of maize quality in 1995. Ni B. etc.(Ni B et al.,1997)graded corn kernel according to its size in 1998. H. ENg. etc.(Ng et al.,1998) identified damaged kernel and moldy kernel of maize kernel in 1998. Zhou Yiming and Wang Fengyuan.etc.(Wang et al.,1995) measured and verified the basic shape of corn seeds in 1995. Song Tao and Zeng Dechao(Song et al.,1996) identified integrity and damage of maize kernel by BP network in 1996. Ning Jifeng and He Dongjian.etc.(Ning et al.,2004) identified the tip cap and germ surface of corn kernel in 2002.

The paper puts forward a method of maize seeds varieties identification based on multi-object feature extraction and identification algorithm, and explores a new method of identification and detection for corn seeds based on machine vision.

2. MAIZE IDENTIFICATION SYSTEM AND MAIZE VARIETY

2.1 Maize identification system

Maize seeds varieties identification system is mainly composed of hardware system and software system. The hardware system shown by figure 1 is composed of the image acquisition equipment, computer information processing equipment, image information storage devices, image information and results output devices and so on. The software system is developed using VC++ 6.0.

2.2 Maize varieties

Four varieties of maize seeds including Nongda 108, Ludan 981, Zhengdan 958 and Wuyue 18, which are provided by Taian Wuyue Mount Tai Co. Ltd, are selected as test samples in this study. We remove the corn seeds of the two ends of the corn ear and select seeds of morphological characteristics relative standard for research.

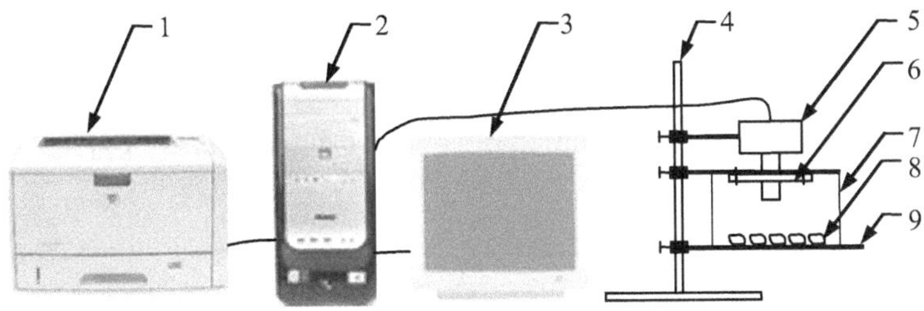

1.Printer 2.Computer 3. Monitor 4. Acquisition shelf

5.Camera 6.Lighting source with ring shape 7.Hood 8.Sample 9.Object stage

Fig.1 The diagram of hardware system composition

3. MULTI-OBJECT CONTOUR EXTRACTION

3.1 Image preprocessing

The images of corn seeds acquired are pretreated before varieties identification to get the target images up to the requirements. Image pretreatment process is shown in figure 2. Firstly, JPG images acquired are transformed into BMP images. Then these BMP images are preprocessed including gray scale transformation, median filtering, binary processing, and morphological transformation. Finally, the necessary images are successfully obtained.

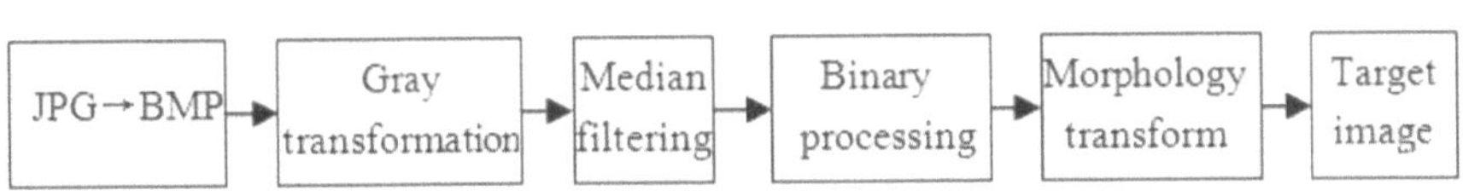

Fig.2 Maize seeds image preprocessing

3.2 Multi-object contour extraction

The traditional contour extraction of images is generally single-object contour extraction and only the single-object contour is tagged and extracted. However, contour extraction and marking of multiple seeds must be simultaneously resolved in this maize seeds varieties identification system, namely multi-object contour extraction.

The key problem of multi-object contour extraction is contour decision of objects. That is, multi-object contour extraction.

The basic idea of multi-object contour extraction is:

1) Traversing image. Enter contour tracking if a target point is found, which is shown in figure 3, and then transfer to step 2) after the contour tracking is completed; If the images traversing complete and does not find target point, then transfer to Step 3);

2) The contour tracked completely is vector tagged. Then transfer to Step 1);

3) End.

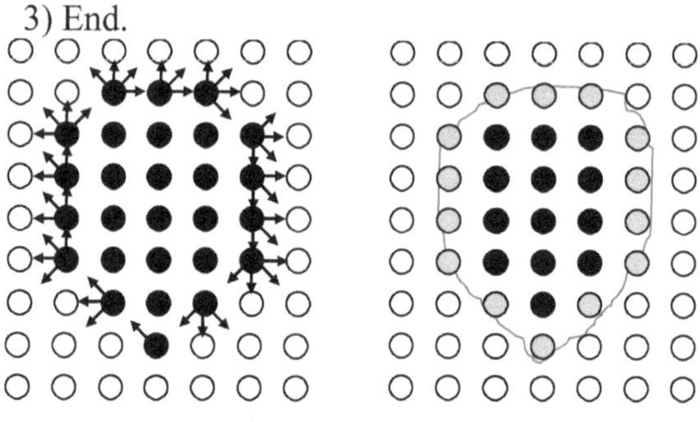
(a) (b)

Fig.3 Diagram of contour following algorithm

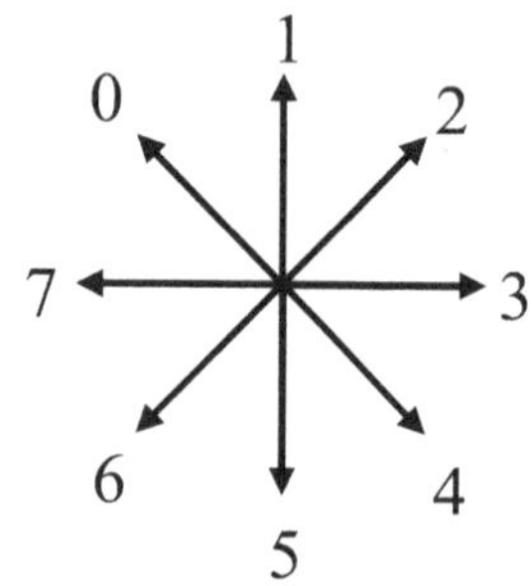

Fig.4 8-connected chain code

In order to convenient for marking the contour, we define 8-connected chain code shown in Figure 4 and establish the xy coordinate deviation of center point and bordering points according to 8-connected chain code in this paper, as shown in table 1.

Table.1 The coordinate deviation of center point and bordering points

Chain code value	0	1	2	3	4	5	6	7
X coordinate deviation	-1	0	1	1	1	0	- 1	- 1
Y coordinate deviation	1	1	1	0	- 1	- 1	- 1	0

To complex contour, it's difficult to accurately determine the region within their borders only using the value of chain code. We solve the problem using the algorithms of vector tagging in this study. The direction of the vector used is from left to right in horizontal direction. We transform the value of chain code of border points according to this direction. Method of conversion is:

If the boundary point is intermediate point of the vector, its value is 0.

If the boundary point is left endpoint of the vector, its value is 1.

If the boundary point is right endpoint of the vector, its value is 2.

If the boundary point is singular point of the vector, its value is 3.

For the contour of figure 5 (a), if it is represented as 00111233455566 beginning from the low-end clockwise using chain code, only the borders of the target region rather than the region within the borders can be expressed and marked. If it is represented as 31111110222222 beginning from the low-end clockwise using the algorithm of vector tagging from left to right, the target area can be represented with seven vector line segments. The target border and the target region within the border which is illustrated in figure 5(b) can be accurately determined. When this method applied to complex contours, the target border and the target region within the border can also be accurately and quickly determined, as shown in figure 5(b).

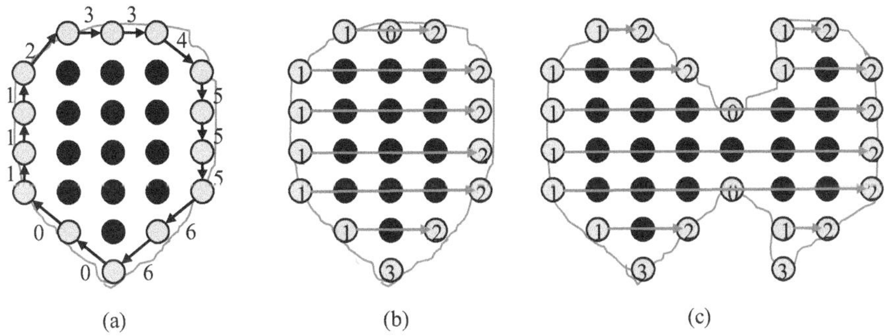

Fig.5 Description of regional contour

For the realization of multi-object contour extraction, a maize seed data linked list is built in this system. The data of each maize seed is stored in the seed data linked list including contour serial number, contour points, coordinate and serial number of each point, coordinates of starting point, adjacent 8-connected chain code, geometric parameters and color parameters of the contour. Multi-object contour extraction is better realized and the implementation of procedures is improved more efficiently by using this data structure. It makes preparation to multi-object feature extraction algorithms.

4. MULTI-OBJECT FEATURE EXTRACTION

4.1 Geometric features extraction

This paper puts forward an algorithm of multi-object feature extraction on the basis of multi-object contour extraction. This algorithm can accomplish feature extraction of many maize seeds on the image by one time without segmenting a group of seeds into many single seeds and improve the speed and efficiency of feature extraction of maize seeds.

Table 2 shows parts of the geometric feature parameters of two varieties maize seeds Nongda 108 and Zhengdan 958.

Table.2 Geometric feature parameters of maize seeds

Geometrical feature	Nongda 108 mean value	Nongda 108 mean-variance	Zhengdan 958 mean value	Zhengdan 958 mean-variance
Contour point number	469.5	21.9	573.2	25.3
Circumference	540.1244	23.57045	664.3753	28.65179
Area	19307.53	1570.919	28368.15	2527.147
The length of long-axis	170.8197	10.97938	226.2442	12.68391
The length of minor-axis	149.7795	11.73738	160.4947	11.31505
Maximum inscribed circle radius	70.28177	3.791605	80.40652	4.961622
Minimum circumscribed circle radius	87.96963	4.807754	115.2072	6.192502
Largest span	173.4634	10.37519	228.6176	12.27242
Equivalent diameter	156.6637	6.324111	189.8631	8.496643
Shape parameter	1.204634	0.039336	1.24148	0.04112
Roundness	1.254113	0.080681	1.436738	0.098116
Elongation	1.148843	0.131217	1.41609	0.122134
Compact ratio	0.91934	0.045075	0.840439	0.035888

4.2 Color feature extraction

The color of maize seeds is an important index to reflect the development and variety of maize seeds, meanwhile, the specific gravity difference of maize seed color components can provide references for maize seed identification and detection. In order to extract and describe the color information in the maize seed images more precisely, this paper also studies H (Hue), S (Saturation), I (Intensity) and other information in the HSI color space.

In order to quantitatively describe the color information of maize seeds, the paper defined 18 color features namely ΣR (Red), R-mean, R-standard deviation, ΣG (Green), G-mean, G-standard deviation, ΣB (Blue), B-mean, B-standard deviation, ΣH (Hue), H-mean, H-standard deviation, ΣS (Saturation), S-mean, S-standard deviation, ΣI (Intensity), I-mean and I-standard deviation. Based on many data analysis, the paper chooses the color features of the most stable part in the maize seed. Shown in figure 6, the areas which are extracted color features are filled with white. Table 3 shows the color feature parameters of Nongda 108, Zhengdan 958 and Ludan 981.

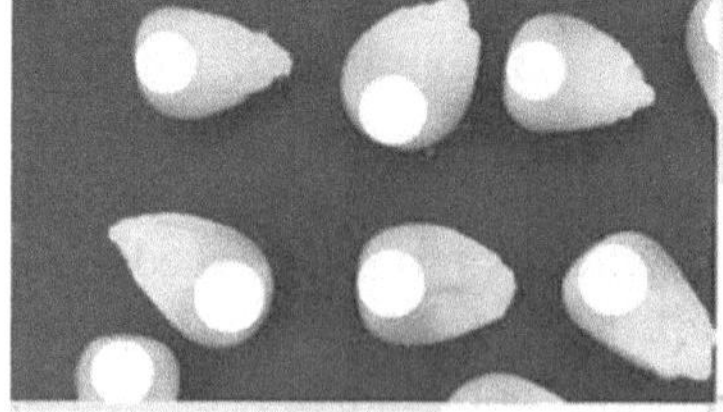

Fig.6 The color feature extraction of Zhengdan 958

Table.3 Color features of maize seeds

Color feature	Nongda 108 mean value	Nongda 108 variance	Zhengdan 958 mean value	Zhengdan 958 variance	Ludan 981 mean value	Ludan 981 variance
$\sum R$	785766.67	120701.33	1394376.45	187810.48	887345.55	130202.78
$\overline{R}$	223.46	7.277	216.88	10.143	209.42	6.717
S_R	5.771	2.762	7.176	3.845	6.324	2.132
$\sum G$	677409.62	104040.56	1145201.25	159049.50	756739.1	108232.05
$\overline{G}$	192.82	11.890	178.06	9.720	178.67	8.543
S_G	7.979	3.368	10.213	4.096	8.756	2.710
$\sum B$	448421.55	118367.10	841963.54	162325.84	556435.08	117621.73
$\overline{B}$	128.03	30.943	130.84	20.136	131.42	22.444
S_B	32.928	10.874	23.344	8.468	23.702	6.353
$\sum H$	2874.57	1534.86	4174.06	1874.997	3252.41	1965.24
$\overline{H}$	0.830	0.493	0.658	0.328	0.782	0.534
S_H	0.393	0.634	0.362	0.586	0.529	0.727
$\sum S$	1111.05	476.21	1685.89	542.443	1066.48	397.29
$\overline{S}$	0.315	0.123	0.263	0.080	0.252	0.088
S_s	0.138	0.0880	0.0357	0.0316	0.0633	0.0878
$\sum I$	636033.77	100661.19	1125079.91	157328.54	732112.75	109220.28
$\overline{I}$	181.605	14.069	175.432	10.981	173.323	10.270
S_I	14.393	12.573	11.978	4.9998	4.985	3.132

5. VARIETIES IDENTIFICATION

As the individual differences of corn seeds and the complexity of its own form, there are certain cross-cutting issues in the demographic characteristics of the different species of seeds. it is difficult to use conventional mathematical model or method to describe the relationship of characteristics and parameters between different species. Therefore, in order to identify varieties of corn seed, the weighted processing needs to be made to the different characteristics of various seeds, and then the contributions of the different characteristics to different varieties of seeds need counting and analyzing. Finally, the characteristics data and statistical data of various corn seeds will be saved to the feature library. After analysis of test, the species identification classifier of maize seed is established, to identify seed through the classifier, as shown in Figure 7.

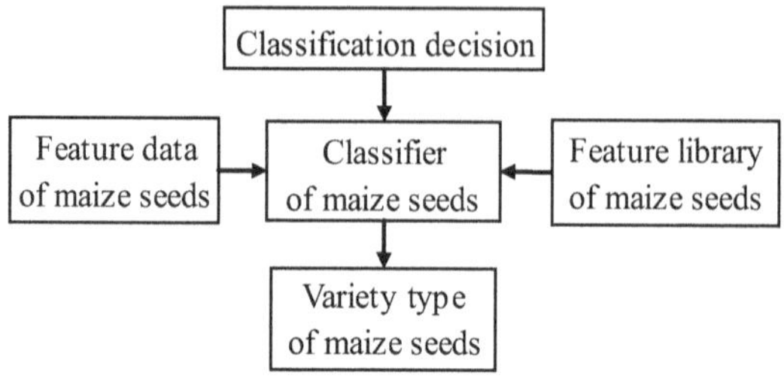

Fig.7 Diagram of maize seeds varieties identification

(1) Normalization processing is done to the various feature parameters of maize seeds. That is, the ratio of various parameters and the average of maize seeds.

(2) After analyzing the test, the weights of various feature parameters are determined. We identify varieties of maize seeds with the following formula:

$$K= k_1Cp^* + k_2P^*+ k_3A^* + \ldots + k_{32}\Sigma I^* + k_{33}\bar{I}^* + k_{34}SI^* \qquad (1)$$

Cp^*, P^*, A^*, ..., ΣI^*, $\bar{I}^*$, SI^* are the normalized parameters of a certain maize seed , k_1 , k_2 , k_3 ,..., k_{33} , k_{34} are the weights of corresponding parameters, and $\Sigma k_i = 1$. When meeting $K \geqslant 0.8$ and meeting the analysis of figure 8, it is the maize seed of this variety.

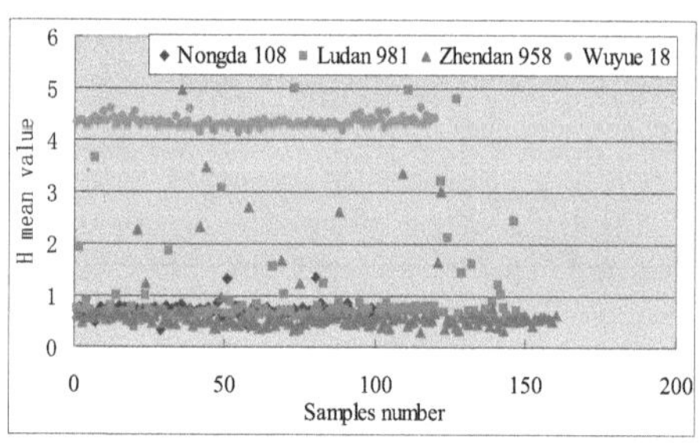

(a) The relationship of H mean value of four varieties maize seeds

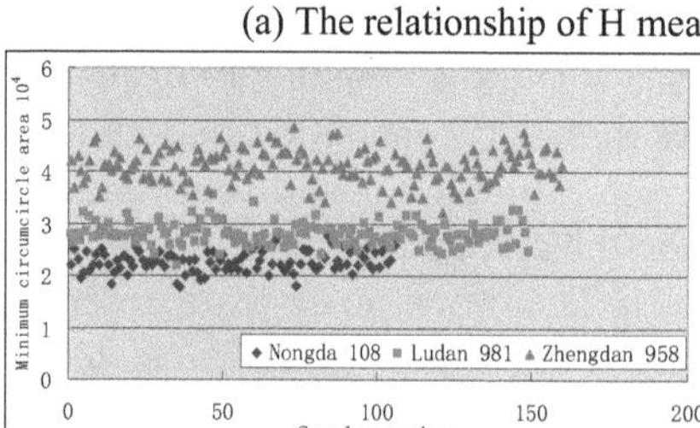

(b) The relationship of the minimum circumcircle area of three varieties maize seeds

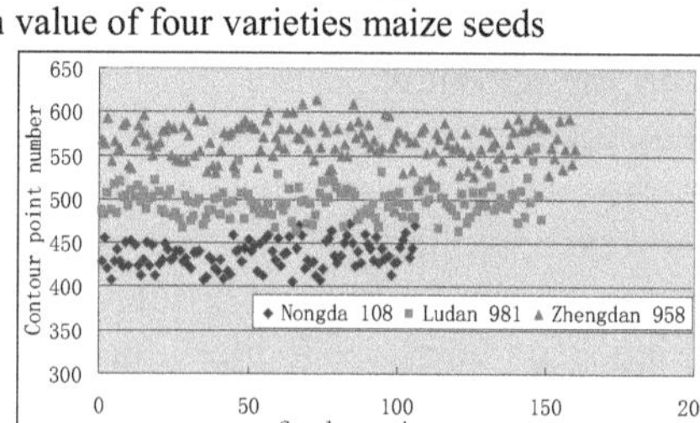

(c) The relationship of contour point number of three varieties maize seeds

Fig.8 Classification foundation of four kinds of maize seeds

Using the varieties identification classifier, the test of single variety identification and mixed varieties identification are done to four varieties maize seeds including 106 tablets of Nongda108, 150 tablets of Ludan981, 160 tablets of Zhengdan958 and 120 tablets of Wuyue18, the identification accuracy are 96.23%, 95.33%, 98.13% and 100% separately, as shown in table 4.

Table.4 Results of varieties identification of 4 varieties maize seeds

Samples	Nongda 108	Ludan 981	Zhengdan 958	Wuyue 18
Sample sizes (grain)	106	150	160	120
Classification features	Contour point number	Minimum circumscribed circle area		H mean value
Threshold values	462	340000		4.00~4.76
Correct identification (grain)	102	148	159	120
Error identification (grain)	4	2	1	0
Recognition accuracy (%)	96.23	98.67	99.38	100
Comprehensive correct identification (grain)	102	143	157	120
Comprehensive Error identification (grain)	4	7	3	0
Comprehensive Recognition accuracy (%)	96.23	95.33	98.13	100

6. CONCLUSION

The results show that, the proposed multi-object contour extraction algorithm and multi-object feature extraction algorithm of maize seeds based on machine vision are very effective in identifying maize seeds varieties. It is feasible to identify and detect maize seeds varieties using machine vision, meanwhile, the method can be extended to other granular agricultural products in the fields of online identification and testing.

ACKNOWLEDGEMENTS

Funding for this research was provided by Shandong Provincial Department of Education (P. R. China)(NO:J05G16).

REFERENCES

Chengfang, Ying Yibin. Application of Machine Vision to Quality Evaluation of Agricultural Seed Resources [J]. Transactions of The Chinese Society of Agricultural Engineering. 2001,(06):175-179.

Liao K, Paulsen M R,Reid J F ,et al. Corn kernel breakage classification by machine vision using a neural network classifier[J]. Trans of the ASAE.1994,36(6):1949-1953.

Ng, H. F., W. F. Wilcke, R. V. Morey, and J. P. Lang. Machine vision evaluation of corn kernel mechanical and mold damage[J]. Trans of the ASAE,1998,41(2): 415-420.

Ni B,Paulsen M R,Reid J F.Corn Kernel Shape identification Using Image Processing[J].Trans of the ASAE,1997,40(3): 833-838.

Ning Jifeng, He Dongjian, Yang Shuqin. Identification of Tip Cap and Germ Surface of Corn Kernel Using Computer Vision. Transactions of The Chinese Society of Agricultural Engineering. 2004,(03):117-119

Songtao, Zeng Dechao. An Investigation on Morphological Discrimination of Corn Kernels Based on Neural Network [J]. Transactions of The Chinese Society of Agricultural Engineering. 1996,12(1):177-181.

Wang Fengyuan, Zhou Yiming. Seed Shape Detection and Measurement with Computer Image Processing. Transactions of The Chinese Society of Agricultural Engineering. 1995,11(2): 52-57

Wang Guiqin,Yang Zibiao,Zheng Limin. etc. Application of Computer Vision Technology in Detection of Agricultural Products [J]. Review of China Agricultural Science and Technology. 2003,5(3):52-55.

STUDY ON DETECTION TECHNOLOGY OF PESTICIDE RESIDUES IN VEGETABLES BASED ON NIR

Jingzhu Wu[1,*], Cuiling Liu[1], Yuanyuan Chen[1], Yan Chen[1], Yun Xu[2]

[1] *School of Information Engineering, Beijing Technology & Business University, Beijing, P. R. China 100037*

[2] *College of Information and Electrical Engineering, China Agricultural University, Beijing, P. R. China 100083*

[*] *Corresponding author, Address: School of Information Engineering, Beijing Technology & Business University, Beijing, 100037, P. R. China, Tel: +86-10-68985506, Email: pubwu@163.com*

Abstract: A quantitative technology based on near infrared spectroscopy (NIR) is proposed to detect pesticide (chlorpyrifos) residues in vegetables in the paper. There are three groups of samples (totally 100 unit, concentration from 0.1mg/kg to 4mg/kg) used in the experiment. The first group sample is confected to simulate vegetable juice to validate feasibility of the NIR quantitative method. As shown in this work, the first NIR model has good indexes (R2 is 0.9807, RMSEP is 0.149). Next the second group sample is confected by adding material standards of 1 mg/ml of chlorpyrifos to spinage juice. The second NIR model also has good performance (R2 is 0.9858, RMSEP is 0.207). The third group sample is confected as the second group to validate repeatability of the detection method based on NIR. The third model is also excellent (R2 is 0.9986, RMSEP is 0.0485). The exploratory experiment results indicate the detection technology of pesticide residue in vegetables based on NIR is feasible in the given conditions. But supposed to the natural vegetable samples that include pesticide residue, the method should to be validated more.

Keywords: NIR, Pesticide Residue, Detection, Spinage

Please use the following format when citing this chapter:

Wu, J., Liu, C., Chen, Y., Chen, Y. and Xu, Y., 2009, in IFIP International Federation for Information Processing, Volume 295, *Computer and Computing Technologies in Agriculture II, Volume 3*, eds. D. Li, Z. Chunjiang, (Boston: Springer), pp. 2217–2222.

1. INTRODUCTION

The invention and use of pesticides has greatly enhanced the crop yield. However, with the widely and unreasonable use of pesticide the negative impact is increasingly exposed, especially poisoning incidents occur from time to time caused by pesticide residues in vegetables. The detection technology of pesticide residues gets great attention. At present there are some traditional chemical method to detect pesticide residues, such as GC, PLC and CG-MS (Wang, et al, 2006). But these methods are time-consuming and complex. Otherwise chemical methods need consume chemical and destroy the estimate sample. So it's unfit for the national supervise department to execute the inspect task locally.

Near infrared spectrometry is fast developed after ninety in twentieth century in China. The NIR-based (Near Infrared Spectroscopy analysis) technology is recommended for agricultural products detection because it is a fast, lossless and sustainable technology. At present, Pierre Billeen had been using near-infrared detection of pesticides used in seed dressing, and achieved good results (Pierre, et al, 2000). Wen-Xiu Li, such as the use of high pesticide residues in vegetables dichlorvos trichlorfon and the infrared absorption of solvent juice, vegetables directly on the pesticide residue was detected (LI, et al, 2004). Xiang-Yang Zhou, analyzed a variety of vegetable samples near-infrared spectra of the differences in the use of phosphorus-containing group characteristic absorption frequency of the area to identify, and GC-MS method than the right, made with the differential effects (ZHOU ,ct al, 2004).

Pesticide residues detection belongs to trace detection. This paper does exploratory studies about quantitative detection technology of pesticide (chlorpyrifos) residues in vegetables based on near infrared Spectroscopy.

2. THEROY

Near-infrared is the electromagnetic wave between the middle-infrared and visible. Its wavelength range is between 800nm to 2500nm.

Near infrared spectroscopy is the absorption spectrum which is mainly obtained by organic molecules hydrogen group (C-H, N-H, O-H, etc.) stretching at all levels and frequency of these groups of stretching and bending of the frequency of vibration. The spectrum of areas include almost all of the hydrogen-containing organic compounds in the group, contains elements of the structure, composition status, and other information. The amount of information is extremely rich, so as to near-infrared spectral analysis of quantitative samples of the physical nature and chemical composition may be provided.

There are many kinds of pesticide are widely used in China, especially organophosphorus pesticides. Chlorpyrifos is one kind of organophosphorus and it is always detected in leaf vegetable higher than national standard. The Molecular formula of chlorpyrifos is $C_9H_{11}Cl_3NO_3PS$ including hydrogen group, which provide the theoretical basis for Near Infrared Spectroscopy analysis.

3. EXPERIMENT

3.1 Sample preparation

There are three group of samples designed in the experiment listed in table 1. The background solution of first group is simulation of pesticide residues in samples of vegetables. The components are includes sugar, Vitamin C and drinking water mixed solution. The samples with different pesticide residues can be obtained by adding material standards of 1 mg/ml of chlorpyrifos according to design requirements.

To approach natural samples the background solution of the second and third group is spinage juice. The spinage purchased from the supermarket is Organic vegetables without pesticide residues. The spinage juice is prepared in the Chinese Academy of Agricultural Sciences. The samples with different pesticide residues can be obtained like the first group.

Table 1. Sample preparation

Group Label	Preparation time	Background Solution	Range of Concentration (mg/kg)	Number of Sample
1	2007-12-11	simulation vegetable juice	0.1-3	30
2	2008-03-07	spinage juice	0.1-4	35
3	2008-05-07	spinage juice	0.1-4	35

3.2 Spectral data

The spectrums of the first group are collected by fiber Optic transmission from 12500 to 4000 cm^{-1} in 8 cm^{-1} on MPA spectrometer with the liquid fiber probe of 2 m, the wavelength path of 1 mm. The spectrums of the second and third group are collected on MATRIX-F spectrometer. The other parameters are same. Spectrum acquisition and analysis software used OPUS5.0.

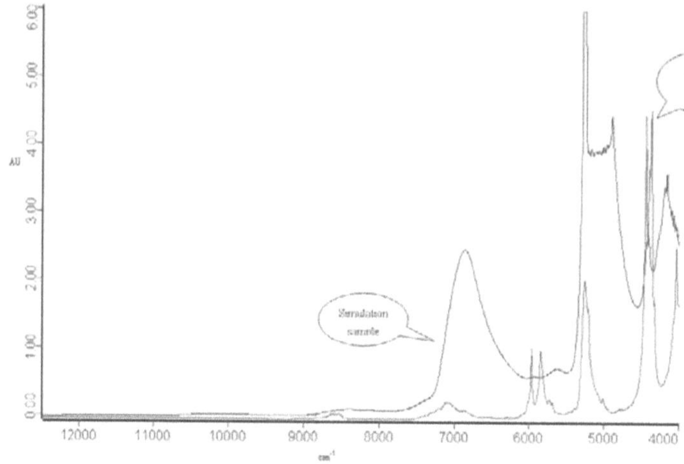

Fig.1: 2 unit spectrums

4. RESULTS AND DISCUSSION

According to the concentration difference, the sample set is divided to train set and test set. The number of the train set and test set is approximately 3. The result lists in the table 2.

From the table 2 the NIR model built by simulation samples is good. The second NIR model built on the natural spinage juice samples also has good performance. The third model is built based on the third group samples to validate the last model's reliability. The result is also satisfied.

Table 2. NIR index of 3 groups of samples

Group Label	Train set	Test set	R^2	RMSECV	RMSEP
1	23	7	0.9807	0.123	0.149
2	27	8	0.9858	0.154	0.207
3	27	8	0.9986	0.0437	0.0485

From the table 2 the NIR model built by simulation samples is good. The second NIR model built on the natural spinage juice samples also has good performance. The third model is built based on the third group samples to validate the last model's reliability. The result is also satisfied. The prediction of test sample of the second and third group list in Table 3 and Table 4.

Pesticide residues detection belongs to trace detection. It is very difficult to detect theoretically. But the result of the experiment in this paper is satisfied unexpectedly. Maybe the reason is that samples are made uniformity solution, which is conducive to collecting spectrum with strong signal, and all samples are ideal samples confected by manual in give some condition. That reduced sample complexity. The Study on quantitative detection technology of pesticide (chlorpyrifos) residue in vegetables based on near infrared Spectroscopy should be continued deeply.

Table 3. Prediction of test sample of Group 2

Filename	True	Prediction	Difference
04.0	0.35	0.3307	0.0193
08.0	0.55	0.7266	-0.177
12.0	0.8	0.9077	-0.108
16.0	1.1	0.8374	0.263
20.0	1.5	1.644	-0.144
24.0	1.9	1.64	0.26
28.0	2.5	2.873	-0.373
32.0	3.2	3.246	-0.0456

Table 4. Prediction of test sample of Group 3

Filename	True	Prediction	Difference
04.0	0.35	0.3099	0.0401
08.0	0.55	0.4504	0.0996
12.0	0.8	0.7574	0.0426
16.0	1.1	1.149	-0.0486
20.0	1.5	1.506	-0.00582
24.0	1.9	1.933	-0.033
28.0	2.6	2.643	-0.043
32.0	3.4	3.39	0.00996

5. CONCLUSION

This paper discusses the quantitative detection technology of pesticide (chlorpyrifos) residue in vegetables based on near infrared Spectroscopy. The experimental results of three groups of samples indicate the detection technology of pesticide residue in vegetables based on NIR is feasible in the given conditions in this paper. But supposed to the natural vegetable samples that include pesticide residue, the method should to be validated more.

ACKNOWLEDGEMENTS

Funding for this research was provided by Beijing Natural Science Foundation, P. R. China (project No. 4073031). The authors would like to thank Doctor Yuan Yuwei from the Chinese Academy of Agricultural Sciences to help us confect the spinage samples. Deeply Tanks for BRUKER OPTICS corp. to supply the NI Spectrograph to help us collect samples' spectrum.

REFERENCES

LI Wen-iu, XU Kexin. Investigation on the Detection of Pesticide Residue in Vegetable Based on Infrared Spectroscopy. Spectroscopy and Spectral Analysis, 2004, 24(10):1202-1204

Pierre Billeen, etal. Testing seeds protected with an insecticide by NIR. Proceeding of the 9th International Conference, 2000, 625-628.

Wang chaojing, Cai qi. Tendency of Detection of Residual Pesticides in Farm Produce. Modern Scientific Instruments, 2006, 01:106-108

ZHOU Xiang-yang, LIN Chung-zhon. Rapid Indentification of Organophosphorus Pesticide Residues in Leaf-vegetable by FT-NIR. Food Science, 2004，25(05):151-154

ON-LINE SORTING MATURITY OF CHERRY TOMATO BY MACHINE VISION

Yinglong Zhang [1], Xiaoping Yin [2], Xiaobo Zou [2,*], Jiewen Zhao [2]

[1] *Industry Center, Jiangsu University, Zhenjiang, Jiangsu 212013, China*

[2] *Agricultural Product Processing and Storage Lab, Jiangsu University, Zhenjiang, Jiangsu 212013, China*

[*] *Corresponding author, Address: Agricultural Product Processing and Storage Lab, Jiangsu University, Zhenjiang, Jiangsu 212013, China, Tel:+86-0511-88780201, Fax:+86-0511-88780201, Email: Zou_xiaobo@ujs.edu.cn*

Abstract: The cherry tomatoes online sorting according to their maturity is an important procedure after harvest. This research proposed an automated cherry tomato grading system base on machine vision. Three images of different angles are obtained from each cherry tomato, allowing the inspection of approximately 90% of the fruit surface. 9 features were extracted from the one cherry tomato images. In order to distinguish into three grades (immature, half ripe and ripe), Principal component analysis (PCA) and linear discrimination analysis (LDA) were used to analyze the features. The PCA results show that ripe cherry tomatoes are distinguished from immature and half ripe ones. 414 cherry tomatoes were tested by the online sorting system. The overall accuracy was up to 94.9%. Furthermore, the grading speed of the sorting line reaches 7 cherry tomatoes per second which meet the actual demand of many farms.

Keywords: Cherry tomato; Machine Vision; On-line Sorting; maturaty

1. INTRODUCTION

A cherry tomato is a smaller garden variety of tomato. With its highly nutritional value and good appearance, cherry tomato had become one of the most popular fruit in the world.

Nowadays cherry tomatos are sorted by hand in many farms. However, the manual inspection process is not only labor-intensive and tedious, but

---

Please use the following format when citing this chapter:

Zhang, Y., Yin, X., Zou, X. and Zhao, J., 2009, in IFIP International Federation for Information Processing, Volume 295, *Computer and Computing Technologies in Agriculture II, Volume 3*, eds. D. Li, Z. Chunjiang, (Boston: Springer), pp. 2223–2229.

also subject to human error which results in poor quality. Farmers want an automated grading device to facilite this work. Considerable effort has been made in the field of machine-vision-based sorting and grading of fruit. Miller took into account the mean fruit colour and a measureof the dispersion (normalised mean squared differences)of the colour, plus a shape parameter to grade citrus fruits according to their external quality. The author compared three different classifiers and had the best results with Bayesian - Gaussian techniques, with between 69 and 86% of the fruit correctly graded into two classes (accepted or rejected) (Miller W M ,1995). Otherwise, V. Leemans proposed a method for apple external quality grading showed correct classification rates of 78 and 72%, for Golden Delicious and Jonagold apples, respectively, according to European standards(V. Leemans etal. ,2002). However, much of the works have not been used in commercial sorting systems because of the constraints in speed, accuracy and flexibility.

Colour grading based on United States Department of Agriculture (USDA) standard colour charts is according to six different maturity stages(United States Department of Agriculture, 1991). Surface colour is considered as a good maturity index.

In this research, a machine-vision system was developed for on-line high-speed cherry tomato maturity sorting according to the surface colour, grading of cherry tomatos into three classes from immature to ripe.

2. MATERIALS AND METHODS

cherry tomato "little angel" was selected for the experiment. The samples were hand harvested on 23th November 2007 from the experimental orchard in "jin rui" Institute of Agricultural, zhenjiang.

Cherry tomatoes were selected completely randomized in the same plant with each fruit as an experimental unit. All fruits of each sample were individually numbered. Without any procedure, five assessors with previous experience in tomato assessment were invited to classified cherry tomatoes into three different maturity states (immature , half-ripe, and full-ripe), each with 30 samples. A total of 90 machine vision measurements were performed.

For validation, the same variety, cherry tomato 'little angel ', was selected for the experiment. A total of 414 cherry tomatoes were ispectioned for validation.

2.1 Hardware

The machine vision system was composed of a three charge coupled device (CCD) colour camera (SenTec STC-1000) and a frame grabber

(GRABLINK Value), connected to a compatible personal computer [Pentium 2.8GHz, 512Mb random access memory (RAM)]. The system provides images of 768 per 576 pixels. The frame grabber digitised and decoded the composite video signal from the camera into three user-defined buffers in red, green and blue colour coordinates (RGB). Lighting system is the most important part of the machine vision system(Sunil Kumar Kopparapu,2006). In this project lighting system was composed of two ring-shaped LEDs inside of a chamber, with a hole in the top to place the camera.

The vision system was part of the robotic system for automatic inspection and sorting. Before entering the inspection chamber the fruit was individualised ,then made the fruit to be presented to the camera in three different, nonoverlapped positions, in order to inspect as much of the fruit surface as possible. Entire system is made of 4 part as follows: ①mechanical conveyer; ②CCD combine with PC; ③ executive mechanism ; ④electronic device(Fig 1).

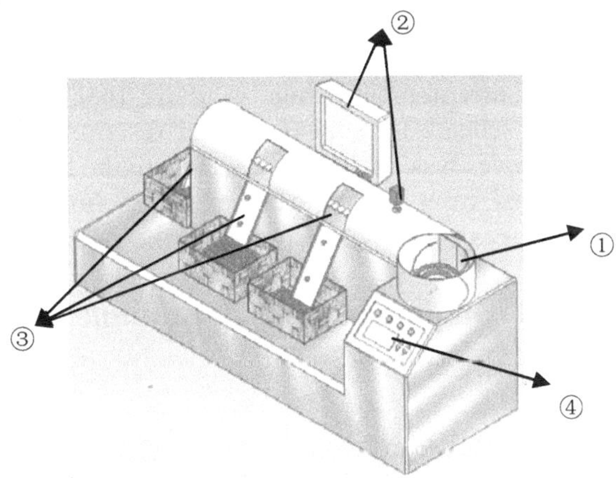

Fig.1: Cherry tomato online sorting device entire map

2.2 Image analysis

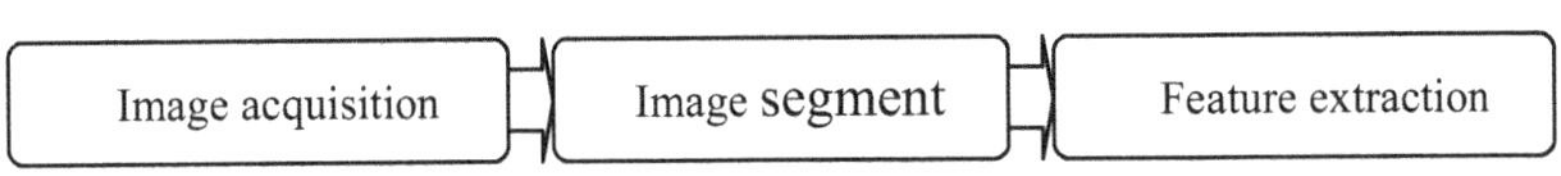

Fig.2: On-line grading software flow chart

On-line operation started with the acquisition of the first image. Three images of different angles are obtained from each cherry tomato, allowing the inspection of approximately 90% of the fruit surface fig3(a).

The second step consisted of image segment use fixed threshold (Rafael C.Gonzelez Richard et al., 2003)as :

$$f_t(x,y)=\begin{cases}0, f(x,y)<T\\255, f(x,y)\geq T\end{cases} \quad (1)$$

Cherry tmatoes was separated from background as shown in fig3(b).

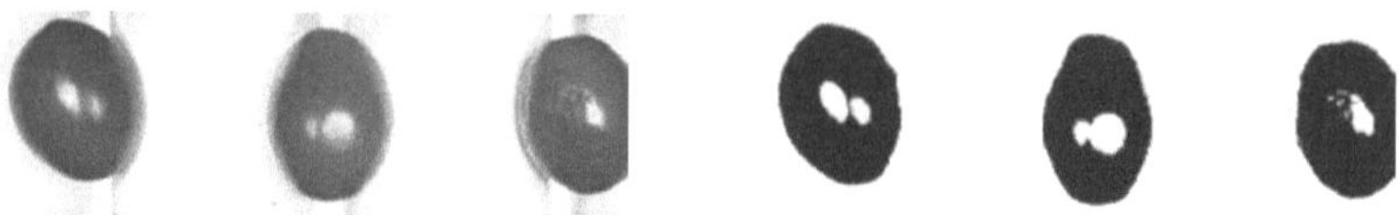

(a)Raw image of cherry tomato (b) Result of image segment

Fig.3: Result of image processing

The third step is feature extracting. Color is one of the most significant inspection criteria related to fruit quality, in that surface color of a fruit indicates maturity. Color representation in RGB (red, green and blue) provides an efficient scheme for statistical color discrimination. Therefore, color evaluation of apples was achieved by analyzing the RGB value of each cherry tomato. A total of 9 feature were extracted from the one cherry tomato, because 3 images are obtained from each cherry tomato.

2.3 Principal component analysis and linear discriminant analysis

Principal component analysis (PCA) is a linear, unsupervised and pattern recognition technique used for analyzing, classifying and reducing the dimensionality of numerical datasets in a multivariate problem. This method permits extraction of useful information from the data, and exploration of the data structure, the relationship between objects, the relationship between objects and variables, and the global correlation of the variables. The main features of PCA are the coordinates of the data in the new base (scores plot) and the contribution to each component of the 9 features(loads plot). The score plot is usually used for studying the classification of the data clusters, while the loads plot can provide information on the relative importance of the feature array to each principal component and their mutual correlation.

The linear discriminant analysis (LDA) calculates the discriminant functions and similar to the PCA—a two- or threedimensional display of the training set data. The difference between PCA and LDA is that PCA does not consider the relation of a data point to the specified classes, while the LDA calculation uses the class information that was given during training. The LDA utilizes information about the distribution within classes and the distances between them. Therefore, the LDA is able to collect information from all sensors in order to improve the resolution of classes (Cheong Hee Park,etal.,2008; Yang, J., Yang, J.-Y,2003)

2.4 Cherry tomato samples

The cherry tomato samples were of the same variety, from the experimental orchard in "jin rui" Institute of Agricultural, zhenjiang. The maturity of cherry tomato is normally sorting into 3 classes which are ripe, half-ripe and immature.

3. RESULTS AND DISCUSSION

To investigate whether the machine vision system was able to distinguish between different ripe state, PCA and LDA analysis were applied to 90 samples. PCA and LDA analysis results are shown in Fig.4 and Fig.5. These figures show the analysis results on a two-dimensional plane, principal component 1 (PC1) and principal component 2 (PC2) in Fig. 4and first and second linear discriminant LD1 and LD2 in Fig.5.

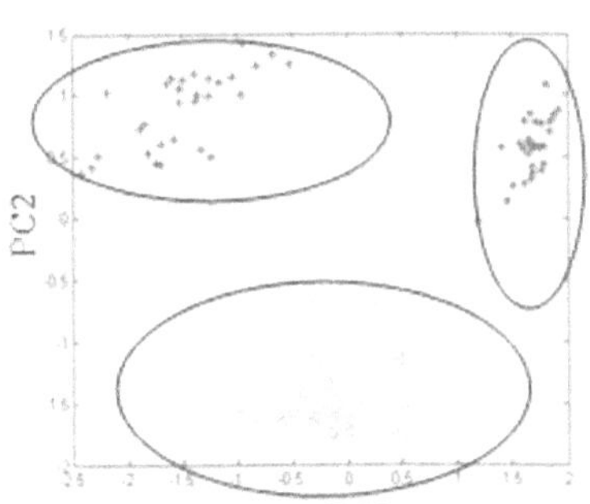

Fig.4: PCA analysis for tomato ripeness

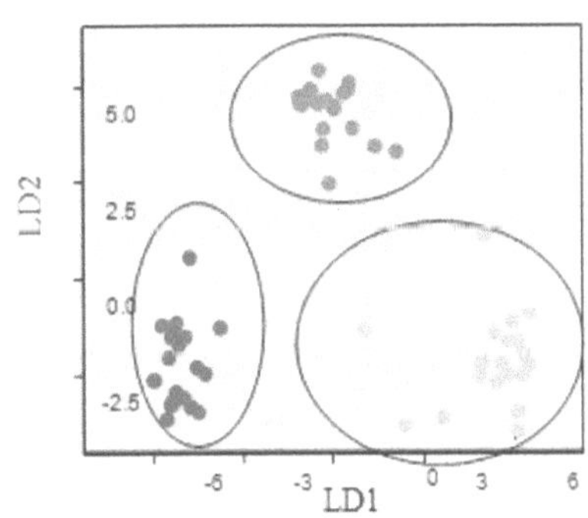

Fig.5: LDA analysis for tomato ripeness.

PCA is a linear combinatorial method, which reduces the complexity of the dataset. The inherent structure of the dataset is preserved while its resulting variance is maximized. Fig.4 shows a clearer discrimination among the various clusters representing the cherry tomato ripe state. Each group was clearly distinguishable from the other groups by using PCA analysis.The processed data show a shift of the different maturity state coinciding with the classification by the trained profile panel. The first principal component, PC1, explains 60.10% of the total variation, while 36.03% of the total variance is explained by PC2 as shown in fig.4. The system has enough resolution to explain the tomato ripe state. PCA analysis showed the variation of each group along the abscissa (PC1) with a trend. The ripe and immature groups show a clear upward and downward displacement in negative and positive direction on the ordinate axis (PC2), respectively, moving these groups away from the other two groups.

The LDA analysis was applied to the same dataset, and it showed a very clear discrimination among the various clusters representing different cherry tomato ripeness state, all cherry tomatoes were perfectly classified (Fig.5). In this plot, about93.3% of the total variance of the data is displayed. LDA function 1 (LD1) and function 2 (LD2) accounted for 84.6% and 8.7% of the variance, respectively as shown in fig.5.

Using PCA and LDA analysis, it is possible to classify the fruit into Three maturity states. When the machine vision system was performed with LDA, better classification rates were observed.

Validation analysis was performed using 414 samples.Tomatoes were of the same variety, from the experimental orchard in "jin rui" Institute of Agricultural, zhenjiang.The Result is as table1. The main error is caused by the half-ripe.

*Table.1:*The detection accuracy rate of cherry tomato

	ripe	Half-ripe	immature
total	211	120	83
correct	205	121	88
error	5	11	5
Repeatability		94.9%	

4. CONCLUSION

This paper has presented a method of cherry tomatos maturaty detecion by machine vision. The main conclusions of this study are as follows:

1.Three images of a cherry tomato by the CCD camera during the motion of the fruit on the grading line. Three images of a cherry tomato could enough cover the 90% of the fruit surface.

2.The cherry tomato is segmented from the background by fixed threshold method, allowed fruits to be precisely distinguished from the background. 9 features (red, green and blue value) were extracted from three image of one cherry tomato.

3.Principal componentanalysis (PCA) and linear discriminant analysis (LDA) were used to investigate whether the 9 features was able to distinguishing among different ripe states. Results indicated that using LDA analysis is possible to differentiate and to classify the different cherry tomato maturity states, and this method was able to classify 94.9% of the total samples in each respective group.

Furthermore, the grading speed of the sorting line reaches 7 cherry tomatoes per second.The sorting line can be used in most of cherry tomato farms, and with a slight change of software it also can be used to sort the other miniature fruit

ACKNOWLEDGEMENTS

This work is supported by "Natural Science Foundation of China" and "863 high-tech fund of China" Fund. We also wish to thank many of our colleagues for many stimulating discussions in this field.

REFERENCES

Cheong Hee Park, Moonhwi Lee. On applying linear discriminant analysis for multi-labeled problems .Pattern Recognition Letters, 2008,29(7): 878-887

Miller W M. Optical defect analysis of Florida citrus. Applied Engineering in Agriculture, ASAE, 1995,11(6):855-860

Rafael C.Gonzelez Richard, Digital Image Processing [M].Beinjing :Publishing house of electronics industry,2003

Sunil Kumar Kopparapu. Lighting design for machine vision application. Image and Vision Computing , 2006,24:720-726

United States Department of Agriculture.United States Standards for Grades of Fresh Tomatoes.1991:4~4

V. Leemansa, H. Mageinb, M.-F. Destain. On-line Fruit Grading according to their External Quality using Machine Vision. Biosystems Engineering,2002, 83 (4): 397-404

Yang, J., Yang, J.-Y. Why can LDA be performed in PCA transformed space?. Pattern Recognition, 2003,36: 563 - 566.

ON-LINE DETECTION OF DEFECTS ON FRUIT BY MACHINE VISION SYSTEMS BASED ON THREE-COLOR-CAMERAS SYSTEMS

Qiaobao Xu[1], Xiaobo Zou[2,*], Jiewen Zhao[2]

[1] *Electro-mechinery General Works Co.Ltd, Jiangsu University, Zhenjiang 2120132*

[2] *School of Food and Biological Engineering, Jiangsu University, Zhenjiang 2120132*

[*] *Corresponding author, Address: School of Food and Biological Engineering, Jiangsu University, Zhenjiang 212013, Jiangsu Province, P. R. China, Tel: +86-511-88780201, Fax: +86-511-88780201, Email: zou_xiaobo@ujs.edu.cn*

Abstract: How to identify apple stem-ends and calyxes from defects is still a challenging project due to the complexity of the process. It is know that the stem-ends and calyxes could not appear at the same image. Therefore, a contaminated apple distinguishing method is developed in this article. That is, if there are two or more doubtful blobs on an apple's image, the apple is contaminated one. There is no complex imaging process and pattern recognition in this method, because it is only need to find how many blobs (including the stem-ends and calyxes) in an apple's image. Machine vision systems which based 3 color cameras are presented in this article regarding the online detection of external defects. On this system, the fruits placed on rollers are rotating while moving, and each camera which placed on the line grabs 3 images from an apple. After the apple segmented from the black background by multi-thresholds method, defect's segmentation and counting is performed on the apple's images. Good separation between normal and contaminated apples was obtained for three-camera system (94.5%), comparing to one-camera system (63.3%), two-camera system (83.7%). The disadvantage of this method is that it could not distinguish different defects types. Defects of apples, such as bruising, scab, fungal growth, and disease, are treated as the same.

Keywords: Apple, defect, segmentation, imaging processing, prediction

Please use the following format when citing this chapter:

Xu, Q., Zou, X. and Zhao, J., 2009, in IFIP International Federation for Information Processing, Volume 295, *Computer and Computing Technologies in Agriculture II, Volume 3*, eds. D. Li, Z. Chunjiang, (Boston: Springer), pp. 2231–2238.

1. INTRODUCTION

China has an annual year apple production over 20 million tons since 2000. After apples are harvested, they are transported to the packing plant to be tested for various quality attributes that determine their price and destination. The external appearance is one of the most important factors in pricing the apples. Machine vision and imaging processing techniques have been increasingly important for the fruit industry, especially when applied to quality inspection and defect sorting applications. Automated inspection of apple quality involves computer recognition of good apples and blemished apples based on geometric or statistical features derived from apple images. Nowadays, mechanical non-destructive devices for on-line measurement of weight, size and color working at high speed (several fruits/s), are common in current packinghouses. However, automating the defect sorting process is still a challenging project due to the complexity of the process. One of the biggest difficulties involved in the technology of automated machine vision inspection of fruit defects is how to distinguish the stem-end (stem cavity) and calyx (bloom bottom) from true defects such as bruises, insect damages, and blemishes.

As it may be seen, most researchers did not consider how to manage several images representing the whole surface of the fruit(Leemans, 2004). It seems that each image was treated separately and that the fruit was classified according to the worse result of the set of representative images. It is know that the stem-ends and calyxes could not appear at the same image. Therefore, if there are two or more doubtful blobs on an apple's image, the apple is contaminated one. There is no complex imaging process and pattern recognition in this method, because it is only need to find how many blobs (including the stem-ends and calyxes) in an apple's image. However, this method needs many images to cover the whole surface of the fruit. The objective of this paper investigates machine vision systems which based on 3 normal-color-cameras regarding the online detection of external defects and get the whole surface of the fruit.

2. MATERIALS AND METHODS

2.1 The hardware

The lighting and image acquisition system were designed to be adapted on an existing single row grading machine (prototype from Jiangsu Univ., China). Four lighting tubes (18 W, type 33 from Philips, Netherlands) were placed at the inner side of a lighting box while three cameras (colour 3CCD uc610 from Uniq, U.S.A.), two inclined at about 60° and one above observed the grading line in the box, as shown in Fig. 1. The standard image treatment functions were based on the Matrox libraries (Matrox, Canada) and the other algorithms were implemented in C++. A local network was build among the three computers in order to communicate results data. The central processing unit of each computer was a Pentium 4 (Intel, USA) clocked at 3.0GHz. The fruits placed on rollers are rotating while moving. The rotational speed of the rollers was adjusted in such a way that a spherical object having a diameter of 80 mm made a rotation in exactly three images. The moving speed in the range 0~15 apples per second could be adjusted by the stepping motor. The fruit grading was achieved in seven steps: image acquisition; image preprocessing; size and shape classification; color classification; defect segmentation and recognition; three computer communication data and finally the fruit classification into quality classes.

2.2 Image acquisition and preprocessing

The Matrox digitizer supports trigger input, this all us to grab a frame upon external trigger. The external trigger that is composed of an emitter and an acceptor was place at the grading line as shown in fig.1. The three frame-grabbers grab images when every roller passes through the trigger. As it mentioned above that there are three apples in the view field of each camera, therefore, nine images were grabbed from am apple as shown in fig.2.

The images in the field of view are not only the apples waiting for measurement but also including the rollers in grading line and other mechanical parts above line. Image preprocessing includes background segmentation, image de-noise, child image segmentation and sequential images processing.

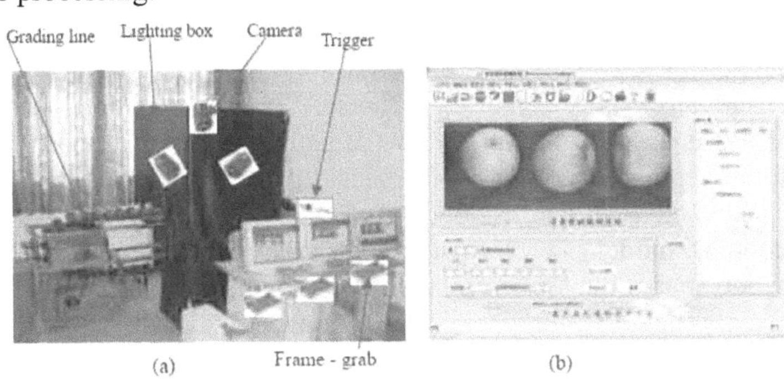

Fig.1: Sketch map of hardware system of apple on-line detection
(a) System hardware (b) Software interface

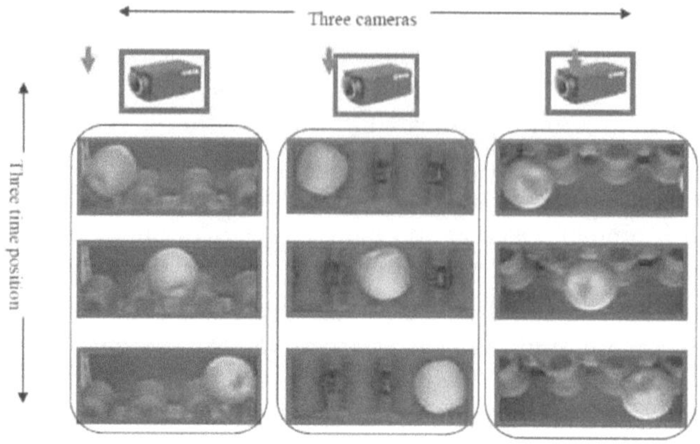

Fig.2: Trigger grab of nine images for an apple by three cameras at three position

The background is relatively complicated. To get rid of the background, multi-thresholds method was put forward. That is, the R value in RGB(red-green-blue) and S value in HIS (hue- intensity - saturation) were taken into account. The segmentation values are as follows

$$p(x,y)=\begin{cases}\text{background pixel:} & R<90 \,\|\, (S<0.20 \,\&\&\, R<200)\\ \text{apple pixel:} & \text{else}\end{cases} \quad (1)$$

There may still be some noises in the image after getting rid of the background, so this paper introduces medial filter to getting rid of the noise.

There are three apples waiting for measurement in the field of view at most. In order to take out one's own information of the individual apple, single apple division has become inevitable operation. The minimum enclosing rectangle of each single apple was used to divide the view image to three child images.

2.3 Blemish (or defects) segmentation and recognition

An image analysis scheme for accurate detection of fruit blemishes proposed by is used in this study. The detection procedure consists of two steps: initial segmentation and refinement. In the first step, blemishes are coarsely segmented out with a flooding algorithm and in the second step an active contour model, i.e. a snake algorithm, is applied to refine the segmentation so that the localization and size accuracy of detected blemishes is improved. The tests for sample apple images have shown very good results than other methods.

The blemish recognition steps are as fallowing:

First, the regions issued from the segmentation process including the defects, calyx and stem ends are hereafter called blobs.

Second, the number of blobs is count in each single child apple image.

Third, logic distinguish method were developed. That is, since calyx and stem ends could not appear in a single child image at the same time, an apple has defects as long as any one of an apple's nine images has more than two blobs. Fig.3 shows an example of apple images has two blobs.

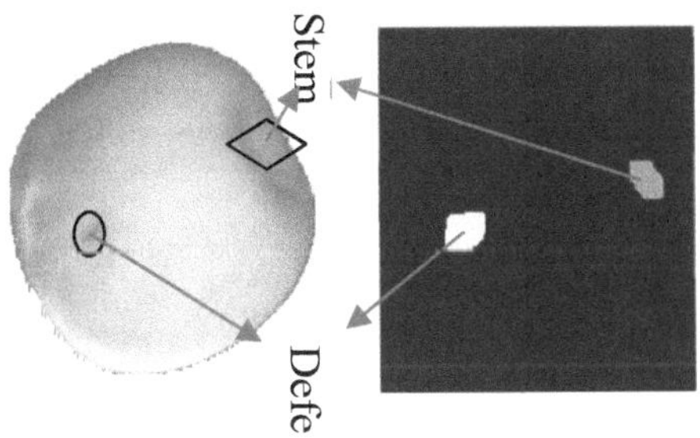

Fig.3 Precise segmentation of blob area

Fourth, three computers are used in the grading line. The apples character parameters of an apple got by each computer should be integrate into one computer. The defect detection mentioned above is all based on the three computers information. One of the three computers is server; the other two computers are customers.

Since nine image of an apple could enough cover the whole surface of the apple, any defects in the apple surface could not be omitted by this method. The disadvantage of this method is that it could not distinguish different defect types. Defects of apples, such as bruising, scab, fungal growth, and disease, are treated as the same. The fruits were then graded to reject apples if there is defect fined in the apple.

2.4 Fruits

All the fruits used in this experiment were selected and came from the same grower. 318 fruits were used in only one experiment and each fruit was thus presented only once to the machine, to avoid any additional bruises. The apples were classified into two classes: accepted (199 apples) and rejected (with blemish, 119 apples).

3. RESULTS AND DISCUSSION

The results obtained by the three color cameras grading line are given in table1. The error rate reached 11% and most of these were accepted fruits which were classified into rejected. When these errors were analyzed, half of the errors were fruits with over-segmentation of healthy tissue and especially the tissue near the boundaries in the defect segmentation processing. The other half was due to two reasons. Firstly, spot blush on the accept apples surface is segment as defect. When there is a spot blush on the accept apple surface, the spot blush is easy to be segment as defect, and the apple is classified into rejected class. Secondly, errors were fruits with defects which were accepted. The defects were always present in the upper part of the fruits. They have almost the same appearance as the russet around the stem end and, because of the proximity in position and appearance, were probably confused with the latter.

Table.1 The results obtained by the three color cameras grading line

Graded in / True groups	Accepted (199 apples)	Rejected(with defects, 119 apples)
I (accepted)	169	5
Rejected	30	114
Classification error	15.07%	4.2%
Global classification error	11%	

As comparing study, the results of sorting line with only one camera (the above camera) and sorting line with the two inclined camera, were shown in table2 and table3. The errors of classification are rising when the cameras decline, especially for the rejected class as 4.2% for three cameras and 21.8% for one camera which are shown in table 1-3. However, at the same time, the error rates for accept class are decline slightly, as 15.07% for three cameras and 11% for one camera. This is mainly caused by the images decline as the cameras decline. A statistic test was carried out for the losing of apple's information when the different number of camera was used in the sorting line. There is 15%-20% of the apple's surface can not see from the 3 images which obtained by the above camera, and there is 5%-10% of the apple surface will be lose in the 6 images which obtained by the two inclined camera. However, the 9 images obtained by the three cameras could cover the whole surface of the apple. For the reject apples, the more images there are more opportunity to find the blemish, and the less error classification of rejected class. For the accept apples, the more images, the more spot blush on the accept apples surface will be find, and the more apples will be misclassified. The global classification error for the online sorting of the three cameras is the highest. There are little reject apples misclassified into accept ones. Compared with many former works in articles(Bennedsen, et al., 2005, Blasco, et al., 2007), several images representing the whole surface of the fruit are considered in this work, and the defect recognition algorithm is simpler and faster.

Table.2 The results obtained by the two inlined color cameras grading line

Graded in / True groups	Accepted(199 apples)	Rejected (with defects, 119 apples)
I (accepted)	172	17
Rejected	25	102
Classification error	12.5%	14.3%
Global classification error	13.2%	

Table.3 The results obtained by the only one camera (the above camera) grading line

Graded in / True groups	Accepted (199 apples)	Rejected (with defects, 119 apples)
I (accepted)	177	26
Rejected	22	93
Correct classification rate	11%	21.8%
Global correct classification rate	15.1%	

4. CONCLUSION

The grading of apples into quality classes is a complex task involving different stages. The prior step is the images acquisition, which was performed by three color CCD cameras during the motion of the fruit on an adapted machine. It was followed by a first segmentation to locate the fruits on the background and a second one to find the possible defects. Once more than one defect was finding, the apple was classified into reject class.

The correct classification rate of 318 “fuji” apples was of 89%, and significantly increased, especially for the reject class. The errors were due to defects difficult to segment such as russet and bruises. Some wounds located near the stem ends were probably confused with these ends. Enhancement of the grading process should come from every stage, and particularly, from the image acquisition stage. Indeed, better light repartition and use of specific wavelengths revealing high contrast between defects and healthy tissue should probably enhance the global machine performances.

ACKNOWLEDGEMENTS

This work is supported by “Natural Science Foundation of China” and “863 high-tech fund of China” Fund. We also wish to thank many of our colleagues for many stimulating discussions in this field

REFERENCES

Abdullah, M. Z., Mohamad-Saleh, J., Fathinul-Syahir, A. S. &Mohd-Azemi, B. M. N. (2006). Discrimination and classification of fresh-cut starfruits (Averrhoa carambola L.) using automated machine vision system. Journal of Food Engineering 76(4): 506-523.

Bennedsen, B. S., Peterson, D. L. &Tabb, A. (2005). Identifying defects in images of rotating apples. Computers and Electronics in Agriculture 48(2): 92-102.

Blasco, J., Aleixos, N. &Molto, E. (2007b). Computer vision detection of peel defects in citrus by means of a region oriented segmentation algorithm. Journal of Food Engineering 81(3): 535-543.

Furfaro, R., Ganapol, B. D., Johnson, L. F. &Herwitz, S. R. (2007). Neural network algorithm for coffee ripeness evaluation using airborne images. Applied Engineering in Agriculture 23(3): 379-387.

Gu, X. (1994). Multi-spectrum imagery techniques quicken the sorting of agricultural products. Hongwai/Infrared (5): 41-42.

Leemans, V. D., M.-F. (2004). A real-time grading method of apples based on features extracted from defects. Journal of Food Engineering 61: 83-89.

Lefcout, A. M., Kim, M. S., Chen, Y. R. &Kang, S. (2006). Systematic approach for using hyperspectral imaging data to develop multispectral imagining systems: Detection of feces on apples. Computers and Electronics in Agriculture 54(1): 22-35.

Mehl, P. M., Chao, K., Kim, M. &Chen, Y. R. (2002). Detection of defects on selected apple cultivars using hyperspectral and multispectral image analysis. Applied Engineering in Agriculture 18(2): 219-226.

Throop, J. A., Aneshansley, D. J., Anger, W. C. &Peterson, D. L. (2005). Quality evaluation of apples based on surface defects: development of an automated inspection system. Postharvest Biology and Technology 36(3): 281-290.

Wen, Z. &Tao, Y. (1998).Dual-wavelength imaging for on-line identification of stem-ends and calyxes. Vol. 3460, 249-253 San Diego, CA, United States: The International Society for Optical Engineering.

Yang, Q. (1994). Approach to apple surface feature detection by machine vision. Computers and Electronics in Agriculture 11(2-3): 249-264.

A COLLECTION SCHEME FOR TRACING INFORMATION OF PIG SAFETY PRODUCTION

Qingyao Luo [1,2] , Benhai Xiong [1,2,*] , Liang Yang [1,2]
[1] *Institute of Animal Sciences, Chinese Academy of Agricultural Sciences, Beijing 100193, China;*
[2] *State Key Laboratory of Animal Nutrition, Beijing 100193, China*
[*] *Corresponding author, Address: Institute of Animal Sciences, Beijing 100193, China, Tel:+86-10-62816017, Fax:+86-10-62815988,Email: bhxiong@iascaas.net.cn*

Abstract: This study takes one main production pattern of smallhold pig farming in Tianjin as a study prototype, deeply analyzes characters of informations about tracing inputs including vaccines,feeds,veterinary drugs and supervision test in pig farming, proposes inputs metadata, criteria for integrating inputs event and interface norms for data transmision, developes and completes identification of 2D ear tags and traceability information collection system of pig safety production based on mobile PDA. The system has implemented functions including setting and invalidate of 2D ear tags, collection of tracing inputs and supervision in the mobile PDA and finally integration of tracing events (the epidemic event,feed event,drug event and supervision event) on the traceability data center (server). The PDA information collection system has been applied for demonstration in Tianjin, the collection is simple, convenient and feasible. It could meet with requirements of traceability information system of pig safety production.

Key words: Pig; PDA; Traceability; Information Collection

1. INTRODUCTION

The establishment of animal identification and disease traceability system is an urgent demand for effectively preventing and controling major animal diseases, insuring animal-derived food security and promoting the sustained and healthy development of animal husbandry.Because mainbodies engaged

---

Please use the following format when citing this chapter:

Luo, Q., Xiong, B. and Yang, L., 2009, in IFIP International Federation for Information Processing, Volume 295, *Computer and Computing Technologies in Agriculture II, Volume 3*, eds. D. Li, Z. Chunjiang, (Boston: Springer), pp. 2239–2247.

in rearing, slaughter, movement of pigs have poor organization, unsubstantial information infrastructure, collect and manage data mainly by hand, and these data collections are characterred by scatterred sites, long average cycles, low rate, small volume data and poor collection environments (Wang,2007), therefore, in order to improve quality and safety of the traceability data collection, to reduce cost and to improve the running efficiency, it is very essential to study optimizing scheme of information collection in pig safety traceability system.Since mad cow disease occurred in Britain in 1986, avian influenza and other zoonoses, Sudan, inferior milk powder and other food safety incidents sometimes occurred, food safety issues have aroused people's great concerns (Chen et al.,2007; Zhang,2006). The EU,Canada, the United States, Japan, Brazil, Australia and New Zealand and other countries have begun to implement animal identification tracking system(Zhang et al.,2008; Xiong et al.,2007; Ammendrup et al.,2001; Jill et al.,2005; N schembti et al.,2007; Caja et al.,2005).Studies for traceability system in China began in 2002 and have gradually constituted a number of relevant standards and guidelines such as “product traceability general specification”, “traceabilty guide for beef products” constituted by Chinese goods code center, “the Administrative Rules on identification and rearing files for animal and poultry” (the Administrative Rules) issued by the Ministry of Agriculture on June 1, 2007(Decree No 67,2006). “Animal traceability system” led by the Ministry of Agriculture has been demonstrating in Beijing, Shanghai, Sichuan and Chongqing, in which main adopted technologies include 2D bar code identification and read technologies.The 2D bar code were used for animal identification, IC card as the interim carrier of animal quarantine and epidemic prevention data(Chen et al.,2007; Zhang,2006).

2. MATIERIALS AND METHOD

2.1 Study Object

This study takes smallhold farmers in Tianjin as a study prototype, takes commercial pigs and their products as study objects, takes rearing files contents stipulated by the Administrative Rules as basic reqiurements of information collection, carrys out the solution of information collection and management for pig safety traceability system.

2.2 Traceability Ear Tag for Pigs

Complied with the animal traceability system led by the Ministry of Agriculture, this study adopts 2D ear tags for pigs which basic function is to track and identify pigs and their products. Ear tags are made of special plastic materials which conform to food contact safety regulations and do not result in product pollution and are set on the left or right ear. As show in Fig 1, on the surface of ear tags there print identification numbers saw by the eyes and engrav 2D bar code image identified by the 2D bar code reader.

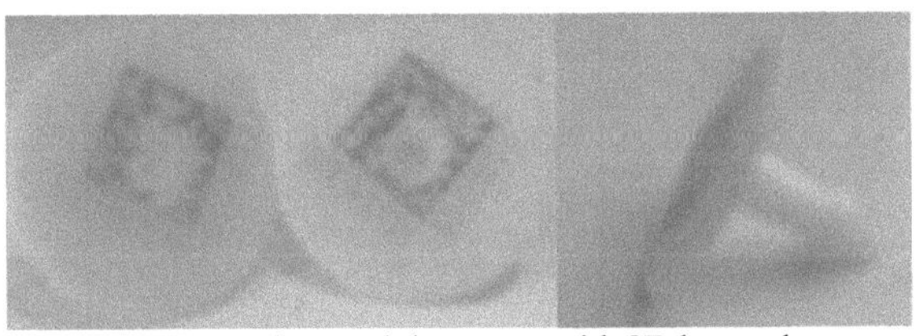

Fig.1 the special ear tags with 2D bar code

Code rules of ear tags is the important basis of achieving general search and identification in all sections of the traceability system.According to identification code rules of animals stipulated by the Administrative Rules, identification code of the pig is consists of 15 digits, the first digit represents the type of livestock ("1" for pigs), the next six digits(2 to 7) the administrative region code of the county (city) in accord with GB T2260-1999 (Nov. 31,2006) (Decree No 67,2006), the final 8 digits is the sequence which is the sequence order number in the same district. For example, the code like "1 120115 00006306" represents that the type is the pig, the farmimg place locates at Baodi District in Tianjin (administrative region code is 120115) and the individual order number is the six thounsands three hundreds and sixth in the scope of Baodi district.

2.3 Identification Device for Ear Tags

In order to strengthen the safety management of pig identification and improve the read efficiency of identification devices, the administrative office of animal traceability proposes the standard norm of traceability 2D ear tag used for pigs. Pig identification codes are converted to relative 2D code image according to the norm and engraved in the upper surface of ear tags through laser.On the surface of ear tags, there have both 15 digits code discerned by naked eyes and the corresponding 2D bar code image automatically identified by identification devices with speacial decode library for 2D bar code by scanning image.The Administrative office takes public bidding and has selected two kinds of smart mobile PDA with automatic identification for 2D pig ear tags (Fig 2). As show in Fig 2, the left smart PDA reader is based on the windows mobile 5.0 operation system and the right based on the Linux operation system.

Fig.2 Intelligent PDA ear tag reader

2.4 Collection and Management of Ttraceabilty Information

2.4.1 Identification of Ear Tags

Based on PDA devices with the windows mobile 5.0, the study carries out to identify traceability ear tags by the P / Invoke quoting the library functions including API functions in coredll.dll, Mobile5DLL.dll, scanning functions in HWCamera.dll for PDA camera and the decoding library for 2D ear tags of pigs. In order to prevent and reduce unanticipated scanning, the system loads functions with scanning and identification just in modules required to identify pig ear tags.The solution can identify both standard 1D or 2D bar code through the camera and pig ear tag with especial 2D bar code and display the identified code number in the ear-tag information textbox. If the code number is repeated, the system gives a warning sound and does not show the code number in the textbox.

2.4.2 Collection of Traceabilty Inputs

Traceability Inputs in the process of pig farming including vaccines, feeds, veterinary drugs are the major information in pig product safety traceability system. Use informations of pig feeds and vaccines are registerred by group (pen) whereas veterinary drug registerred by head of pig. But in some circumstances, feeds and vaccines data can also record every pig and the veterinary drug data records by group.

In order to reduce the labor intensity for PDA operators, there are two ways to register informations of feeds and (or) veterinary drugs: Firstly, after registration of the epidemic prevention immediately to register uses of feeds and (or) veterinary drugs, the system automatically load the information of the required identity inforamtion of pig owner (owner, type of livestock , village, number of ear tags, ID of ear tags) which can not be amended. Secondly the registration of feeds and (or) veterinary drugs separating from registration of the epidemic prevention will need identify ear tags and fill out the required identity inforamtion of pigs owner.

In order to monitor and eradicate use of illicit drugs, veterinary administratives carry out occasional sampling supervision in process of rearing, transport and before-slaughtering of pigs. Test items mainly include clenbuterol hydrochloride, ractopamine, sulfa drugs.Testing methods are respectively test strip for rapid detection, ELISA and GC-MS. If testing result is positive, the sampling pig do not pass the supervision test.Whether positive or not, results record by head of pigs and transmit to the traceability data center. If the sampling result is unqualified, the group pigs including the sampling pig are regarded as ineligible. For inforamtion queries, query results show the last testing result by group. Supervision testing can detect a number of supervision items at one time.

2.4.3 Transmission of Traceabilty Information

All informations will be processed and finally transmitted to the data center.Traceabilty data can be saved to the local SQL Server CE database or text files or XML files in the PDA and text files in smart IC cards connected with the PDA. Once the network works normally, the system timely uploads informations to the traceabilty data center by GPRS. This study proposes metadata standards for traceabilty inputs, integrating norms for traceabilty events, interface specifications for data transmission and realizes the transmission solution of data updating and downloading between PDA and the data center based on the web service (Liu et al.,2006;Fu et al.,2005).

2.5 System Development Tools

System development carrys out on the operation systerm of the Windows 2003 enterprise, the used device is ZX-6000 PDA, main tool softwares include Microsoft Visual Studio .NET, Pocket PC SDK, Microsoft ActiveSync 4.5, Microsoft SQL Server 2005 and so on.

3. RESULT AND ANALYSIS

3.1 Setting and Invalidation of Tracing Ear Tags

According to the Administrative Rules, pigs must be given the unique ID which is weared on one of ears in the first disease immunity. As Fig 3 show, firstly identify ear tags through ear tag reader devices (most continuous ear tags is 15), then enter the information of pig owners and pen No (or group

No).In the system, the default type of livestock is pig, the reaing purpose is fattening and the default address is the township related to the devices. In case of death or slaughter, the relative ear tag must be Invalidated disposal in order to avoid repeated use (as shown in Fig 3).

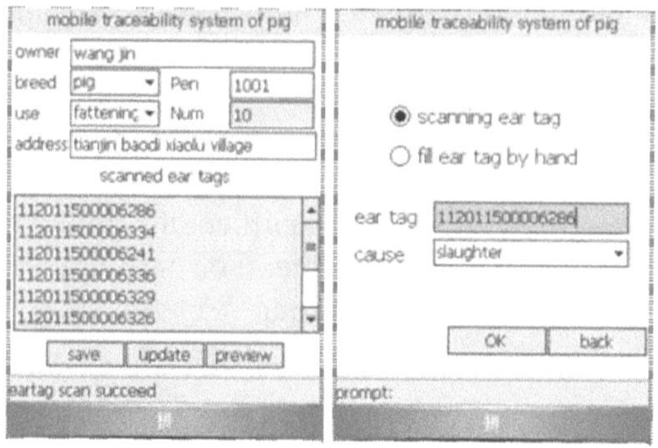

Fig.3 Setting and Invalidation of Pig Ear tags

3.2 Information Collection of Traceability Inputs

Information collection of inputs is to collect use informations of vaccines, feeds and veterinary drugs. In general, registration of the epidemic prevention event for vaccines needs to identify one ear tag or more every pen or group of pigs, then the system will create the same event information for all other pigs in the same pen or group. The field of Note is not required to fill out but the others required. When the data uploads, the system automatically loads information about operating deviec ID and operating date (Fig 4).

After registration of the epidemic prevention event, to register for use inforamtions of feeds and veterinary drugs. Registration of the feed event mainly register name,source and use date of pig feeds. Though the Note is

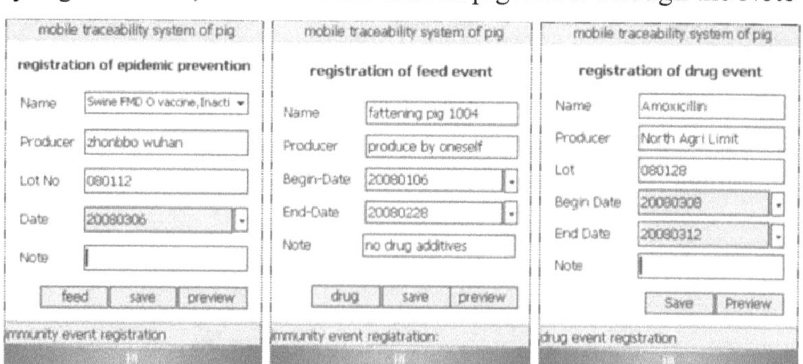

Fig4. Information collection of Input Events

not required to fill out, but reuqired when the pig feeds have added drug addtives requiring the withdrawal time to achieve the level of unidentified drug residues before sold for slaughter.

Whereas registration of the veterinary drug event mainly record use information of drugs treated for diseases, in general official veterinarians examine the drug event information before sold for slaughter or quarantine in producing area in order to ensure no drug residues in pigs. Note is not required to fill out any information.

When traceability data center receives uploading informations from the client PDA, the system automatically conducts integration process of event inforamtions respectively including epidemic prevention event, veterinary drug event and feed event.

3.3 Information Collection of Supervision

The system provides the collection solution for informations produced in the testing process by official veterinarians (Fig 5).In order to simplify operations and avoid operation errors, every event information is divided into detail selection items to fill out. The system automatically updates device ID. Traceability data center will timely receive the data and integrate them into the supervision event.

When a official veterinarian is in the process of monitoring, he can query traceability inforamtion for sampling pigs by his PDA device.

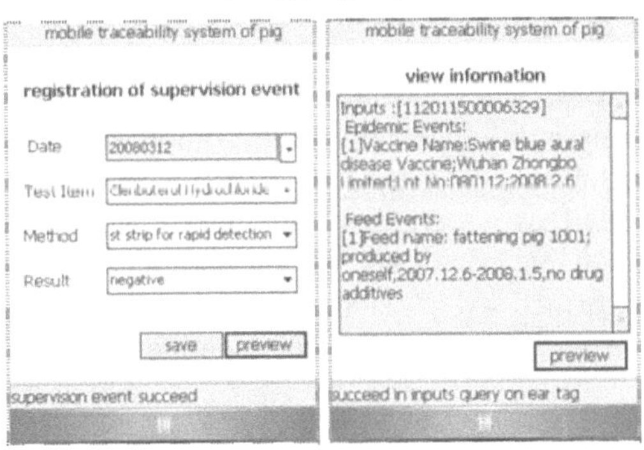

Fig5. Information collection and Query of Superveision Event

4. DISCUSSION

After many years' construction of agricultural information collection system, a number of relatively stable channels of information collection have put in practice in areas of agriculture and animal husbandry, but the major information belongs to statistical information after the event. Majority of systems are vertically deployed with multi-level, high maintenance costs and slow updating rate. Because rural infrastructure is still developing, the national agricultural information collection system can only be deployed to county levels and above nowadays and is very difficult to every town and village in rural areas. Farmers are lack of abilities and means of access information.The delivery methods of agricultural information is not in good harmony with current widespread used information tools and information use abilities of farmers(Wang,2007).Because this study adopts 2D ear tags and PDA readers for pigs appointed by the Ministry of Agriculture, produced in appointed enterprises, managed and used in scopes of country, the information collection solution will is feasible in extension and application.

This study conducts and realizes mobile PDA collection, edition and transmission of traceability system for smallhold pig farming by using PDA with embedded system.In comparing with internet traceability information systems, data collecting and editing is simple and concise.Event information for various inputs is reasonablly distributed to information collection forms where operations are simple and data volume is low, which not only reduces the workload and operation cost but also meets with demands of pork quality traceability, is a practical and reliable information collection solution by the mobile communication.

Apart from information infrastructure, one more important task is continually to perfect principles of information collection and transmission, and to confirm responsible persons in charge of collection and upload of key traceability informations including setting and invalidation of ear tags, inputs event (vaccines, feed, veterinary drugs) and supervision event. Because official veterinarians is in charge of immunization, diseases treatments and supervision of establishing rearing files for every pig from farms or households on behalf of government, all mobile PDA in the solution are assigned to official veterinarians.All traceability information must be examined and confirmed by official veterinarians who are responsible for authenticity of the informations before updating to the traceability data center.

ACKNOWLEDGEMENTS

Appreciation is extended to Ma Xing-jian, Beijing PingZhi Dongfang Sci&Tech Development Limited, for his practical suggestions and helps in process of this system developement. Funding were provided by the national high technology research and development program of China (863 Program) (2006AA10Z266), Tianjin municipal science & technology Innovation fund(06FZZDNC0100) and the national key technology R&D program (2006BAK02A16).

REFERENCES

Ammendrup S, Fussel A-E. Legislative requirements for the identification and traceability of farm animals within the European Union[J].Rev sci tech Off in Epiz,2001,20:437–444.

Caja G, Hernandez M, Conill C et al. Use of ear tags and injectible transponders for the identification and traceability of pigs from birth to the end of the slaughter line[J].J Anim Sci,2005,83:2215–2224.

Chen Hong-hua,Tian Zhi-hong. Comparative study on Farm Products Traceability System at Home and Abroad.Market Modernization[J],2007,(21):5-6.

Decree No 67,the Ministraty of Agrilculture of the People's Republic of China. the Administrative Rules on identification and rearing files for animal and poultry. http://www.agri.gov.cn/blgg/t20060628_638621.htm,June 28,2006.

Fu xi,Qi ning,Xu Jun. Mobile Telephone Application and Development of Windows Mobile [M]. Beijing:posts & telecom press,2005.

Jill E. Hobbs,Deevon bailey,et al..Traceability in the Canadian red meta sector: do consumers care?.Canadian Journal of Agrcultural Economics , 2005,53: 47-65.

Liu Yan-bo,Hu Yan,Ma qi. Application and Developement of Windows Mobile Platform [M]. Beijing:posts & telecom press,2006.

N schembti, Ja toribio.Review of identification and traceability legislation for pigs in Australia[J].Australian Veterinary Journal , 2007,85(7):255-260.

Wang Xin. Mobile communications technology's application in the agriculture instant information collection and service [J].Agriculture Network Information.2007,(12):14-17.

Xiong ben hai et al. A practical web-based tracking and traceability information system for the pork products supply chain[J]. New Zealand Journal of Agricultural Research ,2007,50:725~733.

Zhang Guoqing. The Framework of Forest Product Traceability System. http://www.scichi.com/new/ duzhe/Article/154.html. 1,16,2008.

Zhang Jing-sheng. Methods of Building Animal Identification Traceability System in the World [J].China Animal Husbandry Bulletin,2006,(7):60-62.

A SOLUTION ON IDENTIFICATION AND REARING FILES IN SMALLHOLD PIG FARMING

Benhai Xiong [1,2,*], Runting Fu [3], Zhaohui Lin [3], Qingyao Luo [1,2], Liang Yang [1,2]

[1] *Institute of Animal Sciences, Chinese Academy of Agricultural Sciences, Beijing, 100193*

[2] *State Key Laboratory of Animal Nutrition, Beijing 100193*

[3] *Animal Husbandry and Veterinary Bureau, Tianjin 300210*

[*] *Corresponding author, Address: No. 2 Yuanmingyuan West Road Beijing 100193, China, Tel:+86-10-62811680, Fax:+86-10-62815988,Email: bhxiong@iascaas.net.cn*

Abstract: In order to meet government supervision of pork production safety as well as consumer's right to know what they buy, this study adopts animal identification, mobile PDA reader , GPRS and other information technologies, and put forward a data collection method to set up rearing files of pig in smallhold pig farming, and designs related metadata structures and its mobile database, and develops a mobile PDA embedded system to collect individual information of pig and uploading into the remote central database, and finally realizes mobile links to the a specific website. The embedded PDA can identify both a special pig bar ear tag appointed by the Ministry of Agricultural and a general data matrix bar ear tag designed by this study by mobile reader, and can record all kinds of inputs data including bacterins, feed additives, animal drugs and even some forbidden medicines and submitted them to the center database through GPRS. At the same time, the remote center database can be maintained by mobile PDA and GPRS, and finally reached pork tracking from its origin to consumption and its tracing through turn-over direction. This study has suggested a feasible technology solution how to set up network pig electronic rearing files involved smallhold pig farming based on farmer and the solution is proved practical through its application in the Tianjin's pork quality traceability system construction. Although some individual techniques have some adverse effects on the system running such as GPRS transmitting speed now, these will be resolved with the development of communication technology. The full implementation of the solution around China will supply technical supports in guaranteeing the quality and safety of pork production supervision and meet consumer demand.

Please use the following format when citing this chapter:

Xiong, B., Fu, R., Lin, Z., Luo, Q. and Yang, L., 2009, in IFIP International Federation for Information Processing, Volume 295, *Computer and Computing Technologies in Agriculture II, Volume 3*, eds. D. Li, Z. Chunjiang, (Boston: Springer), pp. 2249–2258.

Keywords: Pig; Smallhold Pig Farming; Mobile PDA; Rearing files; Traceability

1. INTRODUCTION

Since mad cow disease happened in Britain in 1986 (Gollam, E. et al.,2004), the EU takes the lead in study on traceability of beef cattle and calves, and has established identification and tracing system for cattle and beef(F. Madec et al., 2001) . In Canada, identification management for livestock and poultry is taken as an important part of agricultural development strategy called " Brand Canada " and will cover 80 percent of the whole livestock and poultry and their products until the end of 2008 according to this plan(Brand Canada, 2008). After the first case of mad cow disease occurred in the US in 2003, US government makes a plan for the identification and traceability system of cattle and decides to enforce it in the whole country until 2009 so as to track relevant animals within 48 hours in case of epidemic diseases(USDA, 2005). In June 2006 , the Japanese Congress enacted " the beef production traceability" and made certain to establish a national animal traceability information system. as the act stipulates, all cattle produced in Japan must have their own identification number after birth and information including identification number, date of birth, cattle breed, delivery records each cattle are managed by the Livestock Improvement Center(Chinacity,2008). The same are Australia and the New Zealand. On June 1, 2007, the Ministry of Agriculture promulgated the " the Administrative Rules on identification and rearing files for animal and poultry"(Chinaagri,2008), which has put into effect. Study on key traceability technologies for animal diseases and quality and safety and construction of traceability system have be establishing by various project channels.

On the basis of "the Administrative Rules" enacted by the Ministry of Agriculture(MAPRC,2006) and intelligent reader devices, together with identification information technologies, mobile intelligent technology and network database technology, regard to application in practice and economical characteristic of technologies, through integration of technologies and devices, this study proposes a solution for individual electronic rearing files of pig and corresponding system for smallhold pig farming, and is also demonstrated for application.

2. EQUIPMENT AND METHOD

2.1 Equipment

Basis identifying device provided by Beijing PingZhi Dongfang Sci & Tech Development Limited is ZX-6000 mobile intelligent readers (PZDF,2008) in which the operation platform is window mobile 5.0 system, its characteristics are as follows: 1) embedded smart IC card reading and writing device, consistent with "China financial integrated circuit IC card and application standards", can read and write contact smart card IC , 2) identifying bar code: integrate a camera with 300 thousands pixels, one third inch of optical specification, one-dimension or two-dimension bar code can be photographed through the camera and identified.

2.2 Selection of individual identification for pig

Individual identification for pigs adopts integrated identification mode combining special individual barcode for pig with common barcode. One side adopts barcode ear tag recommended by the Ministry of Agriculture(Fig.1,left) which tag code rules comply to "the Administrative Rules" , and detailed contents are described in Xiong benhai et al.(2007) and must select the tag identification devices appointed by the Ministry of Agriculture, including ZhiXing ZX-6000 mobile identification reader(Fig.1,right). The advantage by use of this way is that ear tags could produce and manage in the whole, but its disadvantage is easy to result in monopolization and to affect technologies renovation and to make against application of universal technologies. Therefore, this study develops identification embedded module fro common code on ZX-6000 in order to be able to identify other bar ear tag with other code standards.

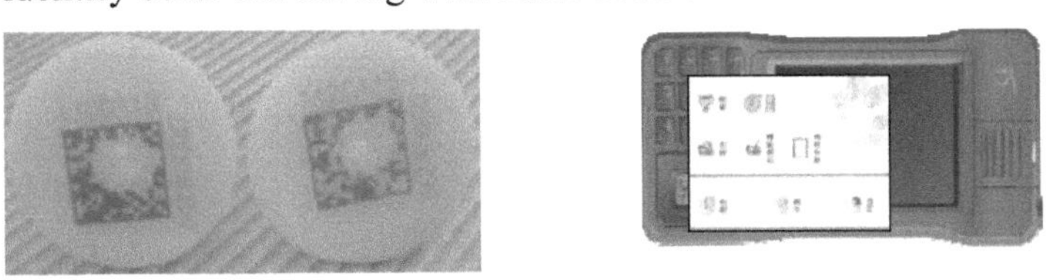

Fig.1 Special pig ear-tag and ZX-6000 PDA

As Fig. 2 shows, the ear tag designed and produced in the study is plastic and conforms to the requirement of the Ministry of agriculture, but its picture bar code uses Data Matrix(DM) code which bears more information in the smaller area(Ec21cn, 2008). Despite its appearance size is the same as the ear tag in Fig 1, it is improved including: Firstly, the latter uses bolder font with legible and concise, and can be identified by naked eyes in case of no reader or disused. Secondly, because the latter adopts short-matrix 2D bar code, area required for bearing the same information can wholly be printed in any side from center dot, and its identification percent could almost achieve 100%, but the ear tag in the Fig.1 must cover the center dot and has spoiled picture when printed resulting in identifying slowly and sometimes un-identified. Thirdly, from view of the reader, the former can only be identified by appointed ZX -6000, but the later can be identified by all-purpose reader on the right of Fig.2. This study adds common decode programs on ZX-6000 with embedding special identification code in Fig 1 so as to identify DM code (also other generic code). Tag code information identified by using the PDA reader is directly sent into modules of immunity, supervision or tracing in PDA system, that is say identification and data treatment accomplish in the same device, but code data identifying through other general readers need transport into the system by line of USB.

Fig.2 DM-bar ear tag(left) and all purpose bar scanner

2.3 Information collection and transmission in rearing section

Identifying pigs and setting electronic rearing files as showed in Fig 3. pigs wearing ear tags in Fig 1 or Fig 2 are identified by the improved mobile intelligent reader, then by the function modules embedded in PDA to collect and to edit information in rearing process, which are directly send to remote traceability database server constructed in the study by GPRS and Internet(Luan Chang-ha et al. 2006) , interconnected with information in process of following pig delivery, slaughter or sale or supplied traceability query.

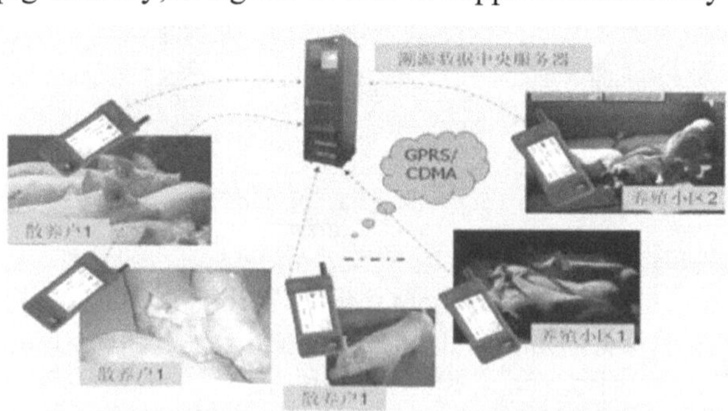

Fig.3 data collection solution of rearing files

2.3.1 Design for mete-data items of pig identification and inputs

Information collection modules in rearing section for swine scattered breeding mainly include setting of ear tag, immunity events, feed events, veterinary events and surveillance events etc.. Design principles of data items reflect primary content of events and obligatory content filled in easily.

As Fig,3, all of rearing information collected by mobile devices from different farmers or breeding districts need to connect with internet through GPRS and then send to traceability center database. But before data storage, information must be reorganized and remove redundant content. Table 1 shows main data fields in the traceability database, of which event fields are used to storage information integrated in the server.

Table.1 The Structure of Tracing Database on Smallhold Farmer mode(tracing database: porktrace.dbo)

Serial No.	Filed Name	Length
1	Ear tag	15
2	Owner	10
3	Breed of pig	12
4	Location of breeding	16
5	Immunity events	128
6	Feed events	128
7	Veterinary events	128
8	Slaughter events**	128
9	Carcass 1*	20
10	Carcass 2*	20
11	Official Vet	64
12	Note	128

**field content offered by slaughter units, others by breeding owners or farms.*

2.3.2 Development of Embedded PDA System based on rearing section in smallhold pig farming mode

The embedded system is developed on ZX-6000 mobile PDA reader according to the design in 2.3.1~2.3.2. Developing platform is windows mobile 5.0, and tool software include Visual Studio 2005, Visual C# 2005, the class libraries run in the mobile device with small memory are .NET Compact Framework(Liu Yan-bo et al. 2006; Fu xi et al.,2005), Database used: is SQL mobile 2005. SQL mobile Client Agent need to be installed on PDA, and Client Agent is communicated with SQL mobile Server Agent through HTTP. Application programming interface for SQL mobile is ADO.net(Fu xi et al.,2005). There are two ways for data storage: Firstly, if signal of wireless GPRS is good, edited data is directly send to traceability center database (porktrace.dbo); Secondly, the data is saved by other storage medium when signal is not good.

3. RESULTS AND DISCUSSION

3.1 Setting of ear tags

As to application of mobile smart traceability system for pig developed in 2.3.2, the first important function is to record ear tag code and to identify ear tags through PDA system as well as recording information about owner, variety and quantity of pigs, dependency etc. In Fig.4, the form shows ear tag ID and basic information of 10 pigs which information is wirelessly send to porktrace.dbo, the traceability database designed according to 2.3.2. Subsequently, the system creates the same quantity of new records in the database making ready for following data processing.

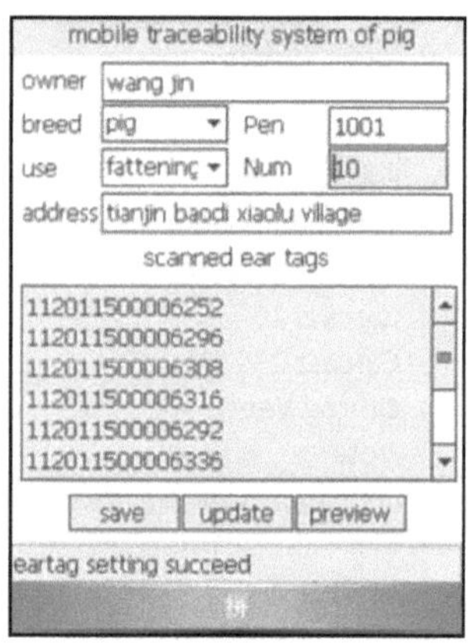

Fig.4 Setting Ear tag by PDA

3.2 Immunity events

Fig.5 shows the record and query result of the pig immunity event. After identifying ear tags through the camera or inputted pig ID, the system displays the Fig.5 form(left), then input basic immunity information which is directly sent to relevant data records by button of the "Save" and update records of the immunity event in the database. Moreover, the event data can also be remotely viewed and edited by wireless. The right in Fig 5 shows the query result of the immunity event for a certain ear tag.

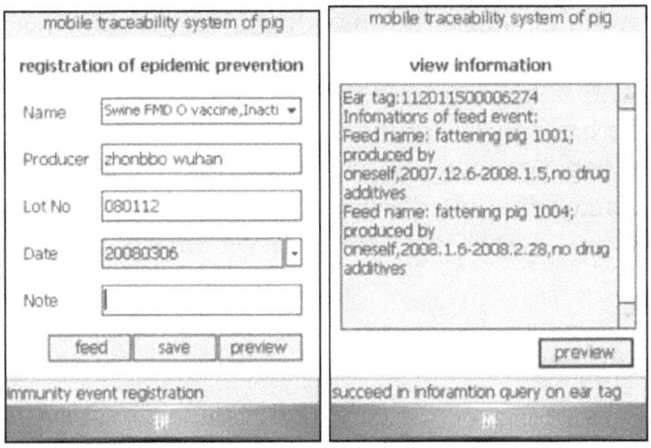

Fig.5 Record immunity event(left) and mobile query (right)

3.3 Feed events

Fig 6 shows record and query result of pig feed event. After identifying ear tags through the camera or inputted pig ID, the system displays Fig 6 form, then input basic feed information which is directly sent to relevant data records by the "save" and update records of the feed event in the database. Moreover, the event data can also be remotely viewed and edited by GPRS. The right in Fig 6 shows the query result of the feed event for a certain ear tag, including used feed additives and some drugs to meet their right to know.

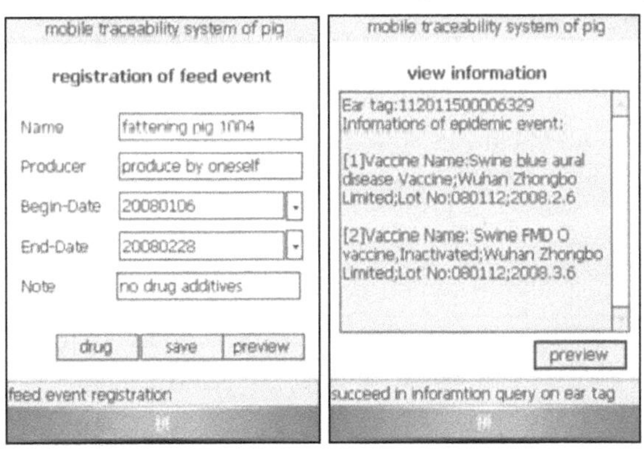

Fig.6 Record Feed event (left) and mobile query(right)

3.4 Veterinary and Surveillance events

When outbreak of disease for swine there need to treat diseased pigs with using provided veterinary drug. According to “safe use standards for feed additives"(Yang Zhen-hai et al., 2003), 79 kinds of veterinary drugs are allowed to use for the swine in China. The mobile system provides function to record information of used veterinary drugs in breeding process for the treatment of illness. In addition , the mobile PDA system also can record test results by official vet in the breeding sector through module of the surveillance event. At present main testing items in the system include clenbuterol, ractopamine, sulfa drugs and other prohibited drugs. Different from 3.1~3.3, the module is generally provided access for official veterinary not for farmers or consumers.

3.5 Query results based on Internet

These ID, inputs and other data collected by the above from 3.1 to 3.4 modules are submitted to remote traceability database, mainly meet to provide information query for consumers after slaughter of pigs with ID and sales in the supermarket. Data query in the study has integrated on traceability query platform that is www.tjporktrace.com or www.porktrace.com.cn built by Xiong Benhai et al. (2007) . This study only offers the client query page in Fig 6 to show integrality of the system, but technical details do not discuss in this study.

Fig. 6 Web traceability platform (www.tjporktrace.com)

3.6 Discussion

Firstly, this solution is a part of whole technical solution in the project of "digital monitor and traceability technologies of pork quality safety in Tianjin city of China", and makes an application in demonstration districts and obtains expected effects. The mobile PDA system can conveniently collect information without restrictions of time and space and provides a feasible technical solution for building electric rearing files of a lots of smallhold farmers(Zhang Jing, 2007), also is applied to those scale farms.

Secondly, the running fee for GPRS is not a problem. At present, China Mobile Company provides a monthly package "GPRS mobile Internet five Yuan package with 10 Mbytes mobile Internet flow by a named dynamic zone of the SIM card (BDZGP, 2008). Data contents in the mobile data collection system are mainly text and digital data, even if a scale of pig-farm achieves its data from several hundreds of heads or thousands, data capacity of rearing information is not more than 10 Mbytes in one month. If the installed SIM card in the ZX-6000 reader is only used to collect and send data without functions of calling, the monthly fee will not exceed five People's Yuan.

The third, this study also provides a solution when GPRS does not work. When GPRS can be connected, the system will send the data to the traceability database, otherwise, it storages the information inside a special IC card owned by the PDA firstly and follows to transmit them when GPRS works good.

ACKNOWLEDGEMENTS

Appreciation is extended to Ma Xing-jian, fromBeijing PingZhi Dongfang Sci & Tech Development Limited, for his practical suggestions and helps in process of this system development.

REFERENCES

Beijing D-Zone GPRS Package(BDZGP). http://www.lanxingkong.net/ shoujishang wang/ article.asp?id=28.2008-07-16

Brand Canada. http://www.aquaculture.ca/bcanada.htm :2008-07-18

Chinaagri. http://www.agri.gov.cn/xxlb/t20070319_787945.htm.2008-07-18(in Chinese)

Chinacity.http://www.chinacity.org.cn/cankao/22233_2.html.2008.-07-18(in Chinese)

EC21CN. http://www.ec21cn.org/Education/UploadFiles/CourseWare/tiaoma/tmjc/2/study/5-3-2.htm.2008-07-18(in Chinese)

F. Madec, R. Geers, P. Vesseur , N. Kjeldsen & T. Blaha. Traceability in the pig production chain. Rev. sci. tech.Off. int. Epiz., 2001, 20 (2), 523-537.

Fu xi,Qi ning,Xu Jun. Mobile Telephone Application and Development of Windows Mobile [M].2005(in Chinese)

Gollam, E. et al. Traceability in the U.S. Food Supply: Economic Theory and Industry Studies. Washington, United States Department of Agriculture,2004. (Agricultural Economic Report Number 830) GS1. Wine Supply Chain Traceability: EAN-UCC Standards Application Guideline. Brussels: 2005.

John D. Lawrence, Daryl Strohbehn, Dan Loy, and Reg Clause. Lessons Learned from the Canadian Cattle Industry:National Animal Identification and The Mad Cow.2003. http://www.card.iastate.edu/publications/DBS/PDFFiles/03mrp7.pdf

Liu Yan-bo,Hu Yan,Ma qi. Application and Developement of Windows Mobile Platform [M].2006(in Chinese)

Luan Chang-hai,Wang Meng. study on embedded internet design based on GPRS [J]. control & automation.2006.(22)17,94-96(in Chinese)

Ministraty of Agrilculture of the People's Republic of China(MAPRC).Decree No 67, the Administrative Rules on identification and rearing files for animal and poultry. http://www.agri.gov.cn/blgg/t20060628_638621.htm,June 28,2006(in Chinese)

Ping Zhi Dong Fang(PZDF). http://www.pda9.com/ZX6000.htm.2008-07-18. (in Chinese)

United States Department of Agriculture(USDA), Marketing and Regulatory Programs Animal and Plant Health Inspection Service. National Animal Identification System (NAIS)- Technical Supplement to Draft Program Standards. July 26, 2005.http://animalid.aphis.usda.gov/nais/

Xiong ben hai et al. A practical web-based tracking and traceability information system for the pork products supply chain. New Zealand Journal of Agricultural Research, 2007,Vol.50:725~733.

Yang Zhen-hai,Cai Hui-yi. Standards for safety application of Feedstuff Addtives[M]. Beijing:china agriculture press,2003(in Chinese)

Yu Linhua,Yang yanlin. status and preventing measures of high-pathogenic swine of blue ear disease [J]. HUNAN journal of a science & veterinary medicine,2007,(3):25-28(in Chinese)

Zhang Jing. Data commentate “force”—Market Analyse for China Swine main production and sale zones in 2006. swine industry science. 2007, 24(1):26~28(in Chinese)

RESEARCH OF UNCERTAINTY REASONING IN PINEAPPLE DISEASE IDENTIFICATION SYSTEM

Liqun Liu[1,*], Haifeng Fan[2]

[1] *Ministry of Personnel, Guangdong Ocean University, Zhanjiang, Guangdong Province, P.R. China 524088*

[2] *Navy south China sea fleet command post, Zhanjiang, Guangdong Province, P.R. China 524055*

[*] *Corresponding author, Address: Ministry of Personnel, Guangdong Ocean University, Zhanjiang 524088, Guangdong Province, P. R. China, Tel: +86-759-2383115, Fax: +86-759-2383115, Email: gdoullq@126.com*

Abstract: In order to deal with the uncertainty of evidences mostly existing in pineapple disease identification system, a reasoning model based on evidence credibility factor was established. The uncertainty reasoning method is discussed, including: uncertain representation of knowledge, uncertain representation of rules, uncertain representation of multi-evidences and update of reasoning rules. The reasoning can fully reflect the uncertainty in disease identification and reduce the influence of subjective factors on the accuracy of the system.

Keywords: pineapple, identification system, certainty factor, uncertainty reasoning

1. INTRODUCTION

Pineapple is one of the important fruits in the tropical and subtropical regions of China. There are large areas of cultivation, mainly concentrated in Guangdong, Guangxi, Fujian and Taiwan. The cultivation of pineapple is attacked by diseases frequently, causing a decline in output and quality. It is particularly important to create a disease identification system for pineapple. However, in the process of establishing the system, the pineapple disease lacks a clear description, because experts often use some ambiguous

---

Please use the following format when citing this chapter:

Liu, L. and Fan, H., 2009, in IFIP International Federation for Information Processing, Volume 295, *Computer and Computing Technologies in Agriculture II, Volume 3*, eds. D. Li, Z. Chunjiang, (Boston: Springer), pp. 2259–2264.

language to describe the symptom of a particular disease, such as: "shorter" and "smaller" etc., which leads to the uncertainty of knowledge reasoning in the system (Zhang Wen et al., 2008). It is difficult to solve such problems if certainty theory and method is used. Therefore, it will adopt uncertainty reasoning method to solve such problems.

2. REASONING MODEL

2.1 Analysis of reasoning problem

Uncertainly reasoning based on conclusions, in practice, cannot avoid some problems as the followings:(1) the premise of application demands the sub conditions in condition set are relatively independent; (2) if the certain degree of general assumption condition is 1 or 0, when a disease performance is atypical or when users' subjective understanding is uncertain, human errors increase, which will increase subjective influence on reasoning results (Zhang Xuenong, 2005).

2.2 Uncertain representation of knowledge

The reasoning that is based on the uncertainty knowledge is defined as uncertainty reasoning. In this system, the production rule with a certainty factor (*CF*) is adopted for the uncertain representation of knowledge (Yang Jian, 2003):

IF *E* THEN *H(CF(H, E))*

E stands for the rules preconditions which can be a simple single precondition, or a complex precondition, i.e. a complex precondition derived from conjunction and extraction. *H* stands for the rules conclusions which can be a single conclusion, or conclusions.

CF (H, E) stands for the rules credibility, also known as the rules strength which reflects the level of reliability of conclusion *H* when precondition *E* is true, $-1 \leq CF(H, E) \leq 1$. The greater the value of *CF (H, E)* is, the higher reliability of the truth of conclusion *H* becomes when precondition *E* is true. On the contrary, the smaller the value of *CF (H, E)* is, the lower reliability of the truth of conclusion *H* becomes when precondition *E* is true.

2.3 Reasoning strategy

The reasoning strategy in the system is mixed reasoning (Chen Zhuangjian et al., 2007). In the process of reasoning, the system conducts the forward reasoning firstly, that is, reasoning the assumed conclusion from the initial symptoms. Lastly it conducts backward reasoning, that reasoning the try-out of the assumed conclusion to get the final identification summary (Lin Peiguang et al., 2008). The detailed reasoning process is shown in Fig.1.

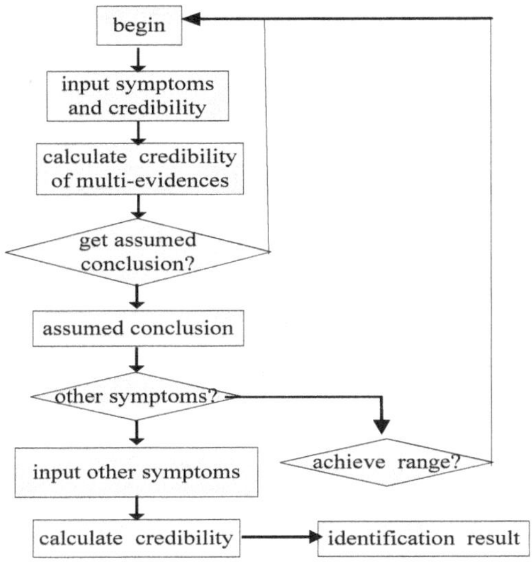

Fig.1: Reasoning process

3. UNCERTAINTY OF REASONING PROCESS

3.1 Uncertain representation of rules

In this system, a rule is known: $E \rightarrow H$. But how to define its credibility: *CF (H, E)*. First, define a prior probability *P (H)*. It expresses the believable degree of conclusion *H* when there is no evidence. Then, conditional probability *P (H / E)*. It expresses the believable degree of conclusion *H* when there is evidence *E* (Bai Yong et al., 2006).

(1)*P (H / E)*: it means that the credibility of the conclusion *H* is increased by the appearance of evidence *E*. We use *MB (H, E)* to express that the evidence *E* works on the growth of trust of conclusions *H*, then:

$$MB(H,E)=\{\frac{\max[P(H/E)^{1},P(H)]-P(H)}{1-P(H)} \quad (1)$$

(2)*P (H / E)*: it means that the credibility of the conclusion *H* is reduced by the appearance of evidence E. We use *MD (H, E)* to express that the evidence E works on the growth of distrust of conclusions *H*, then:

$$MD(H,E)=\{\begin{matrix}\min[P(H/E)^1,P(H)]-P(H)\\ -P(H)\end{matrix} \quad (2)$$

The definition of certainty factor *CF (H, E)* can be got when based on the definition of *MB (H, E)* and *MD (H, E)*:

$$CF(H,E)=MB(H,E)-MD(H,E) \quad (3)$$

Thereinto, their value ranges are: $0 \le MB(H, E) \le 1$, $0 \le MD(H, E) \le 1$, $-1 \le CF(H, E) \le 1$. Certainty factor *CF (H, E)* is given by pineapple experts.

3.2 Uncertain representation of multi-evidences

The certainty factor of evidence is expressed with *CF (E)*, $-1 \le CF(E) \le 1$. *CF (E)* = 1 means evidence *E* is true with assurance; *CF (E)* =-1 means evidence *E* is false with assurance; *CF (E)* = 0 means evidence *E* is beyond our knowledge; 0<*CF (E)* < 1 means the extent of truth of *E*; - 1< *CF (E)* <0 means the extent of falsehood extent of *E*.

(1)When the evidence is the conjunction of evidences, that is, *E* = *E1* and *E2* and… and E_n, and the corresponding certainty factor are: $CF(E_1)$, $CF(E_2)$… $CF(E_n)$, it is resulted: $CF(E) = min(CF(E_i))$, (i = 1, 2…n) (Shi Minghui et al., 2006);

(2)When the evidence is the extraction of evidences, that is, *E* = *E1* or *E2* or… or E_n, and the corresponding certainty factor are: $CF(E_1)$, $CF(E_2)$… $CF(E_n)$, it is resulted: $CF(E) = max(CF(E))$, (i = 1, 2… n);

(3)If the following two rules exist:

R_1: IF E_1 THEN E_2 $(CF(E_2, E_1))$

R_2: IF E_2 THEN H $(CF(H, E_2))$

That is, when the middle result deduced from R_1 acts as the evidence of R_2, certainty factor of E_2 $CF(E_2)$ is the certainty factor deduced from the use of R_1 (Shi Minghui et al., 2007).

3.3 Update of reasoning rules

(1)When the certainty factor of conclusions is calculated with the certainty factor of preconditions and rules, the formula for certainty factor of the conclusions *CF (H)* is as follows:

$$CF(H)=\max\{0,CF(E)\}\times CF(H,E) \quad (4)$$

(2)When two separate evidences and two different rules are used to deduce the certainty factor of a same conclusion, that is,

E1 → *H*, certainty factor of evidence and conclusion are respectively: *CF (E1)*, *CF (H, E1)*;

E2 → *H* certainty factor of evidence and conclusions are respectively: *CF (E2)*, *CF (H, E2)*;

CF1 (H), *CF2 (H)* are used to express the certainty factor of the two rules as:

$$\begin{aligned} CF_1(H) &= \max(0, CF(E_1)) \times CF(H, E_1) \\ CF_2(H) &= \max(0, CF(E_2)) \times CF(H, E_2) \end{aligned} \tag{5}$$

CF (H) is defined as follows:

$$CF(H) = \begin{cases} CF_1(H) + CF_2(H) - CF_1(H) \times CF_2(H) \\ CF_1(H) + CF_2(H) + CF_1(H) \times CF_2(H) \\ CF_1(H) + CF_2(H) / (1 - \min(CF_1(H), CF_2(H)) \end{cases} \tag{6}$$

4. EXPERIMENT

The above reasoning method has been widely applied to pineapple disease identification system. A specific disease, Heart Rot Disease (short as *H1*), a common disease of pineapple cultivation, is quoted as an example to illustrate its experimental reasoning process.

A rule is assumed as:

IF E_1 *and* (E_5 *or* E_4) *and* E_5 *and* E_6 THEN H_1 (0.96)

The certainty factor of the rules is given by experts as 0.96. Contrastingly, and the certainty factor of the evidences is given by users, including: $CF\ (E_1) = 0.\ 75$, $CF\ (E_2) = 0.\ 8$, $CF\ (E_4) = 0.\ 85$, $CF\ (E_5) = 0.\ 9$, $CF\ (E_6) = 0.\ 9$. The meanings of evidences are shown in Table 1.

Table 1. Description of evidence

Evidence	Description
E_1	N,P,K in pineapple are lower than the standard
E_2	the pineapple abdominal shallowness
E_4	the pineapple abdominal height
E_5	the pineapple root slight rot
E_6	the pineapple dingy skin

According to the calculation method of certainty factor of conclusions discussed previously:

$$CF(H) = \max\{0, CF(E)\} \times CF(H, E) \tag{7}$$

A formula can be got:

$$CF(H1)=max\{0,CF(E1 \text{ and } (E2 \text{ or } E4) \text{ and } E5 \text{ and } E6)\}$$
$$=max\{0,min\{CF(E1),max\{CF(E2),CF(E4)\}, CF(E5),CF(E6)\}\}\times 0.96 \quad (8)$$
$$=max\{0,min\{0.75,max\{0.8,0.85\},0.9,0.9\}\}\times 0.96$$
$$=max\{0,0.75\}\times 0.96$$
$$=0.72$$

That is: the credibility factor of Heart Rot Disease is 0.72.

5. CONCLUSION

The reasoning based on the evidence of the credibility fully considers the complexity of the objective world and the inaccuracy and incompleteness when human beings come to know the objective things. At the same time, it reduces the impact of subjective factors on the judgment and reasoning process. The reasoning method introduced by this paper has been used in the pineapple disease identification system. Experiments prove that when the value of CF (E) is set reasonably by users, reasoning result is closer to experts' practical experience and psychological instinct.

REFERENCES

Bai Yong, LanZhangli. Research of individual algorithm based on uncertainty inference, Computer Science, 2006, 33(3): 205-207

Chen Zhuangjian, Xie Wei. Implementation of import and export ontology and its reasoning frame, Computer Engineering, 2007, 33(21): 186-188(in Chinese)

Lin Peiguang. Indefinite knowledge representation based on ontology and cloud theory, Computer Engineering and Applications, 2008, 44(5): 51-54(in Chinese)

Shi Minghui, Zhou Changle, Wu Qingfeng. Uncertain inference based on certainty degree and its implementation by artificial neural network, Application Research of Computers, 2007, 24(1): 241-243(in Chinese)

Shi Minghui, Zhou Changle. Approach to uncertain inference for syndrome differentiation in TCM based on artificial neural network, Computer Engineering and Applications, 2006, 43(7):10-13(in Chinese)

Yang Jian. Research and application of CBR's progression, Computer Engineering and Design, 2003, 29(3): 710-712(in Chinese)

Zhang Wen, Zhang Guixu. Improved rule-based backward fuzzy reasoning algorithm, Journal of Communications, 2008, 29(2): 101-105(in Chinese)

Zhang Xuenong. A gradual approach for model-based diagnosis, Journal of Software, 2005, 19(3): 584-593(in Chinese)

STUDY ON FULL SUPPLY CHAIN QUALITY AND SAFETY TRACEABILITY SYSTEMS FOR CEREAL AND OIL PRODUCTS

Shihong Liu [1,*], Huoguo Zheng [1], Hong Meng [1], Haiyan Hu [1], Jiangshou Wu [1], Chunhua Li [1]

[1] *Agricultural Information Institute, Chinese Academy of Agricultural Sciences, Beijing, P. R. China 100081*

[*] *Corresponding author, Address: Agricultural Information Institute, Chinese Academy of Agricultural Sciences, Beijing 100081, Beijing, P. R. China, Tel: +86-10-68975098, Fax: +86-10-68975098, Email: lius@mail.caas.net.cn*

Abstract: Global food industry and Governments in many countries are putting increasing emphasis on establishment of food traceability systems. Food traceability has become an effective way in food safety management. Aimed at the major quality problems of cereal and oil products existing in the production, processing, warehousing, distribution and other links in the supply chain, this paper firstly proposes a new traceability framework combines the information flow with critical control points and quality indicators. Then it introduces traceability database design and data access mode to realize the framework. In practice, Code design for tracing goods is a challenge thing, so this paper put forward a code system based on UCC/EAN-128 standard. Middleware and Electronic terminal design are also briefly introduced to accomplish traceability system for cereal and oil products.

Keywords: cereal and oil products; full supply chain; traceability; tracking; electronic terminal; RFID

1. INTRODUCTION

Food traceability system, also called food tracking and tracing system, has become an effective way in food safety management. After several food safety related issues, particularly several food sandals, the global food

---

Please use the following format when citing this chapter:

Liu, S., Zheng, H., Meng, H., Hu, H., Wu, J. and Li, C., 2009, in IFIP International Federation for Information Processing, Volume 295, *Computer and Computing Technologies in Agriculture II, Volume 3*, eds. D. Li, Z. Chunjiang, (Boston: Springer), pp. 2265–2273.

industry and governments in many countries have paid increasing attention to systems along the food chain. In order to rebuild the confidence of the consumers and to gain competitive advantages of their brands of agricultural products, governments in many countries compete to pass legislations and regulations to guide developing and implementing of identification systems and traceability systems.

Table1. shows characterized differences between countries all around the world. In June 2002, the Canadian federal government established an ambitious goal that, before 2008, the country would achieve tracing back 80 percent of agricultural products to its source, supporting the "Brand Canada strategy"(Lu Changhua et al., 2007), of which a mandatory identification system for cattle and beef on July 1, 2002 came into operation. In December 2003, the United States developed the statutes of tracking food safety, which required all enterprises involved in food transportation, distribution and import recording their trade information for tracking and tracing back(Cheng Hao. 2007). In addition, the United States also plans to include 70 percent of the cattle in the NAIS (National Animal Identity System) project at the end of 2009 (Xing Wenying. 2006). Traceability system was initially promoted by the EU to control the risks of mad cow disease in 1997. In July 2003, The EU published a White Paper on Food Safety, proposing a new framework for food safety system. Most countries in the EU have implemented mandatory livestock and meat products traceability system (Zhu Haipeng. 2007). The EU also had a plan called TraceFish, the main objective of which is to study the full supply chain traceability of aquatic products (Liu Junrong. 2007). The Japanese Government has passed new legislations on cattle and beef requiring a mandatory traceability system from farm to retail (Wang Lifang et al., 2005). The system allows consumers via the Internet to enter identification numbers of beef on the packaging box getting access to production information of the beef. Australia has plans for general mandatory traceability. NLIS is currently carrying out, which enables trace backward and forward from farm-of-origin to abattoir (Schroeder et al., 2005). At present, most agriculture developed countries are active to implement food.safety traceability systems, but they have different focus.

Table 1. Features in typical traceability systems

Countries or regions	Traceability system	Key technology	Application scope
Canada	Cattle identification	barcode, eartag	beef
US	NAIS	RFID, DDB	beef
EU	TraceFish	barcode, data exchange	aquatic, products
Japan	Food identity	barcode	beef
Austria	Livestock identification	eartag, RFID	beef
Britain	Critter identification	eartag	critter

Table2. shows aspects of traceability systems in china. China has also achieved important progress in food tracing. In April 2004, the State Food and Drug Administration and 7 other departments chose meat industry as a pilot industry, started meat and meat products traceability institution construction and system implementation (General Administration of Quality Supervision. 2002). The main tasks include: developing suitable technical standards and Management norms, publishing guidelines for implementing traceability system including "Meat products tracking and tracing Guide" and "Fresh product tracking and tracing Guide". In June 2004, Administration of national barcode management promoting investigated on vegetable products traceability and started an application project on two vegetable production bases located in Shouguang and Luocheng respectively in Shandong province (Zhou Yingheng et al., 2002). The project was successful i n food quality control in the origin and enforcing standards of market access, product identification and recall. Integrated with electronic auctions and E-commerce, this project established a traceability system for pollution-free vegetables. Shanghai Livestock Bureau legislated to build digital archives for pigs, cattle, sheep and other critters, and the residents can now get access to the egg production information through internet (Shanghai agricultural committee. 2001). In August 2 0 0 8 , Beijing will enforce a food traceability system along the full supply chain for the food supplied for Olympic games to secure food quality and safety (Chang Xiang. 2007).

Table 2. Traceability systems for Agricultural products in china

Project name	Project Implementation Unit	Application scope	Start year	Key technologies
Traceability systems for vegetables supplied to Beijing	Beijing Municipal Bureau of Agriculture, Department of Agriculture of Hebei Province	vegetables	2002	barcode
Traceability institutions and systems for meat and meat products	State Food and Drug Administration	Meat and meat products	2004	eartag, RFID
Egg production information	Shanghai Livestock Bureau	egg	2003	barcode
Traceability systems for vegetables	Administration of national barcode	vegetables	2004	barcode

2. KEY TECHNOLOGIES

Aiming at the major quality problems of cereal and oil products existing in the production (cultivation), inspection, storage, processing, circulation and other links, this paper starts with systematical analysis of the source, channel and the kind of contamination, which forms the traceability system framework.

2.1 The framework

In this paper, integrated with the information flow analysis, FMECA（Failure Mode Effect and Criticality Analysis）(Kang Ri. 2006) approach is used as a tool to detect the possible critical control points. HACCP (Bao Dayue . 2007)（Hazard Analysis and Critical Control Point）quality system and some related national standards on cereal and oil products are referred to determine the key quality indicators in the supply chain. Then, the traceability framework and an optimal set of quality indicators come into appearance.

2.2 Coding, identification, and digital files management

The information flow along with the food chain can be constructed by the following steps and technologies: first, use electronic tags to track the physical flow and give each batch/item a unique number. Second, design coding standard for materials，semi-products and products in the food chain. Third, Integrating with TQM (Total Quality Management) system, full supply chain digital quality profile management can be realized.

2.3 Middleware for communication

Middleware is a kind of software between the operating system and application software. In traceability system, Middleware is needed to support data exchange between electronic terminals and the traceability platform.

2.4 Integrated Platform

This paper now has briefly introduced principal things including the framework, identification system, coding system, full supply chain digital file management and middleware for communication, to build a traceability system. Integrate all of the technologies introduced above based on the framework, we can successfully construct the traceability platform for agriculture products.

3. RESEARCH ACHIEVEMENTS

3.1 Determine critical control points and key quality indicators

In our research, through HACCP (Hazard Analysis and Critical Control Point) and FMECA (Failure Mode Effect and Criticality Analysis), we deduced the critical control points and determined the key quality indicators of the full supply chain.The model is shown in Fig.1.Determination of critical control points and quality indicators also provides fundamental data in the traceability system. Coupled with the information flow analysis, this paper proposed an information tracking and tracing back model, which combine the information flow with the critical control points, therefore, endowing the traceability system the function of quality and safety control.

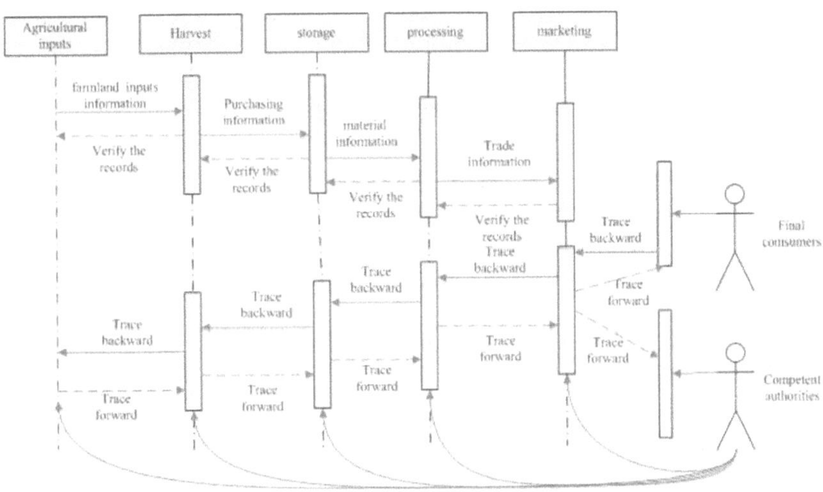

Figure 1: Tracking down and tracing up in the information flow

Now take Peanut oil as an example, after the analysis of production process using the FMECA approach, this paper gives the critical control points, which includes agricultural input, materials test, primary oil-filtering, detection of semi-finished oil, secondary oil-filtering and detection of refined peanut oil.Table3. shows the critical control points and the corresponding key quality indicators.If any of these critical control points missed, it would result in quality information loss and broken of the quality information flow.

Table 3: The critical control points and the key quality indicators of peanut oil production

Critical Control Points	Quality indicators
agricultural input	Seed, pesticide, fertilizer
material test	rate of oil, PH value, percentage of moldy peanuts, impunities in peanuts, humidity
primary oil-filtering	Temperature, additives
detection of semi-finished oil	PH value, color and luster, humidity, volatile
secondary oil-filtering	Temperature, additives
detection of refined peanut oil	color and luster, PH value, humidity, volatile, infusible precipitate, peroxide value, residual menstruum, aflatoxin

3.2 Central traceability database design and data access mode

Based on research above and the peanut oil supply chain, we developed a central traceability database to manage the information flow linked with "retail products code--finished products batch code--semi-finished products batch code--material batch code--source of origin code" ,which were "Primary Keys" of Critical Control Points information. The central traceability database gathered all the necessary quality related information in the supply chain and made the chain transparent.

The system of central traceability database and data access pattern is illustrated in Fig.2.This paper also designed a pattern for data access. Base on RBAC（Role-Based Access Control）,we designed three different roles including consumers, enterprises and quality supervision departments with different rights to use the central database. In the help of the middleware for communication, all users can get access to the central traceability database through three kinds of devices, including fixed terminals in the supermarket, mobile terminals and computers.

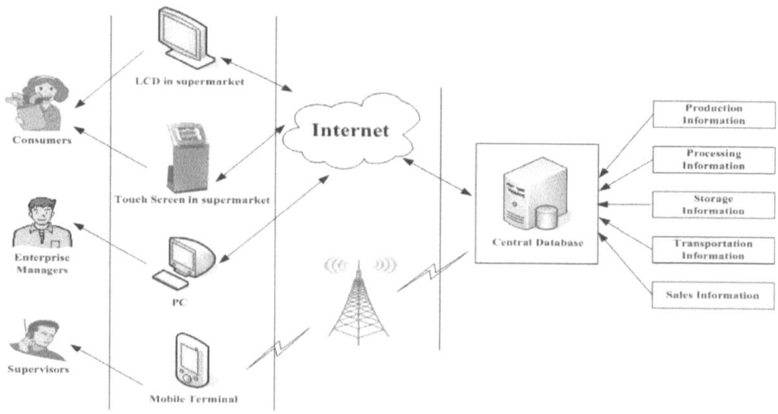

Figure 2: The central traceability database and the pattern for data access

3.3 Traceability code design

Our encoding system follows the UCC/EAN-128 standard (Lin Ling et al., 2004), which is a worldwide standard for exchanging data between different companies. We put forward a unique identification number for encode the data concerning cereal and oil products. The identification number consists of five parts, including enterprise identification number, commodity identification number, batch number and two kinds of verifying number. The first two items in fact form the Global Trade Item Number (short for GTIN). The code structure for cereal and oil products is shown in Fig.3, and an example is shown in Fig.4.

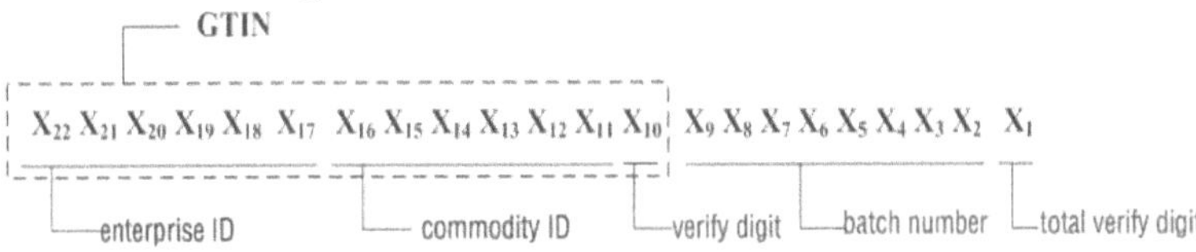

Figure 3: The code structure

Figure 4: Id number for cereal and oil products

3.4 Electr`onic terminal design based on GPRS network

Since the wireless application widely spread and for the consumers convenience, we developed a kind of mobile electronic terminal based on GPRS data transmission, which was consisted of three important parts: the motherboard, GPRS communication module and a micro Liquid-crystal display (short for LCD). The mobile terminal is shown in Fig.5.The mother board is used for processing information, while the GPRS communication module used for receive the information and send the results back. The micro LCD is simply for displaying the results in a friendly way.

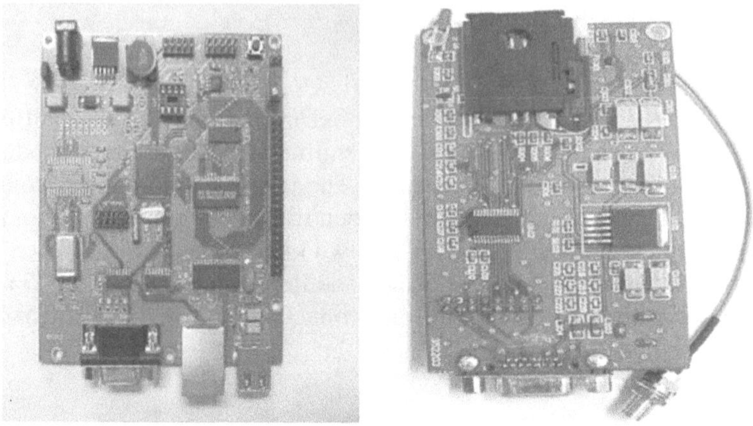

Figure 5: The motherboard and the GPRS communication module of the mobile terminal

4. DISCUSSIONS

1、Cereal and oil products take a large account in agriculture products. Therefore, cereal and oil product quality management has an overall significance in china's food safety, and the research on quality and safety traceability systems for cereal and oil products potentially yields enormous social and economic benefits.

2、The methodology, application system and the pilot project of our traceability systems for cereal and oil products would help the related enterprises to establish traceability systems and rebuild the consumer confidence of Chinese food products, promoting the enforcement of the food safety law of agricultural products in a practical way.

3、All users including consumers, enterprises and the competent authorities can track down and trace up along the information flow easily.

ACKNOWLEDGEMENTS

Funding for this research was provided by National High Technology Research and Development Program of China called as "863" (2006AA10Z268).

REFERENCES

Bao Dayue. Guide to implementation of HACCP. Chemical Industry press, 2007

C. T. Schroeder, T.G.Tonsorb. Australia' S Live—stock Identification Systems: Implications for United States Programs[EB / OL]. http: / / www. agmanager. info / events / risk_ profit / 2004 / Schroeder. 2005-04-27

Chang Xiang. Three food quality and safety traceability systems for Beijing Olympic games will be put into use in August. http://www.chinanews.com.cn/ 2007.07.09

Cheng Hao. Animal product safety control and traceability technologies. Modern Agriculture science and technology, 2007(13), 169~170

Fan Hongping, Feng Zhongze, Yang Ling, Ren Aisheng. Appliance and Discussion of Traceability System in Food Chain. Ecological Economy. 2007, 17(4): 30~33.

General Administration of Quality Supervision, Inspection and Quarantine of the People's Republic of China. GB/T 15425-2002, 2002

Kang Ri. FMECA approach and its application. National Defense Industry Press, 2006

Lin Ling, Zhou Deyi. On the Construction of Food Quality and Safety Traceability System. Commercial research, 2005, (21)

Liu Junrong. An New Legal Requirement on International Fish Market in the Near Future--EU TraceFish Project. Fisheries science, 2007(4), 42~43

Lu Changhua, Wang Changjiang, Hu Sinong etc. Identification and Traceability System for Animals and Animal Products. Chinese Agricultural science and technology press, 2007, 35~36

Shanghai agricultural committee. Notification in carrying out animal identification in Shanghai. Shanghai agricultural committee (2001) NO:126, 2001-08-24

Wang Lifang, Lu Changhua, Xie Jufang, Hu Yinong. Review of traceability system for domestic animals and livestock products. Transactions of the Chinese Society of Agricultural Engineering. 2005(07)

Xing Wenying. The Policy of Traceability in Quality and Safety of Agricultural Products in USA. World Agriculture, 2006, (04)

Xu Jin. Agricultural product supply chain——Guarantee food safety. China Logistics & Purchasing, 2005, (07)

Yu Hui, An Yufa. Theoretical discussion of implementation Traceability System in food supply chain. Agricultural Quality and Standards, 2005, (03)

Zhang Gumin, Chen Gongyu. Food safety and Traceability System. China Logistics & Purchasing, 2005, (14)

Zhou Yingheng, Geng Xianhui. Application of Traceability in Food Safety. Research of Agricultural Modernization, 2002, (06)

Zhu Haipeng. Research and implementation on traceability system of key cereal enterprise. Chinese Academy of Agriculture Science, 2007, 10~11

RFID BASED TRACEABILITY SYSTEM FOR SHEEP BREEDING

Yan Xing[1], Hui Li[1], Jian Zhang[1], Zetian Fu[1,*]
[1] *College of Engineering, China Agricultural University, Beijing, China, 100083*
[*] *Corresponding author, Address: P. O. Box 121, China Agricultural University, 17 Tsinghua East Road, Beijing, 100083, P. R. China, Tel:+86-10-62736717, Fax:+86-10-62736717, Email: fzt@cau.edu.cn*

Abstract: As China's meat production is growing, the quality safety of meat food as an important human food sources has increasingly become a major issue which related to the national economy and people's livelihood. Using early warning technique, tracking and tracing techniques and so on in the key link of meat food production in order to fulfill quality control is an important method which can effectively guarantee the food quality and safety. The use of RFID techniques and handhelds can automatically collect information about sheep breeding. Meanwhile this information could be inquired conveniently through Internet. This study uses C/S (Client/Server) model structure, and builds the sheep breeding traceability system based on RFID.

Keywords: sheep breeding; radio frequency identification (RFID); traceability

1. INTRODUCTION

With the development of the economy and the improvement of people's standard of living, sheep breeding has been developing rapidly. Especially during lasted 10 years, sheep breeding has gained a big achievement, as well its share in gross national product (GNP) increases tremendously (MU Shu-qin, 2005). However, sheep breeding in China is still in its infancy at present, mutton production system has not been completely established, the information of breeding, raising, migration, epidemic prevention and treatment are incomplete. Because of above reasons, mutton production in China has met the challenge of safety, quality, efficiency and continuation.

Please use the following format when citing this chapter:

Xing, Y., Li, H., Zhang, J. and Fu, Z., 2009, in IFIP International Federation for Information Processing, Volume 295, *Computer and Computing Technologies in Agriculture II, Volume 3*, eds. D. Li, Z. Chunjiang, (Boston: Springer), pp. 2275–2281.

In addition international marketplace constantly has built green and technical trade barrier on animal products from China for food safety (ZAN Lin-Sen, 2006), therefore animal breeding which was predominant before is in a very hot spot now. Meanwhile, because of trade globalization and Industrial production of livestock product, production and consumption of mutton have separated with each other. On the other hand, it is difficult to acquire safety information of consumption for consumers because of complicated supply system, so consumers can't easily realize the safety of consumption through termination products if production information can't be passed effectively. Traceability has become an important issue in the international food sector as result of the higher profile of food safety problems (Montanari and Coleman, 1995; Clarke, 1986; 1992; Pilz, 2000; Chanet and Eynard, 2000)

There is need, for both economic and quality assurance reasons, for an efficient and cost-effective method for identifying and tracking livestock product, and for the monitoring of the processing of those livestock (Eggen, 2004& Verbeke 2006). China has recognized an urgent need to increase mutton quality and realized the importance of fast tracing of beef products (Wang LF 2005).By using the sheep breeding traceability system; it is easily to transmit information and resolve the problem of quality control and information dissymmetry. So this measure can meet the home market and international market requirement, it also provide technical guarantee for production standard and production behavior, meanwhile it may be as an essential livestock products' measure to break the trade barrier in international marketplace. This paper had done the research by use of advanced equipments and established traceability system based on the actual research. So not only inaccuracy for manual production was largely avoided, but also efficiency raised, production cost was saved and great economic achievement was gained.

2. THE SCHEME

2.1 Overview

The information records in the traditional sheep breeding and the identification of sheep depend on artificial means. Construction of sheep breeding management information system by using RFID tags and electronic hardware devices such as handhelds can record detailed and accurate breeding information timely and raise the traceability of breeding information. Sheep breeding management information system includes eight functional modules: Routine management, Feed storage, Disease prevention,

Veterinary drug management,Breeding management, Market management, System management.

2.2 Functional modules

(1)Routine management

Feeders feed sheep with forages in accordance with the feed formulation and record detailed information on specific sheep cot after getting forages from forages management department. When the sheep grow to a certain age, sheep need to be transferred to different sheep cot, and detailed records should be kept. In the process of feeding, feeders should conduct regular medical examinations to sheep, and fill out the sheep table. In the case of disqualification during medical examination, the sheep have to be washed out, and relevant record should be filled out.

(2)Feed storage

Forage management department contacts feed suppliers for procurement in determining the need to purchase feed varieties. Before forage entry warehouse, forage officers examine and verify forage, then determine to accept or not. Forage officers record forage entry information as soon as forage entry.

(3)Disease prevention

In the course of sheep farming, as soon as sheep show undesirable symptoms, the veterinary should be immediately notified, then detect the sheep and fill out detection tables. If epidemic situation was found, treatment should be taken immediately, and treatment measurement and drug usage should be kept. At the same time feeders will conduct regular desensitization and disinfection.

(4)Veterinary drug management

Veterinary Drug Management department contact suppliers for the procurement of veterinary drugs after determining which kind veterinary drugs should be procured. Before veterinary drugs entry, veterinary drugs officers should examine and verify veterinary drugs, then determine to accept or not. Veterinary drugs officers record veterinary drugs entry information as soon as veterinary drugs entry.

(5)Breeding management

when ewe are in estrum, breeding manager will test semen from ram, and fill out test report, then determine which rams should be chosen for breeding. During breeding, the breeding ewe, breeding ram, breeding method, the sequence and date information and so on must be recorded.

To improve the efficiency of breeding, embryo transplantation has also been brought into sheep breeding. In sheep embryo transplantation, the embryo's donator, embryo receptor, date of embryo transfer, and other information should be recorded in detail.

When ewes are parturient, feeders and breeding managers transfer the parturient ewes to the special cots which are used to lambing. Lamb number, lamb sex, lamb varieties, lambing date and lamb weight must be recorded off grade and aborted lambs would be washed out and sheep washout sheet should be filled out.

(6)Market management

Market managers must register and audit the information of all the suppliers and customers, unqualified entities would be rejected. Market managers should register order information, and identify pickup ways and pickup date. Meanwhile supplies of farm should be able to meet the order, if not, further consultations are required to take so that the final order would be determined. On delivery date, feeders submit the list of the sheep sale, market managers register records of the sheep sale, and write-off the corresponding orders and store.

(7) System management

The main function of system management module is the user's rights management and parameters set of plug-in equipment. Design thinking of user rights management system is: different enterprise employee has different rights in the system, so corporate information can be managed in safety.

3. SYSTEM IMPLEMENT

3.1 Development platform

This system uses Visual Studio2005 + SQL Server 2005 database as the development platform and c# as server-side programming language. C# language was chosen mainly because: C# (pronounced "See Sharp") is a simple, modern, object-oriented, and type-safe programming language. C# has its roots in the C family of languages and will be immediately familiar to C, C++, and Java programmers. C# full has the advantages provided by the common language runtime, and can be easily integrated and interactive with other applications (XIE Ju-fang, 2006).

3.2 Key Technology

(1) Readers, RFID technology and handhelds

The traditional sheep breeding process management adopts the artificial forms, so the information is serious loss, efficiency is not high, and it's difficult to trace the information. RFID is beginning to be widely used for tracking cattle from birth to the processing plant (Conill et al., 2002). It uses tags affixed to assets to transmit accurate, real-time information to users' information systems (Gerdeman, 1995). Using reader can give individual sheep an only RFID identification; information is read by handheld for the purpose of routine management and then passed to the computer in the processing plant.

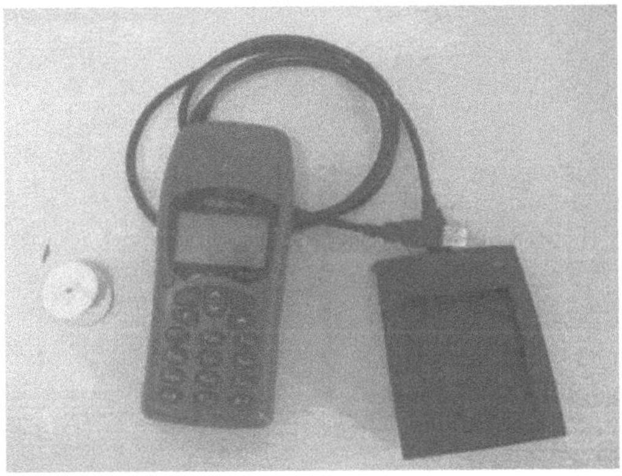

Fig.1 Desktop readers, RFID technology and handhelds

(2)Database access method

SQL Server 2005 provides a stable support and easy maintenance for the system. Database access technology adopts ADO.NET which is a rich data access technology, and one of the key features of the .NET Framework. ADO.NET still has a slew of advanced features and tricks that can be quite useful when you are dealing with data in a real-world application.

3.3 Characteristics of the system

(1)Maternal warning

The system can forecast maternal of the breeding sheep and give a date of childbirth which is based on different species of sheep, and do the appropriate warning.

(2)Sheep genealogical query

Traditional sheep genealogical is mainly managed by manual. It is difficult to manage the breeding sheep when the amount of information is extremely large .This system adopts recursive algorithm dynamic family tree, so the growth of genealogical information is also increased by time and the information is accurate.

(3)The traceability of the information

The system records the various segments of sheep breeding information which is started from birth and achieves the raising stage traceability of the information in order to provide this information for the sale of sheep, sheep slaughter, and so on, so the information is passed.

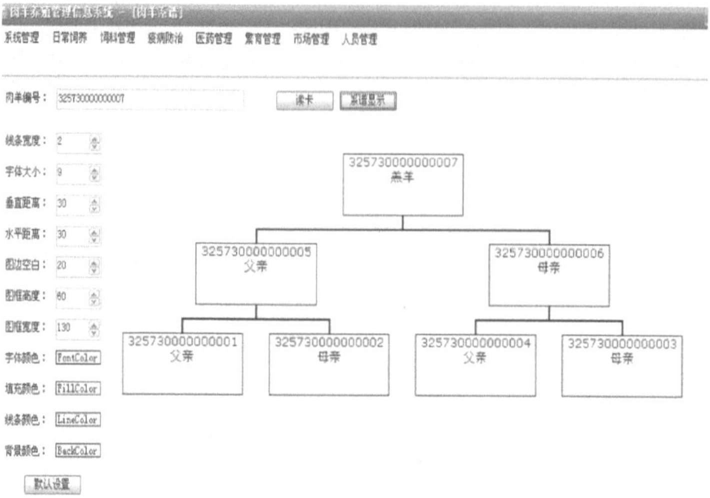

Fig.2 Sheep genealogical query

4. CONCLUSION

It is drawn all attention to inducing livestock product tracing system into food safety and quality control system. And the research of livestock product traceability system is underway in China. The information stored in livestock breed process could not be transferred effectively. It may cause information losses during the tracing processing. A well-designed traceability system may benefit many in the animal food industry (Dickinson and Bailey, 2002), and traceability could become a valued public good (Xie Jufang et al., 2004), especially for food safety. This paper does the preliminary research of the key technology (RFID, computer technology, database and so on) and designs a traceability system for sheep breeding. The introduction of traceability can enhance overall effectiveness and efficiency.

ACKNOWLEDGEMENTS

Thanks are due to National Science & Technology Support Program (2006BAD10A07-02), and National 863 Program (2006AA10Z267) from Ministry of Science & Technology of China for financial support.

REFERENCES

Bailey DV, Jones E, Dickinson DL. 2002. Knowledge management and comparative international strategies on vertical information flow in the global food system. Am J Agric Econ 84(5):1337–44.

Chanet, J.P. and Eynard, P., Technologies for the traceability in the meat industry. Technical Report: Adaption Control & Information Tracing, Proceedings of the First International Meat Automation Congress MAC 2000, pp. 1-6.

Conill C, Caja G, Nehring R, et al 2002. The use of passive injectable transponders in fattening lambs from birth to slaughter: Effects of injection position, age, and breed. Journal of Animal Science 80(4): 919-925.

Eggen A, Hocquette JF 2004. Genomic approaches to economic trait loci and tissue expression profiling: application to muscle biochemistry and beef quality.

Gerdeman JD 1995. Radio Frequency Identification Application 2000: a guide to understanding and using radio frequency identification. North Carolina: Research Triangle Consultants.

Montanari, D.J. and Coleman, G.M., Inventors of Methods for Tracking the Production History of Food Products, a US registered patent system, 1995.

MU Shu-qin,LU Wen-long,LIQian-jun,et al. The Sheep Breeding Situation and Sheep Breeding Expert System Development and Application .Tianjin Agricultural Sciences, 2005，11(4)：61-64(in Chinese)

Pilz, P., Traceability of beef-practical solutions, Proceedings of the First International Meat Automation Congress MAC 2000, pp. 49-52.

Wang LF, Lu CH, Xie JF, et al. 2005. Review of traceability system for domestic animals and livestock products. Transactions of the Chinese Society of Agricultural Engineering. 21(7): 168-174 (in Chinese with English abstract).

Xie Jufang, Lu Changhua, Li Baoming, Wang Lifang.,2004. Development of Monitoring and Traceability System for Prok Production. World Enginers Convention. November 2-6,Shanghai, China.

XIE Ju-fang,LU Chang-hua,LI Bao-ming. Implementation of pork traceability system based on .NET framework . Transactions of the Chinese Society of Agricultural Engineering, 2006，22(5)：218-220(in Chinese).

ZAN Lin-Sen, ZHENG Tong-Chao, SHEN Guang-lei,et al. Design and Development of Quality Traceability Information Management System and Safety of the Beef Production' s Entire Processes. Scientia Agricultura Sinica, 2006，39(10)：2083-2088(in Chinese).

STUDY ON CARRIER MATERIAL OF IMMOBILIZATION ACETYLCHOLINESTERASE FOR BIOSENSOR IN DETECTION OF ORGANOPHOSPHORUS PESTICIDE RESIDUES

Xia Sun[1,2], Xiangyou Wang[1,*], Chuandong Jia[3]

[1] *College of Light Industry & Agricultural Engineering, Shandong University of Technology, 12 Zhangzhou Lu, Zibo, Shandong P. R. China 255049*

[2] *College of Engineering, China Agricultural University, 17 Qinghua Donglu, Beijing, P. R. China. 100083*

[3] *School of Mingshui, Zhoucun Zibo Shandong, P.R. China.255318*

[*] *Corresponding author: Professor Xiang-You Wang, College of Light Industry & Agricultural Engineering, Shandong University of Technology, P. O. Box 50, 12 Zhangzhou Lu, Zibo 255049, Shandong P. R. China; Fax: +86-533-2786558; E-mail: wxy@sdut.edu.cn.*

Abstract: A comparison between several immobilization materials of AChE on surface of glassy carbon electrode(GCE) was presented. The immobilization methods employed crosslinking method with glutaraldehyde as a cross-linking agent and bovine serum albumin(BSA) as a protectant, AChE was immobilized on different membranes including nylon membrane, cellulose nitrate membrane and chitosan membrane respectively. The enzyme membrane was then fixed on the surface of glassy carbon electrode(GCE) with O-ring to prepare an amperometric biosensor for the detection of organophosphorus pesticides. The activity of immobilization AChE was detected by measuring the oxidation current of thiocholine, the results showed that the activity of immobilization AChE were all different with different membrane as carrier material. Compared with nylon membrane and cellulose nitrate membrane, chitosan membrane was obviously good. So chitosan membrane can be selected as immobilized AChE carrier material.

Keywords: Acetylcholinesterase, pesticides residue, biosensor, immobilization, carrier material

---

Please use the following format when citing this chapter:

Sun, X., Wang, X. and Jia, C., 2009, in IFIP International Federation for Information Processing, Volume 295, *Computer and Computing Technologies in Agriculture II, Volume 3*, eds. D. Li, Z. Chunjiang, (Boston: Springer), pp. 2283–2292.

1. INTRODUCTION

Organophosphorus pesticides (OP) are very toxic compounds, they can lead to a severe impairment of nerve functions of human or even death because they inhibit main metabolic pathways (Ivanov et al., 2002, Neufeld et al., 2000). The intensive use of organophosphorus pesticides in agriculture requires further progress in the development of rapid and efficient monitoring methods. Nowadays the conventional analytical techniques mainly used for detecting organophosphorous and carbamate pesticides residues, the chromatographic ones, such as HPLC and GC,has been most commonly used as criterion methods due to their high sensitivity, reliability and precision. However these analytical techniques are time-consuming and expensive and above all, require pretreatment of the sample and highly qualified technicians (Kim et al., 2007, Schulze et al., 2002).

Biosensor techniques based on enzymatic inhibition have gained considerable attention in the past decade due to the advantages of simplicity, rapidity, reliability and low cost devices compared with the conventional analytical techniques (Andreescu et al., 2005). According to the difference of transducers, biosensors mainly included in three types: potentiometric (Marek et al. 2005), amperometric (Dan Du et al.,2007), and optical biosnsors(A.L. Simonian et al.,2005). Among these methods, amperometric biosensor is the most frequently reported because of the simple configuration, close contact between enzyme layer and transducers, high sensitivity.

Enzymes immobilization is a crucial step for the design of the biosensor. it's methods include that enzyme is immobilized on the working electrode surface directly(Arduini et al.,2006), and enzyme is immobilized on membranes, the enzyme membrane is then fixed on the surface of electrode (Fuxiang Wei et al.,2007). Appropriate selection of the electrode, membranes material and immobilization methods is essential for a reliable biosensor, because it will strongly affects the performance of the biosensor in term of sensitivity, stability, response time and reproducibility.

In this work, a comparison of the AchE activity after immobilization with different materials in biosensors was presented. The immobilization methods employed crosslinking method with glutaraldehyde, and under the immobilization methods, the carrier materials were explored, AChE was immobilized on different membranes that were nylon membrane, cellulose nitrate membrane and chitosan membrane respectively. The enzyme membrane was then fixed on the surface of GCE to prepare an amperometric biosensor. The results were evaluated by electrochemical measurements, the results showed that the rate recovery activity of enzyme immobilized with different membrane was different, the activity of enzyme immobilized with chitosan membrane was the highest, and the activity of enzyme immobilized with nylon membrane was the worst.

2. MATERIALS AND METHODS

2.1 Apparatus

Specrophotometric measurements were performed on UV-Vis spectrophotometer purchased from DaoJin Japan, high-speed freezing centrifuge was supplied from Sigma(3K15,Germany); Acidometer was the product of Vastell (230A+,USA); JJ-2 tissue triturator from Ronghua instrument manufacture Co. (Jiangsu,China), DK-98- I galvanothermy constant temperature boiler was obtained from Taisite instrument Co. (Tianjin,China). Electrochemical measurements were performed on an CHI600 Electrochemical Workstation from Shanghai Chenhua instrument Ltd.(Shanghai,China). 3- electrode was purchased from Aida Hengsheng technology Co.(Tianjin, China). glassy-carbon electrode with 3mm diameter as working electrode, platinum electrode as counter electrode, saturated calomelelectrode(SCE)as reference electrode.

2.2 Chemicals and reagents

Chicken brain-esterase was extracted from living chicken bought from market. Acetylthiocholine chloride (ATCh) as substrates was purchased from Nuoyawei Biology Tech.Co.(Shanghai,China). 5,5′-dithiobis(2′-nitrobenzoic acid) (DTNB) as color react of reagent provided by Sigma. Glutaraldehyde and BSA were provided by Sigma. Dichlorvos(40% milk grease)was the product of Dacheng pesticide(Shandong,China). All other chemicals were of analytical grade and were used without further purification. Distilled water was used throughout for the preparation of solutions.

2.3 Preparation and extraction of the chicken brain esterase

Chicken brain-esterase was extracted using a method, which was from a previously reported technique (Sunxia et al.,2008). The fresh chicken brain with phosphate buffer (1%Triton X-100,pH8.0, 0.1mol/l) in the ratio of 1g/3 ml is homogenized for 40 min at 8000 rpm at 4℃ with a tissue triturator. And the chicken brain-esterase solution is obtained from the final supernatant after centrifugation and then preserved in refrigerator at 4℃.

For measuring AChE activity was modified from the assay described by Ellman (Ellman et al.,1961). The Ellman method has found widespread application for determination of cholinesterase (ChE) activities using acetylthiocholine (ACTC) as substrates (T. Hamers et al.,2000). Ellman method is based on the increase of yellow colour nitrobenzoate (TNB) ion produced from thiocholine when it reacts with dithiobisnitrobenzoate (DTNB) ion (Eq. 1 and 2, below) (Zou et al.,2006).

$$\text{acetylcholine} + H_2O \xrightarrow{\text{AChE}} \text{thiocholine} + \text{acetic acid} \quad (1)$$

$$\text{Thiocholine+DTNB} \longrightarrow \text{Formation of TNB [yellow colour (412nm)]} \quad (2)$$

The calculation of enzyme activity of formation is described as followed:

$$E\ (\mu mol \cdot min^{-1} \cdot mL) = V \times A \times 10^6 / (v \times K \times L)$$

where E is the activity of enzyme, V is the volume of enzyme liquid (mL), v=0.02. K is extinction coefficient (L/mmol·mm), K=1.36. L is light length of enzyme liquid (10mm). v is total volume of reaction of liquid (mL), v=3.14. Every experiment was done replicatively in triplicate. After calculated the activity of chicken brain-esterase, recorded and stocked at 4℃for use.

2.4 Preparation of immobilization enzyme membrane

2.4.1 Pretreatment of carrier material

Carrier material was pretreated using a method, which was slightly modified from a previously reported technique (Fuxiang Wei et al.,2007). Commercial carrier material were made into circular sheet about 10 mm diameter, and employed the following method of activation respectively. nylon carrier was immersed in anhydrous methanol for 30 min, then washed 5 min with phosphate buffer (0. 1mol/l, pH8.0), naturally dried and stock for use. Cellulose acetate porous membrane and cellulose nitrate porous membrane were immersed in phosphate buffer (0.01mol/l, pH7.0) for 48 h at 4℃.

2.4.2 Preparation of chitosan porous membrane

The preparation method of chitosan porous membrane was modified from the assay described by Zhibin Qiang (Zhibin Qiang et al.,2007).Weighing 0.1 g chitosan powder,then added 10ml of acetate solution (1%,m/m), and mixed round for 10 min to yellow sol.Then centrifugal for 5min in high-speed freezing centrifuge at 3000r/min at room temperature to remove insoluble particles. Then the pretreated cellulose nitrate porous membrane was immersed in the sol for 12h, and then immersed in phosphate buffer (0.1mol/L,pH8.0) for 12, dried and stock for use.

2.4.3 Immobilization of enzyme

Chicken brain-AChEs, diluted at a concentration of 100 U mL^{-1} in phosphate buffer (0.1mol/L,pH8.0), were immobilized on sensitive membrane. The immobilization method of enzyme was modified from the assay described by Fuxiang Wei (Fuxiang Wei et al.,2007), the protocol were performed according to the following procedure: 10U of AchE liquid and 30.0μL of BSA (1.0%), 2.0μL of glutaraldehyde (5.0%) in 0.50 mL of centrifuge tube, were blended in vortex mixer. Then nylon membrane, cellulose nitrate membrane and chitosan membrane were immersed in 0.50 mL of centrifuge tube which contained those enzyme liquid for 8h at 4℃ respectively. And enzyme membranes was washed with phosphate buffer (0.1mol/L,pH8.0), and then dried at room temperature, and determined enzyme activity by electrochemical measurements.

2.5 Amperometric measurement of AChE enzyme activity

2.5.1 Amperometric measurement principle

The detecting principle is that pesticides can be measured by comparing the residual activity of the enzyme after exposure to OP compounds with the initial activity. After exposure to OP compounds, enzyme activity is decreased, depending on the concentration and the time of exposure. The enzyme activity is estimated amperometrically at about between 100 mV and 700mV versus reference electrode by the anodic oxidation of the thiocholine (TCh) produced by hydrolysis of the acetylthiocholine (ATCh) substrate. (Eq. 3 and 4, below) (Lihong Wang et al.,2006.)

$$\text{acetylcholine} + H_2O \xrightarrow{\text{AChE}} \text{thiocholine} + \text{acetic acid} \quad (3)$$

$$\text{2thiocholine} \longrightarrow \text{dithio-bis-choline} + 2H^+ + 2e^- \quad (4)$$

The following general expression of enzyme-substrate-inhibitor is used to obtain a quantitative estimate of the relation between dichlorvos concentration and the degree of inhibition. The remaining activity (I%) is determined according to the following formula: I%=((I0−I1)/I0)×100%

Where I% is the degree of inhibition related to the inhibitor concentration. I0 is the initial current (I0) of the biosensor is measured without inhibitor in phosphate buffer. I1 is the current after the incubation in the phosphate buffer with different concentrations of inhibitor(Dan Du et al., 2007).

2.5.2 Amperometric measurement of free AChE enzyme activity

Firstly, the surface of GCE was polished by 0.05μm alumina suspension to a mirror finish. Secondly, 3-electrode connected with electrochemical workstation was immersed in 10ml of electrochemical reaction cell in which 9ml of phosphate buffer (0.1mol/L,pH7.5) as electrolyte was added, and then different concentration AChE liquid was added respectively. Then electrochemical workstation was run, the electrodes were tested with electrical current–time at a working potential of +600 mV versus SCE. After the current was stable, 0.5ml of ATCh was injected, the change of current produced by the oxidation of thiocholine was observed. The cell was washed with distilled water between measurements.

2.5.3 Amperometric measurement of immobilization AChE enzyme activity

The procedure amperometric measurement of immobilization AChE enzyme activity was similar to the amperometric measurement of free AChE enzyme activity basically, only one was difference, substituting AChE liquid for the enzyme membrane fixed on the surface of glassy carbon working electrode with O-ring.

3. RESULTS

3.1 The current of catalytic reaction under free AChE

The current which corresponds to the oxidation of thiocholine, the product of the enzymatic hydrolysis of ATCh substrate under different concentration AChE liquid was recorded after plateau. The relation curve between AChE concentration and electrical current volume was drawn. The curve shown in Fig1 indicated that the relation between the current value and enzyme activity was obvious linear correlation. And its linear correlation coefficient was 0.9945. So the current value of catalytic reaction can be used to measure enzyme activity.

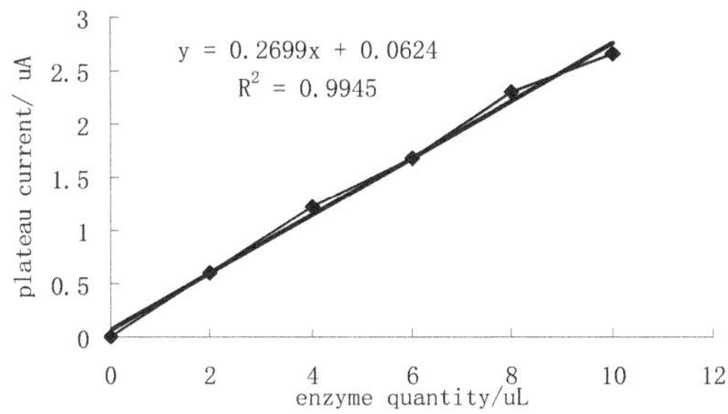

Fig.1 The relation curve between AchE concentration and electrical current volume

3.2 The current of catalytic reaction under immobilization AchE employed different carrier materials

The enzyme membrane fixed on the surface of glassy carbon working electrode with O-ring. 3-electrode connected with electrochemical workstation was immersed in 10ml of electrochemical reaction cell in which 9ml of phosphate buffer (0.1mol/L,pH7.5) as electrolyte was added. Then electrochemical workstation was run, the electrodes were tested with electrical current –time at a working potential of +600 mV versus saturated calomelelectrode. After about 50 second, electrical current reached a plateau, 0.5ml of ATCh was injected at this time.

3.2.1 The current of catalytic reaction employed nylon porous membrane

The curves shown in Fig.2 indicated that 0.5ml of ATCh was injected at 50 second, and at 120 second the current began descend. At 300 second, the current achieved to valley value that was 1.643uA.

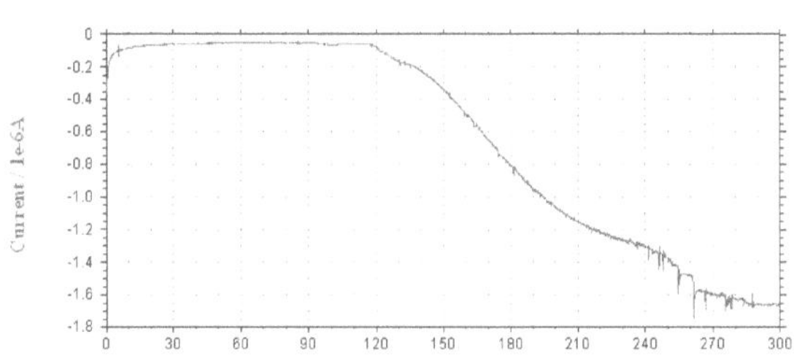

Fig.2 The current curve of catalytic reaction employed nylon membrane

3.2.2 The current of catalytic reaction employed cellulose nitrate porous membrane

The curves shown in Fig.3 indicated that 0.5ml of ATCh was injected at 50 second, and at 60 second the current began descend. At 240 second, the current achieved to valley value that was 1.12uA.

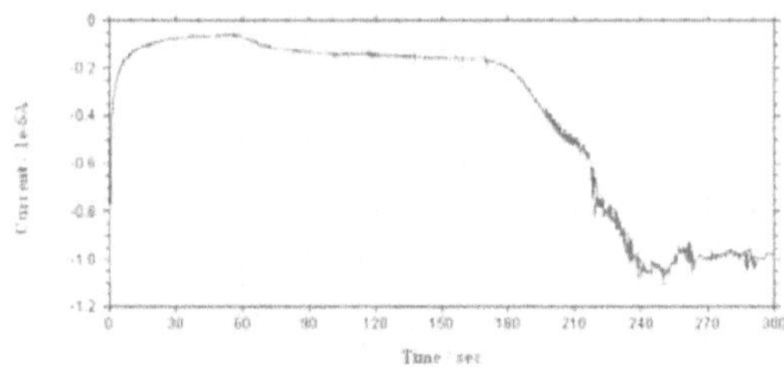

Fig.3 The current curve of catalytic reaction employed cellulose nitrate membrane

3.2.3 The current of catalytic reaction employed chitosan porous membrane

The response curve for the current of catalytic reaction employed chitosan membrane was shown in Figure4. The curves showed that at 50 second, 0.5ml of ATCh was injected, and at 120 second the current began descend. At 180 second, the current achieved to valley value that was 2.4 uA.

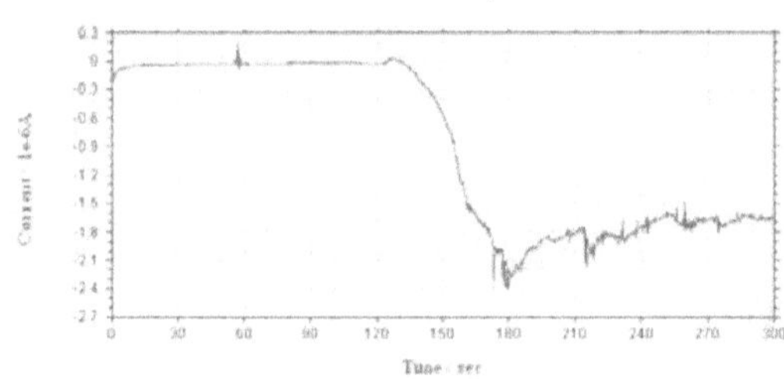

Fig.4 The current curve of catalytic reaction employed chitosan porous membrane

3.3 Discussion

The results showed that the current of response time, stability time and the current stability values were all different in fig.2, fig.3, fig.4 respectively. In other words, the selection of carrier materials directly effect the activity, stabilty of immobilization enzyme. Compared with nylon membrane, the time of the current achieved to valley value immobilized with cellulose nitrate membrane was different, it showed that substrate diffusion speed in cellulose nitrate membrane was quicker than nylon membrane. But the valley value in fig2 was bigger than fig3, it showed that cellulose nitrate membrane osmosis was better than nylon membrane, so cellulose nitrate membrane absorption enzmye quantity less than nylon membrane, after the cellulose nitrate membrane modified by chitosan, the situation improved greatly in fig4. The reason is that there is amino structure in chitosan molecular structure, that have high affinity, that can easy to fix the enzmye.

4. CONCLUSION

The results were evaluated by catalytic reaction showed that the rate recovery activity of enzyme employed different membrane as carrier material was different, chitosan membrane was the highest compared with cellulose nitrate membrane and nylon membrane. And the results also showed that chitosan membrane prevented leakage of the enzmye, improved the activity of immobilization enzmye. So the AChE immobilized with chitosan membrane used in biosensor can improve the performance of the biosensor in term of sensitivity, stability, response time and reproducibility. we also can conclude that promote the application of immobilized enzymes in pesticide residue rapid detecting fields by improving traditional materials, design and exploit new carrier materials. In the future, the design and preparation of carrier materials with excellent integration performances will be one of the most significant researches for the development of immobilized enzymes.

ACKNOWLEDGEMENTS

The authors are gratefully acknowledging the financial support of the Science and Technology Research Key Program of Shandong province(No.2008GG10009027), and the Technology Program of Shandong Provincial Education Department(No. J07YF11).

REFERENCES

A.L. Simonian, T.A. Good, S.-S. Wang, & J.R. WildNanoparticle-based optical biosensors for the direct detection of organophosphate chemical warfare agents and pesticides. Analytica Chimica Acta, 2005, 534:69-77.

Andreescu, D., Andreescu, S., Sadik, O. A., & Gorton, L. Chapter 7 New materials for biosensors, biochips and molecular bioelectronics. In Comprehensive Analytical Chemistry, 2005, 44: 285-327.

Avramescu, A., Andreescu, S., Noguer, T., Bala, C., Andreescu, D., & Marty, J.-L. Biosensors designed for environmental and food quality control based on screen-printed graphite electrodes with different configurations. Analytical and Bioanalytical Chemistry, 2002,374(1):25-32.

Dan Du, Xi Huang, Jie Cai, & Aidong Zhang. Amperometric detection of triazophos pesticide using acetylcholinesterase biosensor based on multiwall carbon nanotube–chitosan matrix Sensors and Actuators B: Chemical, 2007, 127(2): 531-535.

Ellman, G. L., Courtney, K. D., Andres, V., & Featherstone, R. M. A new and rapid colorimetric determination of acetylcholinesterase activity. Biochemical Pharmacology, 1961,7(2):88-90.

Fuxiang Wei, Zhenchuan Wang, & Jinmei Wang, Determination of Organophosphorus Pesticide Residues by AnAcetylcholinesterase Biosensor in Vegetables and Fruits, Food science, 2007,28(2):229-231.

Ivanov, A. N., Lukachova, L. V., Evtugyn, G. A., Karyakina, E. E., Kiseleva, S. G., Budnikov, H. C., et al. Polyaniline-modified cholinesterase sensor for pesticide determination. Bioelectrochemistry, 2002,55(1-2):75-77.

Kim, N., Park, I.-S., & Kim, D.-K. High-sensitivity detection for model organophosphorus and carbamate pesticide with quartz crystal microbalance-precipitation sensor. Biosensors and Bioelectronics, 2007, 22(8):1593-1599.

Lihong Wang, Lin Zhang, & Huanlin Chen Enzymatic Biosensors for Detection of Organophosphorus Pesticides. Progress in chemistry, 2006,18(4): 440-451.

Marek Trojanowicz, Michael L. Hitchman. Determination of pesticides using electrochemical biosensors. TrAC Trends in Analytical Chemistry, 1996,15 (1):38-45.

Neufeld, T., Eshkenazi, I., Cohen, E., & Rishpon, J. A micro flow injection electrochemical biosensor for organophosphorus pesticides. Biosensors and Bioelectronics, 2000, 15(5-6):323-329.

Schulze, H., Schmid, R., & Bachmann, T. Rapid detection of neurotoxic insecticides in food using disposable acetylcholinesterase-biosensors and simple solvent extraction. Analytical and Bioanalytical Chemistry, 2002,372(2):268-272.

Siriwan Suwansa-ard, Proespichaya Kanatharana, Punnee Asawatreratanakul, Chusak Limsakul, Booncharoen Wongkittisuksa, & Panote Thavarungkul. Semi disposable reactor biosensors for detecting carbamate pesticides in water.Biosensors and Bioelectronics, 2005, 21, (3):445-454.

Sunxia, Xiangyou Wang, Xiao yu Wang, & Zhe Liu. A Comparative Study of Sensitivity of Acetylcholinesterase in Detection of Organophosphorus Pesticide Residues. International Journal of Food Engineering, 2008,4(3), article 7.

T. Hamers, K.R.J. Molin, J.H. Koeman and A.J. Murk, A small-volume bioassay for quantification of the esterase inhibiting potency of mixtures of organophosphate and carbamate insecticide in rainwater: development and optimization, Toxicological Science 2000,58:60-67.

Vangelis G. Andreou, Yannis D. Clonis A portable fiber-optic pesticide biosensor based on immobilized cholinesterase and sol–gel entrapped bromcresol purple for in-field use. Biosensors & Bioelectronics, 2002,17:61–69.

Zhibin Qiang, Yinsheng Chen, Huawen Guo, & Jianqi Liu, Property of Different Molecular Weight and Degree of Deacetylation Chitosan Membranes, Journal of Dong Hua University, 2007, 33(2):212-215.

Zou, M.-Q., Yang, R., Wang, D.-N., Li, J.-F., & Jin, Q.-H. A novel immobilized cholinesterase for on-site screening of organophosphate and carbamate compounds. Pesticide Biochemistry and Physiology, 2006,86(3): 162-166.

TRACEABILITY SYSTEM FOR AGRICULTURAL PRODUCTS BASED ON RFID AND MOBILE TECHNOLOGY

Koji Sugahara
National Agricultural Research Center, National Agriculture and Food Research Organization, 3-1-1 Kannondai, Tsukuba,305-8666, Japan

Abstract: In agriculture, it is required to establish and integrate food traceability systems and risk management systems in order to improve food safety in the entire food chain. The integrated traceability system for agricultural products was developed, based on innovative technology of RFID and mobile computing. In order to identify individual products on the distribution process efficiently, small RFID tags with unique ID and handy RFID readers were applied. On the distribution process, the RFID tags are checked by using the readers, and transit records of the products are stored to the database via wireless LAN. Regarding agricultural production, the recent issues of pesticides misuse affect consumer confidence in food safety. The Navigation System for Appropriate Pesticide Use (Nouyaku-navi) was developed, which is available in the fields by Internet cell-phones. Based on it, agricultural risk management systems have been developed. These systems collaborate with traceability systems and they can be applied for process control and risk management in agriculture.

Key words: food safety, traceability system, risk management, RFID, mobile devices

1. INRTODUCTION

In recent years, Japanese people have been increasingly aware and concerned about food safety issues. This is a result of various incidents relating to the food safety since 1996. Examples of such incidents are food poisoning by bacteria of *E. coli* O157 and *Vibrio parahaemolyticus*, large-scale food poisoning on the processed milk, mixing of genetically modified

Please use the following format when citing this chapter:

Sugahara, K., 2009, in IFIP International Federation for Information Processing, Volume 295, *Computer and Computing Technologies in Agriculture II, Volume 3*, eds. D. Li, Z. Chunjiang, (Boston: Springer), pp. 2293–2301.

corn (GMO) in the processed food, pesticide residue on the imported frozen vegetable products, uses of the illegal agricultural chemicals, and product recalls caused by pesticide misapplications. The first cow infected with BSE (bovine spongiform encephalopathy) in Japan was discovered in 2001, and the import of beef from USA was stopped in 2003 because the BSE cow had discovered there. Many other issues on meat products have occurred such as the false labeling. Moreover, the outbreak of bird flu in Japan and East Asia was reported in 2004. Therefore, many consumers feel insecure in the safety of agricultural products including imported ones.

In this paper, current state and perspective of the researches on food safety, especially ICT-based food traceability system in Japan are presented. The policy that has been promoted by the Japanese Government to improve food safety and to establish food traceability systems is outlined. Moreover, the integrated traceability system for agricultural products and the agricultural risk management system are introduced as the research works that the author has been involved in.

2. POLICY FOR FOOD SAFETY

2.1 Policy of the Japanese Government

The Japanese Government enacted the Food Safety Basic Law in May 2003. The purpose of this law is to promote the measure concerning food safety. The Food Safety Commission was organized in July 2003, based on the law. This organization aims to undertake scientific risk assessment about foods and their effects on human health, and it is independent from the organizations that implement risk management.

The Ministry of Agriculture, Forestry and Fisheries (MAFF) announced "regeneration plan of food and agriculture" in April 2002 and "policy outline of food safety and security" in June 2003. Securing food safety is positioned as an important theme of the policy. MAFF are implementing the measures such as support system for producers, information provision to consumers, and risk management on food production. Promotion of development and introduction of “food traceability systems (FTS)” has been carried out as one of those measures (Tomiyama, T., 2005).

2.2 System Development

Table 1 shows the institutional FTS in Japan. The beef traceability system based on the Beef Traceability Law (December 2003) is mainly for domestic beef as the measure for the BSE problem. The food traceability systems are for various food products to support independent activities, however there is not legal obligation to use the systems and to store the products' information. In the campaign to record production processes of whole food products for the producers around Japan, there is effort obligation to record the items based on the Food Sanitation Law. The system of production information disclosure as JAS (Japan Agricultural Standards) is for beef, pork and farm products, in order to provide production information of the products with the JAS marks to consumers.

MAFF is promoting development of FTS to establish food traceability and safety. The budgets for development, demonstration and introduction of FTS have been offered to the private sector since 2001. The feature of these projects is to utilize ICT, especially "ubiquitous computing" by the u-Japan Strategy, and to develop not only FTS but risk management systems for agriculture and food industry. "Ubiquitous computing" technology enables anyone easily to access various information anytime and anywhere, and it has shown remarkable progress with the spread of RFID and mobile devices. MAFF has required the FTS development with the following features: a) speed-up of callback and cause unfolding at food accidents, b) efficiency improvement of food production and distribution, c) enhancement of various food information that is provided to consumers.

This measure has prompted the establishment of FTS and systems to disclose information about food production and distribution processes. The local governments and the agricultural cooperatives (JA) as well as the Japanese government have been actively promoting development and practical introduction of FTS in the national projects.

Table 1. Institutional food traceability systems in Japan.

System Name	Target	Record items
The beef traceability system	Domestic beef	Date of birth, sex, type, breeding location, breeding manager
The food traceability systems	Foods other than domestic beef	The producers or the corporations independently set recorded data
The campaign to record production processes	whole foods	The items based on Food Sanitation Law (effort obligation)
The system of production information disclosure (JAS)	beef, pork and farm products	Animal medicines, agricultural chemicals, fertilizers, and so on

3. BASIC TECHNOLOGIES

3.1 Scheme of Food Traceability

As the basic scheme of ICT-based FTS, "Virtually Identified Produce System (VIPS)" (http://www.vips.gr.jp/) was developed (Sugiyama, 2001).

National Agriculture and Food Research Organization (NARO) owns the business model patent of this system (Japanese patent No. 3355366). Regarding VIPS, an identification data (ID) is assigned to an identification unit (lot) of food products and it is printed on the products' labels or packages. The producers input production data of their products and send these data to the Internet-accessible database server. The consumers can browse the products' data by accessing the website of VIPS and entering the products' ID. Based on VIPS, a practical information disclosure system for farm products, "SEICA" (http://seica.info/) was developed and open to the public in 2002 (Sugiyama, 2004).

3.2 RFID and Mobile Technology

For the practical traceability systems, it is important to apply automatic identification technology such as barcode system. RFID (Radio Frequency Identification) is innovative noncontact automatic identification technology to read and write the data on electronic media (IC chips) more quickly and accurately. It has been applied mainly to logistics of industrial products in order to identify the products and check their distribution processes. RFID has recently shown promise for more advanced and effective FTS.

Mobile technology is based on wireless networking as cell-phone service and wireless LAN (mainly Wi-Fi). In Japan, recent models of cell-phones have increased range of functions such as camera and Internet access, and such "Internet cell-phones" have been very familiar ICT tools. On the other hand, high-speed wireless networks have been easily available around the Wi-Fi access points (hotspots) in various places and facilities, using mobile devices such as laptop PC, PDA, and portable game consoles.

4. TRACABILITY SYSTEM

4.1 Integrated Traceability System

In the FTS research project that the author has been involved in since 2003, an integrated traceability system for agricultural products in the food chain have been developed, based on the technologies of network computing, cell-phones, and RFID (Sugahara, 2005). Figure 1 shows the concept of this system. It consists of three subsystems: a production record system using Internet cell-phones, a distribution record system using RFID devices, and an information disclosure system for consumers on WWW (World Wide Web). The production record system has been improved as an agricultural risk management system, as described later.

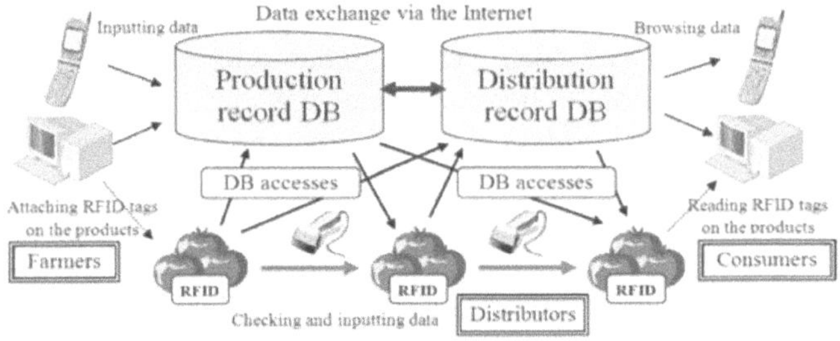

Figure 1. The concept of the integrated traceability system for agricultural products based on RFID and mobile technology, to record and check the data on the production and distribution processes.

4.2 Distribution Record System

A key concept to record distribution processes as the food chain is that a temporal identification unit is an "event" which is recorded at each specific point of the distribution process, such as incoming, outgoing, combination, and division. Data of an event record consist of an event ID assigned to each event, a point ID, event time, and involved product lot IDs. Based on this concept, a database and a data-inputting system were designed.

The on-site data-inputting system based on RFID was developed in order to identify the product lots and record the "event" at each point of the distribution process efficiently and quickly. RFID tags which have unique ID to identify individual lots and handy RFID reading devices (readers) compatible with Wi-Fi to read the ID were applied (Figure 2). Additionally, an Internet accessible database server was built. It can store the data which are sent from the readers and it enables to retrieve and refer the distribution process records on WWW.

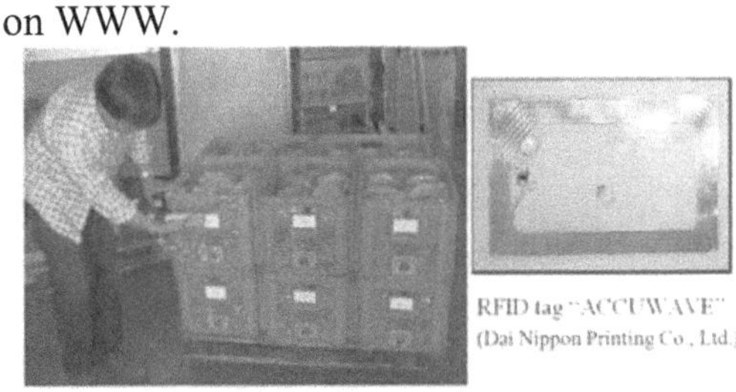

Figure 2. Reading the RFID tags attached to the container cases by using a handy reader in the experiment on lettuce products.

4.3 Information Disclosure System

In order to disclose proper information about agricultural products to consumers, information disclosure systems has been developed based on

VIPS and additionally applying "QR code" (Denso Wave Inc.) which is a kind of two-dimensional (2D) barcode system. In Japan, camera-equipped Internet cell-phones have been very popular and most of them have a barcode-reading function. Therefore, the consumers that purchase the products with the QR codes can immediately browse and check the information about the products' production and distribution, by reading the QR codes using their own cell-phones (Figure 3).

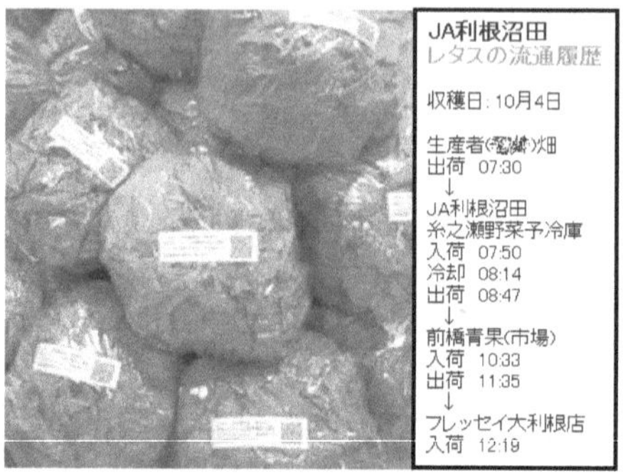

Figure 3. Lettuce products for sale which the "QR code" labels are pasted on (left) and a display of the distribution records on a cell-phone (right).

4.4 Field Experiments

In order to verify the effectiveness of the integrated traceability system in the actual food chain of vegetable and fruit products, the field experiments of the system were carried out in Yamagata, Saitama, Shizuoka, and Gunma from 2003 to 2005, in cooperation with farmers, agricultural cooperatives, wholesale markets, cooperative societies (co-op), supermarkets and the local governments. Regarding RFID, "mu-chip" (Hitachi, Ltd.) as microwave-type tags (2.45GHz) and "ACCUWAVE" (Dai Nippon Printing Co., Ltd.) as electromagnetic-type tags (13.56MHz) were applied. In these experiments, the information disclosure on the products was well received by consumers. However it was required to improve workability of data inputting and RFID handling.

5. AGRICULTURAL RISK MANAGEMENT

5.1 Nouyaku-Navi

The recent issues on agricultural chemicals (AC) misuse and residue on the agricultural products have affected consumer confidence in food safety.

It is required for farmers to make proper uses of AC to their products. In the research project supported by MAFF, Navigation System for Appropriate Pesticide Use was developed as a basic system for agricultural risk management (Nanseki et al., 2005). It was called "Nouyaku-navi" (Nouyaku is the Japanese word for AC including pesticide). Its main part is composed of a database of the legal standard for the registered AC uses and a server application to judge whether each AC use is appropriate for the standard or not, as a Web-based system which is accessible by Internet-enabled PC and cell-phones. This system has the following functions: creation of guidelines and rules on pest control, creation and judgment of AC application plans, on-site warning and recording of AC applications, and so on. Figure 4 shows a flow of using the application of Nouyaku-navi.

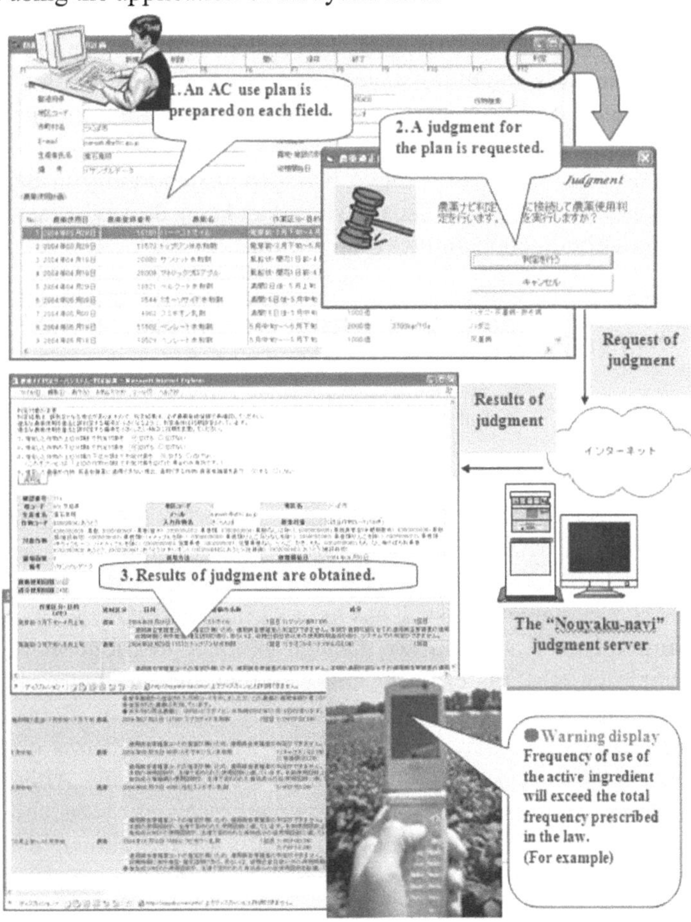

Figure 4. A flow of using the application of "Nouyaku-navi".

5.2 Risk Management System

Some of existing "food traceability systems" also have function to check AC application records on the products. Although they can check whether the AC applications are appropriate after the application or the harvest, they can not prevent AC misapplication due to carelessness before the application.

Based on Nouyaku-navi, a risk management system on AC applications was developed in part of the national project by MAFF (Nanseki et al., 2006). The main feature of this system is that it is possible to check whether AC applications on individual products are appropriate for multiple standards and at multiple stages such as planning, application, harvest, distribution and sale. This system can prevent AC misapplication by checking before the application, and it can record the detail data (5W1H) of the AC application while checking it. The system was tested and evaluated practically in the agricultural cooperatives in Yamagata. Effectiveness of the method to check on the multi-standards and multi-stages has been confirmed.

Moreover based on this system, an integrated ICT system for agricultural risk management and traceability on vegetable and fruit production were developed and released by the Nogyo Navigation Laboratory in September 2007 (http://www.nnavi.org/). This system has been utilized practically in farms and agricultural cooperatives around Japan.

6. CONCLUDING REMARKS

In Japan, various ICT-based food traceability systems (FTS) have been developed and introduction of them into agriculture and food industry has been promoted, in order to deal with consumers' interest in food safety and security. However, FTS have not spread enough because the benefits to introduce those ICT systems are not clear for the food industry. They would need to be applied not only for traceability but for process control, quality control and risk management, such as Good Agricultural Practice (GAP).

Regarding the ICT systems to manage product distribution and sale, although applying RFID is efficient and effective, the problems are cost to utilize RFID tags and devices, and workability of RFID handling such as tagging and reading. To solve these problems, the researchers are studying how to reduce the cost and how to automate RFID handling. Although the information disclosure on the products was well received by the consumers, it was difficult for them to understand details of the shown data. It is important to research what kind of information the consumers require.

A goal of the agricultural risk management system is to enable workers automatically to prevent mistakes in various agricultural operations such as AC applications. Moreover, by collaboration of the risk management system and FTS, it is expected that safety and security of agricultural products will be further improved.

ACKNOWLEDGEMENTS

The author would like to thank Professor Teruaki Nanseki, Kyushu University, Dr. Kazunori Sato, Dr. Masayuki Hirafuji and Dr. Seishi Ninomiya, National Agricultural Research Center, for their cooperation and helpful advices. This work was supported in part by a research project for utilizing advanced technologies in agricultural, forestry and fisheries of Agricultural, Forestry and Fisheries Research Council, MAFF.

REFERENCES

Nanseki, T., 2005. A navigation system for appropriate pesticide use and food safety, Proceedings of International Seminar on Technology Development for Good Agriculture Practice in Asia and Oceania: 134-152

Nanseki, T., 2007. Status and Perspective of Food Traceability in Japan, Proceedings of International Food Traceability Symposium -Korea, Japan, EU-

Nanseki, T., et al., 2005. A navigation system for appropriate pesticide use: design and implementation, Agricultural Information Research, 14(3): 207-226 (in Japanese)

Nanseki, T., et al., 2006. Development of a risk management system for agricultural chemical use, Agricultural Information Research, 15(4): 359-372 (in Japanese)

Sugahara, K., 2005. Traceability system for agricultural products using RFID and mobile technology. Proceedings of International Seminar on Technology Development for Good Agriculture Practice in Asia and Oceania: 204-212

Sugahara, K., et al., 2006. Effective distribution traceability system for agricultural products based on RFID technology. Proceedings of AFITA 2006: 424-431

Sugiyama, J., 2001. Virtually Identified Produce System (VIPS) using the Internet. Agriculture and Horticulture 76(8): 845-854 (in Japanese)

Sugiyama, J., 2004. Traceability and accountability for produce. Journal of Agricultural Science 59(5): 193-197 (in Japanese)

Tomiyama, T., 2005. Introduction spread of food traceability, Y. Niiyama [ed.] Traceability in Food Chain, Showado: 114-126 (in Japanese)

Kolibri 360 GmbH
Ferdinand-Jühlke-Straße 7
99095 Erfurt, Deutschland
produktsicherheit@kolibri360.de